ICAME 2005

ICAME 2005

Proceedings of the 28th International Conference on the Applications of the Mössbauer Effect (ICAME 2005) held in Montpellier, France, 4–9 September 2005

Volume I *(Part I–II/V)*

Edited by

P.-E. LIPPENS

Université Montpellier II - CNRS, Montpellier, France

J.-C. JUMAS

Université Montpellier II - CNRS, Montpellier, France

and

J.-M. R. GÉNIN

Université Henri Poincaré-Nancy I, Villers-lès-Nancy, France

Reprinted from *Hyperfine Interactions*
Volume 165, Nos. 1–4 (2005)
Volume 166, Nos. 1–4 (2005)

Springer

A C.I.P. Catalogue record for this book is available from the Library of Congress.

ISBN 3-540-49848-3

Published by Springer
P.O. Box 990, 3300 AZ Dordrecht, The Netherlands

Sold and distributed in North, Central and South America
by Springer
101 Philip Drive, Norwell, MA 02061, U.S.A.

In all other countries, sold and distributed
by Springer
P.O. Box 990, 3300 AZ Dordrecht, The Netherlands

Printed on acid-free paper

Table of Contents

Preface 1–2

G. SHENOY, G. SMIRNOV and J. ARTHUR / Stanley Ruby 1924–2004 3–4

J. D. CASHION and T. L. GREAVES / *In Situ* Mössbauer Studies of Electrochemical Processes 5–16

H. PAULSEN, P. WEGNER, H. WINKLER, J. A. WOLNY, L. H. BÖTTGER, A. X. TRAUTWEIN, C. SCHMIDT, V. SCHÜNEMANN, G. BARONE, A. SILVESTRI, G. LA MANNA, A. I. CHUMAKOV, I. SERGUEEV and R. RÜFFER / Dynamics of Metal Centers Monitored by Nuclear Inelastic Scattering 17–24

I. P. SUZDALEV and YU. V. MAKSIMOV / Critical Sizes and Characteristics of Nanoclusters and Nanostructures 25–35

F. VARRET, K. BOUKHEDDADEN, E. CODJOVI and A. GOUJON / Molecular Switchable Solids: Towards Photo-controlled Magnetism 37–47

J. BALOGH, D. KAPTÁS, L. F. KISS, T. KEMÉNY, I. VINCZE, K. TEMST and C. VAN HAESENDONCK / Fe–Ag Granular Multilayers and Heterostructures Studied in Applied Magnetic Field 49–53

J.-P. CANDY, E. ROISIN, J.-M. BASSET, D. UZIO, S. MORIN, L. FISCHER, J. OLIVIER-FOURCADE and J.-C. JUMAS / Evidence for Direct Observation by Mössbauer Spectroscopy of Surface Tin Atoms in Platinum–tin Particles 55–60

C. ESTOURNÈS, C. D'ORLÉANS, J.-L. REHSPRINGER, E. MANOVA, B. KUNEV, D. PANEVA, I. MITOV, L. PETROV and M. KURMOO / Mössbauer and Magnetic Study of $Co_xFe_{3-x}O_4$ Nanoparticles 61–67

R. MANTOVAN, S. SPIGA and M. FANCIULLI / Low Temperature CEMS of Sn-implanted SiO_2 69–73

M. MIGLIERINI, T. KAŇUCH, Y. JIRÁSKOVÁ, M. MASHLAN, R. ZBOŘIL and M. VŮJTEK / Surface Properties of $Fe_{76}Mo_8Cu_1B_{15}$ Alloy After Annealing 75–80

V. ŠEPELÁK, I. BERGMANN, A. FELDHOFF, P. HEITJANS, F. J. LITTERST and K. D. BECKER / Mechanosynthesis of Spinel Ferrite Nanoparticles Followed by Mössbauer Spectroscopy 81–87

S. HASLBECK, K.-P. MARTINEK, L. STIEVANO and F. E. WAGNER / Formation of Gold Nanoparticles in Gold Ruby Glass: The Influence of Tin 89–94

S. BAN, Y. TAKANO, N. OHKUBO, H. HOYA, Y. TAKAHASHI, K. TAKASE and K. SEKIZAWA / Mössbauer Study of the Transition Metal Phosphorus Trichalcogenide $FePS_3$ and Spin Glass $Mn_{0.5}Fe_{0.5}PS_3$ 95–99

L. BALAN, A. DAILLY, R. SCHNEIDER, D. BILLAUD, P. WILLMANN, J. OLIVIER-FOURCADE and J.-C. JUMAS / Synthesis of Nanoscale Antimony Particles 101–106

B. F. O. COSTA, J. M. LOUREIRO and G. LE CAËR / Phase Transformations of σ-FeCr Induced by Ball Milling 107–112

V. G. DE RESENDE, G. M. DA COSTA, E. DE GRAVE and L. DATAS / Chemical Reduction of Hematite by Sodium Borohydride 113–119

J. SITEK and J. DEGMOVÁ / Aluminium Based Amorphous and Nanocrystalline Structure 121–125

J. FEDOTOVA, J. KALININ, A. FEDOTOV, A. SITNIKOV, I. SVITO and A. ZALESKI and A. JABLONSKA / The Effect of the Sputtering Process Ambient on the Magnetic State and Phase Composition of the Film Nanocomposites $(Fe_{0.45}Co_{0.45}Zr_{0.10})_x(Al_2O_3)_{1-x}$ 127–134

P. A. BINGHAM, S. D. FORDER, R. J. HAND and A. LAVAYSIERRE / Mössbauer Studies of Phosphate Glasses for the Immobilisation of Toxic and Nuclear Wastes 135–140

G. WELLENREUTHER, H. FRANZ, U. V. BÜRCK and I. SERGUEEV / Rotational Dynamics in Anisotropic Confinement 141–146

N. S. GAJBHIYE and S. BHATTACHARYYA / Mössbauer Studies of ε-$Fe_{3x-x}Ni_xN$ and γ'-$Fe_{4-y}Ni_yN$ Nanoparticles 147–151

N. S. GAJBHIYE, S. BHATTACHARYYA, G. BALAJI, R. S. NINGTHOUJAM, R. K. DAS, S. BASAK and J. WEISSMÜLLER / Mössbauer and Magnetic Studies of MFe_2O_4 (M=Co, Ni) Nanoparticles 153–159

J. S. GARITAONANDIA, K. SUZUKI, J. D. CASHION, E. GOIKOLEA, M. INSAUSTI, F. PLAZAOLA, R. PIZARRO, P. GORRIA and P. MINGUEZ / Mössbauer Study of the Crystallization Products of a $Fe_{75}Zr_{25}$ Amorphous Alloy 161–165

S. H. KILCOYNE, D. GREIG and M. N. GONA / ^{57}Fe Mössbauer Studies of Amorphous $Pd_{40}(NiFe)_{40}P_{20}$ Alloys 167–174

B. SAHOO, V. KUNCSER, W. KEUNE and J. PIEPER / Mössbauer Spectroscopical Investigation of Amorphous Fe–Y Alloy Ribbons Prepared by Melt Spinning 175–181

M. KOPCEWICZ, J. KALINOWSKA and T. KULIK / Magnetic Properties of Nanocrystalline $FeNiCo_xZrB$ Alloys (x = 0, 10, 20) 183–188

M. KOPCEWICZ, J. KALINOWSKA and J. LATUCH / Crystallization of Amorphous FeNiZrB and FeNiCoZrB Alloys 189–194

A. MIKÓ, E. KUZMANN, M. LAKATOS-VARSÁNYI, A. KÁKAY, F. NAGY and L. K. VARGA / Mössbauer and XRD Study of Pulse Plated Fe–P and Fe–Ni Thin Layers 195–201

A. LANČOK, K. ZÁVĚTA, M. POPOVICI, C. SAVII, M. GICH, A. ROIG, E. MOLINS and K. BARČOVÁ / Mössbauer Studies on Ultraporous Fe-Oxide/SiO_2 Aerogel 203–208

D. R. MABE, A. M. KHASANOV and J. G. STEVENS / A Novel Investigation of Tin-doped Ferrihydrite Nanoparticles 209–213

E. MANOVA, C. ESTOURNÈS, D. PANEVA, J.-L. REHSPRINGER, T. TSONCHEVA, B. KUNEV and I. MITOV / Mössbauer Study of Nanodimensional Nickel Ferrite – Mechanochemical Synthesis and Catalytic Properties 215–220

H. BARTONKOVA, M. MASHLAN, R. ZBORIL, J. PECHOUSEK, O. SCHNEEWEISS and P. MARTINEC / Mössbauer Study of Iron Oxide Modified Montmorillonite 221–225

J. OLENIACZ, P. DUDA, W. ZYCH and J. GRABSKI / Magnetic Properties of Metallic Glasses of the Type $Fe_{80-x}Pd_xB_{20}$ and $Fe_{80-x}Pt_xB_{20}$ 227–230

R. KALAI SELVAN, C. O. AUGUSTIN, M. I. OSHTRAKH, O. B. MILDER and V. A. SEMIONKIN / Mössbauer and D.C. Magnetization Studies of $(CuFe_2O_4)_{1-x}$ $(SnO_2)_x$ (x=0 and 5 wt.%) Nanocomposites 231–237

G. C. PAPAEFTHYMIOU, F. X. REDL, C. T. BLACK, R. L. SANDSTROM, M. YIN, C. B. MURRAY and S. P. O'BRIEN / Hybrid Magnetic Nanoparticles Derived from Wüstite Disproportionation Reactions at the Nanoscale 239–245

H.-D. PFANNES, D. S. VIEIRA JUNIOR, R. PANIAGO and J. H. DIAS FILHO / Superparamagnetic Relaxation Mössbauer Spectra by Phonon–magnon Interaction Using the Multilevel Method 247–252

J. M. FLOREZ, J. MAZO-ZULUAGA and J. RESTREPO / Local Structural Order in Nanostructured Hematite 253–259

R. K. SHARMA, O. SUWALKA, N. LAKSHMI and K. VENUGOPALAN / Mössbauer Studies of Cr Substituted Bulk and Nano Co–Zn Ferrites 261–266

R. E. VANDENBERGHE, I. NEDKOV, T. MERODIISKA and L. SLAVOV / Surface Oxidation Control of Nanosized Magnetite and Mössbauer Measurements 267–271

S. ZHU, W. LIU, C. FAN and Y. LI / Mössbauer Study of Nano-TiO_2 Doped with Fe 273–278

S. CROCI, O. J. BABALOLA, S. BETTATI, C. VALENTI, I. ORTALLI and F. PARAK / Quadrupole Splitting Temperature Dependence of High and Low Affinity Deoxyhemoglobin Encapsulated in Wet Silica Gel 279–283

J. GALAZKA-FRIEDMAN, E. R. BAUMINGER, A. FRIEDMAN, D. KOZIOROWSKI and K. SZLACHTA / Human Nigral and Liver Iron – Comparison by Mössbauer Spectroscopy, Electron Microscopy and ELISA 285–288

K. KOVÁCS, E. KUZMANN, F. FODOR, A. VÉRTES and A. A. KAMNEV / Mössbauer Study of Iron Uptake in Cucumber Root 289–294

A. X. TRAUTWEIN, P. WEGNER, H. WINKLER, H. PAULSEN, V. SCHÜNEMANN, C. SCHMIDT, A. I. CHUMAKOV and R. RÜFFER / Nuclear Resonance Vibrational Spectroscopy of Iron–sulfur Proteins 295–298

G. BARONE, L. H. BÖTTGER, J. A. WOLNY, H. PAULSEN, A. X. TRAUTWEIN, A. SILVESTRI, I. SERGUEEV and G. LAMANNA / Tin–DNA Complexes Investigated by Nuclear Inelastic Scattering of Synchrotron Radiation 299–302

A. A. KAMNEV, L. A. KULIKOV, YU. D. PERFILIEV, L. P. ANTONYUK, E. KUZMANN, A. VÉRTES / Application of ^{57}Co Emission Mössbauer Spectroscopy to Studying Biocomplexes in Frozen Solutions 303–308

T. OHYA, J. TAKEDA and M. SATO / Is the Hematin Derived from Dodecaphenylporphyrin a μ-oxo Dimer? 309–312

M. I. OSHTRAKH / Mössbauer spectroscopy: Application in Biomedical Research 313–320

M. I. OSHTRAKH, O. B. MILDER, V. A. SEMIONKIN, L. I. MALAKHEEVA and P. G. PROKOPENKO / Study of Chicken Liver and Spleen by Mössbauer Spectroscopy 321–326

E. I. YURYEVA and M. I. OSHTRAKH / Ab Initio Study of the ^{57}Fe Quadrupole Splitting in the Heme Models of α- and β-subunits in Tetrameric Deoxyhemoglobin 327–331

G. C. PAPAEFTHYMIOU, A. J. VIESCAS, R. HORN, E. CARNEY, G. ZHAO, N. D. CHASTEEN, J. LEE and S. M. GORUN / Deuterium Isotope Effects on Iron Core Formation in Ferritin 333–338

S. C. BHATIA, B. H. CARDELINO and N. RAVI / A ^{57}Fe Mössbauer Characterization of Fe-biopolymer Complexes and their Relevance to Biological Molecules 339–344

G. DÉNÈS, A. MUNTASAR, S. BOUFAS and H. MERAZIG / Search for the stannous ion in a chloride/fluoride matrix: cases of $Ba_{1-x}Sn_xCl_{1+y}F_{1-y}$ and of Ba_2SnCl_6 345–349

R. H. HERBER and I. NOWIK / Exploring metal atom dynamics in organometallics: methodology and results for C_{60} ferrocenes 351–355

Y. KOBAYASHI, H. NONAKA, J. MIYAZAKI, M. K. KUBO, H. UENO, A. YOSHIMI, H. MIYOSHI, D. KAMEDA, K. SHIMADA, D. NAGAE, K. ASAHI and Y. YAMADA / Reactions of ^{57}Mn implanted into solid oxygen 357–361

F. J. BERRY, G. R. LUMPKIN, G. OATES and K. R. WHITTLE / Iron-57 Mössbauer spectroscopy study of phases in the $CaZrTi_{2-2x}Nb_xFe_xO_7$ zirconolite system 363–366

S. K. DEDUSHENKO, M. G. ZHIZHIN and YU. D. PERFILIEV / X-ray and Mössbauer study of structural changes in $K_3Na(FeO_4)_2$ 367–371

J. HANTASH, A. BARTLETT, G. DÉNÈS, A. MUNTASAR and P. OLD-FIELD / $BaSnF_4$ fast ion conductor: Variations versus the method of preparation and anomalous temperature variation of the quadrupole splitting 373–378

J. KOCHUPARAMPIL, G. DÉNÈS and A. MUNTASAR / Phase transition of the doubly disordered $Ba_{1-x}Sn_xCl_{1+y}F_{1-y}$ solid solution at high temperature 379–384

M. SEREDYUK, A. B. GASPAR, V. KSENOFONTOV, S. REIMAN, Y. GALYAMETDINOV, W. HAASE, E. RENTSCHLER and P. GÜTLICH / Multifunctional materials exhibiting spin crossover and liquid-crystalline properties. Interplay between spin crossover and liquid-crystal properties in iron(II) coordination complexes 385–390

J.-M. R. GÉNIN, M. ABDELMOULA, R. AÏSSA and C. RUBY / Ordering in Fe^{II-III} Hydroxysalt green rusts from XRD and Mössbauer analysis (chloride, carbonate, sulphate, oxalate...); about the structure of hydrotalcite-like compounds 391–396

S. IIJIMA, F. MIZUTANI, O. NIWA, N. MATSUMOTO, Y. SUNATSUKI and M. KOJIMA / Mössbauer studies of mixed-valence spin-crossover iron complexes with a hexadentate tripod ligand 397–402

K. IKEDA, N. KOJIMA, Y. KOBAYASHI and M. SETO / Study on charge transfer interaction and valence state of layered perovskite-type mixed valence complex $[NH_3(CH_2)_nNH_3]_2[(Au^{I}I_2)(Au^{III}I_4)(I_3)_2]$ (n = 7, 8), by means of ^{197}Au Mössbauer spectroscopy 403–407

D. AFROJ and M. KATADA / Variable temperature Mössbauer spectroscopic studies of oxo-centered trinuclear mixed-valence iron succinate, mesaconate and isophthalate complexes 409–416

T. KAWASAKI, T. KITAZAWA, T. NISHIMURA, M. NAKADA and M. SAEKI / Relationship between ^{237}Np Mössbauer parameters and bond distances in nitrogen coordinated $[NpO_2(acac)_2py]$ 417–423

M. K. KUBO, Y. KOBAYASHI, H. NONAKA, Y. YAMADA, Y. SAKAI, H. SHOJI and H. MATSUE / Neutron in-beam Mössbauer spectroscopic study of iron disulfide at room temperature 425–428

S. KUBUKI, N. KAWAKAMI, T. KAMIKAWA, M. FUKAGAWA, T. NISHIZUMI, T. NISHIDA, Z. HOMONNAY and E. KUZMANN / Corelationship between local structure and water purifying ability of iron-containing waste glasses 429–436

V. M. BYAKOV, L. A. KULIKOV, Y. D. PERFIL'EV and S. V. STEPANOV / Investigation of primary chemical reactions of electrons by emission Mössbauer and positron spectroscopies 437–442

J. LADRIERE and V. SENTERRE / Mössbauer study on the gamma radiolysis of lithium trisoxalatoferrate (III) pentahydrate 443–447

F. J. BERRY, J. R. GANCEDO, J. F. MARCO and X. REN / Characterisation of the reduction properties of $La_{0.5}Sr_{0.5}Fe_{0.5}Co_{0.5}O_3$ 449–453

T. M. MEAZ and C. B. KOCH / An investigation of trivalent substituted M-type hexagonal ferrite using X-ray and Mössbauer spectroscopy 455–463

A. LINDNER, M. MENZEL, F. RENZ, D. G. KURTH and A. F. THÜNEMANN / Metallosupramolecular coordination polyelectrolytes investigated by Mössbauer spectroscopy 465–468

S. NAKASHIMA, T. MORITA, K. YAMADA and K. INOUE / Spin-crossover phenomena of the assembled iron complexes by enclathrating an organic molecule 469–473

J.-L. REHSPRINGER, S. VILMINOT, D. NIZNANSKY, K. ZAVETA, C. ESTOURNES and M. KURMOO / A temperature and magnetic field dependence Mössbauer study of ϵ-Fe_2O_3 475–481

Y. SAKAI, R. OMATSUZAWA, I. SAKAZAKI, S. SUZUKI, H. HASHINO, T. SAITO and T. IRIYAMA / Structure comparison between Th_2Zn_{17}-type and $TbCu_7$-type Sm–Fe intermetallic compounds and their nitrides by means of ^{57}Fe-Mössbauer spectroscopy 483–487

N. NGUYEN, A. DUCOURET, F. STUDER, V. AUBIN, D. CAURANT, D. GOURIERN and J.-M. COSTANTINI / Defects induced by electron irradiation in hollandite ceramics, specific radioactive cesium-host wasteforms: a ^{57}Fe Mössbauer study 489–493

J. A. WOLNY, H. PAULSEN, H. WINKLER, A. X. TRAUTWEIN and J.-P. TUCHAGUES / DFT calculations as a tool to analyse quadrupole splittings of spin crossover Fe(II) complexes 495–498

C. I. WYNTER, D. H. RYAN, S. P. TANEJA, L. MAY, F. W. OLIVER, D. E. BROWN and M. IWUNZIE / Mössbauer studies of ^{151}Eu in europium oxalate, europium bissalen ammonium and europium benzoate 499–503

C. BENDER KOCH and E. A. FERROW / Impact related changes in mineral properties: Ti-magnetite from basaltic rocks at Lonar crater, India 505–509

H. P. GUNNLAUGSSON, P. BERTELSEN, C. BUTZ-JØRGENSEN, G. KLINGELHÖFER, I. KUVVETLI, M. B. MADSEN, J. P. MERRISON, P. NØRNBERG, H. RASMUSSEN and G. WEYER / On the merits of conversion electron Mössbauer spectroscopy in geosciences 511–516

A. L. VENDELBOE, H. P. GUNNLAUGSSON, Ö. HELGASON and P. NØRNBERG / Characterization of burned soil profiles by Mössbauer spectroscopy 517–522

P. BERTELSEN, M. B. MADSEN, C. S. BINAU, W. GOETZ, H. P. GUNNLAUGSSON, S. F. HVIID, K. M. KINCH, G. KLINGELHÖFER, K. LEER, D. E. MADSEN, J. MERRISON, M. OLSEN, S. W. SQUYRES and THE ATHENA SCIENCE TEAM / Backscattering Mössbauer spectroscopy of Martian dust 523–527

D. MALCZEWSKI, J. E. FRĄCKOWIAK and E. V. GALUSKIN / ^{57}Fe Mössbauer spectroscopy and x-ray diffraction study of some complex metamict minerals 529–536

B. PANDEY, A. TRIPATHI, R. P. TRIPATHI and H. C. VERMA / KTB iron mineralogy at iridium deficient layers revealed by Mössbauer spectroscopy 537–542

Y. A. ABDU, I. SOUZA AZEVEDO, S. J. STEWART, A. LÓPEZ, M. E. VARELA, G. KURAT and R. B. SCORZELLI / Mössbauer study of glasses in meteorites: the D'Orbigny angrite and Cachari eucrite 543–547

G. KLINGELHÖFER, E. DEGRAVE, R. V. MORRIS, A. VAN ALBOOM, V. G. DE RESENDE, P. A. DE SOUZA JR, D. RODIONOV, C. SCHRÖDER, D. W. MING and A. YEN / Mössbauer spectroscopy on Mars: goethite in the Columbia Hills at Gusev crater 549–554

H. P. GUNNLAUGSSON, E. S. WORM, P. BERTELSEN, W. GOETZ, K. KINCH, M. B. MADSEN, J. P. MERRISON and P. NØRNBERG / Simulations of the magnetic properties experiment on Mars Exploration Rovers 555–560

H. RASMUSSEN, H. P. GUNNLAUGSSON, C. TEGNER, L. KRISTJÁNSSON / Magnetic properties of Martian olivine basalts studied by terrestrial analogues 561–566

C. SCHRÖDER, G. KLINGELHÖFER, B. E. BAILEY and H. STAUDIGEL / Mössbauer spectroscopy as a tool in astrobiology 567–571

M. MOROZOV, C. BRINKMANN, M. GRODZICKI, W. LOTTERMOSER, G. TIPPELT, G. AMTHAUER and H. KROLL / Octahedral cation partitioning in Mg,Fe^{2+}-olivine. Mössbauer spectroscopic study of synthetic $(Mg_{0.5}\ Fe^{2+}{}_{0.5})_2SiO_4$ (Fa_{50}) 573–578

C. BENDER KOCH, O. K. BORGGAARD and A. GAFUR / Formation of iron oxides in soils developed under natural fires and slash-and-burn based agriculture in a monsoonal climate (Chittagong Hill Tracts, Bangladesh) 579–584

A. BUSTAMANTE, S. ESPINOZA, G. MORALES and R. B. SCORZELLI / Mössbauer studies of impactites from Huamalies province in Huanuco Region 585–591

A. BUSTAMANTE, J. CABRERA, V. GARCIA, E. URDAY, Y. A. ABDU and R. B. SCORZELLI / Mössbauer spectroscopy description of limonite from Taraco, in the Huancane Province of the Puno Region, Peru 593–597

A. BUSTAMANTE, E. MATTIEVICH, H. S. DE AMORIM, I. VENCATO and M. M. SILVEIRA / The Mössbauer spectrum of synthetic hureaulite: $Fe_5{}^{2+}(H_2O)_4(PO_4H)_2(PO_4)_2$ 599–603

H. BUSTOS RODRIGUEZ, D. OYOLA LOZANO, Y. A. ROJAS MARTÍNEZ, G. A. PÉREZ ALCÁZAR and A. G. BALOGH / Invisible gold in Colombian auriferous soils 605–611

N. I. CHISTYAKOVA, V. S. RUSAKOV, T. V. GUBAIDULINA and S. V. KOZERENKO / Investigations of sulfide minerals with layered structure by Mössbauer spectroscopy methods 613–617

A. S. ANASTACIO, J. D. FABRIS, J. W. STUCKI, F. S. COELHO, I. V. PINTO and J. H. M. VIANA / Clay fraction mineralogy of a Cambisol in Brazil 619–624

V. RUSANOV, V. GUSHTEROV, H. WINKLER and A. X. TRAUTWEIN / Iron-containing atmospheric aerosols in the Chernobyl fallout 625–630

B. KOPCEWICZ, M. KOPCEWICZ, M. JELENSKA and A. HASSO-AGOPSOWICZ / Mössbauer study of chemical transformations in soil samples during thermomagnetic measurements 631–636

A. KUNO, M. MATSUO, Q. WANG and C. J. ZHANG / Iron speciation of sediments from Kuhai Lake in China by Mössbauer spectroscopy 637–642

A. KOMLÓSI, E. KUZMANN, Z. HOMONNAY, N. M. NAGY, S. KUBUKI and J. KÓNYA / Effect of $FeCl_3$ and acetone on the structure of Na–montmorillonite studied by Mössbauer and XRD measurements 643–649

M. MIGLIERINI, J. DEKAN and D. STAŠŠÍKOVÁ-ŠTUKOVSKÁ / Analysis of surface layers of an iron axe from Early Medieval Age 651–655

K. NOMURA, M. TAKEDA, T. IIYAMA and H. SAKAI / Mössbauer studies of Jarosite, Mikasaite and Yavapaiite, and implication to their Martian counterparts 657–664

E. V. ZHIGANOVA, M. I. OSHTRAKH, O. B. MILDER, V. I. GROKHOVSKY, V. A. SEMIONKIN and A. V. MEZENTSEV / Mössbauer spectroscopy of ordinary chondrites: an analysis of the metal phases 665–670

V. I. GROKHOVSKY, M. I. OSHTRAKH, O. B. MILDER and V. A. SEMIONKIN / Mössbauer spectroscopy of iron meteorite Dronino and products of its corrosion 671–677

A. VAN ALBOOM and E. DE GRAVE / Mössbauer effect in natural strunzite, ferristrunzite and ferrostrunzite 679–685

F. B. WAANDERS and A. GOVENDER / Mineral associations in coal and their transformation during gasification 687–691

F. B. WAANDERS, M. C. BRINK and A. A. BISSCHOFF / Mössbauer study of two different aged rock types in the Vredefort structure, South Africa 693–698

D. J. M. BURKHARD and T. SCHERER / Volume and surface oxidation of basalt 699–703

G. R. FERN, T. A. MATHER and D. M. PYLE / Investigation of near-source basaltic glasses using ^{57}Fe Mössbauer spectroscopy 705–708

Author Index

Hyperfine Interact (2005) 165: 1–2
DOI 10.1007/s10751-006-9403-5

Preface

P.-E. Lippens · J.-C. Jumas · J.-M. R. Génin

Published online: 14 November 2006

The 28th International Conference on the Applications of the Mössbauer Effect (ICAME 2005) was held at the conference center "Le Corum" in Montpellier (France) from Sunday 4 to Friday 9 September 2005. ICAME 2005 was very successful with attractive scientific and social programs. Only the weather was not as sunny as expected, but this helped the participants to attend the conference.

The scientific program of ICAME 2005 was comprised of invited lectures, plenary oral talks and poster sessions. The 19 invited lectures were selected based on propositions of IBAME members by the scientific committee: F.J. Berry, M. Carbucicchio, R. Gancedo, J.M. Génin, J.M. Grenèche, P. Gütlich, J. Ladrière, P.E. Lippens, G. Long, J.F. Marco, R. Rüffer, E.Tronc. More than 500 abstracts were received and most of them were of high scientific level. The selection of 58 oral talks by the members of the Groupe Francophone de Spectroscopie Mössbauer (GFSM) was really a challenge. Both invited lectures and oral talks were given in the comfortable Pasteur room of the conference center. The posters were shown during all the week in the same large room (Antigone) in order to make easier scientific discussions between the participants. The stands of our partners: Mössbauer Effect Data Center, MIMOS, AMETEK, LEA-CERCA, RITVERC and WISSEL were also located in the Antigone room, as well as the IBAME meetings, the coffee breaks and the very sought-after Internet connections.

Tutorials were organized for the first time in an ICAME conference on Sunday 4 September 2005. About 60 participants learnt a lot about the Mössbauer effect (P. Bonville), the analysis of experiments and data (J.M. Grenèche), the nuclear resonant scattering of synchrotron radiation (R. Coussement), the applications of the Mössbauer effect in industry (F.J. Berry) and the molecular magnetism (P. Gütlich). In addition, a

P.-E. Lippens (✉) · J.-C. Jumas
LAMMI, UMR 5072 CNRS, Université Montpellier II, F-34095 Montpellier Cedex 5, France
e-mail: icame05@univ-montp2.fr

J.-M. R. Génin
LCPME, UMR 7564 CNRS, Université Henri Poincaré Nancy 1, 405 rue de Vandœuvre,
F-54600 Villers-Lès-Nancy, France

FORUM on the applications of the Mössbauer effect in Energy, Environment and Health was open to the public.

ICAME 2005 was organised by the Mössbauer group of Montpellier (LAMMI-UMR 5072 CNRS-UMII), the CNRS for the financial management and the team of Le Corum for the logistic aspects. The conference was attended by 324 participants from 45 different countries including France (56), Germany (43), Japan (30), Russia (20), Poland (18) and the USA (15). More than 25 participants from Eastern Europe countries were financially supported by the French ministry of education and research. Reduced registration fees were proposed to about 80 students and some of them obtained financial supports from a generous IBAME member (eight students) and the GFSM (eight students). The organising committee also received financial supports from: Université Montpellier II, Pôle Universitaire Européen de Montpellier et du Languedoc-Roussillon, Ville de Montpellier, Communauté d'Agglomération de Montpellier, Département Hérault, Région Languedoc-Roussillon, Institut Gerhardt, Société Française de Chimie, Commissariat à l'Energie Atomique, Centre National de la Recherche Scientifique, Ministère de l'Education Nationale de l'Enseignement Supérieur et de la Recherche.

The ICAME 2005 proceedings contain all the accepted papers reviewed by two referees. They begin with the contributions of the invited speakers followed by the nine sections corresponding to the topics of the conference. Each section contains the contributions of the oral and poster presentations in the same order as in the conference abstract book.

Hyperfine Interact (2005) 165: 3–4
DOI 10.1007/s10751-006-9348-8

OBITUARY

Stanley Ruby 1924–2004

Gopal Shenoy · Gennadii Smirnov · John Arthur

Published online: 26 October 2006

Stanley L. Ruby, who made major contributions to Mössbauer spectroscopy and who inspired the community with the idea of observing the Mössbauer effect using synchrotron radiation, passed away on October 18, 2004, in Los Gatos, California. His boundless intellectual curiosity and passion for life was an inspiration to all around him, especially his scientific colleagues.

Born in New York City in 1924, Stan served in the US Army Signal Corps during World War II in New Guinea and the Philippines. He performed his undergraduate and graduate studies at Columbia University, guided in his graduate work primarily by Professor Madame C.-S. Wu. He briefly worked at the IBM Watson Laboratory before starting his work on Mössbauer spectroscopy with Fe^{57} in 1960 at Westinghouse Electric Corporation. He collaborated with Paul Flinn (Stanford University) and Gen Shirane (Brookhaven National Laboratory), studying a large class of magnetic compounds. His most recognized work from this period was on the acoustic modulation of the wavelength of resonant gamma radiation, detected using the Mössbauer effect. During a visit to the Weizmann Institute in 1962, he observed the 9.3 keV Mössbauer resonance in Kr^{83}.

Stan joined Argonne National Laboratory in 1964, where he was involved in the discovery of many Mössbauer resonances, including K^{40}. Since this isotope has no radioactive parent, Stan and R. E. Holland (Argonne) observed it by populating the

G. Shenoy (✉)
Argonne National Laboratory, Argonne, IL, USA
e-mail: gks@aps.anl.gov

G. Smirnov
Russian Research Center "Kurchatov Institute", Moscow, Russia

J. Arthur
Stanford Linear Accelerator Center, Stanford, CA, USA

29.4 keV excited nuclear state by means of deuteron bombardment ($K^{39}(d,p)K^{40}$). Later, Stan and D. H. Vincent (University of Michigan) excited the K^{40} resonance through the neutron capture reaction ($K^{39}(n, \gamma)K^{40}$).

Argonne provided an ideal setting for Stan. He could incubate his ideas with colleagues who helped them blossom into successful experiments. He actively worked with a large number of scientists from different divisions at Argonne, with backgrounds in nuclear physics, materials science, and solid state physics. This was essential for establishing new Mössbauer resonances and finding the best techniques to unravel nuclear, chemical and solid state properties. In particular, his collaboration with Michael Kalvius, Bobby Dunlap and Gopal Shenoy led to many publications dealing with resonances in Sn^{119}, Sb^{121}, Te^{125}, $I^{127,129}$, U^{238}, Np^{237}, Am^{243}.

Stan will be best remembered for his proposal in 1974 to excite the 14.4 keV Mössbauer resonance in Fe^{57} using synchrotron radiation rather than a radioactive source to populate the nuclear excited state. Stan struggled for years to develop techniques for separating the nuclear resonant X-rays from the overwhelming background. In the early 1980s, Stan moved from Argonne to the Stanford Synchrotron Radiation Laboratory with the sole desire to succeed with his idea. Although it was the group of Erich Gerdau that first achieved this goal in 1985 at DESY, Stan and the SSRL group followed in 1987 and proceeded to make important contributions to this new field. During the period 1990–1995 Stan and the Stanford group collaborated closely with scientists from Russia (Kurchatov Institute) and Germany (Munich Technical University) in a series of experiments at SSRL and CHESS. They observed and explained several surprising features of the coherent nuclear exciton created by a synchrotron pulse in a nuclear sample, such as the extreme speed-up of the nuclear exciton decay in scattering from a multilayer, and the nuclear exciton echo induced by vibrating a portion of an excited sample. They were also the first to use synchrotron radiation to excite the narrow 6.2 keV Mössbauer resonance in Ta^{181}. Stan was particularly concerned with the conceptual problem of understanding "when and where" the interaction of X-ray photons with nuclei occurs during the propagation of radiation pulses through a target.

Fundamental physics was central to Stan's life, though his interests ranged from cosmology to biology. While he allowed that quantum mechanics was useful for calculations, he found it very unsatisfactory on a philosophical level. During his later career, and especially after his retirement, he worked hard to find simple, alternative explanations for quantum phenomena. Stan was young in soul, always individualistic and passionate about his scientific ideas, which sometimes touched very exotic fields. His mind was most acute when he was talking, so he was eager to sit and talk about his latest ideas. These discussions were always interesting, involving physics, history, astronomy, and politics. Stan's lifelong concern with the impact of science on society was expressed in his leading role in the campaign against anti-ballistic missiles during the 1970s. Beyond his consuming interest in physical science, his many pursuits included international travel, outdoor recreation, marine life, and observing the human parade.

He was truly a family man and took an avid interest in the lives of all his colleagues. He is survived by his wife of 57 years, three children, and four grandchildren.

His friends on several continents will miss him. We remember Stan for his never-fading enthusiasm to discuss physics with anyone who would listen, his friendship, and above all his curiosity for new ideas.

Hyperfine Interact (2005) 165: 5–16
DOI 10.1007/s10751-006-9241-5

In situ Mössbauer studies of electrochemical processes

J. D. Cashion · T. L. Greaves

Published online: 26 October 2006

Abstract Recent electrochemical studies involving *in situ* Mössbauer spectroscopy are reviewed. The principal application has been the development of new electrodes for lithium-ion batteries, where Mössbauer spectroscopy plays an important complementary role to XRD in characterizing, and usually identifying, nanocrystalline and intermediate phases of iron and tin. Other applications include the crystallography of Prussian Blue, taken using a recently developed spectrometer for acquiring multiple spectra during continuous cyclic voltammetry, valence interconversion in clay minerals and electrochemical oxygen reduction.

Key words Mössbauer spectroscopy · *in situ* electrochemistry · cyclic voltammetry · lithium-ion batteries · clay minerals · Prussian Blue

1 Introduction

Although the use of Mössbauer spectroscopy to study electrochemical processes stretches back for 30 years, it has never been widely used, and indeed has probably not been used as widely as it deserves to be. The uses can be divided into *ex situ* (removed from the cell), in situ (usually at constant potential) and battery type cells (charging and discharging). After a burst of *in situ* and *ex situ* Mössbauer studies around 20 years ago, progress was extensively reviewed at the end of the 1980s [1–3] after which the impetus seemed to wane. At that stage the principal applications had been to lithium batteries, iron oxides and corrosion, passivation layers and pitting in steels, protective films such as iron phthalocyanine and tin plating. Since then, interest in its use for *in situ* corrosion and protection studies seems to have almost disappeared, while studies of lithium-ion batteries have grown considerably, in

J. D. Cashion (✉) · T. L. Greaves
School of Physics, Building 27, Monash University, Clayton, Vic 3800, Australia
e-mail: john.cashion@sci.monash.edu.au

Present address:
T. L. Greaves
CSIRO Division of Molecular and Health Technologies, Clayton, Vic 3680, Australia

conjunction with the rapid expansion of this technology. This review covers the principal *in situ* work carried out over approximately the past six years.

2 *In situ* Mössbauer electrochemical cells

The use of Mössbauer spectroscopy for electrochemical investigations has to take several different constraints into consideration. The first is the obvious one that Mössbauer spectroscopy can only study solids so the material of interest must be incorporated into an electrode. Secondly, there is the necessity to have reasonable loadings of the active nucleus in the working electrode. If the material is conducting, this is not usually a problem. However, insulating materials must be crushed and mixed with carbon or some other conductor, in order to make the electrode sufficiently conducting. With the addition of a binder, this means that typical loadings may be only 20–40% of the total material. Consequently, it is often advantageous to use isotopically enriched samples, although this severely limits the ability to study industrial samples. Incorporation of all this material into the electrode is only useful if all the resonant nuclei are available to take part in the electrochemical reaction. This places a limit on the thickness of the electrode such that all of the electrode is available for diffusion of the relevant species, otherwise the unavailable material contributes an unchanging background spectrum from which you must extract the results of interest.

In the case of cells with a liquid electrolyte, some other considerations in the design of *in situ* Mössbauer electrochemical cells include the relative area and positioning of the working and counter electrodes, getting the reference electrode close to the working electrode without interfering with the gamma ray transmission and minimizing the gamma path length in the electrolyte. A schematic diagram of the cell used in our laboratory is shown in Figure 1.

In the case of cells with solid electrolytes, such as those for lithium-ion batteries, it is most likely that the gamma path length will include all of the components of the cell. Figure 2 shows a schematic of the cell used by Dunlap and co-workers at Dalhousie University [4], which is based on the 2325 coin-type cell hardware in which circular holes have been cut in the body and lid of the can and covered with beryllium windows. It is a fortunate accident that the mass absorption coefficient of lithium is very small at the ^{57}Fe and ^{119}Sn gamma energies so that, even though all the components are in the gamma path, this design suffers less from electronic absorption than ones with an aqueous electrolyte. A modification to allow low temperature spectra of the cell without opening has also been described which enabled identification of superparamagnetic iron [5].

There are two alternative designs which are also suitable for Mössbauer studies. The first is the X-ray cell, PLIon™, designed by Bellcore [6] which uses a flat, plastic cell with three electrodes and separators, sandwiched between Al and Cu grids and the whole being hermetically sealed. The second is the pouch cell designed for EXAFS [7] which uses a single positive electrode sandwiched between two Li metal negative electrodes. The whole pouch is sealed after pumping out any residual gas.

3 Lithium-ion batteries

The principal area of recent *in situ*, and most of the *ex situ*, Mössbauer studies involving electrochemistry is in the development of lithium-ion batteries, which are widely used as

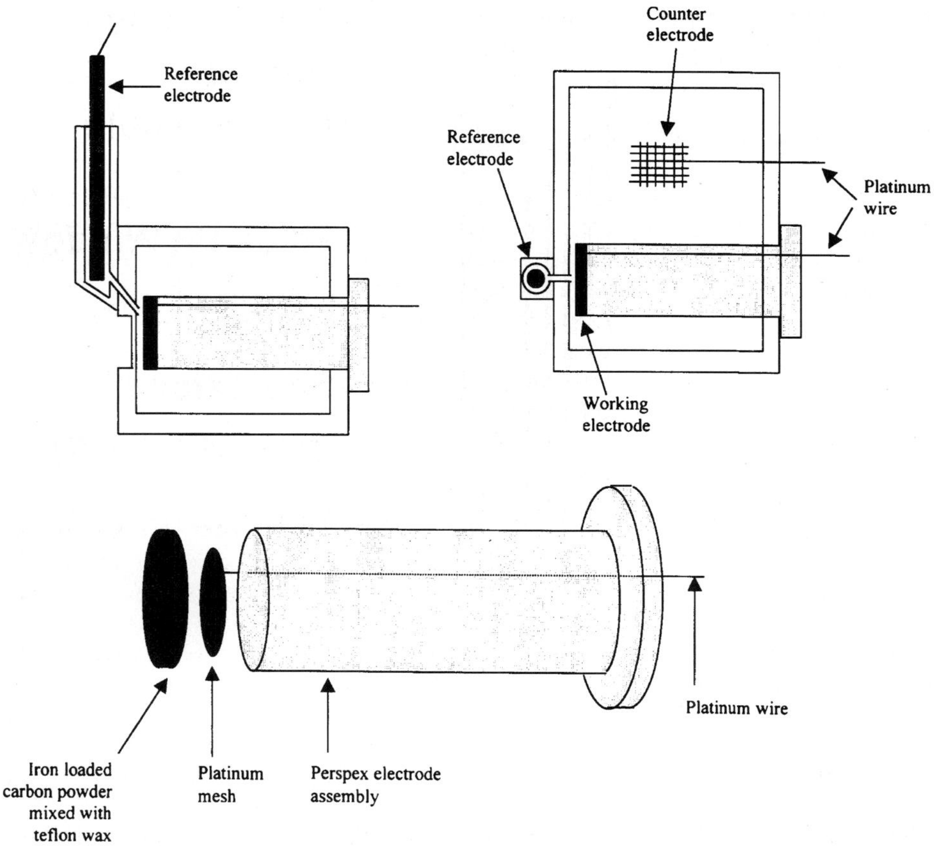

Figure 1 Schematic diagram of the *in situ* Mössbauer electrochemical cell used in our laboratory.

rechargeable power packs for laptop computers, cameras and cell phones because of their high energy density. However, many improvements are needed before they can take over the high power market (power tools) from the environmentally-unfriendly Ni-Cd batteries and the high energy density (hybrid vehicle) market from the Ni/metal hydride batteries. This section does not claim to be a balanced view of the development of electrodes for Li-ion batteries for which there is an excellent recent review by Whittingham [8]. We hope that the present examples may encourage other groups to include Mössbauer spectroscopy as an additional complementary technique.

Both electrodes in Li-ion batteries are intercalation materials. The discharge cycle consists of the extraction of lithium from the negative electrode, which is usually a carbon composite, by the reaction:

$$LiC_6 \quad \rightarrow \quad 6C + Li^+ + e^-.$$

The Li^+ is transferred to the positive electrode, where it becomes intercalated, while the electron travels around the external circuit. For ease of intercalation, the positive electrode materials are either layered materials with planes of cations between close packed anions or the open or tunnel compounds such as MnO_2, vanadates and phosphates.

Figure 2 Schematic diagram of the *in situ* Mössbauer electrochemical cell used by Dunlap and co-workers for studying lithium-ion batteries.

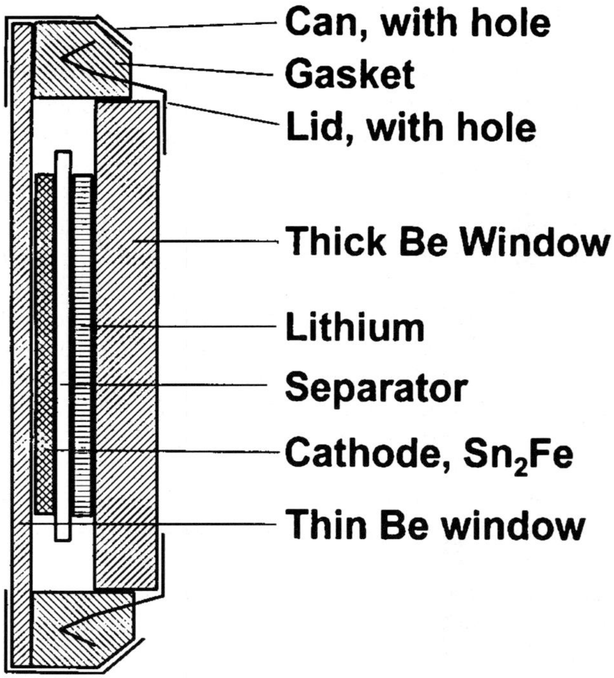

3.1 Positive electrode materials

Current batteries principally use the layered $LiCoO_2$ as the positive electrode material, in preference to the competing double oxides such as $LiNiO_2$ and $LiMn_2O_4$, despite its disadvantages, which include the cost and limited availability of Co, the possibility of oxygen production upon accidental heating and the increased reactivity as the lithium concentration decreases. The others also have disadvantages. There has been considerable research to try and find replacements for the $LiCoO_2$ positive electrodes, with various iron-containing materials being strong contenders. These include various modified oxides, double oxides and intermetallic compounds. The high potential of the Fe^{3+}/Fe^{4+} couple of around 4 V (vs Li/Li^+) makes it an attractive choice for new systems and Mössbauer analysis has been employed in many of these investigations. However, conventionally prepared $LiFeO_2$ is electrochemically inactive, so research has covered both other iron oxides and modifications to the preparation and composition of $LiFeO_2$. For example, $Li_{1-z}(Ni_{1-y}Fe_y)_{1+z}O_2$ has been studied by several groups because the mixing of Fe with either Ni or Co in the $LiFeO_2$ structure puts crystallographic pressure on the Fe-O octahedra and enables the Fe^{3+}/Fe^{4+} couple to be accessed. The strong dependence of the ease of lithium intercalation on the metal–oxygen bond distance has been summarized by Delmas *et al.* [9].

Most of these electrodes suffer from the so-called irreversible capacity of the material. Much of the problem arises because of cation disorder in the alternating layers of metal ions, which affects the lithium intercalation. In the case of $Li_{1-z}(Ni_{1-y}Fe_y)_{1+z}O_2$, a possible mechanism for the irreversibility, which involves two different diffusion rates and hence creates a Li-rich surface layer, has been proposed [10] and experiments show that it can be reversed by cycling to a much lower potential.

In $LiFe_xMn_{2-x}O_4$ spinel, with a small concentration of Fe, two plateaus have been observed with the well-known Mn^{3+}/Mn^{4+} couple at 4 V and new one at nearly 5 V, which the Mössbauer measurements showed to be due to Fe^{3+}/Fe^{4+} [11]. The development of a 5 V battery would be a boon for high power and high energy uses but will require the development of a new electrolyte as current ones are unstable at this potential.

In the borates $FeBO_3$ and Fe_3BO_6, the structures do not allow intercalation and the Mössbauer results [12] confirmed that the first cycle destroyed the crystalline borates. However the materials were then able to continue to cycle reversibly between Fe^0 and $Fe^{2+/3+}$ states, with the latter phases being probably nanocrystalline or amorphous.

Although not collected by *in situ* Mössbauer measurements, the variations in the spectra of Denis *et al.* [13] as lithium was reacted with triclinic $FeVO_4$ up to Li_8FeVO_4 are a very good illustration of how informative Mössbauer spectra can be, particularly in interpreting complex arrangements, in this case as the Fe^{III} reduces to Fe^{II} and Fe^0. Other recent *ex situ* studies have been on $Li_x(Ni_{0.70}Fe_{0.15}Co_{0.15})_{1.03}O_2$ [14] which showed well-defined Fe^{4+} and "$Fe^{3+\delta}$" doublets, $Li(Co_{1-x}Fe_x)O_2$ [15], Li_2FeS_2 [16] which had superparamagnetic Fe^0, $Li_2Fe_xMn_{1-x}O_3$ [17], $Li/NiFe_2O_4$ [18], $Li_3Fe_2(PO_4)_3$ [19] and $Li_{2-2x}Fe_{1+x}Cl_4$ [20]. Some of the experiments show excellent resolution of phases, but for some with poorer resolution, the fitting is not unambiguous, casting doubt on the consequent interpretations. The most promising iron-containing material for future studies, with and without Mössbauer spectroscopy, appears to be the olivine structure $LiFePO_4$ because of its low cost, availability and reversibility. Electrical conductivity is a problem with all of these materials and a recent development has been the discovery that an improvement of eight orders of magnitude when $LiFePO_4$ was doped with small amounts of elements such as Nb, due to the formation of a percolating nano-network of metal rich iron phosphides [21].

3.2 Negative electrode materials

A new direction in lithium-ion batteries started when Fuji Photo Film [22] suggested the use of an amorphous tin-based composite oxide in place of the carbon-based negative material. Such tin-based materials are, of course, just as amenable to Mössbauer effect studies as iron-based materials are. An essential background study of the seven intermetallic phases in the Li-Sn phase diagram was carried out by Dunlap *et al.* [23, see also 24].

Dahn, Dunlap and co-workers [4, 25, 26] have studied ball milled $FeSn_2$ and FeSn electrodes versus lithium with a combination of ^{57}Fe and ^{119}Sn Mössbauer spectroscopy. The slow discharge–recharge cycles took approximately 160 h each, with spectra recorded every 3 h. In the ^{57}Fe spectra of the $FeSn_2$ system [4, 25], the $FeSn_2$ was initially a superparamagnetic singlet and the first discharge converted this to a doublet attributed to superparamagnetic iron metal on the surfaces of the Li-Sn alloy grains. Allowing the system to stand for two weeks converted the doublet to the more conventional superparamagnetic iron singlet with an estimated size of 3 nm. After five cycles, the singlet was principally converted to a sextet at room temperature showing that the particles had grown to approximately 10 nm. By this stage, the iron was no longer taking part in the reactions, and did not form a supporting matrix, so that the cell was behaving essentially as a Li/Sn cell as studied by Chouvin *et al.* [24]. In the ^{119}Sn spectra, the initial $FeSn_2$ exhibited a superparamagnetic doublet and, on discharge, gradually converted to a singlet whose isomer shift indicated it to be heavily lithiated metallic tin. On the first recharge, a singlet of lightly lithiated tin appears, and some reformation of $FeSn_2$. Similar results were observed for the FeSn sample [26], except that in this case the iron formed a skin over the FeSn. It is noteworthy that none of the superparamagnetic nanometre-sized components of iron and tin, nor most of the reformed $FeSn_2$ in these experiments were observable in XRD spectra.

Performance of the cells was improved by ball-milling carbon into the electrodes [27, 28] and forming an inert matrix of $SnFe_3C$ such that the lithium could migrate along the grain boundaries. Cycle life was considerably improved if the voltage was restricted to 0–0.55 V. A recent alternative approach to achieving an inactive matrix and reducing the effects of intergrain cracking was the preparation of ball-milled nanocomposites of Al_2O_3 with $FeSn_2$ [29].

The other tin-based system studied recently by *ex situ* Mössbauer spectroscopy was the sulphide spinel, $Li_yCu_2CoTi_xSn_{3-x}S_8$, which can take up to 12 lithium atoms per formula weight at which stage the spectra had contributions from Sn^{4+}, Sn^{2+} and Sn^0 [30].

4 Spectra under cyclic voltammetry conditions

The long times involved in the discharge and recharge of lithium-ion batteries make it possible to take many spectra during the two parts of the cycle. We have recently developed two spectrometers, which are capable of taking multiple spectra, separated by uniform time intervals, of changes in the state of materials which are varying with time, either cyclically or following some frequent triggering event. The second of these is being used to study pulsed RF collapse of the hyperfine field in soft magnetic materials [31], while the former is used for systems traversing reasonably fast cyclic voltammetry loops. This requires that the redox processes be reproducible over the many days that it takes to collect the multiple spectra (typically eight), which are uniformly arranged around the voltammetry loop. Apart from enabling identification of the species, the system also enables quantitative analysis of the proportion of the material taking part in the reaction. The cycling rate can be chosen independent of the number of spectra taken per cycle, so this enables study of the charge transfer rate.

Prussian Blue is a highly insoluble iron cyanide based complex, which contains high spin ferric iron and low spin ferrocyanide. It has been extensively investigated electrochemically, where due to its insolubility it is usually incorporated into a carbon based electrode, or as a thin film on a working electrode. We have used the cyclic system to study a Prussian Blue-modified carbon electrode [32, 33 and to be published]. Although Prussian Blue has been well-known for over 200 years, its chemistry and crystal structure are still the subject of dispute. Figure 3 shows a set of eight spectra taken by our system as it cycled between 0.25 V and +0.85 V. Strong reduction of the high spin ferric iron was achieved, and the change in the areas of the different components are shown in Figure 4. No change was observed in the Mössbauer parameters of the low spin ferrocyanide site, though this could be either due to the small difference in Mössbauer parameters for ferricyanide and ferrocyanide, or because the applied potential was in insufficient to oxidize this site.

5 Spectra of clay minerals

Mössbauer spectroscopy has been a long-standing method for the study and characterization of clay minerals, particularly regarding the site occupancy of the iron atoms. The octahedral sites can be described as $M(H_2O)_4(OH)_2$, with $M=Fe^{3+}$, Mg^{2+} or Al^{3+}, and there is the possibility of cis and trans coordination of the M ions by the OH^- ions. There has been a long-running dispute in the literature as to whether these cis and trans sites can be

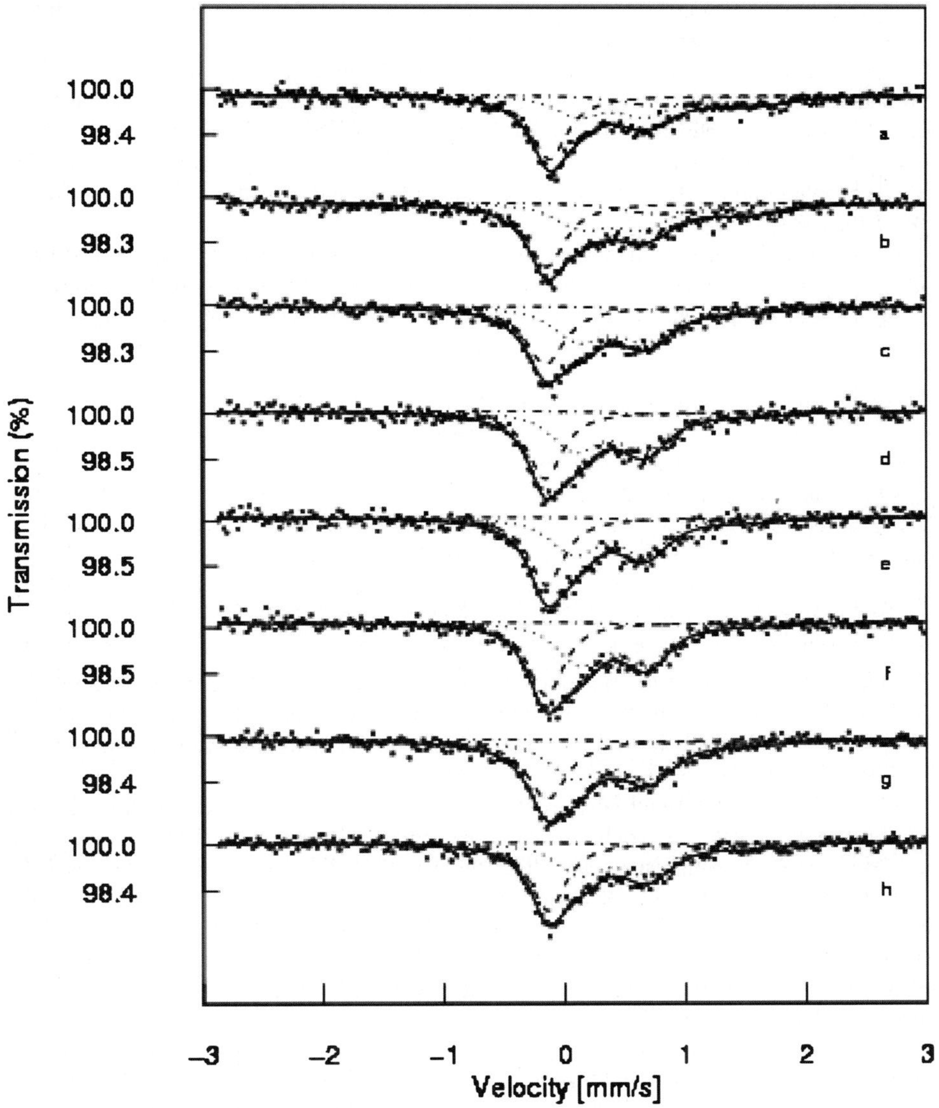

Figure 3 A set of eight spectra taken during continuous cyclic voltammetry of Prussian Blue. The potential ranges are (a) −0.250 to 0.025 V, (b) 0.025 to 0.300 V, (c) 0.300 to 0.575 V, (d) 0.575 to 0.850 V, (e) 0.850 to 0.575 V, (f) 0.575 to 0.300 V, (g) 0.300 to 0.025 V, (h) 0.025 to −0.250 V.

differentiated in the Mössbauer spectra. Fitting of the spectra to two doublets was common for many years, with the doublets frequently being ascribed to the cis and trans sites. However, the legitimacy of this assignment was questioned by Cardile and Johnston [34] and Rancourt [35, 36] who suggested that a larger change in the quadrupole splitting was caused by variations in the local structural and chemical arrangements than is caused by the change from cis to trans in the coordination of the OH groups.

We hope to throw light on this, and other ambiguities which exist in the interpretation of Mössbauer spectra of clay minerals, by taking spectra under either oxidizing or reducing

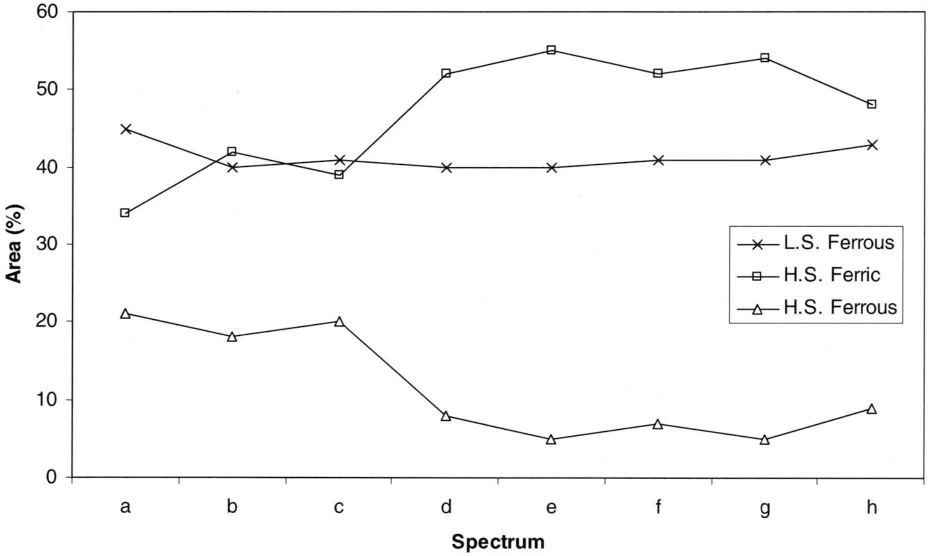

Figure 4 The areas of the Prussian Blue sites in the different potential ranges shown in Figure 3.

conditions. By changing the valency of some of the iron selectively, for example by reduction, we hope to provide information which will link the parameters of the disappearing ferric site with the appearing ferrous site, knowing that the iron cannot move in the process.

Montmorillonite and nontronite are both di-octahedral smectites, with the former being iron-poor and the latter iron-rich. Smectites are quite susceptible to the redox environment and even washing with water can affect the Fe^{2+}/Fe^{3+} ratio of montmorillonites [37]. Furthermore, reoxidation of chemically reduced smectites does not always restore the original state. It was hoped that electrochemical treatment should produce relatively large changes in valence in some samples and that they should be reversible. Electrodes were made of sodium saturated samples of several clay minerals and placed in the *in situ* cell using a NaCl electrolyte.

The cyclic voltammogram of Otay Na^+ montmorillonite is shown in Figure 5, taken at cycling rates of 10 mV/s and 1 mV/s. Clear peaks are observable in both traces, which correspond to the reduction of Fe^{3+} (lower branch) and oxidation of Fe^{2+} (upper branch). The voltage difference between the two peaks is a measure of the reversibility of the reaction and the difference between the traces taken at the two different, but modest rates is indicative of the slow diffusion rate in the sample. Preliminary Mössbauer spectra have been taken of Cloisite Na^+ montmorillonite, where Figure 6a shows the spectrum of the initial sample and Figure 6b shows the spectrum under a reducing potential of approximately −0.6 V. The enhancement of the ferrous doublet by approximately 15% is quite clear, with the parameters agreeing well with the most commonly observed ferrous site in montmorillonite (IS=1.2 mm/s w.r.t. α-Fe at room temperature, QS=2.8 mm/s). Both have been fitted with a distribution of quadrupole splittings.

In contrast, no reduction to ferrous species was observed in the nontronite spectra, but changes in the ferric quadrupole splitting distribution were observed. These appear to be due to a change in the interlayer spacing similar to those observed by Johnston and Cardile

Figure 5 Cyclic voltammograms of Na Otay montmorillonite taken at speeds of (a) 10 mV/s and (b) 1 mV/s.

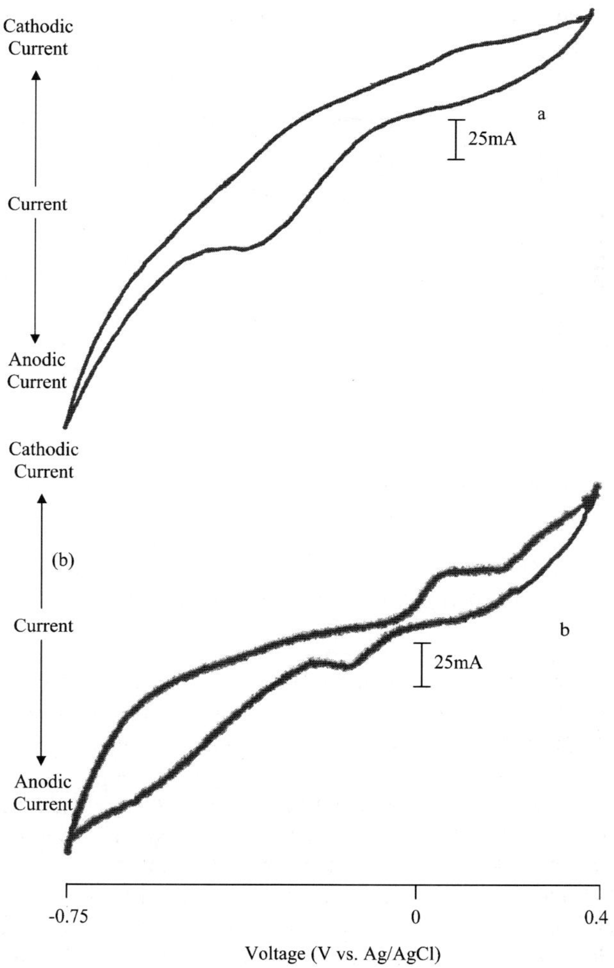

[38]. Mössbauer studies have observed that nontronites are predominately to exclusively trivalent [39], and hence the failure to produce any Fe^{2+} could just be indicative of the strong preference of iron in the nontronites for the trivalent state. However, it should be noted that electrochemical reduction has previously been used to produce appreciable ferrous iron for a nontronite sample [37].

6 Catalysis

The electrochemical reduction of oxygen can be catalysed by materials such as carbon-supported transition metal-N4 macrocyclics, for example, iron porphyrins. These systems have a potential application in direct methanol fuel cells with acid electrolytes. Bouwkamp-Wijnoltz *et al.* [40] have studied the reactions of isotopically-enriched iron tetraphenyl porphyrin (FeTTP) under different conditions of preparation and operation. They found that four doublets were produced, which were difficult to assign and which may contain iron in

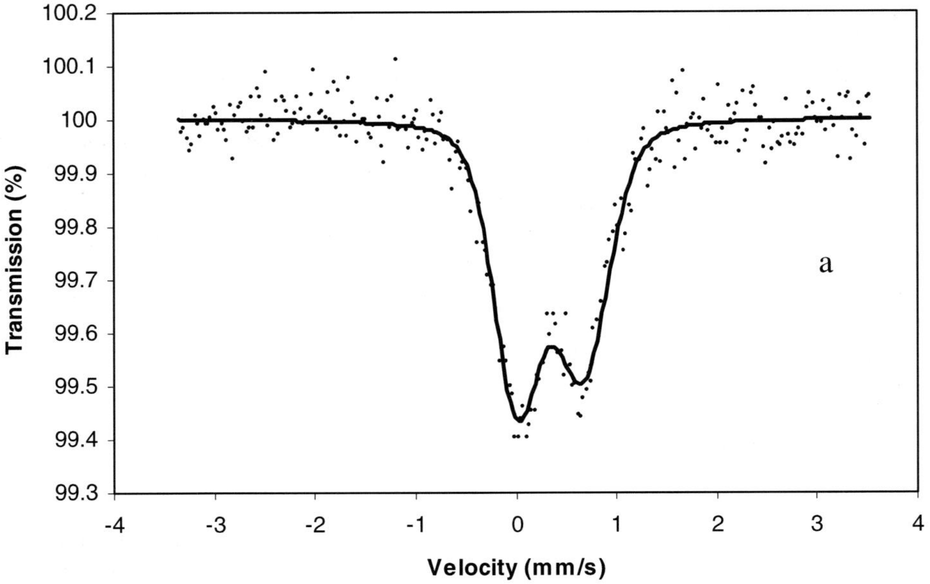

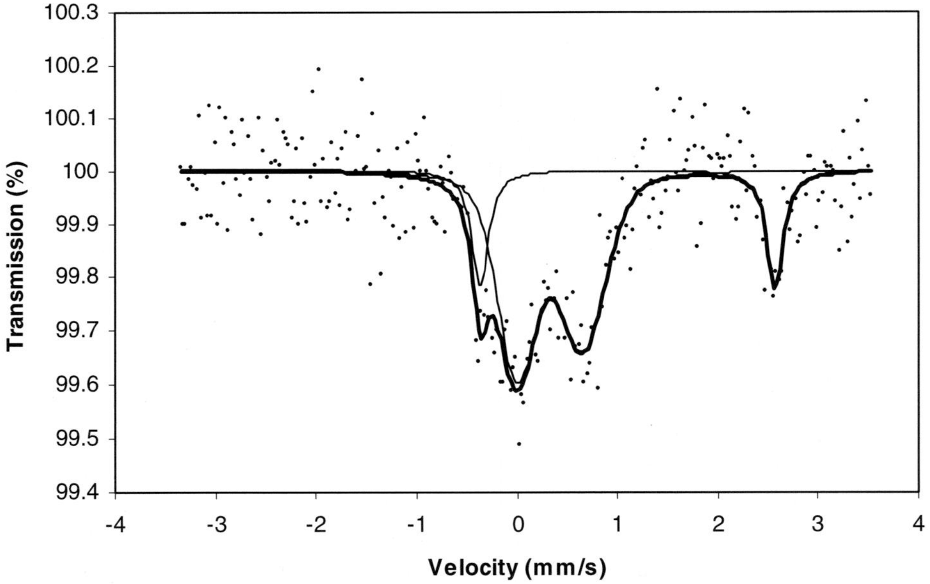

Figure 6 Room temperature Mössbauer spectra of Cloisite montmorillonite taken at (a) zero potential and at (b) −0.6 V, showing the creation of a ferrous doublet.

all the valence states from Fe^{+} to Fe^{4+}. The iron phases were well-defined and approximately two-thirds was electro-active, although it was not clear which were playing a significant role in the reaction of interest since the cyclic voltammetry indicated only about a 10% contribution. The Fe-N4 moiety was still regarded as the most likely usefully active species, but this was not proven.

7 Conclusions

In situ electrochemical Mössbauer studies have been shown to have a useful role to play in many areas, with the development of electrodes for lithium-ion batteries being currently the most important. Its particular niche in analysis is in the observation and identification of phases which do not show up in other techniques. For example, several samples were X-ray amorphous and some had multiple phases, which would not have otherwise been distinguishable. The resolution available for distinguishing phases is quite variable for both ^{57}Fe and ^{119}Sn in the materials studied to date. Although some of the observed subspectra did not necessarily correspond to currently known species, their approximate composition and valence were identifiable and provided important information on the redox processes. A literature search shows that the most attractive, Mössbauer-sensitive material currently being researched, $LiFePO_4$, is rapidly overtaking $LoCoO_2$ as the most studied electrode material. Other recent electrochemical work surveyed included the crystallography of Prussian Blue, the interconversion of ferric and ferrous sites in clay minerals and the electrochemical reduction of oxygen.

Acknowledgements This research has been supported by the Australian Research Council and the Monash Research Fund. We are grateful to W.P. Gates, M. Forsyth and P.C. Howlett for helpful discussions, and to L. Bohni for technical assistance.

References

1. Scherson, D.A., Corrigan, D.A., Fierro, C., Carbonio, R.: In: Electrochemical Surface Science, ACS Symposium Series 378. Amer. Chem. Soc., p. 257. Washington, DC (1988)
2. Ananth, M.V., Parthasaradhy, N.V.: J. Sci. Ind. Res. **47**, 28 (1988)
3. Vertes, A.: Hyp. Int. **45**, 105 (1989)
4. Mao, Ou, Dunlap, R.A., Courtney, I.A., Dahn, J.R.: J. Electrochem. Soc. **145**, 4195 (1998)
5. Obrovac, M.N., Dunlap, R.A., Sanderson, R.J., Dahn, J.R.: J. Electrochem. Soc. **148**, A576 (2001)
6. Tarascon, J.-M., Gozdz, A.S., Schmutz, C., Shokoohi, F., Warren, P.C.: Solid State Ion. **86–88**, 49 (1996)
7. Johnson, C.S., Kropf, A.J.: Electrochim. Acta **47**, 3187 (2002)
8. Whittingham, M.S.: Chem. Rev. **104**, 4271 (2004)
9. Delmas, C., Prado, G., Rougier, A., Suard, E., Fournès, L.: Solid State Ion. **135**, 71 (2000)
10. Mueller-Neuhaus, J.R., Dunlap, R.A., Dahn, J.R.: J. Electrochem. Soc. **147**, 3598 (2000)
11. Shigemura, H., Sakaebe, H., Kageyama, H., Kobayashi, H., West, A.R., Kanno, R., Morimoto, S., Tabuchi, M.: J. Electrochem. Soc. **148**, A730 (2001)
12. Ibarra-Palos, A., Darie, C., Hazemann, J.L., Aldon, L., Jumas, J.C., Morcrette, M., Strobel, P.: Chem. Mater. **14**, 1166 (2002)
13. Denis, S., Dedryvère, R., Baudrin, E., Laruelle, S., Touboul, M., Olivier-Fourcade, J., Jumas, J.C., Tarascon, J.M.: Chem. Mater. **12**, 3733 (2000)
14. Prado, G., Fournès, L., Delmas, C.: J. Solid State Chem. **159**, 103 (2001)
15. Kobayashi, H., Shigemura, H., Tabuchi, M., Sakaeba, H., Ado, K., Kageyama, H., Hirano, A., Kanno, R., Wakita, M., Morimoto, S., Nasu, S.: J. Electrochem. Soc. **147**, 960 (2000)
16. Takada, K., Kitami, Y., Inada, T., Kajiyama, A., Kouguchi, M., Kondo, S., Watanabe, M., Tabuchi, M.: J. Electrochem. Soc. **148**, A1085 (2001)
17. Tabuchi, M., Nakashima, A., Shigemura, H., Ado, K., Kobayashi, H., Sakaebe, H., Kageyama, H., Nakamura, T., Kohzaki, M., Hirano, A., Kanno, R.: J. Electrochem. Soc. **149**, A509 (2002)
18. Alcántara, R., Jaraba, M., Lavela, P., Tirado, J.L., Dumas, J.C., Olivier-Fourcade, J.: Electrochem. Commun. **5**, 16 (2003)
19. Zhu, S., Zhou, H., Miyoshi, T., Hibino, M., Honma, I., Ichihara, M.: Adv. Mater. **16**, 2012 (2004)
20. Kajiyama, A., Takada, K., Inada, T., Kouguchi, M., Kondo, S., Watanabe, M., Tabuchi, M.: Solid State Ion. **152–153**, 295 (2002)
21. Herle, P.S., Ellis, B., Coombs, C., Nazar, L.F.: Nat. Materials **3**, 147 (2004)
22. Fuji Photo Film.: European patent, EP 0 704 921 A1 (1995)

23. Dunlap, R.A., Small, D.A., MacNeil, D.D., Obrovac, M.N., Dahn, J.R.: J. Alloys Compd. **289**, 135 (1999)
24. Chouvin, J., Olivier-Fourcade, J., Dumas, J.C., Simon, B., Godiveau, O.: Chem. Phys. Lett. **308**, 413 (1999)
25. Dunlap, R.A., Mao, Ou, Dahn, J.R.: Phys. Rev. **B59**, 3494 (1999)
26. Mao, Ou, Dunlap, R.A., Dahn, J.R.: Solid State Ion. **118**, 99 (1999)
27. Beaulieu, L.Y., Larcher, D., Dunlap, R.A., Dahn, J.R.: J. Electrochem. Soc. **147**, 3206 (2000)
28. Mao, Ou, Dunlap, R.A., Dahn, J.R.: J. Electrochem. Soc. **146**, 405, 414, 423 (1999)
29. Dunlap, R.A., Kehoe, S.W., Dahn, J.R., O'Brien, J.W.: J. Alloys Compd. **400**(1–2), 252 (2005)
30. Lavela, P., Pérez-Vicente, C., Tirado, J.L., Branci, C., Olivier-Fourcade, J., Jumas, J.C.: Chem. Mater. **11**, 2687 (1999)
31. Davis, J.A., Cashion, J.D., Kopcewicz, M.: Industrial Applications of the Mössbauer Effect, AIP Conf. Proc., vol. 765, Amer. Inst. Phys. 185 (2005) (New York)
32. Greaves, T.L.: PhD thesis, Monash Univ., (2004)
33. Greaves, T.L., Cashion, J.D.: Proc. 27th Annual A&NZIP Cond. Matter and Mater. Meeting, Wagga Wagga, http://www.aip.org.au/wagga_2003/ww03_32 (2003)
34. Cardile, C.M., Johnston, J.H.: Clays Clay Miner. **34**, 307 (1986)
35. Rancourt, D.G.: Phys. Chem. Miner. **21**, 250 (1994)
36. Rancourt, D.G.: Hyp. Int. **117**, 3 (1998)
37. Rozenson, I., Heller-Kallai, L.: Clays Clay Miner. **26**, 88 (1978)
38. Johnston, J.H., Cardile, C.M.: Clays Clay Miner. **33**, 21 (1985)
39. Murad, E., Cashion, J.D.: Mössbauer Spectroscopy of Environmental Materials and their Industrial Utilization, p. 152. Kluwer, Boston (2004)
40. Bouwkamp-Wijnoltz, A.L., Visscher, W., van Veen, J.A.R., Boellard, E., van der Kraan, A.M., Tang, S. C.: J. Phys. Chem. B **106**, 12993 (2002)

Hyperfine Interact (2005) 165: 17–24
DOI 10.1007/s10751-006-9242-4

Dynamics of metal centers monitored by nuclear inelastic scattering

H. Paulsen · P. Wegner · H. Winkler · J. A. Wolny · L. H. Böttger · A. X. Trautwein · C. Schmidt · V. Schünemann · G. Barone · A. Silvestri · G. La Manna · A. I. Chumakov · I. Sergueev · R. Rüffer

Published online: 26 October 2006

Abstract Nuclear inelastic scattering of synchrotron radiation has been used now since 10 years as a tool for vibrational spectroscopy. This method has turned out especially useful in case of large molecules that contain a Mössbauer active metal center. Recent applications to iron–sulfur proteins, to iron(II) spin crossover complexes and to tin-DNA complexes are discussed. Special emphasis is given to the combination of nuclear inelastic scattering and density functional calculations.

Key words nuclear inelastic scattering · vibrational spectroscopy · iron–sulfur proteins · spin crossover complexes · density functional theory calculations

H. Paulsen (✉) · P. Wegner · H. Winkler · J. A. Wolny · L. H. Böttger · A. X. Trautwein
Institut für Physik, Universität zu Lübeck, D-23538 Lübeck, Germany
e-mail: paulsen@physik.uni-luebeck.de

C. Schmidt
Institut für Biochemie, Universität zu Lübeck, D-23538 Lübeck, Germany

V. Schünemann
Fachbereich Physik, Technische Universität Kaiserslautern, D-67663 Kaiserslautern, Germany

G. Barone · A. Silvestri
Dipartimento di Chimica Inorganica e Analitica, Universitá degli Studi di Palermo, I-90128 Palermo, Italy

G. La Manna
Dipartimento di Chimica Fisica, Universitá degli Studi di Palermo, I-90128 Palermo, Italy

A. I. Chumakov · I. Sergueev · R. Rüffer
ESRF, F-38043 Grenoble Cedex 9, France

Abbreviations

NIS	nuclear inelastic scattering
SCO	spin crossover
DFT	density functional theory
EXAFS	extended X-ray absorption fine structure
msd	mean square displacement
HS	high-spin
LS	low-spin

1 Introduction

Nuclear inelastic scattering (NIS) of synchrotron radiation (also known as nuclear resonant vibrational spectroscopy, NRVS) can be regarded as an extension of the conventional, energy-resolved Mössbauer spectroscopy to energies of the order of molecular vibrations [1, 2]. NIS has similarity with resonance Raman spectroscopy: the energy of monochromatized synchrotron radiation is detuned from nuclear resonance by a certain amount of energy (several meV) that is sufficient to create or annihilate a phonon, or in terms of molecular spectroscopy, to excite or deexcite a molecular vibration. While the Raman intensity for a given molecular vibration depends on the change of polarizability with respect to the motion of the nuclei, the NIS intensity depends on the contribution of this particular vibration to the mean-square displacement (msd) of the Mössbauer nucleus. NIS is, therefore, a highly site-selective tool and it is as vibrational spectroscopy complementary to Raman or IR spectroscopy. The interpretation of measured NIS spectra is greatly facilitated due to the fact that the NIS intensity depends only on mechanical and not on electric properties of the investigated molecules. For example, iron-ligand bond stretching vibrations of an iron-containing molecule are normally easy to identify because of their relatively large contribution to the msd. The dependence on purely mechanical properties also facilitates the simulation of measured NIS spectra with the help of quantum chemical calculations, which are usually based on density functional theory (DFT). The calculation of IR or Raman intensities, which depend on electric properties like changes of dipole moment or polarizability tensor, is far less accurate. A quantitative analysis of this fact with the example of the ferrocene complex is given in Section 2.

The comparison of experimental and simulated spectra provides the possibility to assign peaks in the measured NIS spectrum with calculated normal modes of the molecule under study. This procedure has been applied first to small and medium-sized metal complexes (up to 100 atoms) [3, 4] but is now applied also to larger biological molecules like rubredoxin [5] or DNA with chelated metal ions [6] (in the latter two cases DFT calculations are restricted to a cluster of about 100 atoms around the iron center). First applications of NIS to problems of biological relevance were related to model hemes like tetraphenylporphyrins [7, 8] and to heme-based proteins like myoglobin [7–9] (see also references in [10]). Recently NIS measurements have been performed on model complexes of iron–sulfur proteins [11], on a rubredoxin-type iron–sulfur protein in the reduced and in the oxidized state [5], and on Benzoyl-CoA reductase [5], which contains $[4Fe–4S]^{2+}$ centers. The NIS spectra for the rubredoxin-type protein allowed for the first time to investigate the change of the metal dynamics between the reduced and the oxidized form (see also Sections 3.1 and 3.3).

Another class of metal-containing molecules, which is being studied by NIS, is formed by the so called spin crossover (SCO) complexes. For these molecules NIS has turned out to be a very valuable method, since the dynamics of the metal center governs the crossover from the low-spin to the high-spin isomer (Section 3.4).

2 Density functional theory calculations

A detailed analysis of the dynamics of metal centers in large molecules requires combination of experimental investigations and theoretical calculations. In case of molecules that contain transition metals, density functional theory (DFT) calculations are currently the method of choice. They have been shown to reproduce the frequencies and vibrational modes of transition metal compounds with reasonable accuracy and can be applied also to large molecules containing hundreds of atoms. Frequencies and vibrational modes are used to simulate a NIS spectrum for comparison with experiment [3, 7]. Additionally, DFT calculations provide the partial vibrational density of states (PVDOS) for the Mössbauer nucleus which can be compared with a PVDOS that has been extracted from the measured spectrum. A detailed description how the calculated vibrational modes can be used to simulate NIS spectra can be found for instance in [3, 7].

In a low-temperature approximation the absorption probability S that is measured in a NIS experiment is proportional to the spectral contribution to the mean-square displacement (msd) of the Mössbauer nucleus:

$$S(E,\mathbf{k}) \propto \sum_i \delta(E - h\nu_i)\left\langle(\mathbf{k}\cdot\mathbf{u}_i)^2\right\rangle \tag{1}$$

Here E and $\mathbf{k}$ denote energy and wavevector, respectively, of the absorbed photon, index i enumerates calculated normal modes of vibrations. ν_i is the frequency of mode i and the vector $\mathbf{u}_i$ contains those three components of the normal coordinate vector that refer to the motion of the Mössbauer nucleus in x, y and z direction. The sum over all terms $\left\langle(\mathbf{k}\cdot\mathbf{u}_i)^2\right\rangle$ yields the total measured msd of the Mössbauer nucleus. In order to simulate the NIS spectrum of a molecule it is, therefore, sufficient to know the frequencies ν_i and to which degree the Mössbauer nucleus is participating in the individual vibrations. The latter information is given by the vectors $\mathbf{u}_i$. The frequencies and normal modes of molecular vibrations may be regarded as mechanical properties of the molecule, although they are based on the molecule's electronic structure.

With the example of the ferrocene molecule we demonstrate here that these mechanical properties are more easily calculated with sufficient accuracy than those electric properties (changes of dipole moment or polarizability tensor) that are needed for the simulation of IR or Raman spectra. The dynamics of the iron center in ferrocene has been studied intensively by various spectroscopic techniques including NIS [12] and quasielastic nuclear forward scattering [13]. We have performed DFT calculations using the hybrid functional B3LYP [14] together with five different basis sets, 3-21G [15–20], CEP-31G [21–23], 6-311G [24, 25] (with the Wachters–

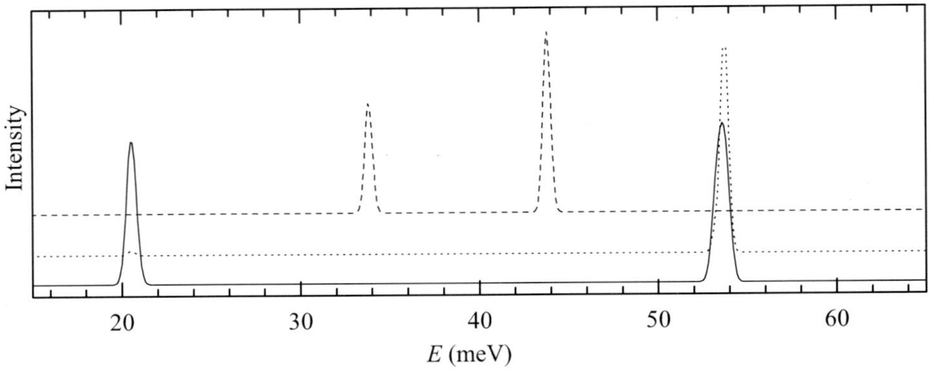

Figure 1 NIS, IR and Raman intensities (*solid, dotted* and *dashed line,* respectively) of ferrocene in arbitrary units calculated with B3LYP/TZVP.

Hay double zeta basis for Fe [26, 27]), TZV and TZVP [28, 29] as included in the Gaussian 03 program package [30]. The simulated spectra (Figure 1) are in reasonable agreement with the experimental spectra given in [12]. The comparison of calculated IR, Raman and NIS spectra in the region of metal-ligand vibrations exemplifies the complementarity of NIS and Raman spectra. In the region from 0 to 60 meV nine vibrational modes are obtained from the calculation. Modes 1 and 7 are practically invisible for all three spectroscopic techniques, modes 2 and 3, modes 5 and 6 and modes 8 and 9 are in pairs nearly degenerate. In this region mode 4 has the strongest Raman intensity, while the almost degenerate pair of modes 8 and 9 exhibits the largest IR and NIS intensities. We have examined the influence of the basis set on the calculated IR, Raman and NIS intensities. In Figure 2 the basis sets are ordered on the horizontal axis with increasing quality. It turns out that the calculated NIS intensity is almost independent of the employed basis set whereas the calculated IR and Raman intensities exhibit large oscillations for small and medium sized basis sets. This result is even more striking if one takes into account that, according to Lambert–Beer law, there is an exponential relation between the calculated IR and Raman activities and the measured spectral intensities. In summary, a reasonable theoretical simulation of IR and, especially, of Raman spectra is computational much more demanding than a simulation of NIS spectra. This is an important aspect if large molecules are investigated where demanding computational methods are prohibitive.

3 Recent examples

3.1 Rubredoxin-type iron–sulfur proteins

Iron–sulfur proteins are widely used for electron transport in cells. For a deeper understanding of the reorganisation of these proteins upon electron uptake or release the knowledge of the vibrational modes of the iron–sulfur core is of special importance. For the oxidized and reduced forms of a rubredoxin-type iron–sulfur protein from *Pyrococcus abyssi*, which contains a single Fe–S_4 core [31], we have measured NIS spectra at the nuclear resonance beamline ID 18 of ESRF in Grenoble,

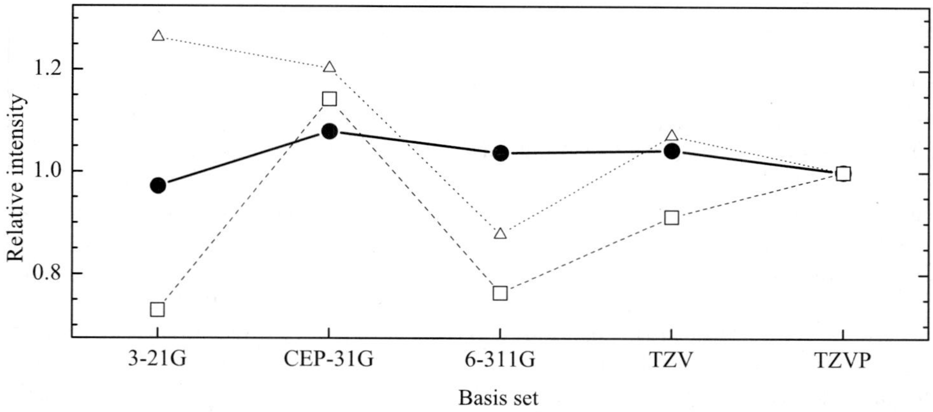

Figure 2 Relative NIS (•), IR (□) and Raman (△) intensities of mode 4 (Raman) and the nearly degenerate modes 8 and 9 (IR, NIS) of ferrocene (see text) calculated with B3LYP and different basis sets as indicated. For each spectroscopic technique the calculated intensities are divided by the value obtained with the TZVP basis set. The *lines* serve as guides to the eyes.

France. In the NIS spectrum of the oxidized form Fe–S bond-stretching modes could be identified which are in agreement with resonance Raman studies. The spectrum of the reduced form, where no Raman data are available, shows a down shift of these bond-stretching modes by about 5 meV. This observation is in line with an increase of the Fe–S bond length upon reduction as observed by EXAFS. In addition to the peaks that are assigned to the Fe–S bond stretching modes, a second group of prominent peaks was observed in the NIS spectra, which hardly change their position upon reduction. In order to assign these peaks to molecular vibrations, DFT calculations have been performed for a series of models of the protein under study, $Fe(SH)_4$, $Fe(SCH_3)_4$ and $FeCys_4$, i.e., containing 9, 21 and 49 atoms, respectively. The simulated NIS spectra for these three models gain approach to the experimental spectra (oxidized and reduced) with growing model size and allow to assign the observed second group of peaks to Fe–S–C bending modes. The most simple model, $Fe(SH)_4$, although far less realistic than the larger models, is nevertheless useful since it provides a qualitatively correct picture of the experimentally observed vibrations of the FeS_4 core. More details of this study are described in [5].

3.2 Iron corroles

Corroles are aromatic macrocycles that are closely related to porphyrins and have unique chemical properties such as stabilizing high oxidation states of coordinated metal ions, e.g., Fe(IV) [32]. We have performed NIS measurements on iron corroles with two different axial ligands at the nuclear resonance beamline ID 18 of ESRF. Currently, a detailed analysis of the observed molecular vibrations of these molecules is performed that exploits the complementarity of IR, Raman and NIS measurements in combination with DFT calculations [33].

3.3 Tin-DNA complexes

NIS is a highly site-selective tool, focussed on Mössbauer nuclides. Therefore, most applications of NIS in biology and bioinorganic chemistry have so far been performed with molecules containing iron centers. Recently, we have investigated tin-DNA complexes at the nuclear resonance beamline ID 22 of ESRF; this is, to our knowledge, the first example of a NIS study on a biological sample that contains a Mössbauer nuclide other than ^{57}Fe. This study has been performed to elucidate the binding site of organotin cations, Sn(II) and Sn(IV), chelated to DNA. The binding site of such tin-DNA complexes could not be determined by conventional structural methods like X-ray diffraction. Spectroscopic techniques like IR or Raman are hampered by the complexity of the spectra which are dominated by the vibrations of the large organic ligands. The site-selectivity of NIS focusses on those vibrations that are connected with a considerable msd of the tin nucleus. Our NIS investigation in combination with DFT calculations have corroborated the former hypothesis about the binding sites of tin that was based on conventional Mössbauer studies and semiempirical molecular-orbital calculations [34]. More details can be found in [6, 35].

3.4 Spin crossover complexes

These molecules exhibit a transition from a low-spin (LS) to a high-spin (HS) isomer that is induced by change of temperature or pressure or by irradiation with light [36]. Spin crossover complexes are not only very promising materials for technical applications, they also serve as models for the transition of the metal spin in biomolecules, observed for instance in a catalytic cycle. The driving force for the spin transition is the entropy difference between the two spin isomers. The major contribution to this entropy difference arises from the dynamics of the metal ion: in the LS state the *d* electrons of the metal center occupy non-bonding or weakly π-anti-bonding molecular orbitals. Upon transition to the HS isomer part of the *d* electrons are transferred to σ-anti-bonding molecular orbitals. Consequently, in the HS isomer the metal-ligand bonds are weaker and longer and also softer, i.e., the force constants of the metal-ligand bond stretching vibrations are lower. This leads to a significant down-shift of the corresponding frequencies upon crossover from LS to the HS state (about 10–20 meV in case of typical Fe(II) SCO complexes). For this reason, the HS isomer has – in comparison to the LS isomer – many more vibrational states that are thermally accessible, and the vibrational entropy contribution of the HS isomer is significantly higher than the corresponding value of the LS isomer (some 10 J mol^{-1} K^{-1} in case of Fe(II) SCO complexes). First NIS experiments have been performed on samples of $[Fe(tpa)(NCS)_2]$ (tpa = tris(2-pyridylmethyl)amine) and $[Fe(bpp)_2](BF_4)_2$ (bpp = 2,6-bis(pyrazol-3-yl)pyridine) [3, 37, 38]. Subsequent measurements on a single crystalline sample of $[Fe(tptMetame)](ClO_4)_2$ (tptMetame = 1,1,1-tris((N-(2-pyridylmethyl)-N-methylamino)methyl)ethane) yielded angular resolved NIS spectra [4]. With the same sample the influence of a pressure induced SCO on the metal dynamics has been investigated with NIS [39]. Studying $[Fe(phen)_2(NCS)_2]$ (phen = 1,10-Phenanthroline) the complementarity of NIS and IR and Raman spectroscopy has been demonstrated. Combining these three vibrational spectroscopic techniques with DFT calculations allows an almost complete assignment of the observed vibrational modes [40]. One of the most recent NIS experiments on SCO complexes [41] has been used to investigate five different phases of the complex $[Fe(ptz)_6](BF_4)_2$ (ptz = 1-*n*-Propyl-tetrazole) that can be reached by combination of temperature changes and irradiation with light of different frequencies [42].

4 Conclusion

In summary we conclude that the comparison of complementary experimental (i.e., NIS, IR, Raman) and theoretical spectra provides the possibility to assign measured peaks with calculated normal modes of the metal complex under study. This procedure has been applied to small and medium-sized iron complexes and has recently been applied also to iron-containing proteins and to DNA being chelated with tin.

Acknowledgement Financial support by the Deutsche Forschungsgemeinschaft (DFG) is gratefully acknowledged (Tr 97/31 and Tr 97/32).

References

1. Seto, M., Yoda, Y., Kikuta, S., Zhang, X., Ando, M.: Phys. Rev. Lett. **74**, 3828–3831 (1995)
2. Stuhrhahn, W., Toellner, T., Alp, E.E., Zhang, X., Ando, M., Yoda, Y., Kikuta, S., Seto, M., Kimball, C., Dabrowski, B.: Phys. Rev. Lett. **74**, 3832–3835 (1995)
3. Paulsen, H., Winkler, H., Trautwein, A.X., Grünsteudel, H., Rusanov, V., Toftlund, H.: Phys. Rev. B **59**, 975–984 (1999)
4. Paulsen, H., Benda, R., Herta, C., Schünemann, V., Chumakov, A.I., Duelund, L., Winkler, H., Toftlund, H., Trautwein, A.X.: Phys. Rev. Lett. **86**, 1351–1354 (2001)
5. Trautwein, A.X., Schünemann, V., Wegner, P., Winkler, H., Paulsen, H., Schmidt, C., Chumakov, A.I.: Hyp. Interact (in press)
6. Barone, G., Silvestri, A., Böttger, L.H., Wolny, J.A., Paulsen, H., Sergueev, I., Rüffer, R., Trautwein, A.X., La Manna, G.: Hyp. Interact (in press)
7. Sage, J.T., Paxson, C., Wyllie, G.R., Stuhrhahn, W., Durbin, S.M., Champion, P., Alp, E.E., Scheidt, W.R.: J. Phys.: Condens. Matter **13**, 7707–7722 (2001)
8. Leu, B.M., Zgierski, M.Z., Wyllie, G.R., Scheidt, W.R., Stuhrhahn, W., Alp, E.E., Durbin, S.M., Sage, J.T.: J. Am. Chem. Soc. **126**, 4211–4227 (2004)
9. Achterhold, K., Keppler, C., van Bürck, U., Potzel, W., Schindelmann, P., Knapp, E.-W., Melchers, B., Chumakov, A.I., Baron, A., Rüffer, R., et al.: Biophys. Lett. **25**, 43–46 (1996)
10. Paulsen, H., Schünemann, V., Trautwein, A.X., Winkler, H.: Coord. Chem. Rev. **249**, 255–272 (2005)
11. Oganesyan, V.S., Barclay, J.E., Hardy, S.M., Evans, D.J., Pickett, C.J., Jayasooriya, U.A.: Chem. Commun. **2**, 214–215 (2004)
12. Jayasooriya, U.A., Malone, S.A., Chumakov, A.I., Rüffer, R., Overweg, A.R., Nicklin, C.R.: Chem. Phys. Chem. **3**, 177–180 (2001)
13. Asthalter, T., Franz, H., van Bürck, U., Messel, K., Schreier, E., Dinnebier, R.: J. Phys. Chem. Solids **64**, 677 (2003)
14. Becke, A.: J. Chem. Phys. **98**, 5648 (1993)
15. Binkley, J., Pople, J., Hehre, W.: J. Am. Chem. Soc. **102**, 939 (1980)
16. Gordon, M., Binkley, J., Pople, J., Pietro, W., Hehre, W.: J. Am. Chem. Soc. **104**, 2797 (1982)
17. Pietro, W., Francl, M., Hehre, W., Defrees, D., Pople, J., Binkley, J.: J. Am. Chem. Soc. **104**, 5039 (1982)
18. Dobbs, K., Hehre, W.: J. Chem. Phys. **7**, 359 (1986)
19. Dobbs, K., Hehre, W.: J. Chem. Phys. **8**, 861 (1987)
20. Dobbs, K., Hehre, W.: J. Chem. Phys. **8**, 880 (1987)
21. Stevens, W., Basch, H., Krauss, J.: J. Chem. Phys. **81**, 6026 (1984)
22. Stevens, W., Krauss, M., Basch, H., Jasien, P.: Can. J. Chem. **70**, 612 (1992)

23. Cundari, T., Stevens, W.: J. Chem. Phys. **98**, 5555 (1993)
24. McLean, A., Chandler, G.: J. Chem. Phys. **72**, 5639 (1980)
25. Krishnan, R., Binkley, J., Seeger, R., Pople, J.: J. Chem. Phys. **72**, 650 (1980)
26. Wachters, A.: J. Chem. Phys. **52**, 1033 (1970)
27. Hay, P.: J. Chem. Phys. **66**, 4377 (1977)
28. Schaefer, A., Horn, H., Ahlrichs, R.: J. Chem. Phys. **97**, 2571 (1992)
29. Schaefer, A., Huber, C., Ahlrichs, R.: J. Chem. Phys. **100**, 5829 (1994)
30. Frisch, M.J., Trucks, G.W., Schlegel, H.B., Scuseria, G.E., Robb, M.A., Cheeseman, J.R., Montgomery, Jr. J.A., Vreven, T., Kudin, K.N., Burant, J.C. et al.: Gaussian 03, Revision C.02, Gaussian, Wallingford, CT (2004)
31. Czernuszewicz, R., LeGall, J., Moura, I., Spiro, T.G.: Inorg. Chem. **25**, 696 (1986)
32. Zakharieva, O., Schünemann, V., Gerdan, M., Licoccia, S., Cai, S., Walker, F.A., Trautwein, A.X.: J. Am. Chem. Soc. **124**, 6636–6648 (2002)
33. Tuczek, F., Schünemann, V., Paulsen, H., Licoccia, S., Walker, F.A., Trautwein, A.X., Zakharieva, O., Veeger, C.: J. Mol. Struct. Theochem **723**, 171–182 (2005)
34. Barone, G., Ramusino, M.C., Barbieri, R., La Manna, G.: J. Mol. Struct. Theochem **469**, 143–149 (1999)
35. Barone, G., Böttger, L.H., Wolny, J.A., Paulsen, H., Sergueev, I., Rüffer, R., Silvestri, A., Trautwein, A.X., La Manna, G.: (to be published)
36. Gütlich, P., Goodwin, H.A. (eds.): Spin Crossover in Transition Metal Compounds I–III, Topics in Current Chemistry, 1st edition. Springer, Berlin Heidelberg New York (2004)
37. Paulsen, H., Grünsteudel, H., Meyer-Klaucke, W., Gerdan, M., Winkler, H., Toftlund, H., Trautwein, A.X.: Eur. Phys. J. B **23**, 463–472 (2001)
38. Chumakov, A.I., Rüffer, R., Leupold, O., Grünsteudel, H., Barla, A., Asthalter, T. In: ESRF Highlights 1999. ESRF, Grenoble, p. 50 (2000)
39. Averseng, F., Giefers, H., Leupold, O., Paulsen, H., Rüffer, R., Toftlund, H., Trautwein, A.X., Wegner, P., Wolny, J.A., Wortmann, G.: (to be published)
40. Ronayne, K.L., Paulsen, H., Höfer, A., Dennis, A.C., Wolny, J.A., Chumakov, A.I., Schünemann, V., Winkler, H., Spiering, H., Bousseksou, A., et al.: Phys. Chem. Chem. Phys., (2006) DOI:10.1039/6610634j
41. Böttger, L.H., Chumakov, A.I., Grunert, C.M., Gütlich, P., Kusz, J., Paulsen, H., Ponkratz, U., Rusanov, V., Trautwein, A.X., Wolny, J.: Chem. Phys. Lett. **429**, 189–193 DOI:10.10161j.cplett.2006.08.004
42. Kusz, J., Gütlich, P., Spiering, H.: In: Spin Crossover in Transition Metal Compounds II, pp. 129-153. Springer, Berlin Heidelberg New York (2004)

Hyperfine Interact (2005) 165: 25–35
DOI 10.1007/s10751-006-9243-3

Critical sizes and characteristics of nanoclusters and nanostructures

I. P. Suzdalev · Yu. V. Maksimov

Published online: 26 October 2006

Abstract Critical sizes and characteristics of nanoclusters and nanostructures are introduced as parameters of nanosystems. The following critical characteristics are under consideration: atomic and electronic magic numbers, critical size of cluster nucleation, critical size of melting–freezing of cluster, critical size of the first order magnetic phase transition. The critical characteristics are treated in terms of thermodynamics and tested by Mössbauer spectroscopy, Atomic Force Microscopy, heat capacity and other methods.

Key words nanocluster · nanostructure · critical characteristics · "magic numbers" · atomic motion · melting points · magnetic phase transitions · Mössbauer spectroscopy · AFM

1 Introduction

Nano-scale clusters and nanostructures possess the structure and properties that strongly differ from the bulk materials and single molecules. They show size dependent structure, stability, atomic dynamic, electronic, optic and magnetic effects [1]. New peculiarities of nanosystem arise when isolated clusters are assembled to form a large nanostructure. The properties of isolated nanoclusters and nanostructures can be described in terms of critical sizes and characteristics. These critical parameters are determined by the conditions of cluster formation and nucleation, cluster structure, cluster melting, cluster atomic dynamics, cluster nonlinear optical effects, single electron conductivity, magnetic transitions etc. For isolated and interacted clusters these properties can be treated in terms of structure, electronic, quantum and thermodynamic models [2].

This paper was supported by RFBR (06-03 32006) and INTAS (01-204).

I. P. Suzdalev (✉) · Y. V. Maksimov
Semenov Institute of Chemical Physics, Russian Academy of Sciences,
ul. Kosygina 4, Moscow 119991, Russia
e-mail: suzdalev@chph.ras.ru

This paper deals with the review and study of the following critical parameters: (1) “Magic numbers” for critical cluster sizes. (2) Critical parameters for cluster nucleation. (3) Critical characteristic of cluster atomic dynamics. (4) Critical characteristics of magnetic phase transitions.

2 Atomic “magic numbers” for critical cluster sizes

“Magic numbers” for most stable composition of molecular clusters, noble gas clusters, carbon clusters have been introduced in [3]. Molecular clusters consist of metal or metal oxide core stabilized by the ligand shell. Most stable core structure is governed by the layer rule [4]

$$N = 1/3\left(10n^3 + 15n^2 + 11n + 3\right), \tag{1}$$

where N is the number of metal atoms in the core, n is the number of layers. In accordance with (1) the most stable metallic core of molecular cluster should have N=13, 55, 147, 309, 561, 923, 1415, 2057, 2869, while the critical cluster size will change within ~0.5–5 nm. The part of metallic atoms is connected with ligands, for example, phenantrolins, phosfins, carbonyls, etc.

In palladium cluster $Pd_{561}phen_{60}$ $(OAc)_{180}$, surface part of metallic core is stabilized by ligands and therefore has chemical (covalent or coordination) bonds, but inner part of core possesses metallic bonds and provides most density packing – icosahedron [5] (Figure 1). Full icosahedron gives maximum bonds on atoms, maximal bond energy and maximum stability of cluster.

For noble gas clusters produced in atomic and molecular beams of supersonic jets the expression (1) for “magic number” is also valid. They are “naked” clusters and icosahedron atomic packing with van der Waals’ forces provides its stability.

Carbon cluster also have critical sizes and “magic number” of carbon atoms. There are two different regions of stability – the small carbon clusters C_n, with $n \leq 24$ and the large carbon clusters, with $n \geq 24$, known as the fullerenes [6]. Small carbon clusters are formed by high frequency discharge or by laser vaporization and display “magic numbers” (Figure 2). For fullerene an even number of carbon is more favorable for maximum stability. Besides very famous C_{60}, a lot of fullerenes, for example C_{70} and up to C_{180}, C_{240}, C_{540}, were observed and predicted. Contrary to the previous two cases, the most important reasons for formation of “magic numbers” and critical sizes for carbon clusters are strong covalent bond and sp, sp^2 hybridization for one- and two-dimensional carbon structures.

Alkali clusters with free metal s-electrons also possess “magic numbers” and critical sizes. As a factor of stability the ionization potential (IP) can be used (Figure 3). A maximum IP for clusters Na_n, K_n and also for Ag_n exists [7]. “Magic numbers” and cluster critical sizes can be understood in terms of the shell structure or jellylike model. In this model the cluster core (positive ions) forms a classical liquid medium, while the electrons at the same temperature form quantized medium in its ground state. Similar to atomic shell structure, cluster electronic shell is formed from the free s-electrons of every atom-entering cluster. As a result, the rule for “magic numbers” takes a view

$$n = 2(2l + 1), \tag{2}$$

where l is the orbital momentum. Thus, for completed electronic shell the “magic numbers” are: n=2, 8, 20, 40, 70 etc., providing critical sizes for clusters Na_n, K_n and Ag_n.

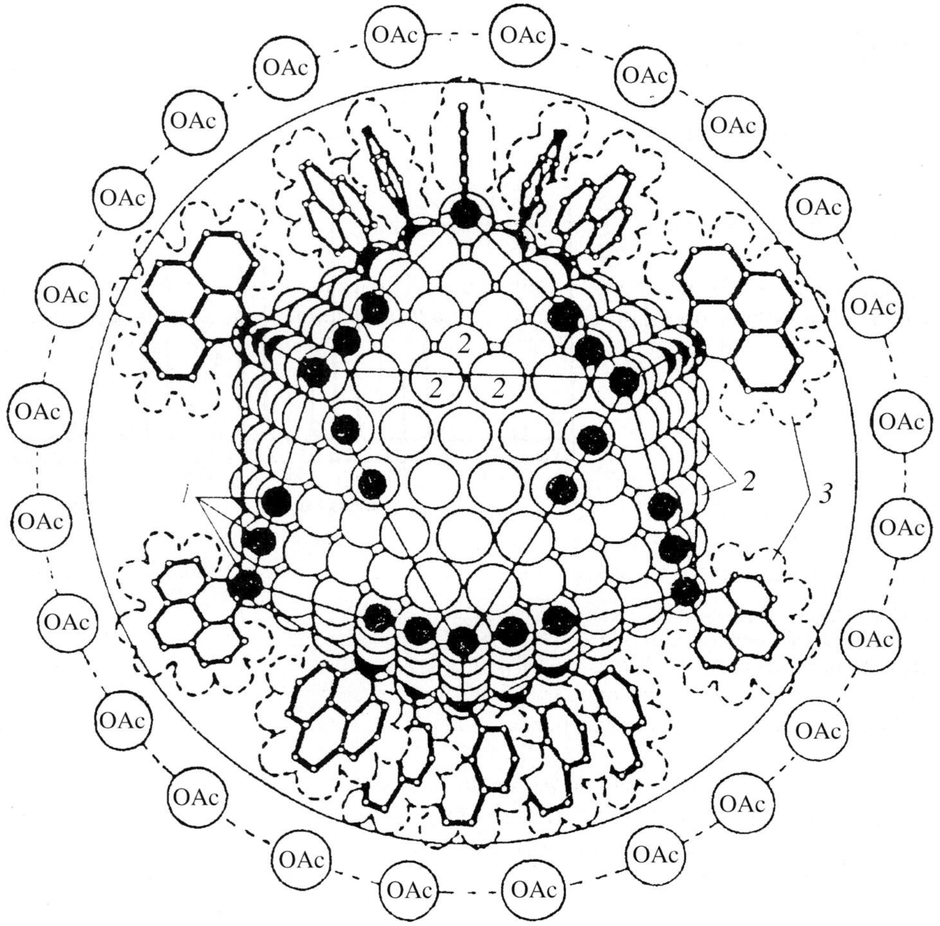

Figure 1 Idealized structure of Pd cluster $Pd_{561}phen_{60}\,(OAc)_{180}$.

3 Critical parameters for cluster nucleation

Cluster nucleation can take place in gas, liquid and solid media [8]. The existence of critical cluster sizes for cluster nucleation and cluster sintering have been predicted and proved experimentally on iron oxide nanoclusters [9]. The formation of clusters from solution in a single pore under weak cluster–matrix interaction can be treated as [9]

$$\Delta G = \alpha S - [\Delta\mu + \ln{(n_0 - n)}/N]\rho V, \tag{3}$$

where ΔG is the change in the Gibbs' free energy, α is the density of the surface energy of the cluster (α=const), ρ is the substance density in the cluster, $\Delta\mu$ is the change in the chemical potential of the system caused by the transfer of a single iron atom from the medium to the cluster, N are atoms of medium, n_0 are atoms of a dissolved substance, e.g., iron, n are iron atoms entering cluster, R is cluster radius. Upon nucleation of nanoclusters in a single (closed) pore the function $\Delta G(n)$ possesses both maximum and minimum with two critical parameters $n_{\min}$ and $n_{\max}$, respectively. The limited concentration of the solute inside the pore entails the existence of limiting volume of the cluster, $n_{\max}$. Once this

Figure 2 Mass-spectra of positively charged carbon clusters.

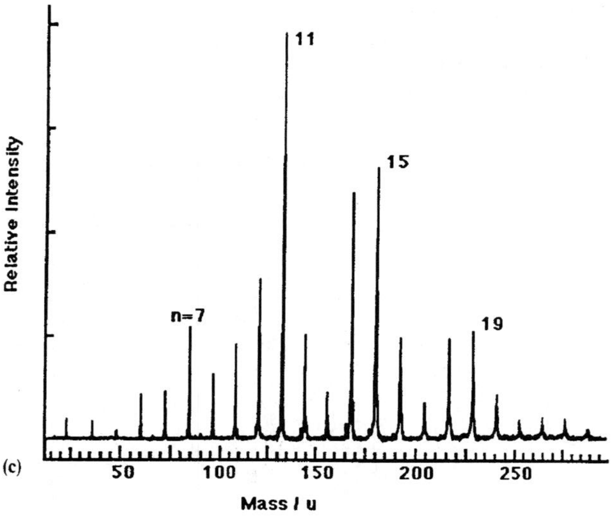

Figure 3 Ionization potensial of silver, sodium and potassium clusters as a function of size.

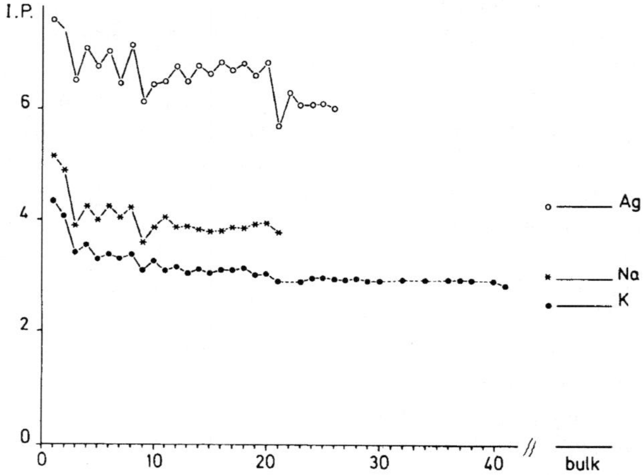

volume has been reached, cluster growth becomes energetically unfavorable. Heterogeneity decreases energy barrier for formation of a stable cluster seed of a cluster having critical volume [9].

For the nucleation in pore $n_{\min}=32\alpha^3/3\rho^2\Delta\mu^3$, energy barrier corresponding $n_{\min}$ is $\Delta G_{\mathrm{m}}=16\pi\alpha^3 K/3\rho^2\Delta\mu^2$ and $n_{\max}=n_0\Delta\mu/2+\mu$, where $K<1$ is a heterogeneity factor.

For nanoclusters of iron oxides obtained by thermal decomposition of ferric oxalate $Fe_2(C_2O_4)_3$ $5H_2O$ the nucleation critical parameters are introduced as in the following. In this case the role of pore confinement fulfills diffusion restricting nucleation time. For nucleation of small clusters showing weak cluster–cluster interactions the dependence of Gibbs' energy (ΔG_{f}) on cluster radius (R) for the single cluster, can be described by (1) with $S_{\mathrm{f}}=4\pi R^2$, $V_{\mathrm{f}}=4/3\pi R^3$. As is seen in Figure 4, the Gibbs' function passes through a maximum at $R=R_{\mathrm{CR}}$ (R_{CR} is the nucleation critical radius) and a minimum at $R=R_{\max}$ ($R_{\max}$ is the cluster radius corresponding to a stable equilibrium state). Upon further growth the

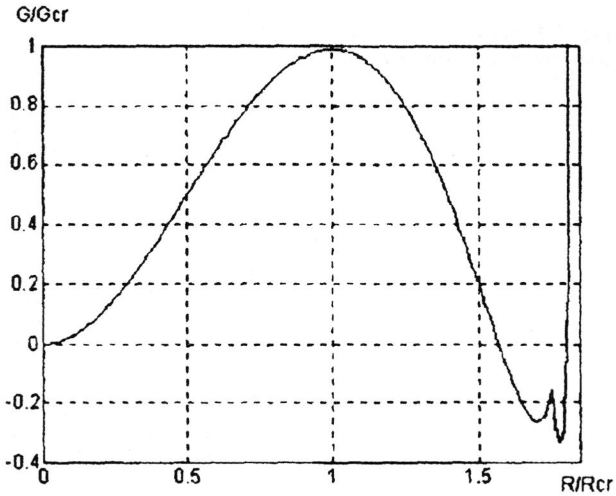

Figure 4 Change in the Gibbs' free energy (G/G_{CR}) against R/R_{CR}: nucleation, cluster growth and cluster sintering.

clusters come into contact with each other and boundaries between clusters vanish (sintering). The onset of cluster sintering corresponds to transition from isolated or weekly interacting centimeter.

Thus, there are two ranges in nanostructure formation: (1) $R<R_t$ and (2) $R>R_t$. Thus, for a system densely populated by clusters, at $k=6$ and $R_t \leq 1.7R_{CR}$ sintering occurs without a barrier. At $R_t=1.75R_{CR}$ the process of sintering requires overcoming a potential barrier. At least, for a system less densely populated by clusters, at $R=1.8R_{CR}$ sintering does not occur at all.

The DTA and DTG curves recorded during thermal decomposition of $Fe_2(C_2O_4)_3$ $5H_2O$ in air exhibit two minima at $T=200$ and 260°C [2]. The first minimum corresponds to dehydration accompanied by formation of mobile medium where iron oxide cluster nucleation and growth proceeds. The second minimum seems to correspond to origin of sintering. The cluster sizes were defined by BET, X-ray analysis, atomic force microscopy (AFM) and Mössbauer spectroscopy (MS). Mössbauer spectra of iron oxide nanoclusters prepared by thermal decomposition of iron oxalate $T_D=215$°C are shown on Figure 5. Analysis of typical superparamagnetic relaxation spectra allowed estimating the nanocluster size ranging 2–3 nm that is in full agreement with BET measurements. These nanoclusters correspond to the region of cluster sizes at $R<R_t$ (Figure 4) as well as to isolated week interacting clusters.

AFM image and AFM profiles for weak interacted nanoclusters ($T_D=215$°C) are shown in Figure 6. One can see that a nanocluster system consists of rather big (~80–90 nm) diffusive agglomerates that in turn are gathered from small (~2–3 nm) or larger nanoclusters. Indeed, the AFM images (Figure 6, bottom) show 30–40 small reflections all over the main peak belonging to cluster ~80–90 nm.

An AFM image of iron oxide nanoclusters at the beginning of sintering ($T_D=265$°C) is shown in Figure 7. The scale of the image (500×500 nm) permits to estimate the cluster sizes in the range ~30–50 nm. X-ray data also testify to the presence of α-Fe_2O_3–γ-Fe_2O_3 clusters with the same sizes [10]. Figure 8 shows Mössbauer spectra of this nanostructure that is characterized by strong cluster–cluster interaction at $R>R_t$. No superparamagnetic behavior is visible for these big nanoclusters. In fact, Mössbauer spectra are composed from magnetic hyperfine structure lines arising from α-Fe_2O_3–γ-Fe_2O_3 clusters in magnetic state and quadrupole doublet arising from smaller cluster in "paramagnetic" state. The origin of these phenomena will be interpreted in chapter 4.

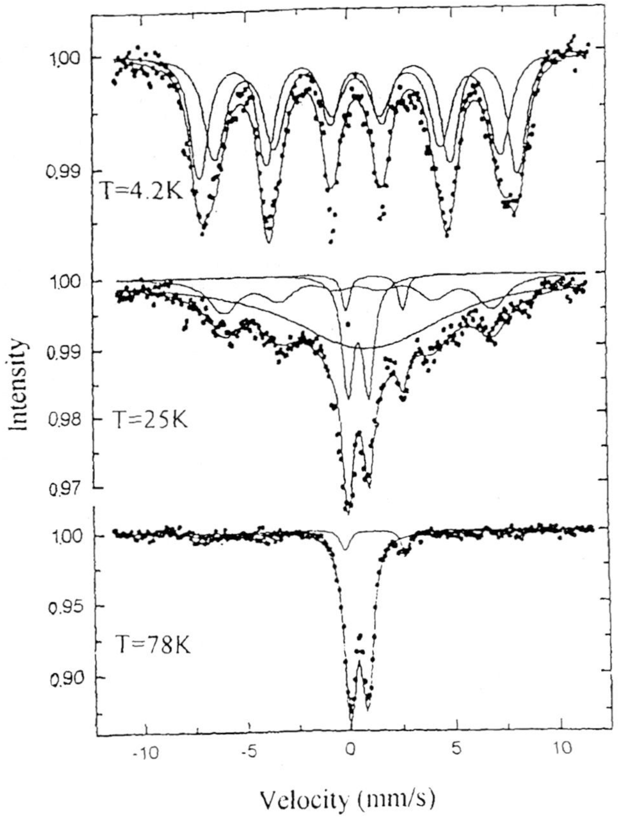

Figure 5 Mössbauer spectra of iron oxide nanoclusters obtained at T_D=215°C at different temperatures. a – 4.2 K, b – 25 K, c – 78 K.

4 Critical characteristic of cluster atomic dynamics

Intracluster atomic dynamics defines properties of nanoclusters such as, for example, heat capacity and melting. One of the most important characteristics is the mean square displacement of atoms. The increase of mean square displacements for surface atoms of cluster leads to the increase of heat capacity and the decrease of melting point. Thermodynamic formalism permits to predict cluster melting and to bind it with the values of mean square displacement [11–15]. The difference between the free energies of the cluster in liquid and solid states may be expressed by

$$\Delta G = 4\pi R^2 \Delta\alpha + 4/3\pi R^3 \rho\Delta\mu, \tag{4}$$

where $\Delta\alpha=\alpha_l-\alpha_s$ is the difference between surface energy and $\Delta\mu=\mu_l-\mu_s$ is the difference between the chemical potentials for liquid and solid states of cluster. It is obvious that, for melting, $\Delta\alpha<0$ and $\Delta\mu>0$. Typical $\Delta G(R)$ curves are shown on Figure 9. The critical cluster size corresponding to the transition from solid to liquid state ($\Delta G=0$) is calculated from the formula

$$R_{CR} = 3\Delta\alpha/\rho\Delta\mu. \tag{5}$$

When $R<R_{CR}$ a cluster occurs in liquid state. As it follows from (5), solid–liquid transition state cannot be realized for separate atoms in matrix ($\Delta\alpha=0$) when cluster still is not exists.

Figure 6 Atomic Force Microscopy image (*top*) and profiles (*bottom*) for weak interacted iron oxide nanoclusters organized in aggregates obtained at T_D=215°C.

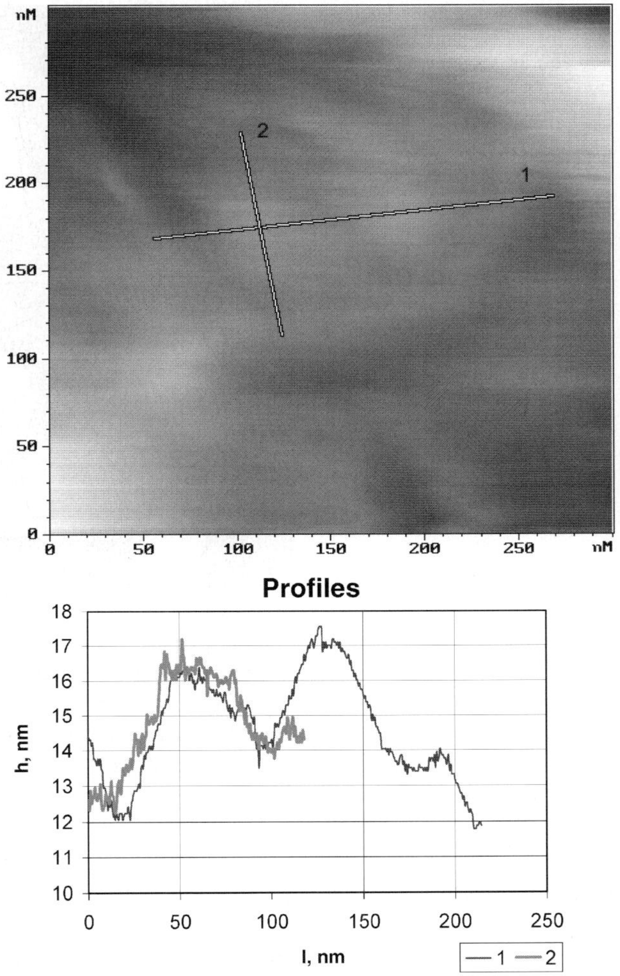

The $\Delta\alpha$ value should be the greatest for individual cluster. Any intercluster interactions would decrease this value and, hence, the value of R_{CR} that determines the region of possible fluctuations of solid and liquid states. The decrease of melting point of cluster is

$$\delta T / T_o \approx 3\Delta\alpha / \rho q R, \tag{6}$$

where q is the latent heat of the phase transition, T_o is the melting point of the bulk material.

Expression (6) indicates that the melting point of cluster decreases with a decrease of cluster size. A decrease in the melting point (down to room temperature) with a decrease in the cluster size has been observed for gold, tin and CdS nanoclusters [14–16]. A decrease of critical cluster size for nucleation and sharp increase of mean square displacements for iron hydroxide nanoclusters in polymer pore can be qualified as melting of nanocluster [17–19]. The intensity of MS lines for iron hydroxide clusters in polysorbs abruptly decrease upon a decrease of polysorb pore size and of nanocluster size. This indicates an increase intracluster atomic mobility corresponding to melting processes.

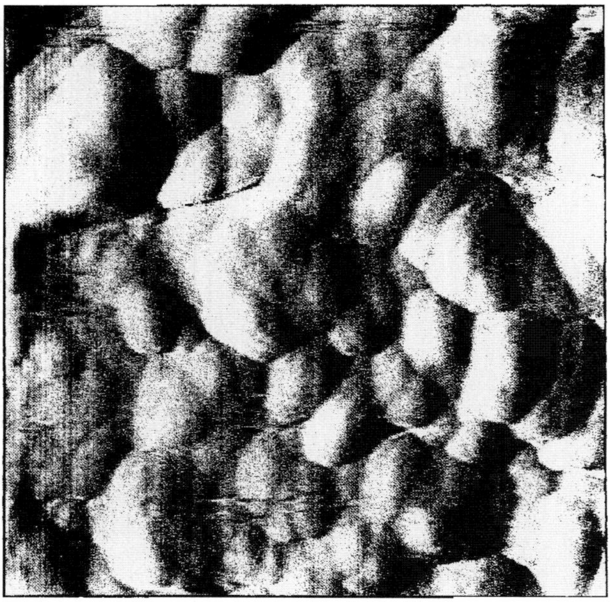

Figure 7 Atomic Force Microscopy image (resolution 500× 500 nm) for strongly interacting iron oxide nanoclusters obtained at T_D=265°C.

For nanoclusters of γ-Fe_2O_3 obtained by thermal decomposition of iron oxalate the dependence of spectral area on T_D has a minimum for smallest clusters [10]. Relation (6) can be used to estimate the value of R_{CR} corresponding to the onset of cluster melting. Using in (6) $\rho=5.18\times10^6$ g/m^3, $\Delta\alpha\cong1$ J/m^2, $q\cong2\times10^{-19}$ J [20], we obtain $R=aT_0/\delta T$, where a=0.1–1.0 nm. For T_m=300 K, R is equal to 0.1–1.2 nm, that is in agreement with the estimate $d\sim2$ nm found from Mössbauer spectra.

Temperature dependences of heat capacity for cluster 2–3, 7–8 and 150–300 nm are shown on Figure 10. The increase of heat capacity with the decrease of cluster size is caused by the influence of large surface area of clusters and by the increase of mean square displacement, supporting the idea of solid–liquid state for small clusters. Thus, the decrease of melting point and the existence of solid–liquid state for nanoclusters were found in accordance with thermodynamic predictions and can be described in terms of critical cluster sizes.

5 Critical characteristics of magnetic phase transitions

A thermodynamic model for magnetic phase transitions in nanocluster has been proposed in [21]. A cluster is assumed to exhibit a pronounced magnetostriction effect and the Curie point of a substance depends on its volume

$$T_c = T_o(1+\beta)(V-V_o)/V_o, \tag{7}$$

where β is the magnetostriction constant of the substance, V and V_0 are molecular volumes of the substance in the magnetically ordered and paramagnetic states. Minimization of the free energy gives

$$T\Big/T_{co} = m\big(1/3\gamma m^2+1\big)[1/2\ln(1+m)/(1-m)]^{-1}, \tag{8}$$

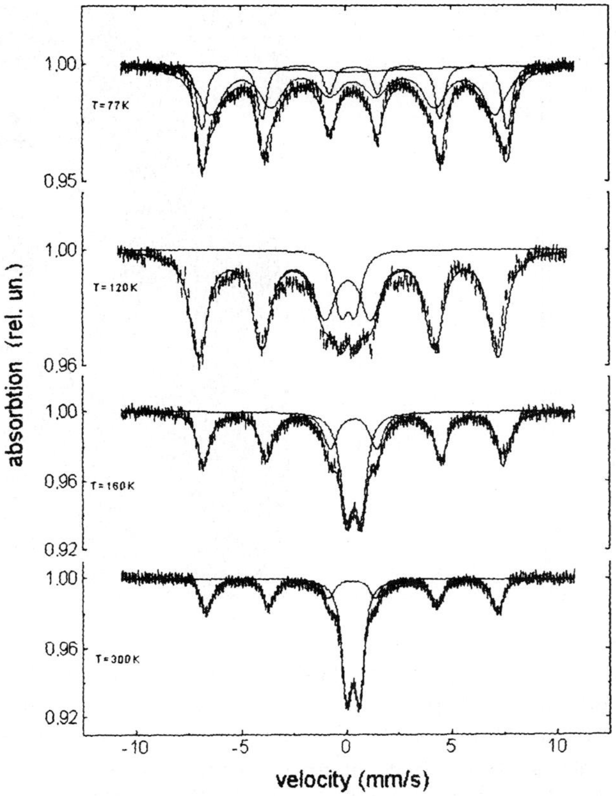

Figure 8 Mössbauer spectra of α-Fe_2O_3–γ-Fe_2O_3 nanostructure obtained at T_D=265°C at different temperatures.

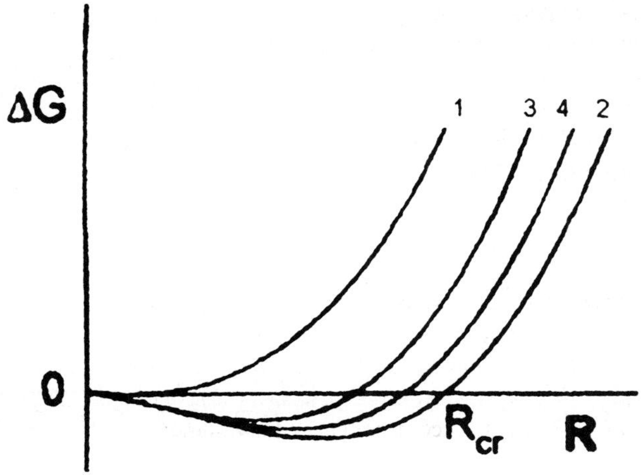

Figure 9 Change in the Gibbs' Free Energy versus cluster size: an ensemble of separate atoms in a solid (1) (α_1=0), isolated clusters (2) (α_2=α_{max}), systems of interacting clusters (3,4) occurring in the solid and liquid state (α_3<α_4).

where T_{co}=T_o(1-$\beta\eta P$) shows the change in T_o due to pressure $P = 2\alpha/R, \gamma = 3/2NkT_o\eta\beta^2$ $(1 - \beta\eta P)^{-1}, \eta$ is the compressibility constant, α is the surface tension, T_o is the Curie point for the bulk, m is the magnetization.

Figure 11 presents the temperature dependence m=f(T/T_{co}) for different values of the parameter γ, that determines the character of magnetic phase transition. For γ<1, the phase

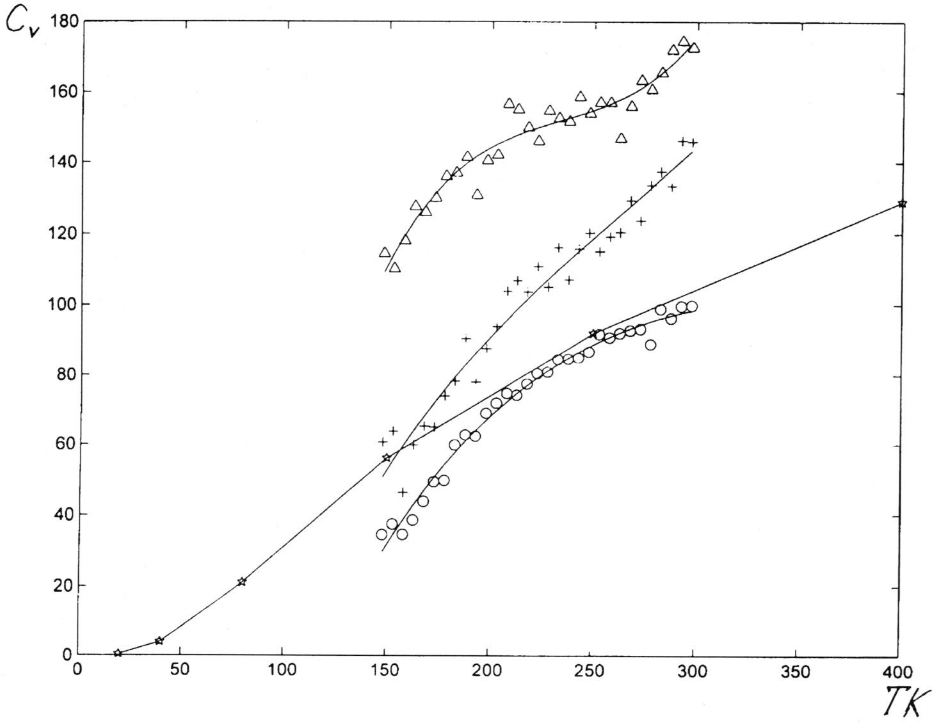

Figure 10 Temperature dependence of heat capacity for iron oxide clusters: *triangle*-clusters 2–3 nm, *plus*-clusters 7–8 nm, *circle*-clusters 150–300 nm, *star*-bulk hematite.

Figure 11 Temperature dependence of cluster magnetization for different γ: (1) $\gamma<1$, (2) $\gamma=1$, (3) $\gamma>1$.

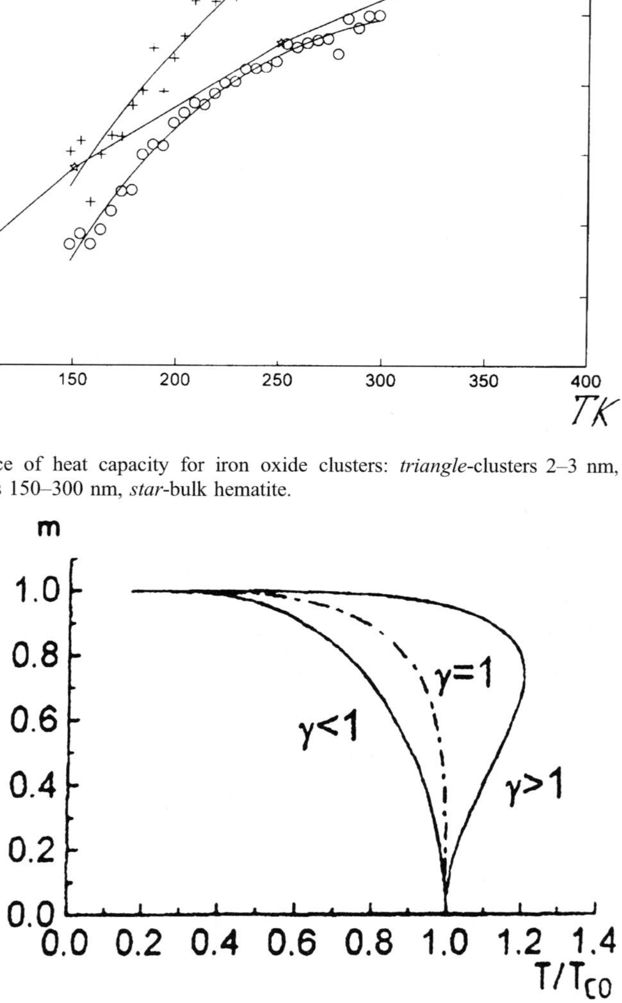

transition corresponds to a second-order magnetic transition, while at $\gamma\geq1$ this is a first-order magnetic transition. If $\gamma=T_{cc}/T_{co}$, where $T_{cc} = 3/2NKT_0^2\eta\beta^2$, for the cluster size $R \leq R_{CR}\{R_{CR} = 2\alpha\beta\eta/1-(T_{cc}/T_{co})\}$, at $\gamma\geq1$, cluster passes from the magnetically ordered to the paramagnetic state by jump, i.e., the first-order magnetic phase is realized. For larger clusters, only the second-order magnetic phase transition takes place. For $kT>K_aV$ (K_a is the anisotropy constant, V is the cluster volume) it manifests itself as phenomenon of superparamagnetism. The first and the second-order magnetic phase transitions were observed for the clusters $Fe_5HO_8\ 4H_2O$ (ferrihydrite) in the polysorb pores [21]. The presence of the first-order magnetic phase transition is clearly seen in Figure 8.

References

1. Cava, R.J.: Future directions in solid state chemistry. Prog. Solid State Chem. **30**, 1 (2002)
2. Suzdalev, I.P., Suzdalev, P.I.: Russ. Chem. Rev. **70**:177 (2001)
3. Haberland, H. (ed): Clusters of Atoms and Molecules. Springer, Berlin Heidelberg New York (1994)
4. Chini, P.: Gazz. Chim. Ital. **109**:225 (1979)
5. Vargaftik, M.N., Kozitsyna, Yu.N., Cherkashina, N.V.: In: Braunstein, P., Oro, L.A., Raithly, P.R. (eds.) Metal Clusters in Chemistry, vol. 3, p. 1364. Wiley-VCH, Weinheim (1999)
6. Campbell, E.E.: In: Haberland, H. (ed.) Clusters of Atoms and Molecules, p. 331. Springer, Berlin Heidelberg New York (1994)
7. Brechignac, C.: In: Haberland, H. (ed.) Clusters of Atoms and Molecules, p. 250. Springer, Berlin Heidelberg New York (1994)
8. Oxtoby, D.W.: In: Prigogin, I.I. (ed.) Advances in Chemical Physics, vol.70, p. 140. Wiley, New York (1988)
9. Suzdalev, I.P., Maksimov, Yu.V., Buravtsev, V.N., Imshennik, V.K., Kazakevitch, A.G., Novochikhin, S.V.: Kolloid. Zh. **62**:257 (2000)
10. Suzdalev, I.P., Maksimov, Yu.V., Novichkhin, S.V., Buravtsev, V.N., Imshennik, V.K.: Khim. Fiz. **19**:105 (2000)
11. Suzdalev, I.P., Buravtsev, V.N., Imshennik, V.K., Maksimov, Yu.V., Matveev, V.V., Volynskaya, A.V., Trautwein, A.X., Winkler, H.: Z. Phys. D **36**:163 (1996)
12. Suzdalev, I.P., Buravtsev, V.N., Imshennik, V.K., Maksimov, Yu.V., Trautwein, A.X., Winkler, H.: Hyp. Int. **C1**:290 (1996)
13. Suzdalev, I.P., Shklovskaya, N.I.: Colloid. Surf. **A108**:315 (1996)
14. Buffet, Ph., Borel, J.P.: Phys. Rev. A. **13**:22 (1976)
15. Osima, Y., Takaynagi, K.: Melting in tin small particles. Z. Phys. D. **27**:287 (1993)
16. Goldstein, A.M., Echer, C.M., Alivisatos, A.P.: Science **256**:1425 (1992)
17. Suzdalev, I.P., Buravtsev, V.N., Imshennik, V.K., Maksimov, Yu.V., Novichikhin, S.V.: Khim. Fiz. **12**:1255 (1993)
18. Grigorov, I.R., Meilichov, V.S. (eds.): Fizicheskie Velichiny, Spravochnik (Physical Values, Handbook). Energoatomizdat, Moscow (1991)
19. Ormont, B.F.: Dokl. Akad. Nauk SSSR **124**:132 (1959)
20. Rehbinder, P.A.: In: Surface effects in dispersed systems. Physicochemical Mechanics, p.142. Nauka, Moscow (1979)
21. Suzdalev, I.P., Buravtsev, V.N., Imshennik, V.K., Maksimov, Yu.V., Novichikhin, S.V., Trauwein, A.X., Winkler, H.: Z. Phys. D **37**:55 (1996)

Hyperfine Interact (2005) 165: 37–47
DOI 10.1007/s10751-006-9244-2

Molecular Switchable Solids: towards photo-controlled magnetism

François Varret · Kamel Boukheddaden · Epiphane Codjovi · Antoine Goujon

Published online: 26 October 2006

Abstract Molecular Switchable Materials are considered here in view of their potentialities towards the photo-control of magnetic properties. Major results obtained in the past few years are reviewed here, viz. generic features of photo-induced magnetism (external control of magnetic dilution, Raw Photo-Induced magnetic state), various photo-magnetic functionalities and recent successes in single-shot photo-excitation. The main challenges for the future clearly are : understanding the nature of the so-called Domino effect, reaching room-temperature photo-magnets.

Key words Molecular Switchable Solids · Raw Photo-Induced magnetic state · photo-induced magnetization

1 Introduction

The switching properties of Spin Transition Solids, mostly discovered over the last two decades, are shared by a growing number of Molecular Solids [1–5], including valence tautomeric complexes, bimetallic Co–Fe Prussian Blue analogues, and others involving electron transfers and/or Jahn–Teller cooperative distortions. Such materials are termed Molecular Switchable Solids, according to the generic property of molecular bi- (or multi-) switchability by external stimuli such as pressure or temperature variations, intense magnetic field and light irradiation. Interesting photo-magnetic properties require that the photo-induced electronic state acquires (or looses) both magnetic moments and interactions, which is hardly conceivable in the case of spin-crossover compounds. The first example of

F. Varret (✉) · K. Boukheddaden · E. Codjovi
Laboratoire de Magnétisme et d'Optique, CNRS/UVSQ, 45, Avenue des Etats-Unis, 78035 Versailles Cedex, France
e-mail: fvarret@physique.uvsq.fr

A. Goujon
Laboratoire Léon Brillouin (CEA), Saclay, France

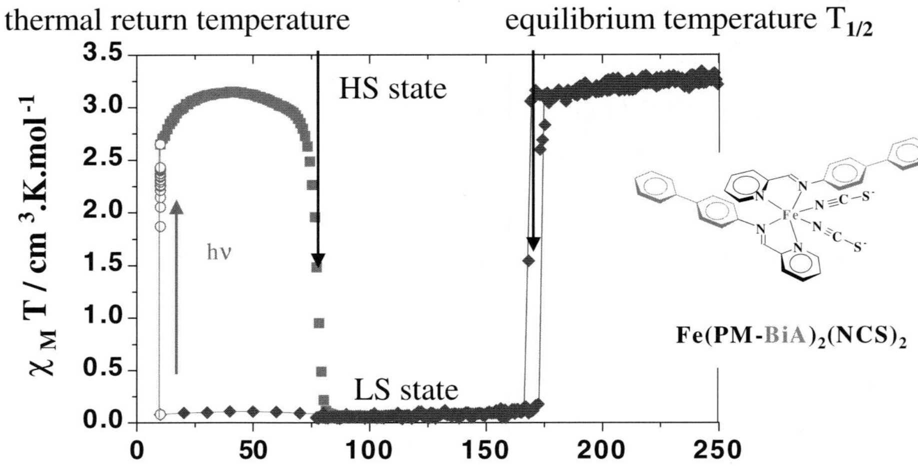

Figure 1 Recent example of thermal and optical switching of spin states (after [23]) Thermal switching spontaneously occurs at $T_{1/2}$, the "spin transition," a phase transition driven by the entropy difference between the competing molecular electronic states (low spin LS, high-spin HS). Thermal hysteresis at $T_{1/2}$ results from elastic interactions in the solid [24]. The thermal return temperature, somewhat dependent on the scan rate (typically 0.3 K/min) is associated with the thermal activation of the tunneling process of relaxation of the metastable HS state.

Figure 2 Low-temperature optical switching of $[Fe_{0.64}Co_{0.36}(btr)_2(NCS)_2]$. H_2O, at 8 K, 50 Oe, 550 nm, 25 mW/cm^2 (after [25]). Magnetization and reflected intensity were simultaneously recorded. Both photo-magnetic and photochromic effects are evidenced. The jumps of the magnetic signal when the light is switched on and off are due to the warming effect of light on the sample. The different kinetics of the magnetic and reflectivity data illustrate the bulk absorption of light in the sample: reflectivity is a "surface" response sensitive only to the top layers of the sample.

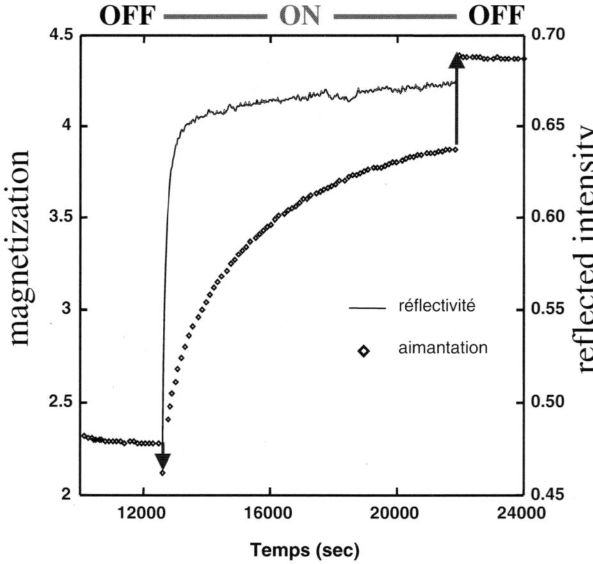

light-induced magnetic order was provided by the bi-metallic Co–Fe bi-metallic cyanides of Prussian blue structure (often designed as Prussian blue analogues, PBA) [6–8], allowing the first generic features of "ordered" photo-induced magnetization to be observed [9]. The photo-control of magnetic order by light [10] became reality.

On another hand, efficient photo-switching requires that the photo-excitation process overcomes the relaxation of the photo-induced metastable electronic state, which

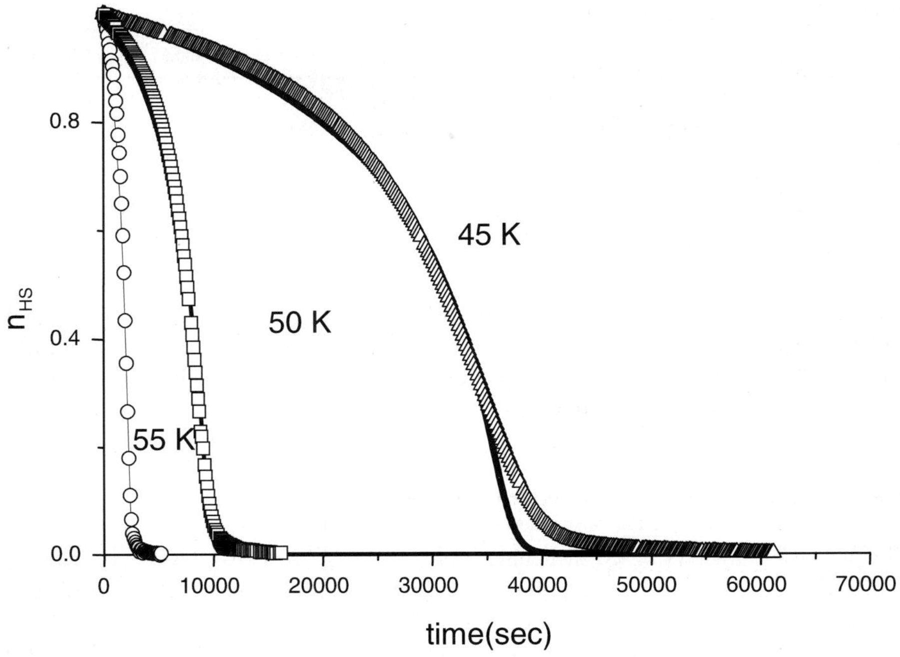

Figure 3 Relaxation of the metastable HS state of $[Fe_{0.5}Zn_{0.5}(btr)_2(NCS)_2]\times H_2O$, in the dark (after [26]). The self-acceleration effect, responsible for the sigmoidal shape of the curves, is explained by the building-up of internal pressure due to the elastic effect of the increasing number of low-volume LS units [24]. The departure from mean-field simulations performed using Hauser's factor (*full lines*), was denoted "relaxation tail" and attributed to short-range correlations created at long times by short-range interactions [27–29]. Note the strong dependence of relaxation rates upon temperature, due to the thermal activation of the tunneling process.

Figure 4 Creation of the metastable Raw Photo-Induced state of $Rb_{0.52}$ Co $[Fe(CN)_6]_{0.84}$, 2.31 H_2O, and subsequent magnetic mesurements in the dark, under constant magnetic field (after [7]).

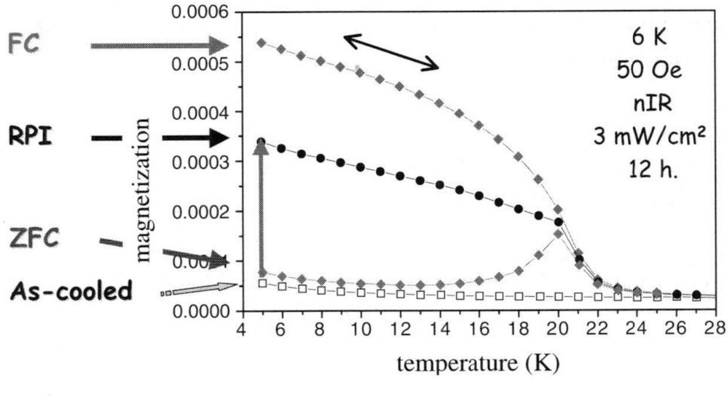

dramatically increases as a function of temperature, reaching the nanosecond range for molecular spin-crossover compounds. The challenge of spin-crossover-based optical switches working at room-temperature faced the problem of a very high irradiation intensity, only acceptable in a transient regime such as that delivered by one-pulse laser sources. Fortunately, recent examples [11–14] of successful single shot pulse-laser excitations gave some evidence [11, 12] of an increase in the quantum yield with respect to the expectations based on low-temperature data. Such an increase was attributed to the

Table I Spin crossover-based switchable solids

Type	Low-temp. state	High-temp. or photo-excited state	Typical examples
Spin crossover	Fe^{II} $t_{2g}6$ LS (S=0)	Fe^{II} $t_{2g}4$ $e_{2g}2$ HS (S=2)	[17–19]
	Fe^{III} $t_{2g}5$ LS (S=1/2)	Fe^{III} $t_{2g}3$ $e_{2g}2$ HS (S=5/2)	[30]
Bimetallic cyanides	$Fe^{II}{}_{LS}(S=0)$ $Co^{III}{}_{LS}(S=0)$	$Fe^{III}{}_{LS}(S=1/2)$ $Co^{II}{}_{HS}(S=3/2)$	[6–8]
	$W^{IV}(S=0)$ $Co^{III}{}_{LS}(S=0)$	$W^{V}(S=1/2)$ $Co^{II}{}_{HS}(S=3/2)$	[31]
Valence tautomerism	$Co^{III}{}_{LS}(S=0)$ (sq)(cat)	$Co^{II}{}_{HS}(S=3/2)$ (sq) (sq)	[32]

so-called Domino effect [15, 16] which describes the propagation of a cascade of single-molecule excitations in the solid. The present report will mainly focus on experimental aspects, including a brief review of the various photo-magnetic functionalities reported so far, to the best authors' knowledge.

2 The essentials of thermal and optical switching of spin states

Comprehensive reviews of the switching properties of spin-crossover systems have been published [17–19]. We just remind the essential features, illustrated by typical examples in Figures 1, 2, 3 and 4. The crucial Jablonski diagram of the photo-excitation process (usually called LIESST=Light Induced Excited Spin State Trapping [20]) is reproduced in all the previously quoted references. However, it is worth recalling that careful continuous laser experiments [21] have confirmed at low-temperature the single-molecule character of the photo-induced process, so far admitted [22].

3 Further switchable systems

Other systems exhibit switching properties based on a spin crossover triggered by (or associated with) an electron transfer, which was termed "charge transfer induced spin transitions" (CTIST) [6], see Table I.

Further switchable systems are not based on the spin-crossover instability, but on other examples of vibronic lability [33] (a crucial concept for molecular multistability), e.g., cooperative Jahn–Teller instability. As a matter of fact the field of photo-switchable materials is extremely large and every year further materials join the list (see for instance recent reviews of photomagnetic materials by Epstein [34] and Sato [35]). Among these materials, those presenting extended charge-transfer phase transitions, viz. neutral-to-ionic [36], spin-Peierls [37] or metal-to-insulator [38], might be very promising in terms of ultra-fast optical switches [39]. However, they are not central here due to their (so far) poor magnetic properties. At last, switchable nitroprussides (see [2] for an extensive review of this controversial system), novel high spin photo-magnetic molecules [40], and other switchable systems which do not involve a spin crossover [41, 42] are worth mentioning.

4 Generic features of photo-induced magnetism

During previous investigations of the dia-to-ferrimagnetic photo-transformation of Co–Fe based Prussian Blue derivatives [7–9] we were looking for specific magnetic features

Figure 5 Room temperature photo-switching of $Na_{0.68}Co_{1.20}Fe(CN)_6$ 3.7 H_2O, using single-shot irradiation (532 nm, 6 ns, after [11]). Remark: the photo-induced phase transition was monitored by the analysis of (very) broad absorption, so that the indication of a complete transition is very approximative.

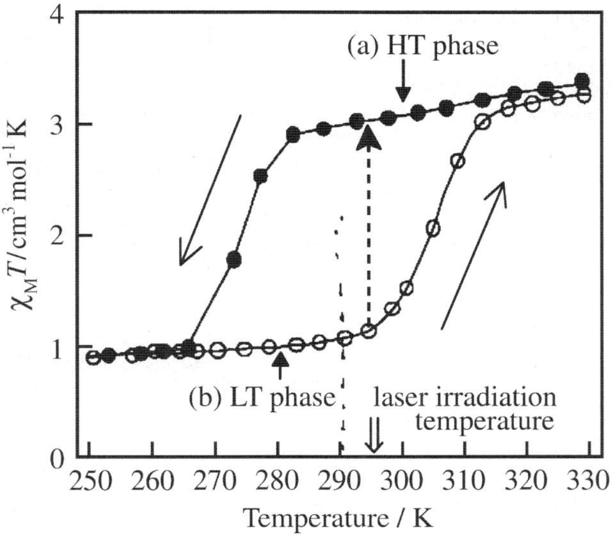

associated with the lack of thermal fluctuations during the build-up of the magnetic system. For example the so-called Raw Photo Induced (RPI) magnetic state obtained at sufficiently low field and temperature, see Figure 4, irreversibly evolves under thermal treatments far below the thermal return temperature which is ~110 K.

A systematic investigation, reported in [4] showed that the RPI state behaves as any state inside the B–H hysteresis loop of the photo-induced material. Actually, the RPI state is governed by the previous photo-thermo-magnetic history of the sample. In addition, after thermal treatment above Tc or application of a magnetic field in excess of the hysteretic range, the system does not retain any memory of the magnetic field which was applied during photo-excitation. Systematic repetition of the same experimental sequence (thermal annealing at 150 K for re-initialization of the electronic state, photo-excitation at 10 K, RPI, FC, ZFC curves) have shown that the RPI curve spans all the space between the ZFC and FC curves when the external field is varied from zero to the coercive field value of the final photo-induced magnetic state. We further reported in [43] that the RPI magnetization curve was well reproduced by a linear combination of field-cooled (FC) and zero-field cooled (ZFC) curves, and proposed a simple description of the photo-process in terms of successive equilibrium and out-of-equilibrium processes. Such a qualitative model, formerly proposed in terms of ferrimagnetic domains, could as well be expressed in terms of dynamic processes better suited to the actual cluster glass state established by [44]. It is finally worth noting that the absence of any thermal fluctation would lead to a totally disordered magnetic structure (~spin glass). Such a limit situation has not been obtained, and even seems hardly plausible, because the non-adiabatic decay involved in the photo-processes *locally* delivers sizable excitation energies.

A second generic character of the photo-induced magnetism is the ability of controlling the concentration of a magnetic component of a given solid, by external means. The trivial strategy of merely controlling the irradiation time has led so far to an inhomogeneous state of photo-transformation, because of bulk absorption of light in the sample. We tried an alternative way, viz. saturating the sample at low temperature by a long irradiation stage, and then trying to induce a controlled thermal relaxation of the metastable electronic state.

Figure 6 Reversible photo-switching of $Na_{0.36}Co_{1.32}Fe(CN)_6$ $5.6H_2O$, using single-shot irradiation at 2.38 eV, after [12]. Same remark as for previous figure.

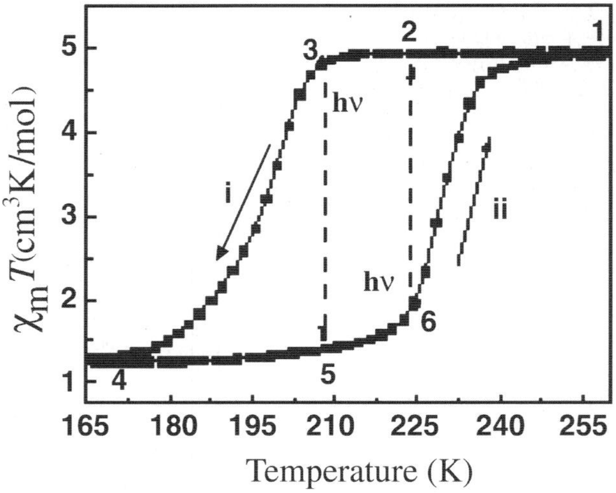

The first results have been encouraging [45]; however an open problem is the occurrence of non-random dilution, due to the eventual onset of short-range correlations. In addition the eventuality of a nucleation and growth process cannot be discarded, according to recent structural investigations of thermal or photo-induced phase transition in spin-crossover solids [46–53].

5 One-shot pulsed laser experiments in the thermal hysteresis loop

For real applications, the temperature range at which the photo-excitation process dominates over relaxation has to be raised up to room temperature. The thermal return temperatures of a large number of spin-crossover or charge-transfer-induced spincrossover materials have been collected by J.-F. Létard in a database (T_{LIESST}, [53]), which points out that progress is continuously reported on the way to room temperature, particularly thanks to the Co–Fe based Prussian Blue analogues (now reaching up to ~170 K). It however seems that the relaxation rates at room temperature remain in the sub-microsecond range still for a long time. The molybdenum-based photo-magnetic molecule quoted in [40] was reported to have a return temperature above 280 K, but so far there is no clear way to using it in 3D magnetic compounds. The same problem holds for the novel ligand-driven light-induced spin crossover solids (LD-LISC) [54] which reversibly operate at room temperature.

Because of rapid relaxation, a very large intensity is needed for achieving photo-excitation at high temperatures. For example using the low-temperature experimental data of $Na_{0.26}Co\ [Fe^{3+}(CN)_6]_{0.74}\cdot 3.3H_2O$, we estimated by mean-field macroscopic Master Equation approach [55] a threshold intensity value ~300 mW cm^{-2} at the equilibrium temperature of 230 K. The estimated value reaches some W/mm^2 at room temperature. Such a high intensity would merely destroy the sample. An obvious solution to this problem was the recourse to short pulses of laser irradiation, and successful experiments [11–14] were reported as soon as one-shot lasers in the ns range became available. The second problem on the way to optical memories at room temperature is the short lifetime of the photo-induced state, and the proposed solution consisted in operating in the thermal

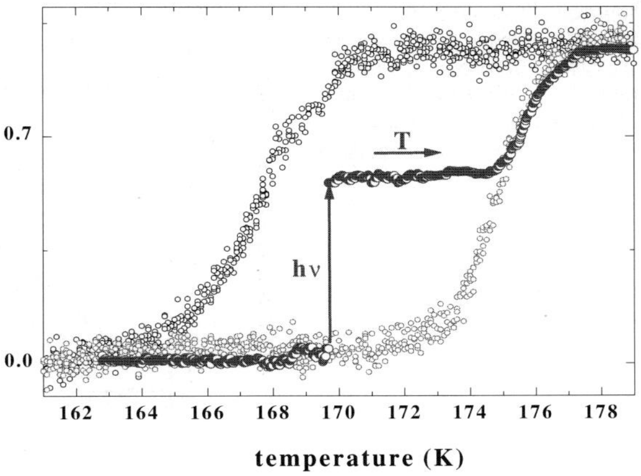

Figure 7 One-shot photo-switching of the spin-crossover system Fe(PM-BiA)$_2$(NCS)$_2$ from optical reflectivity measurements (pump-probe) using pulsed laser (532 nm, 6 ns, 10 mJ), after [13]. Accurate measurements indicate the incomplete character of the transition.

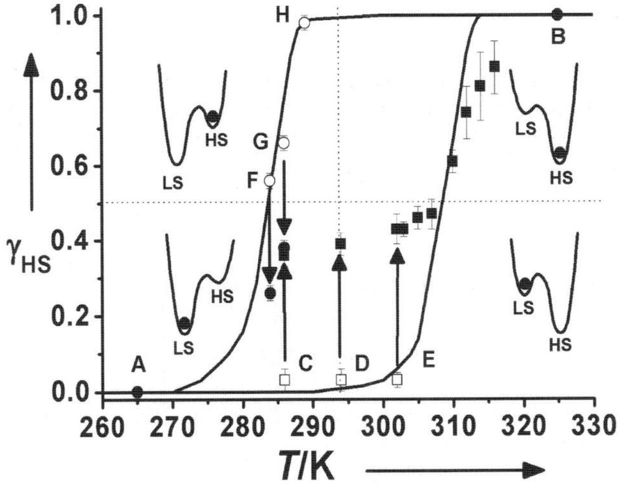

Figure 8 One-shot photo-switching of the spin-crossover system {Fe(C$_4$H$_4$N$_2$)[Pt(CN)$_4$]} from Raman measurements after pulsed laser excitation (532 nm, 6 ns, 10 mJ), after [14]. Note (1) the reversible character of the process which at 285 K may lead to a final state irrespective of the initial state and (2) the very incomplete character of the photo-induced transition.

hysteresis range, because of the onset of the macroscopic barrier energy associated with the long-lifetimes of the macroscopic states. The four examples [11–14] are now discussed, in the light of Figures 5, 6, 7 and 8.

6 Discussion

Both experiments shown in Figures 5, 6, dealing with similar Na-based Co–Fe Prussian Blue analogues, report an intensity threshold effect typical for a non-linear process. This should hardly be attributed to cooperative relaxation, since the relaxation process is not expected to take place in a very short time (a few nanoseconds). In addition the measured quantum efficiency (ratio of transformed Fe–Co pairs/absorbed photons), close to 1, is roughly 20 times larger than the values measured at low temperature. Such an enhancement of the quantum yield was attributed to the "Domino effect" describing the propagation of cascades of single-molecule excitations through the lattice [15]. The occurrence of the

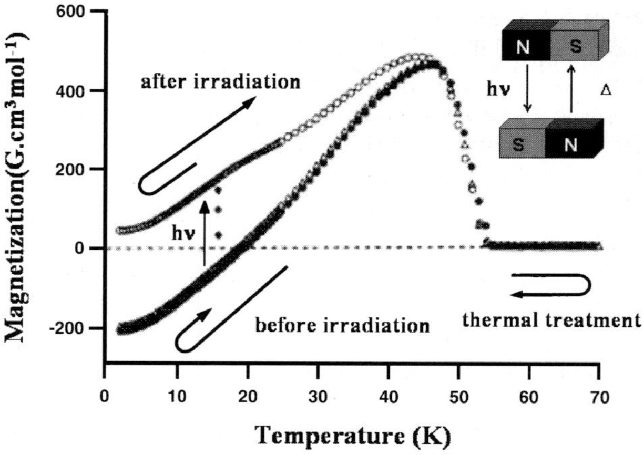

Figure 9 Photo-induced magnetization reversal in the ferrimagnet $(Fe_{0.4} Mn_{0.6})Cr(CN)_6$, $7.5H_2O$, (after [60]), just below the compensation temperature. Photo-induced electron transfer results in the completion (depletion) of the sublattice parallel (antiparallel) to the total magnetization. The systems can be re-initialized by thermal annealing above the electronic return temperature.

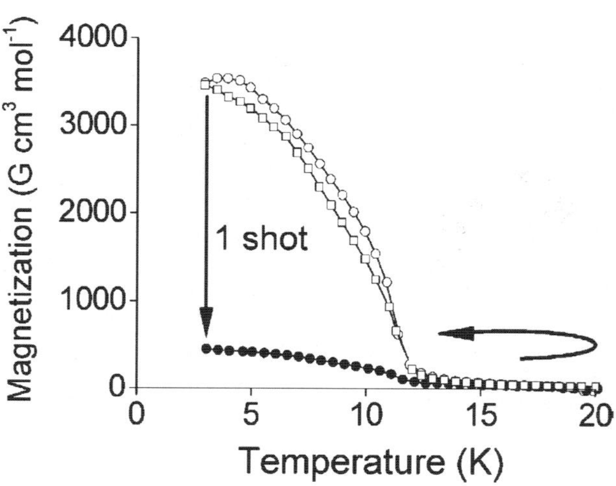

Figure 10 Photo-extinction of magnetization in the ferrimagnet $RbMn^{II} Fe^{III} (CN)_6$ $0.6 H_2O$, after [40], ascribed to the creation of an antiferro-magnetic state.

Domino effect in the photo-excitation of a spin transition solid was reported since the 1990s [56], but later on it was ruled out by the low-temperature accurate measurements by Hauser's group [21]. However, the occurrence of the Domino effect in the *high-temperature* photo-excitation of Prussian Blue analogues cannot be a priori excluded. Even the propagation of metastable spin states through the lattice has to be inferred from the recent observations of nucleation and growth processes in various types of materials (spin transition, charge transfer induced spin transition), during both the spontaneous and the light-induced phase transitions, through refined structural investigations [46–53]. These direct observations are consistent with the demixing effect (that is, the phase separation) we previously observed in the light-induced instability regime of a spin transition solid [57]. All existing models (up to e.g., [58]) should probably be revised from the viewpoint of *mobile* metastable states, i.e., through the excitonic formalism.

A surprising feature is the incomplete character of the spin-crossover photo-induced transitions, at least in the LS$\rightarrow$HS sense, because low-temperature irradiations usually lead to complete transitions. On the reverse, a complete photo-induced transition it not be expected in the case of Prussian blue analogues, due to the broad overlap between the absorption bands of the LS and HS forms. Such an overlap is indeed the mechanism which explains the reversible effect observed in Figure 6, where the same wavelength induces opposite processes at different temperatures. As an illustrative approach, the mean-field dynamic simulation [55] showed that the breaking of the balance between opposite relaxation terms, induced by the temperature variation may drive in opposite ways the phase transitions on the two sides of the equilibrium temperature. Due to the weak overlap of absorption lines in spin-crossover solids, the same mechanism should not hold for the example reported in Figure 8, which therefore remains unexplained.

7 Perspectives and conclusions

Various examples in literature show that the net magnetization can be as well created (Figure 4), reversed (Figure 9) or even suppressed (Figure 10). In most cases, reversibility is only provided by the thermal return of the metastable electronic state, excepted for the Co–Fe photo-magnetic Prussian blues. Also, transient effects have been observed since the very beginning of photo-magnetic investigations [59], and more recently reported in terms of optical activation of the Magnetic After Effect [4]. Due to the short lifetime of the photo-excited species, these effects were not discussed here.

No doubt the optical control of these functions would enrich the versatility of current devices for spintronics, e.g., the bias effect in a multi-layer system. However, any future application at room temperature will require improvement of both the switchability and the Curie temperature of the system. The first challenge is understanding the Domino effect, the second is designing novel materials, e.g., based on Prussian Blue derivatives which are ferrimagnets above room temperature [61].

Acknowledgements We are indebted to Network of Excellence MAGMANET (FP6-515767-2), Conseil Régional d'Ile de France, CNRS, Université de Versailles Saint Quentin and Commissariat à l'Energie Atomique for financial support.

References

1. Varret, F., Constant-Machado, H., Dormann, J.L., Goujon, A., Jeftic, J., Noguès, M., Boussekou, A., Klokishner, S., Dolbecq, A., Verdaguer, M.: ICAME'97 invited lecture, Proceedings in Hyperfine Interactions **113**, 37–46 (1998)
2. Gütlich, P., Garcia, Y., Woike, T.: Coord. Chem. Rev. **839**, 219–221 (2001)
3. Varret, F., Bleuzen, A., Boukheddaden, K., Boussekou, A., Codjovi, E., Enachescu, C., Goujon, A., Linares, J., Menendez, N., Verdaguer, M.: Pure Appl. Chem. **74**, 2093–2102 (2002)
4. Varret, F., Nogues, M., Goujon, A. In: Miller, J. Drillon, M. (eds) Magnetism: Molecules to Materials, vol. 2, pp. 257–295. Wiley, New York (2001)
5. Soraï, M.: Bull. Chem. Soc. Jpn. **74**, 2223–2253 (2001)
6. Sato, O., Iyoda, T., Fujushima, A., Hashimoto, K.: Science **272**, 704–705 (1996)
7. Goujon, A., Roubeau, O., Noguès, M., Varret, F., Dolbecq, A., Verdaguer, M.: Eur. Phys. J., B Cond. Matter Phys. **14**, 115–124 (2000)
8. Bleuzen, A., Lomenech, C., Escax, V., Villain, F., Varret, F., Cartier dit Moulin, C., Verdaguer, M.: J. Am. Chem. Soc. **122**, 6648 (2000)
9. Goujon, A., Varret, F., Escax, V., Bleuzen, A., Verdaguer, M.: Polyhedron **20**, 1347 (2001)
10. Pekajovic, D.A., Manson, J.L., Kitamura, C., Miller, J.S., Epstein, A.J.: Mol. Cryst. Liq. Cryst. **374**, 289–302 (2002)

11. Shimamoto, N., Ohkoshi, S., Sato, O., Hashimoto, K.: Chem. Lett. **4**, 486–487 (2002)
12. Liu, H.W., Matsuda, K., Gu, Z.Z., Takahashi, K., Cui, A.L., Nakajima, R., Fujishima, A., Sato, O.: Phys. Rev. Lett. **90**, 167403 (2003)
13. Freysz, E., Montant, S., Létard, S., Létard, J.F.: Chem. Phys. Lett. **394**, 318 (2004)
14. Bonhommeau, S., Molnár, G., Galet, A., Zwick, A., Real, J.-A., McGarvey, J. G., Bousseksou, A.: Angew. Chem. **117**, 4137 (2005)
15. Koshino, K., Ogawa, T.: J. Phys. Soc. Jpn. **68**, 2164 (1999)
16. Koshino, K., Ogawa, T.: J. Phys. Soc. Jpn. **67**, 2174 (1998)
17. Kahn, O.: Molecular Magnetism. VCH, New York (1993)
18. Gütlich, P., Hauser, A., Spiering, H.: Angew. Chem. Engl. **33**, 2024 (1994)
19. Gütlich, P., Goodwin, H.A. (eds.): Spin Crossover in Transition Metal Compounds. Top. Curr. Chem., vols. 234, 235, 236 (2004)
20. Decurtins, S., Gütlich, P., Köhler, C.P., Spiering, H., Hauser, A.: Inorg. Chem. **24**, 2174 (1985)
21. Enachescu, C., Oetliker, U., Hauser, A.: J. Phys.Chem. **37**, 9540 (2002)
22. Hauser, A.: Comments Inorg. Chem. **17**, 17 (1995)
23. Létard, J.F., Chastanet, G., Nguyen, O., Marcen, S., Marchivie, M., Guionneau, P., Chasseau, D., Gütlich, P.: Montashefte für Chemie, **134**, 165 (2003)
24. Hauser, A., Gütlich, P., Spiering, H.: Inorg. Chem. **25**, 4245 (1986)
25. Codjovi, E., Morscheidt, W., Jeftic, J., Linares, J., Nogues, M., Goujon, A., Roubeau, O., Constant-Machado, H., Desaix, A., Bousseksou, A., Verdaguer, M., Varret, F.: Mol. Cryst. Liq. Cryst. **335**, 1295–1304 (1999)
26. Enachescu, C., Constant, H., Codjovi, E., Linares, J., Boukheddaden, K., Varret, F.: J. Phys. Chem. Solids **62**, 1409–1422 (2001)
27. Spiering, H., Kohlhaas, T., Romstedt, H., Hauser, A., Bruns-Yilmas, C., Kusz, J., Gütlich, P.: Coord. Chem. Rev. **629**, 190–192 (1999)
28. Hôo, B., Boukheddaden, K., Varret, F.: Eur. Phys. J., B Cond. Matter Phys. **17**, 449–457 (2000)
29. Boukheddaden, K., Shteto, I., Hôo, B., Varret, F.: Phys. Rev., B Condens. Matter **62**, 14796–14805 and 14806–14817 (2000)
30. Hayami, S., Gu, Z.Z., Shiro, M., Einaga, Y., Fujishima, A., Sato, O.: J. Amer. Chem. Soc. **122**, 7126 (2000)
31. Arimoto, Y., Ohkoshi, S., Hozumi, T., Zhang, Z.J., Seino, H., Mizobe, Y., Hashimoto, K.: J. Amer. Chem. Soc. **125**, 9240 (2003)
32. Adams, D.M., Dei, A., Rheingold, A.L., Hendrickson, D.N.: J. Amer. Chem. Soc. **115**, 8221 (1993)
33. Ammeter, J.H.: Nouv. J. Chim. **4**, 631 (1980)
34. Epstein, A.: Mat Res Bull. 492–499 (July) (2003)
35. Sato, O.: J. Phys. Photochem. Photobiol. Rev. **5**, 203–223 (2004)
36. Collet, E., Lemée-Cailleau, M-H., Buron-Le Cointe, M., Cailleau, H., Wulff, M., Luty, T., Koshihara, S., Meyer, M., Toupet, L., Rabiller, P., Techert, S.: Science **300**, 612 (2003)
37. Matzusaki, H., Fujita, W., Awaga, K., Okamoto, H.: Phys. Rev. Lett. **91**, 7403 (2003)
38. Liu, X.J., Moritomo, Y., Machida, A., Nakamura, A., Asano, H.: J Phys. Chem. Solids **63**, 921–924 (2002)
39. Chollet, M., Guérin, L., Uchida, N., Fukaya, S., Shimoda, H., Ishikawa, T., Matsuda, K., Hasegawa, T., Ota, A., Yamochi, H., Saito, G., Tazaki, R., Adachi, S., Koshihara, S.: Science **307**, 86–89 (2005)
40. Herrera, J.M., Marvaud, V., Verdaguer, M., Marrot, J., Kalisz, M., Mathonière, C.: Angew. Chem. **43**, 5468 (VIP) (2004)
41. Tokoro, H., Ohkoshi, S., Matsuda, T., Hashimoto, K.: Inorg. Chem. **43**, 5231 (2004)
42. Kojima, N., Aoki, W., Itoi, M., Ono, Y., Seto, M., Kobayashi, Y., Maeda, Y.: Solid State Commun. **120**, 165–170 (2001)
43. Varret, F., Goujon, A., Bleuzen, A.: Hyperfine Interact. **134**, 69–80 (2001)
44. Pekajovic, D.A., Manson, J.L., Miller, J.S., Epstein, A.J.: Synth. Met. **122**, 529–533 (2001)
45. Varret, F., Goujon, A., Boukheddaden, K., Noguès, M., Bleuzen, A., Verdaguer, M.: Mol. Cryst. Liq. Cryst. **379**, 333–340 (2002)
46. Morimoto, Y., Kato, K., Kuriki, A., Nakamoto, A., Kojima, N., Takata, M., Sakata, M.: J. Phys. Soc. Jpn. **71**, 2609 (2002)
47. Hanawa, M., Morimoto, Y., Kuriki, A., Tateishi, J., Kato, K., Takata, M., Sakata, M.: J. Phys. Soc. Jpn. **72**, 987 (2003)
48. Kunz, J., Gütlich, P., Spiering, H. In: Gütlich, P., Goodwin, H.A. (eds) Top. Curr. Chem., vol. 234, p. 129. Springer, Berlin Heidelberg New York (2004)
49. Pillet, S., Hubsch, J., Lecomte, C.: Eur. Phys. J., B Cond. Matter Phys. **38**, 541–552 (2004)

50. Huby, N., Guérin, L., Collet, E., Toupet, L., Ameline, J-C., Cailleau, H., Roisnel T, Tayagaki, T., Tanaka, K.: Phys. Rev B **69**, 020101 (2004) (R)
51. Legrand V. PhD thesis, Nov 2005, Nancy, France
52. Ichiyanagi, K., Herbert, J., Toupet, L., Cailleau, H., Guionneau, P., Létard, J.-F., Collet, E.: Phys. Rev B **73**, 060408 (2006)(R).
53. Létard, J.F., Guionneau, P., N'Guyen, O., Sanchez, J., Marcen, S., Chastanet, G., Marchivie, M., Capes-Goux, L.: Chem Eur. J. **11**, 4582 (2005)
54. Boillot, M.L., Zarembowitch, J., Sour, A. Spin-Crossover in Transition Metal Compounds (see ref. 18), pp 261–276 (2004)
55. Varret, F., Boukheddaden, K., Codjovi, E., Maurin, I., Ohkoshi, S., Tokoro, H., Hashimoto, K.:Int Conf. Molecular Magnetism (ICMM04 Tsukuba, Japan, 5–8 Oct 2004) Polyhedron **24**, 2857 (2005)
56. Koshihara, S., Tokura, Y., Takeda, K., Koda, T.: Phys. Rev. Letters **68**, 1148 (1992)
57. Desaix, A., Roubeau, O., Jeftic, J., Haasnoot, J.G., Boukheddaden, K., Codjovi, E., Linares, J., Nogues, M.,Varret, F.: Eur. Phys. J. B **6**, 183–193 (1998)
58. Nishino, M., Boukheddaden, K., Miyashita, S., Varret, F.: Phys Rev., B Condens. Matter **72**, 064452 (2005)
59. Teale, R.W., Temple, D.W.: Phys. Rev. Lett. **19**, 904 (1967)
60. Ohkoshi, S., Yorozu, S., Sato, O., Iyoda, T., Fujishima, A., Hashimoto, K.: Appl. Phys. Lett. **70**, 1040 (1997)
61. Ferlay, S., Mallah, T., Ouahès, R., Veillet, P., Verdaguer, M.: Nature **378**, 701 (1995)

Hyperfine Interact (2005) 165: 49–53
DOI 10.1007/s10751-006-9245-1

Fe–Ag granular multilayers and heterostructures studied in applied magnetic field

J. Balogh · D. Kaptás · L. F. Kiss · T. Kemény ·
I. Vincze · K. Temst · C. Van Haesendonck

Published online: 26 October 2006

Abstract An (0.2 nm $^{57}Fe/2.6$ nm $Ag)_{75}$ granular multilayer sample and heterostructures with additional continuous Fe layers in different sequences were studied in magnetic field applied at different temperatures. The broadening of the superparamagnetic lines was found to be very similar for the three samples in applied fields both parallel and perpendicular to the sample plane. While the layer sequence has no significant effect on the superparamagnetic properties, the continuous magnetic layers follow a different approach to saturation in perpendicular magnetic fields.

Key words granular alloys · multilayers · magnetic properties · in field Mössbauer spectroscopy

Thin films of nano-scale granular composites can be prepared by alternate deposition of the constituents in the form of discontinuous multilayers or, as these are frequently called, granular multilayers. Recently, it was shown [1] by Mössbauer spectroscopy, that continuous magnetic layers can be inserted into Fe/Ag granular multilayer structures [2] without modifying the average blocking temperature (i.e., the average magnetic grain size) and the almost perpendicular magnetic alignment of the granular layers. It was also shown [1], that the heterostructures containing both granular and continuous magnetic layers exhibit a difference of the low field magnetization after field cooling (FC) and zero field cooling (ZFC), which is mainly due to the continuous layers and depends on the layer sequence. Here we report on the investigation of the Mössbauer spectra of these samples above the blocking temperature (T_B=40 K) in applied magnetic field.

J. Balogh · D. Kaptás · L. F. Kiss (✉) · T. Kemény · I. Vincze
Research Institute for Solid State Physics and Optics, 1525 Budapest, P. O. Box 49, Hungary
e-mail: kissl@szfki.hu

K. Temst · C. V. Haesendonck
Laboratorium voor Vaste-Stoffysica en Magnetisme,
K.U.Leuven, Celestijnenlaan 200 D, 3001 Leuven, Belgium

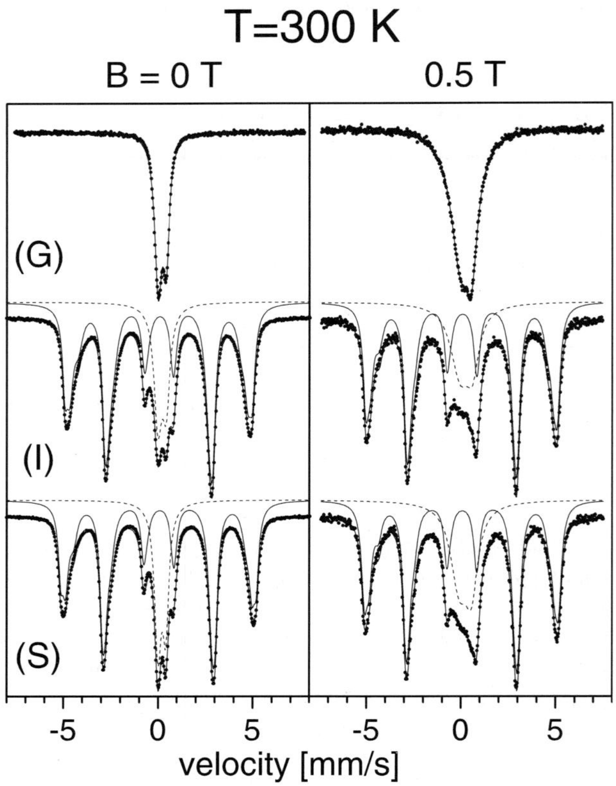

Figure 1 Mössbauer spectra of the granular Fe/Ag multilayer (*G*) and the heterostructures containing interleaved (*I*), and separated (S) continuous Fe layers, measured at 300 K with and without the application of 0.5 T magnetic field lying in the sample plane. The granular components (*the sum of two single lines*) and the layer components (*the sum of three sextets*) of (*I*) and (*S*) are indicated by *dotted and continuous lines*, respectively.

The layer sequence of the granular sample (G), the interleaved (I), and separated (S) heterostructures was the following:

G [0.2 nm ^{57}Fe/2.6 nm Ag]$_{75}$
I [(0.2 nm ^{57}Fe/2.6 nm Ag)$_3$/(1.5 nm ^{57}Fe/2.6 nm Ag)]$_{25}$
S [1.5 nm ^{57}Fe/2.6 nm Ag]$_{25}$/[0.2 nm ^{57}Fe/2.6 nm Ag]$_{75}$

Details of the sample preparation and structural characterization can be found in [1]. For the Mössbauer measurements the samples were removed from the Si substrate. The transmission Mössbauer spectra were measured by a standard constant acceleration spectrometer with ^{57}Co(Rh) source. At room temperature the magnetic field was applied in the sample plane (parallel field) with a conventional electromagnet of 0.55 T maximum fields. The homogeneity of the magnetic field was tested by a Hall probe and by measurements of the Fe calibration foil. The Fe hyperfine field was found decreasing by the value of the applied field and no line broadening was detected. At low temperature the samples were measured in a Janis cryostat with a 7 T superconducting magnet and the magnetic field was perpendicular to the sample plain.

The spectra of the three samples measured at 300 K in zero and 0.5 T parallel field are shown in Figure 1. The parameters of the fitting of the granular component by two single lines and the continuous layers by three sextets [3] are shown in Table I. The average hyperfine field of the layer components are slightly different [4] for (I) and (S), but the parameters of the granular component agree well within the experimental errors for the

Table I Parameters derived from fitting the Mössbauer spectra of Figure 1 by two single lines and three sextets: the line positions and width of the two single lines (P1, P2, W1, W2), the relative area under the two single lines ($I_{granular}$), the average hyperfine field and isomer shift (HF_{av} and IS_{av}) of the magnetic layer

External magnetic field	Sample	P1 mm/s	P2 mm/s	W1 mm/s	W2 mm/s	$I_{granular}$ %	HF_{av} T	IS_{av} mm/s
B=0 T	(G)	−0.01 (1)	0.41 (1)	0.46 (2)	0.40 (2)	100	–	–
	(I)	−0.02 (1)	0.41 (1)	0.50 (4)	0.48 (4)	21 (5)	29.0(2)	0.06(1)
	(S)	−0.01 (1)	0.40 (1)	0.44 (4)	0.39 (4)	24 (5)	30.4(2)	0.06(1)
B=0.5 T	(G)	−0.05 (2)	0.55 (2)	1.26 (4)	0.79 (4)	100	–	–
	(I)	−0.04 (2)	0.61 (2)	1.24 (8)	0.80 (8)	26 (5)	30.0(3)	0.06(2)
	(S)	−0.08 (2)	0.57 (2)	1.11 (8)	0.80 (8)	30 (5)	30.6(3)	0.06(2)

three samples at both applied field values. The applied parallel field does not influence significantly the magnetic alignment of the continuous layers (the intensities of the second and fifth lines of the sextets are 3.5 and 3.6 for (I) and (S), respectively), but causes an appreciable broadening of the superparamagnetic component. The similar broadening observed in the three samples supports our conclusion derived from the blocking temperatures [1], that the inserted continuous magnetic layers have negligible effect on the superparamagnetic properties of the granular layers.

The spectra measured at 150 K in zero, 0.25 T and 0.75 T perpendicular applied field are shown in Figure 2. The granular components of the 0.25 T spectra were fitted with a single broadened line because of the larger statistical error compared to that of the spectra measured at 300 K. The linewidth of this component was 1.6, 1.7, and 1.7 mm/s for samples (G), (I), and (S), respectively. A slightly smaller value, 1.4 (±0.05) mm/s, was obtained when the granular component of the 300 K, 0.5 T spectra was also fitted with a single line neglecting the quadrupole splitting (0.42 mm/s). In case of interaction-free superparamagnetic particles [5] the magnetization measured at different temperatures and fields follow a single curve when scaled as B/T, the applied field divided by the temperature. Magnetization of the granular sample was indeed found [1] to follow this scaling above the blocking temperature. The similar broadening of the granular sample in 0.25 T at 150 K and in 0.5 T at 300 K is in line with this observation and it seems that the ferromagnetic layers do not significantly perturb this behavior, as well.

While the broadening of the granular component is quite similar for the three samples both in parallel and perpendicular applied fields, the spectra of the continuous layers of samples (I) and (S) show a difference in perpendicular applied fields. Although the hyperfine parameters of the layer components for samples (I) and (S) agree well, the intensity of the second and fifth lines (I_{2-5}) is quite different, especially in 0.75 T. This is well seen in the spectra of Figure 2, where I_{2-5} changes from 3.5 and 3.6 at 0 T to 0.8 and 1.3 at 0.75 T for (I) and (S), respectively. (We note that in the evaluation of the 0.75 T spectra the line positions for the granular component were fixed to those of sample (G) because of the strong overlap between the lines of the granular and the layer components.) Magnetization measurements performed up to 5 T magnetic field also indicate that the saturation of sample (S) is more sluggish than that of sample (I), when the magnetic field is

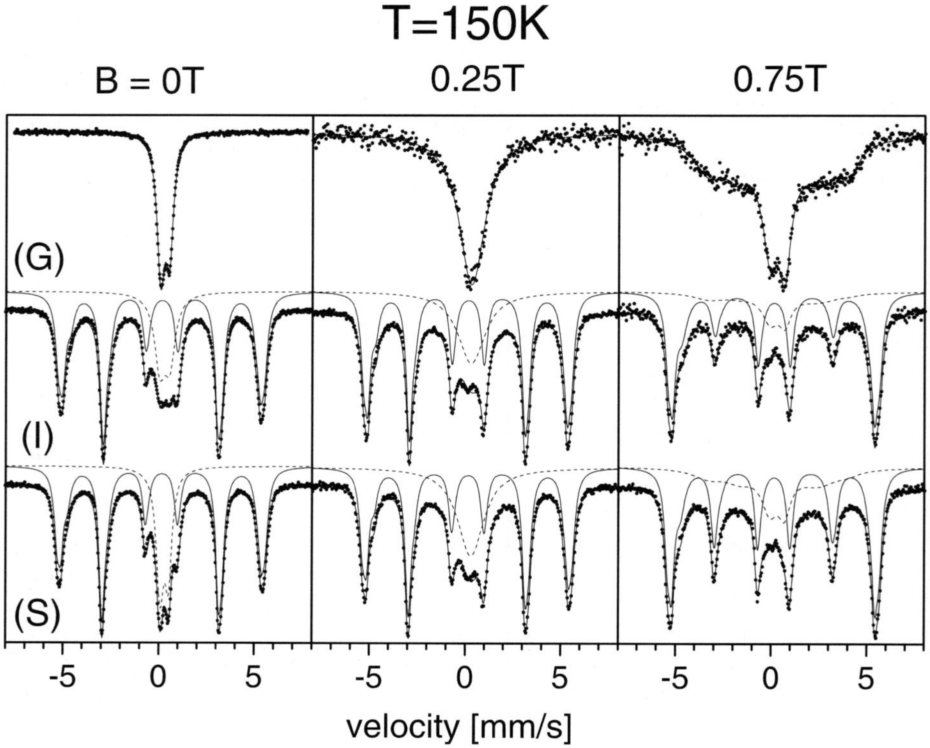

Figure 2 Mössbauer spectra of the samples (*G*), (*I*), and (*S*) measured at 150 K in magnetic fields indicated in the figure. The field was applied perpendicularly to the sample plane. The granular and the layer components of (*I*) and (*S*) are indicated by *dotted and continuous lines*, respectively.

perpendicular. The Mössbauer results undoubtedly show that it is due to the different behavior of the continuous layers and most probably reflects differences of the micro-structure due to the different layer sequence.

In conclusion the broadening of the superparamagnetic lines due to small applied fields were studied in an Fe/Ag granular multilayer and in multilayer hetrostructures with additional continuous Fe layers deposited in two different sequences. The superparamagnetic component showed similar broadening in the different samples both in parallel and perpendicular applied fields and the broadening can be scaled to B/T. The magnetic saturation process of the continuous Fe layers in perpendicular applied field was found to depend on the layer sequence. The results demonstrate that in a magnetic composite, containing both superparamagnetic and ferromagnetic particles, the microstructure can influence more significantly the ferromagnetic properties than the superparamagnetic ones.

Acknowledgements This work was supported by the Hungarian Research Found OTKA T048965 and T046795.

References

1. Balogh, J., Kaptás, D., Kiss, L.F., Pusztai, T., Szilágyi, E., Tunyogi, Á., Swerts, J., Vandezande, S., Temst, K., Van Haesendonck, C.: Appl. Phys. Lett. **87**, 102501 (2005)
2. Balogh, J., Kaptás, D., Kemény, T., Kiss, L.F., Pusztai, T., Vincze, I.: Hyperfine Interact. **141/142**, 13 (2002)
3. Schurer, P.J., Celinski, Z., Heinrich, B.: Phys. Rev. B **51**, 2506 (1995)
4. Gutierrez, C.J., Qiu, Z.Q., Tang, H., Wieczorek, M.D., Mayer, S.H., Walker, J.C.: Phys. Rev. B **44**, 2190 (1991)
5. Dormann, J.L., Fiorani D., Tronc, E.: In: Rice, P.I., Stuart, A. (eds.): Adv. Chem. Phys. **XCVIII**, 283 (1997)

Hyperfine Interact (2005) 165: 55–60
DOI 10.1007/s10751-006-9247-z

Evidence for direct observation by Mössbauer spectroscopy of surface tin atoms in platinum–tin particles

J.-P. Candy · E. Roisin · J.-M. Basset · D. Uzio · S. Morin · L. Fischer · J. Olivier-Fourcade · J.-C. Jumas

Published online: 26 October 2006

Abstract ^{119}Sn Mössbauer spectroscopy at 77 K was used to observe surface tin atoms on $PtSn_x/SiO_2$ catalysts. Two families of $PtSn_x/SiO_2$ catalysts with fully reduced $Pt^{(0)}$ and $Sn^{(0)}$ atoms were prepared following the Surface Organo-metallic Chemistry on Metal strategy. These two families of catalysts exhibit about the same Sn/Pt composition, but very different metallic particle diameters, each with narrow particle size distribution, so that, the ratio between surface tin atoms and "bulk" tin atoms was significantly different for the two families. In addition to the well known "Pt–Sn" bulk alloy with small quadripole splitting, a new Pt–Sn species with large quadripole splitting and isomer shift in the range of $Sn^{(0)}$ was observed. The relative proportion of this species with respect to "Pt–Sn" bulk alloy is greater for small particles than for the larger ones, which suggests that this species is a surface tin atom present in a "Pt–Sn" alloy.

Key words Mössbauer spectroscopy · surface organo-metallic chemistry on metal route · platinum–tin catalysts

1 Introduction

Platinum–tin catalysts are widely used in catalytic processes such as alkane dehydrogenation or reforming. The presence of tin has two clear benefit effects. It greatly improves the selectivity of the catalyst toward unsaturated corresponding olefinds, by decreasing the

J.-P. Candy (✉) · E. Roisin · J.-M. Basset
LCOMS-CPE, 43 bd du 11 Novembre 1918, 69616 Villeurbanne, France
e-mail: candy@cpe.fr

D. Uzio · S. Morin · L. Fischer
IFP, CEDI Solaize, BP 3, 69390 Vernaison, France

J. Olivier-Fourcade · J.-C. Jumas
LAMMI, Université de Montpellier II, pl. E. Bataillon, 34095 Montpellier, France

Table I Metal loading and metallic particle size of $PtSn_x/SiO_2$ catalysts

Ref	Pt (%)	Sn (%)	Sn/Pt	*d* (nm)	M_s/M_t
ER016	1.4	0.7	0.3	1.7	0.61
ER017	1.4	2.3	1.0	1.9	0.56
ER022	5.5	2.7	0.3	3.9	0.32
ER023	5.5	4.5	0.5	3.9	0.32

catalytic activity toward hydrogenolysis into lower alkanes [1–6]. It considerably reduces the deactivation of the catalyst which occurs by coke formation and by coalescence of the metallic particles [2, 7, 8]. While these facts are now generally known, the exact mechanism leading to these improvements is still under some controversy. Two effects are associated with the presence of tin: (a) a site isolation effect [9–11] and (b) an electronic effect, caused by the tin influencing the electronic density on platinum and thereby modifying their bonding properties [12, 13].

Numerous studies have been undertaken to characterize these metallic particles, and among them, Mössbauer spectroscopy of ^{119}Sn has appeared to be very reliable with respect to oxidation state. It has been demonstrated that the nature of the support (alumina, silica) and the route followed to prepare the catalyst greatly influence the oxidation state of the tin atoms [14–19].

Regarding the use of Mössbauer in the characterisation of Pt–Sn catalysts, numerous tin species were evidenced, but, to our knowledge, only Bowles and Cranshaw, in 1965, [20] suggested that fully reduced tin atom located on the surface of a metallic particle could present a quadripole splitting greater than zero, due to the anisotropy of the electronic cloud.

In this work, we will demonstrate that surface tin atoms can be clearly observed by ^{119}Sn Mössbauer spectroscopy, on the surface of $PtSn_x/SiO_2$ catalysts. Two families of $PtSn_x/SiO_2$ catalysts with fully reduced $Pt^{(0)}$ and $Sn^{(0)}$ atoms were prepared following the Surface Organo-metallic Chemistry on Metal route [21]. These two catalysts exhibit about the same Sn/Pt composition, but very different metallic particle diameter, with narrow particle size distribution, so that, the ratio between surface tin atoms and "bulk" tin atoms was largely different for the two families.

2 Experimental

The starting Pt/SiO_2 catalyst (AG65) was prepared from $Pt(NH_3)_4(OH)_2$ (Aldrich) and silica Aerosil from Degussa (200 m^2/g), following the procedure already described [22, 23]. The metal loading was 1.35% wt and the metal particle size was close to 1.5 nm with a very narrow size distribution.

The metal loading and the metallic particle size of the AG65 catalyst was increased by selective reaction of $Pt(acac)_2$ on the reduced surface of the Pt/SiO_2 sample, in toluene as solvent, according to the procedure already described [24]. Four successive reaction were undertaken, to obtain the ER18 catalyst, with a metal loading of 5.5% wt and a metallic particle size of about 3.9 nm with a narrow size distribution.

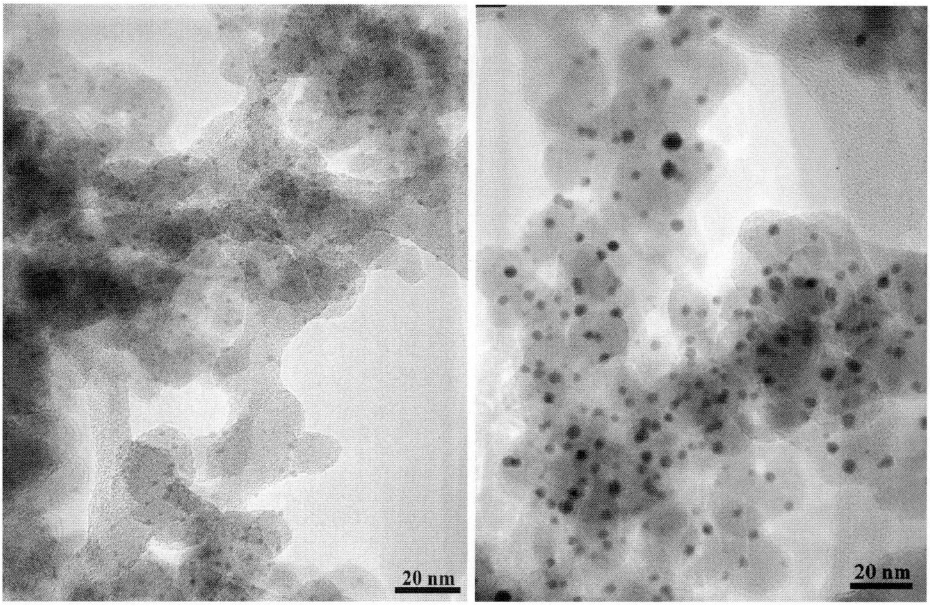

Figure 1 TEM pictures of $PtSn_x/SiO_2$ ER016 and ER023 catalysts.

$PtSn_x/SiO_2$ catalysts have been prepared following the Surface Organo-metallic Chemistry on Metal route [21, 25] by reaction, at 300 K under H_2 (atm. pressure) and in *n*-heptane as solvent, of given amounts of tetrabutyltin ($Sn(n\text{-}C_4H_9)_4$) on the reduced metallic surface of well defined Pt/SiO_2 (AG65 and ER18) "host" catalysts. Two families of $PtSn/SiO_2$ catalysts have been thus prepared with various Sn/Pt ratio and very different sizes.

The transmission electron microscopy (TEM) examination of the samples was carried out on a JEOL 2010 microscope with a high resolution pole piece. The microscope was operated at an accelerating voltage of 200 kV and gave a nominal structural reolution of 0.19 nm. Samples were sonically dispersed in ethanol and deposited on a holey carbon copper grid before examination. The size of the particles were determined by observation of more than 200 particles from about five TEM pictures.

^{119}Sn Mössbauer spectra were recorded at 77 K in transmission geometry on a standard EG&G spectrometer in the constant acceleration mode, by using a $Ba^{119m}SnO_3$ source of 10 mCi nominal activity. The velocity scale was calibrated by means of a room temperature spectrum of α-Fe recorded with a $^{57}Co(Rh)$ source. The hyperfine parameters δ (isomer shift) and Δ (quadrupole splitting) were determined by fitting Lorentzian lines to the experimental data, by using the ISO programme [19]. All isomer shifts are given with respect to the spectrum of $BaSnO_3$. The effective thickness of all absorbers was in the range 0.5–1.0 mg ^{119}Sn cm^{-2}.

Specific glass (Pyrex) Mössbauer cells have been developed for these experiments. These glass cells were sealed on the reducing reactor (Pyrex). The $PtSn_x/SiO_2$ samples were treated under flowing H_2 at 450°C during 4 h in the reducing reactor and then introduced into the Mössbauer cells, under hydrogen without any contact with air. The Mössbauer cells was then sealed under hydrogen and separate from the reducing reactor. Mössbauer spectra of these samples have been taken at 77 K.

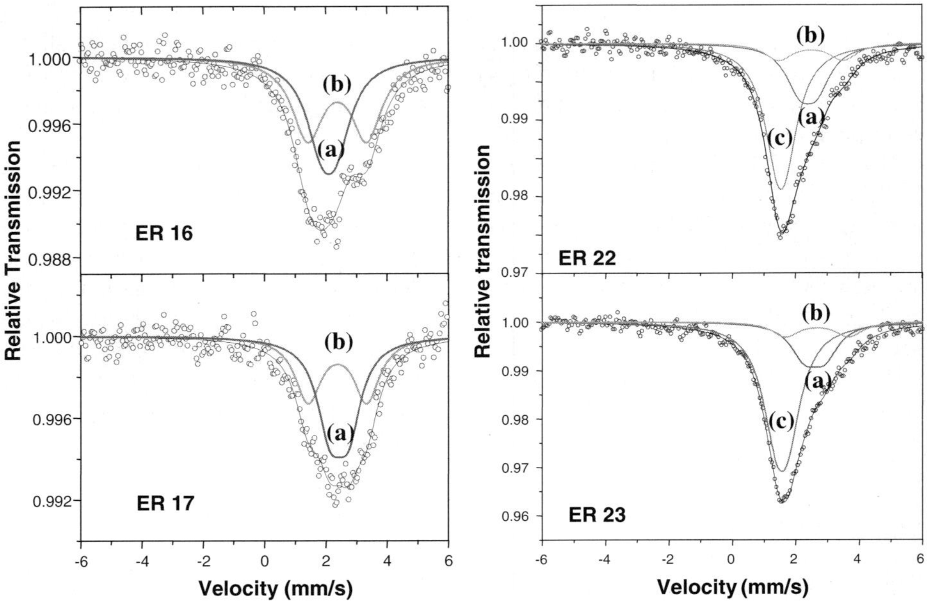

Figure 2 ^{119}Sn Mössbauer spectra recorded at 77 K of $PtSn_x/SiO_2$.

3 Results and discussion

The metal loading and the metallic particle size of the various $PtSn_x/SiO_2$ catalysts were reported on Table I. Electron microscopy pictures of these samples are shown in Figure 1.

Assuming that the particles possess a cubo–octahedral shape, the number of surface metal atoms (M_s) and total metal atoms (M_t) for $PtSn_x$ particles of about 1.7, 1.9 and 3.9 nm are reported on Table I, [26].

Mössbauer spectra of these samples taken at 77 K are reported in Figure 2. The fitted parameters are reported on Table II.

Four samples have been analysed (ER016, ER017, ER022, ER023). Two different families of species have been identified from the experimental spectra (Figure 2).

Table II Mössbauer parameters computed from spectra of the samples, recorded at 78 K

Ech.	*d* (nm)	Species	δ ($mm\cdot s^{-1}$)	Δ ($mm\cdot s^{-1}$)	*Γ* ($mm\cdot s^{-1}$)	Percent (%)
ER016	1.7	(a) $PtSn_x$	2.01	0.48	1.24	46
		(b) Sn_s	2.29	1.93	1.24	54
ER017	1.9	(a) $PtSn_x$	2.34	0.60	1.03	56
		(b) Sn_s	2.29	1.92	1.03	44
ER022	4.0	(a) $PtSn_x$	2.44	0.4	0.99	21
		(b) Sn_s	2.36	1.9	0.99	14
		(c) Pt_3Sn	1.51	0.4	0.99	65
ER023	4.0	(a) $PtSn_x$	2.51	0.64	1.07	24
		(b) Sn_s	2.60	1.99	1.07	11
		(c) Pt_3Sn	1.48	0.38	1.07	65

The first family contains several species with small Δ (less than 0.6 mm·s^{-1}) and δ of 1.51, 2.01 2.34, 2.44 and 2.51 mm·s^{-1}. These species have been attributed to $PtSn_x$ alloys. From [19], δ lower than ca. 1.7 mm·s^{-1} correspond to platinum rich particles (ca. Pt_3Sn), δ larger than ca. 2.3 mm·s^{-1} correspond to tin rich particles and δ of about 2 mm·s^{-1} correspond to approximately equal Pt and Sn proportion.

The second family contains species with large Δ (1.9 mm/s) and δ of 2.29 and 2.36 mm·s^{-1}. This species cannot be attributed to either $Sn^{(II)}O$ (δ expected at values greater than 2.6 mm·s^{-1}) or $Sn^{(IV)}O$ (δ expected at values lower than 1.0 mm·s^{-1}) [27]. According to the hypothesis of Bowles and Cranshaw [20], this species could be attributed to surface tin atoms (Sn_s), the electronic environment of which is asymmetrical. If this is true, the relative intensity of this species must be lower on large bimetallic particles than on the smaller. This is the case as shown in Table II. Therefore, we can suggest that the Mössbauer signal with high Δ (1.9 mm·s^{-1}) and δ of 2.36 mm·s^{-1} to surface tin atoms. To our knowledge, this is the first time that the differentiation between surface and bulk atoms of supported particles can be observed by Mössbauer spectroscopy.

4 Conclusion

^{119}Sn Mössbauer spectroscopy at 77 K of two families of $PtSn_x/SiO_2$ catalysts with fully reduced $Pt^{(0)}$ and $Sn^{(0)}$ and very different metallic particle size shows the well known $PtSn_x$ alloy species with small Δ and a new species with large Δ and δ in the range of $Sn^{(0)}$. The correlation between the relative proportion of this new species and the relative proportion of surface tin atoms suggests that this new species is surface tin atom.

Acknowledgment We greatly thank M. Aouine from IRC (CNRS, Villeurbanne, France) for electron microscopy pictures.

References

1. Clippinger, E., Rafael, S., Mulaskey, F.: U.S.A. Patent 3 531 5431970
2. Humblot, F., Candy, J.P., Le Peltier, F., Didillon, B., Basset, J.M.: J. Catal. **179**, 459 (1998)
3. Loc, L.K., Gaidai, N.A., Gudkov, B.S., Kiperman, S.L., Kogan, S.B.: Kinet. Catal. **27**, 1184 (1987)
4. Loc, L.C., Gaidai, N.A., Kiperman, S.L.: In: Proc. 9th Internat. Congr. Catal., vol. 3, p. 1261. Calgari, Canada (1988)
5. Siri, G.J., Casella, M.L., Santori, G.F., Ferretti, O.A.: Ind. Eng. Chem. Res. **36**, 4821 (1997)
6. Stagg, S.M., Querini, C.A., Alvarez, W.E., Resasco, D.E.: J. Catal. **168**, 75 (1997)
7. Fan, Y., Lin, L., Zang, J., Xu, Z.: Stud. Surf. Sci. Catal. **75**, 2507 (1993)
8. Seleznev, V.N., Fomichev, Y.V., Levinter, M.E.: Neftekhimiya **14**, 205 (1974)
9. Ichikawa, M., Lang, A.J., Shriver, D.T., Sachtler, W.M.H.: J. Am. Chem. Soc. **107**, 7216 (1985)
10. Dautzenberg, F.M., Helle, J.N., Biloen, P., Sachtler, W.M.H.: J. Catal. **63**, 119 (1980)
11. Biloen, P., Helle, J.N., Verbeek, H., Dautzenberg, F.M., Schtler, W.M.: J. Catal. **63**, 112 (1980)
12. Burch, R.: J. Catal. **71**, 348 (1981)
13. Burch, R., Garla, L.A.: J. Catal. **71**, 360 (1981)
14. Bacaud, R., Bussière, P., Figuéras, F.: J. Catal. **69**, 399 (1981)
15. Kuznetsov, V.I., Belyi, A.S., Yutchenko, E.N., Smolikov, M.D., Protasova, M.T., Zatolokina, E.V., Duplyakin, V.K.: J. Catal. **99**, 159 (1986)
16. Vértes, C., Tálas, E., Czakó-Nagy, I., Ryczkowski, J., Göbölös, S., Vértes, A., Margitfalvi, J.: Appl. Catal. **68**, 149 (1991)
17. Hobson, M.C., Goresh, S.L., Khare, G.P.: J. Catal. **142**, 641 (1993)
18. Sharma, L.D., Kumar, M., Saxena, A.K., Rawat, D.S., Prasada Rao, T.S.R.: Appl. Catal., A Gen. **168**, 251 (1998)

19. Olivier-Fourcade, J., Womes, M., Jumas, J.-C., Le Peltier, F., Morin, S., Didillon, B.: ChemPhysChem **5**, 1734 (2004)
20. Bowles, B.J., Cranshaw, T.E.: Phys. Lett. **17**, 258 (1965)
21. Candy, J.P., Didillon, B., Smith, E.L., Shay, T.B., Basset, J.M.: J. Mol. Catal. **86**, 179 (1994)
22. Goguet, A., Aouine, M.A., Cadete Santos Aires, F.J., De Mallmann, A., Schweich, D., Candy, J.P.: J. Catal. **209**, 135 (2002)
23. Goguet, A., Schweich, D., Candy, J.P.: J Catal **220**, 280 (2003)
24. Womes, M., Cholley, T., Le Peltier, F., Morin, S., Didillon, B., Szydlowski-Schildknecht, N.: Appl. Catal., A Gen. **283**, 9 (2005)
25. Bentahar, F.Z., Bayard, F., Candy, J.P., Basset, J.M., Didillon, F.: Catal. Today **66**, 303 (2001)
26. Van Hardeveld, R., Hartog, F.: Surf. Sci. **15**, 189 (1969)
27. Choi, J.S., Maugé, F., Pichon, C., Olivier-Fourcade, J., Jumas, J.-C., Petit-Clair, C., Uzio, D.: Appl. Catal., A Gen. **267**, 203 (2004)

Hyperfine Interact (2005) 165: 61–67
DOI 10.1007/s10751-006-9248-y

Mössbauer and magnetic study of $Co_xFe_{3-x}O_4$ nanoparticles

C. Estournès · C. D'Orléans · J.-L. Rehspringer · E. Manova · B. Kunev · D. Paneva · I. Mitov · L. Petrov · M. Kurmoo

Published online: 7 November 2006

Abstract Magnetic nanoparticles of cobalt ferrites $Co_xFe_{3-x}O_4$ (x=1 or 2) have been obtained either by mechanical milling or thermal treatment of pre-prepared layered double hydroxide carbonate x-LDH–CO_3. Mechanical milling of the 1-LDH–CO_3 leads to the large-scale preparation of nearly spherical nanoparticles of $CoFe_2O_4$, the size of which (5 to 20 nm) is controlled by the treatment time. Core-shell structure with surface spin-canting has been considered for the nanoparticles formed to explain the observed hysteresis loop shift (from ZFC–FC) in the magnetic properties. Annealing treatment of the 2-LDH–CO_3 below 673 K results in the formation of nearly spherical pure Co_2FeO_4 nanoparticles. At 673 K and above, the LDH decomposition leads to the formation of a mixture of both spinels phases Co_2FeO_4 and $CoFe_2O_4$, the amount of the latter increases with annealing temperature. Unusually high magnetic hardness characterized by a 22 kOe coercive field at 1.8 K has been observed, which reflects the high intrinsic anisotropy for Co_2FeO_4.

Key words $Co_xFe_{3-x}O_4$ (x=1, 2) nanoparticles · ferrite spinel · high energy ball-milling · high coercivity · Mössbauer

C. Estournès (✉)
Plateforme Nationale de Frittage Flash and CIRIMAT, UMR5085 CNRS – Université Paul Sabatier, 118 route de Narbonne, 31062 Toulouse Cedex 04, France
e-mail: estournes@chimie.ups-tlse.fr

C. Estournès · C. D'Orléans · J.-L. Rehspringer
I.P.C.M.S. Groupe des Matériaux Inorganiques, UMR7504 CNRS, ULP 23 rue du Loess, BP 43, 67034 Strasbourg Cedex 02, France

E. Manova · B. Kunev · D. Paneva · I. Mitov · L. Petrov
Institute of Catalysis, Bulgarian Academy of Sciences, Acad. G. Bonchev Str., bl. 11, 1113 Sofia, Bulgaria

M. Kurmoo
Laboratoire de Chimie de Coordination Organique, CNRS-UMR7140, Tectonique Moléculaire de solide, Université Louis Pasteur, Institut Le Bel, 4 rue Blaise Pascal, 67000 Strasbourg, France

1 Introduction

Nano-sized materials are currently of great technological importance and consequently, one of the research subjects under intense investigation for their mechanical, electrical or chemical properties due to the fact that the properties of the nanophases being very different from those of the bulk material. Nanoparticles of materials are likely to improve dramatically the properties of the resulting material. Consequently, the synthesis of spinel ferrite nanoparticles has been intensively studied in recent years due to their potential application in high-density magnetic recording, microwave devices and magnetic fluids.

2 Experimental

Cobalt ferrite powders ($Co_xFe_{3-x}O_4$, $x=1$ or 2) were prepared from co-precipitation of layered iron–cobalt hydroxide carbonates x-LDH–CO_3, by high-energy ball-milling or annealing thermal treatment. The starting materials used were $Fe(NO_3)_3{\cdot}9H_2O$ (purity 99%), $Co(NO_3)_2{\cdot}6H_2O$ (purity 96%) and Na_2CO_3.

The powder X-ray diffraction (XRD) patterns were recorded by use of a TUR M62 diffractometer with Co-Kα radiation ($\lambda=1.789$ Å). The observed patterns were cross-matched with those in the JCPDS database. Transmission Electron Microscopy (TEM) investigations were made using a Topcon 002B electron microscope operating at 200 kV with a point-to-point resolution $r=1.8$ Å. The samples were sonicated in ethanol and deposited on a pre-deposited polymer on a copper grid. Mössbauer spectra of nano-size cobalt ferrite particles were recorded at 295 and 77 K on a Wissel electromechanical Mössbauer spectrometer (Wissenschaftliche Elektronik GmbH, Germany) working in a constant acceleration mode. A $^{57}Co/Cr$ (activity$\cong$10 mCi) source and an α-Fe standard were used. The experimental spectra were treated using the least squares method. The parameters of hyperfine interaction such as isomer shift (IS), quadrupole splitting (QS) and effective internal magnetic field (H_{eff}) as well as the line widths (FWHM) and the relative weight (G) of the partial components of the spectra were determined. Temperature and field dependences of the magnetization of the cobalt ferrite were measured on a Quantum Design MPMS-XL SQUID operating in the temperature range 2–400 K and an applied field up to 5 T. In order to avoid sample rotation in the applied field the nanoparticles were embedded in a polymer matrix, PMMA (poly methyl methacrylate).

3 Results and discussion

Nearly spherical $CoFe_2O_4$ nanoparticles were obtained by high energy ball milling of the 1-LDH-CO_3. X-ray powder diffraction and Transmission Electron Microscopy show that the average size of the particles increases (ca. 5, 8 and 20 nm) with milling time (ca. 5, 10 and 15 h). The increase in size is also evidenced by the reduction of the microstrain that is commonly associated with a diminution of structural defects. It is interesting to note that the process of crystallization continues as a function of treatment time and at later stages is accompanied by the reduction to cobalt. Similarly, Menzel et al. have shown that high-energy milling of nickel ferrite spinels induces the formation of metallic iron [1].

Cobalt ferrite bulk material ($CoFe_2O_4$) is known to be a ferrimagnetic material with very high cubic magnetocrystalline anisotropy leading to high theoretical coercivity: 25.2 kOe at

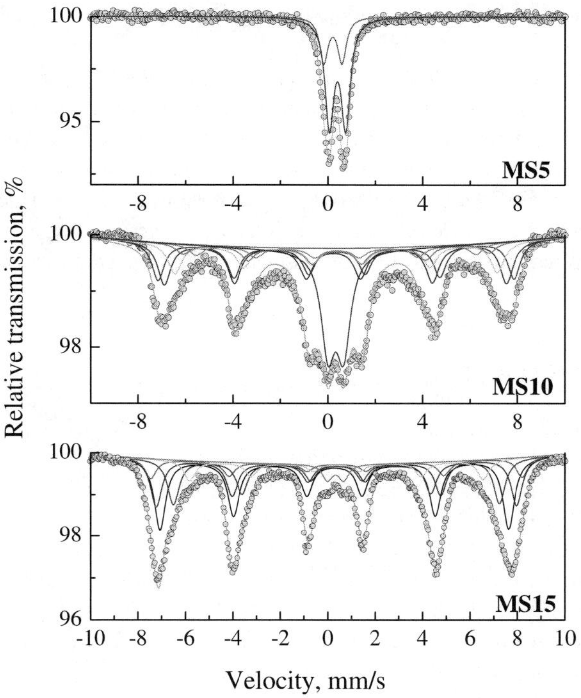

Figure 1 Mössbauer spectra of $CoFe_2O_4$ at 295 K for different milling time.

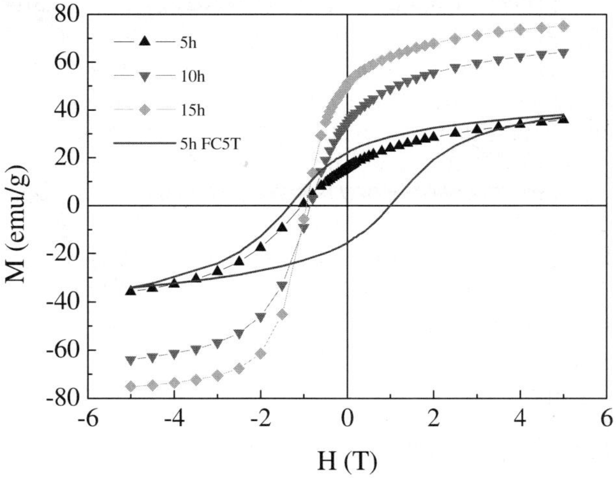

Figure 2 ZFC–isothermal magnetization (5 K) of the milled $CoFe_2O_4$ and FC (5 T) for the 5 h milled sample.

5 K (maximum observed at 2 K is 22 kOe) and 5.4 kOe at 300 K, and a saturation magnetization of 93.9 emu/g at 5 K and 80.8 emu/g at 300 K [2–4].

Concerning our samples, the absence of coercivity, remanence and saturation at 295 K for the sample milled during 5 h suggests that this material behaves as a superparamagnet at room temperature. This superparamagnetic character is confirmed by Mössbauer spectrometry, two doublets arising from the two possible iron environments in spinels are observed at room temperature. At lower measuring temperatures, these doublets tend to disappear to

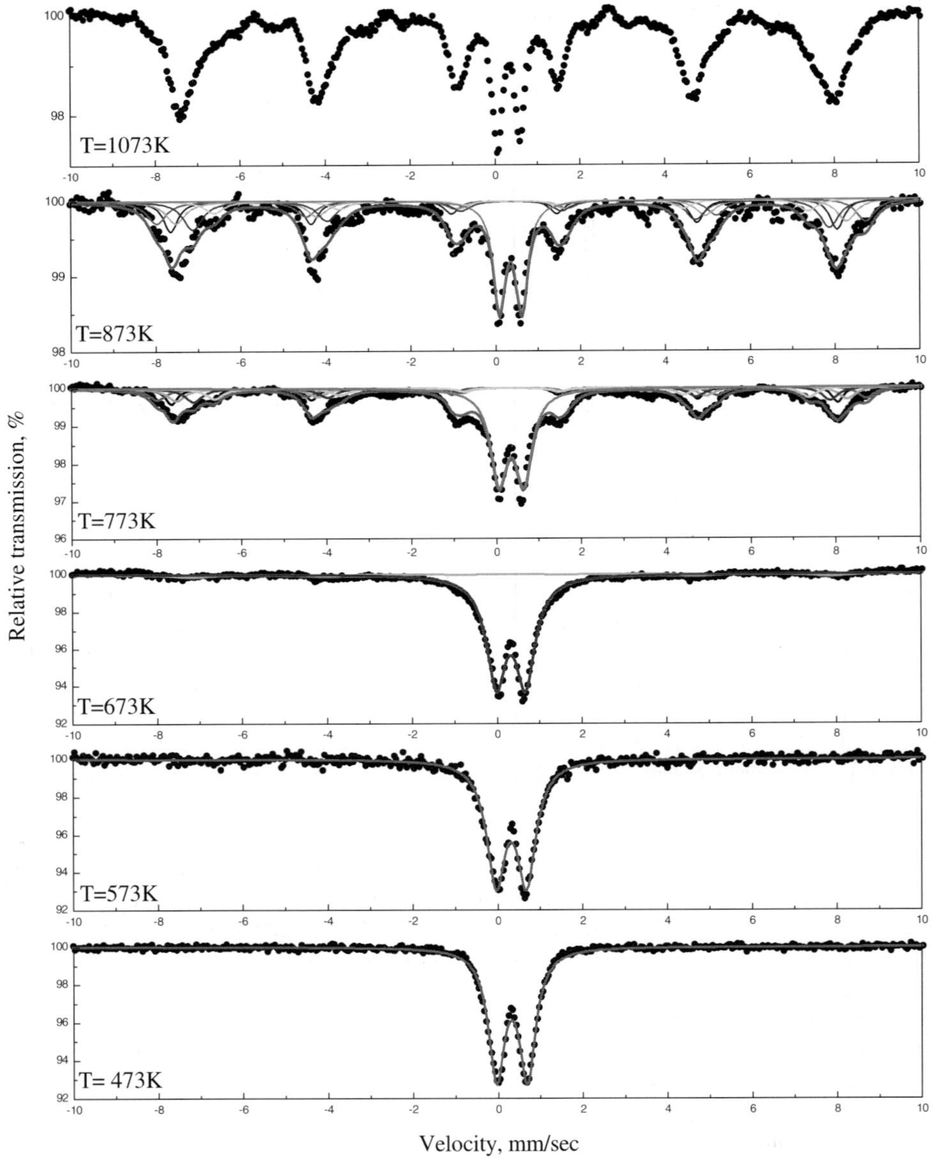

Figure 3 RT Mössbauer spectra of all synthesised Co_2FeO_4 samples.

the benefit of sextets. The blocking temperature T_B associated to the maximum in the Zero Field Cooling curve increases (ca. 200, 310 and 390 K) with milling time (ca. 5, 10 and 15 h) indicating that the samples milled for 10 or 15 h are blocked at room temperature. Magnetic splitting is evidenced on room temperature. Mössbauer spectra of the samples milled for 10 and 15 h confirm the presence of ferrimagnetic species (Figure 1). Saturation of the magnetization is never reached (in the magnetic field range explored (±5 T) for all the samples and measuring temperatures (Figure 2). However, saturation magnetization

Figure 4 Isothermal magnetization of the Co_2FeO_4 annealed at 573 K.

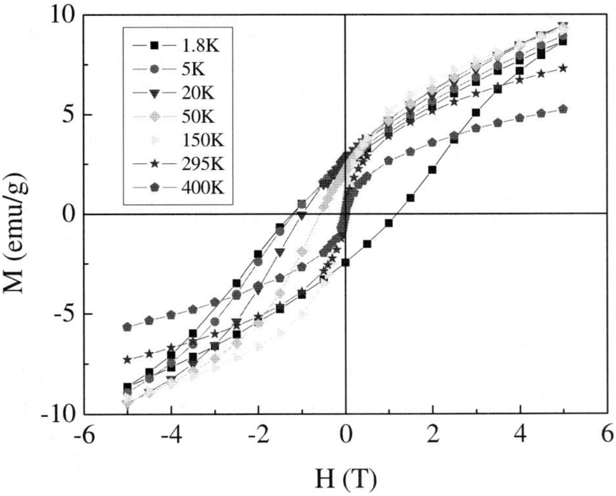

(M_s) can be estimated from zero extrapolation of M versus $1/H$ and the ratio M_r/M_s can be calculated [5]. Both calculated parameters increase with milling time but never reach the theoretical values [6]. This indicates that even for such long treatment time some of the ultrafine particles remain or that the objects formed are core/shell particles with a spin-glass like surface layer [5]. The later hypothesis is confirmed by the observed hysteresis loop shift (from ZFC to FC) that has been attributed to canting of the surface spins. Furthermore, Mössbauer spectra simulations show that the degree of inversion of cobalt ferrite increases with increasing milling time and approaches that of thermochemically synthesized samples.

Heating the 2-LDH up to 673 K leads to the formation of pure cubic [a=8.19±0.01 Å] spinel Co_2FeO_4 nanoparticles while above this temperature a second phase ($CoFe_2O_4$) is formed [a=8.37±0.01 Å]. The amount of the latter increases at the expense of the Co_2FeO_4 phase upon increasing the treatment temperature. Our results surprisingly disagree with those reported by Fereira et al. [7], where the existence of two spinel phases in the temperature range 570–1070 K is discussed. The average size of the nanoparticles of Co_2FeO_4 increases from 3 to 22 nm by increasing treatment temperature from 473 to 1073 K. The Co_2FeO_4 spinel phase has not been extensively reported in the literature. According to Takahashi et al. [8] bulk Co_2FeO_4 has a Curie temperature around T_C=425 K, saturation magnetization around M_s=15 emu/g and coercive force H_c of the order of 2 T.

For annealing temperature below 673 K, the nanoparticles formed behave as superparamagnets at room temperature. Mössbauer spectrometry indicates that only Fe^{3+} iron are present. The blocking temperature T_B associated to the maximum in the Zero Field Cooling curve increases (ca. 45, 220 and 250 K) with annealing temperature (ca. 473, 573 and 673 K). However, for the latter temperature coercivity is observed at room temperature on magnetization curve which can be associated to the presence of small quantity of $CoFe_2O_4$ particles. Room temperature Mössbauer spectroscopy was applied to all synthesized products. Figure 3 shows the RT Mössbauer spectra of the samples obtained after treatment at different temperatures. The doublets observed in all samples can be attributed to cobalt ferrite phase rich in cobalt (Co_2FeO_4), while the sextet component can be related with the presence of $CoFe_2O_4$. These results agree with the XRD data, where for samples annealed at temperatures higher than 673 K a second phase ($CoFe_2O_4$) is observed. When the

Table I Magnetic parameters for the Co_2FeO_4 samples

Samples	Temperature (K)	Hc (T)	M_r (emu/g)	Estimated M_s (emu/g)	M_r/M_s	T_B (K)
473 K	295	0	0	6	0	45
	5	0.68	0.93	9	0.10	
573 K	295	0	0	12	0	220
	5	1.20	2.57	13	0.19	
673 K	295	0.02	0.69	23	0.03	250
	5	1.50	4.71	30	0.16	

Mössbauer measuring temperature is decreased these doublets tend to disappear to the benefit of sextets.

Magnetization measurements have been performed, on pure Co_2FeO_4 samples immobilized in a polymer matrix (PMMA), as a function of applied magnetic field and temperature (Figure 4). Several important observations can be made from these experiments. First is the large opening of the hysteresis loop at low temperature. The coercive force can reach an unprecedented value of 2.2 T at 1.8 K for field-cooled sample. The second point is that saturation magnetization is far from being reached in the magnetic field explored. These two observations reflect the high intrinsic anisotropy for Co_2FeO_4 and one can expect higher coercivity in extended applied magnetic field. However, saturation magnetization can be estimated from zero extrapolation of M versus $1/H$ and the ratio M_r/M_s can be calculated. It is observed that these two parameters (Table 1) increase with annealing temperature in agreement with the increase of the average size of the particles. The estimated magnetization saturation (30 emu/g at 5 K) for the sample annealed at 673 K is higher than the value given by Takahashi et al. [8]. Our over-estimation is due to the fact that at this temperature the sample contains a small quantity of $CoFe_2O_4$. The ratio M_r/M_s obtained are far below the expected values for randomly oriented nanoparticles [9]. This is probably due to the fact that the hysteresis loops obtained in the magnetic field range explored is in fact only an inner loop. Further experiments in magnetic field up to 60 Tesla will be necessary.

4 Conclusion

Large-scale preparation of cobalt ferrite $Co_xFe_{3-x}O_4$ (x=1 or 2) nanoparticles has been achieved by combining chemical precipitation of layered double hydroxide carbonate (x-LDH–CO_3, x=1 or 2) and their treatments by high-energy mechanical milling or thermal decomposition. The size of the particles increases with increasing milling time or temperature. Controlled temperature treatment of 2-LDH–CO_3 results in the thermodynamically less stable Co_2FeO_4 which is progressively transformed to the more stable $CoFe_2O_4$ at higher temperatures. Magnetic properties display finite size effects and in the case of $CoFe_2O_4$, core-shell structure with a canting of surface spins has been considered to explain the observed hysteresis loop shift (from ZFC to FC). For Co_2FeO_4 nanoparticles with diameter of 6 nm, unprecedented coercivity of 2.2 T has been measured at 1.8 K.

Acknowledgments The authors thank the National Science Fund of the Bulgarian Ministry of Education and Science for the financial support of Project X-1504/05.

References

1. Menzel, M., Šepelák, V., Becker, K.D.: Solid State Ion. **141–142**, 663 (2001)
2. Grigova, M., Blythe, H.J., Blaskov, V., Rusanov, V., Petkov, V., Masheva, V., Nihtianova, D., Martinez, Ll.M., Munõz, J.S., Mikhov, M.: J. Magn. Magn. Mater. **183**, 163 (1998)
3. Chinnasamy, C.N., Jeyadevan, B., Shioda, K., Tohji, K., Djayaprawira, D.J., Takahashi, M., Joseyphus, R.J., Narayanasamy, A.: Appl. Phys. Lett. **83**, 2862 (2003)
4. Hutlova, A., Nyznansky, D., Rehspringer, J.L., Estournès, C., Kurmoo, M.: Adv. Mater. **15**, 1622 (2003)
5. Tung, L.D., Kolesnicheko, V., Caruntu, D., Chou, N.H., O'Connor, C.J., Spinu, L.: J. Appl. Phys. **93**, 7486 (2003)
6. Manova, E., Kunev, B., Paneva, D., Mitov, I., Petrov, L., Estournès, C., d'Orléans, C., Rehspringer, J.L., Kurmoo, M.: Chem. Mater. **16**, 5689 (2004)
7. Ferreira, T.A.S., Waerenborgh, J.C., Mendoca, M.H.R.M., Nunes, M.R., Costa, F.M.: Solid State Sci. **5**, 383 (2003)
8. Takahashi, M., Fine, M.: J. Appl. Phys. **43**, 4205 (1972)
9. Berkowitz, A.E., Kneller, E.: Magnetism and Metallurgy. Academic, New York (1, chap 8) (1969)

Hyperfine Interact (2005) 165: 69–73
DOI 10.1007/s10751-006-9249-x

Low temperature CEMS of Sn-implanted SiO_2

R. Mantovan · S. Spiga · M. Fanciulli

Published online: 31 October 2006

Abstract Conversion electron Mössbauer spectroscopy (CEMS) at room and low temperature has been used to study thin SiO_2 films implanted with Sn atoms and annealed at 900°C. This work focuses on the determination of the Debye temperature (θ_D) and Debye–Waller factors (f) of the Sn oxidized phases formed in this system. The Sn^{2+} oxidation state is the predominant one, even if a small percentage of the Sn atoms is in the Sn^{4+} oxidation state. The real Sn-oxides fractions are calculated by normalizing the resonant areas to the f values, as calculated from the temperature dependence of the related resonant areas within a Debye model. The Sn^{4+} oxidation state, possibly related to Sn atoms close to the SiO_2 surface, represents less than 20% of the Sn atoms. For the Sn^{2+} oxidation state, two different electronics configurations *a* and *b*, having different Debye temperature and hyperfine parameters are identified. The component *a*, with a lower θ_D (137 K), is the predominant one and might be related to small (2–3 nm) amorphous SnO_x clusters in the SiO_2 matrix. The component *b* could be related to substitutional Sn atoms in the SiO_2 network forming a local Sn environment similar to the SnO amorphous compound.

Key words Sn nanocrystals · low-temperature conversion electron Mössbauer spectroscopy · Debye temperature

1 Introduction

Metallic and semiconducting nanocrystals embedded in dielectric materials are attracting great interest due to their novel electrical and optical properties. In particular, for Sn nanocrystals formed in SiO_2 by ion implantation, single electron charging effects [1] and new luminescence properties [2] have been observed. Memory devices using nanocrystals as charge storage elements have been also demonstrated [3]. In a previous work, we performed a detailed study on the local structure of Sn-implanted in thin SiO_2 films by means of conversion electron Mössbauer spectroscopy (CEMS) at room temperature (RT),

R. Mantovan (✉) · S. Spiga · M. Fanciulli
Laboratorio Nazionale MDM-CNR-INFM, via C. Olivetti 2, Agrate Brianza 20041, Italy
e-mail: roberto.mantovan@mdm.infm.it

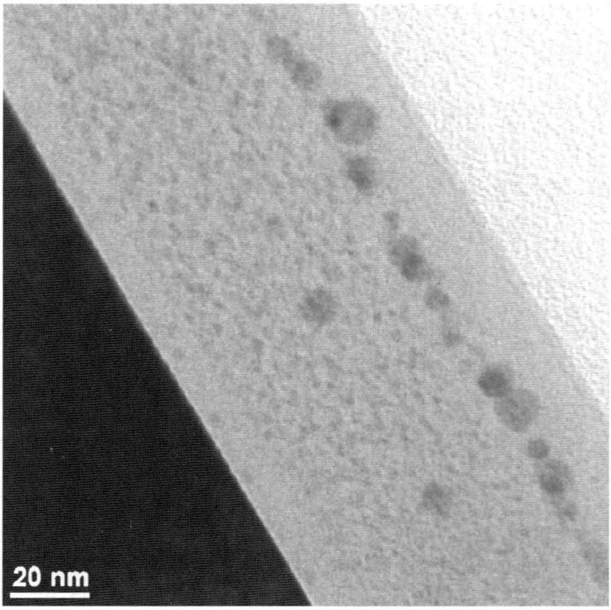

Figure 1 TEM image of the sample annealed at 900°C, 30 s.

extended X-ray absorption spectroscopy (EXAFS), and transmission electron microscopy (TEM) [4]. The results there obtained show that in the as-implanted samples all the Sn atoms are oxidized (with both Sn^{2+} and Sn^{4+} oxidation states present), while annealing induces the formation of metallic β-Sn nanocrystals, leaving a fraction of Sn atoms oxidized. Figure 1 shows a TEM image of the sample annealed at 900°C for 30 s. It is evident the presence of clusters of different diameter, in the range 2–10 nm.

High-resolution TEM (not shown) evidences that the biggest clusters are crystalline and associated with β-Sn. Small amorphous cluster (2–3 nm) are also shown by TEM either in the as-implanted and annealed samples. The CEM-spectra at RT of the implanted samples show the presence of a component related to Sn^{4+} and two doublets related to different Sn^{2+} configurations. The latter contributions were attributed to small amorphous SnO_x clusters and to Sn atoms dissolved in the SiO_2 matrix [4].

In the present work, we report on low temperature (LT) CEMS performed on the Sn-implanted SiO_2 samples, with the aim to gain more knowledge on the nature of the oxidized phases. Indeed, CEMS performed at different temperatures allows the determination of the Debye–Waller factors f and a better identification of the different phases. The corresponding Debye temperatures (θ_D) are used to make a comparison between the different phases.

2 Experimental

Thin SiO_2 films (85 nm thick) are implanted with $^{119}Sn^+$ ions at RT, with energy of 80 keV and fluence 1×10^{16} cm^{-2}. Rapid thermal annealing follows ion implantation, in order to both recover the implantation damage and to obtain a thermally activated and controlled redistribution of the implanted ions by phase separation.

Table I Hyperfine parameters for the phases Sn^{2+}(a), Sn^{2+}(b), and Sn^{4+}

Sample	Parameters	Sn^{2+}(a)		Sn^{2+}(b)		Sn^{4+}	
	Temperature (K)	295	123	295	123	295	123
900°C	δ	2.92	2.98	3.10	3.10	−0.10	0.10
30 s	Δ	2.26	2.30	2.05	2.00	0.70	0.70
	Γ	0.98(5)	0.98(4)	0.85	0.85	0.85	0.85
	A	40(3)		47(3)		12(1)	
	θ_D (K)	137(7)		168(13)		210(37)	
	f(295 K)	0.060(17)		0.153(44)		0.294(90)	
	A_{real}	61(14)		30(20)		9(9)	

δ is the isomer shift, Δ is the quadrupole splitting, Γ is the line width, and A is the spectral fraction as obtained from the fit. θ_D is the Debye temperature calculated with the Debye model from the T-dependent resonant absorption area. The recoilless fraction at RT is indicated with f, and A_{real} is the spectral contribution obtained by normalizing A to the real f value.

CEMS measurements have been performed at RT (295 K) and LT (123 K) on the Sn-implanted samples annealed at 900°C for 30 s, 900°C for 120 s, and 1,100°C for 30 s. We use a 15 mCi ^{119m}Sn source in a $CaSnO_3$ matrix, which is moved by a standard constant acceleration drive. The samples are incorporated as electrodes in a specially developed parallel-plate avalanche counter, which allows to measure down to 120 K. The detector is filled with a $He-CH_4$ gas. Details on the LT-CEMS experimental set-up will be given elsewhere (Mantovan and Fanciulli, unpublished). The isomer shifts are given relative to $CaSnO_3$ at RT.

3 Results and discussion

Figure 2 shows the RT and LT spectra for the sample annealed at 900°C for 30 s.

CEM-spectra are fitted in terms of four contributions, related to the oxidized Sn^{2+}(a), Sn^{2+}(b), Sn^{4+} phases, and to the metallic β-Sn phase [4]. We use the least-squares Normos-90 fitting program [5] to fit the experimental data. The sample annealed at 900°C for 30 s shows the lowest contribution from the β-Sn phase, therefore it is well suited to investigate the oxidized phases. The discussion of the β-Sn phase will be reported elsewhere (Mantovan et al., unpublished). Table I summarizes the hyperfine parameters (for the oxidized phases), i.e., the isomer shift (δ), the quadrupole splitting (Δ), and the line width (Γ), as obtained from the fit of the CEM-spectra shown in Figure 2. The new RT measurements reveal, within the experimental errors, the same hyperfine parameters obtained previously on the same samples [4]. Table I reports also the calculated θ_D and the corresponding f values at RT. θ_D is calculated according to the expression of f in the Debye model

$$f = \exp\left\{-\frac{3E_R}{2k\Theta_D}\left[1+4\left(\frac{T}{\Theta_D}\right)^2\int_0^{\frac{\Theta_D}{T}}\frac{x}{e^x-1}\mathrm{d}x\right]\right\} \tag{1}$$

where E_R is the recoil energy of the γ-ray, k is the Boltzman constant, and T is the samples temperature. Since, for each component, f is proportional to the resonant absorption area A, we determine the θ_D for the oxidized phases from the temperature dependence of A. Finally,

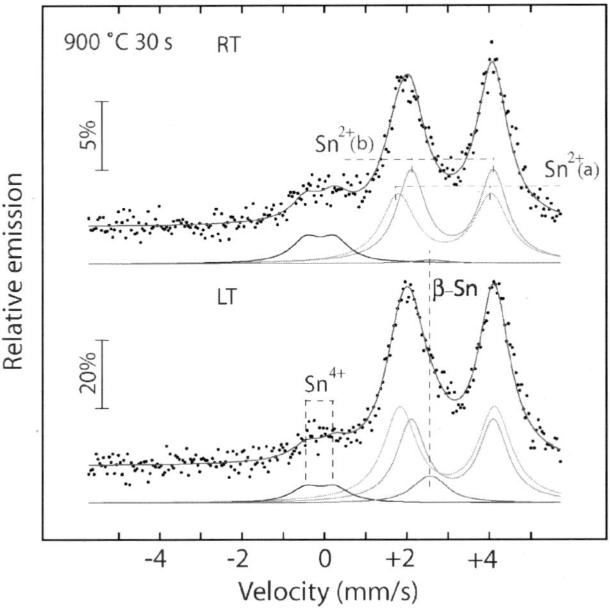

Figure 2 CEM-spectra at RT and LT for the sample annealed at 900°C for 30 s. The spectra are fitted using four components: Sn^{2+}(a), Sn^{2+}(b), Sn^{4+}, and β-Sn.

we calculate the real site-populations (A_{real}) by normalizing the spectral fractions A to the measured f factors for all the Sn-related phases.

In contrast to the results based on CEMS at RT [4] for the sample annealed at 900°C for 30 s, it is found that the Sn^{2+}(a) phase is predominant with respect to the Sn^{2+}(b) one (Table I). For the other samples (annealed at 900°C for 120 s and 1,100°C for 30 s), the considerable increase of the β-Sn fraction prevents to calculate A_{real} for the oxides, with reasonable errors. Less than 20% of Sn atoms are in the Sn^{4+} local electronic configuration. This oxidized phase shows a lower θ_D, compared to the reported value for crystalline and amorphous SnO_2 compounds [6]. A lowering of θ_D has been observed in the case of Sn^{4+} close to the surface of float glasses [7]. Moreover, for a bulk reference β-Sn sample, which has a fraction of SnO_2 on its surface, we calculate θ_D=215(18) K for SnO_2, which corresponds to the value measured for the Sn^{4+} oxidation state (Table I) in this work. We might, therefore, relate the Sn^{4+} oxidation state detected in this work to oxidized Sn atoms close to the SiO_2 surface.

The two different Sn^{2+} local configurations have distinct θ_D values. θ_D for Sn^{2+}(a) is well below the values reported for crystalline and amorphous SnO, which are 203(1) and 181(2) K, respectively [6]. On the other hand, for Sn^{2+}(b) the calculated θ_D is very close to that of the amorphous SnO compound (Table I). For Sn^{2+}(a) and Sn^{2+}(b), the hyperfine parameters δ and Δ are higher than those characterizing SnO, either crystalline or amorphous [6]. We suggest that Sn^{2+}(a) contribution in the CEM-spectra is related to the small (2–3 nm) amorphous "SnO" clusters, also detected by TEM (Figure 1), in which the Sn atoms are located in a highly disordered local environment. This is supported by the high line width measured for this phase (Table I), together with the very high Δ value, when compared to the values reported in literature for SnO [6]. The component b might be related to Sn atoms dispersed in the SiO_2 matrix. After ion implantation and annealing Sn might be present into the SiO_2 matrix in substitutional position. The Sn local environment is similar to that one of the SnO amorphous compound, as confirmed by the measured θ_D (168 K) and hyperfine parameters. Another possible explanation is that the two distinct components *a* and *b* are due to different SnO_x mean cluster size. Following annealing processes at higher temperatures or for longer time, CEMS revealed a decrease of the spectral intensity related to component Sn^{2+}(b) [4]. This is possibly due to the thermal instability of SnO, which may dissociate forming β-Sn clusters.

4 Conclusions

Temperature-dependent CEMS is used to calculate the Debye–Waller factors for the oxidized phases in the Sn-implanted samples. The temperature dependence of the resonant absorption area, within the Debye model, allows the determination of θ_D. The SnO_2 phase shows a lower θ_D, when compared to the bulk value, and we conclude that this is due to SnO_2 distributed near the sample surface. The Debye temperature calculation confirms the presence of two different Sn^{2+} local configurations, since they show two well distinct θ_D values. We calculate the real site-populations for the oxidized phases by normalizing the resonant absorption area to the calculated *f* values.

References

1. Nakajima, A., Futatsagi, T., Nakao, H., Usoki, T., Horiguchi, N., Yokoyama, N.: J. Appl. Phys. **84**, 1316 (1998)
2. Lopes, J.M., Zawislak, F.C., Fichter, P.F.P., Lovey, F.C., Condò, A.M.: Appl. Phys. Lett. **86**, 23101 (2005)
3. Futatsagi, T., Nakajima, A., Nakao, H.: FUJITSU Sci. Tech. J. **34** (1998)
4. Spiga, S., Mantovan, R., Fanciulli, M., Ferretti, N., Boscherini, F., d'Acapito, F., Schmidt, B., Gröetzschel, R., Mücklich, A.: Phys. Rev., B **68**, 205419 (2003)
5. Brandt, R.A.: Wissenschaftliche GmBH. Starubey, Germany (1994)
6. Collins, G.S., Kachnowski, T., Benczer-Koller, N., Pasternak, M.: Phys. Rev., B **19**, 1369 (1979)
7. Williams, K.F.E., Johnson, C.E., Nikolov, O., Thomas, M.F., Johnson, J.A., Greengrass, J.: J. Non-Cryst. Solids **242**, 183 (1998)

Hyperfine Interact (2005) 165: 75–80
DOI 10.1007/s10751-006-9250-4

Surface properties of $Fe_{76}Mo_8Cu_1B_{15}$ alloy after annealing

M. Miglierini · T. Kaňuch · Y. Jirásková · M. Mashlan · R. Zbořil · M. Vůjtek

Published online: 26 October 2006

Abstract NANOPERM-type alloy $Fe_{76}Mo_8Cu_1B_{15}$ is investigated in amorphous and in partially crystallized state. Samples were prepared by 1 h isothermal annealing in vacuum at temperatures ranging from 330°C up to 700°C. Bulk and surface microstructural characteristics were studied using transmission and conversion electron Mössbauer effect techniques, respectively. Surface features were checked by the help of atomic force microscopy. Presence of nanocrystalline bcc-Fe phase was detected during the first crystallization stage. The crystallization process starts at 450°C and it is more pronounced in surface regions than in the bulk. With progressing crystallization, hyperfine parameters especially of the amorphous residual phase are altered. Distinctions in surface morphology are revealed between wheel and air side of the ribbon-shaped samples.

Key words Mössbauer effect · nanoscale materials · magnetic properties of nanostructures

PACS 76.80.+y · 61.46.+w · 75.75.+a

1 Introduction

$Fe_{76}Mo_8Cu_1B_{15}$ belongs to the NANOPERM-type nanocrystalline alloys. Recently we have studied the correlation between magnetic properties and structural arrangement of this alloy which was in amorphous [1] and in nanocrystalline state. The latter was examined

M. Miglierini (✉) · T. Kaňuch
Department of Nuclear Physics and Technology, Slovak University of Technology, Ilkovičova 3, 812 19 Bratislava, Slovakia
e-mail: marcel.miglierini@stuba.sk

Y. Jirásková
Institute of Physics of Materials, Academy of Sciences of the Czech Republic, Žižkova 22, 616 62 Brno, Czech Republic

M. Miglierini · M. Mashlan · R. Zbořil · M. Vůjtek
Departments of Experimental Physics and Physical Chemistry, Palacky University, Svobody 26, 771 46 Olomouc, Czech Republic

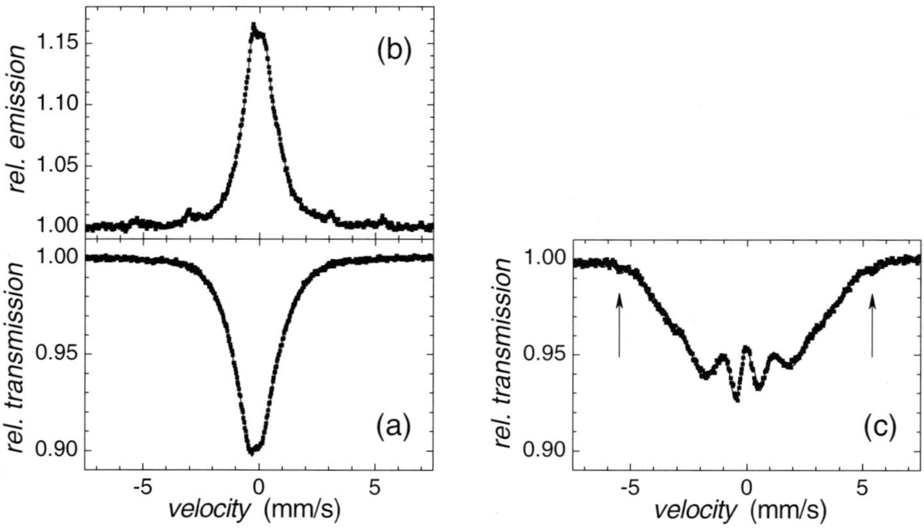

Figure 1 RT transmission (a), CEMS (b), and LNT TMS (c) Mössbauer spectra of annealed (450°C) $Fe_{76}Mo_8Cu_1B_{15}$ alloy. The *arrows* mark traces of crystalline component.

both from macroscopic [2] and from microscopic [3] viewpoints. Methods of structural analyses such as X-ray diffraction, transmission electron microscopy, high resolution electron microscopy, and differential scanning calorimetry were applied along with transmission Mössbauer effect studies [3]. Evolution of magnetic microstructure with temperature was reported in [4].

Nevertheless, all experiments performed in [1–4] provided information on the bulk of the studied samples employing transmission Mössbauer spectrometry (TMS) as a principal method of investigation. In the present work, we report on conversion electron Mössbauer spectrometry (CEMS) measurements aimed to elucidate surface properties of this system. Additional characterization of surface features is performed by atomic force microscopy (AFM).

2 Experimental details

An originally amorphous $Fe_{76}Mo_8Cu_1B_{15}$ alloy prepared by planar flow casting in a form of about 6 mm wide and 20 μm thick ribbon was subjected to 1-h isothermal treatment at annealing temperatures 330°C<T_a<700°C in a vacuum. ^{57}Fe Mössbauer effect experiments with a ^{57}Co(Rh) radiation source were performed in transmission mode at room (RT) and liquid nitrogen (LNT) temperatures and using the detection of conversion electrons at room temperature. AFM measurements were executed on the AFM Explorer (Thermomicroscopes, USA) in the non-contact mode. No special treatment was applied to the sample surface prior to inspection by the tips with a nominal radius of 10 nm.

3 Results and discussion

The samples annealed at temperatures below 450°C are amorphous and show broadened V-shaped Mössbauer patterns like the spectrum in Figure 1a. Such Mössbauer spectra are

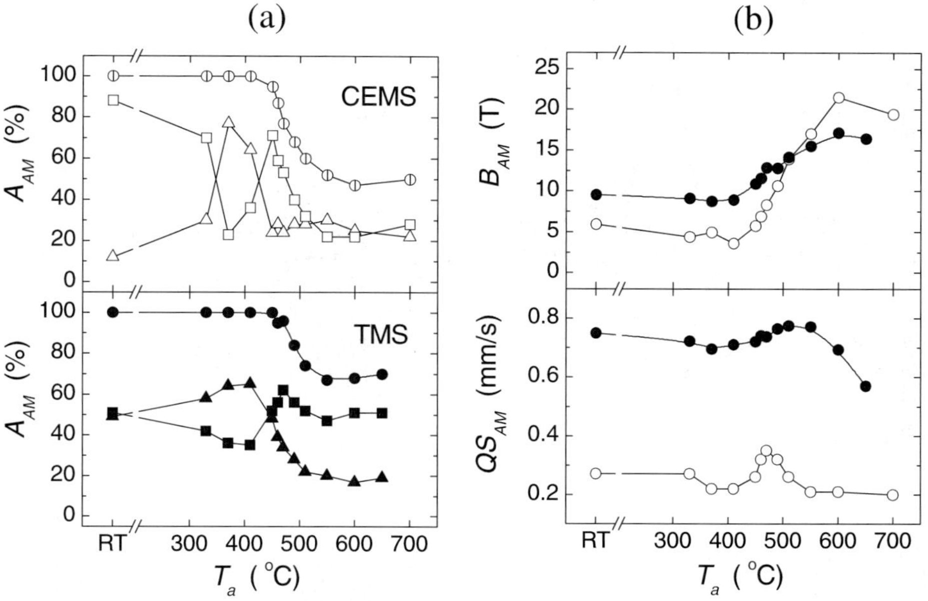

Figure 2 Relative fraction of amorphous matrix A_{AM} (a) and corresponding hyperfine parameters of magnetic B_{AM} and non-magnetic QS_{AM} regions (b) plotted against T_a as derived from TMS and CEMS. *Full symbols* – TMS, *open symbols* – CEMS, *circles* – amorphous rest, *squares* – magnetic AM regions, *triangles* – non-magnetic AM regions.

characteristic for combined electric quadrupole and magnetic dipole interactions in the amorphous phase (AM), which correspond to non-magnetic (triangles in Figure 2a) and magnetic (squares) regions, respectively. While in the bulk (TMS: full symbols) their mutual ratio is nearly 50:50, a significantly higher contribution of magnetic regions is observed on their surface (CEMS: open symbols) as demonstrated in Figure 2a. The total relative fraction of the amorphous matrix (circles) shows a decrease after the onset of crystallization for T_a>450°C. An inverse feature is observed for the crystalline phase contribution (not shown).

Notable decrease in the fraction of magnetic regions (open squares in Figure 2a) in AM is associated with structural relaxation, which takes place after annealing at moderate T_a. Its subsequent increase indicates the beginning of crystallization which starts earlier on the surface. Non-magnetic regions exhibit an opposite behaviour. Decrease of both types of AM regions for T_a>450°C is a result of progressive crystallization. It should be noted that from a qualitative point of view the same T_a vs. A_{AM} dependence is observed for TMS and CEMS data.

After annealing at T_a=450°C, traces of bcc-Fe were found. Spectral component corresponding to CR is clearly visible in the CEMS pattern in Figure 1b even though it was not revealed in the RT TMS spectrum in Figure 1a. Nevertheless, it can be identified in the LNT spectrum in Figure 1c.

With rising T_a, contribution of bcc-Fe nanograins becomes more pronounced. The amount of nanocrystallites affects the evolution of hyperfine interactions as demonstrated in Figure 2b. It introduces relatively strong magnetic interactions in the AM remainder caused by enhanced exchange interactions among nanocrystals, which grow in size and number. This results in an increase in B_{AM}. A decrease in B_{AM} and strong reduction of QS_{AM} as

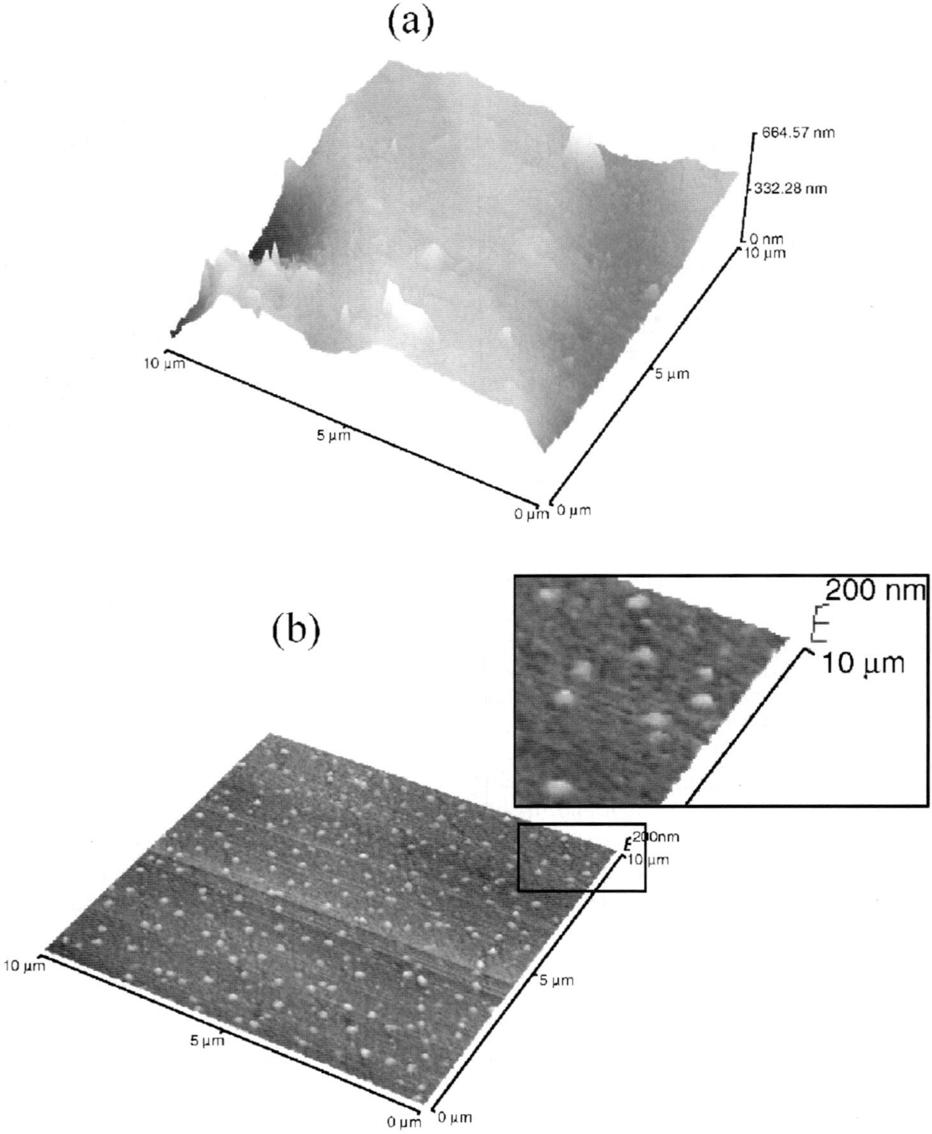

Figure 3 AFM scans taken from (a) wheel side and (b) air side of the $Fe_{76}Mo_8Cu_1B_{15}$ ribbon after annealing at 450°C. The inset shows a detailed view of the air side.

observed in Figure 2b is due to new crystalline phase that appears for T_a>600°C and also due to grain growth of the original bcc phase. Changes in both spectral parameters are caused by an increase of the relative ratio of nonmagnetic atoms (Mo, Cu, B) as compared to the magnetic ones (Fe) in AM residual matrix.

AFM scans performed upon the wheel side of the ribbon (dull) exhibit wave-like bending of the surface with variety of features different in size as seen in Figure 3a. The height of the "waves" is approximately 600 nm. On the other hand, the air surface (shiny)

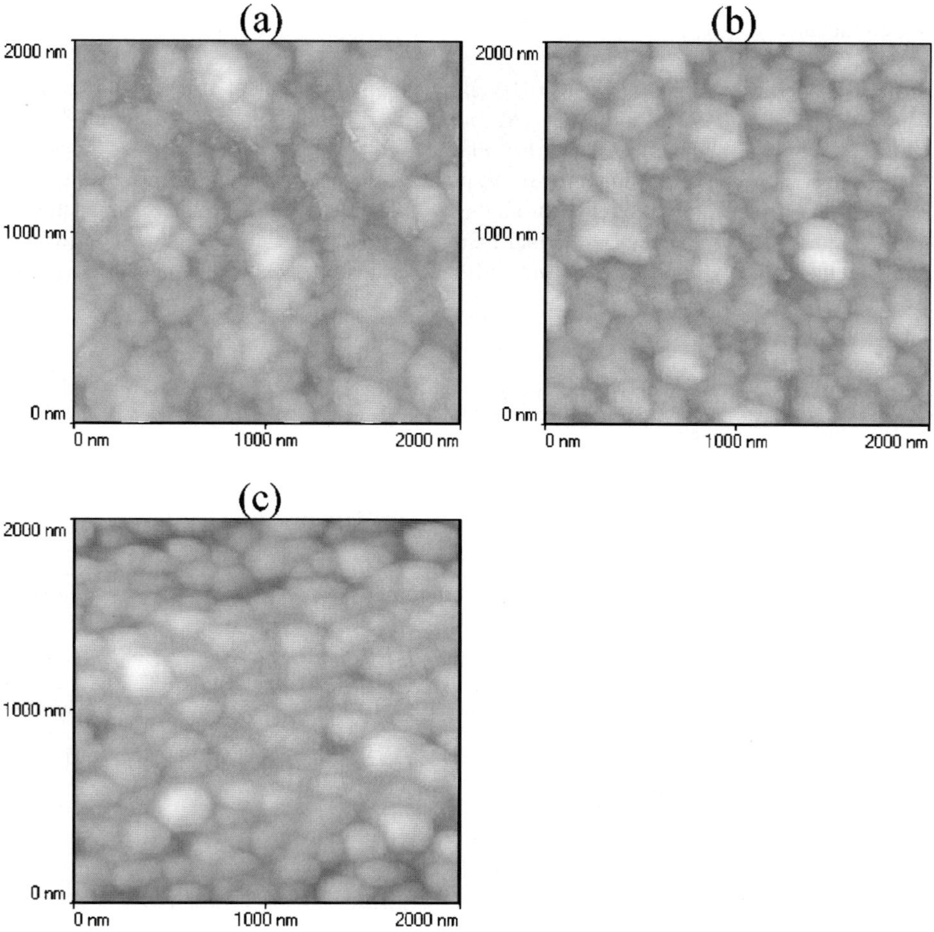

Figure 4 AFM scans of $Fe_{76}Mo_8Cu_1B_{15}$ alloy annealed at: (a) 450, (b) 510, and (c) 600°C.

in Figure 3b is flat and it is covered with homogeneously scattered peaks about 30 nm in height. The "waves" and tall peaks on the wheel side of the ribbon are most probably caused by scratches and dips which were produced by surface finishing tools during the fabrication of the quenching copper wheel. That is why we have concentrated on the inspection of the air surface of the ribbons.

Examples of AFM scans taken from the air side of $Fe_{76}Mo_8Cu_1B_{15}$ nanocrystalline ribbons are illustrated in Figure 4. It is clearly seen from this figure that the density of bright spots is directly related to the temperature of annealing, i.e., to the relative fraction of nanocrystallites formed in the alloy. The vertical scale is the same for all images (150 nm). One can speculate that the observed peak features are somehow related to nanocrystalline grains. The size of the grains was estimated from the results of XRD, TEM, and HREM experiments [3] to be of about 10 nm for $T_a < 550$°C. That is why the features in Figure 4a and b hardly represent the nanocrystals directly. It is assumed that the nanocrystalline grains form centres of surface contamination (fine dust particles, air humidity, etc.) in the places where they are located.

4 Conclusions

Surface properties of the $Fe_{80}Mo_8Cu_1B_{12}$ alloy in amorphous and nanocrystalline state were monitored by CEMS and AFM. An onset of crystallization at T_a of about 430°C, which is by about 20° lower than that in the bulk detected by TMS, was revealed. Moreover, variations in the amount of magnetic and non-magnetic regions in the AM residual phase are more pronounced on the surface than in the bulk. Compositional modifications of AM induced by thermal treatments at different T_a either due to structural rearrangement and/or crystallization are reflected in quantitative distinctions in QS_{AM} as well as in B_{AM}.

Acknowledgements This work was performed under the projects: CZ/SK/057, FR/SK/23, VEGA 1/1014/04, APVT-20-008404, Czech Science Foundation 202/05/2111, and Ministry of Education of the Czech Republic MSM6198959218.

References

1. Jirásková, Y., Maryško, M., Miglierini, M.: Czechoslov. J. Phys. **54**, D77 (2004)
2. Miglierini, M., Kaňuch, T., Krenický, T., Škorvánek, I.: Czechoslov. J. Phys. **54**, D73 (2004)
3. Miglierini, M., Degmová, J., Kaňuch, T., Švec, P., Illeková, E., Janičkovič, D.: Czechoslov. J. Phys. **54**, D161 (2004)
4. Miglierini, M., Degmová, J., Kaňuch, T., Greneche, J.M.: Phys. Status Solidi(a) **201**, 3280 (2004)

Hyperfine Interact (2005) 165: 81–87
DOI 10.1007/s10751-006-9251-3

Mechanosynthesis of spinel ferrite nanoparticles followed by Mössbauer spectroscopy

V. Šepelák · I. Bergmann · A. Feldhoff · P. Heitjans · F. J. Litterst · K. D. Becker

Published online: 26 October 2006

Abstract The single-step synthesis of nanosized $MgFe_2O_4$ and $NiFe_2O_4$ via mechanochemical processing of binary oxide precursors is followed by ^{57}Fe Mössbauer spectroscopy. Quantitative information is provided on both ionic and spin configurations in mechanosynthesized spinels. The response of the mechanosynthesized ferrite nanoparticles to changes in temperature is also studied.

Key words mechanosynthesis · spinel ferrite · nanocrystalline material · cation distribution · spin canting

1 Introduction

Spinel ferrites of the type $M\mathrm{Fe}_2\mathrm{O}_4$ (M is a divalent metal cation) are currently key materials for advancements in electronics, magnetic storage, ferrofluid technology, and in many bioinspired applications [1]. Conventional ceramic processing of ferrites requires a number

V. Šepelák (✉) · P. Heitjans
Center for Solid State Chemistry and New Materials,
University of Hannover, Callinstraße 3-3A, 30167 Hannover, Germany
e-mail: v.sepelak@tu-bs.de

V. Šepelák · I. Bergmann · K. D. Becker
Institute of Physical and Theoretical Chemistry, Technical University of Braunschweig,
Hans-Sommer-Straße 10, 38106 Braunschweig, Germany

A. Feldhoff · P. Heitjans
Institute of Physical Chemistry and Electrochemistry,
University of Hannover, Callinstraße 3-3A, 30167 Hannover, Germany

F. J. Litterst
Institute of Condensed Matter Physics, Technical University of Braunschweig,
Mendelssohnstraße 3, 38106 Braunschweig, Germany

V. Šepelák
Slovak Academy of Sciences, Watsonova 45, 04353 Košice, Slovakia

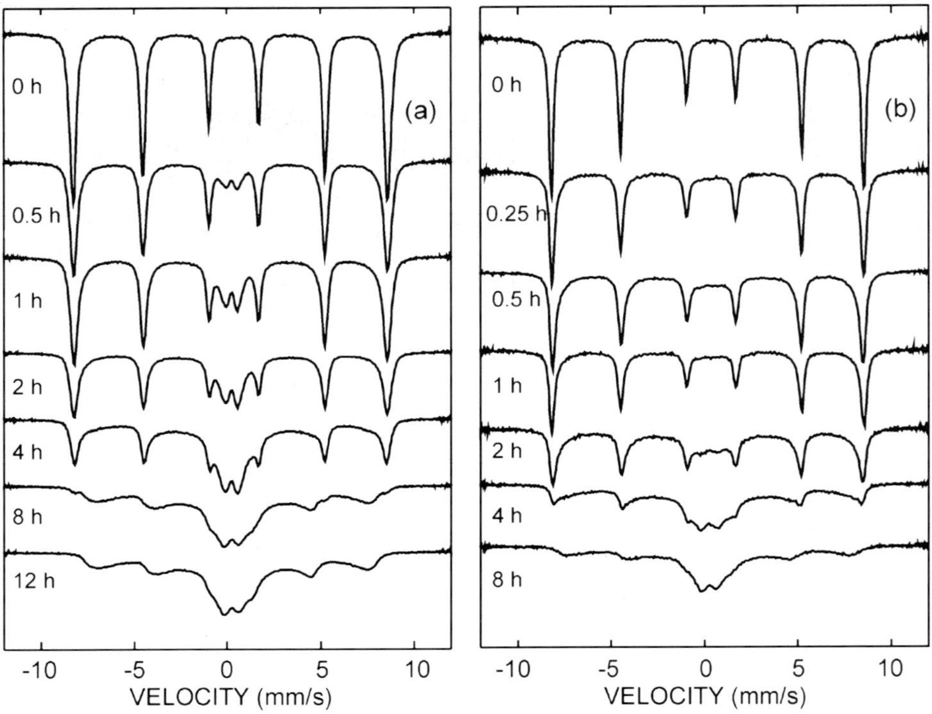

Figure 1 Room-temperature Mössbauer spectra of the milled α-Fe_2O_3/MgO (a) and α-Fe_2O_3/NiO (b) mixtures showing the mechanosynthesis of $MgFe_2O_4$ and $NiFe_2O_4$, respectively. The milling times are indicated in the figure.

of stages, including homogenization of the powder precursors, compaction of the reactants, and finally prolonged heat treatment at considerably elevated temperatures. One goal of modern ferrite research and developments has been to identify simpler processing schemes that do not rely upon high temperature treatments for inducing solid-state reactions.

In this article, the single-step synthesis of nanosized $MgFe_2O_4$ and $NiFe_2O_4$ via mechanochemical route is followed by ^{57}Fe Mössbauer spectroscopy. Information is obtained on the nonequilibrium cation distribution, the noncollinear spin arrangement, and the thermal stability of mechanosynthesized spinel ferrite nanoparticles. To emphasize the site occupancy on the atomic level, the structural formula of MFe_2O_4 spinel may be written as $(M_{1-\lambda}Fe_{\lambda})[M_{\lambda}Fe_{2-\lambda}]O_4$, where round and square brackets denote cation sites of tetrahedral (A) and octahedral [B] coordination, respectively, and where λ represents the degree of inversion defined as the fraction of the (A) sites occupied by Fe^{3+} cations.

2 Experimental

High-purity powders of MO (M=Mg, Ni) and α-Fe_2O_3 (Merck, Darmstadt) were used as precursors for the mechanosynthesis of $MgFe_2O_4$ and $NiFe_2O_4$. Stoichiometric α-Fe_2O_3/MO mixtures (10 g) were milled for various uninterrupted times (up to 12 h) in a Pulverisette 6 planetary ball mill (Fritsch, Idar-Oberstein) at room temperature. The grinding chamber (250 cm^3 in volume) and balls (5 mm in diameter) made of tungsten

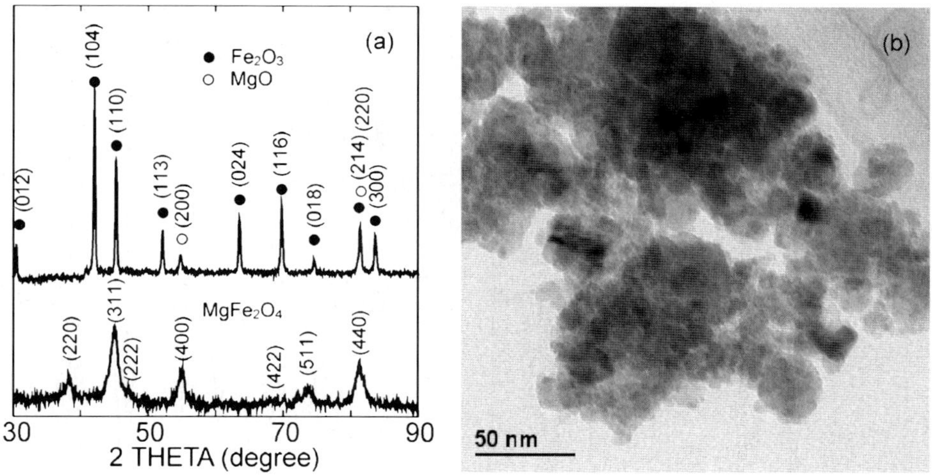

Figure 2 (a) XRD pattern of the nonactivated α-Fe_2O_3/MgO mixture (*top*) and of the mixture milled for 12 h (*bottom*). Diffraction lines of the reaction precursors and of the mechanosynthesized spinel phase are denoted by Miller indices. (b) TEM bright-field image showing the nanoscale structure of mechanosynthesized $MgFe_2O_4$.

carbide were used. The ball-to-powder weight ratio was 20:1. Milling experiments were performed in air at 600 rpm.

Mössbauer spectroscopy measurements were performed in transmission geometry using a ^{57}Co/Rh γ-ray source. High-field Mössbauer measurements were carried out at an external magnetic field of 5.5 T applied perpendicular to the γ-ray direction. The velocity scale was calibrated relative to ^{57}Fe in Rh. 'RECOIL' spectral analysis software [2] was used for the quantitative evaluation of the spectra. The degree of inversion was calculated from the subspectral intensities $\left(I_{(A)}/I_{[B]} = \lambda/(2-\lambda)\right)$. The average canting angle, Ψ, was calculated from the ratio of the intensities of lines 2 and 1, $I_{2,1}$, according to $\Psi = 90^\circ - \arcsin\left[(3I_{2,1}/2)/(1+3I_{2,1}/4)\right]^{1/2}$.

The X-ray diffraction (XRD) patterns were collected using a URD 6 diffractometer (Seifert-FPM, Freiberg) with Fe Kα radiation. The morphology of powders and the sizes of individual crystallites were studied using a transmission electron microscope (Jeol JEM-2100F).

3 Results and discussion

The mechanically induced evolution of the α-Fe_2O_3/MgO mixture and of the α-Fe_2O_3/NiO mixture is shown in Figure 1a and b, respectively. In both cases, with increasing milling time, the sextet corresponding to α-Fe_2O_3 becomes asymmetric towards the inside of each line, decreases in intensity, and is gradually replaced by a central doublet and a new broadened sextet. The Mössbauer spectrum of the α-Fe_2O_3/MgO sample milled for 12 h and of the α-Fe_2O_3/NiO sample milled for 8 h does not reveal any magnetically split lines corresponding to α-Fe_2O_3 (see Figure 1a and b, bottom). The spectra of both milled mixtures are dominated by a doublet with an isomer shift of about 0.22 mm/s characteristic of ferrite nanoparticles in the superparamagnetic state [3, 4].

To determine the phase evolution of the α-Fe_2O_3/*M*O (*M*=Mg, Ni) mixtures during high-energy milling by an independently supporting method, the mechanochemical reactions

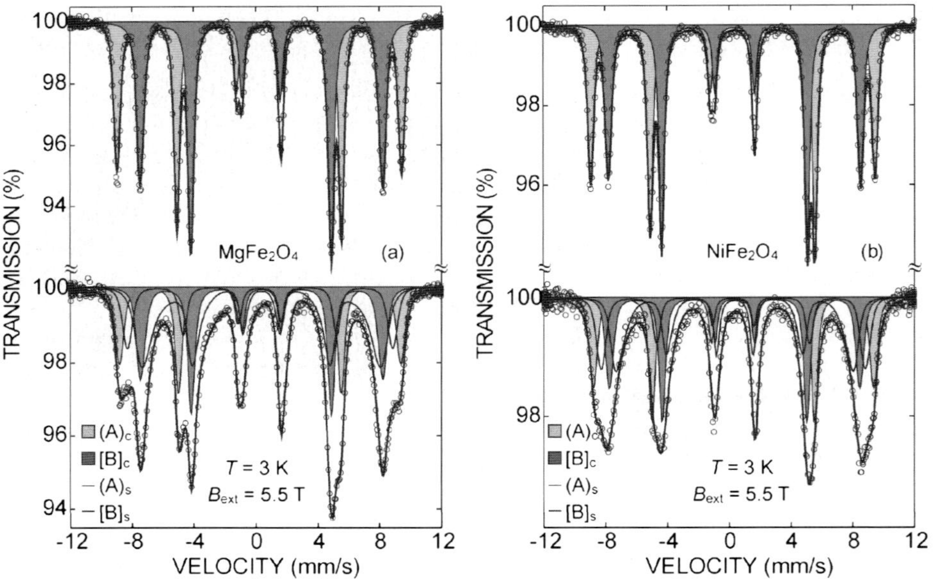

Figure 3 Mössbauer spectra of bulk (*top*) and nanosized mechanosynthesized (*bottom*) spinels $MgFe_2O_4$ (a) and $NiFe_2O_4$ (b). Spectra were taken at 3 K at an external magnetic field of 5.5 T applied perpendicular to the γ-ray direction. $(A)_c$, $[B]_c$ and $(A)_s$, $[B]_s$ denote cation sites of tetrahedral and octahedral coordination in the inner core and in the surface shell of ferrite nanoparticles, respectively.

were also followed by XRD technique. As demonstrated for the case of the α-Fe_2O_3/MgO mixture in Figure 2a, the XRD pattern of the starting mixture is characterized by sharp diffraction peaks corresponding to the simple oxides. After 12 h of milling, all peaks detected above background are due to $MgFe_2O_4$, see Figure 2a bottom. It has been found that in the case of the α-Fe_2O_3/NiO mixture, the mechanochemical route to $NiFe_2O_4$ takes 8 h. Compared to the traditional ferrite fabrication processes, the mechanochemical route used in the present work is a one-step, low-temperature, high-yielding and low-cost procedure for the synthesis of spinel ferrite nanoparticles. The line-broadening analysis of the XRD data reveals an average crystallite size of 8.5(2) and 8.6(3) nm for mechanosynthesized $MgFe_2O_4$ and $NiFe_2O_4$, respectively.

TEM revealed that both mechanosynthesized ferrites consists of crystallites mostly in the 5–15 nm size range, consistent with the average crystallite size determined from the XRD data. The shape of the majority of the crystallites appears spherical. As shown in the TEM bright-field micrograph (Figure 2b), nanoscale crystallites tend to agglomerate and to form aggregates with sizes mostly up to 30 nm.

The high-field Mössbauer spectra of nanosized mechanosynthesized $MgFe_2O_4$ and $NiFe_2O_4$ and of their bulk counterparts (prepared by a conventional ceramic method) taken at 3 K are compared in Figure 3. For the bulk $MgFe_2O_4$ and $NiFe_2O_4$ materials, two sextets were necessary to fit the absorption due to (A)- and [B]-site Fe^{3+} ions (Figure 3, top). The degree of inversion of the bulk $MgFe_2O_4$ and $NiFe_2O_4$ was found to be $\lambda \approx 0.91$ and $\lambda \approx 1$, respectively. The intensity ratio $I_{2,1}=4/3$ indicates that both bulk materials exhibit a Néel-type collinear spin arrangement ($\Psi_{(A)} = \Psi_{[B]} \approx 0°$). The spectra of mechanosynthesized $MgFe_2O_4$ and $NiFe_2O_4$ nanoparticles (Figure 3, bottom) were fitted by assuming an ordered particle core surrounded by disordered grain boundary (surface) regions. In order to separate the surface effects from the bulk effects in the Mössbauer spectra of mechanosynthesized nanoparticles, the following assumptions were taken into account:

(a) The core of nanoparticles possesses the same structure as the bulk material. Thus, the fits to the spectra of nanoscale $MgFe_2O_4$ and $NiFe_2O_4$ were made by imposing constraints on the Mössbauer parameters of two subspectra corresponding to (A)- and [B]-site ferric ions of the particle core (denoted by $(A)_c$ and $[B]_c$ in Figure 3), such that these parameters were fixed and equal to those obtained from the fit of the spectra of the bulk materials. The rest of the spectral intensity was fitted by two additional subspectra (denoted by $(A)_s$ and $[B]_s$), which were ascribed to Fe^{3+} ions at (A) and [B] sites in the grain boundary (surface) regions of nanoparticles.
(b) The relaxation effects (superparamagnetic relaxation, collective magnetic excitations) are suppressed at the temperature of Mössbauer measurements (3 K), i.e., they do not contribute to a reduction of the hyperfine fields.
(c) The mechanosynthesized material consists of non-interacting nanoparticles with a narrow size distribution.

It is clear that a more realistic model should include various effects involving a reduction of the hyperfine fields, such as particle size distribution, cluster aggregation, magnetic interactions between nanoparticles, influence of relaxation effects, etc. Therefore, the results obtained using the fitting strategy taking into account the above-given assumptions represent a simplified case and should be taken as an approximation.

For both mechanosynthesized materials, the intensity ratio of the $(A)_s$ and $[B]_s$ sextets to the total spectral intensity indicates that about 50% of magnetic cations are located in the surface shell of nanoparticles. Considering a spherical shape of the mechanosynthesized nanoparticles with an average diameter of about 9 nm, the thickness of the disordered surface shell is estimated to be about 0.9 nm. This value is comparable to the lattice constant of the spinel ferrites.

While the core (c) of the mechanosynthesized nanoparticle possesses an ordered spinel structure ($\lambda_c \approx 0.91$ for $MgFe_2O_4$, $\lambda_c \approx 1$ for $NiFe_2O_4$) with a Néel-type collinear spin arrangement ($\Psi_{(A)c} = \Psi_{[B]c} \approx 0°$), the shell region of nanoparticles is structurally and magnetically disordered. The major feature of the atomic configuration in the shell (s) of the $MgFe_2O_4$ and $NiFe_2O_4$ nanoparticles is a nonequilibrium cation distribution characterized by degrees of inversion of $\lambda_s \approx 0.69$ and $\lambda_s \approx 0.67$, respectively. These values correspond nearly to the random distribution of cations ($\lambda = 2/3$). The average canting angles were found to be $\Psi_{(A)s} \approx 29°$, $\Psi_{[B]s} \approx 39°$ and $\Psi_{(A)s} \approx 28°$, $\Psi_{[B]s} \approx 40°$ for mechanosynthesized $MgFe_2O_4$ and $NiFe_2O_4$, respectively. As a consequence of frustrated superexchange interactions, the average sublattice magnetic fields experienced by Fe^{3+} ions located in the near-surface layers are reduced ($B_{(A)s}=47.75(3)$ T, $B_{[B]s}=50.76(2)$ T for $MgFe_2O_4$; $B_{(A)s}=48.44(2)$ T, $B_{[B]s}=52.91(5)$ T for $NiFe_2O_4$) in comparison with those acting on iron nuclei in the inner core ($B_{(A)c}=50.99(6)$ T, $B_{[B]c}=53.47(1)$ T for $MgFe_2O_4$; $B_{(A)c}=51.62(2)$ T, $B_{[B]c}=55.83(1)$ T for $NiFe_2O_4$). These results are consistent with those of previous studies on nanosized spinel ferrites [4–6], where both different spin canting in (A) and [B] sublattices and reduced sublattice magnetic fields were observed.

The nanostructured spinel ferrites prepared by high-energy milling are metastable with respect to structural changes at elevated temperatures. As it is shown for the case of mechanosynthesized $NiFe_2O_4$ (Figure 4), the shape of the Mössbauer spectra of the nanosized ferrite does not change significantly in the temperature range 373–673 K. This

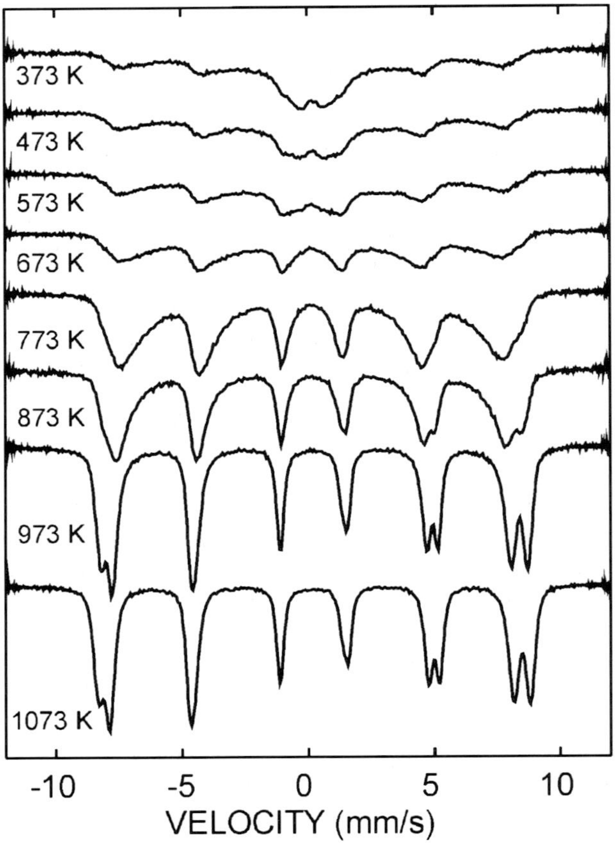

Figure 4 Room-temperature Mössbauer spectra of mechanosynthesized $NiFe_2O_4$ after annealing at various temperatures for 30 min.

demonstrates that the range of the thermal stability of the mechanosynthesized product extends up to about 673 K. However, at temperatures above 673 K, a gradual crystallization of the nanoscale powders takes place. As to be expected, the superparamagnetic doublet component gradually vanishes due to the particle growth of the spinel phase. Simultaneously, the sextet structure, typical for the long-range ferrimagnetic state, develops due to the thermally induced changes in the spin configuration. In conclusion, during the annealing process, the nonequilibrium cation distribution and the disordered spin arrangement, resulting from the mechanochemical preparation route, relax towards their equilibrium configurations characteristic of the bulk material.

4 Conclusions

The mechanosynthesis of nanosized $MgFe_2O_4$ and $NiFe_2O_4$ with average crystallite sizes of about 9 nm has been followed by ^{57}Fe Mössbauer spectroscopy. Based on a simplified model that does not include effects involving a reduction of the hyperfine fields, the comparative in-field Mössbauer study of the bulk and nanosized ferrites enables us to separate the surface effects from the bulk effects in the spectra of mechanosynthesized $MgFe_2O_4$ and $NiFe_2O_4$ nanoparticles. It is found that the fraction of magnetic cations located in the grain boundary (surface) regions of mechanosynthesized nanoparticles is about 50%. The disorder in the grain boundary regions is characterized by both a considerable degree of spin canting (up to≈40°) and a nearly random distribution of cations (λ≈2/3). On heating to elevated temperatures, the mechanosynthesized spinel ferrites relax to a structural and magnetic state that is similar to the bulk one.

Acknowledgments The work was supported by the Deutsche Forschungsgemeinschaft (DFG). One of the authors (V. Šepelák) thanks the DFG for supporting his work at the Center for Solid State Chemistry and New Materials, University of Hannover, within a Mercator Visiting Professorship. Partial support by the Grant Agency of the Ministry of Education of the Slovak Republic and of the Slovak Academy of Sciences (Grant 2/5146/25) and by the Alexander von Humboldt Foundation is gratefully acknowledged.

References

1. Wang, Z.L., Liu, Y., Zhang, Z.: Handbook of Nanophase and Nanostructured Materials, vol. 3. Plenum, New York (2002)
2. Lagarec, K., Rancourt, D.G.: RECOIL – Mössbauer Spectral Analysis Software for Windows, Version 1.02. Department of Physics, University of Ottawa, Ottawa (1998)
3. Šepelák, V., Baabe, D., Mienert, D., Litterst, F.J., Becker, K.D.: Scr. Mater. **48**, 961 (2003)
4. Šepelák, V., Baabe, D., Mienert, D., Schultze, D., Krumeich, F., Litterst, F.J., Becker, K.D.: J. Magn. Magn. Mater. **257**, 377 (2003)
5. Goya, G.F., Leite, E.R.: J. Phys., Condens. Matter **15**, 641 (2003)
6. Ponpandian, N., Narayanasamy, A., Chinnasamy, C.N., Sivakumar, N., Greneche, J.-M., Chattopadhyay, K., Shinoda, K. Jeyadevan, B., Tohji, K.: Appl. Phys. Lett. **86**, 192510 (2005)

Hyperfine Interact (2005) 165: 89–94
DOI 10.1007/s10751-006-9246-0

Formation of gold nanoparticles in gold ruby glass: The influence of tin

S. Haslbeck · K.-P. Martinek · L. Stievano · F. E. Wagner

Published online: 14 November 2006

Abstract ^{197}Au and ^{119}Sn Mössbauer spectroscopy was used to study the effect of tin on the formation of gold nanoparticles in gold ruby glasses containing 200 ppm of gold and up to 2 wt.% of tin. The tin was found to speed up the formation of the metallic nanoparticles during annealing at temperatures around 600°C and to cause very small particles with sizes of only a few nanometres to be formed, whereas fewer but larger particles are formed in the absence of tin. This has consequences for the colour of the glasses, which changes from purple red for virtually tin-free to brown for tin-rich glasses.

Key words gold · tin · ruby glass · Mössbauer spectroscopy

1 Introduction

Gold ruby glass is a silicate glass containing nanometre sized metallic gold metal particles imparting the glass its distinctive red colour. Gold ruby glass is mainly used for making decorative glassware, but recently glasses of this type have also been suggested for technical applications [1]. Gold ruby glass seems to have been known already in antiquity, as is demonstrated by the so-called Lycurgus cup, a cage cup probably dating from Late Roman times and now in the collections of the British Museum [2]. The modern use of gold ruby glass began in the seventeenth century A.D., when the Italian Antonio Neri gave a recipe for making it and the German alchemist and glassmaker Johann Kunckel started its mass production in good quality [3].

S. Haslbeck · F. E. Wagner
Physik-Department, Technische Universität München, 85748 Garching, Germany

K.-P. Martinek
F. X. Nachtmann Bleikristallwerke GmbH, 94566 Riedlhütte, Germany

L. Stievano (✉)
Laboratoire de Réactivité de Surface, Université Paris VI, 4 Place Jussieu, 75252 Paris, France
e-mail: lorenzo.stievano@upmc.fr

Gold ruby glass is usually made by adding several hundred parts per million of a gold precursor to the melt of a silicate glass. At the usual melting temperatures around 1,400°C the gold dissolves in the melt and remains in this state on quenching to ambient temperature. Only when the quenched glass is annealed at about 500°C for times of the order of minutes or hours does it strike red owing to the formation of metallic gold nanoparticles with sizes between a few and about 50 nm [4–6]. The typical ruby colour of the glass arises from a strong light absorption of the gold particles in the green region of the visible spectrum caused by surface plasmon excitation and from the interband absorption of metallic gold in the blue. The hue of the colour strongly depends on the size distribution of the gold particles.

This size distribution depends on the duration and temperature of the annealing step, but also on the composition of the base glass and particularly on the presence of characteristic elements influencing the precipitation of the gold nanoparticles. Tin is known to increase the speed with which the glass strikes red [7], but the mechanism by which it acts is not yet well understood. In order to elucidate the role of tin in the formation of gold ruby glass, we have undertaken Mössbauer experiments with both ^{197}Au and ^{119}Sn.

^{197}Au Mössbauer spectroscopy has shown that the gold in the quenched, colourless glass yields a quadrupole doublet with a splitting of about 6.5 mm/s and an isomer shift of about 1.3 mm/s, which is typical for Au(I) linearly coordinated to two nearest oxygen neighbours [8]. The formation of the metallic gold nanoparticles during the annealing of the glass can be followed by Mössbauer spectroscopy. Tin in silicate glasses has been studied in some detail by Mössbauer spectroscopy [9–12], showing varying amounts of Sn(IV) and Sn(II), depending on the redox conditions in the glass melt.

2 Experimental

Two different base glasses were used: one a simple sodium silicate glass (Glass I: 26 wt.% Na_2O; 74 wt.% SiO_2) and the other a soda lime glass (Glass II: 22 wt.% Na_2O; 8 wt.% CaO; 70 wt.% SiO_2). To all glasses 200 ppm of Gold was added either as $KAu(CN)_2$ (Glass I) or as $HAuCl_4$ xH_2O (Glass II). Different amounts of tin were added as SnO_2. The glasses were melted at 1,400°C and then quenched by pouring them into water or onto a cold steel surface.

All Mössbauer measurements were performed with both the source and the absorber cooled to 4.2 K in a liquid He bath cryostat. All isomer shifts are given with respect to the source. The sources of ^{197}Pt were made by neutron irradiation of isotopically enriched ^{196}Pt metal. The absorbers for the ^{197}Au measurements had thicknesses of 6 to 8 g/cm^2 of the ruby glass. The ^{119}Sn measurements were made with a commercial source of $^{119\,m}$Sn in $CaSnO_3$.

3 Results and discussion

Figure 1 shows the ^{197}Au Mössbauer spectra of type II glasses containing between 20 ppm and 2 wt.% of tin in the quenched state and on annealing at 590°C for 10 h. The quenched glasses all show a quadrupole doublet with a slight asymmetry due to a distribution of the quadrupole splittings and correlated isomer shifts [8], the mean shift being 0.92(3) mm/s and the mean splitting 6.19(6) mm/s. This doublet is attributable to the Au(I) in the glass matrix [8]. An additional weak single line at −1.23 mm/s is due to metallic gold. X-ray

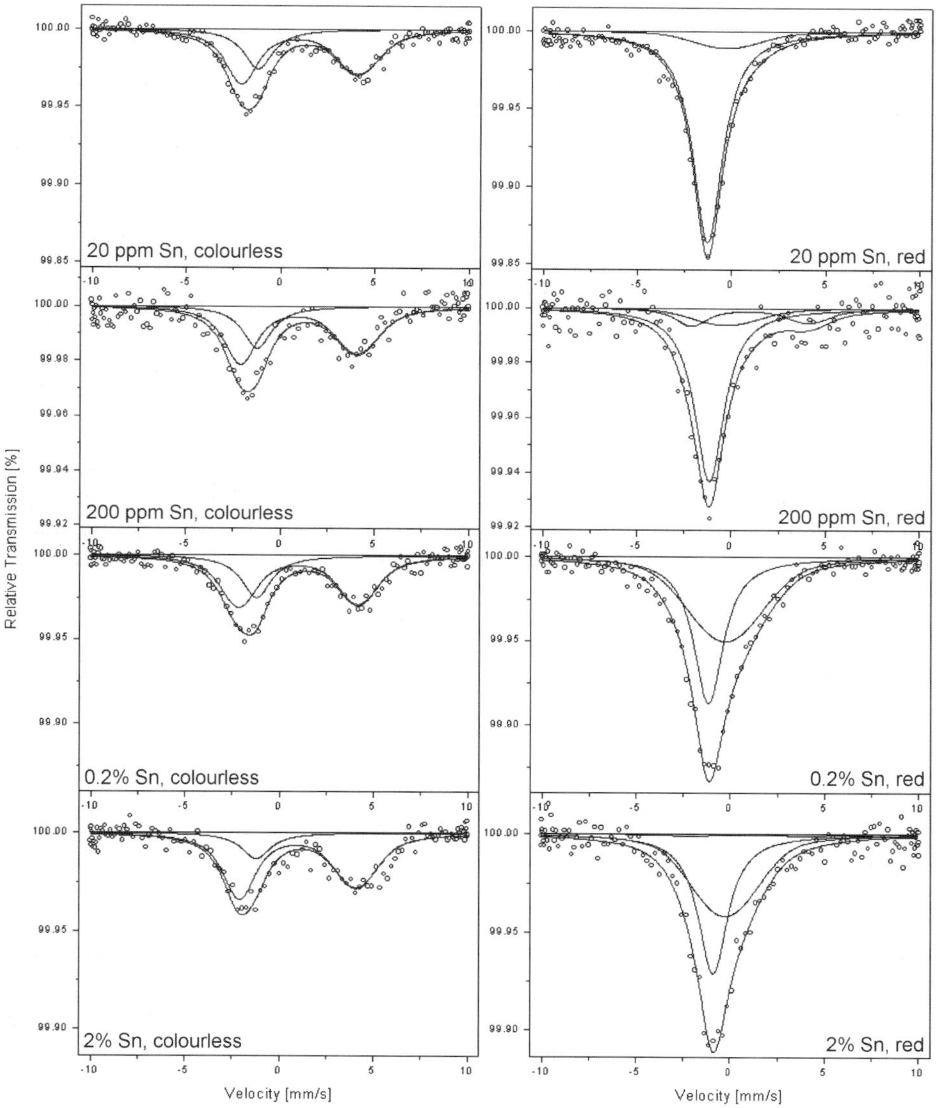

Figure 1 ^{197}Au Mössbauer spectra of gold ruby glasses (type II) containing different amounts of tin in the quenched, colourless state (*left*) and after striking red on annealing at 590°C for 10 h (*right*). Note that the intensity scales are the same for the quenched and the respective annealed glasses in order to show the increase of the total spectral area occurring when the metallic gold forms. From the total spectral areas, and taking into account the slightly different absorber thicknesses, one concludes that the Lamb–Mössbauer factor of the monovalent gold in the glass matrix is only a bit more than half of that of the metallic gold, and that it is not much different for the surface and core atoms of the nanoparticles.

diffraction studies indicate that this is metallic gold as micrometer-size particles which did not dissolve properly during the melting of the glass.

The annealed glasses containing 20 and 200 ppm of tin show practically only the narrow single line of metallic gold, but in those with 0.2 and 2% of tin an additional, broad component appears on the right side of the line of metallic gold. This component could be

Figure 2 Fractional area attributable to metallic gold (including the surface gold component) as a function of the annealing time for a type I glass containing no tin and 2.4 wt.% of tin. Note that the annealing temperatures are substantially different.

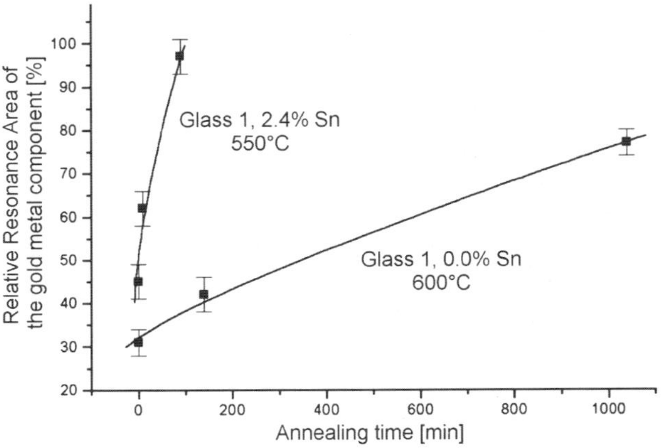

fitted by a quadrupole doublet at about −0.3 mm/s with a strong Gaussian broadening and unresolved splitting. The relative intensity of this component is about 50% in both glasses. A similar component has previously been found for gold nanoparticles in a Mylar matrix, where it was attributed to surface gold atoms on nanoparticles smaller than a few nanometres [13]. We therefore attribute the broad component to surface atoms on gold particles with sizes of a few nanometres only. From the relative intensities of the core and the surface components, one estimates that the size of the gold nanoparticles in the tin-rich glasses is between 2 and 3 nm. For the glasses containing 200 and 20 ppm of tin the broad component is at best very weakly visible, showing that there the particles are so large that the fraction of surface atoms is too small to contribute noticeably to the spectra.

The Mössbauer spectra thus show that, already in glasses with tin contents as small as a few tenth of a percent, much smaller gold particles form than in glasses with less or no tin. This is also reflected in the optical absorption spectra of the glasses. In fact, at a tin content of 20 ppm the glass still showed a pronounced absorption peak with its maximum at 534 nm, while in the glass with 0.2% of tin the peak is much less pronounced and at 503 nm, and for 2% of tin the peak has virtually vanished, the absorption decreasing monotonically from the blue to the red region of the spectrum. The colour changes concomitantly from purple red at 20 ppm of tin to brown at 2% of tin.

Tin in gold ruby glass thus seems to lead to the formation a larger number of smaller metallic particles. It also accelerates the formation of the metallic particles. This was demonstrated by Mössbauer measurements on two samples of type I glass with no tin and 2.4 wt.% of tin added. These were annealed for increasing times at a constant temperature of 650°C (no tin) and 550°C (2.4 wt.% tin), respectively. In Figure 2 the relative spectral intensity attributable to metallic gold (including the broad surface component, which is very intense for the tin-containing glass) is shown as a function of the annealing time. In the tin-containing glass, the formation of the metal particles turns out to be more than an order of magnitude faster than in the tin-free one, although the annealing temperature was lower in the former than in the latter.

In order to better understand the chemistry of tin during the reduction of gold, we have taken also ^{119}Sn Mössbauer spectra of all tin-containing glasses (not shown). These consist mainly of a Sn(IV) component with an unresolved quadrupole splitting of about 0.5 mm/s and an isomer shift of about −0.2 mm/s with respect to the $CaSnO_3$ source. These values of the hyperfine parameters are typical for tin in octahedral coordination dissolved in the glass matrix [14]. The quenched glasses also exhibit a weak Sn(II) component with a shift of

2.9 mm/s, a quadrupole splitting of 2.0 mm/s, and a fractional area of up to 4%. The Sn(II) was found to oxidise to Sn(IV) during annealing except for the highest tin content, but in the glasses with 20 and 200 ppm of gold a weak indication of a broad single line at about 2.4 mm/s appears. On the other hand, the single line in the ^{197}Au spectra attributed to gold in the cores of the metallic nanoparticles shifts from −1.23 mm/s for glasses without tin to −0.9 mm/s in the type II glass with 2 wt.% Sn. This and the single line at 2.4 mm/s in the ^{119}Sn spectra suggest that an alloy forms between tin and gold. Using Mössbauer data for Au–Sn alloys [15, 16], one estimates that the metallic gold particles in the glass with a tin content of 2 wt.% contain between 10 and 15 at.% of tin.

^{197}Au and ^{119}Sn Mössbauer spectroscopy together indicate that there are close interactions between tin and gold in the metallic gold particles. In particular, the Mössbauer results give indications on the mechanisms by which tin speeds up the kinetics of formation of the metallic gold particles and results in a reduction of the average gold particle size:

a. Highly dispersed tin seems to provide condensation nuclei for the nanoparticles. The simultaneous presence of many condensation nuclei and the low total amount of gold in the glass will then prevent the formation of gold particles of bigger sizes.
b. Tin may act a "surfactant" at the surface of the gold particles. Tin species bonding to the gold atoms on the surface of the metal particles might decrease their surface tension in the glass matrix. Such a mechanism would thermodynamically stabilise small gold metal particles compared to large ones having a lower surface-to-bulk atomic ratio. This view agrees well with the formation of a gold–tin alloy. In fact, in Au–Sn alloys, tin is expected to be segregated on the surface in thermodynamic equilibrium [17].
c. Finally, the acceleration of the kinetics of the formation of gold metal might be explained by the following redox mechanism, which usually occurs during the formation of "Purple of Cassius" [3], a mixed colloid of Sn(IV) hydroxide and gold metal:

$$Sn^{2+} + 2\,Au^{+} \rightarrow Sn^{4+} + 2\,Au^{0} \quad (1)$$

The equilibrium concentration of Sn(IV) and Sn(II) depends on oxygen partial pressure and, at the annealing temperature, is strongly displaced towards Sn(IV), as ^{119}Sn Mössbauer spectroscopy shows. For the samples containing large tin contents, the total amount of Sn (II) would be large enough to reduce all of the cationic gold. In the case of small tin contents, however, there would not be enough Sn(II) to reduce all the gold. Nevertheless, at the annealing temperature, the Sn(II) consumed by the reaction might reform from Sn(IV), to attain again its equilibrium concentration. The reformed Sn(II) could thus react with more cationic gold, according to the following reaction cycle:

If the redox equilibrium between Sn(II) and Sn(IV) is fast enough, the tin can be effectively considered as a catalyst.

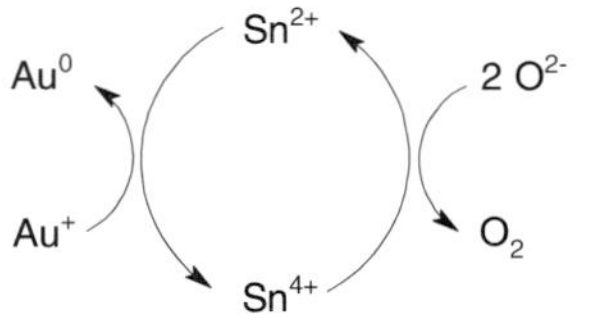

Further work is necessary to figure out which, among the proposed mechanisms, takes place or is predominant.

4 Conclusions

^{197}Au and ^{119}Sn Mössbauer spectroscopy together show that tin (a) accelerates the formation of gold clusters in silicate glasses, confirming previous findings [7, 18], and (b) induces a decrease of the average size the gold nanoparticles, compared to the size of particles formed in glasses with no or very little tin. On the basis of the Mössbauer results, three different possible mechanisms explaining the action of tin, namely the formation of condensation nuclei, the action as a surfactant, or the action as a catalyst, are suggested.

References

1. Eichelbaum, M., Rademann, K., Müller, R., Radtke, M., Riesemeier, H., Görner, W.: Angew. Chem. Int. Ed. **44**, 7905 (2005)
2. Barber, D.J., Freestone, I.C.: Archaeometry **32**, 33 (1990)
3. Frank, S.: Glass Technol. **25**, 47 (1984)
4. Badger, A.E., Weyl, W., Rudow, H.: Glass Ind. **20**, 407 (1939)
5. Doremus, R.H.: J. Chem. Phys. **40**, 2389 (1964)
6. Kreibig, U.: J. Physique, Colloque C2-97 (1977)
7. Williams, J.A., Rindone, G.E., McKinstry, A.: J. Am. Ceram. Soc. **64**, 709 (1981)
8. Wagner, F.E., Haslbeck, S., Stievano, L., Calogero, S., Pankhurst, Q. A., Martinek, K.-P.: Nature **407**, 691 (2000)
9. Dannheim, H., Oel, H.J., Tomandl, G.: Glastech. Ber. **49**, 170 (1976)
10. Johnson, J.A., Johnson, C.E., Williams, K.F.E., Holland, D., Karim, M.M.: Hyperfine Interact. **95**, 41 (1995)
11. Johnson, J.A., Johnson, C.E., Holland, D., Sears, A., Bent, J.F., Appleyard, P.G., Thomas, M.F., Hannon, A.C.: J. Phys. Condens. Matter **12**, 213 (2000)
12. Benne, D., Rüssel, Ch., Lehmann, O., Menzel, M., Niemeier, D., Becker, K.D.: J. Non-Cryst. Solids **318**, 202 (2003)
13. Stievano, L., Santucci, S., Lozzi, L., Calogero, S., Wagner, F.E.: J. Non-Cryst. Solids **232–234**, 644 (1998)
14. Casagrande, M., Storaro, L., Lenarda, M., Gersich, J., Stievano, L., Wagner, F.E.: J. Mater. Chem. **14**, 1010 (2004)
15. Charlton, J.S., Harris, I.R.: Phys. Status Solidi. **39**:K1 (1970)
16. Charlton, J.S., Hayes, M.C., Harris, I.R.: J. Less-Common Metals **20**, 105 (1970)
17. Overbury, S.H., Somorjai, G.A.: J. Chem. Phys. **66**, 3181 (1977)
18. Baozhong, Y., Changlin, G., Changwei, L.: J. Non-Cryst. Solids **95 & 96**, 725 (1987)

Hyperfine Interact (2005) 165: 95–99
DOI 10.1007/s10751-006-9290-9

Mössbauer study of the transition metal phosphorus trichalcogenide $FePS_3$ and spin glass $Mn_{0.5}Fe_{0.5}PS_3$

Shuichi Ban · Yoshiki Takano · Naoki Ohkubo · Hisanobu Hoya · Yumiko Takahashi · Kouichi Takase · Kazuko Sekizawa

Published online: 31 October 2006

Abstract The transition metal phosphorus trichalcogenides $MnPS_3$ and the $FePS_3$ are $CdCl_2$ type layered compounds, where the transition metal ions form a hexagonal lattice. While these compounds order anti-ferromagnetically at low temperature, the magnetic structures are different. We have reported that these mixtures $Mn_{0.5}Fe_{0.5}PS_3$ is a spin glass with a glass transition temperature T_g=33.7 K. Then, in this work, we report that the results of the temperature variation of the ^{57}Fe Mössbauer spectra of $FePS_3$ and $Mn_{0.5}Fe_{0.5}PS_3$, in detail. In the anti-ferromagnetic state of $FePS_3$, the hyperfine magnetic field H_{int} increases with decreasing temperature and the Isomer shift I. S. increases slightly with decreasing temperature. However in $Mn_{0.5}Fe_{0.5}PS_3$, the two broadened peaks are observed and the two peaks became a single peak with decreasing temperature at about 50.0 K, which is higher than T_g=33.7 K. In the spin glass $Mn_{0.5}Fe_{0.5}PS_3$, the Mössbauer spectra suggest that the magnetic interactions exist far above T_g.

Key words spin glasses · Mössbauer effect · remanence · manganese compounds · iron compounds · transition metal phosphorus trichalcogenides · magnetic interactions

1 Introduction

The transition metal phosphorus trichalcogenides $MnPS_3$ and the $FePS_3$ are $CdCl_2$ type layered compounds, where the transition metal ions form a hexagonal lattice. While these compounds order anti-ferromagnetically at low temperature, there are

S. Ban (✉) · N. Ohkubo
Physics Laboratory, College of Science and Technology, Nihon University,
7-24-1 Narashinodai, Funabashi-City, Chiba 274-8501, Japan
e-mail: van@phys.ge.cst.nihon-u.ac.jp

Y. Takano · H. Hoya · Y. Takahashi · K. Takase · K. Sekizawa
Department of Physics, College of Science and Technology, Nihon University,
1-8-14 Kanda Surugadai, Chiyoda-ku, Tokyo 101-8308, Japan

reported that $MnPS_3$ is a two-dimensional Heisenberg anti-ferromagnet and $FePS_3$ is a two-dimensional Ising anti-ferromagnet [1].

The direction of the ordered magnetic moment is parallel to the c axis in both compounds, but the magnetic structures are different. The magnetic ions (Mn) have an anti-ferromagnetic coupling with the nearest neighbour magnetic atoms (Mn) in the *c* plane in the $MnPS_3$ with the Néel temperature T_N of 78 K.

In the $FePS_3$, the magnetic ions (Fe) form ferromagnetic chains and these chains order anti-ferromagnetically in the *c* plane with T_N of 118 K. We have found spin glass phenomena in these mixtures $Mn_{0.5}Fe_{0.5}PS_3$ compound.

We reported that its magnetic properties, specific heat and the spin glass transition temperature as T_g=33.7 K [2]. The remarkable feature of this spin glass phase is that the spin freezing is not suppressed up to a rather high magnetic field compared with similar spin glass (Mn, Fe)TiO_3 [3–5].

In $Mn_{0.5}Fe_{0.5}PS_3$, the cusp temperature of $\chi'(\omega)$ does not show magnetic field amplitude dependence below 10 Oe and increases with an increase in frequency, obeying the Vogel–Fulcher rule. At below T_g, the temperature dependence of $\chi''(\omega)$ shows two extra peaks. The time dependence of thermoremanent magnetization shows power law dependence below T_g. The value of the molar specific heat is about an order of magnitude larger than that of $MnPS_3$ and $FePS_3$. The T-linear term characteristic of spin glass is observed below about 10 K. The temperature dependence of the magnetic specific heat shows no anomaly at T_g [2, 6–8].

Then, in this work, we report that the results of the temperature variation of the ^{57}Fe Mössbauer spectra of $FePS_3$ and $Mn_{0.5}Fe_{0.5}PS_3$, in detail.

2 Experimental procedure

Single crystal of $FePS_3$ and $Mn_{0.5}Fe_{0.5}PS_3$ were grown by chemical vapor transport. The proper weights of starting materials, transition metal (Mn and Fe), phosphorus and sulfur powders were thoroughly mixed with iodine in an Ar atmosphere. They were sealed in an evacuated quartz tube (1×10^{-3} Pa). The chemical vapor transport reactions were carried out in a two-zone electric furnace. The distance between the charge zone and the growth zone is about 10 cm. The temperature of the charge zone and that of the growth zone are 730 and 630°C for $FePS_3$ and $Mn_{0.5}Fe_{0.5}PS_3$, respectively. Transport was carried out for 120 h.

The Mössbauer measurements were performed by a conventional constant acceleration type spectrometer with ^{57}Co in Rh source. We calibrated the velocity with pure Fe metal absorber at room temperature. The single crystal samples were used for the measurements, and the gamma ray was irradiated parallel to the c axis (parallel to the direction of the ordered magnetic moment) in zero field.

3 Result and discussion

Figure 1 shows the temperature variation of the ^{57}Fe Mössbauer spectra of $FePS_3$ and $Mn_{0.5}Fe_{0.5}PS_3$. The Mössbauer spectrum of single crystal of $FePS_3$ was composed of a paramagnetic quadrupole doublet of asymmetric shape at room temperature [9]. As the Néel temperature T_N of $FePS_3$ is 118 K, two paramagnetic

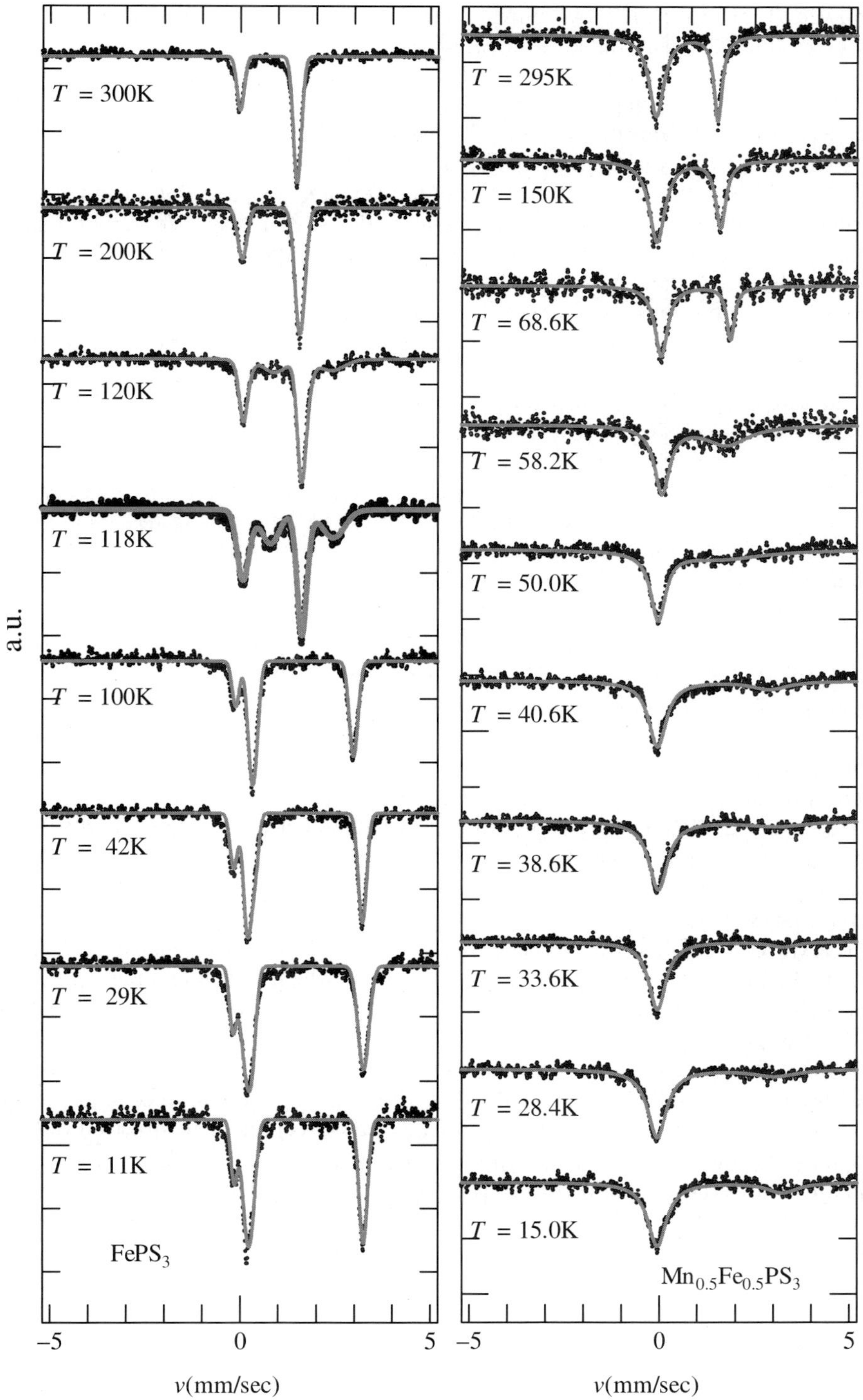

Figure 1 The temperature variation of ^{57}Fe Mössbauer spectra of the single crystals $FePS_3$ and $Mn_{0.5}Fe_{0.5}PS_3$. The gamma ray was irradiated parallel to the c axis in zero field.

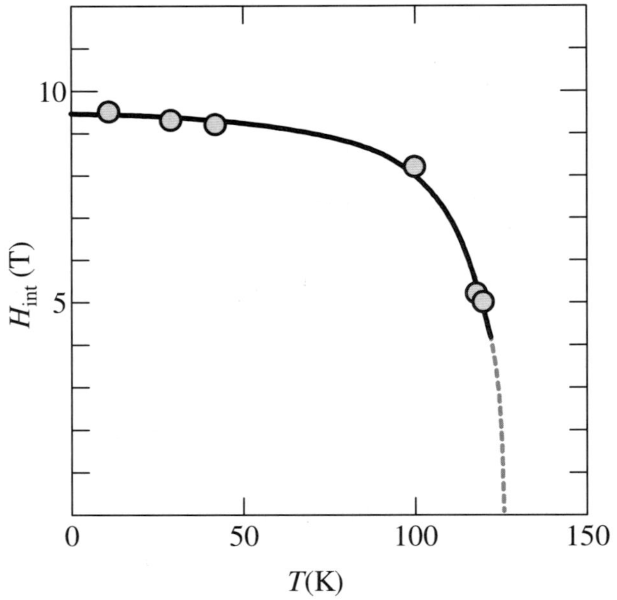

Figure 2 The temperature dependence of Hyperfine field H_{int} of $FePS_3$ in the anti-ferromagnetic state.

peaks are observed above 120 K [10, 11]. In the anti-ferromagnetic state, the hyperfine magnetic field H_{int} increases with decreasing temperature, H_{int} becomes 9.5 T at T=11 K shown Figure 2. The Isomer shift splitting I. S. increases slightly with decreasing temperature, as I. S.=1.71_2 mm/s at 11 K, I. S.=1.67_0 mm/s at 100 K.

However, in $Mn_{0.5}Fe_{0.5}PS_3$, two rather broadened peaks compared with $FePS_3$ are observed even at room temperature (Figure 1). The intensity of these peaks is almost same. The width of these peaks increases slightly and the intensity of the off-center peak which is observed about 1.7 mm/s decreases with decreasing temperature. The off-center peak is collapsed and broadened at about 58 K and two peaks became a single peak at about 50.0 K, which is higher than T_g=33.7 K. And below about 40.6 K, the very broad peak near 3 mm/s present as like a peak of $FePS_3$ (suggested Fe^{2+}).

As the above results are very interesting, we consider the magnetic interactions exist far above T_g in the spin glass $Mn_{0.5}Fe_{0.5}PS_3$. As the fluctuation of the hyperfine fields increases with decreasing temperature, the off-center peak smeared out. We consider that these results support the result of the temperature dependence of the specific heat at the low temperature [2]. Although the Mössbauer spectra far above T_g is similar to that of the spin glass (Mn, Fe)TiO_3, it is quite different near and below T_g [5].

References

1. Takano, Y., Arai, A., Takahashi, Y., Takase, K., Sekizawa, K.: J. Appl. Phys. **93**, 8197 (2003)
2. Takano, Y., Arai, N., Arai, A., Takahashi, Y., Takase, K., Sekizawa, K.: J. Magn. Magn. Mater. **E593**, 272–276 (2004)

3. Ito, A., Aruga, H., Torikai, E., Kikuchi, M., Syono, Y., Takei, H.: Phys. Rev. Lett. **57**, 483 (1986)
4. Aruga Katori, H., Ito, A.: J. Phys. Soc. Jpn. **63**, 3122 (1994)
5. Ito, A., Morimoto, S., Aruga Katori, H.: J. Phys. Soc. Jpn. **65**, 3331 (1996)
6. Sherrington, D., Kirkpatrick, S.: Phys. Rev. Lett. **35**, 1792 (1975)
7. de Almeida, J.R.L., Thouless, D.J.: J. Phys. **A11**, 983 (1978)
8. Thouless, D.J.: Phys. Rev. Lett. **56**, 1082 (1986)
9. Tsuji, T., Watanabe, H.: J. Alloys Compd. **383**, 259–264 (2004)
10. Taylor, B.E., Steger, J., Wold, A.: J. Solid State Chem. **7**, 461 (1973)
11. Jernberg, P., Bjarman, S., Wäppling, R.: J. Magn. Magn. Mater. **46**, 178 (1984)

Hyperfine Interact (2005) 165: 101–106
DOI 10.1007/s10751-006-9252-2

Synthesis of nanoscale antimony particles

L. Balan · A. Dailly · R. Schneider · D. Billaud · P. Willmann · J. Olivier-Fourcade · J.-C. Jumas

Published online: 26 October 2006

Abstract For the search of new negative electrodes of Li-ion batteries, a low-temperature method has been developed for the preparation of nanoscale antimony particles which uses an alkoxide-activated sodium hydride as reducing agent of antimony pentachloride. X-ray diffraction and TEM studies confirm the obtaining of amorphous Sb nanoparticles dispersed in an organic matrix. ^{121}Sb Mössbauer spectroscopy gives evidence for the occurrence of interactions between antimony and the matrix. These interactions are modified by the washing treatments.

Key words nanosized antimony particles · amorphous antimony · ^{121}Sb Mössbauer spectroscopy

1 Introduction

The increasing demand of electric energy for many applications motivates an intensive research for Lithium-ion batteries. Improvement of their electrochemical performances to obtain higher specific capacity, better cycleability, large reversibility and good safety requires the search of new electrode materials. In particular the research of new negative

L. Balan · A. Dailly · D. Billaud (✉)
Laboratoire de Chimie du Solide Minéral (UMR 7555 CNRS)Université Henri Poincaré Nancy I, BP 239, 54506 Vandoeuvre les Nancy Cedex, France
e-mail: Denis.Billaud@lcsm.uhp-nancy.fr

R. Schneider
Laboratoire de Synthèse organométallique et Réactivité (UMR 7565 CNRS)Université Henri Poincaré Nancy I, BP 239, 54506 Vandoeuvre les Nancy Cedex, France

P. Willmann
Centre National d'Etudes Spatiales, 18 Avenue Edouard Belin, 31401 Toulouse Cedex 9, France

J. Olivier-Fourcade · J.-C. Jumas
Laboratoire des Agrégats Moléculaires et Matériaux Inorganiques (UMR 5072 CNRS), Université Montpellier II, Place Eugène Bataillon, 34095 Montpellier Cedex 5, France

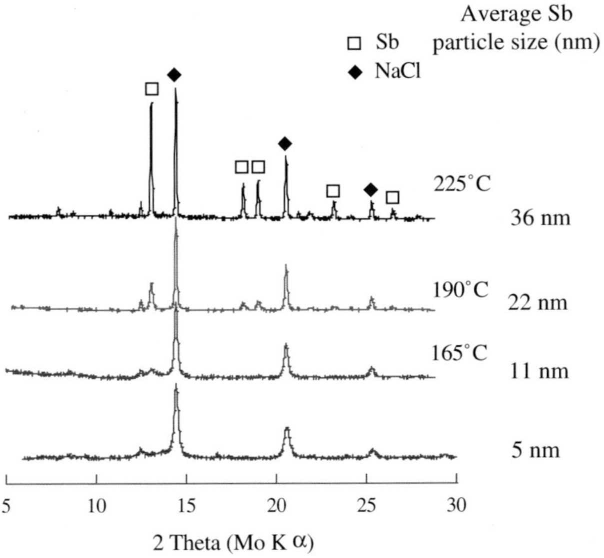

Figure 1 XRD patterns of Sb-based systems at different thermal treatments. Note the intensity increase of the Sb reflections with the temperature increase.

electrodes knows an increasing interest since the announcements of the high performance of tin composite [1] confirmed Sony Corporation and the industry's first hybrid lithium ion rechargeable battery [2].

With the new methods of nanomaterial syntheses, extensive research about lithium metal alloys is now in progress. A low-temperature method has been developed for the preparation of nanoscale metal (or semi-metal) particles which uses an alkoxide-activated alkaline hydride as reducing agent.

This communication is related to the Sb-containing systems; the obtained materials have been characterized by X-ray diffraction (XRD), transmission electron microscopy (TEM) and ^{121}Sb Mössbauer spectroscopy.

2 Experimental

The chemical reaction that we employed for the preparation of M (Sb, Sn) nanoparticles [3, 4] using $M\mathrm{Cl}_X$ ($\mathrm{SbCl_5}$, $\mathrm{SnCl_4}$) and t-BuONa-activated NaH in tetrahydrofuran (THF) can be written as follows:

$$M\mathrm{Cl}_x + \mathrm{x(NaH)/y}(t-\mathrm{BuONa}) \xrightarrow[65^\circ\mathrm{C}]{\mathrm{THF}} M^{(0)}\Big/\mathrm{y}\,(t-\mathrm{BuOna}) + \mathrm{x}/2\,\mathrm{H_2} + \mathrm{x\,NaCl}$$

t-BuONa activated NaH is used both as reducing and stabilizing agent. The method is technically simple, affordable and gives access under mild conditions to nanometer-sized metal particles.

The synthesis was handled under standard airless techniques in a nitrogen atmosphere. A slight excess of NaH was employed to ensure a rapid and complete reduction of $M\mathrm{Cl}_x$ into $M^{(0)}$. In a typical synthesis, a Schlenk tube was loaded with degreased NaH and THF and the mixture was heated to 65°C. A solution of t-BuOH in THF was then injected into the tube and the mixture was held at 65°C for 5 min. After cooling to room temperature, $M\mathrm{Cl}_x$

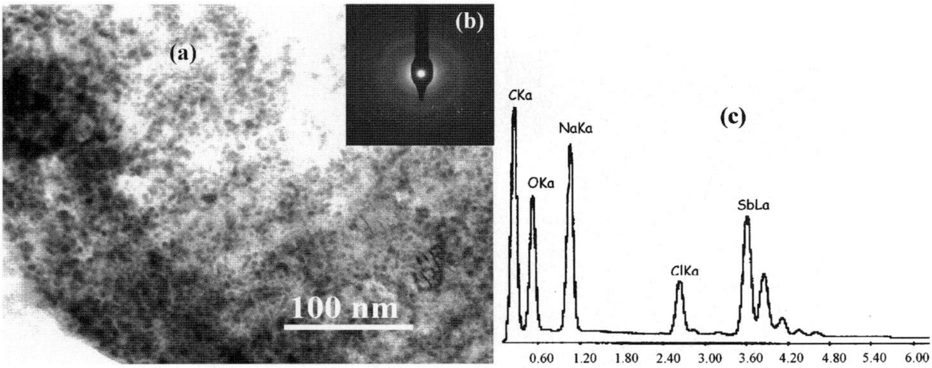

Figure 2 TEM studies of the as-synthesized material: (a) bright field micrograph, (b) SAED and (c) EDXS spectrum.

was added in one portion. The Schlenk contents were further stirred at 65°C for 30 min. After completion of the reaction, the solution was rotary evaporated under air to remove volatile organics, which led to a grey powder of $M^{(0)}$ nanoparticles stabilized by *t*-BuONa. Excess of NaH and NaCl was removed by subsequent washings with ethanol and water, respectively.

Powder X-ray diffraction (XRD) analyses were obtained using an automated powder diffractometer with MoKα radiation (Rotaflex RU-200B, RIGAKU generator and CPS 120 INEL detector, transmission assembly). Powders were also characterized by transmission electron microscopy (TEM PHILIPS CM 20 instrument operated at 200 kV) and energy dispersive X-ray spectrometry (EDXS). ^{121}Sb Mössbauer measurements were performed in standard transmission geometry using a ^{121m}Sn in $BaSnO_3$ source of nominal activity 0.5 mCi on a constant acceleration spectrometer. During the measurements, both source and absorber were held at liquid helium temperature to increase the fraction of recoil-free emission and absorption processes. Samples were prepared in a glove box by carefully mixing the powders with A-Piezon grease and then were placed on a specific sample holder transparent for γ rays. The velocity scale was calibrated with the standard spectrum of an iron absorber using a ^{57}Co(Rh) source. Zero isomer shift was defined from the spectrum of InSb at 4 K. The data was processed using a transmission integral analysis software GM5SIT [5] with the Mössbauer source fraction f_s and source line width Γ_s held constant at 0.62 and 1.45 mm/s, respectively.

3 Results and discussion

Figure 1 shows the XRD patterns of the as-synthesized material at different thermal treatments. At room temperature, the XRD patterns do not present any reflections related to antimony particles which are either amorphous or smaller than the X-ray coherence length. After heating, reflections of crystalline antimony are present and one can observe an increase of the Sb particle size with the temperature rise. Washing with EtOH and with H_2O results in the removal of excess NaH and NaCl without modification of the Sb particle size.

Figure 2 presents the TEM bright field micrograph of the as-synthesized material obtained after a few seconds of irradiation to obtain a partial coalescence of

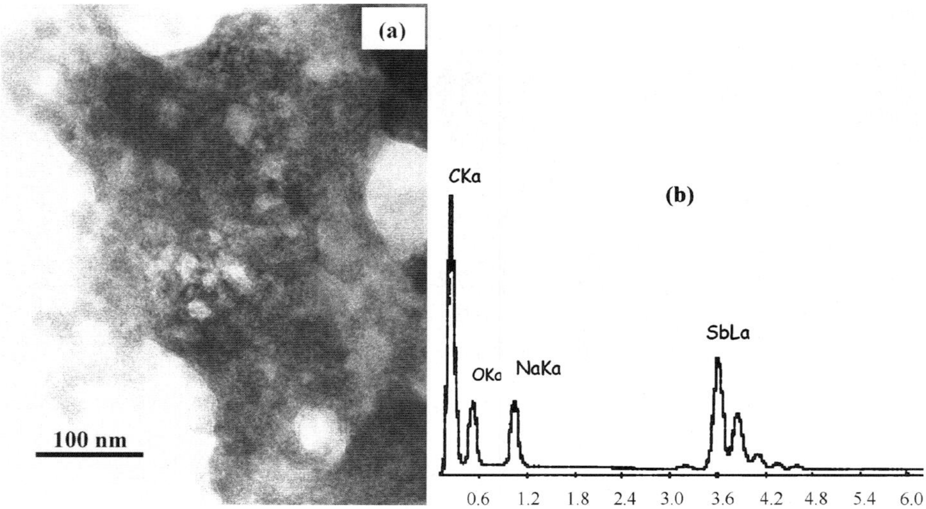

Figure 3 TEM bright field micrograph of the $Sb^{(0)}$ particles after washing with ethanol and with water. Associated EDXS spectrum.

subnanoparticles present in the material. Thus, Sb nanoparticles with an average diameter of 5 nm appear dispersed fairly homogeneously in an amorphous organic matrix (Figure 2a). Some aggregates are also present. The associated EDXS spectrum (Figure 2c) gives evidence for the elements present in the material (C, O, Na, Cl, Sb) and the selected area electron diffraction (SAED) pattern (Figure 2b) displays a low intensity and diffuse ring characteristic of amorphous Sb.

Figure 3 shows the bright field micrograph of the materials washed with EtOH and then with H_2O. The Sb particle size is not modified significantly by the washing treatments (Figure 3a). In the associated EDXS spectrum (Figure 3b), lines attributed to chlorine have disappeared while the intensity of that of Na has decreased, in agreement with excess NaH and NaCl elimination by EtOH and H_2O washings, respectively. Moreover, the decrease of the O/Sb ratio and the presence of Na gives evidence for a partial removal and/or a chemical modification of the amorphous organic matrix.

Figure 4 presents the ^{121}Sb Mössbauer spectra of Sb nanoparticles obtained from $SbCl_5$ and t-BuONa activated NaH (a), after washing with EtOH (b) and after washing with EtOH then H_2O (c). In Table I, the corresponding Mössbauer parameters δ (isomer shift relative to InSb), Δ (quadrupole splitting) and Γ (line width) are summarized.

The three Mössbauer spectra exhibit two absorption bands which have been analyzed as two sub-spectra:

– The first one which is the more intense and corresponds to a Sb contribution of 74 to 82% is characteristic of neutral $Sb^{(0)}$. The isomer shift δ (−2.57 to −2.96 mm/s vs. InSb) even close to that of elemental Sb (−3.25 mm/s) [6, 7] would be related to an electron density increase around Sb due to a ligand effect. As previously discussed [8] for the $Sb^{(0)}$ compounds, these variations can be explained from an increase of the screening effect due to the weak electronic transfer between Sb and the organic matrix. The quadrupole splittings Δ (+5.8 to 10.7 mm/s) are however significantly larger than that of $Sb^{(0)}$ (+3.2 mm/s). These observations give evidence for the $SbCl_5$ reduction

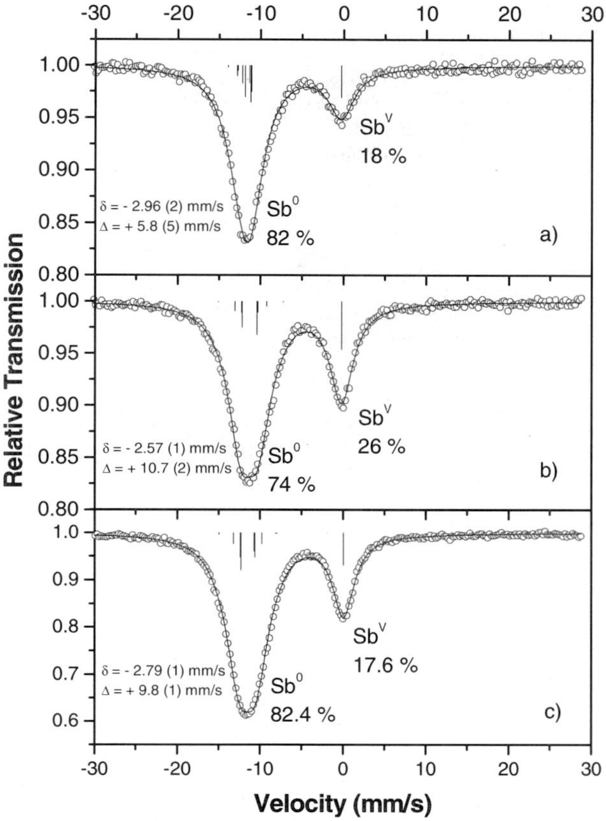

Figure 4 ^{121}Sb Mössbauer spectra of Sb nanoparticles (a) obtained from $SbCl_5$ and *t*-BuONa activated NaH, (b) after washing with EtOH, (c) after washing with EtOH then H_2O.

from Sb^V to $Sb^{(0)}$ and that $Sb^{(0)}$ interacts with the organic matrix. These interactions are modified by the washing treatments and are related to an electron density increase around Sb due to a ligand effect. Such an interaction is depending on the washing treatments and therefore on a chemical modification of the matrix induced by washings as evidenced for the quadrupole splittings which strongly become larger after washings.

– The second one corresponding to a Sb abundance of 18 to 26% can be attributed to the presence of Sb^V likely in the oxide Sb_2O_5 form (δ=+9.2 mm/s) [9, 10] even if some interactions with the organic matrix cannot be totally excluded.

4 Conclusion

The reduction of $SbCl_5$ by *t*-BuONa activated NaH leads to the formation of amorphous Sb nanoparticles with a diameter size lower than 5 nm. These particles are dispersed fairly homogeneously in an organic matrix. Subsequent washings with EtOH and H_2O results in the removal of NaH in excess of NaCl and of a partial elimination or chemical modification of the organic matrix. ^{121}Sb Mössbauer spectroscopy gives evidence for the occurrence of interactions between elemental $Sb^{(0)}$ and the matrix.

Table I ^{121}Sb Mössbauer data of Sb nanoparticles obtained from $SbCl_5$ and *t*-BuONa activated NaH, (a) as synthesized, (b) after washing with ethanol and (c) after washing with ethanol and with water

Sample	δ (mm/s)	Δ (mm/s)	Γ (mm/s)	Contribution (%)
a	−2.96 (2)	+5.8 (5)	2.23 (7)	82.0 (3)
	+8.44 (5)	–	2.23 (7)	18.0 (3)
b	−2.57 (1)	+10.7 (2)	1.78 (4)	74.0 (2)
	+8.52 (2)	–	1.78 (4)	26.0 (2)
c	−2.79 (1)	+9.8 (1)	1.42 (2)	82.4 (1)
	+8.78 (1)		1.42 (2)	17.6 (1)

δ=isomer shift relative to InSb at 4 K (−8.70 (4) mm/s relative to the $Ba^{121m}SnO_3$ source); Δ=quadrupole splitting; Γ=absorber line width.

Acknowledgments The authors acknowledge CNES, Toulouse, France, for financial support (Contract N° 04/1756/00, EOTP N°: R-S04/PF-0005-001) and Drs. J. Ghanbaja and G. Medjadi for TEM and X-ray characterizations.

References

1. Fuji Photo Film Co. Ltd. European Patent, EP 0 704 921 A1, 1995
2. Sony, internet: http://www.sony.net/SonyInfo/News/Press/200502/05-006E/ (2005)
3. Dailly, A., Schneider, R. Billaud, D., Fort, Y., Ghanbaja, J.: J. Nanopart. Res. **5**, 389 (2003)
4. Balan, L., Schneider, R., Billaud, D., Ghanbaja, J.: Mat. Lett. **59**(8–9), 1080 (2005)
5. Rubenbauer, K., Birshall, T.: Hyperfine Interact. **7**, 125 (1979)
6. Berry, F.J., Gogarty, P.M., Jenkins, J.W.: J. Chem. Soc., Dalton Trans. 2161 (1988)
7. Aldon, L., Garcia, A., Olivier-Fourcade, J., Jumas, J.C., Fernandez-Madrigal, F.J., Lavela, P., Perez-Vicente, C., Tirado, J.L.: J. Power Sources **119–121**, 585 (2003)
8. Lippens, P.E.: Solid State Commun. **113**, 399 (2000)
9. Birchall, T., Della Valle, B.: Chem. Commun. 675 (1970)
10. Jansen, M., Pebler, J., Dehnicke, K.: Z. Anorg. Allg. Chem. **495**, 120 (1982)

Hyperfine Interact (2005) 165: 107–112
DOI 10.1007/s10751-006-9253-1

Phase transformations of σ-FeCr induced by ball milling

B. F. O. Costa · J. M. Loureiro · G. Le Caër

Published online: 9 November 2006

Abstract A tetragonal sigma phase σ-$Fe_{0.54}Cr_{0.46}$ transforms gradually into a bcc phase by ball-milling in argon. Mössbauer spectroscopy shows further the presence of an amorphous phase. It starts to form after 25 h of milling in our experimental conditions.

Key words Fe–Cr · sigma phase · ball-milling · amorphization · Mössbauer spectroscopy

1 Introduction

Nearly equiatomic binary Fe–Cr alloys and the multinary alloys derived from them are technologically important materials because of their high-temperature corrosion and mechanical resistance. At high temperature, these body centered cubic (bcc) alloys are however not stable. The precipitation of a sigma Fe–Cr phase [1, 2] makes for instance the steel very brittle.

Phase transformations in nanocrystalline materials (grain size $d<100$ nm) can condition some of their applications, as reported for instance for Fe–Cr and Fe–Cr–Sn nanostructured alloys [3, 4]. Recently, a study of phase transformations occurring during long-time annealing in Fe–Cr films showed that an amorphous phase is formed along with the crystalline phases expected for Cr-rich alloys [5].

The behaviour under milling in vacuum of a sigma Fe–Cr phase was studied earlier by Bakker et al. [6–8]. The sigma phase is not stable and transforms into a bcc nanostructured phase which is the sole phase identified from X-ray diffraction patterns. However, this

B. F. O. Costa (✉) · J. M. Loureiro
Departamento de Física, Universidade de Coimbra, P3000-516 Coimbra, Portugal
e-mail: benilde@ci.uc.pt

G. Le Caër
GMCM UMR CNRS 6626, Universite de Rennes I, Campus de Beaulieu, Bat. 11A, F35042 Rennes Cedex, France

Figure 1 X-ray diffraction patterns of $Fe_{0.54}Cr_{0.46}$ milled for different times. The *first pattern* corresponds to the starting tetragonal sigma phase. The *bottom pattern* is that of the as-cast bcc alloy.

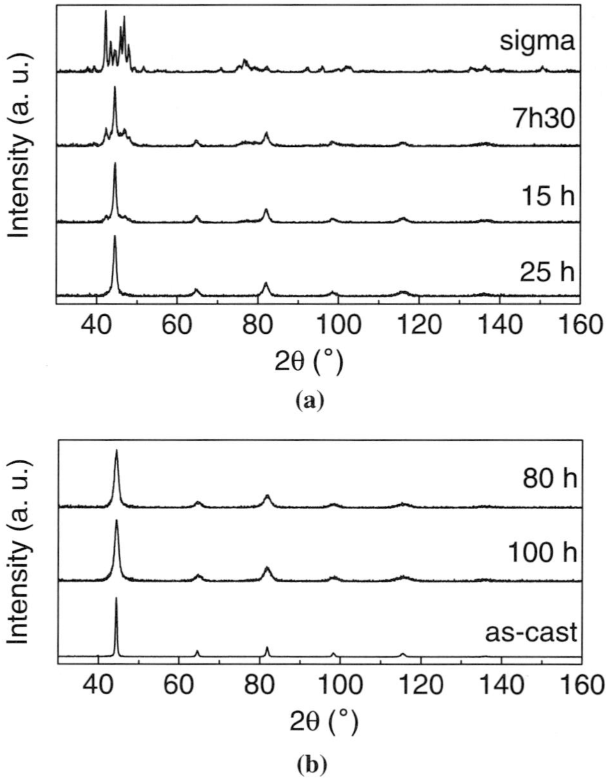

interpretation does not account for the variation of the low-temperature magnetisation with milling time. The magnetisation expected for a coarse-grained bcc Fe–Cr alloy with the same composition [9, 10] is about 2 μ_B/at. Fe instead of the reported value of 1.2 μ_B/at. Fe. This large difference cannot be explained by the effect on magnetic properties of the decrease of the bcc grain size into the nanometer range as proposed. For instance the magnetisation of alpha iron remains practically unchanged in such conditions. Similarly, the mean magnetic hyperfine fields of coarse-grained $Fe_{0.51}Cr_{0.49}$ are essentially identical ($\langle B \rangle = 16$ T) when the alloy is filed ($d = 250$ nm) or when it is ball-milled ($d = 25$ nm) despite coherent domain sizes d which differ by an order of magnitude [11]. Further, a contamination by steel coming from milling tools would contribute to a Fe-enrichment of the milled powders and thus to an increase of the measured magnetic moment. At least two assumptions can be put forward to explain a magnetisation decrease: (a) the formation of an amorphous phase [12, 13], which is non-magnetic in the studied concentration range and is possibly favoured by the presence of oxygen in ball-milled alloys [14]; (b) a phase separation with a significant amplitude which is difficult to evidence by X-ray diffraction in such alloys.

As a significant amount of an amorphous phase is however not seen on the X-ray diffraction patterns, we started an investigation of the stability of a sigma-FeCr alloy under high-energy ball-milling in argon, using Mössbauer spectroscopy as the main characterization technique of the structural changes.

Figure 2 RT ^{57}Fe Mössbauer spectra as a function of milling time. The top spectrum is that of the starting sigma phase.

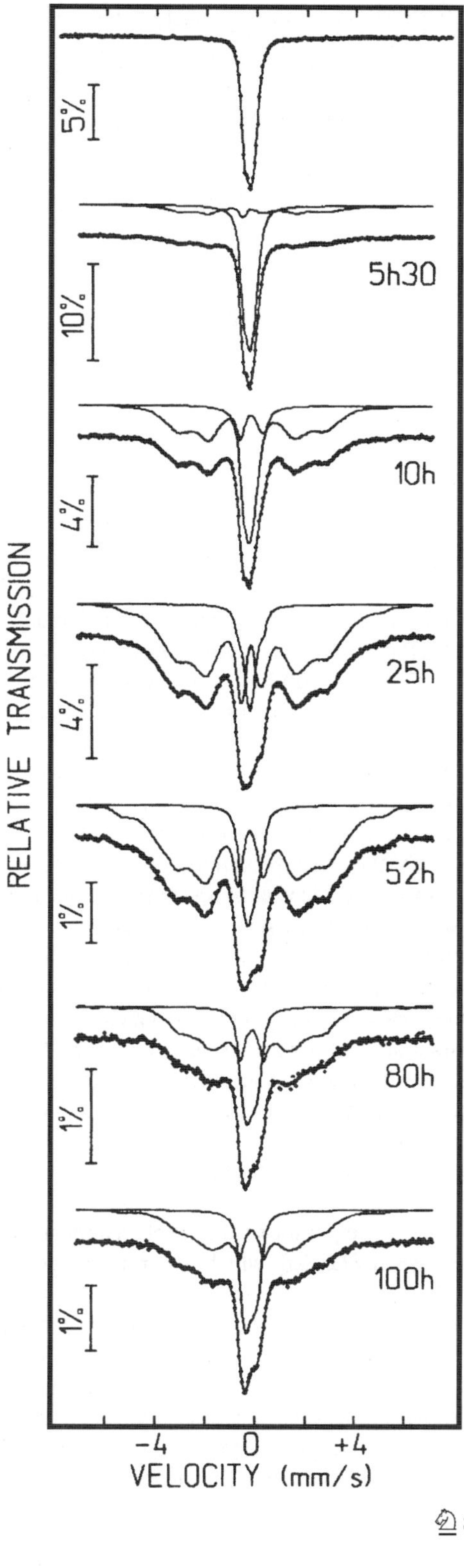

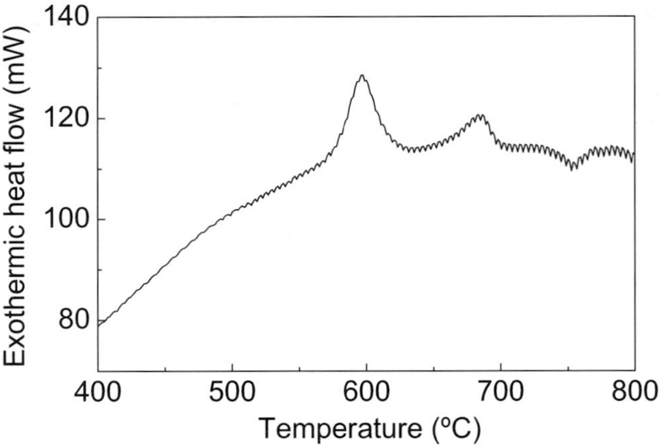

Figure 3 DSC trace of the $Fe_{0.54}Cr_{0.46}$ sample milled for 100 h.

2 Experimental details

A bcc $Fe_{0.536}Cr_{0.464}$ alloy was prepared by melting together appropriate amounts of Fe (99.9+% purity) and Cr (99.995% purity) under argon in an induction furnace. The composition, near the centre of the σ phase domain, was chosen to be close to that studied by Bakker et al. [6–8]. The sigma phase was formed by annealing the as-cast bcc alloy in vacuum at 700°C for 100 h. It was powdered in an agate mortar into particles with a mean size of about 90 μm. A mass of 5 g of σ-FeCr was then ball-milled in a Fritsch P0 vibratory mill with a hardened steel vial and ball. The latter has a diameter of 5 cm and a mass of 500 g. The vial was sealed under argon atmosphere at the start and after every 100 min of milling. The mill was working at its maximum amplitude of vibration.

Microprobe analysis was used to determine the composition of the alloy. X-ray diffraction (XRD) was performed at room temperature using Cu K_α radiation (λ= 0.154184 nm). ^{57}Fe Mössbauer spectra were recorded at room temperature (RT) in a transmission geometry using a standard constant acceleration spectrometer. A ^{57}Co source in Rh matrix with a strength of ≈35 mCi was used. The experimental spectra were analysed by a constrained Hesse–Rübartsch method [15], which yields a hyperfine magnetic field distribution, P(B). Lorentzian line-shapes were employed in this procedure. The ^{57}Fe isomer shifts are given with respect to α-Fe at RT.

Differential Scanning Calorimetry (DSC) in an argon flow was used to study the transformations of the final as-milled samples during heating from RT to 800°C at a rate of 40°C/min. The saturation magnetization of the final sample was determined with a SQUID magnetometer at a temperature of 4.2 K.

3 Results and discussion

Figure 1 shows X-ray diffraction patterns (XRD) recorded from powders milled for different times and for comparison, that of the bcc coarse-grained as-cast alloy. The XRD pattern of the starting material is characteristic of the σ-phase. The intensities of the σ-phase diffraction peaks decrease steadily with milling time t_m and the peaks of the α-phase appear progressively. After 25 h of milling and for longer milling times, only the bcc peaks are seen (Figure 1b). The peaks are broadened due to the decrease of the mean

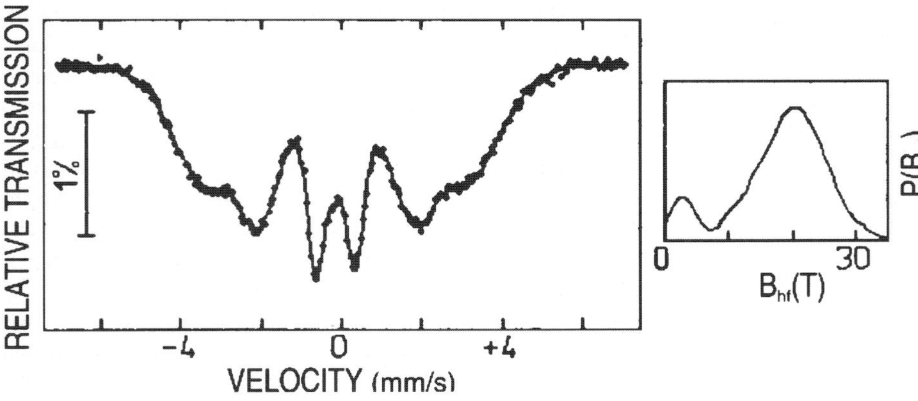

Figure 4 RT ^{57}Fe Mössbauer spectrum and hyperfine field distribution of the sample milled for 100 h and then annealed at 600°C for 6 min.

crystallite size and to strain. As expected, the peaks of the XRD pattern of the as-cast alloy are much narrower than those of ball-milled samples. The mean grain size, determined from the widths of the XRD peaks by the Williamson–Hall method, is 7 nm for the sample milled during 100 h and the microstrain is 1.3%.

Figure 2 shows RT ^{57}Fe Mössbauer spectra of the FeCr sample milled for the indicated periods. The first spectrum corresponds to the starting sigma phase which is paramagnetic at RT with an average isomer shift of −0.17(2) mm/s. A broad magnetic component, associated with the magnetic alpha phase, appears with increasing t_m. The contribution of this phase increases until about 25 h of milling. At this stage, the shape of the central peak clearly changes and its relative contribution increases with t_m. The paramagnetic contribution can no more be attributed to the sole sigma phase. Another phase transformation is clearly occurring while XRD patterns show only a broadening of the peaks of the bcc phase. The average hyperfine field is 18.5(2) T and the average isomer shift is −0.02(2) mm/s for the sample milled for 100 h. The fraction of 'magnetic' Fe is about 66%. A DSC trace of the latter sample is shown in Figure 3. The two exothermic peaks, located at about 600°C and 680°C, are attributed to a two-step crystallization of an amorphous phase as advocated in [16].

Thus, we attribute the central peak of the Mössbauer spectra to an amorphous phase with an average isomer shift of −0.08(2) mm/s in agreement with the results of Xia et al. [13].

The sample milled for 100 h was annealed at 600°C in vacuum for 6 min and a Mössbauer spectrum and an XRD pattern were subsequently recorded at RT. The Mössbauer spectrum (Figure 4) shows a magnetic phase corresponding to the bcc Fe–Cr phase with an average hyperfine magnetic field $\langle H \rangle = 18.2(2)$ T and an average isomer shift $\langle IS \rangle = -0.05(2)$ mm/s while the XRD peaks are narrower than they are before annealing. The mean crystallite size and the microstrain are found to be 13 nm and 0.66%, respectively.

The X-ray diffraction results that we obtain are similar to those of Bakker et al. [6–8]. The broad peaks were attributed by them to a bcc solid solution with a small grain size. They also attributed the small saturation magnetization values to the nanocrystalline size of domains.

The saturation magnetization of the sample milled for 100 h is measured to be of 1.130 μ_B/at. Fe, a value which is similar to that obtained by Bakker et al. [8]. It is 1.871 μ_B/at. Fe for the same sample annealed at 600°C for 6 min while the value given by Aldred [9] for a bulk coarse-grained bcc alloy of the same composition is close to 2 μ_B/at. Fe. The amorphous phase is thus not fully crystallized as suggested by the small contribution seen in the central part of the spectrum of Figure 4a and reflected by a peak at low field in the corresponding P (B) distribution (see also [16]).

The sample milled for 100 h contains 1.2 at. % O as determined by its combustion in an oxygen determinator equipped with a solid state infrared detector. Fnidiki et al. [17] showed recently how the presence of significant amounts of oxygen in the grain boundaries of ball-milled FeCr alloys favours the formation of a 'disordered' phase. The oxygen content found here is however smaller than that investigated in the latter work. As discussed in more detail in [16], the formation of an amorphous phase at grain boundaries favoured by the presence of oxygen, or even that of an amorphous-bcc composite with amorphous layers surrounding bcc grains, appears to be a reasonable assumption both consistent with the X-ray diffraction pattern of a bcc phase and the change of magnetic properties. In a dynamical equilibrium phase diagram of near-equiatomic Fe–Cr alloys milled under vacuum, the domain of amorphization is expected to be shifted downwards to smaller injected powers with respect to that of Fe–Cr alloys milled in the presence of residual gases which stabilize the amorphous phase [16]. An amorphous Fe–Cr phase can indeed be formed in experimental conditions in which oxygen does not play a significant role [5, 12].

4 Conclusions

The sigma phase, whose composition is here $Fe_{0.54}Cr_{0.46}$, is not stable during high-energy ball milling either in argon or in vacuum [6–8]. The transformation does not yield a sole bcc Fe–Cr phase as concluded from the X-ray diffraction patterns. Mössbauer spectroscopy shows that the formation of the bcc phase occurs together with a partial amorphization of the ball-milled sample. Amorphization, which is known to be favoured by the presence of oxygen, might however be largely of intrinsic origin as discussed further in [16].

References

1. Kubachewski, O.: Iron-Binary Phase Diagrams, p. 185. Springer, Berlin (1982)
2. Dubiel, S.M., Inden, G., Metalkd, Z.: **78**, (1987)
3. Costa, B.F.O., Le Caër, G., Ayres de Campos, N.: Phys. Status Solidi (a) **183**, 235 (2001)
4. Costa, B.F.O., Le Caër, G., Luyssaert, B.: J. Alloy Compd. **350**, 36 (2003)
5. Levin, A.A., Meyer, D.C., Gorbunov, A., Mensch, A., Pompe, W., Paufler, P.: J. Alloy Compd. **360**, 107–117 (2003)
6. Bakker, H., Zhou, G.F., Yang, H.: Prog. Mater. Sci. **39**, 159 (1995)
7. Yang, H., Di, L.M., Bakker, H.: Intermetallics **1**, 29 (1993)
8. Yang, H., Bakker, H.: Math. Sci. Eng. **A181/A182**, 1207 (1994)
9. Aldred, A.T.: Phys. Rev., B **14**, 219 (1976)
10. Aldred, A.T., Rainford, B.D., Kouvel, J.S., Hicks, T.J.: Phys. Rev., B **14**, 228 (1976)
11. Delcroix, P., Le Caër, G., Costa, B.F.O.: J. Alloy Compd. ISMANAM 2005, Paris 4–7 July 2005 (in press)
12. Xia, S.K., Baggio-Saitovitch, E., Larica, C.: Phys. Rev., B **49**, 927 (1994)
13. Xia, S.K., Baggio-Saitovitch, E., Rizzo Assunçao, F.C., Pena Rodriguez, V.A.: J. Phys., Condens. Matter **5**, 2729 (1993)
14. Lemoine, C., Fnidiki, A., Lemarchand, D., Teillet, M J.: J. Phys. Condens., Matter **11**, 8341 (1999)
15. Le Caër, G., Dubois, J.M.: J. Phys. E **12**, 1083 (1979)
16. Costa, B.F.O., Le Caër, G., Loureiro, J.M., Amaral, V.S.: J. Alloy Compd. **424**, 131–140 (2006)
17. Fnidiki, A., Lemoine, C., Teillet, J.: Physica. B **363**, 271 (2005)

Hyperfine Interact (2005) 165: 113–119
DOI 10.1007/s10751-006-9254-0

Chemical reduction of hematite by sodium borohydride

V. G. de Resende · G. M. da Costa ·
E. De Grave · L. Datas

Published online: 26 October 2006

Abstract Well-crystallized hematite was suspended in water and treated at room-temperature (RT) with sodium borohydride. The product of the reaction is a highly magnetic black powder, which is stable at RT. The $NaBH_4$ treatment converts about half of the hematite to an amorphous Fe–B alloy and to a small fraction of sub-micron sized, amorphous metallic-Fe nodules. Heating at 400°C of this composite has resulted in the crystallization and/or oxidation of more than half of the amorphous Fe–B phase to α-Fe and Fe_3O_4 and B_2O_3, respectively. After treatment at 800°C, the metallic Fe and the amorphous Fe–B have completely vanished, and the resulting product consists of hematite and $FeBO_3$ embedded in the matrix of α-Fe_2O_3.

Key words Fe–B alloy · hematite · sodium borohydride · Mössbauer · spectroscopy

1 Introduction

The use of sodium borohydride for the production of amorphous transition-metal boron alloys (TM-B) has been the subject of intense research in recent years [1–3]. The soft route using $NaBH_4$ offers several advantages with respect to other, more traditional methods (vapour deposition, melt spinning, sputtering, etc.) such as ease and promptness of the synthesis procedures and the possibility of preparing TM-B alloys with a broad variety in composition [4]. As far as it is known, all preparations methods reported so far in literature involve the addition of $NaBH_4$ to a solution containing the metallic ions. The effect of the

V. G. de Resende · G. M. da Costa
Dep. de Química, Universidade Federal de Ouro Preto, 35400 Ouro Preto (MG), Brazil

V. G. de Resende · E. D. Grave (✉)
Department of Subatomic and Radiation Physics, University of Ghent, 9000 Gent, Belgium
e-mail: Eddy.DeGrave@UGent.be

L. Datas
CIRIMAT (LCMIE), Université Paul Sabatier, 31062 Toulouse, France

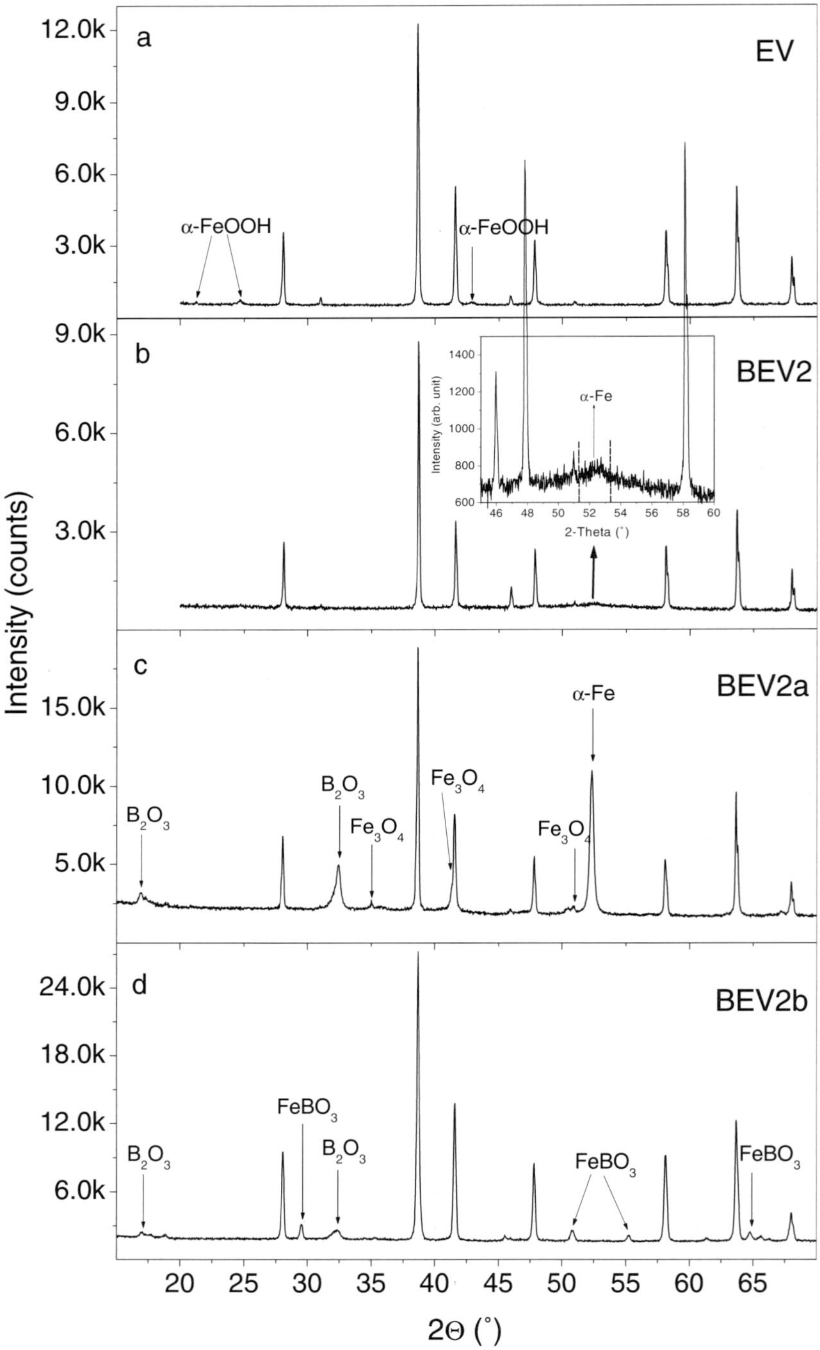

Figure 1 X-ray diffraction patterns of the original sample (EV), and after treatment with borohydride (BEV2). Samples BEV2a and BEV2b were obtained by heating BEV2 at 400 and at 800°C, respectively.

pH, reaction time, and even of the sequence of the addition of the reactants on the composition and on the physicochemical properties of the obtained products have been examined by a number of authors [4–6]. On the other hand, the reaction of $NaBH_4$ with solid particles has not been investigated, and the main objective of this work is to report for the first time the formation of amorphous Fe–B alloy in aqueous medium by the reduction of hematite particles.

2 Experimental

The parent sample (hereafter called EV) involved in this study is an iron ore concentrate containing mainly hematite, with a negligible fraction of goethite. About 1.2 g of the sample was dispersed in 200 ml of distilled water in which 4.0 g of $NaBH_4$ had been dissolved prior to the dispersion. Hydrochloric acid (100 ml, 0.20 M) was added dropwise under vigorous stirring. Four portions of 0.8 g each of $NaBH_4$ were added at regular intervals together with 100 ml of HCl. After 30 min of reaction, the solid was washed several times with distilled water and finally dried at 100°C (sample BEV2). Fractions of this sample were heated at 400°C (sample BEV2a) and 800°C (BEV2b) in static air during 1 h.

X-ray diffractograms (XRD) were recorded in the range 15–70° (2θ) with goniometer speed of 2°/min using a XRD-6000 Shimadzu diffractometer equipped with a Co tube and Fe filter. Counts were registered every 0.02°.

Mössbauer spectra were collected with spectrometers operating at constant acceleration mode with triangular reference signals. ^{57}Co(Rh) sources were used. At room-temperature, both conventional transmission Mössbauer spectroscopy (TMS spectra) and integral low-energy electron Mössbauer spectroscopy (ILEEMS spectra) have been applied. The latter variant probes the very top monolayers of the crystallites and gives information about those Fe species lying at distances of a few nanometers below the physical surfaces of the particles [7]. For samples BEV2 and BEV2a, TMS spectra were also recorded with the absorbers at a number of different temperatures between 12 K and RT. All Mössbauer spectra were computer-analysed in terms of model-independent distributions of hyperfine-parameter values and numerical data quoted hereafter refer to maximum probability values. Isomer shifts are referenced with respect to α-Fe at RT.

3 Results and interpretation

Sample EV has the typical macroscopic appearance of an iron ore with specularite being the dominant Fe phase. Specularite is a variety of hematite characterized by aggregates of silvery, specular hematite flakes or tabular crystals and as such shows a metallic brilliance. Hematite is weakly ferromagnetic at RT and for this reason, it is only slightly attracted by a hand magnet. On the other hand, sample BEV2 is black and highly magnetic. At first glance, the XRD pattern of BEV2 only shows the peaks corresponding to well-crystallised hematite (Figure 1b), but a closer examination of the diffractogram (inset Figure 1b) reveals the appearance of a shallow bump near 52° (2θ), a position that corresponds to the principal diffraction line of α-Fe. In addition, it can be noticed that the peaks corresponding to goethite (arrows in Figure 1a) are not present in BEV2, indicating that this phase has completely reacted with the borohydride.

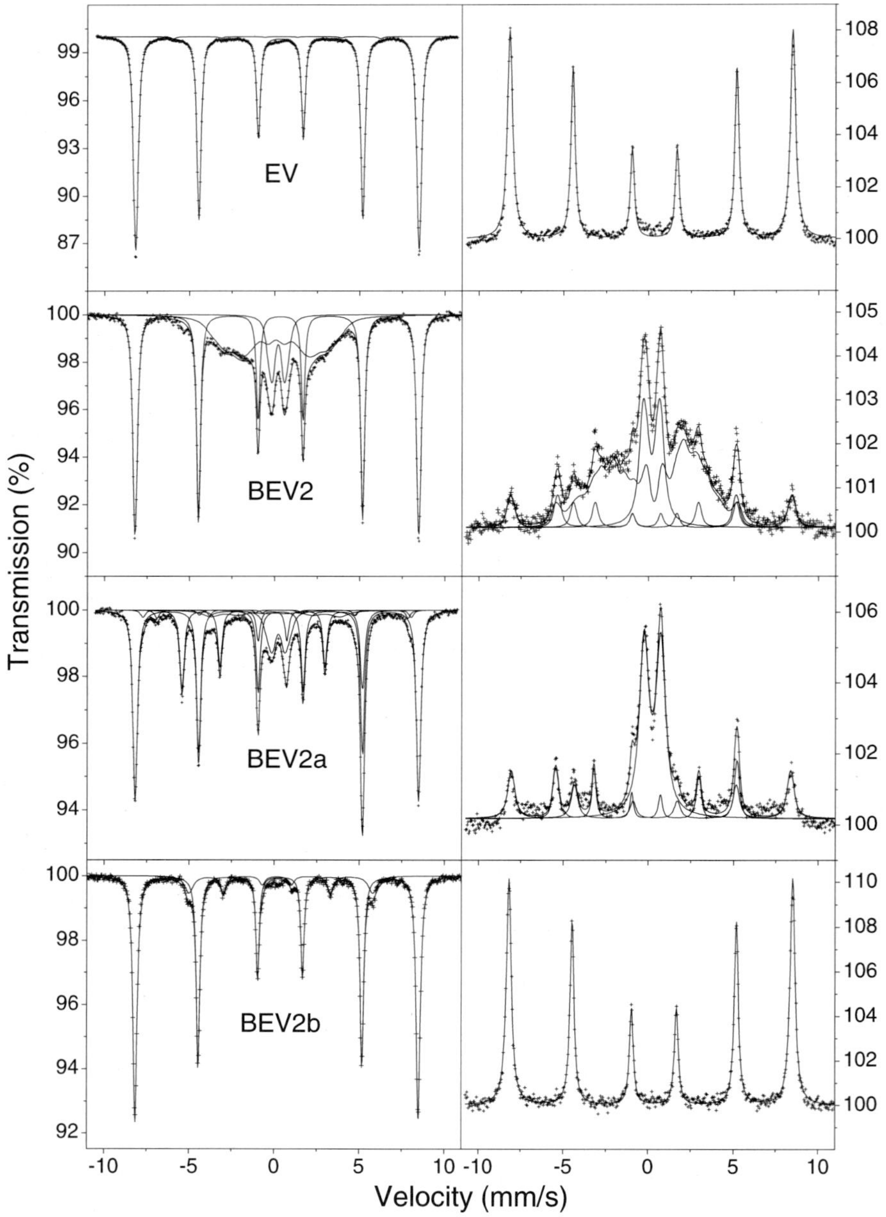

Figure 2 Room temperature transmission (*left*) and ILEEMS (*right*) Mössbauer spectra of the studied samples.

TMS and ILEEMS of sample EV at RT (Figure 2) yield one sharp sextet of which the relevant Mössbauer parameters, i.e., hyperfine field H_{hf} and quadrupole shift $2\varepsilon_{\mathrm{Q}}$ (see Table I) are diagnostic of hematite. A very weak second sextet (area S=1% relative to total spectral area) could be resolved from the TMS spectrum. Its hyperfine field and quadrupole shift at RT are 378 kOe and −0.26 mm/s, respectively, and can therefore be assigned to well-crystallized goethite, α-FeOOH [8].

Table I Mössbauer results of the samples measured at room temperature

	EV				BEV2				BEV2a				BEV2b			
	H_{hf}	$2\varepsilon_Q$	δ	S	H_{hf}	$2\varepsilon_Q{}^{b}$	δ	S	H_{hf}	$2\varepsilon_Q{}^{b}$	δ	S	H_{hf}	$2\varepsilon_Q$	δ	S
Sextet I																
Trans.	517	−0.20	0.37	100	517	−0.21	0.35	54	517	−0.22	0.37	56	516	−0.20	0.37	87
ILEEMS	516	−0.19	0.37	100	515	−0.20[a]	0.37[a]	11	511	−0.25	0.39	23	518	−0.19	0.37	100
Sextet II																
Trans.	–	–	–	–	185	0.00[a]	0.19	35	237	0.00[a]	0.16	7	–	–	–	–
ILEEMS	–	–	–	–	208	0.00[a]	0.24	53	–	–	–	–	–	–	–	–
Sextet III																
Trans.	–	–	–	–	–	–	–	–	329	−0.01	−0.01	22	–	–	–	–
ILEEMS	–	–	–	–	327	0.00[a]	0.00[a]	12	331	0.00	0.00	20	–	–	–	–
Sextet IV/V																
Trans.	–	–	–	–	–	–	–	–	487	0.00[a]	0.26[a]	2	–	–	–	–
ILEEMS	–	–	–	–	–	–	–	–	457	0.00[a]	0.58	2	–	–	–	–
Sextet VI																
Trans.	–	–	–	–	–	–	–	–	–	–	–	–	335	0.24	0.38	13
ILEEMS	–	–	–	–	–	–	–	–	–	–	–	–	–	–	–	–
Doublet																
Trans.	–	–	–	–	–	0.63	0.31	11	–	0.64	0.35	11	–	–	–	–
ILEEMS	–	–	–	–	–	0.93	0.36	24	–	1.00	0.35	57	–	–	–	–

For the doublet, ΔE_Q means the quadrupole splitting (mm/s). The values of isomer shifts are with reference to metallic iron.

H_{hf} maximum-probability hyperfine fields (kOe), $2\varepsilon_Q$ quadrupole shifts (mm/s), δ isomer shifts (mm/s) and S relative areas (%) of the various spectral components resolved from the room-temperature transmission and ILEEMS Mössbauer spectra

[a] Fixed parameter

[b] ΔE_Q for the doublet component

The RT TMS of sample BEV2 presents, in addition to the sharp hematite sextet (relative spectral area S=54%), a very broad sextet (35%, henceforward termed sextet II) and a doublet (11%). In fitting this spectrum, the area ratio of outer lines to middle lines to inner lines for the broad sextet was forced to equal 3:2:1. From the results of the effect of temperature on the TMS of BEV2 (not shown) and the numerical results derived from these spectra (also not shown), it is obvious that sextet II remains considerably broadened, even at 12 K, and that its contribution slightly increases with lowering of temperature. As for the doublet, it seems to persist down to temperatures as low as 12 K, but it is quite possible that it would completely vanish on further lowering the temperature towards 0 K. Due to the weak contribution of this doublet, its Mössbauer parameters are ill defined and the adjusted hyperfine-parameter values inevitably are uncertain, thus explaining the unreal fluctuations in the obtained temperature variations of these values. It is believed, though, that the consistent decrease of the doublet's relative area with decreasing temperature, the amount by which fully accounting for the increase of the area of sextet II, is not unreal.

The above mentioned findings suggest that sextet II and doublet, as resolved from the TMS spectra of BEV2, both are due to one and the same structural phase of which the magnetic order–disorder-transition temperature is spread out over a very broad temperature range. Such behaviour one may expect to be observed for magnetic small-particle or amorphous systems with a broad distribution in particle size or composition, respectively. Based upon published data referring to amorphous Fe–B alloys [4], upon the type of elements present in the starting mixture and upon the observation that no diffraction lines other than those from α-Fe_2O_3 and α-Fe are obvious, one may assign sextet II and doublet to an amorphous Fe–B alloy, henceforward referred to as the a-(Fe,B) phase. The global composition of this phase is $Fe_{0.9}B_{0.1}$, as determined by wet chemical analysis and ICP.

Finally, the $2\varepsilon_Q$ data for the hematite sextet indicate that the Morin transition of the hematite phase occurs at a temperature well above 220 K, implying a high degree of crystallinity of the parent ferric oxide with minor or no substitutions of the ferric cations by other species such as Al, Mn, Ti, etc. [9], which are common in natural hematites.

The ILEEMS spectrum of BEV2 at RT shows that the amount of hematite in the surface layers, as quantified by the relative spectral area S=11% of the hematite component appearing in the ILEEMS, is much less than in the bulk, where S=54%. On the other hand, the area of doublet has nearly doubled from 11% (bulk) to 24% (surface). Two other components are clearly identified in this ILEEMS spectrum: the broad sextet (53%) corresponding to sextet II in the TMS spectra and a third sextet, designated as sextet III in Table I, which obviously is due to metallic iron (12%). The observation that both sextet II and doublet components in the ILEEMS spectrum are enhanced with respect to their respective counterparts in the TMS spectrum, is consistent with the suggestion that they are indeed attributable to the same constituent, which thus is most likely initially formed in the surface layers of the hematite grains.

The XRD patterns of the samples obtained by heating BEV2 at 400°C (BEV2a, highly magnetic) and 800°C (BEV2b, non-magnetic) are reproduced in Figure 1c and d, respectively. The most noticeable aspect is the presence of a sharp peak corresponding to metallic iron in BEV2a, and the absence of this peak upon heating at 800°C. Additional diffraction lines, due to magnetite, Fe_3O_4, and B_2O_3 in sample BEV2a and to $FeBO_3$ and B_2O_3 in BEV2b, are readily recognized.

The TMS spectra of BEV2a (see Figure 2) clearly show the presence of α-Fe and Fe_3O_4, (sextets III and IV/V, respectively, in Table I) along with hematite and the iron phases that give rise to the broad sextet II and a weak doublet. The fraction of the total Fe atoms that forms the hematite phase has remained unchanged from that in the parent sample BEV2 prior to heating. In contrast, the contribution of sextet II to the total spectrum of BEV2a, i.e., ~10%, is considerably lower as compared to BEV2. Its Mössbauer parameters are ill defined and consequently, as one can notice from the numerical data, they bear a large scatter for the various temperatures. Remarkably, the relative area of the doublet has remained unaffected by the heating at 400°C, and as for BEV2 tends to zero when the temperature lowers towards 0 K. Further, the amount of metallic Fe has doubled as a result of the heating.

The ILEEMS spectrum of BEV2a (Figure 2) shows a predominant contribution from the doublet. As for BEV2, the hematite fraction in the surface layers is less than that in the bulk, whilst no difference in that respect is observed for the metallic-Fe fraction. An additional component that would correspond to the sextet II in the TMS spectra and a magnetite contribution are not resolved. Both the transmission and ILEEMS spectra of BEV2b show the predominant sharp sextet for hematite, but in the transmission data the presence of a second sextet is clearly recognized. Its parameters are similar to those of $FeBO_3$ [10], which are consistent with the XRD pattern of this sample. The borate phase does not appear in the ILEEMS spectrum, implying that the borate grains are embedded in the hematite matrix.

More detailed results concerning this work will be published in a forthcoming paper.

Acknowledgements This work was partially funded by CNPq and Fapemig (Brazil), and by the Fund for Scientific Research – Flanders, Belgium.

One of us (V. G. de Resende) thanks the Program Alban, the European Union Programme of High Level Scholarships for Latin America, scholarship no. E04M034189BR.

References

1. Dragieva, L.D., Stoynov, Z.B., Klabunde, K.J.: Scr. Mater. **44**, 2187 (2001)
2. Foster, G.L., Barquín, L.F., Cohen, N.S., Pankhurst, Q.A., Parkin, I.P.: J. Mater. Process. Technol. **92–93**, 525 (1999)
3. Petit, S., David, K., Doumerc, J.P., Grenier, J.C., Seguelong, T., Pouchard, M.: C. R. Acad. Sci. Paris Série IIc **517**, (1998)
4. Linderoth, S., Mørup, S.: J. Appl. Phys. **69**, 5256 (1991)
5. Saida, J., Ghafari, M., Nakamura, Y., Inoue, A., Masumoto, T.: Nucl. Instrum. Methods Phys. Res. **B76**, 223 (1993)
6. Song, X., Yusheng, X., Huali, J., Qing, X.: Nucl. Instrum. Methods Phys. Res. **B76**, 260 (1993)
7. De Grave, E., Vandenberghe, R.E., Dauwe, C.: Hyperfine Interact. **161**, 147 (2005)
8. De Grave, E., Vandenberghe, R.E.: Hyperfine Interact. **28**, 643 (1986)
9. De Grave, E., Bowen, L.H., Vochten, R., Vandenberghe, R.E.: J. Magn. Magn. Mater. **72**, 141 (1988)
10. Bayukov, O.A., Abd-Elmeguid, M.M., Ivanova, N.B., Kazak, N.V., Ovchinnikov, S.G., Rudenko, V.V.: Phys. Solid State **46**(6), 1088 (2004)

Hyperfine Interact (2005) 165: 121–125
DOI 10.1007/s10751-006-9401-7

Aluminium based amorphous and nanocrystalline structure

Jozef Sitek · Jarmila Degmová

Published online: 3 November 2006

Abstract Aluminium based rapidly quenched alloys of nominal composition $Al_{90}Fe_7Nb_3$ and $Al_{94}Fe_2V_4$ were studied by Mössbauer spectroscopy. Samples annealed up to 573 K showed amorphous structure represented by quadrupole doublet. From corrected spectra areas the values of *f*-factor were calculated. In the case of AlFeNb samples *f*-factor was estimated as $f = 0.26$ and for AlFeV $f = 0.31$. The corresponding Debye temperatures were also calculated. Higher temperature annealing at 773 and 873 K induced deformation of nano- and microcrystalline state. Mössbauer spectra of samples of both compositions (with vanadium as well as with niobium) annealed at 773 K showed superposition of crystalline phases with dominant role of Al_3Fe alloy. During annealing at 873 K, phases with large grains and small amount of FeAl metastable phase were developed (Das et al. *Mat. Sci. Eng.*, A304–A306, 159, 2001; Illeková et al. *Mat. Sci. Eng.*, A375–A377, 946, 2004).

Key words Al-based alloys · amorphous structure · crystallization

1 Introduction

Recently Al-based systems have attracted much attention due to their outstanding properties. Al-rich glasses containing more than 80% of Al with combination of transition and rare earth element additions have very good mechanical properties and corrosion resistance [1, 3]. It is known that Al and Fe atoms may form strong Al–Fe association in the melt. These two elements in combination with transition metals as Nb or V promote the glass-forming and α-Al nanocrystallization in the alloy [4]. By rapid solidification or in a more controlled manner through the annealing of precursor, amorphous phase and traces of

J. Sitek · J. Degmová (✉)
Department of Nuclear Physics and Technology, Faculty of Electrical Engineering and Information Technology, Slovak University of Technology, Ilkovičova 3, 81219 Bratislava, Slovakia
e-mail: jarmila.degmova@stuba.sk

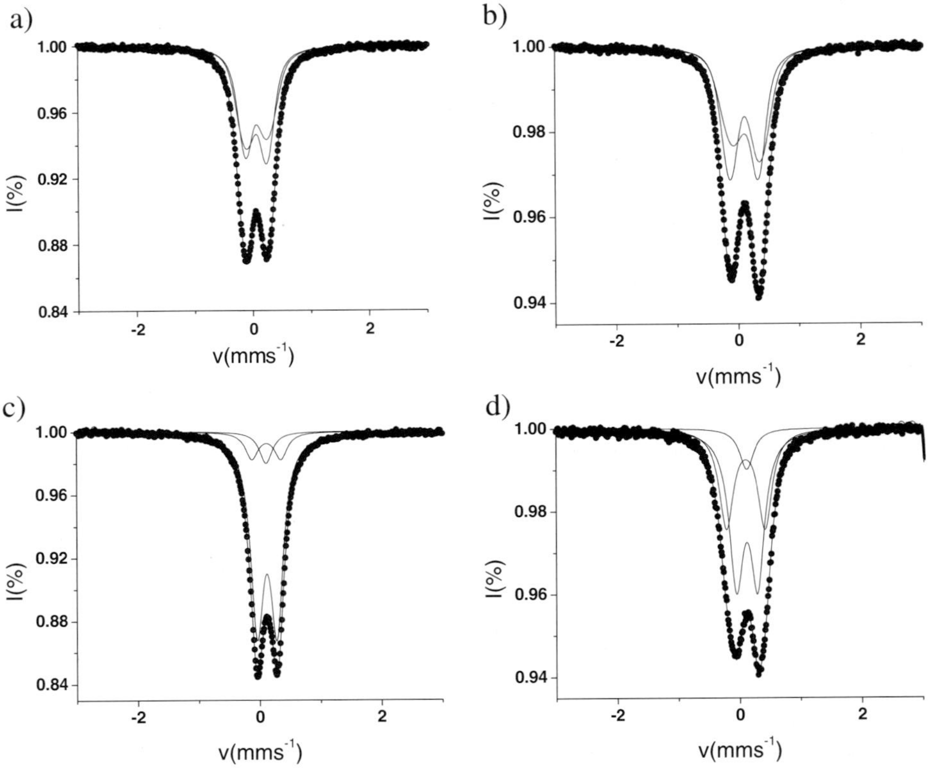

Figure 1 Mössbauer spectra of AlFeNb and AlFeV alloys: **a** amorphous AlFeNb, **b** amorphous AlFeV, **c** AlFeNb after annealing at 773 K for 1 h and **d** AlFeV after annealing at 773 K for 1 h.

nanocrystalline Al are created. During further annealing of Al based ribbon, due to primary and secondary crystallization, α-Al crystalline phase is developed and the remaining matrix starts to crystallize in Al-intermetallic phases.

In the present work, Mössbauer spectroscopy was applied on Al–Fe–Nb and Al–Fe–V samples in order to determinate Lamb–Mössbauer factor of amorphous ribbons and to study process of crystallization kinetics.

2 Experimental

Aluminium based alloys in the form of ribbons (25 μm × 10 mm) were prepared by planar flow casting. The nominal compositions were $Al_{90}Fe_7Nb_3$ and $Al_{94}Fe_2V_4$. Mössbauer spectra were collected in transmission geometry by a conventional constant-acceleration spectrometer with a $^{57}Co(Rh)$ source at room temperature. The spectra were evaluated by the NORMOS (Brand, unpublished, 1989) program which allows simultaneous treatment of crystalline, residual amorphous matrix and interface using set of single individual lines and a distribution of hyperfine components. Lamb–Mössbauer factor was calculated using area method and the Debye temperature was also determined. All compositions were annealed in vacuum in temperature range from 473 up to 873 K for 1 h.

3 Results and discussion

3.1 Amorphous state

Experimental Mössbauer spectra taken on the amorphous samples exhibit quadrupole doublet with broadened line shape (see Figure 1a, b). Spectra of amorphous samples were fitted with two doublet distributions and the area of the spectra was calculated with respect to non-resonant background. The cluster structure for rapidly quenched metals having a well-developed medium range order, namely for a pure Al, was theoretically predicted [5]. The results in [4] confirmed that amorphous structure of the $Al_{90}Fe_7Nb_3$ alloy can be interpreted by two-type clusters model, namely Al clusters and Al–Fe–Nb clusters. This cluster model can be applied also on Al–Fe–V system. Because of strong Fe–Al bounding in Fe–Al alloys, we suppose that iron atoms are situated correspondingly in two different neighbourhoods of Al atoms. Therefore, Mössbauer spectra were fitted with two broad doublets.

Area method was used to determinate the Lamb–Mössbauer factor. The effective Debye temperature was calculated using formula [6] (see Eq. 1):

$$f = \exp\left\{\frac{-3}{2} \times \frac{E_R}{k_B\theta}\left[1 + \left(\frac{T}{\theta}\right)^2 \int_0^{\theta/T} \frac{x\mathrm{d}x}{e^x - 1}\right]\right\} \tag{1}$$

Where E_R is recoilless energy, k_B is Boltzman constant, θ is Debye temperature and T is temperature.

In the case of AlFeNb alloy, the Lamb–Mössbauer factor was determined as $f = 0.26 \pm 0.02$ and for AlFeV as $f = 0.31 \pm 0.02$. The corresponding Debye temperatures were calculated as $\Theta_D = 296 \pm 5$ K and $\Theta_D = 318 \pm 5$ K, respectively.

In the case of first sample the Debye temperature was derived from second order Doppler shift as $\Theta_D = 325 \pm 5$ K.

Slight amount of nanocrystalline Al was found in AlFeV as-prepared ribbons by the XRD technique [2]. However, Mössbauer spectroscopy did not reveal any Fe-containing nanocrystalline phase. An only difference was found: Mössbauer spectrum of amorphous AlFeV sample shows small asymmetry in comparison with the spectrum of AlFeNb alloy.

The values of Debye temperature calculated from the spectra are low in comparison with the value for pure aluminium i.e. $\Theta_D(\mathrm{Al}) = 428$ K. One has to take into account that the measured samples contain two kinds of impurities (Fe and Nb or V). In this case, the effective Debye temperature for binary system can be expressed by a relation (see Eq. 2) as:

$$\Theta_{\mathrm{eff}} = \Theta_{\mathrm{host}}\sqrt{\frac{M_{\mathrm{host}}}{M_{\mathrm{im}}}} \times \sqrt{\frac{\lambda'}{\lambda}} \tag{2}$$

Where Θ is the Debye temperature, M is the mass of host or impurities, λ and λ' are force constants for host and impurities and $\lambda'/\lambda \cong 1$.

According to this formula, Debye temperatures were determined as Θ_D=295 K. Both values are close to those achieved experimentally.

Table I Selected Mössbauer parameters of amorphous and annealed Al–Fe–Nb and AlFeV ribbons: T_a is annealing temperature, Δ describes the quadrupole splitting, IS is isomer shift and *A* is the relative area of component in the spectra

Sample	T_a (K)	Δ_1 (mm/s)	IS_1 (mm/s)	A_1 (%)	Δ_2 (mm/s)	IS_2 (mm/s)	A_2 (%)	IS_3 (mm/s)	A_3 (%)
$Al_{90}Fe_7Nb_3$	0	0.42	0.07	49	0.40	0.06	51		
	623	0.42	0.09	49	0.45	0.06	51		
	773	0.54	0.08	15	0.34	0.12	68	0.10	17
	873	0.47	0.12	12	0.33	0.11	81	0.11	7
$Al_{94}Fe_2V_4$	0	0.46	0.12	49	0.51	0.11	51		
	623	0.49	0.11	51	0.48	0.11	49		
	773	0.64	0.10	37	0.35	0.12	54	0.12	9
	873	0.42	0.20	93	0.14	0.06	7		

3.2 Crystallization process

The linear heating of as-quenched samples of both compositions exhibits four distinct transformation stages [4]. The first stage has onset at 544.5 K. Under this temperature no significant changes in Mössbauer parameters (as isomer shift, quadrupole splitting and spectrum area) was found. The second transformation stage was achieved at 681.7 K. After annealing for 1 h at 623 K, which is the temperature below limit of the second transformation stage, a slight increase of the quadrupole splitting of the second spectral component was observed (see Table I). During this heating, the amount of α-Al nanocrystalline phase also increased, which can be connected with the quadrupole splitting change.

The third transformation stage was achieved at 724.7 K. Above this temperature two intermetalic phases (Al_3Fe and Al_3Nb) and α-Al are formed. After annealing at 773 K, the appearance of the spectra changed significantly and a new singlet appeared corresponding to the Al-phase with a small amount of Fe (see Figure 1 c, d). This phase has Fe atoms in symmetric environment of eight non-magnetic Al-atoms.

The first doublet corresponds to Al_3Nb containing small amount of Fe-atoms. This can by also labelled as $(AlFe)_3Nb$. The second, more intensive, doublet corresponds to Al_3Fe. This phase was also identified by XRD [2].Similar interpretation could be presented also for AlFeV alloy because the second doublet has the same parameters as in the previous case of AlFeNb alloys. The first doublet is solid solution of Al–Fe–V.

The fourth transformation stage starts at 841 K. Mössbauer spectra of samples annealed at 873 K showed one intensive and one quite weak doublet. The more intensive doublet corresponds to the rest coarsely grained crystalline phase which may be mixture of (FeAl) Nb or (FeAl)V surrounded by AlFe (see Table I). These results supported XRD measurements [2] where was found that after annealing at 830 K for 1 h the alloy structure consist of polycrystalline Al and intermetallic phases of FeAlNb and FeAlV, respectively. The same authors confirmed existence of these phases also by TEM measurements [2].

4 Conclusions

Analysis by Mössbauer spectroscopy showed that Debye temperature of amorphous Al-based alloys is deep below the Debye temperature for pure Al. Kinetic of crystallization is strongly dependent on the constituent elements. The short-range ordered structure, especially that related to Al–Fe bond which is responsible for transformation from amorphous to crystalline state is modified by the third (Nb or V) element.

Acknowledgement The master ribbons were supplied by P. Švec from Slovak Academy of Sciences. This work was supported partly by grants Vega 1/0284/03 and French–Slovak Grant Fr/Sk/23.

References

1. Das, S., Peperezko, J., Wu, R., Wilde, G: Mat. Sci. Eng. **A304–A306**, 159 (2001)
2. Illeková, E., Janičkovič, D., Kubečka, P., Švec, P., Gachon, J.C.: Mat. Sci. Eng. **A375–A377**, 946 (2004)
3. Masumoto, T.: Mater. Sci. Eng. **A179–A180**, 57 (1994)
4. Audebert, F., Arcondo, B., Rodriguez, D., Sirkin, H.: Mater. Sci. Eng. Forum **360–363**, 98 (2001)
5. Rangsu, Z., Jiyong, L., Kejun, D., Cuixing, Z., Hairong, L.: Mater. Sci. Eng. **B94**, 141 (2002)
6. Grenwood, N.N., Gibb, T.C.: Mössbauer Spectroscopy, Chapman & Hall, London (1971)

Hyperfine Interact (2005) 165: 127–134
DOI 10.1007/s10751-006-9255-z

The effect of the sputtering process ambient on the magnetic state and phase composition of the film nanocomposites $(Fe_{0.45}Co_{0.45}Zr_{0.10})_x(Al_2O_3)_{1-x}$

J. Fedotova · J. Kalinin · A. Fedotov · A. Sitnikov · I. Svito · A. Zaleski · A. Jablonska

Published online: 26 October 2006

Abstract The effect of oxygen-containing ambient arising at sputtering of granular nanocomposites $(Fe_{0.45}Co_{0.45}Zr_{0.10})_x(Al_2O_3)_{1-x}$ (30 at.%$\leq x \leq$65 at.%) on their magnetic state and phase composition has been investigated. It was shown that the presence of oxygen resulted in the formation of oxide shells preventing the ferromagnetic interaction between $Co_{0.45}Fe_{0.45}Zr_{0.10}$ nanoparticles and also the formation of metallic percolative net beyond the percolation threshold (as opposed to the films prepared in pure argon atmosphere).

Key words granular nanocomposites · magnetization · Mössbauer spectroscopy

1 Introduction

Granular films containing soft ferromagnetic FeCo-based nanoparticles in dielectric matrix are potential candidates for ultrahigh density magnetic recording media [1–3] though superparamagnetism places a lower limit on the size of useable nanoparticles. Such nanoparticle composites have also the added benefits of higher resistivity controlled by the interparticle distances and hence reduced eddy current losses, and the ability to tailor the magnetic properties of the composite system by the control of such physical properties of the nanoparticles as size, shape, orientation, volume fraction, etc.

J. Fedotova (✉)
NC PHEP of Belarusian State University, Minsk, Belarus
e-mail: julia@hep.by

J. Kalinin · A. Sitnikov
Voronez State University, Voronez, Russia

A. Fedotov · I. Svito · A. Zaleski
Belarusian State University, Minsk, Belarus

A. Jablonska
Institute of Atomic Energy, Otwock-Swierk, Poland

However, FeCo-based nanoparticles can be surrounded by oxides due to the influence of residual oxygen during the film deposition. These oxides show different magnetic properties (ferrimagnetic for γ-Fe_2O_3 or Fe_3O_4; antiferromagnetic for CoO; ferromagnetic, superparamagnetic or ferrimagnetic for $CoFe_2O_4$, depending on the temperature, matrix type, etc.) and oxidation of ferromagnetic nanoparticles can alter their magnetic behavior, namely magnetization, coercivity, remanence [1, 2].

The aim of the present study is to investigate the effect of oxygen in the sputtering ambient on the magnetic state and phase structure in $(Co_{0.45}Fe_{0.45}Zr_{0.10})_x(Al_2O_3)_{100-x}$ (30 at.%$\leq x \leq$65 at.%) composites.

2 Experimental procedures

The studied film samples with thickness of 3 to 5 μm in the form of $Co_{0.45}Fe_{0.45}Zr_{0.10}$ nano-particles embedded into Al_2O_3 matrix were sputtered either in pure argon (Ar) at a pressure of 8.0×10^{-4} Pa (samples of series 1) or in argon–oxygen (Ar+O) gas mixture at the total pressure of 9.6×10^{-2} Pa and partial pressure of oxygen of about 4.4×10^{-2} Pa (samples of series 2) [4]. The composites have been investigated by means of TEM, Mössbauer spectroscopy and magnetization curves measurements [4, 5]. TEM microscopy of as-deposited films revealed their granular structure with randomly distributed amorphous metallic alloy particles (with the dimensions between 6 and 10 nm) and amorphous alumina [5]. Mössbauer spectra have been recorded at room temperature in transmission geometry using ^{57}Co/Rh source (25 mCi). The fitting procedure has been performed with the use of MOSMOD program assuming the distribution of hyperfine magnetic fields (H_{hf}) and quadrupole splitting (QS). All isomer shifts (IS) were given with respect to α-Fe. Lorentzian width (FWHM) of fitted spectral lines was equal to 0.15 mm/s.

The magneto-response measurements was carried out by alternation grads magnetometer at 300 K with the applied magnetic induction $B\leq600$ mT. The pick-up coil signal recorded at different B was proportional to the value of magnetization M and reduced to the volume of the film.

3 Mössbauer spectroscopy and magnetization measurements

The results of Mössbauer measurements are presented in Figures 1a and 2a for the samples of series 1 and 2, respectively. The magnetization curves *M*(*B*) for both series are shown in Figures 1b and 2b. Note that all signals, detected by the pick-up coil and presented in Figures 1b and 2b, were additionally normalized on the signal measured for the sample of series 2 with $x=64$ at.% which had maximal value of M at $B=0.5$ T among the samples of both series 1 and 2. The derivatives $(dM/dB)_{B\to0}$, that is proportional to magnetic susceptibility of the sample, as a function of metallic element content for both series of samples are presented in Figure 3.

3.1 Granular films sputtered in pure argon (series 1)

The Mössbauer spectra recorded for the granular films of series 1 strongly depend on the atomic fraction (x) of $Co_{0.45}Fe_{0.45}Zr_{0.10}$. As evidences from Figure 1a, the spectra look like totally non-ferromagnetic at $x\leq42$ at.%. They could be fitted with two Fe^{3+} doublets (D1:

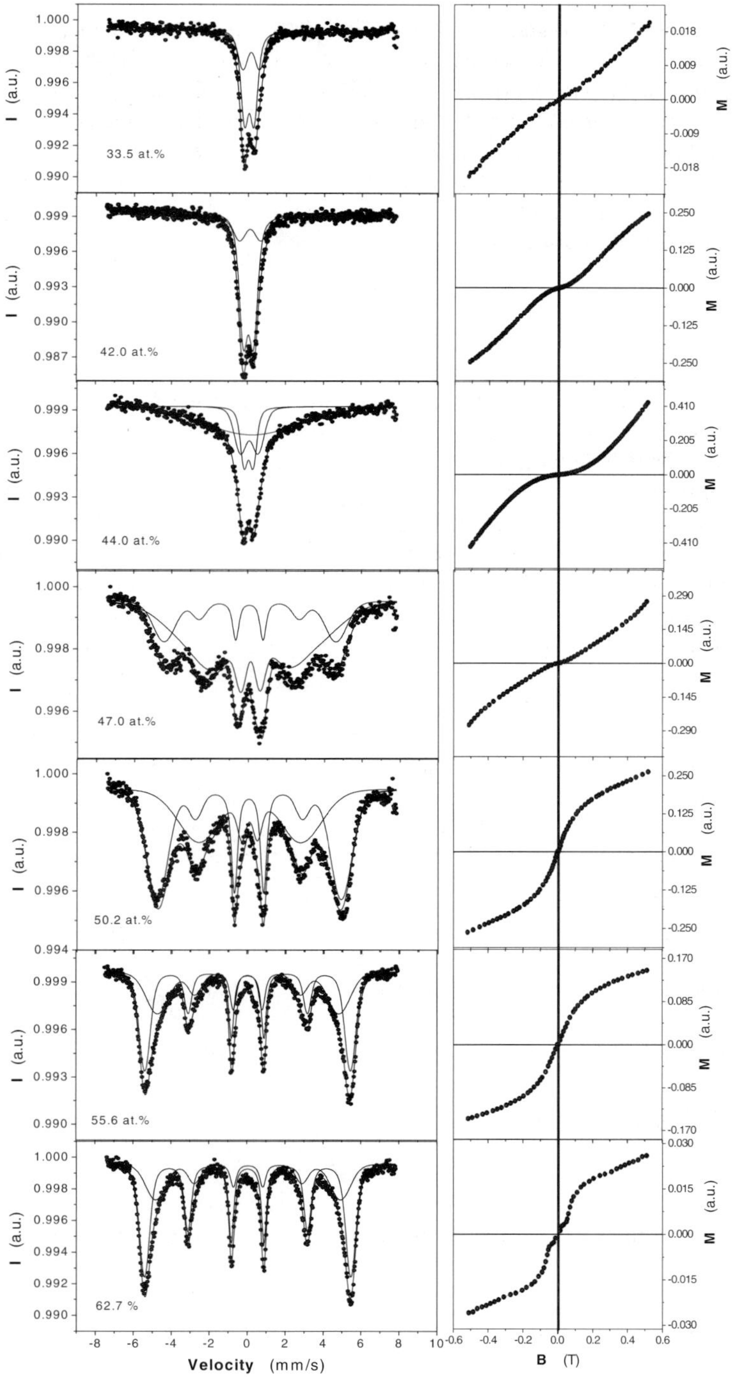

(a) (b)

Figure 1 Mössbauer spectra (a) and normalized magnetization curves (b) for different concentrations of metallic components in the samples of series 1.

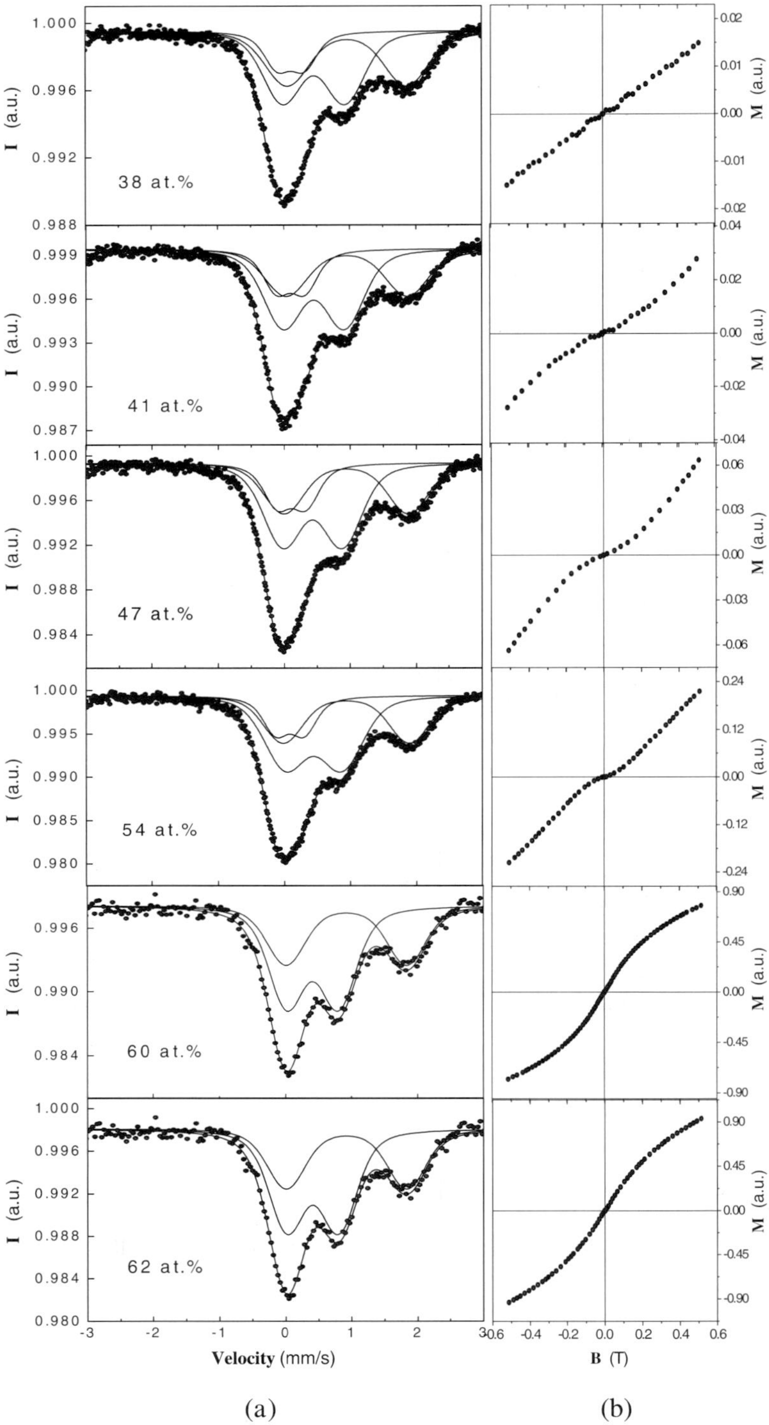

Figure 2 Mössbauer spectra (a) and normalized magnetization curves (b) for different concentrations of metallic components in the samples of series 2.

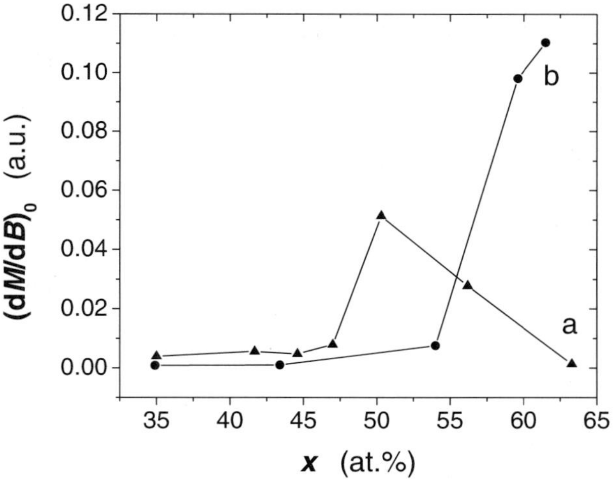

Figure 3 The derivative $(dM/dB)_{B\to 0}$, estimated from M (B) curves in Figures 1b and 2b, as a function of the metallic components concentration for the samples of series 1 (a) and series 2 (b).

IS=0.13–0.14 mm/s, QS=0.88–1.12 mm/s; D2: IS=0.04–0.05 mm/s, QS=0.49–0.52 mm/s). At x approaching the percolation threshold $x_c \approx 43$–47 at.% (estimated from the electric properties of the composites [6]), the appearance of a wide partially collapsed magnetic subspectrum with IS close to 0.08 mm/s has been observed in addition to the above mentioned two doublets (see Figure 1a). Spectra of the films with $x > x_c$ are magnetically split and could be evaluated in the assumption of two magnetic sextets (H_{hf1}=16.0–30.0 T; H_{hf2}=28.2–33.8 T). It is noteworthy that D2 doublet could be eventually fitted in the spectra of films with $x > x_c$ but its relative contribution was found to be about 3% (i.e., within experimental error).

As a whole, the behavior of the magnetization curves $M(B)$ shown in Figure 1b for the series 1 samples correlated with the Mössbauer data. As has been found, dielectric samples far from the percolation threshold ($x \leq 35$ at.%) revealed a nearly linear increase of M with B growth, which looks like a paramagnetic behavior. At higher concentrations of the metallic phase up to the percolation threshold $x_c \approx 43$–47 at.% the room temperature magnetization curves show no saturation and hysteresis features to confirm the Mössbauer data concerning the fact that the metallic alloy nanoparticles are in the superparamagnetic state. Both the remanence and coercivity in this case are zero at 300 K.

Note also that M values measured at constant B increase with x [6] although $(dM/dB)_{B\to 0}$ is practically invariable up to x_c, where it increases sharply (see Figure 3a) to its maximal value for this series of samples

Beyond the percolation threshold x_c the shape of magnetization curves is changed (Figure 1b) approaching saturation without any hysteresis that is characteristic for soft amorphous ferromagnetic materials and confirmed by the magnetically split Mössbauer spectra. At the same time both M at constant B (Figure 1b) and $(dM/dB)_{B\to 0}$ (Figure 3a) values decrease with x.

3.2 Granular films sputtered in Ar+O ambient (series 2)

Contrary to the previous case, Mössbauer spectra of the samples deposited in Ar+O mixture show nonmagnetic behavior independently on the $Co_{0.45}Fe_{0.45}Zr_{0.10}$ content (both before and beyond percolation threshold $x_c \approx 55$ at.% as estimated in [6]). However, the local Fe states observed in this case and *M*(*B*) curves also demonstrated a threshold character. In particular, at $x \leq 55$ at.% Mössbauer spectra were evaluated assuming three local Fe states in the films: two doublets (D3: IS=0.08–0.11 mm/s, QS=0.41–0.49 mm/s; D4: IS=0.40–0.46 mm/s, QS=0.77–0.94 mm/s) associated with Fe^{3+} ions and one doublet assigned to Fe^{2+} state (D5: IS=0.90–0.96 mm/s, QS=1.81–1.91 mm/s). Starting from *x* values above 59 at.% (i.e., far beyond x_c), the Mössbauer spectra could be fitted with only D4 and D5 doublets, while D3 disappeared. It should be noticed that a relative contribution of D4 subspectrum increased with the value of *x* at the expense of the D5 contribution. A fraction of D3 before x_c was constant within the experimental error.

As a whole, the behavior of magnetization curves *M*(*B*) for the series 2 samples with oxidized nanoparticles correlated with Mössbauer data also. As is seen from Figure 2b, formally *M*(*B*) curves for this series are similar to those for the samples of series 1. Also at $x << 55$ at.% (far from the percolation threshold x_c) the series 2 films exhibit the paramagnetic behavior of *M*(*B*) but with lower values of $(dM/dB)_{B\to 0}$ (Figure 3b) and *M* measured at constant *B* (Figure 2b) than for the series 1 films. As seen from Figure 2b, before the percolation threshold the magnetization curves show neither saturation nor hysteresis, indicating the superparamagnetic behavior of the oxidized alloy nanoparticles.

Note also that $(dM/dB)_{B\to 0}$ and *M* values measured at constant *B* beyond x_c increase (see Figures 3b and 2b) with *x* however *M* values for *B*=0.5 T are twice as high as for the samples of series 1.

Beyond the percolation threshold x_c the shape of magnetization curves is changed from non-saturating to the saturating one, also showing no features of hysteresis (Figure 1b).

4 Discussion

The Mössbauer spectra of the films sputtered in pure Ar with $Co_{0.45}Fe_{0.45}Zr_{0.10}$ concentrations $x < x_c \approx 43$–47 at.%, i.e., before the percolation threshold, have been fitted with two doublets (D1 and D2). Superposition of these doublets could be associated with non-interacting aggregation of superparamagnetic nanoparticles with two different mean sizes isolated in the dielectric Al_2O_3 matrix. The Mössbauer measurements performed at liquid nitrogen temperature (to be published elsewhere) have demonstrated that blocking temperature of these $Co_{0.45}Fe_{0.45}Zr_{0.10}$ nanoparticles lies far below 77 K. The assumption concerning the coexistence of nanoparticles with two different mean sizes was also confirmed by the temperature dependences of DC resistance and real part of impedance, demonstrating Mott law for the hopping carrier transport with two different activation energies observed in these films before percolation threshold in the temperature range 77–350 K [4, 6].

As follows from electric measurements [4, 6], the occurrence of two magnetically split subspectra with two different hyperfine magnetic field values at $x_c \geq 47$ at.% most probably originated from the interaction between these aggregations due to joining of the nanoparticles in the percolative current-conducting net. The step-like form of *M*(*B*) curves far beyond the percolation threshold characteristic for multiple-phase magnetic substances (see curve for *x*=63 at.% in Figure 1b) can be regarded as an indirect support for the presence of nanoparticles with two different mean sizes.

Mössbauer spectra of the films sputtered with Ar+O mixture have revealed the superparamagnetic or non-magnetic phases over the whole range of $Co_{0.45}Fe_{0.45}Zr_{0.10}$

concentrations. It is noteworthy that hyperfine parameters of D3 doublet are close to those extracted for D2 doublet and observed in the films sintered in pure Ar. So far D3 doublet could also be assigned to the non-interacting superparamagnetic nanosized $Co_{0.45}Fe_{0.45}Zr_{0.10}$ assemblies embedded into Al_2O_3 matrix. Two other Fe states (D4 and D5) may respectively be attributed to the oxide shells of $Co_{0.45}Fe_{0.45}Zr_{0.10}$ nanoparticles and $FeAl_2O_4$ compound (hercinite) nanoparticles [7] formed at the interface CoFeZr alloy – alumina.

Such an illogical behavior of the magnetization curves and Mössbauer spectra for the films from series 2 with increase of $Co_{0.45}Fe_{0.45}Zr_{0.10}$ concentration may be explained as follows. Firstly, due to the presence of oxygen in the sputtering ambient, the metal-containing nanoparticles may form at least three different phases – (à) residual $Co_{0.45}Fe_{0.45}Zr_{0.10}$ nanoparticles, (b) oxide shells of $Co_{0.45}Fe_{0.45}Zr_{0.10}$ nanoparticles, and (c) $FeAl_2O_4$, formed at the $Co_{0.45}Fe_{0.45}Zr_{0.10}$–$Al_2O_3$ interface. Secondly, Fe- and (or) Co-containing oxides are semiconducting (but not metallic) substances and may not be ferromagnetic. This means that the ferromagnetic interaction between the metallic $Co_{0.45}Fe_{0.45}Zr_{0.10}$ nanoparticles separated by oxide shells cannot occur even at their direct electric contact and these nanoparticles can remain in the superparamagnetic state. This can explain the absence of magnetic splitting in the Mössbauer spectra and the non-hysteresis form of magnetization curves as well beyond the percolation threshold.

Moreover, a threshold character of the magnetic and structural transformations in composite films was found to be in the evident correlation with the same character of the earlier performed carrier transport and magnetotransport measurements [4, 6]. Specifically, the occurrence of the interacting $Co_{0.45}Fe_{0.45}Zr_{0.10}$ particle aggregations for the samples of series 1 deposited in pure Ar atmosphere is in agreement with a change of the electronic transport mechanisms from the Mott-like hopping to the power-like metallic one beyond the percolation threshold concentration $x_c \sim 45$–47 at.% [6]. At the same time, the formation of partly or totally oxidized metallic nanoparticles in the films of series 2 at x values higher than 55–59 at.% may explain the Shklovski-Efros hopping mechanism of the electronic transport observed at the temperatures 160–350 K, that is characteristic for semiconductor materials (like Fe- and/or Co-containing oxides).

5 Summary

Summarizing, we can conclude that the Mössbauer and magnetization measurements performed confirm the fact that the magnetic and structural states of the studied granular films are under the strong influence of oxygen addition into the sputtering ambient. In particular, the presence of oxygen caused the formation of oxide shells around the $Co_{0.45}Fe_{0.45}Zr_{0.10}$ nanoparticles, preventing the ferromagnetic interaction between residual metallic nanoparticles, and also the formation of the metallic percolative net beyond the percolation threshold (as opposed to the films prepared in pure argon atmosphere).

Acknowledgements The financial support of this project was received from the Belarus state program "Nanotechnology."

References

1. Skomski, R.: J. Phys.; Phys. Rev., B Condens. Matter **15**, R841 (2003)
2. Jiles, D.S.: Acta Mater. **51**, 5907 (2003)
3. Bader, S.D.: Surf. Sci. **500**, 172 (2002)

4. Saad, A.M., Mazanik, A.V., Kalinin, Yu.E., Fedotova, J.A., Fedotov, A.K., Wrotek, S., Sitnikov A.V., Svito, I.A.: Rev. Adv. Mater. Sci. **8**, 34 (2004)
5. Kalinin, Yu.E., et al.: Bulletin of Voronezh State Technical University: Material Science **N113**, 43 (2003)
6. Saad, A., Fedotov, F., Fedotova, J., Svito, I., Andrievsky, B., Kalinin, Yu., Fedetova, V., Malyutina-Bronskaya, V., Patryn, A., Mazanik, A., Sitnikov, A.: Phys. Status Solidi **C3**, 1283 (2006)
7. Paesano A. Jr., Matsuda, C.K., et al.: J. Magn. Magn. Mater. **264**, (2003)

Hyperfine Interact (2005) 165: 135–140
DOI 10.1007/s10751-006-9256-y

Mössbauer studies of phosphate glasses for the immobilisation of toxic and nuclear wastes

P. A. Bingham · S. D. Forder · R. J. Hand · A. Lavaysierre

Published online: 26 October 2006

Abstract Various iron-containing phosphate glasses were investigated by Mössbauer spectroscopy. Iron was found to occur predominantly as Fe^{3+} in all glasses, and largely occupied sites with distorted octahedral coordination for both redox states. Using a base glass of nominal composition 60 P_2O_5–40 Fe_2O_3 (mol%), stepwise molar replacement of Fe_2O_3 by (0.67 $Na_2O \times 0.33$ Al_2O_3) increased the redox ratio, $Fe^{2+}/\Sigma Fe$, from 0.13 at 40% Fe_2O_3 to 0.25 at 10% Fe_2O_3. The centre shift increased and quadrupole splitting decreased by up to ~20% over this range, interpreted as a decrease in the average distortion of Fe sites from cubic symmetry, and an increase in average iron coordination. Literature revealed that recoil-free fraction ratio $f\,(Fe^{3+})/f\,(Fe^{2+}) \approx 1.3$ in iron phosphate glasses, and this was considered when assessing redox. Mössbauer parameters of these and other glasses demonstrated a combination of structural stability and compositional flexibility which makes them so suitable for waste immobilisation.

Key words Mössbauer · glass · phosphate · iron

1 Introduction

Vitrification is amongst the safest and most proven waste immobilisation technologies. Nuclear and toxic wastes must be safely immobilised so dangerous components cannot leach out into the environment [1]. The majority of high level nuclear waste (HLW) which is highly radioactive and heat-generating is currently immobilised by vitrification in borosilicate glasses. However these glasses have some shortcomings which include limited capacity to immobilise certain elements, e.g., P, S, Mo, noble metals and halogens [1].

P. A. Bingham (✉) · R. J. Hand
Immobilisation Science Laboratory, Department of Engineering Materials, University of Sheffield, Mappin Street, Sheffield S1 3JD, UK
e-mail: p.a.bingham@sheffield.ac.uk

S. D. Forder · A. Lavaysierre
Materials and Engineering Research Institute, Sheffield Hallam University, Howard Street, Sheffield S1 1WB, UK

Russian HLW has been immobilised in P_2O_5–Al_2O_3–Na_2O glasses [2], and substantial work has been carried out on immobilisation of certain wastes in glasses in the P_2O_5–Fe_2O_3 system [3, 4]. Several differences in properties arise between borosilicate and phosphate glass systems, including chemical durability, thermal properties, waste loading and elemental solubility.

^{57}Fe Mössbauer spectroscopy has been widely used to investigate the environment, redox and structure of iron-containing phosphate glasses [3–8]. For high-spin Fe^{3+} (as occurs in glasses), the electrons are spherically symmetrical, so only distortions from cubic symmetry contribute to quadrupole splitting (QS) and this parameter is therefore a direct measure of Fe site distortion. Values of QS (Fe^{2+}) are larger than QS (Fe^{3+}) owing to lack of symmetry caused by the additional electron of Fe^{2+}. Linewidth (LW) can be a measure of the range of site distortions, as opposed to the average site distortion which can be assessed from QS.

2 Experimental

Glasses were prepared from analytical grade >99% $NH_4H_2PO_4$, Fe_2O_3, Na_2CO_3, Na_2SO_4, $Al(OH)_3$, SiO_2, and H_3BO_3. Batches to produce 200 g of glass were melted for 3 h under conditions shown in. Table I and detailed in [5]. The calculated and analysed glass compositions, together with fitted Mössbauer parameters, are given in Table I. Glass PFe-1 was used as the base glass.

Mössbauer spectroscopy was carried out at 293 K using a constant acceleration spectrometer with a 25 mCi$\times^{57}$Co source in a Rh matrix. Absorbers were prepared from ground glass mixed with graphite powder to ensure a Mössbauer thickness $t<1$. Spectra were measured in the velocity range ±5 mm s^{-1} relative to α-Fe and were fitted using Recoil analysis software [9]. Eight broadened Lorentzian paramagnetic doublets were fitted to all spectra, a method used elsewhere [4, 5]. Mössbauer parameters shown in Table I are weighted averages based on fitted doublet areas. Figure 1 shows selected fitted spectra. Compositions were analysed using a Link EDS unit fitted to a Philips 500 scanning electron microscope. The composition of sample B-5 was measured by ICP-OES using an ARL 3410 spectrometer.

3 Results and discussion

For many phosphate glasses, the recoil-free fraction ratio $f(Fe^{3+})/f(Fe^{2+})\approx 1.3$ [6, 7], and therefore we have assumed that $f(Fe^{3+})/f(Fe^{2+})=1.30$. Whilst critical to absolute measurements of redox, $f(Fe^{3+})/f(Fe^{2+})$ was expected to vary little for the glasses measured and hence the redox differences were confirmed. Glasses PFe-1, H-10, H-20 and H-30 constituted stepwise shifts between the P_2O_5–Fe_2O_3 and P_2O_5–Al_2O_3–Na_2O systems whilst maintaining a constant nominal P_2O_5 content. The Al_2O_3:Na_2O ratio was 1:2 as used in Russian HLW immobilisation [2]. Batches to produce glasses H-10, H-20 and H-30 contained Na_2SO_4 in order to study the effects of composition upon sulphate solubility; this will be discussed in a future publication. Sodium sulphate is an oxidising agent, therefore a decrease in $Fe^{2+}/\Sigma Fe$ might normally be expected. However this did not occur. It can therefore be assumed that any effects of sulphate on redox in these glasses, whilst not necessarily negligible, were small by comparison with the effects of glass composition.

Fitted hyperfine parameters demonstrated clear trends with increasing replacement of Fe_2O_3 by (0.67 $Na_2O\times$0.33 Al_2O_3) through the series PFe-1 to H-30. For both redox states, centre

Table I Sample compositions (by batch and analysed) and fitted Mössbauer parameters

	Sample PFe-1	Sample H-10	Sample H-20	Sample H-30	Sample Si-5	Sample Al-5(b)	Sample B-5
P_2O_5 mol% (analysed)	60 (60.9)	60 (57.7)	60 (55.7)	60 (57.3)	57 (58.3)	60 (63.8)	57 (58.1)
Fe_2O_3 mol% (analysed)	40 (38.4)	30 (27.9)	20 (19.4)	10 (9.6)	38 (36.5)	35 (30.5)	38 (35.8)
Na_2O mol% (analysed)	0 (<0.1)	6.7 (10.7)	13.3 (17.3)	20 (22.8)	0 (<0.1)	0 (<0.1)	0 (n/m)
Al_2O_3 mol% (analysed)	0 (0.7)	3.3 (3.3)	6.6 (6.9)	10 (10.1)	0 (0.4)	5 (5.4)	0 (0.5)
SiO_2 mol% (analysed)	0 (<0.1)	0 (<0.1)	0 (<0.1)	0 (<0.1)	5 (4.8)	0 (0.2)	0 (0.2)
B_2O_3 mol% (analysed)	0 (n/m)	0 (n/m)	0 (n/m)	0 (n/m)	0 (n/m)	0 (n/m)	5 (5.3)
SO_3 mol% (analysed)	0 (<0.1)	5.0 (<0.1)	5.0 (<0.1)	5.0 (<0.1)	0 (<0.1)	0 (<0.1)	0 (<0.1)
K_2O mol% (analysed)	0 (0)	0 (0.4)	0 (0.7)	0 (0.3)	0 (0)	0 (0)	0 (0)
Melting Temp./°C	1,150	1,150	1,150	1,150	1,150	1,150	1,150
Crucible Material	Al_2O_3	Al_2O_3	Al_2O_3	Al_2O_3	3 $Al_2O_3 \times 2$ SiO_2	3 $Al_2O_3 \times 2$ SiO_2	3 $Al_2O_3 \times 2$ SiO_2
CS (Fe^{2+})/mm s^{-1} (±0.02)	1.25	1.25	1.32	1.35	1.23	1.25	1.23
QS (Fe^{2+})/mm s^{-1} (±0.02)	2.13	2.15	2.01	1.98	2.13	2.11	2.15
LW (Fe^{2+})/mm s^{-1} (±0.02)	0.15	0.13	0.15	0.14	0.14	0.15	0.17
CS (Fe^{3+})/mm s^{-1} (±0.02)	0.37	0.40	0.39	0.41	0.36	0.36	0.35
QS (Fe^{3+})/mm s^{-1} (±0.02)	0.89	0.84	0.79	0.74	0.89	0.88	0.91
LW (Fe^{3+})/mm s^{-1} (±0.02)	0.15	0.15	0.15	0.18	0.14	0.15	0.15
($Fe^{2+}/\Sigma Fe) \times 100\%$ (±1)	14	17	20	25	20	17	25

n/m not measured

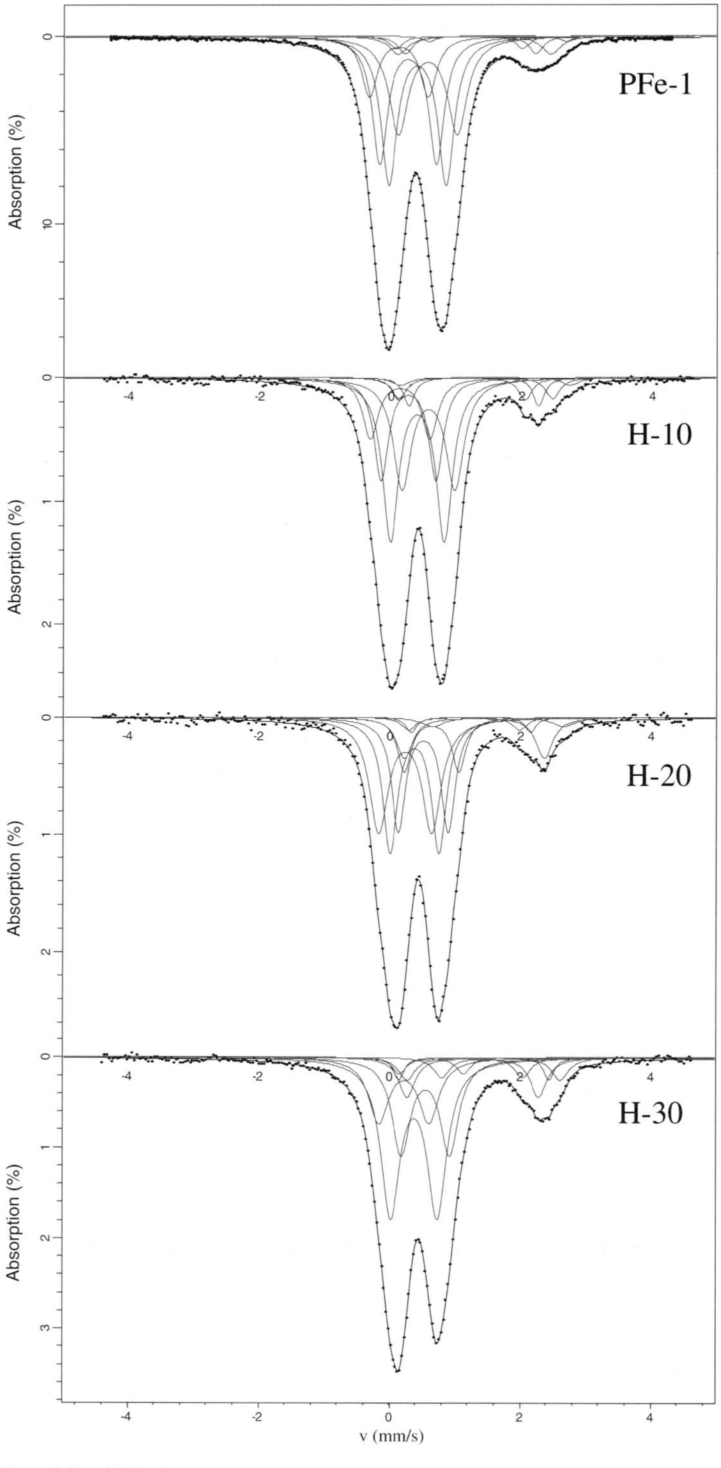

Figure 1 Selected fitted Mössbauer spectra.

shift (CS) increased whilst QS decreased, with changes of up to 20% from base glass (PFe-1) values. Using established parameter ranges to deduce iron coordination geometries, Fe^{2+} and Fe^{3+} ions largely occupied distorted octahedral sites in all glasses. However some sites with lower coordination such as tetrahedral may also have occurred. This was particularly likely in glasses with the highest Fe_2O_3 contents, where CS (Fe^{3+}) was closer to the value of 0.31 mm s^{-1} found for tetrahedral Fe^{3+} in crystalline $FePO_4$, as discussed by Yu et al. [3]. The systematic changes in CS and QS indicated decreasing distortion from cubic symmetry and a shift towards octahedral iron environments. LW values indicated that this was not associated with changes in the *range* of site distortions, therefore it was attributed to a shift in the average iron coordination geometry towards more octahedral, indicative of an increase in the octahedral/tetrahedral Fe ratio. It was estimated that for all glasses this ratio was >>1.

The addition of Na_2O to 60 P_2O_5–40 Fe_2O_3 (mol%) glasses or their equivalent has been carried out in a number of previous studies [3, 4, 6, 7]. Addition of up to 15 mol% Na_2O on a pro rata basis had little effect on fitted Mössbauer parameters, although $Fe^{2+}/\Sigma Fe$ redox ratio increased slightly [4]. Larger changes occurred when Na_2O replaced Fe_2O_3:CS (Fe^{3+}) increased from 0.40 to 0.45 mm s^{-1} upon adding 20% Na_2O, but decreased to 0.43 mm s^{-1} at 40% Na_2O [7]. CS (Fe^{2+}) decreased from 1.02 to 0.97 mm s^{-1} upon adding 15% Na_2O, then increased to 1.13 mm s^{-1} at 40% Na_2O [7]. QS (Fe^{3+}) decreased linearly from 0.85 to 0.55 mm s^{-1} upon adding 35% Na_2O, whilst QS (Fe^{2+}) remained approximately constant up to 25 mol% Na_2O, above which it decreased rapidly [7]. These results were interpreted by Takahashi et al. [7] as indicating that the hyperfine parameters of Fe^{3+} ions were influenced by Na_2O, even at low Na_2O contents. They claimed this was related to the formation of non-bridging oxygens (NBOs). The behaviour of the Mössbauer parameters from these references [4, 7] were consistent with one another and with those shown in this study, although the changes in CS and QS were stronger in those glasses where Na_2O replaced Fe_2O_3 [7] rather than pro rata replacement. However, it has also been demonstrated that P and Fe contents of iron phosphate glasses can themselves influence CS, QS and the $Fe^{2+}/\Sigma Fe$ redox ratio [3]. The magnitudes of the changes observed here were similar to those for systems containing added Na_2O [3, 4, 6, 7]. It was therefore not possible to differentiate between the effects on hyperfine parameters of P/Fe ratio and (Na_2O+Al_2O_3) additions. This may indicate that iron environments were not influenced per se by the addition of Al_2O_3 and Na_2O, but that iron environments may largely be determined in these systems by the P/Fe ratio. Such behaviour would be consistent with the assertion of Marasinghe et al. [4] that the near-neighbour environments of Fe ions are not appreciably altered by waste components (including Na_2O).

The addition of 5% Si, Al or B to the base glass had little measurable effect on hyperfine parameters, although the $Fe^{2+}/\Sigma Fe$ ratio did increase markedly. Comparison of the results discussed in this paper with a wide range of published data [3–8] showed that fitted Mössbauer hyperfine parameters were relatively invariable over a wide range of compositions. This was consistent with the view that in iron phosphate glasses, near-neighbour environments of Fe ions are not greatly affected by additional components. This supports their suitability as hosts for the vitrification of certain wastes.

4 Conclusions

Iron was found to occur predominantly as Fe^{3+} in a range of phosphate glasses, largely occupying sites with distorted octahedral coordination for both redox states. Stepwise molar replacement of Fe_2O_3 by ($0.33\ Al_2O_3 \times 0.67\ Al_2O_3$) in 60% P_2O_5–40% Fe_2O_3 glasses substantially increased the redox ratio. Accompanying changes in centre shift and quadrupole splitting were interpreted as a decrease in the average distortion of Fe sites from cubic symmetry and an increase in average iron coordination. Literature revealed that recoil-free fraction ratio $f\ (Fe^{3+})/f\ (Fe^{2+}) \approx 1.3$ in iron phosphate glasses, and this was considered when assessing redox.

Acknowledgements The authors wish to thank the UK's Engineering and Physical Sciences Research Council (EPSRC) for funding this work.

References

1. Pegg, I.L., Joseph, I.: In: Oh, C.H. (ed.) Hazardous and Radioactive Waste Treatment Technologies Handbook, p. 4.2.1. CRC, London (2001)
2. Kushnikov, V.V., Matyunin, Y.I., Smelova, T.V., Demin, A.V.: Mater. Res. Soc. Symp. Proc. **465**, 55 (1997)
3. Yu, X., Day, D.E., Long, G.J., Brow, R.K.: J. Non-Cryst. Solids **215**, 21 (1997)
4. Marasinghe, G.K., Karabulut, M., Ray, C.S., Day, D.E., Allen, P.G., Bucher, J.J., Edelstein, N.M., Shuh, D.K., Badyal, Y.S., Saboungi, M.L., Grimsditch, M., Shastri, S.D., Haeffner, D.: Ceram. Trans. **93**, 195 (1999)
5. Bingham, P.A., Hand, R.J., Forder, S.D., Lavaysierre, A., Kilcoyne, S.H., Yasin, I.: Mater. Lett. **60**, 844 (2005)
6. Itoh, H., Inamura, T., Wakabayashi, H., Toriyama, T., Iijima, H.: Proc. ICAME-95, 445 (1996)
7. Takahashi, T., Inamura, T., Toriyama, T., Ijima, H.: Nucl. Instrum. Methods Phys. Res. **B76**, 103 (1993)
8. Nishida, T., Shotsuki, T., Takashima, Y.: J. Non-Cryst. Solids **43**, 115 (1981)
9. Lagarec, K., Rancourt, D.G.: Recoil: Mössbauer spectral analysis software for windows, http://www.isapps.ca/recoil/ (1998)

Hyperfine Interact (2005) 165: 141–146
DOI 10.1007/s10751-006-9257-x

Rotational dynamics in anisotropic confinement

G. Wellenreuther · H. Franz · U. v. Bürck · I. Sergueev

Published online: 15 November 2006

Abstract Quasi-elastic nuclear resonant forward scattering experiments of the model glass former DBP:FC (Dibutylphthalate:Ferrocene) have been performed in anisotropic confinement, yielding complex spectra. A model for the rotational relaxation of Ferrocene based upon the Kubo–Anderson process is presented, which explains the observed spectra as superposition of three components: Static Ferrocene, anisotropic rotating Ferrocene and isotropic rotating Ferrocene.

Key words rotational dynamics · glass transition · confinement

1 Introduction

In the late 1960s the theoretical base for the understanding of the influence of rotational or spin dynamics on hyperfine interaction was laid by a series of papers (see e.g. [1] and references in [2]). Their approach was utilized to explain many experiments, e.g. [3]. Especially the rotational dynamics of Ferrocene was observed by Gibb in the restricting geometry of thiourea-clathrates [4], and Sergueev examined the rotational dynamics of Ferrocene in bulk DBP:FC [5]. In the case of no or isotropic confinement only small motional narrowing and an enhanced decay were observed with QNFS (Quasi-elastic nuclear resonant forward scattering), while in the case of anisotropic confinement strong perturba-

G. Wellenreuther (✉) · H. Franz
Hamburger Synchrotronstrahlungslabor, EMBL Hamburg Outstation, Hamburg, Germany
e-mail: Gerd.Wellenreuther@embl-hamburg.de

H. Franz
e-mail: Hemann.Franz@desy.de

U. v. Bürck
Physik Department, Institut für Experimentalphysik E13, TU München, Germany

I. Sergueev
Nuclear Resonance Group, ESRF, Grenoble, France

tions of the hyperfine states occur; we attribute this to an increase of anisotropic rotational dynamics.

To reduce the difficulty in the calculation of the line shapes observed from stochastically rotating molecules only the powder average was determined in the works cited above. The resulting formalism can be used to describe isotropic dynamics as well as anisotropic rotation in samples missing any orientational alignment.

Recently experiments on anisotropic samples were reported showing the need of a theory taking a macroscopic alignment of the quantization axis into account, e.g. by Asthalter in AlPO4–5 [6]. To ensure comparability of experiments in this work we used the same model glass former as already studied in bulk and isotropic confinement.

2 Experimental

We performed nuclear resonant scattering experiments at the ESRF on the model glass former DBP:FC [5, 7] in pores of mesoporous silicon [8] with 9.6 nm mean pore diameter, which have been prepared by Victor Timoshenko [9]. We used the beamline ID18 in 16-bunch mode to measure time spectra in two different orientations and at 14 temperatures in a range from 20 to 220 K (far above the bulk calorimetric glass transition temperature of T_g = 178 K [7]). Hereby we observed pronounced rotational dynamics, which was in addition partially anisotropic [10]. Due to the anisotropic confinement we observed (in contrast to [5]) rather complex spectra at temperatures where relaxation sets in.

3 Phenomenological analysis of experimental spectra

Experimental spectra at four exemplary temperatures are displayed as symbols in Figure 1. At temperatures well below the calorimetric glass transition temperature, all spectra can be fitted with QNFS in the kinematical approximation [11, 12] using the well known quadrupole splitting value of Ferrocene ($\Omega \approx 24\Gamma$, Γ = natural linewidth). For higher temperatures a coherent sum of two QNFS amplitudes is necessary, in this case the value of the second quadrupole splitting is $\approx 8 - 10\Gamma(\approx 35 - 40\%$ of the magnitude of the static quadrupole splitting of Ferrocene). In addition a fast exponential decaying line is observed at short times.

Translational dynamics on a timescale comparable to the nuclear lifetime would cause a decrease of QNFS-signal. Unlike in experiments in bulk or isotropic confinement we did not observe a breakdown of the QNFS-signal above T_g, which we attribute to reduced translational dynamics as compared to bulk. Additionally, a decrease of the quadrupole splitting value by more than 50% can not be explained by translational dynamics. On the other hand the three spectral components may be explained on the basis of different environments of rotating Ferrocene molecules.

4 Modelling

We modeled the three different spectral components observed in our spectra by three relaxation centers containing a Ferrocene molecule which can jump between three perpendicular states of the electric-field gradient (EFG) orientation. The basic equations describing spin–relaxation based on the Kubo–Anderson process have been summarized by

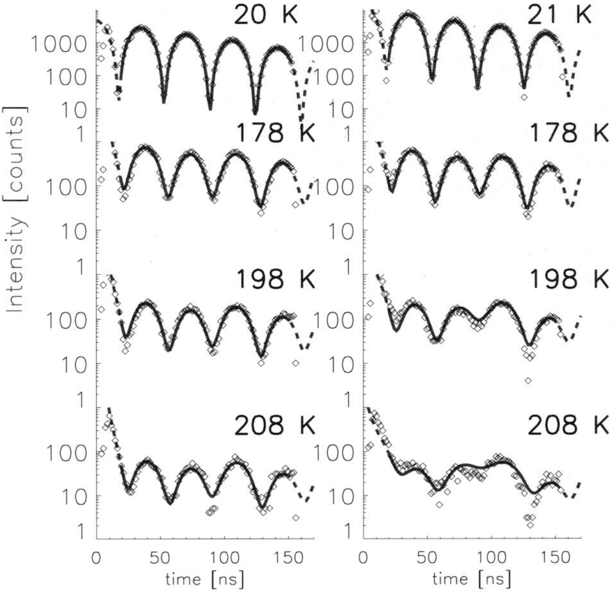

Figure 1 QNFS time spectra. *Symbols*: Time spectra, *Left*: Parallel orientation of pore axis and incident beam, *Right*: Perpendicular orientation. *Solid line*: Least-squares fit to the data according to our model. Jumprates, asymmetry parameters and principal quadrupole splitting as stated in the text. Fraction of first center dropped from ≈100% at 20 K downto 45%, fractions of second and third center increased to 25 and 30%, respectively.

Dattagupta in [2] (we use his conventions regarding variables etc.). As the analytical calculation of the lineshape and even more the time-domain spectrum is getting rather involved we used the SYNFOS software package [13]. This algorithm allowing to calculate a coherent superposition of several relaxation centers was provided by Mati Haas (Institute of Physics at the University of Tartu) and Heiner Winkler (Institute of Physics of the Medical University of Lübeck). We further added a least-squares fit routine to this algorithm. Fitting the time spectra with three relaxation centers was successful as displayed by the solid line in Figure 1, while attempts to fit the time spectra at high temperatures using only one or two different relaxation centers failed.

Figure 2 displays simulated spectra of the three relaxation centers: Static Ferrocene (top), an isotropic relaxation in the motional narrowing regime (bottom, quadrupole splitting is averaged out) and an intermediate state relaxing anisotropically (middle); the inset shows schematically the corresponding spectra of Mössbauer transition lines. Two orientations are displayed (solid/dashed line); obviously the orientation dependence manifests itself predominantly in the intermediate state. The amplitudes from these three types of scattering centers are added up *coherently* to yield the QNFS-signal, as shown in Figure 1. It has to be stressed that we did not apply any distribution function to jump rates, quadrupole splittings or asymmetry parameters. We expect the fit quality to further increase if distribution functions are applied.

The *first center* models the amount of static Ferrocene dissolved in DBP. Thereby the jump rates and the asymmetry parameters were fixed at 0.

The *second center* corresponds to anisotropically rotating Ferrocene, causing an incomplete reduction of the quadrupole splitting [4]. Values for the jump rates were obtained from the fit as $W_{xy} = 0.15\ \hbar\Omega$, $W_{xz} = 0.075\ \hbar\Omega$ and $W_{yz} = 25.0\ \hbar\Omega$. This center displays a strong reduced quadrupole splitting (see inset in middle of Figure 2).

Figure 2 Spectral components corresponding to the relaxation centers calculated with hyperfine parameters as stated in the text. *Top*: Static case, *Mid*: Anisotropic rotation, *Bottom*: Isotropic rotation. *Solid lines*: Parallel orientation of incident beam and pore axis, *dashed lines*: Perpendicular orientation. *Insets*: Schematic representation in the energy domain, disregarding orientation.

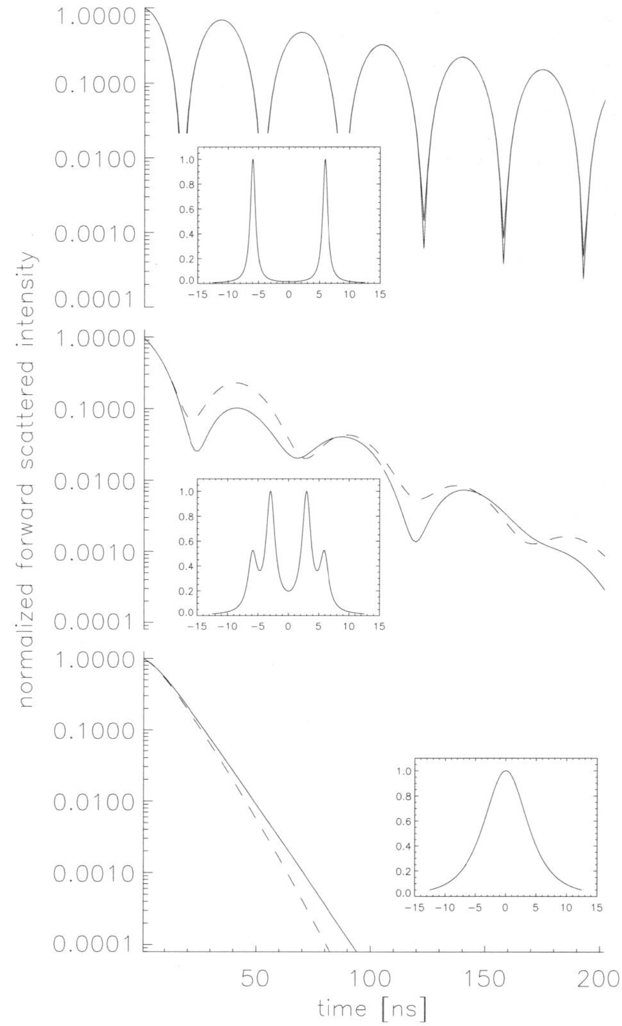

To model quadrupole splittings below 50% of the principal quadrupole splitting, the asymmetry parameters had to be allowed to take non-zero values. Fitting resulted in values of $\eta_x = 0.5$, $\eta_y = 0.0$ and $\eta_z = 0.25$.

The *third center* corresponds to a fraction of Ferrocene molecules rotating isotropically, thereby leading to a fast decaying exponential line and pronounced motional narrowing. The jump rates were set to fixed values ($W_{xy} = W_{xz} = W_{yz} = 1.0\ \hbar\Omega$), and $\eta_x = \eta_y = \eta_z = 0$.

In summary we understand these different centers as being caused by different degrees of influence (not necessarily pair-interaction) of the pore walls. Closest to the wall the environments are frozen in due to the interdiction of cooperative modes (first center) resulting in spectra resembling static Ferrocene as known from experiments in bulk. In the middle of the pore isotropically rotating Ferrocene (third center) causes the fast decaying line; the second center is understood as an in-between state in an environment allowing

Figure 3 Temperature dependence of the population of the three relaxation centers. *Squares*: first center, *Triangles*: second center, *Diamonds*: third center, *Black symbols*: parallel case, *Grey symbols*: perpendicular case.

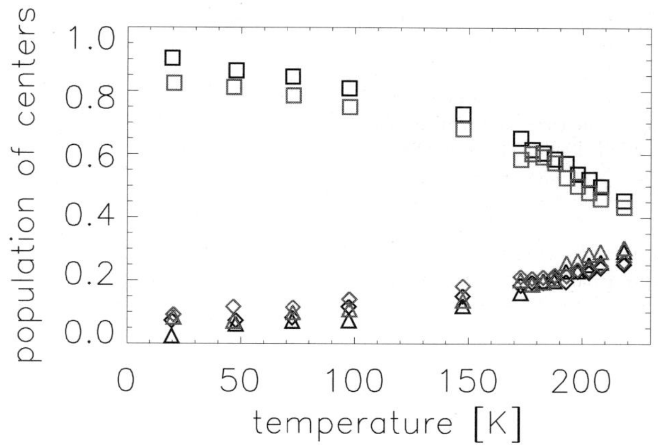

only anisotropic relaxation. As can be seen in Figure 3 the occupancy of the second and third center increases significantly during the glass transition.

We also examined whether Ferrocene in our sample oxidized, which could also lead to reduced quadrupole splittings [14]. In XANES experiments performed at beamline E4 (HASYLAB) no shift of the absorption edge was observed relative to crystalline Ferrocene and an aged solution of DBP:FC, which means that no change of the oxidation state has taken place.

5 Conclusions

We have applied the theory of rotational relaxation to describe EFG relaxation of Ferrocene in an amorphous environment. Here the anisotropic nature of the relaxation spectra had to be computed numerically. We used the computational results to model the time spectra of our QNFS experiment performed on DBP:FC. The results point out the dominant role of rotational relaxation in this kind of system.

Acknowledgements We would like to thank Victor Y. Timoshenko for the preparation of the mesoporous silicon, Aleksandr I. Chumakov and Rudolf Rüffer for their continuous and excellent support at beamline ID 18 (ESRF), as well as Heiner Winkler from the University of Lübeck and Mati Haas from the University of Tartu for their aid to establish the dynamic model.

References

1. Blume, M.: Phys. Rev. **174**, 351 (1968)
2. Dattagupta, S.: Hyperfine Interact. **11**, 77 (1981)
3. Dunlap, B.D., Shenoy, G.K., Asch, L.: Phys. Rev., B **13**, 18 (1976)
4. Gibb, T.: J. Phys., C. Solid State Phys. **9**, 2627 (1976)
5. Sergueev. I.: Nuclear Resonant Scattering for the Study of Dynamics of Viscous Liquids and Glasses. PhD thesis, http://tumb1.biblio.tu-muenchen.de/publ/diss/ph/2004/sergueev.html, Technische Universität München (2004)

6. Asthalter, T., Villanueva Garibay, J., Olszowka, V., Kornatowski, J.: J. Chem. Phys. **122**, 014508 (2005)
7. Meyer, A., Franz, H., Wuttke, J., Petry, W., Wiele, N., Rüffer, R., Hübsch, C.: Z. Phys., B **103**, 479 (1997)
8. Lehmann, V.: Electrochemistry of Silicon. Wiley, Weinheim (2002)
9. Valiullin, R., Kortunov, P., Kärger, J., Timoshenko, V.: J. Chem. Phys. **120**(24), 11804 (2004)
10. Wellenreuther, G.: Glass Dynamics in Confinement Examined with Quasi-Elastic Nuclear Resonant Scattering. PhD thesis, University of Hamburg, published in DESY Berichte 29 (2005)
11. Smirnov, G., Kohn, V.: Phys. Rev., B **52**, 3356 (1995)
12. Kohn,V., Smirnov, G.: Phys. Rev., B **57**, 5788 (1998)
13. Haas, M., Realo, E., Winkler, H., Meyer-Klaucke, W., Trautwein, A.: Hyperfine Interact. **125**, 189 (2000)
14. Toda, Y., Ishimaru, S., Ikeda, R., Mitani, T., Kitao, S., Seto, M.: J. Phys. Chem. Solids **65**, 471 (2004)

Hyperfine Interact (2005) 165: 147–151
DOI 10.1007/s10751-006-9258-9

Mössbauer studies of ε-$Fe_{3-x}Ni_xN$ and γ'-$Fe_{4-y}Ni_yN$ nanoparticles

N. S. Gajbhiye · Sayan Bhattacharyya

Published online: 7 November 2006

Abstract Nanocrystalline ε-$Fe_{3-x}Ni_xN$ ($0.0 \leq x \leq 0.8$) particles are synthesized by precursor technique and nitridation of decomposed products in NH_3 (g) in the temperature range 673 K-823 K. For x=0.1–0.4 compositions, single phase ε-$Fe_{3-x}Ni_xN$ hexagonal structure with space group $P6_3$/mmc is formed, while for x=0.5–0.8, fcc γ'-$Fe_{4-y}Ni_yN$ phase is also precipitated. The room temperature Mössbauer spectrum for all the compositions shows the presence of superparamagnetic doublet, which is attributed to ε-$Fe_{3-x}Ni_xN$ phase. For x=0.5–0.8 compositions, two additional sextets are observed corresponding to two different iron sites, the corner position (Fe^c) and the fcc position (Fe^f), in γ'-$Fe_{4-y}Ni_yN$. The added Ni atoms preferentially substitute the corner Fe^c positions. The isomer shift, quadrupole splitting and hyperfine field values are found to change with the Ni content.

Key words nanoparticle · Mössbauer · iron–nickel nitride · atomic ordering

1 Introduction

Magnetic nanoparticles have great potential applications in various technological areas [1]. Iron nitrides (Fe–N) represent a rich class of ferromagnetic materials. Although reports of substituting various elements in fcc γ'-Fe_4N lattice are available in the literature [2, 3], but very few reports are available for substitution of Co and Ni in hexagonal ε-Fe_3N lattice [4, 5]. In cubic γ'-Fe_4N, preferential occupation of the substituents is reported whereas substitution in hexagonal ε-Fe_3N occur in a random fashion. The Fe–N ferromagnetic nanoparticles below a certain critical size range exhibit superparamagnetic relaxation at room temperature. In ε-Fe_3N, the intermixing of N-2p states with Fe-3d states occur because of the similarities in their density of states or comparable in energy, thus reducing

N. S. Gajbhiye · S. Bhattacharyya
Department of Chemistry, Indian Institute of Technology, Kanpur 208 016 UP, India

N. S. Gajbhiye (✉)
Institute of Nanotechnology, Forschungszentrum Karlsruhe, P.O. Box 3640, 76021 Karlsruhe, Germany
e-mail: nsg@iitk.ac.in

the number of unpaired d-electrons which results in the magnetic moments lower than α-Fe [6]. The motivation of this work is to investigate the change of the intrinsic Mössbauer parameters viz. magnetic hyperfine field (ΔH_f), isomer shift (δ), quadrupole splitting (Δ) and line width (Γ) with the variation of Ni content in ε-Fe_3N. Also, the Ni occupation in fcc γ'-$Fe_{4-y}Ni_yN$ phase (for higher Ni substitutions, x=0.5–0.8) is studied by ^{57}Fe Mössbauer spectroscopy at room temperature.

2 Experimental section

ε-$Fe_{3-x}Ni_xN$ ($0.0 \leq x \leq 0.8$) nanoparticles were synthesized by wet chemical method. The aqueous solution of 0.22 (N) $Fe(NO_3)_3 \cdot 9H_2O$ was added to 0.35 (N) $Ni(NO_3)_2 \cdot 6H_2O$ solution, in desired proportion corresponding to the stoichiometric ratios of ε-$Fe_{3-x}Ni_xN$. The resulting solution was mixed with citric acid solution and refluxed at 343 K for 12 h. The viscous slurry was dried and then decomposition of the citrate precursors in air at 773 K for 4 h, gave finely divided Fe–Ni-oxide particles. The nitridation is carried out under flowing NH_3 (g) at 673 K–823 K for 12 h to get the desired Fe–Ni–N nanostructured phases. The stoichiometry of the final products was confirmed by chemical analysis. X-ray diffractograms (Rich Seifert model Isodebyeflex 2002) were recorded using CuKα radiation. The average particle size was calculated from the X-ray line broadening of (101) plane reflection using Scherrer's formula. The particle size and morphology were characterized by TEM (Philips model E-301). Mössbauer spectra of the samples under study were recorded by using conventional spectrometer and ^{57}Co source embedded in Rh matrix. The spectrum was then analyzed by using software PC-MOS supplied by CMTE Electronik, Germany. The values are presented relative to natural iron. The uncertainty in isomer shift and quadrupole splitting is 0.03 and 0.04 mm/s, respectively and the hyperfine field values are correct to 0.1 T.

3 Results and discussion

Analysis of the XRD patterns, with the help of Rietveld analysis program, confirms the formation of ε-Fe_3N hexagonal structure, with space group $P6_3$/mmc, for the x=0.1–0.4 compositions. However, for x=0.5–0.8 compositions, a small amount of fcc γ'-$Fe_{4-y}Ni_yN$ phase is also precipitated. In ε phase, the lattice parameter a varies from 2.758(1) Å to 2.668(1) Å and c from 4.403(2) Å to 4.259(2) Å with increasing Ni content. TEM confirms the particles to be spherical in shape with a size in the range 28–30 nm.

Figure 1 shows the room temperature Mössbauer spectra for the representative x=0.0, 0.4 and 0.8 systems. The Mössbauer spectra, recorded at 300 K with zero external magnetic field, show one doublet for x=0.0–0.4 compositions, indicating superparamagnetic nature of the ε-$Fe_{3-x}Ni_xN$ nanoparticles. For x=0.5–0.8 systems, the Mössbauer spectra are fitted with one doublet and two sextets, indicating the co-existence of superparamagnetic and ferromagnetic phases.

In ε-$Fe_{3-x}Ni_xN$ system, it has been shown from Rietveld analysis that the Fe atoms are randomly replaced by Ni atoms and N occupies the regular positions of the ε-Fe_3N structure. Due to the existence of partial covalent bonding between Fe and N atoms, ε-Fe_3N deviates from the octahedral coordination around the Fe nucleus, as seen from the quadrupole splitting, Δ=0.47 mm/s and line width Γ=0.87 mm/s. The random distribution of Ni at the Fe sites for ε-$Fe_{3-x}Ni_xN$ is attributed to the slight increase of Δ values for

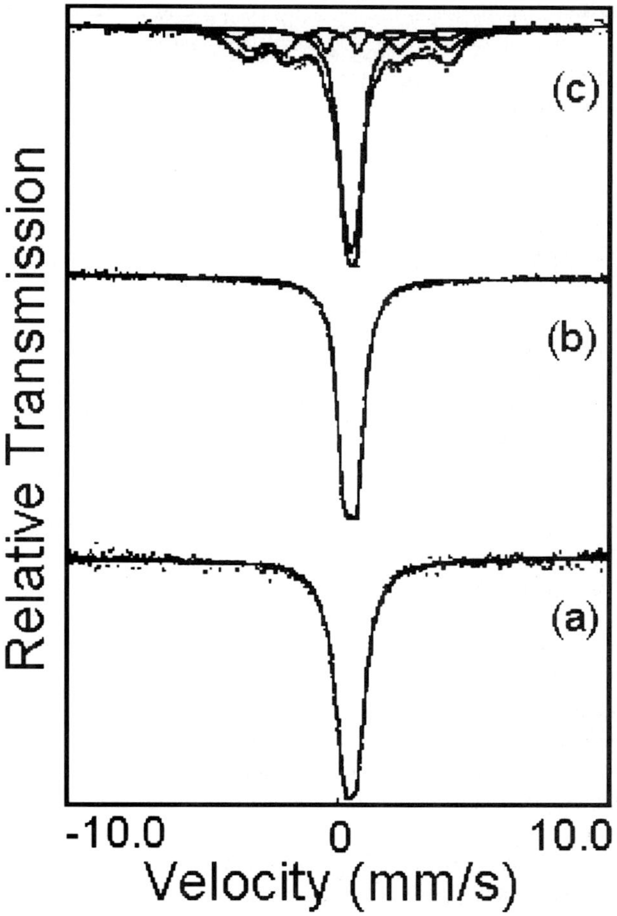

Figure 1 Room temperature Mössbauer spectrum of ε-$Fe_{3-x}Ni_xN$; x is (a) 0.0 (b) 0.4 and (c) ε-$Fe_{3-x}Ni_xN$ (x=0.8) and γ′-$Fe_{4-y}Ni_y$ (y=1.1).

x=0.0–0.4 compositions. Similar trend is also observed for δ and Γ values, for the superparamagnetic doublet part of the ε-$Fe_{3-x}Ni_xN$ system (Table I). The δ values indicate smaller s- or greater p- and d-electron densities of Fe in ε-$Fe_{3-x}Ni_xN$ system. The density of states at the Fermi level (E_f) arise from the unbalanced spins between Fe 3d and Ni 3d orbital and N 2p orbital electrons.

For x=0.5 to 0.8 systems, Mössbauer spectra display one doublet and two sextets (Figure 1). The doublet corresponds to the ε-$Fe_{3-x}Ni_xN$ phase, while the two sextets are due to two different sites, the corner position (Fe^c) and the fcc position (Fe^f) of the fcc lattice of the γ′-$Fe_{4-y}Ni_yN$ phase (y=0.7, 0.8, 0.9 and 1.1 for x=0.5, 0.6, 0.7 and 0.8, respectively). The x and y values are determined separately using XRD-Rietveld analysis using the best fit with calculated parameters to the experimental XRD spectra. In the Mössbauer spectra of γ′-$Fe_{4-y}Ni_yN$ phase, these two sextets corresponding to Fe^c and Fe^f atoms are due to the distribution of internal magnetic fields. The Mössbauer spectra of γ′-$Fe_{4-y}Ni_yN$ have been deconvoluted into five sextets indicating iron atoms in five different environments elsewhere in the literature [7]. However, in this study, such kind of resolution is not

Table I ^{57}Fe Mössbauer resonance parameters for ε-$Fe_{3-x}Ni_xN$ and γ'-$Fe_{4-y}Ni_yN$ phases

Ni (y)	Ni (x)	Iron site	Hyperfine field ΔH_f (T)	Isomer shift δ (mm/s)	Quadrupole splitting Δ (mm/s)	Line widths Γ (mm/s)	Percent obs
–	0.0	Fe_ε	–	0.44	0.46	0.87	100
–	0.1	Fe_ε	–	0.45	0.45	0.64	100
–	0.2	Fe_ε	–	0.44	0.43	0.63	100
–	0.3	Fe_ε	–	0.43	0.53	0.88	100
–	0.4	Fe_ε	–	0.44	0.45	0.67	100
0.7	0.5	$Fe_{\gamma'}{}^f$	21.12	0.13	0.30	2.40	10.8
		$Fe_{\gamma'}{}^c$	26.09	0.21	–	1.11	0.5
		Fe_ε	–	0.39	0.43	0.74	88.7
0.8	0.6	$Fe_{\gamma'}{}^f$	21.71	0.30	0.38	1.27	10.5
		$Fe_{\gamma'}{}^c$	24.78	0.43	−0.85	0.61	3.6
		Fe_ε	–	0.39	0.41	0.68	85.9
0.9	0.7	$Fe_{\gamma'}{}^f$	22.52	0.24	0.29	1.10	10.1
		$Fe_{\gamma'}{}^c$	24.35	0.34	−0.91	0.66	9.0
		Fe_ε	–	0.39	0.38	0.68	80.9
1.1	0.8	$Fe_{\gamma'}{}^f$	22.56	0.23	0.31	1.26	8.2
		$Fe_{\gamma'}{}^c$	24.11	0.40	−0.68	1.06	12.4
		Fe_ε	–	0.38	0.37	0.67	79.4

attainable because of the central peak corresponding to the superparamagnetic ε-$Fe_{3-x}Ni_xN$ phase. Similar kind of results are also observed for γ'-Fe_4N by Saegusa et. al. [8]. For the x=0.5–0.8 systems, the sextet having the larger hyperfine field corresponds to Fe^c sites and the other sextet with smaller hyperfine field corresponds to Fe^f sites. The sextet corresponding to the Fe^c sites have lower intensities compared to the sextet for the Fe^f sites and this is because the added Ni atoms preferentially substitute for the Fe^c sites.

The preference of Ni occupation in γ'-$Fe_{4-y}Ni_yN$ has been discussed in detail in the literature [9]. Assuming the recoil-free fraction to be the same for Fe atoms at different crystallographic sites of the γ'-$Fe_{4-y}Ni_yN$ phase, the Ni occupation is given as,

$$A_c/A_f = \left(1 - N_{Ni}{}^c\right)/\left(3 - N_{Ni}^f\right) \tag{1}$$

where, A_c and A_f are the corresponding peak areas, $N_{Ni}{}^c$ and N_{Ni}^f are the fractional number of Ni atoms replacing Fe^c sites and Fe^f sites, respectively. Using the estimated values of A_c and A_f and the relation, $N_{Ni}{}^c + N_{Ni}{}^f = y$, the percentage of Ni atoms occupying the corner positions are determined. However, some fraction of the Ni atoms, are still present in the face centered site of the γ'-$Fe_{4-y}Ni_yN$ phase. The percentage of Ni atoms occupying the corner positions ($P_{Ni}{}^c$) in γ'-$Fe_{4-y}Ni_yN$ phase are calculated as: 90.2, 88.2, 69.1 and 37.9% for various Ni content, y=0.7, 0.8, 0.9 and 1.1, respectively. $P_{Ni}{}^c$, for y=0.8 is close to the proposed value of 80% [3]. $P_{Ni}{}^c$ decreases from y=0.8 to 1.1, which is in good agreement with the earlier report [9]. The Δ values have different signs for the Fe^f and Fe^c atoms, due to different geometrical orientations. The variation of Γ values is due to fluctuation of the exchange field due to Ni atom substitutions. The ΔH_f for Fe^c atoms is low as compared to γ'-Fe_4N (ΔH_f=34.5 T) and this is attributed to the local inhomogeneities in γ'-$Fe_{4-y}Ni_yN$ phase [3]. The ΔH_f for Fe^f sites are comparable to 21.9 T for γ'-Fe_4N [6]. The Mössbauer data corroborate the magnetization studies, which were investigated in detail [10].

4 Conclusion

The Mössbauer spectra for ε-$Fe_{3-x}Ni_xN$ ($0.0 \leq x \leq 0.8$) nanoparticles are studied at 300 K. For $x=0.0$–0.4 compositions, the spectra shows the presence of superparamagnetic doublet corresponding to single ε-$Fe_{3-x}Ni_xN$ phase. For $x=0.5$–0.8 systems, two additional sextets are observed corresponding to two different Fe sites in γ′-$Fe_{4-y}Ni_yN$ additional phase. The high δ values indicate smaller *s*-electron density or greater *p*- and *d*-electron densities at Fe sites. When Ni content is, $y=0.7$, 0.8 and 0.9, the Ni atoms preferentially occupy Fe^c sites and for higher nickel amount, $y=1.1$, Ni atoms occupy Fe^f sites.

Acknowledgement The authors acknowledge CSIR, New Delhi for the financial support in carrying out this research work.

References

1. Dormann, J.L., Fiorani, D. (eds.): Magnetic Properties of Fine Particles, North-Holland, Amsterdam (1992)
2. Chen, S.K., Jin, S., Tiefel, T.H., Hsieh, Y.F., Gyordy, E.M., Johnson, D.W. Jr.: J. Appl. Phys. **70**, 6247 (1991)
3. Panda, R.N., Gajbhiye, N.S.: J. Magn. Magn. Mater. **195**, 396 (1999)
4. Takahashi, S., Kume, M., Matsuura, K.: IEEE Trans. Magn. **26**, 1632 (1990)
5. Gajbhiye, N.S., Ningthoujam, R.S., Weissmüller, J.: Hyperfine Interact. **156/157**, 51 (2004)
6. Panda, R.N., Gajbhiye, N.S.: IEEE Trans. Magn. **34**, 542 (1998)
7. Bond, G.C.: Catalysis by Metals, p. 371. Academic, London (1994)
8. Saegusa, N., Morrish, A.H., Tasaki, A., Tagawa, K., Kita, E.: J. Magn. Magn. Mater. **35**, 123 (1983)
9. Li, F., Yang, J., Xue, D., Zhou, R.: Appl. Phys. Lett. **66**, 2343 (1995)
10. Gajbhiye N.S., Bhattacharyya Sayan: Phys. Stat. Sol. (c) **1**, 3764 (2004)

Hyperfine Interact (2005) 165: 153–159
DOI 10.1007/s10751-006-9259-8

Mössbauer and magnetic studies of MFe_2O_4 (M=Co, Ni) nanoparticles

N. S. Gajbhiye · Sayan Bhattacharyya · G. Balaji · R. S. Ningthoujam · Raj Kumar Das · Susmita Basak · J. Weissmüller

Published online: 26 October 2006

Abstract Nanocrystalline MFe_2O_4 (M=Co, Ni) particles are synthesized by citrate precursor technique. Mössbauer and magnetic studies are carried out with the $CoFe_2O_4$ samples having particle sizes of 9, 14 and 30 nm and the $NiFe_2O_4$ samples having particle sizes of 9, 21 and 30 nm. The intrinsic magnetic parameters are found to vary with the particle size. The magnetic interactions and cation distribution present in these systems influence the room temperature Mössbauer parameters. Ferrimagnetic sextets are observed for all the different particle sizes. The observed reduction of the magnetic hyperfine field values with the decrease in the size of MFe_2O_4 particles are attributed to the intrinsic size effect and the canted spin structure at the surface of the nanoparticles.

Key words ferrite · nanocrystalline · ferrimagnetism · cation distribution · surface effect

1 Introduction

Ferrite materials encompass a large family of magnetic oxides, which exhibit different structures with soft as well as hard magnetic properties [1]. These materials have the potential applications such as magnetic recording media, ferrofluids, magnetic refrigeration, geomagnetism, catalysis, sensor and biomagnetic applications [2]. The chemical order in nanophase ferrites is different from that in bulk material. The state of chemical order, especially the cation size occupancy has a remarkable effect on the magnetic state of ferrites [3]. The magnetic structure of MFe_2O_4 (M=Co, Ni, Cu) system is ferrimagnetic [4, 5]. It contains two sublattices, A and B, where A and B refer to the tetrahedral site and octahedral site, respectively. Cation distribution occurs between the two sublattices. The cation

N. S. Gajbhiye (✉) · S. Bhattacharyya · G. Balaji · R. S. Ningthoujam · R. K. Das · S. Basak
Department of Chemistry, Indian Institute of Technology, Kanpur 208 016 UP, India
e-mail: nsg@iitk.ac.in

N. S. Gajbhiye · J. Weissmüller
Institute of Nanotechnology, Forschungszentrum Karlsruhe, P.O. Box 3640, 76021 Karlsruhe, Germany

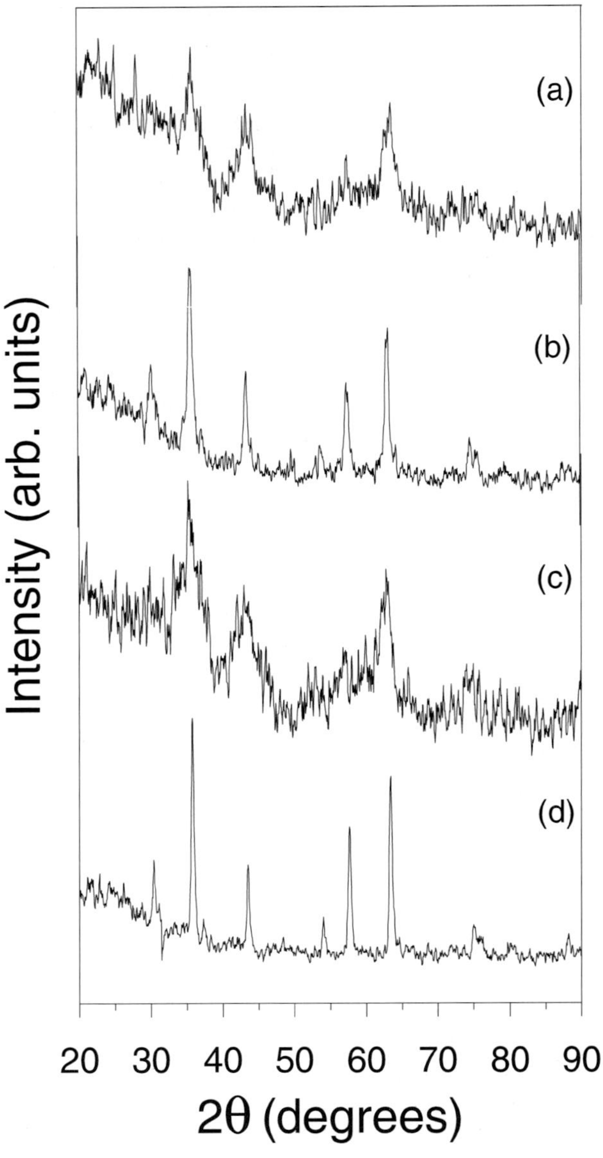

Figure 1 X-ray diffraction pattern for $CoFe_2O_4$ particles having sizes (a) 9 nm and (b) 30 nm and $NiFe_2O_4$ particles having sizes (c) 9 nm and (d) 30 nm.

distribution largely depends on the synthetic method, heat treatment temperature, heat-duration time and cooling rate. The magnetic properties of MFe_2O_4 system are governed by the degree of cation distribution. Moreover, in the nanophase materials the surface to volume ratio is more than the bulk materials. The magnetic ordering at the surface of the nanoparticles is different from that at the core and hence the surface influences the overall magnetic properties of the system. Thus, the motivation of the present work is to investigate the magnetic behavior of the nanostructured $CoFe_2O_4$ and $NiFe_2O_4$ systems having different particle size and the effect of cation distribution and surface effects on Mössbauer and magnetic parameters.

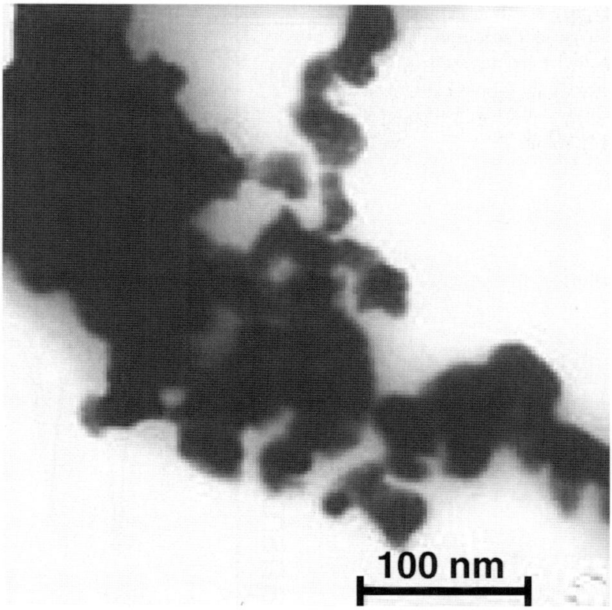

Figure 2 Transmission electron micrograph of 9 nm $CoFe_2O_4$ particles.

2 Experimental section

Nanostructured $CoFe_2O_4$ and $NiFe_2O_4$ particles were synthesized by using precursor technique [6]. The stoichiometric amount of metal salts (cobalt/nickel nitrate, iron nitrate) and citric acid were mixed in the molar ratio 1:2:2.7. The molarity of the solution was maintained at 0.2 M and the pH-value was controlled in the range 1.25–2.5. The final solution was refluxed for 12 h at 333 K with vigorous stirring. The metal citrate precursor was obtained by heating on a water bath at 363 K. The precursor was decomposed to finely divided oxide particles of various sizes by further heat treatment in the temperature range 623–973 K. X-ray diffractograms (Rich Seifert model Isodebyeflex 2002) for these samples were recorded using Cu-Kα radiation to ascertain the formation of $CoFe_2O_4$ and $NiFe_2O_4$ nanoparticles. The average particle size was calculated from the X-ray line broadening of (311) plane reflection using Scherrer's formula. Also, the particle size and morphology were characterized by TEM (Philips model E-301). Mössbauer spectra of the samples under study were recorded by using conventional spectrometer and ^{57}Co source embedded in Rh matrix. The spectrum was then analyzed by using software PC-MOS supplied by CMTE Electronik, Riemerling, Germany. The Mössbauer parameters are presented relative to metallic iron. The uncertainty in isomer shift and quadrupole splitting is 0.03 and 0.04 mm/s, respectively. The hyperfine field values are accurate to 0.1 T.

3 Results and discussion

X-ray diffraction (XRD) patterns indicate the formation of pure and monophasic nanostructured $CoFe_2O_4$ and $NiFe_2O_4$ particles (Figure 1). The cubic crystal structure is stabilized for the different particle sizes. For $CoFe_2O_4$, the particle size obtained was 9, 14 and 30 nm and for $NiFe_2O_4$, size was 9, 21 and 30 nm. The electron micrographs (TEM)

Figure 3 Room temperature Mössbauer spectrum of $CoFe_2O_4$ particles having sizes (a) 9 nm and (b) 30 nm and $NiFe_2O_4$ particles having sizes (c) 9 nm and (d) 30 nm.

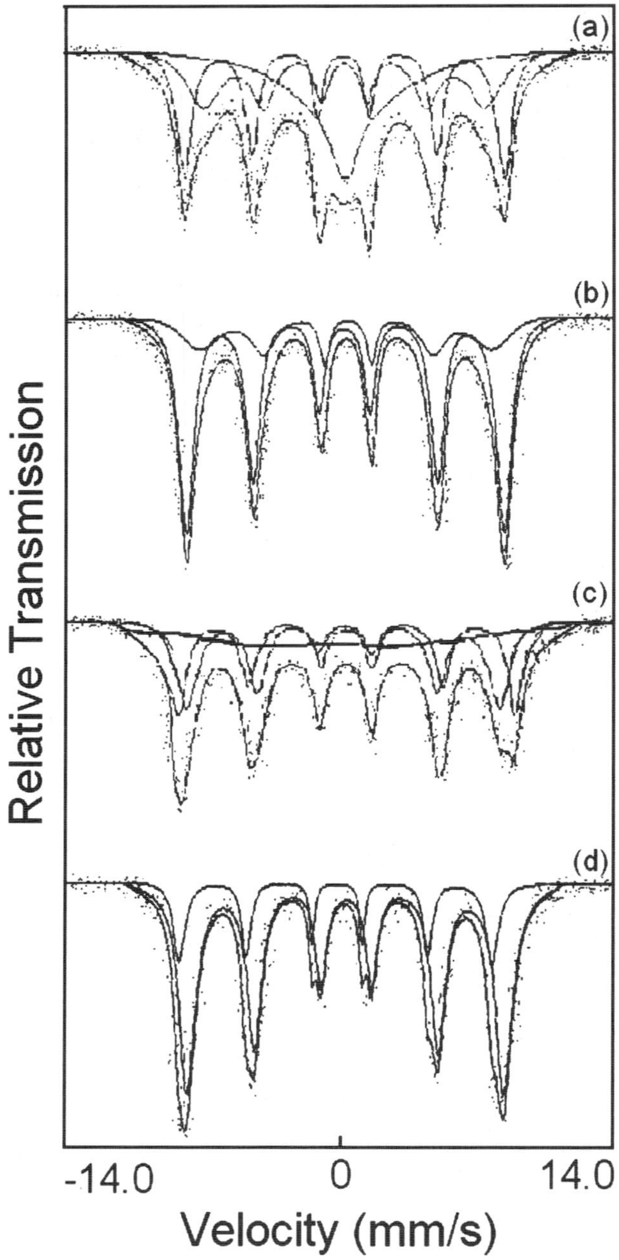

show the aggregation of particles and indicate that the particles are spherical in shape. Figure 2 shows the TEM photograph of 9 nm $CoFe_2O_4$ particles. The distribution of the particle size has been estimated with a variance $\sigma=\pm1.0$ nm.

Figure 3 shows the representative room temperature zero-field Mössbauer spectrum of $CoFe_2O_4$ and $NiFe_2O_4$ particles having sizes of 9 and 30 nm each. The spectra are composed of sextets only without the presence of superparamagnetic doublets. For the 9 nm

Table I Room temperature Mössbauer spectral parameters for nanosize MFe_2O_4 (M=Co, Ni) particles having different sizes

Mössbauer parameters	Subsites	$CoFe_2O_4$				$NiFe_2O_4$			
		9 nm	14 nm	30 nm	[a]	9 nm	21 nm	30 nm	[b]
Hyperfine field (H_{hf}) (T)	A	46.8	43.8	46.4	48.8	46.3	48.6	49.6	48.6
	B	49.9	50.2	50.4	51.5	49.9	49.7	50.2	52.4
	C	18.3	–	–	–	22.2	–	–	–
Quadrupole splitting (ΔE_Q) (mm/s)	A	−0.11	−0.02	−0.10	−0.01	−0.19	0.01	−0.07	–
	B	0.07	−0.01	0.00	0.01	0.10	0.00	−0.02	–
	C	0.25	–	–	–	0.53	–	–	–
Isomer shift (δ) (mm/s)	A	0.20	0.20	0.28	0.28	0.17	−0.17	−0.16	0.26
	B	0.22	0.28	0.30	0.37	0.24	0.29	0.34	0.34
	C	0.41	–	–	–	−2.7	–	–	–
Line widths (Γ) (mm/s)	A	2.3	2.1	2.4	0.51	1.2	1.2	1.1	–
	B	0.82	0.97	0.94	0.40	0.91	1.0	1.4	–
	C	10.3	–	–	–	8.00	–	–	–

[a] [7]; 200–500 nm $CoFe_2O_4$ particles.

[b] [8]; 55 nm $NiFe_2O_4$ particles.

particles, the Mössbauer parameters are obtained by fitting the spectra with two sextets corresponding to A and B sublattices and a third broad subspectra (C) corresponding to the relaxation effects (Table I). For isolated magnetic particles, the subspectra (C) would have been a sharp superparamagnetic doublet at the center but due to the presence of strong interparticle interactions, the relaxation effects give rise to a broad featureless absorption. The strong interparticle interactions are also evident form the large aggregation of particles as observed from TEM picture (Figure 2). For the higher particle sizes, the Mössbauer spectra are fitted with only two sextets corresponding to A and B sublattices. The isomer shift (δ) values for both tetrahedral (δ_{tet}) and octahedral (δ_{oct}) sites indicate that the Fe^{3+} is in high spin state. For both $CoFe_2O_4$ and $NiFe_2O_4$ nanoparticles, no particular trend is observed for the δ_{tet} and δ_{oct} values with increasing particle size. The change in δ values relative to the bulk is attributed to the change in Fe^{3+}–O^{2-} distance correspondingly at A and B sites due to crystal distortion. When the distance of the ligand oxygen to iron is shorter in the oxygen polyhedra, greater is the *s*-electron density at the Fe-nucleus, which results in lowering of isomer shift and vice versa. The quadrupole splitting (ΔE_Q) values are found to be low for 14, 21 and 30 nm particles indicating the overall presence of cubic symmetry at both the sites. However, the larger ΔE_Q values for 9 nm particles are due to the slight distortion from cubic symmetry due to the nanocrystalline nature.

The magnetic hyperfine field (H_{hf}) values for both A and B sites are close to each other. This is because the A and B subsites are magnetically coupled and hence the spins at these sites fluctuate in unison. In contrast, the H_{hf} values at the C site for the 9 nm particles are much lower as compared to the A and B sites, which is due to a much faster relaxation process. Also the H_{hf} values for both A and B sites are found to increase with increasing particle size. The observed decrease in H_{hf} with the decrease in particle size could be explained on the basis of the surface effect and intrinsic size effect models [9, 10]. The probable distribution in H_{hf} is expected over all the nanocrystalline particles. Also, due to the collective magnetic excitations where the magnetic vectors fluctuate near an energy minimum, the H_{hf} values decrease with decreasing particle size. Moreover the increasing concentration of Co^{2+}/Ni^{2+} ions at the octahedral (B) sites, has an effect in reducing the H_{hf}

values. The migration of Co^{2+}/Ni^{2+} ions to octahedral sites induces crystal distortion and this fact is reported in the literature for $CuFe_2O_4$ nanoparticles [11]. This is corroborated by the fact that $NiFe_2O_4$ has an inverse spinel structure and shows ferrimagnetism, that originate from the magnetic moment of the antiparallel spins between Fe^{3+} ions at tetrahedral sites and Ni^{2+} at the octahedral sites [12]. However, $CoFe_2O_4$ is not completely inverse and the degree of inversion depends on the heat treatment temperature.

The Mössbauer parameters are corroborated by the magnetization measurements at 300 K. For $CoFe_2O_4$ nanoparticles, the saturation magnetization (σ_s) are 15.7, 64.5 and 75.6 emu/g for 9, 14 and 30 nm particles, respectively. For $NiFe_2O_4$ nanoparticles, the σ_s values are 33.5, 38.2 and 43.1 emu/g for 9, 21 and 30 nm particles, respectively. These values are little lower than the bulk σ_s values of 93.9 emu/g and 50.0 emu/g for $CoFe_2O_4$ and $NiFe_2O_4$, respectively, measured at room temperature. The decrease may be attributed to the canted spin structure and spin non-collinearity at the surface of these nanoparticles. The coercivity (H_c) values at 300 K for $NiFe_2O_4$ nanoparticles are 0.0005, 0.0065 and 0.0115 T for 9, 21 and 30 nm particles, respectively. For $CoFe_2O_4$ nanoparticles, the H_c values at 300 K are 0.042, 0.109 and 0.137 T for 9, 14 and 30 nm particles, respectively. However, recently a coercivity value of 1.4 T at 5 K is reported for $CoFe_2O_4$ nanoparticles in porous anodic alumina template [13]. Thus, the surface effect, size effect and cation distribution largely influence the Mössbauer and magnetic parameters.

4 Conclusion

In the present study, nanostructured $CoFe_2O_4$ and $NiFe_2O_4$ particles are synthesized by wet chemical method. At 300 K, the Mössbauer spectra are fitted with two sextets corresponding to the tetrahedral (A) and octahedral (B) sites. For the 9 nm particles, an additional broad featureless component is observed because of the relaxation effects. The δ values indicate Fe^{3+} in high spin state and the change in δ values relative to the bulk is attributed to the change in Fe^{3+}–O^{2-} distance correspondingly at A and B sites due to crystal distortion. For the 9 nm particles, the hyperfine field (H_{hf}) is lower than that for the larger particles. The lowering of H_{hf} values is attributed to the surface spin canting effects, nanocrystalline size effects and the crystal distortion due to the migration of Co^{2+}/Ni^{2+} ions to octahedral sites. These observations corroborate the magnetization data measured at 300 K.

Acknowledgement The authors acknowledge MHRD, New Delhi for the financial support in carrying out this research work.

References

1. Lisfi, A., Lodder, J.C., Keim, E.G., Williams, C.M.: Appl. Phys. Lett. **82**, 76 (2003)
2. Dormann, J.L., Fiorani, D. (eds.): Magnetic Properties of Fine Particles. North-Holland, Amsterdam (1992)
3. Hamdeh, H.M., Hikal, W.M., Taher, S.M., Ho, J.C., Thuy, N.P., Quy, O.K., Hanh, N.: J. Appl. Phys. **97**, 064310 (2005)
4. Cullity, B.D.: Introduction to Magnetic Materials. Addison-Wesley, Massachusetts (1978)
5. Gajbhiye, N.S., Balaji, G., Bhattacharyya, S., Ghafari, M.: Hyperfine Interact. **156/157**, 57 (2004)
6. Prasad, S., Vijayalakshmi, A., Gajbhiye, N.S.: J. Therm. Anal. **52**, 597 (1998)

7. Medina-Boudri, A., Bueno-Baqués, D., Fuentes-Cobas, L., Miki-Yoshida, M.: J. Appl. Phys. **87**, 6235 (2000)
8. De Marco, M., Wang, X.W., Snyder, R.L., Simmins, J., Bayya, S., White, M., Naughton, M.J.: J. Appl. Phys. **73**, 6287 (1993)
9. Govaert, A., Dauwe, C., De Grave, E., De Sitter, J.: Solid State Commun. **18**, 389 (1976)
10. Morup, S., Topsoe, H.: Appl. Phys. **11**, 63 (1976)
11. Goya, G.F., Rechenberg, H.R., Jiang, J.Z.: J. Magn. Magn. Mater. **218**, 221 (2000)
12. Kim, C.S., Lee, S.W., Park, S.l., Park, J.Y., Oh, Y.J.: J. Appl. Phys. **79**, 5428 (1996)
13. Jung, J.-S., Lim, J.-H., Choi, K.-H., Oh, S.-L., Kim, Y.-R., Lee, S.-H., Smith, D.A., Stokes, K.L., Malkinski, L., O'Connor, C.J.: J. Appl. Phys. **97**, 10F305 (2005)

Hyperfine Interact (2005) 165: 161–165
DOI 10.1007/s10751-006-9260-2

Mössbauer study of the crystallization products of a $Fe_{75}Zr_{25}$ amorphous alloy

J. S. Garitaonandia · K. Suzuki · J. D. Cashion · E. Goikolea · M. Insausti · F. Plazaola · R. Pizarro · P. Gorria · P. Minguez

Published online: 7 November 2006

Abstract The crystallization of the $Fe_{75}Zr_{25}$ as-milled amorphous alloy has been studied by Mössbauer spectroscopy. The process begins at around 880 K with the complete and fast crystallization of the amorphous phase into the crystalline α-Fe and Fe_2Zr ones but, from around 915 K, there is a sudden phase transformation into the $Fe_{23}Zr_6$ phase, which remains stable after cooling. A fitting for the Mössbauer contribution of the $Fe_{23}Zr_6$ phase, consistent with its crystalline structural features, has been obtained.

Key words FeZr alloys · crystallization process · Mössbauer spectroscopy

1 Introduction

Amorphous Fe-Zr alloys have attracted attention in the recent years because of the great number of different magnetic arrangements and properties which they present [1]. They are the compositional base of the FeZr-based nanocrystalline soft magnetic materials [2]. These are obtained after an adequate thermal treatment of the amorphous precursors, usually produced by the melt-spinning technique, and are basically composed of α-Fe crystallites of 10–20 nm embedded in a remaining amorphous phase. The magnetic softness in nanocrystalline materials is intimately related with the microstructure of the alloy and parameters like the crystallized volume fraction or the grain size of the α-Fe crystallites play a crucial role [3]. So, the study of the crystallization of the amorphous precursors has been, for that reason, the important aim of a great quantity of prior work.

J. S. Garitaonandia (✉) · K. Suzuki · J. D. Cashion
School of Physics and Materials Engineering, Monash University, Building 27, Victoria 3800, Australia
e-mail: garita@we.lc.ehu.es

J. S. Garitaonandia · E. Goikolea · M. Insausti · F. Plazaola · R. Pizarro · P. Minguez
Zientzia eta Teknologia Fakultatea, Euskal Herriko Unibertsitatea, P.O. Box 644, 48080 Bilbao, Spain

P. Gorria
Departamento de. Física, Universidad de Oviedo, Calvo Sotelo s/n, 33007 Oviedo, Spain

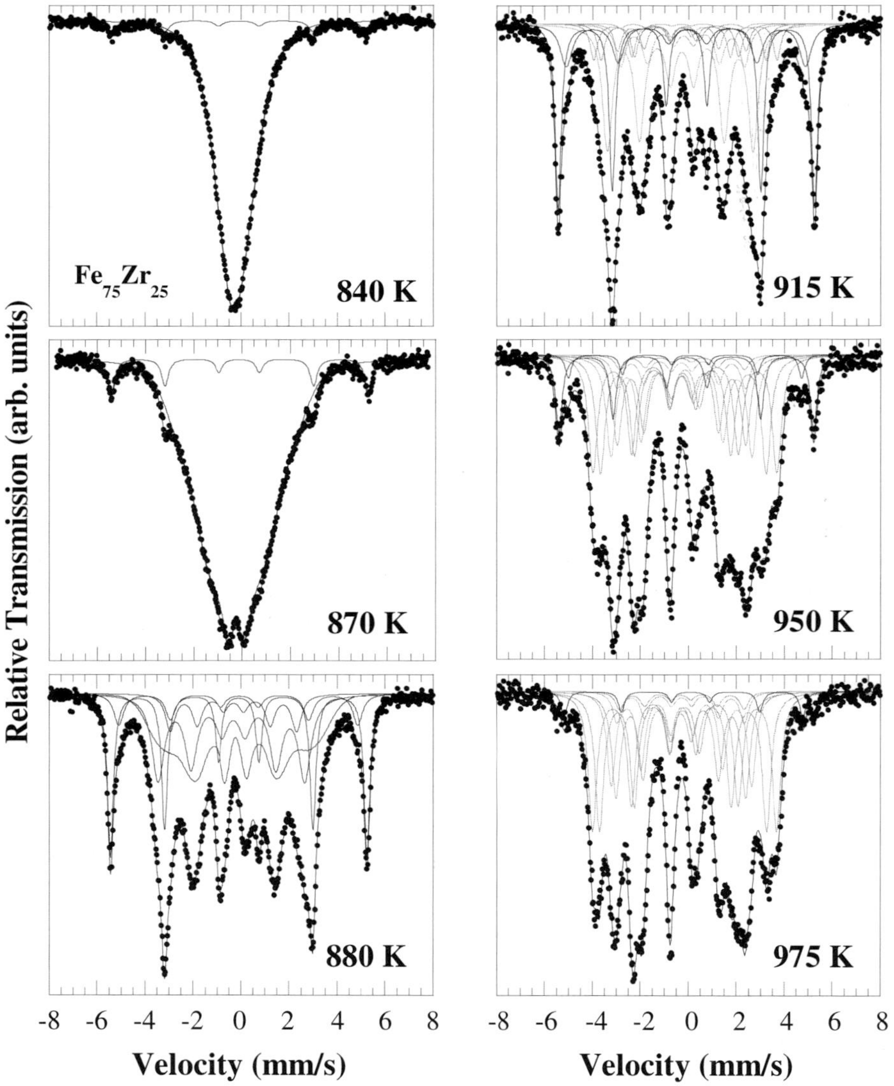

Figure 1 Room temperature Mössbauer spectra of the $Fe_{75}Zr_{25}$ studied alloy after annealing up to the marked temperatures.

As representative of these alloys, the crystallization of the binary $Fe_{90}Zr_{10}$ amorphous ribbon has been extensively studied [4, 5]. However, out of the vicinity of this concentration, the melt-spinning technique cannot provide amorphous ribbons and, so, crystallization of the FeZr amorphous alloys with <88 at.% of Fe has not been deeply studied.

With the objective of complementing a systematic study on the magnetic properties and the crystallization behaviour in Fe-Zr amorphous alloys, we have employed mechanical alloying in order to produce Fe-Zr amorphous alloys out of the eutectic composition range [6]. In this paper we show some preliminary results of such study of the crystallization,

Table I Symmetry, quantity and kind of the different next neighbours for the crystallographic sites in the $Fe_{23}Zr_6$ Fm-3m phase

		Zr	Fe1	Fe2	Fe3	Fe4	Total Fe	Symmetry
Fe1	4b	0	0	0	8	0	8	m-3m
Fe2	24d	4	0	0	4	4	8	m mm
Fe3	32f	3	1	3	3	3	10	0.3 m
Fe4	32f	3	0	3	3	3	9	0.3 m
Zr	24e	4	0	4	4	4	12	4 m m

centering our attention on the Mössbauer spectra of the crystallization products especially on the $Fe_{23}Zr_6$ phase.

2 Experimental section

Some quantities of the $Fe_{75}Zr_{25}$ as-milled amorphous alloy were annealed under high vacuum (10^{-5} Torr) in a MILA 3000 infrared furnace. The annealing procedure consisted of two steps: from room temperature (RT) to 673 K at a heating rate of 4 K/min, and from 673 K to the selected temperatures at a rate of 0.5 K/min. Mössbauer spectra were collected at room temperature in transmission geometry using a conventional constant-acceleration spectrometer with a ^{57}Co-Rh source. The experimental spectra have been fitted by the NORMOS program developed by Brand et al., which allows a simultaneous fit of several crystalline spectra with the possible addition of an amorphous phase [7].

3 Results and discussion

Figure 1 shows room temperature Mössbauer spectra of the sample annealed up to the indicated temperatures. The spectra reflect the different stages of the crystallization, which can be divided into two steps. The first one extends from the beginning of the crystallization up to 915 K, and is characterized by the fast and complete crystallization of the amorphous phase. The second one, located between 915 and 1,100 K consists of a phase transformation of the crystallization products of the first step into a new and different crystalline phase.

According to previous neutron diffraction results, there are only two crystalline phases: α-Fe and Fe_2Zr [8]. Fe_2Zr is a Fd3m cubic Laves phase with two magnetically different Fe sites with 3:1 population ratio when the phase is in the magnetic state. So, the Mössbauer spectra of the samples corresponding to this stage ought to be composed of only three crystalline subspectra besides the amorphous one. However, we have been forced to add a fourth crystalline subspectrum to fit a small, 30 T contribution located close to the α-Fe one.

The second step shows a phase transformation from the α-Fe and Fe_2Zr phases to a new crystalline one which was identified as $Fe_{23}Zr_6$ [9]. This phase becomes dominant as the temperature increases and the phase transformation advances. This process can be also visualized in the Mössbauer spectra by following the decrease of the peaks around ±6 mm/s belonging to the α-Fe phase and the appearance and increase of the peak around −4 mm/s with the increased annealing temperature.

Table II Values of the hyperfine parameters of the Fe_2Zr, $Fe_{23}Zr_6$ and α-Fe crystalline phases obtained from the fittings of the Mössbauer spectra at the different stages of the crystallization process of the $Fe_{75}Zr_{25}$ as-milled alloy

	Fe_2Zr		$Fe_{23}Zr_6$				α-Fe	
	Fe1	Fe2	Fe1	Fe2	Fe3	Fe4	Fe1	Fe2
Bhf (T)	18.3(1)	16.5(1)	16.0(1)	16.6(2)	23.8(1)	21.5(2)	33.0(1)	30.5(2)
QS (mm/s)	−0.03(2)	−0.04(3)	0.03(1)	−0.01(1)	−0.01(1)	0.02(2)	0.00(1)	−0.1(1)
IS (mm/s)	−0.18(2)	−0.19(2)	−0.18(1)	−0.19(2)	−0.04(1)	−0.13(1)	0.00(1)	0.03(3)
HWHM (mm/s)	0.46(4)	0.46(4)	0.47(3)	0.47(3)	0.47(3)	0.47(3)	0.31(2)	0.63(2)

The figure between brackets represents the fluctuation of the value among eight different spectra (see text).

The published crystallographic data describe the structure of the $Fe_{23}Zr_6$ phase as Fm-3m cubic with a lattice parameter of 11.69 Å and with one crystallographic site for Zr and four different ones for Fe atoms [10]. We have not found any reference in the literature related to the Mössbauer spectrum of the $Fe_{23}Zr_6$ phase with the Fm-3m cubic structure. In Table I we show the main crystallographic characteristics of the different Fe sites in this structure. A previous knowledge of the characteristics of the Fe sites is crucial for a correct interpretation of the fitting results. For instance, from Table I we notice that positions Fe3 and Fe4 have the same population (32) and, therefore, their subspectra might present the same intensity. Besides they are the positions with the largest number of Fe atoms as nearest neighbours, 10 and 9, respectively, what suggests that the subspectra with the highest hyperfine field values will correspond to these sites. On the contrary, the Fe1 and Fe2 sites both present only eight Fe atoms in their nearest neighbourhoods and, following the same arguments, their subspectra will present similar, and smaller, magnetic field values. Thus, the Mössbauer spectra of the samples at this stage of the crystallization are composed of at least seven subspectra (four for $Fe_{23}Zr_6$, two for Fe_2Zr and one for α-Fe) and, despite the basic help from the structural parameters, to avoid any misinterpretation of the fitting results as a consequence of the overlap between subspectral components of different phases, other constraints must be used. In the fittings, we have established the following conditions:

(a) The intensity ratio of the subspectra in each phase has been kept proportional to the ideal site occupancy of Fe atoms, that is, 32:32:24:4 for the components of the $Fe_{23}Zr_6$ phase and 3:1 for the subspectra corresponding to Fe_2Zr phase.
(b) For each phase, the widths of the subspectra have been linked to the same value.
(c) We have not taken into consideration any subspectrum with a contribution of less than 2% of the intensity of the total spectrum.

The values of the hyperfine parameters of the different phases obtained after the fittings of the spectra are summarized in Table II.

With respect to the $Fe_{23}Zr_6$ phase, we have assigned the spectra to the different Fe sites in accordance with the structural data shown in Table I, as was explained above. The feature which best supports the assignments and the reliability of the fitting results is the repeatability of the values obtained for the hyperfine parameters of the different crystalline phases in each spectrum in Figure 1. We must emphasize that the figures between brackets in Table II correspond to the fluctuation of the value from the results of the fittings of the Mössbauer spectra of the samples at different stages of the crystallization process. So, even when the contributions of the different phases to the spectra change as the transformation process advances, the spectra corresponding to this step are always very well fitted by the same values of the hyperfine parameters. Therefore, we can conclude that the values shown in Table II are the characteristic Mössbauer parameters of $Fe_{23}Zr_6$ phase. In the same way, we have not observed any noticeable change between the values of the hyperfine parameters of the Fe_2Zr phase. The values shown in Table II are representative for this phase in both steps of the crystallization. There is a great similarity between the values of the hyperfine parameters obtained from the fittings and those found in the literature though, in our case, the values of hyperfine field are around 2–3 T lower [11].

Acknowledgments This work was undertaken under projects MAT2002-4087-C02-01 and UPV 224.310-14553/2002 and support from the Australian Research Council. JSG thanks MECD for support under the program Movilidad de Profesores de Universidad.

References

1. Unruh, K.M., Chien, C.L.: Phys. Rev., B **30**, 4968 (1984)
2. Suzuki, K., Kataoka, N., Inoue, A., Makino, A., Masumoto, T.: Mater. Trans., JIM **31**, 743 (1990)
3. Herzer, G.: IEEE Trans. Magn. **MAG-25**, 3327 (1989)
4. Zarubova, N., Moser, N., Kronmüller, H.: Mat. Sci. Eng. **A151**, 205 (1992)
5. Fujinami, M., Ujihira, Y.: J. Appl. Phys. **59**, 2387 (1986)
6. Pizarro, R., Garitaonandia, J.S., Plazaola, F., Barandiaran, J.M., Grenèche, J.M.: J. Phys., Condens. Matter **12**, 3101 (2000)
7. Brand, R.A., Lauer, J., Herlach, D.M.: J. Phys. F. Met. Phys. **14**, 555 (1984)
8. Gorria, P., Garitaonandia, J.S., Pizarro, R., Campo, J.: In: Cicognani, G., Vettier, C. (eds.) ILL 2001 Annual Report, p. 48. (2002)
9. Gorria, P., Garitaonandia, J.S., Pizarro, R., Martinez-Blanco, D., Campo, J., Plazaola, F.: Physica B **350**, e1075 (2004)
10. Kripyakevich, P.I., Protasov, V.S., Cherkashin, E.E.: Zh. Neorg. Khi. **10**, 151 (1965)
11. Wertheim, G.K., Jaccarino, V., Wernick, J.H.: Phys. Rev. **135**, A151 (1964)

Hyperfine Interact (2005) 165: 167–174
DOI 10.1007/s10751-006-9261-1

^{57}Fe Mössbauer studies of amorphous $Pd_{40}(NiFe)_{40}P_{20}$ alloys

Susan H. Kilcoyne · Denis Greig · Michael N. Gona

Published online: 31 October 2006

Abstract Zero-field ^{57}Fe Mössbauer spectroscopy has been used to study the magnetic properties of an amorphous melt–spun $Pd_{40}Ni_{40-x}Fe_xP_{20}$ (x=5, 10, 17.5, 20) series. The alloys show quadrupole split doublets at room temperature indicating paramagnetic behaviour. At low temperatures the Mössbauer spectra are sextets due to the onset of long range order. The transition temperatures increase with Fe concentration from 12±2 K for $Pd_{40}Ni_{35}Fe_5P_{20}$ to 42.0±2.0 K for $Pd_{40}Ni_{20}Fe_{20}P_{20}$. The hyperfine field was found to follow $\mathrm{Hf}(0)(1-\mathrm{T}/\mathrm{T_c})^{\beta}$ over the critical region with the β exponent being close to 0.4.

Key words Mössbauer spectroscopy · amorphous magnetism · metallic glasses · spin glasses

1 Introduction

Metallic glass systems lack the long-range lattice periodicity found in crystalline alloys. Their solid state properties, which are largely determined by their structure, atomic electron states and interactions between atoms are uniquely different from their crystalline counterparts. Magnetism without long-range crystalline ordering (amorphous magnetism) was realized in the 1960s following the successful production of amorphous alloys of Transition Metal, TM, (Fe, Ni, Co, etc.), Metalloid, M, (P, B, C, Si, etc.) and in some cases with Rare-Earth, RE, elements [1]. Since then it has been a subject of growing scientific

S. H. Kilcoyne (✉)
Institute for Materials Research, University of Salford,
Salford M5 4WT, UK
e-mail: s.h.kilcoyne@salford.ac.uk

D. Greig · M. N. Gona
School of Physics and Astronomy,
University of Leeds, Leeds LS2 9JT, UK

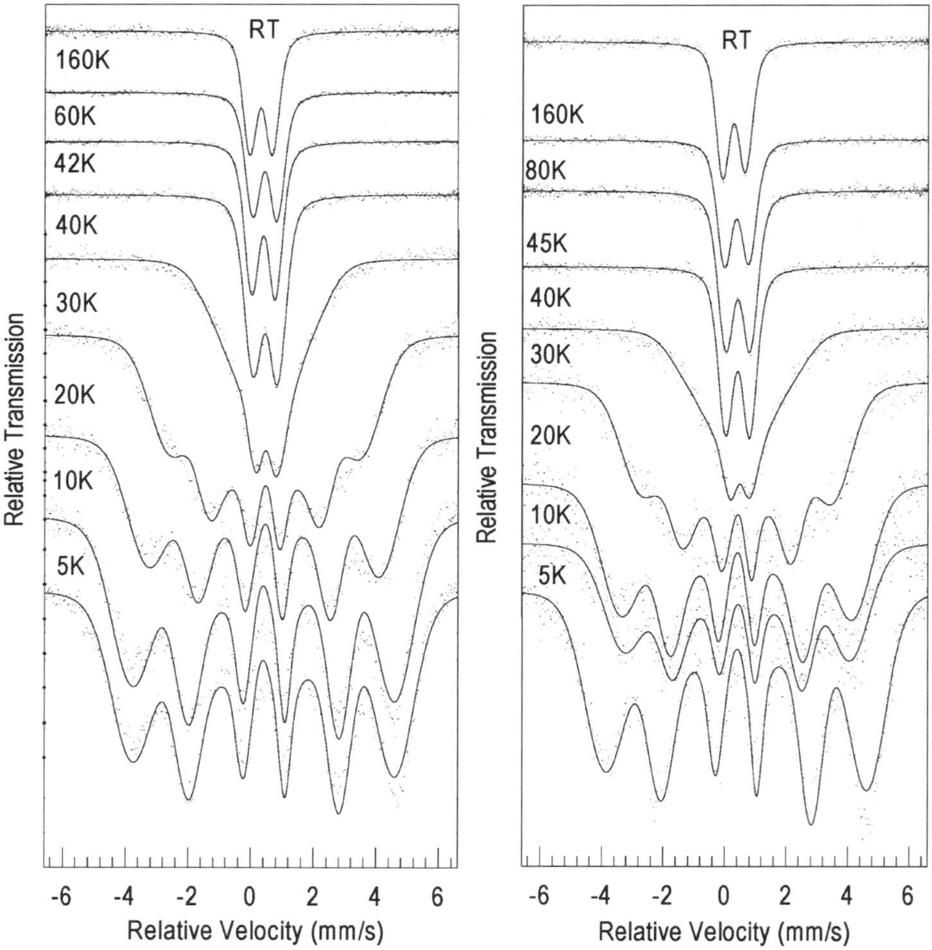

Figure 1 ZF Mössbauer spectra of amorphous $Pd_{40}Ni_{40\text{-}x}P_{20}$ (x=17.5 (right), 20 (left)) alloys.

interest, especially the magnetic properties of TM–M alloys [2, 3], which can be made magnetically soft. In addition to their superior physical properties compared to conventional crystalline alloys these metallic glasses have also become important for magnetic applications such as in transformer cores and motors.

A number of ternary and higher-order alloy compositions have been formed as bulk metallic glasses. These multi-component alloys, for example, Pd-Ni-P [4], Au-Pd-Sb [5], Zr-Ti-Cu-Ni-Be [6], Fe-(Al, Ga, Ge, Co, Cr, Nb, Mo)-(P, C, B, Si) [7] have low critical cooling rates sufficient to suppress crystallization in an under-cooled melt enabling the synthesis of bulk metallic glasses. The metallic alloy $Pd_{40}Ni_{40}P_{20}$ forms close to a low eutectic point in the temperature phase diagram and hence can readily be liquid-quenched into a metastable amorphous state. Although amorphous $Pd_{40}Ni_{40}P_{20}$ is paramagnetic at all temperatures, substituting Ni-Fe atomic species the amorphous $Pd_{40}(NiFe)_{40}P_{20}$ alloy exhibits temperature and field dependent magnetic phase transitions. Shen et al. have

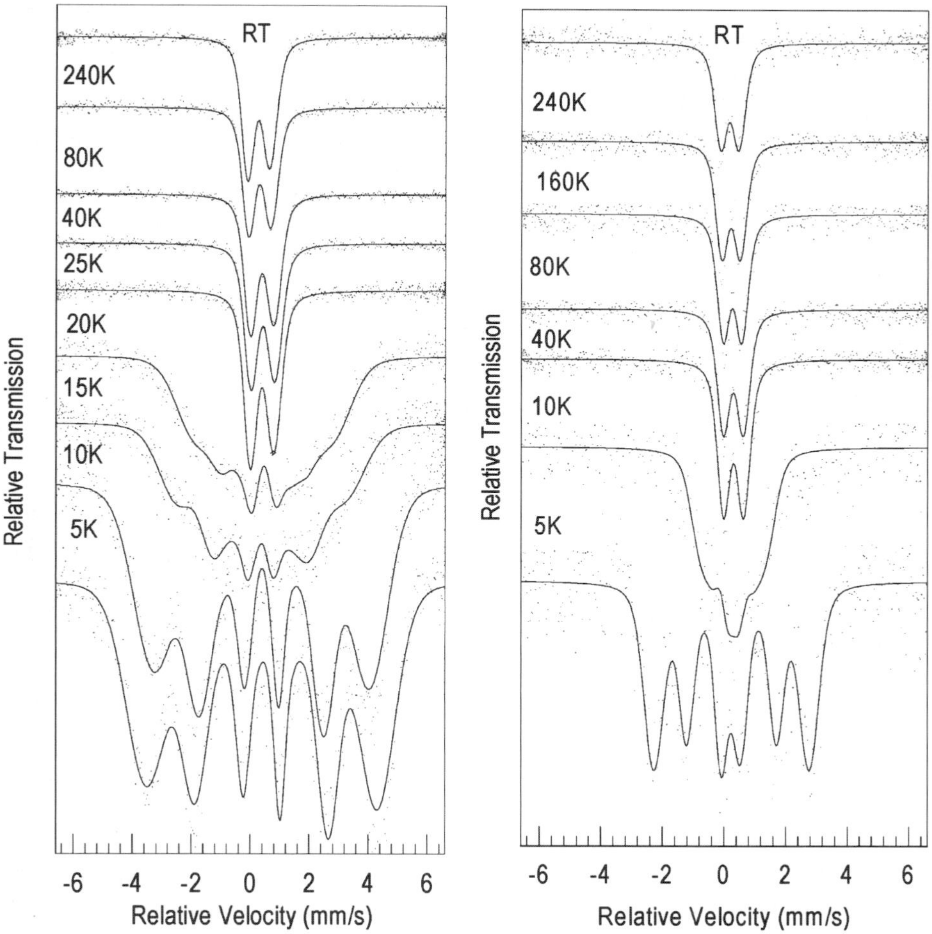

Figure 2 ZF Mössbauer spectra of amorphous $Pd_{40}Ni_{40-x}Fe_xP20$ (x=5 (right), 10 (left)) alloys.

studied the magnetic properties of bulk amorphous $Pd_{40}Ni_{22.5}Fe_{17.5}P_{20}$ as a function of temperature and applied magnetic field [8, 9]. However, the alloy has been investigated almost exclusively using high field magnetization techniques and there are few, if any, studies of this sample in zero-field. Also, there are no systematic studies of the amorphous $Pd_{40}Ni_{40-x}Fe_xP_{20}$ series as a function of Fe concentration.

Here we present the results of a zero-field ^{57}Fe Mössbauer spectroscopy and an AC magnetic susceptibility study of melt–spun amorphous $Pd_{40}Ni_{40-x}Fe_xP_{20}$ in the Fe concentration range $5 \leq x$ atm.%≤ 20.

2 Experimental

Elemental ingots of Pd(99.95%), Ni(99.95%) and Fe(99.98%) were mixed with powders of FeP(99.95%), Fe_2P(99.95%) and Ni_2P(99.98%) in order to prepare four compositions of the

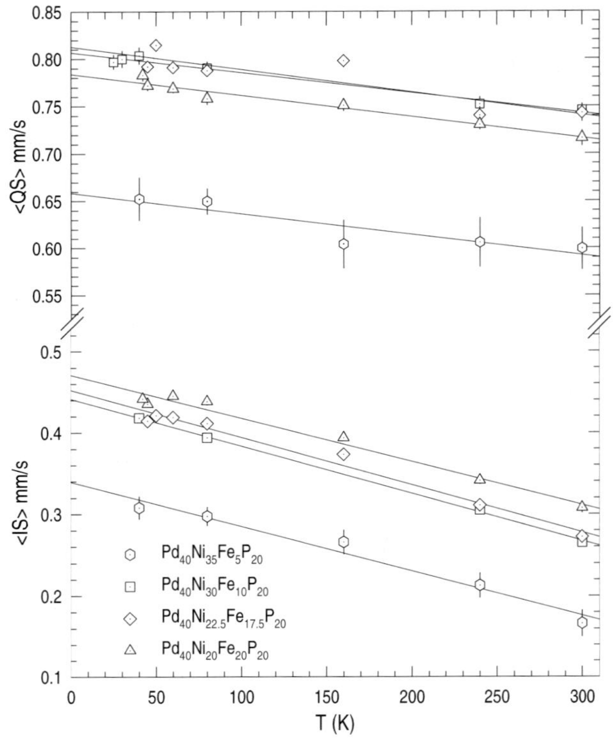

Figure 3 Temperature dependence of the isomer shift <*IS*> and quadrupole splitting <*QS*> (*solid lines* are fits described in the text).

$Pd_{40}Ni_{40-x}Fe_xP_{20}$ series; x=5, 10, 17.5, 20. The mixture was placed in an 18 mm diameter silica tube and heated in an Ar atmosphere to form ingots. These ingots were then re-heated several times in an Ar-arc furnace to ensure uniform alloying of all the constituents. They were subsequently melt–spun on a high speed rotating copper wheel to form amorphous ribbons measuring 1 mm and ~25 μm in width and thickness, respectively. X-ray and high intensity neutron diffraction (D20 at ILL, Grenoble) verified the amorphous structure of the alloys. AC magnetic susceptibility measurement were carried out using a MagLab (Oxford Instruments) system. Approximately 0.3 g of amorphous ribbon was cooled from 300 K in zero field to ~2 K, well below the magnetic transition. The susceptibility data was collected on warming with an AC excitation field of 10 Oe applied along the length of the ribbons. Mössbauer absorbers were prepared by cutting the ribbons and clamping them parallel to each other covering an area of 18 × 18 mm. The samples were mounted in a conventional constant-acceleration spectrometer with a 50 mCi ^{57}Co–Rh source and the Mössbauer experiment was done in transmission geometry. The temperature of the samples was controlled using a Variox (Oxford Instruments) cryostat with a temperature stability of <0.2 K using flow rate adjustment and feedback heating. The samples were cooled to 5 K in zero-field conditions, and Mössbauer spectra were collected at various temperatures between 5 and 300 K in zero-field during warming. The room temperature (~300 K) calibration spectrum was collected from a standard α-Fe foil absorber (18 × 18 mm). The hyperfine interaction parameters derived in the analysis are quoted relative to the α-Fe foil parameters at 300 K.

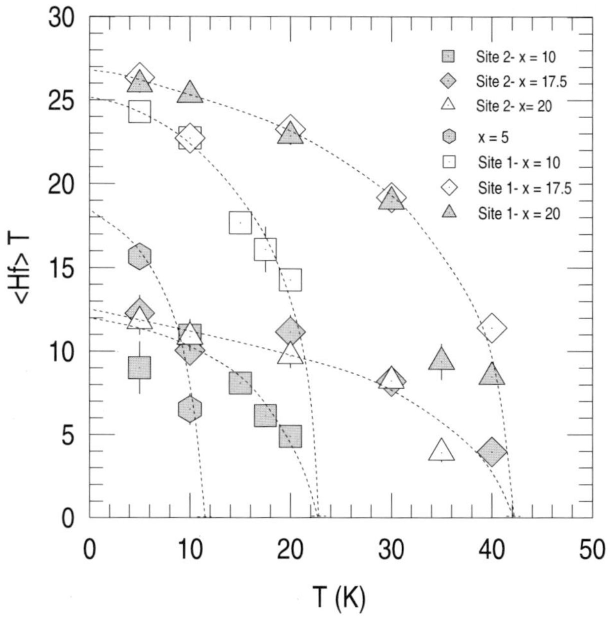

Figure 4 Temperature dependence of the hyperfine field of amorphous $Pd_{40}Ni_{40-x}Fe_xP_{20}$ alloys.

3 Results and discussion

The zero-field Mössbauer spectra (experimental data and fits) collected between 300 and 5 K for the four amorphous $Pd_{40}Ni_{40-x}Fe_xP_{20}$ alloys are shown in Figures 1 and 2. At 300 K the Mössbauer spectra of all the samples are broad paramagnetic doublets. As the temperature is lowered the Mössbauer spectra gradually broaden and change from paramagnetic doublets into sextets, consistent with the appearance of long range magnetic order. The phase transition in these metallic glasses is gradual compared to related crystalline alloys [12, 13]. This could be due to (1) a large number of structural inequivalent Fe sites [14–18] and (2) competing interactions between isolated magnetic moments or clusters.

The spectra were fitted using the extended Voigtian profile option in RECOIL; a generalised analytic line shape calculation where the parameter distributions are represented as a sum of independent Gaussian components [10, 11]. This method has been applied with the assumption that, for the quadrupole splitting and hyperfine field distributions, there is linear coupling between primary and secondary hyperfine parameters. Each parameter is expressed by an average value and a finite probability distribution.

At high temperatures all the spectra were well fitted with a single broadened quadrupole split doublet. At 300 K the isomer shift of this doublet increases as the Fe concentration, x, increases from 0.17(2) mm/s for x=5 to 0.307(6) mm/s for x=20. As the isomer shift is a direct measure of the local electronic charge density at an Fe site the changes observed upon alloying provides valuable information about the corresponding changes in electronic properties of the alloy. The quadrupole splitting also increases with x ranging from 0.60(2) mm/s for x=5 to 0.746(6) mm/s for x=20. The temperature dependence of the isomer shift

Table I Transition temperatures and critical parameters of $Pd_{40}Ni_{40-x}Fe_xP_{20}$ alloys

Sample	T_g (K)	T_c (K)		β	D
$Pd_{40}Ni_{35}Fe_5P_{20}$	8.8(0.5)	12.2(2.0)		–	–
$Pd_{40}Ni_{30}Fe_{10}P_{20}$	17.6(0.5)	23.0(2.0)	Site 1	0.33(0.04)	1.09(0.04)
			Site 2	0.37(0.13)	1.26(0.17)
$Pd_{40}Ni_{22.5}Fe_{17.5}P_{20}$	25.8(0.5)	41.0(2.0)	Site 1	0.30(0.05)	1.01(0.04)
			Site 2	0.38(0.09)	1.03(0.07)
$Pd_{40}Ni_{20}Fe_{20}P_{20}$	33.4(0.5)	42.0(2.0)	Site 1	0.38(0.08)	1.08(0.08)
			Site 2	0.46(0.10)	1.08(0.08)

and quadrupole splitting above the magnetic transition together with linear regression fits are shown in Figure 3. Both parameters increase as the temperature decreases.

At low temperatures the spectra were fitted using two sextets; hyperfine splitting arising from two distinct Fe environments with isomer shift, quadrupole splitting and hyperfine field as variable parameters. The temperature dependence of the magnetic hyperfine field is shown in Figure 4. We have defined the magnetic ordering temperature, T_c as the temperature at which the sextets disappear on warming. The variation of T_c as a function of Fe concentration is presented in Table I. T_c varies relatively slowly with x for high Fe concentration ($x \geq 17.5$) but decreases sharply for low x. The temperature T_c approaches 0 K at a finite value of Fe concentration, thus suggesting a critical Fe concentration, x_c' above which magnetic order is observed. In this system x_c appears to be around 0.5% Fe. The rapid decrease in T_c at low x and the appearance of critical concentration x_c are similar to the dilute ferromagnetic systems, for example Fe–Au [19, 20], $(Fe_xNi_{1-x})_{79}P_{14}B_8$ [21] and $Ce(Fe_xNi_{1-x})_2$ [22].

It is clear from the figure that the variation in hyperfine field does not follow a Brillouin curve. However both theory and experiment [23] suggest that the following relationship (1) holds near the critical temperature T_c:

$$\frac{\mathrm{Hf}(T)}{\mathrm{Hf}(0)} = D\left(1 - \frac{T}{T_c}\right)^{\beta} \tag{1}$$

where Hf(T) and Hf(0) are local Fe-site average fields from hyperfine splitting and saturation (T=0 K), respectively, the β and D are the critical constants for a particular material. The parameters β and D are listed in Table I alongside the transition temperatures of the alloys.

The temperature dependences of the real and imaginary components of the AC magnetic susceptibility have been measured for the four amorphous $Pd_{40}Ni_{40-x}Fe_xP_{20}$ alloys. The real χ' and the imaginary χ'' components are shown in Figure 5. The curves are consistent with the material being a cluster glass or spin glass. The spin freezing temperatures are identified by examining the temperature behaviour of χ' curves. T_g increases with the Fe concentration occurring at 8.8 and 33.4 K for the x=5 and x=20 alloys, respectively (Table I).

It is interesting to note that there are no features in the susceptibility at the transition temperatures determined by Mössbauer spectroscopy. One possible explanation for these results is the formation of superparamagnetic clusters as the temperature is decreased, with the clusters forming at T_c and freezing at T_g.

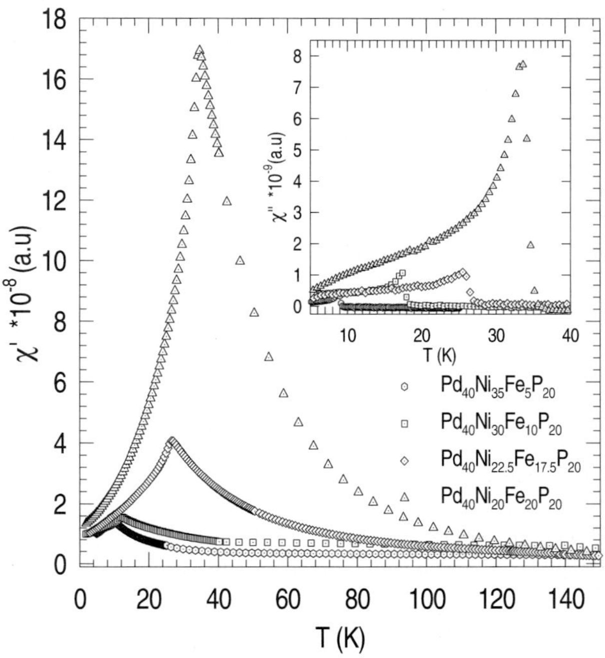

Figure 5 Temperature dependence of the real (χ') and (*inset*) imaginary (χ'') AC susceptibility of amorphous $Pd_{40}Ni_{40\text{-}x}Fe_xP_{20}$ alloys.

4 Conclusion

The magnetic properties of amorphous $Pd_{40}Ni_{40-x}Fe_xP_{20}$ series have been investigated by zero-field Mössbauer spectroscopy and AC susceptibility. The magnetic transition observed in Mössbauer measurements occur at higher temperatures than the spin freezing temperature derived from AC susceptibility. This behaviour, which has also been observed in some dilute Fe concentration alloys (Fe–Au) [19, 20] can be explained by cluster formation. The higher values of the critical exponents β close to 0.4 lie within the range $0.36 \leq \beta \leq 0.41$ predicted from a Heisenberg model.

References

1. Duwez, P.: Trans. Am. Soc. Metals **60**:607 (1967)
2. Bagley, B.G., Turnbull, D.: J. Appl. Phys. **39**:5681 (1968)
3. Davis, H.A.: In: Luborsky, F.E. (ed.) Amorphous Metallic Alloys, p. 8. Butterworth, London (1983)
4. Drehman, A.J., Greer, A.L., Turnbull, D.: Appl. Phys. Lett. **41**:716 (1982)
5. Lee, M.C., Kendall, J.M., Johnson, W.L.: Appl. Phys. Lett. **40**:382 (1982)
6. Peker, A., Johnson, W.L.: Appl. Phys. Lett. **63**:2342 (1993)
7. Inoue, A.: Mater. Sci. Eng. A **226–228**:357 (1997)
8. Shen, T.D., He, Y., Schwarz, R.B.: J. Mater. Res. **14**:2107 (1997)
9. Shen, T.D., Schwarz, R.B., Thompson, J.D.: J. Appl. Phys. **85**:4110 (1999)
10. Lagarec, K., Rancourt, D.G.: Nucl. Instrum. Methods Phys. Res., B**129**:266 (1997)
11. Lagarec, K., Rancourt, D.G.: Recoil Spectral Analysis Software, http://www.science.uottawa.ca/phy/~recoil
12. Longworth, G.: Phys. Rev. **172**:172 (1968)
13. van Deen, J.K., Dijkhuizen, D.J., van der Woude, F.: J. Magn. Magn. Mater. **15–18**:1171 (1980)

14. Lynn, J.W., Shirane, G., Birgeneau, R.J., Chen, H.S.: AIP Conf. Proc. **34**:313 (1976)
15. Rivier, N., Continentino, M.A.: J. Magn. Magn. Mater. **15–18**:1419 (1980)
16. Birgeneau, R.J., Tarvin, J.A., Shirane, G., Gyorgy, E.M., Sherwood, R.C., Chen, H.S., Chien, C.: Phys. Rev. **B18**:2192 (1978)
17. Bhagat, S.M., Spano, M.L., Chen, H.S., Rao, K.V.: Solid State Commun. **33**:303 (1980)
18. Chien, C.L., Musser, D., Gyorgy, E.M., Sherwood, R.C., Chen, H.S., Luborsky, F.E., Walter, J.L.: Phys. Rev. **B20**:283 (1979)
19. Gonser, U., Grant, R.W., Meechan, C.J., Muir, A.H., Wiedersich, H.J.: J. Appl. Phys. **36** (1965)
20. Violet, C.E., Borg, R.J.: Phys. Rev. **149**:540 (1966)
21. Durand, J.: In Amorphous Magnetism II. Plenum, New York (1977)
22. Harris, I.R., Longworth, G.J.: J. Less-common Met. **45**:63 (1976)
23. Kadanoff, L.P., et al.: Rev. Mod. Phys. **39**:395 (1967)

Hyperfine Interact (2005) 165: 175–181
DOI 10.1007/s10751-006-9262-0

Mössbauer spectroscopical investigation of amorphous Fe–Y alloy ribbons prepared by melt spinning

B. Sahoo · V. Kuncser · W. Keune · J. Pieper

Published online: 19 October 2006

Abstract Fe–Y amorphous alloy ribbons were prepared by the melt spinning method and characterized by X-ray diffraction, Mössbauer spectroscopy and inelastic neutron scattering. X-ray diffraction demonstrates that the $Fe_{0.7}Y_{0.3}$ ribbons are completely amorphous, whereas the $Fe_{0.3}Y_{0.7}$ ribbons contain a small fraction of crystalline Y precipitates in the amorphous Fe–Y matrix. Mössbauer spectroscopy between 4.2 to 300 K reveals the amorphous nature of the Fe–Y matrix and the $Fe_{0.7}Y_{0.3}$ ribbons. The preliminary neutron scattering results S(Q, ω) show excess low energy vibrational modes which gives rise to the so called "boson peak" in this amorphous material.

Key words Fe–Y amorphous alloy · Mössbauer spectroscopy · atomic vibrational dynamics · sound velocity · inelastic neutron scattering

1 Introduction

The atomic vibrational dynamics of amorphous binary alloys has attracted a great deal of interest because of the excess vibrational density of states (VDOS), g(E), at low vibrational excitation energies (<10 meV) [1–5]. The excess VDOS, which

B. Sahoo (✉) · V. Kuncser · W. Keune
Laboratory of Applied Physics, University of Duisburg-Essen, D-47048 Duisburg, Germany
e-mail: balaram_sahoo@uni-duisburg.de

V. Kuncser
National Institute of Physics of Materials, RO-76900 Bucharest-Magurele, Romania

J. Pieper
Berlin Neutron Scattering Center (BENSC), HMI, D-14109 Berlin, Germany

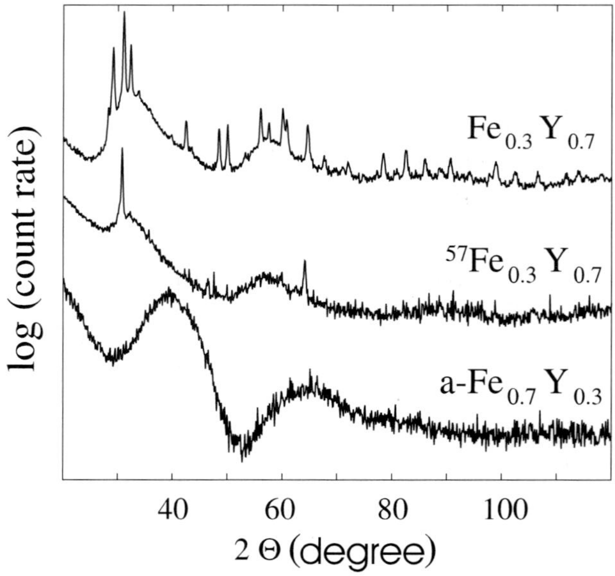

Figure 1 X-ray diffraction pattern of the $^{nat}Fe_{0.3}Y_{0.7}$, $^{57}Fe_{0.3}Y_{0.7}$, and $^{nat}Fe_{0.7}Y_{0.3}$ alloy ribbons. (Cu–K_{α} radiation).

shows up as a peak in the plot of $g(E)/E^2$ versus E, is known as the "boson peak." The boson peak is a characteristics of amorphous materials and is absent in the crystalline state. The unusual excitations do not obey the usual Debye law, $g(E) \propto E^2$. For the study of the behavior of the boson peak, suitable intermetallic amorphous samples are required. Amorphous and/or nanocrystalline intermetallics may be obtained by non-conventional metallurgical methods. Such methods lead to non-equilibrium (metastable) structures due to an additional free energy kept in the system during the preparation procedure. The excess free energy (compared with the equilibrium thermodynamic state) is "frozen" in the metastable system by the fast variation of an intensive thermodynamic parameter, e.g., the temperature in the case of ultrafast solidification of metallic melts. Usual cooling rates during such preparation procedures are about 10^5–10^6 K/s.

The magnetic properties of amorphous (a-) Fe–Y alloys are well investigated [6–11]. The magnetic order in Fe-rich a-Fe–Y alloys is ferromagnetic and of spin-glass type at low temperatures [12]. Although the magnetic properties of a-Fe–Y alloys are well investigated, the study of the vibrational dynamics of amorphous Fe–Y alloys are rare. The present work reports on the preparation and characterization of a-Fe–Y alloys devoted to the future study of the boson peak in the VDOS by inelastic neutron scattering (INS) and nuclear resonant inelastic X-ray scattering (NRIXS). Further, the average sound velocity, V_s, of these materials has been determined. The estimation of the sound velocity is important for a correct characterization of the intensity of the boson peak. Preliminary INS results will also be reported.

2 Sample preparation and characterization

Fe–Y amorphous alloy ribbons were prepared by the melt spinning (single roller) method, in Ar atmosphere. Primary alloy ingots were obtained from high purity

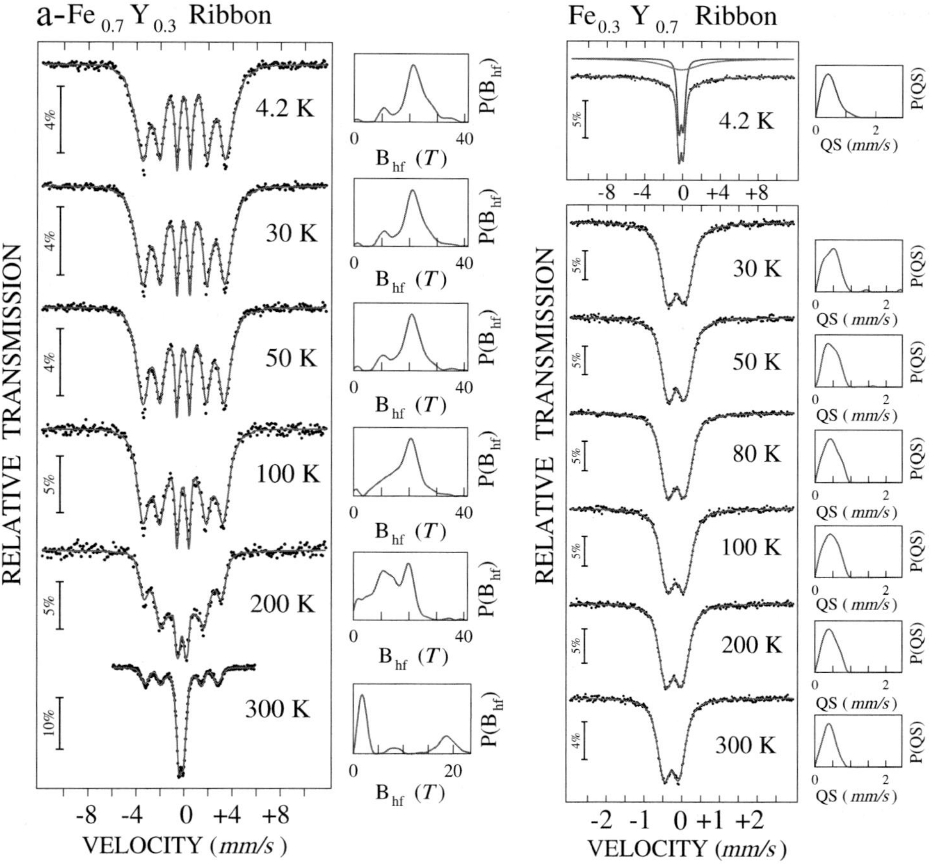

Figure 2 Mössbauer spectra of the $^{nat}Fe_{0.7}Y_{0.3}$ (*right*), and $^{nat}Fe_{0.3}Y_{0.7}$ (*left*) alloy ribbons taken at different temperatures between 4.2 and 300 K. The spectra of the $^{nat}Fe_{0.7}Y_{0.3}$ alloy ribbons (*left*) have been fitted by a distribution of magnetic hyperfine fields, $P(B_{hf})$(*right hand side*). The spectra of $^{nat}Fe_{0.3}Y_{0.7}$ alloy ribbons (*right*) have been fitted by a distribution of quadrupole splittings P(QS). An additional broad singlet has been used for the fitting of the spectrum of $^{nat}Fe_{0.3}Y_{0.7}$ alloy ribbons measured at 4.2 K.

elements (99.99% purity for Y and 99.995% for natural Fe) by induction melting in a special chamber and in Ar protecting atmosphere. Moreover, some ribbons were prepared from ^{57}Fe with 40 % enrichment in ^{57}Fe. The alloys were further induction melted in fused-silica crucibles, and then the melt was ejected through a small hole onto a rotating copper wheel at room temperature. Linear quenching speeds of about 20 m/s were used to obtain rapidly quenched ribbons, tens of centimeters long and with an average cross section of about 2 mm (width) $\times$ 50 μm (thickness).

Amorphous $^{nat}Fe_{0.7}Y_{0.3}$, $^{nat}Fe_{0.3}Y_{0.7}$ and $^{57}Fe_{0.3}Y_{0.7}$ alloy ribbons were characterized by X-ray diffraction (XRD) and transmission Mössbauer spectroscopy at different temperatures between 4.2 K and room temperature (RT). The XRD pattern (Figure 1) for $^{nat}Fe_{0.7}Y_{0.3}$ contains no sharp diffraction peak and indicates the amorphous nature of these ribbons. However, the $^{nat}Fe_{0.3}Y_{0.7}$ and $^{57}Fe_{0.3}Y_{0.7}$

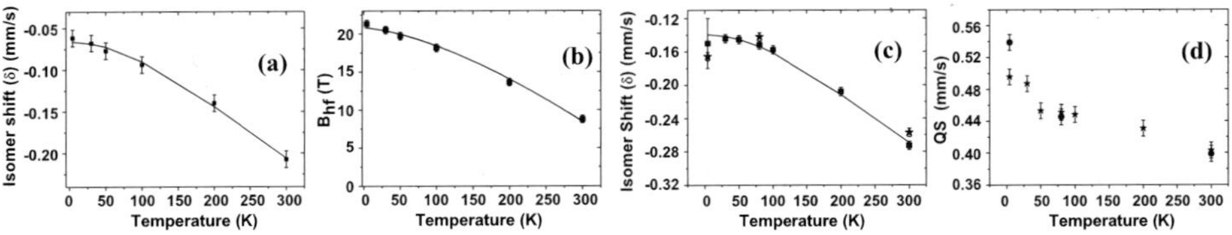

Figure 3 T-dependence of the average isomer shift δ (**a**) and average magnetic hyperfine field B_{hf} (**b**) of the $^{nat}Fe_{0.7}Y_{0.3}$ alloy ribbons; T-dependence of the average isomer shift δ (**c**) and average quadrupole splitting QS (**d**) of the $^{nat}Fe_{0.3}Y_{0.7}$ alloy ribbons. The data for δ(T) are *least-squares* fitted according to Debye model (solid line). The data for B_{hf}(T) are *least-squares* fitted by Bloch's $T^{3/2}$ law (solid line). The *asterisks* in (**c**) and *solid circles* in (**d**) represent the isomer shift and quadrupole splittings, respectively, for the $^{57}Fe_{0.3}Y_{0.7}$ ribbons.

ribbons show a small fraction of crystalline Y precipitates in the amorphous Fe–Y matrix. Hence, XRD suggests that completely amorphous Fe–Y bulk alloys can be prepared only in the higher Fe concentration region. The Mössbauer spectra were least-squares fitted by using the computer program "NORMOS" [13, 14]. The average sound velocity of the $^{nat}Fe_{0.3}Y_{0.7}$ alloy ribbons was measured via a laboratory made set-up based on Hooke's law.

3 Mössbauer spectroscopical results

The Mössbauer spectra of the a-$^{nat}Fe_{0.7}Y_{0.3}$ ribbons (Figure 2 (left)) are fitted by a magnetic hyperfine field distribution P(B_{hf}) shown at the right-hand side of each spectrum. The spectra indicate magnetic ordering at all temperatures; however, at 300 K the spectrum shows a strong paramagnetic central pattern along with a magnetic sextet. The Mössbauer spectra of the $^{nat}Fe_{0.3}Y_{0.7}$ ribbons (Figure 2 (right)) indicate paramagnetism and are fitted by a distribution, P(QS), of quadrupole splittings. These spectra suggest that there is no bcc-Fe segregation, but Fe remains in the amorphous Fe–Y matrix. However, the amorphous matrix does not have the nominal concentration, but has an expected Y composition lower than the nominal composition due to Y precipitation. The spectra indicate paramagnetism at all temperatures between 4.2 and 300 K; however, at 4.2 K the spectrum shows an additional weak contribution from a broad magnetic central pattern. The T-dependence of the obtained hyperfine parameters, namely average isomer shift δ, average quadrupole splitting QS and average hyperfine field B_{hf} are shown in Figure 3. δ is relative to the ^{57}Co (Rh matrix) source. The Mössbauer spectra of the $^{57}Fe_{0.3}Y_{0.7}$ ribbons (Figure 4) are similar to those of $^{nat}Fe_{0.3}Y_{0.7}$ (Figure 2). The T-dependence of the δ (center line shift) for the $^{nat}Fe_{0.3}Y_{0.7}$ and the $^{nat}Fe_{0.7}Y_{0.3}$ ribbons are fitted according to Debye model (Figure 3a and c), yielding Debye temperature $\theta_D = 447$ K and 465 K, respectively. The magnetic hyperfine field data were fitted by assuming Bloch's law ($B_{hf}(T) = B_{hf}(0)[1\text{-}bT^{3/2}]$) (Figure 3b), resulting in $B_{hf}(0) = 20.8$ T and the stiffness constant $b = 1.1 \times 10^{-4}\ K^{-3/2}$.

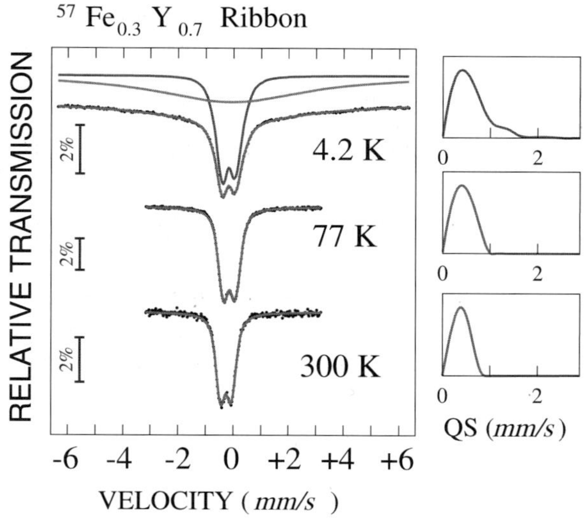

Figure 4 Mössbauer spectra of the $^{57}Fe_{0.3}Y_{0.7}$ alloy ribbons taken at 4.2, 77 and 300 K. The spectra have been fitted by a distribution of quadrupole splittings, P(QS) (*right hand side*). An additional broad singlet has been used for the fitting of the 4.2 K spectrum.

4 The sound velocity

The average (longitudinal) sound velocity, V_s, of the ribbons was calculated from the relation $V_s = \sqrt{\frac{E}{\rho}}$, where ρ = mass density of the ribbons and E = Young's modulus. The density and Young's modulus were measured as described below.

The density of the ribbons was estimated by introducing 2 g of material into a calibrated cylinder with water. Two grams of the ribbons increases the water volume by 0.40 cm^3, which is equal to the volume of the ribbons. There are no gas inclusions in the ribbon material, as expected from the preparation procedure of these intermetallic alloys; further, because water wets the ribbon walls, no air inclusions are expected. The experimentally obtained density of 5,000 kg/m^3 (with ± 5% relative error) agrees well with a 5–10% smaller density (typical for amorphous systems) as compared to the theoretically expected density of 5,500 kg/m^3 that is obtained from linear interpolation, according to the formula $Fe_{0.3}Y_{0.7}$, by using the density of Fe (7,860 kg/m^3) and of Y (4,470 kg/m^3).

The $Fe_{0.3}Y_{0.7}$ ribbons are very brittle and the typical size was 50 mm length, close to 0.6 mm width and 14–25 μm thickness. A stress–strain measurement was performed in order to determine E. The force was applied via the weight with a mass of about 20 g (0.2 N). The ribbon elongation was observed to vary linearly between 12×10^{-6} m and 23×10^{-6} m for different weights. Three such measurements on different ribbons leads to values of Young's modulus E of 5.2×10^{10} N/m^2, 4.6×10^{10} N/m^2 and 5.3×10^{10} N/m^2, which averages to 5.0×10^{10} N/m^2. The relative error is estimated to be about 10%. From the above measurements the sound velocity V_s was calculated to be 3,150 ± 150 m/s for the $Fe_{0.3}Y_{0.7}$ ribbons.

5 Inelastic neutron scattering

Inelastic neutron scattering was performed at Berlin Neutron Scattering Center (BENSC) at the NEAT (Time of Flight) spectrometer (V3 instrument). The binary

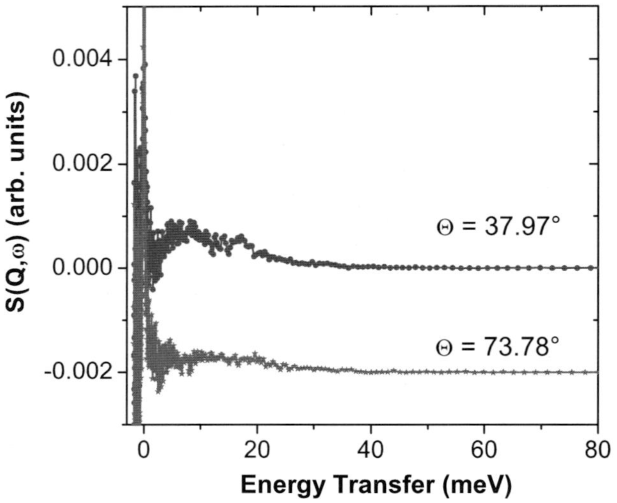

Figure 5 INS spectra of $^{nat}Fe_{0.3}Y_{0.7}$ alloy ribbons obtained with an incident neutron wavelength of 2 Å at 300 K. The spectra obtained for different angles are grouped into two with average angles of 37.97° and 73.78°. For clarity, the bottom spectrum has been shifted downward by −0.002 units.

$Fe_{0.3}Y_{0.7}$ and $Fe_{0.7}Y_{0.3}$ alloy ribbons were studied at different temperatures with two different incident neutron wavelengths of 2.0 Å (incident energy 20 meV) and 5.1 Å (~ 3.2 meV) in order to measure INS spectra for energy gain and energy loss of the neutrons. The corresponding elastic energy resolutions were 921 and 93 μeV, respectively.

The sample was placed inside the cryo-furnace of the V3 instrument by putting the sample into the slab-shaped aluminum sample holder in argon atmosphere. It was sealed airtight by using indium. The sample and background (empty sample holder) spectra were measured, and preliminary analysis has been performed by the standard program available at V3. The details of the measurement procedure is similar to that explained in [15]. Figure 5 shows INS spectra, i.e., S(Q,ω), of $Fe_{0.3}Y_{0.7}$ alloy ribbons obtained with an incident neutron wavelength of 2 Å at 300 K. A maximum is observed below ~10 meV, which provides evidence for excess low-energy vibrational excitations. The detailed analysis will be reported elsewhere.

6 Summary

Amorphous Fe–Y ribbons have been prepared by melt spinning. The XRD results suggest that bulk-like amorphous alloys can be prepared near the eutectic composition. Mössbauer spectra indicate magnetic ordering of the a-$Fe_{0.7}Y_{0.3}$ alloy ribbons, while partly crystalline $Fe_{0.3}Y_{0.7}$ alloy ribbons are paramagnetic, except at 4.2 K, where partial magnetic ordering occurs. The longitudinal sound velocity of $Fe_{0.3}Y_{0.7}$ alloy ribbons is about 3,150 m/s. Preliminary inelastic neutron scattering results indicates the presence of excess low-energy vibrational excitations in this amorphous alloy.

References

1. Chumakov, A.I., Sergueev, I., van Bürck, U., Schirmacher, W., Asthalter, T., Rüffer, R., Leupold, O., Petry, W.: Phys. Rev. Lett. **92**, 245508 (2004)
2. Ruckert, T., Keune, W., Sahoo, B., Sturhahn, W., Toellner, T., Alp, E.E., Röhlsberger, R.: Hyperfine Interact. **144/145**, 65 (2002)
3. Keune, W., Ruckert, T., Sahoo, B., Sturhahn, W., Toellner, T.S., Alp, E.E., Röhlsberger, R.: J. Phys. Condens. Matter **16**, S379 (2004)
4. Walterfang, M., Keune, W., Schuster, E., Zayak, A.T., Entel, P., Sturhahn, W., Toellner, T.S., Alp, E.E., Jochym, P.T., Parlinski, K.: Phys. Rev. B **71**, 035309 (2005)
5. Sahoo, B., Keune, W., Sturhahn, W., Toellner, T.S., Alp, E.E.: J. Phys. Chem. Solids **66**, 2263 (2005)
6. Chappert, J., Coey, J.M.D., Lienard, A., Rebouillat, J.P.: J. Phys. F. Met. Phys. **11**, 2727 (1981)
7. Coey, J.M.D., Givord, D., Lienard, A., Rebouillat, J.P.: J. Phys. F. Met. Phys. **11**, 2707 (1981)
8. Larica, C., Alves, K.M.B., Baggio-Saitovitch, E., Guimarães, A.P.: J. Magn. Magn. Mater. **145**, 306 (1995)
9. Larica, C., Passamani, E.C., Nunes, E., Orlando, M.T.D., Alves, K.M.B., Baggio-Saitovitch, E.: J. Alloys Compd. **274**, 23 (1998)
10. van der Kraan, A.M., Buschow, K.H.J.: Phys. Rev. B **25**, 3311 (1982)
11. Coey, J.M.D., Lienard, A., Rebouillat, J.P., Ryans, D.H., Wang, Z., Yu, B.: J. Phys. F. Met. Phys. **18**, 1299 (1988)
12. Suzuki, T., Fujita, A., Fukamichi, K., Aruga-Katori, H., Goto, T., Hashikura, M., Matsubara, E., Waseda, Y.: J. Phys. Condens. Matter **8**, 2219 (1996)
13. Brand, R.A.: Nucl. Instrum. Methods Phys. Res. **B28**, 398 (1987)
14. Brand, R.A.: Nucl. Instrum. Methods Phys. Res. **B28**, 417 (1987)
15. Pieper, J., Irrgang, K.D., Renger, G., Lechner, R.E.: J. Phys. Chem. B **108**, 10556 (2004)

Hyperfine Interact (2005) 165: 183–188
DOI 10.1007/s10751-006-9263-z

Magnetic properties of nanocrystalline $FeNiCo_xZrB$ alloys ($x=0$, 10, 20)

M. Kopcewicz · J. Kalinowska · T. Kulik

Published online: 26 October 2006

Abstract Magnetic properties of amorphous and nanocrystalline $Fe_{31}Ni_{50}Zr_7B_{12}$, $Fe_{31}Ni_{40}Co_{10}Zr_7B_{12}$ and $Fe_{31}Ni_{30}Co_{20}Zr_7B_{12}$ alloys were studied by an unconventional "rf-Mössbauer" technique. Introduction of Co atoms into FeNiZrB alloy leads to a large increase of anisotropy field that suppresses the rf collapse effect. The rf induced crystallization effect observed in Co-containing alloys was attributed to the rf sidebands effect which induced in the alloys mechanical deformations via the magnetostriction. This effect is particularly strong in amorphous alloys and in nanocrystalline alloys containing significant fraction of amorphous matrix and is absent in Co-free alloy.

Key words nanocrystalline alloys · amorphous alloys · Co-containing alloys · soft magnetic properties · rf-Mössbauer technique · hysteresis loop measurements

1 Introduction

Nanocrystalline alloys are the promising candidates for wide practical applications because of their excellent soft magnetic properties [1]. The main disadvantage is their high brittleness. Recently, the $Fe_{41}Ni_{40}Zr_7B_{12}$ nanocrystalline alloy revealed improved mechanical behaviour [2], however, that was associated with a significant decrease of saturation magnetization (hyperfine field) of nanograins. Here we discuss the experimental results obtained for FeNiZrB alloys to which the Co atoms were added in order to increase the saturation magnetization. Addition of Co atoms increased the hyperfine field to about 35 T (FeCo nanocrystalline phase was formed), however, at the expense of increased coercivity and anisotropy field.

M. Kopcewicz (✉) · J. Kalinowska
Institute of Electronic Materials Technology, Wólczyńska Street 133, 01-919 Warszawa, Poland
e-mail: kopcew_m@itme.edu.pl

T. Kulik
Faculty of Materials Science and Engineering, Warsaw University of Technology,
Wołoska Street 141, 02-507 Warszawa, Poland

Magnetic properties of the alloys: $Fe_{31}Ni_{50}Zr_7B_{12}$, $Fe_{31}Ni_{40}Co_{10}Zr_7B_{12}$ and $Fe_{31}Ni_{30}Co_{20}Zr_7B_{12}$ in amorphous state and after nanocrystallization of amorphous precursors have been studied by an unconventional rf-Mössbauer technique. This technique, which combines the Mössbauer effect with the phenomena induced by the external radio-frequency (rf) magnetic field (the rf-collapse and rf-sideband effects), permits us to study changes of magnetic anisotropy and to follow the changes of magnetostriction associated with the formation of nanocrystalline grains. Recently, the rf-Mössbauer technique was applied successfully for studying nanocrystalline alloys [2, 3] and is discussed in detail in ref. [4, 5].

An unexpected effect – the rf-induced crystallization of nonthermal origin of Co-containing alloys (amorphous and nanocrystalline) was attributed to mechanical deformations introduced by the rf field via magnetostriction (the rf-sidebands effect).

The rf-Mössbauer investigations were accompanied by the conventional measurements of quasi-static hysteresis loops [6].

2 Experimental results

Amorphous alloys: $Fe_{31}Ni_{50}Zr_7B_{12}$ (Co-free alloy), $Fe_{31}Ni_{40}Co_{10}Zr_7B_{12}$ (Co_{10} alloy) and $Fe_{31}Ni_{30}Co_{20}Zr_7B_{12}$ (Co_{20} alloy) were prepared by the melt quenching technique. The ribbons were 3 to 5 mm wide and about 25 μm thick. The as-quenched alloys were fully amorphous as verified by the X-ray diffraction and Mössbauer measurements [7]. The nanocrystalline phases were formed by controlled 1 h annealing in vacuum of amorphous precursors (annealing temperature: T_a = 440–770°C). The rf-Mössbauer measurements were performed, in a similar way as in [3–5], during exposure of the samples to a radio frequency (rf) magnetic field of the frequency 61 MHz which intensity was varied between 0 and 20 Oe. The isomer shifts are related to the α-Fe standard.

The rf-Mössbauer investigations were accompanied by the conventional measurements of quasi-static hysteresis loop [6], which allowed the estimation of the coercive fields and saturation fields for the annealed alloys.

Conventional Mössbauer spectra of $Fe_{31}Ni_{50}Zr_7B_{12}$ (Co-free alloy) are discussed in [7]. The as-quenched alloy is fully amorphous; the nanocrystallization occurs at T_a = 470–670°C. The dominating phase is the bcc-FeNi (hyperfine field $H_{hf} \approx 29$ T, isomer shift $\delta \approx 0.04$ mm/s), accompanied by the $(FeNi)_{23}B_6$ phase [7]. The rf-Mössbauer measurements revealed that the amorphous Co-free alloy is magnetically very soft. A complete collapse of the magnetic hyperfine structure (mhfs) occurs at about 12 Oe at 61 MHz. The rf-Mössbauer spectra of the as-quenched and annealed at 470°C samples (Figure 1a and b) consist of a collapsed doublet accompanied by strong rf sidebands which suggest a substantial magnetostriction of the alloy. The mhfs is averaged to zero due to fast magnetization reversal induced by the rf field, which intensity sufficiently exceeds the local anisotropy field. The shape of the collapsed pattern changes with the increase of annealing temperature. The splitting of the doublet decreases with increasing T_a and the development of nanograins (Figure 1b). For $T_a \approx 520$°C the collapsed pattern consists of a single line originating from a magnetically very soft bcc-FeNi phase. Magnetostriction of the alloy vanished completely; the rf sidebands disappeared (Figure 1c). For higher T_a ($T_a \geq 570$°C, Figure 1d–f) the rf-collapse is not complete. The spectrum consists of a well-resolved doublet, suggesting that the local structural symmetry decreased, accompanied by the magnetic "wings" (Figure 1d) or poorly resolved magnetic hyperfine component (Figure 1e and f). It strongly suggests that the magnetic anisotropy increased and the rf field of 20 Oe

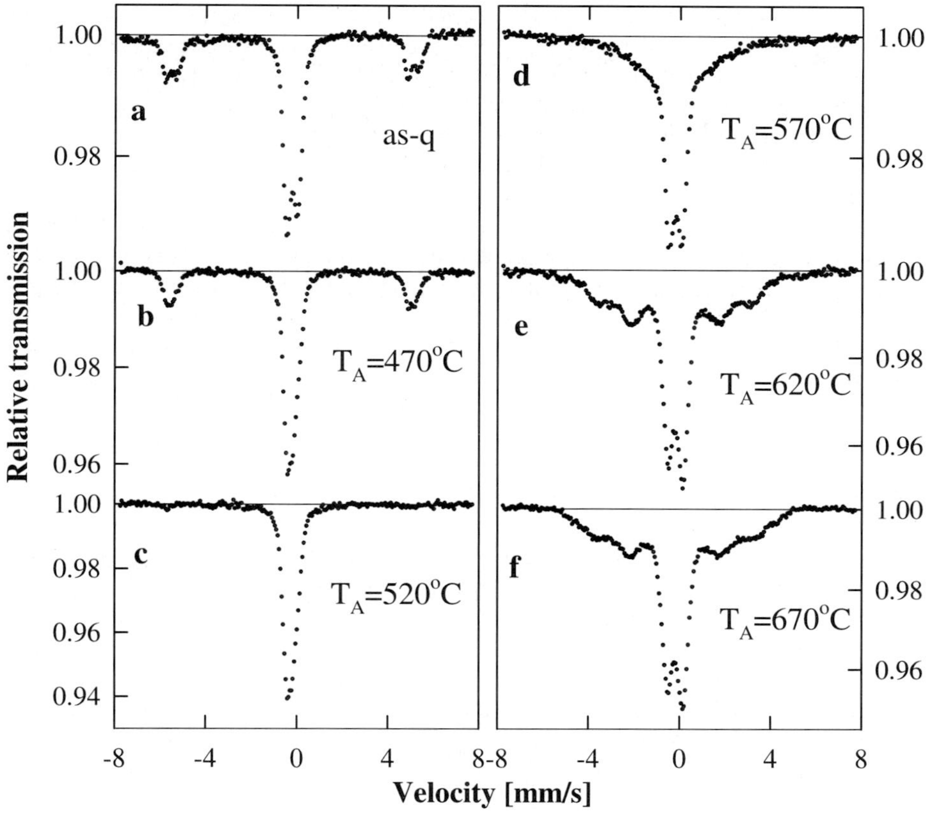

Figure 1 Mössbauer spectra of the annealed at indicated temperatures sample of $Fe_{31}Ni_{50}Zr_7B_{12}$ alloy recorded with the rf field of 20 Oe at 61 MHz.

was unable to average completely the magnetic field to zero. The Mössbauer spectrum recorded after the exposure of the Co-free sample to the rf field is identical to that observed for the as-quenched amorphous alloy before the rf field exposure.

The situation is significantly different in Co-containing alloys. Addition of Co to the amorphous structure caused an increase of the anisotropy fields. The full collapse of the mhfs of the Co_{10}-containing amorphous sample occurs at higher intensity of the rf field. The sidebands, observed in the rf-Mössbauer spectra, reveal very high intensity. However, a larger Co-content in the Co_{20}-alloy causes the instability of the amorphous phase during exposure to the rf field (Figure 2a–e). The amorphous alloy crystallizes partly due to the rf field exposure; the FeCo phase is formed. This phase is magnetically quite hard and the rf field of 20 Oe is unable to collapse the mhfs of this phase (Figure 2d). The rf collapse is not complete and is observed only for the remaining amorphous matrix. The nanocrystalline FeCo and $Fe_{23}B_6$-type phases have larger anisotropy and the rf collapse is suppressed (Figure 2d).

In both Co-containing nanocrystalline alloys, in Co_{20} alloy in particular, the anisotropy field is strongly increased as compared with the Co-free alloy. The rf-Mössbauer spectra of Co_{20} alloy annealed at $T_a \approx 470°C$ (Figure 2a′–d′) show a noncollapsed magnetic component, related to FeCo phase with sufficiently large anisotropy to limit the magnetization

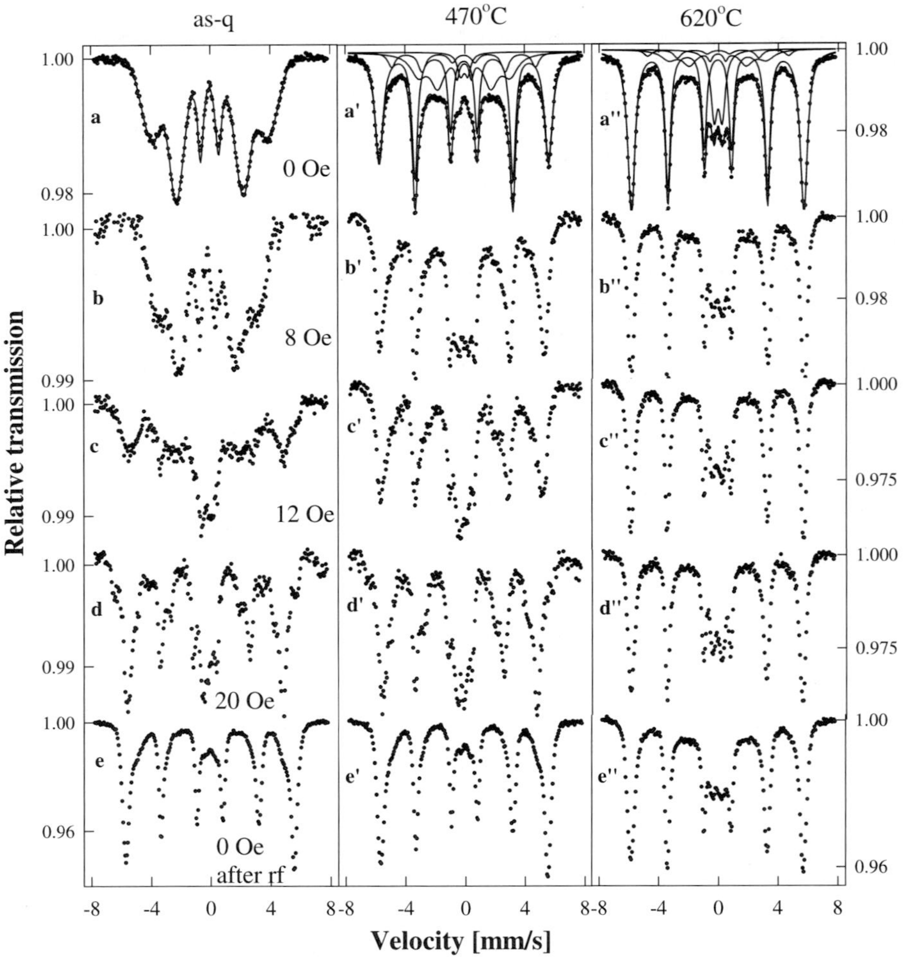

Figure 2 The rf-Mössbauer spectra recorded as a function of the rf field intensity for the as-quenched (*a–e*), annealed at 470°C (*a′–e′*), and 620°C (*a″–e″*) $Fe_{31}Ni_{30}Co_{20}Zr_7B_{12}$ alloy.

reversal in this phase, accompanied by a collapsed doublet of small intensity. The spectral contribution of this doublet, that originates from the collapse of the mhfs of the remaining amorphous matrix, increases with increasing rf field intensity (Figure 2b′–d′). The spectra of the sample annealed at T_a = 620°C measured during the exposure to the rf field (Figure 2b″–d″) do not depend on the rf field intensity. They consist mainly of the mhfs corresponding to the noncollapsed FeCo sextet ($H_{hf} \approx 35$ T). The anisotropy field of this phase exceeds the rf field of 20 Oe that is not sufficient to induce the rf collapse effect. A central quadrupole doublet, related most probably to Fe atoms surrounded by Zr or Zr O_2, is also present.

The Mössbauer spectrum recorded for the as-quenched Co_{20} sample in the absence of the rf field after the exposure to the 20 Oe rf field (Figure 2e) is significantly different than that corresponding to the as-quenched Co_{20} alloy (Figure 2a). The spectrum in Figure 2e corresponds to a sample considerably crystallized due to the rf-exposure; it clearly shows the formation of FeCo crystalline grains. The spectra of the Co-containing samples collected after the rf field exposure (Figure 2e, e′, e″) are quite similar to each other and correspond to the crystallized alloys. The difference between the spectra observed before and after the rf exposure is the largest for the as-quenched alloy (compare Figure 2e and a) and for the nanocrystalline alloy with a large content of amorphous matrix (Figure 2e′ and a′). The rf-induced crystallization is particularly pronounced in the alloys with a substantial fraction of amorphous matrix with large magnetostriction, which suggests that this effect is related to the deformations induced in the samples via magnetostriction (rf sidebands effect). When the rf sidebands disappear in nanocrystalline alloys (magnetostriction vanishes) than the rf induced crystallization does not occur. However, this effect occurs only in Co-containing alloys, so the presence of Co atoms is essential. The rf-induced crystallization of conventional amorphous Co-containing alloys was observed earlier [8]. A crystallization of the Co-containing amorphous phase due to the mechanical deformation induced during the high-energy ball milling was observed also [9].

The rf-Mössbauer studies have been supplemented by the measurements of the quasi-static hysteresis loops from which the magnetization and coercive field were estimated. The coercivity strongly increased in the Co-containing alloys after annealing. The coercivity of the Co_{10}-alloy increased from about 40 A/m (T_a = 440°C) to 278 A/m (T_a = 470°C) and 1130 A/m (T_a = 520°C) and reached the largest value of 8335 A/m (T_a = 620°C). In the Co_{20}-alloy the increase of coercivity was much larger: from about 80 A/m (T_a = 440°C) to 1038 A/m (T_a = 520°C) and 4090 A/m (T_a = 570°C) and finally reached 12072 A/m (T_a = 620°C). At T_a = 670°C coercivity decreased to 4355 A/m. The changes in coercivity reflect the FeCo phase abundance in the alloys [7]. The largest coercivity was observed for the alloys in which the content of FeCo phase reached the saturation. Most of the hysteresis loops had a rectangular shape. The measurements of the hysteresis loops fully support the rf-Mössbauer results. In the Co-containing alloys both coercive field and anisotropy field strongly increase after annealing at elevated temperatures. The rf fields were able to induce the rf-collapse effect only if the rf field intensity was significantly larger than the anisotropy field. The increase of the anisotropy field caused the suppression of the rf collapse.

In summary, the rf-Mössbauer results show that the amorphous precursors are magnetically very soft. The anisotropy field increases with the introduction of Co: in the amorphous alloys this effect is significant and in nanocrystalline alloys this increase is dramatic. The rf-Mössbauer measurements are very useful for the estimation of magnetic anisotropy fields. Presence and the magnitude of rf sidebands provide a good test for the formation of the nanocrystalline alloy: the rf sidebands vanish when the nanocrystalline structure is formed. The nonthermal rf induced crystallization of Co-containing alloys was observed and its origin was attributed to the rf sidebands effect which induced in amorphous phase mechanical deformations via the magnetostriction.

Acknowledgement The financial support from the Grant no. KBN 2 P03B 075 24 from the State Committee for Scientific Research (Poland) is gratefully acknowledged.

References

1. Makino, A., Hanai, T., Inoue, A., Masumoto, T.: Mater. Sci. Eng. A **179/180**, 552 (1994)
2. Kopcewicz, M., Idzikowski, B., Kalinowska, J.: J. Appl. Phys. **94**, 638 (2003)
3. Kopcewicz, M.: J. Alloys. Comp. **382**, 165 (2004)
4. Kopcewicz, M.: In: Long, G.J., Grandjean, F. (eds.) Mössbauer Spectroscopy Applied to Inorganic Chemistry, vol. 3, p. 243. Plenum, New York/London (1989)

5. Kopcewicz, M.: Struct. Chem. **2**, 313 (1991)
6. Kulik, T., Savage, H.T., Hernando, A.: J. Appl. Phys. **73**, 6855 (1993)
7. Kopcewicz, M., Kalinowska, J., Latuch, J.: This conference (2006) (in press)
8. Kopcewicz, M., El Zayat, M., Gonser, U.: J. Magn. Magn. Mater. **72**, 119 (1988)
9. Trudeau, M.L., Schulz, R., Dassault, D., Van Neste, A.: Phys. Rev. Lett. **64**, 99 (1990)

Hyperfine Interact (2005) 165: 189–194
DOI 10.1007/s10751-006-9264-y

Crystallization of amorphous FeNiZrB and FeNiCoZrB alloys

M. Kopcewicz · J. Kalinowska · J. Latuch

Published online: 26 October 2006

Abstract Crystallization of amorphous precursors: $Fe_{31}Ni_{50}Zr_7B_{12}$, $Fe_{31}Ni_{40}Co_{10}Zr_7B_{12}$ and $Fe_{31}Ni_{30}Co_{20}Zr_7B_{12}$ was studied by the Mössbauer spectroscopy and X-ray diffraction. The nanocrystalline phases were formed by annealing the amorphous alloys at temperatures of 440–770°C. The crystallization temperatures were determined by DSC measurements. Addition of Co atoms into FeNiZrB alloy resulted in the formation of FeCo phase with large hyperfine field of 35 T. Formation of FeCo and $(FeNiCo)_{23}B_6$ phases was followed by the Mössbauer measurements and confirmed by XRD measurements. The size of nanograins was estimated from XRD measurements using the Scherrer formula.

Key words nanocrystalline $Fe_{31}Ni_{50}Zr_7B_{12}$, $Fe_{31}Ni_{40}Co_{10}Zr_7B_{12}$ and $Fe_{31}Ni_{30}Co_{20}Zr_7B_{12}$ alloys · crystallization of amorphous precursors · nanograins of FeCo and $(FeNiCo)_{23}B_6$ · Mössbauer spectroscopy · DSC measurements · X-ray diffraction · grain size estimation

1 Introduction

Soft magnetic nanocrystalline alloys are usually produced by partial devitrification of the melt-spun amorphous precursors. They exhibit a two-phase structure: nanocrystalline grains (with the bcc structure) dispersed in the residual amorphous matrix with higher crystallization temperature. The crystallization behaviour of amorphous precursor is crucial both for fundamental research as well as for technological applications. The search continues for soft magnetic nanocrystalline alloys with high saturation magnetization, combined with improved mechanical properties (reduced brittleness). Recently, the $Fe_{41}Ni_{40}Zr_7B_{12}$ nanocrystalline alloy was studied [1], which revealed considerably reduced brittleness as compared with the nanocrystalline bcc-Fe phase observed in NANOPERM alloys.

M. Kopcewicz (✉) · J. Kalinowska
Institute of Electronic Materials Technology, Wólczyńska Street 133, 01-919 Warszawa, Poland
e-mail: kopcew_m@itme.edu.pl

J. Latuch
Department of Materials Science and Engineering, Warsaw University of Technology,
Wołoska Street 141, 02-507 Warszawa, Poland

Table I Crystallization temperatures for the Co-free and Co-containing alloys determined from DSC results

Alloy	First peak		Second peak	
	Onset temperature $T_{\times 1}$ (°C)	Peak position (°C)	Onset temperature $T_{\times 2}$ (°C)	Peak position (°C)
Co-free	469	483	565	581
Co_{10}	486	495	562	600
Co_{20}	484	506	610	634

However, the improved favourable mechanical properties were associated with a significant reduction of saturation magnetization (hyperfine field). In order to increase the saturation magnetization without a significant degradation of mechanical properties a Co was added to FeNiZrB alloys. The addition of Co atoms resulted in a formation of nanocrystalline FeCo phase with high hyperfine field (about 35 T).

We prepared novel amorphous and nanocrystalline alloys: $Fe_{31}Ni_{50}Zr_7B_{12}$ (Co-free alloy), $Fe_{31}Ni_{40}Co_{10}Zr_7B_{12}$ and $Fe_{31}Ni_{30}Co_{20}Zr_7B_{12}$ (Co_{10} and Co_{20} alloys, respectively). The nanocrystalline phase was formed by annealing amorphous precursors. Crystallization of the amorphous precursors was studied by Mössbauer spectroscopy (MS) and supplemented by the differential scanning calorimetry (DSC) and X-ray diffraction (XRD).

2 Experimental results

Amorphous $Fe_{31}Ni_{50}Zr_7B_{12}$, $Fe_{31}Ni_{40}Co_{10}Zr_7B_{12}$ and $Fe_{31}Ni_{30}Co_{20}Zr_7B_{12}$ alloys were prepared by melt-quenching technique. The ribbons were 3 to 5 mm wide and about 25 μm thick. As-quenched alloys were fully amorphous as verified by the XRD and Mössbauer measurements. The nanocrystalline alloys were formed by annealing the amorphous precursors for 1 h in vacuum at temperature ranging from 440°C to 670°C for Co-free alloy, and from 440°C to 770°C for Co-containing alloys. Annealing temperatures were selected basing on the differential scanning calorimetry (DSC) results. DSC measurements revealed that the amorphous alloys exhibit two-stage crystallization behaviour. The first stage, with characteristic onset temperature $T_{\times 1}$, corresponds to a partial crystallization of the amorphous phase and formation of nanocrystalline phases. The second stage ($T_{\times 2}$) corresponds to a complete crystallization of the remaining amorphous phase, increase of nanocrystalline grains and formation of other crystalline phases. The results of the DSC measurements, performed with a heating rate of 20 K/min. are shown in Table I.

The crystallization behaviour of amorphous alloys was studied by the conventional Mössbauer spectroscopy. The measurements were performed at room temperature in transmission geometry. The Mössbauer spectra recorded for $Fe_{31}Ni_{50}Zr_7B_{12}$ and $Fe_{31}Ni_{30}Co_{20}Zr_7B_{12}$ alloys, annealed at indicated temperatures, are shown in Figures 1 and 2, respectively. The first trace of crystallization of the Co-free amorphous alloy was detected at $T_A \approx 440$°C (Figure 1b). At $T_A \approx 470$°C the bcc-FeNi (hyperfine field $H_{hf} \approx 29$ T, isomer shift $\delta \approx 0.04$ mm/s) and $(FeNi)_{23}B_6$ ($H_{hf} \approx 25$ T, $\delta \approx 0.03$ mm/s) phases are clearly formed (Figure 1c). At the annealing temperatures below 520°C these phases are accompanied by the remaining amorphous matrix, the spectral contribution of which decreases with increase of T_A. The remaining amorphous matrix crystallizes at $T_A \approx 520$°C, so well below the $T_{\times 2}$ temperature, and the Fe_3B-type phase is formed. The Mössbauer spectral component assigned to the Fe_3B-type phase consists of two sextets with the hyperfine fields of 22.5 and

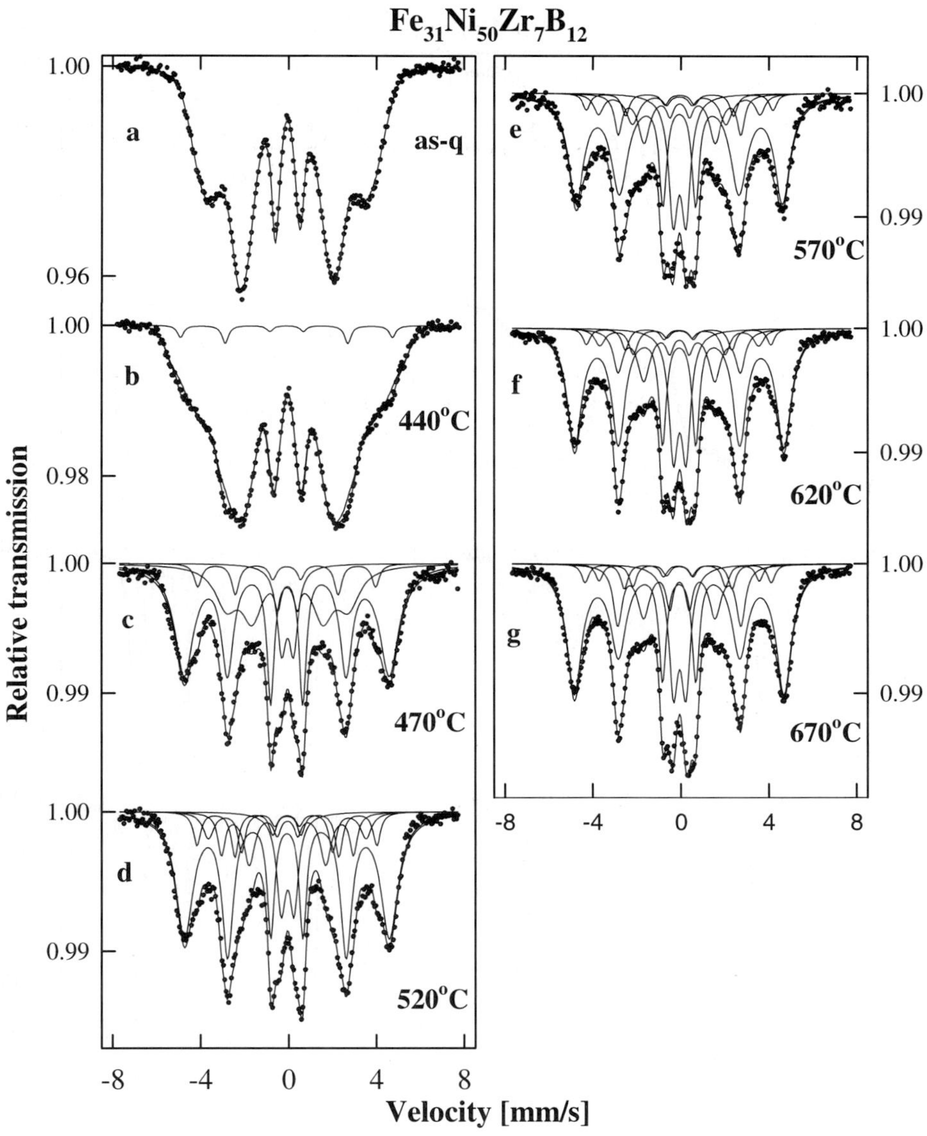

Figure 1 Room temperature Mössbauer spectra of $Fe_{31}Ni_{50}Zr_7B_{12}$ alloy annealed at indicated temperatures.

17.5 T, respectively [2] (Figure 1d, e). In all cases in addition to the magnetically ordered phases discussed above a nonmagnetic fraction with a quadrupole doublet (QS≈0.6 mm/s and $\delta\approx0.05$ mm/s) is observed. The hyperfine parameters of this spectral component do not correspond to any known phase that could be formed as a result of crystallization of amorphous alloys studied. Probably this nonmagnetic fraction is related to Fe atoms surrounded by Zr or ZrO_2. The appearance of the ZrO_2 phase was observed in the XRD results for T_A exceeding 470°C. The QS doublet, however with a larger quadrupole splitting (QS=0.88 mm/s), was observed earlier for $FeZr_2$ compound [3].

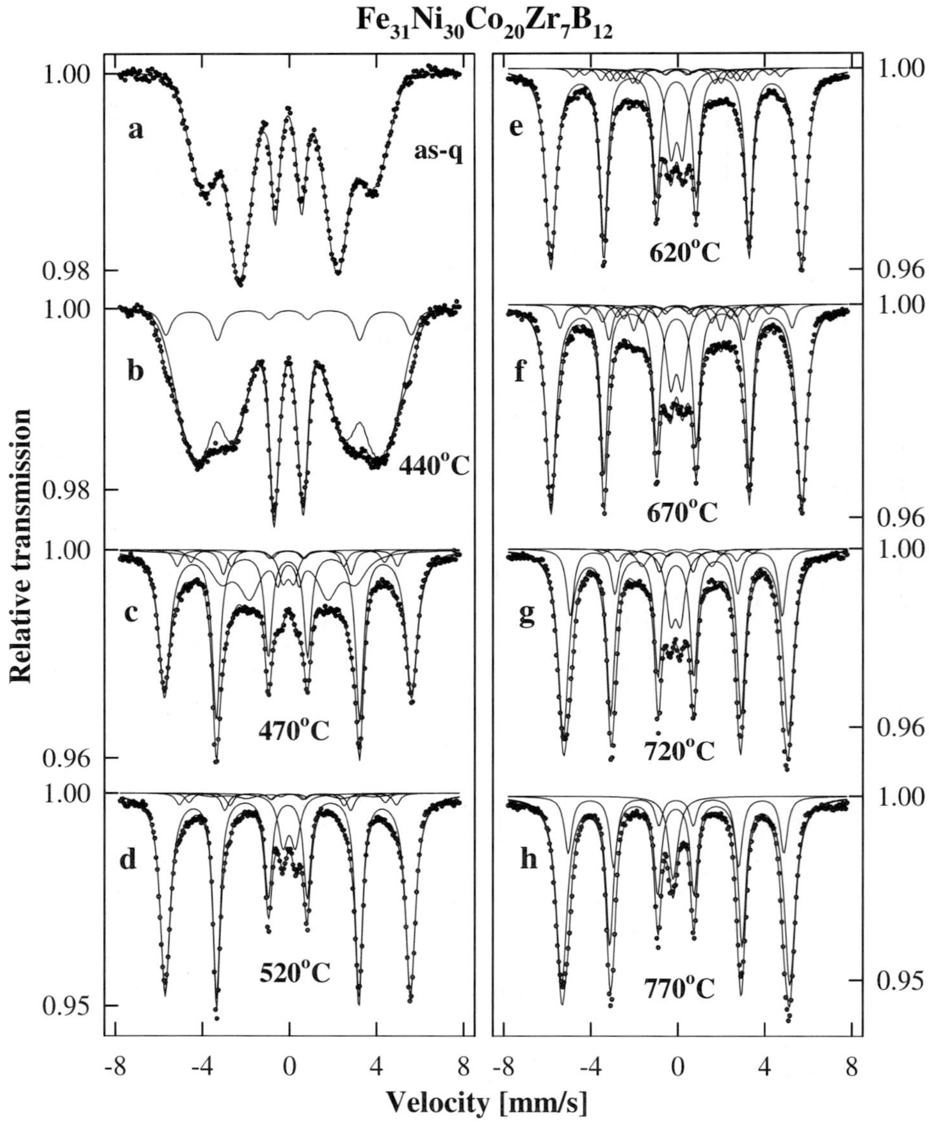

Figure 2 Room temperature Mössbauer spectra of $Fe_{31}Ni_{30}Co_{20}Zr_7B_{12}$ alloy annealed at indicated temperatures.

Crystallization process of both Co-containing alloys (Co_{10} and Co_{20}) proceeds in a different way. The as-quenched Co_{10} and Co_{20} alloys were fully amorphous (Figure 2a). The crystallization starts at $T_A \approx 440°C$ (Figure 2b). Presence of Co in the alloys causes that annealing at temperatures between $T_{\times 1}$ and $T_{\times 2}$ (i.e., between 470°C and 620°C) leads to the formation of nanograins of the FeCo phase ($H_{hf} \approx 35$ T, $\delta \approx 0.04$ mm/s), as identified by the MS and XRD techniques. The relative spectral fraction of the FeCo phase dominates in the MS spectra (Figure 2c, d, e). In addition to the FeCo nanophase less abundant phases: bcc-FeNi ($H_{hf} \approx 29$-31T, $\delta \approx 0.04$ mm/s) and $Fe_{23}B_6$-type phase, most probably the

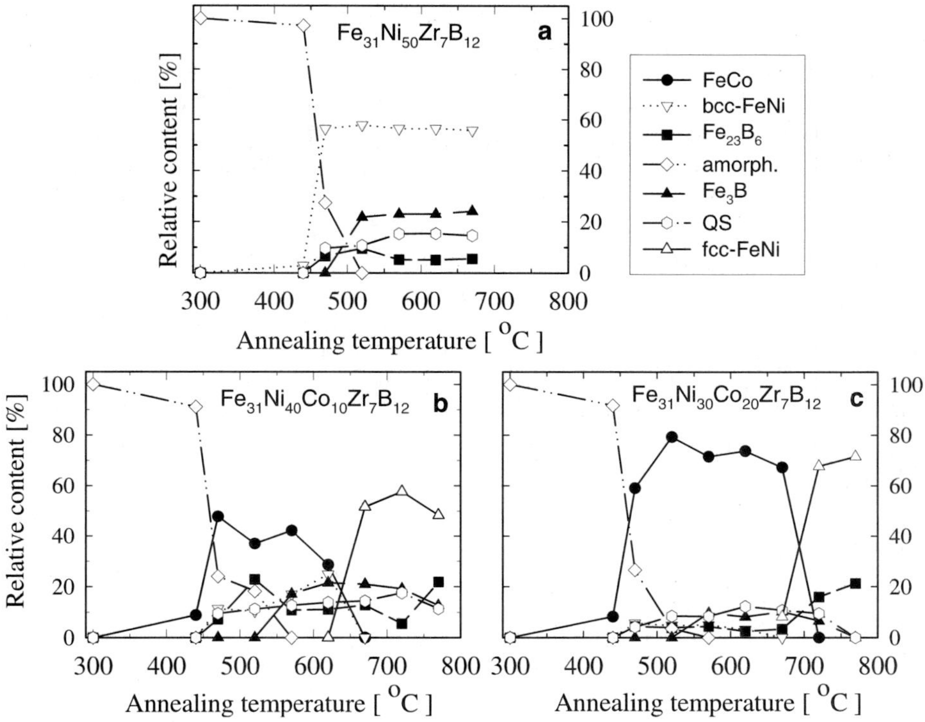

Figure 3 Relative spectral fractions of various phases determined from MS spectra of $Fe_{31}Ni_{50}Zr_7B_{12}$ (a), $Fe_{31}Ni_{40}Co_{10}Zr_7B_{12}$ (b) and $Fe_{31}Ni_{30}Co_{20}Zr_7B_{12}$ (c) alloys vs. annealing temperature.

$(FeNiCo)_{23}B_6$ phase ($H_{hf}\approx 26$ T, $\delta\approx 0.02$–0.04 mm/s) were also formed due to annealing at the same temperatures (Figure 2c, d, e). The remaining amorphous phase was observed in the Mössbauer spectra of the sample annealed at 470°C (Figure 2c). It crystallizes at about $T_a\approx 520$°C (Figure 2d) and disappears at $T_a\approx 570$°C; the Fe_3B-type phase is formed (spectral component consists of two sextets with $H_{hf}\approx 23$ and 18 T) [2]. The hyperfine fields of given phases may vary slightly depending on the Ni and Co content in the crystalline grains. The FeCo phase is stable up to $T_a\approx 620$°C. At higher annealing temperatures (close to $T_{\times 2}$ and above) the bcc-FeNi phase disappears and the phase observed for $T_a>670$°C consists mainly of fcc-FeNi ($H_{hf}\approx 32$T, $\delta\approx 0.02$ mm/s), as suggested by the XRD reults. The $(FeNiCo)_{23}B_6$ and Fe_3B borides remain stable at high annealing temperatures, above $T_{\times 2}$. The nonmagnetic spectral component (QS doublet) related most probably to Fe atoms surrounded by Zr or ZrO_2 is clearly visible in the spectra in Figure 2d–g. The hyperfine parameters (QS, δ) of this doublet are similar to those observed in Co-free alloy. Its spectral contribution of the QS doublet is the largest for Co_{10} alloy and exceeds 15% at $T_a\approx 670$–720°C. The MS spectrum of Co_{20} alloy annealed at the highest $T_a\approx 770$°C (Figure 2h) contained, in addition to the spectral components related to the magnetically ordered fcc-FeNi and boride phases discussed above, a single line component ($\delta\approx 0.06$–0.11 mm/s). This spectral component may be attributed to the fcc-FeNi phase with high Ni-content. Similar single-line component was observed recently for FeNiZrB alloy annealed at $T_a>$ 700°C [1]. The magnetic behaviour of the fcc-FeNi system strongly depends on Ni content and the fcc phase becomes nonmagnetic for Ni content of about 30 at.% [4]. A single line component, corresponding to an unknown Fe–Zr–B phase [3] is rather unlikely. However, this phase has not been identified in our XRD experiments.

The relative spectral contributions of various phases observed of MS spectra of all alloys studied vs. annealing temperature are shown in Figure 3.

The XRD results fully confirm the formation of similar phases formed at the same T_a as observed in the Mössbauer measurements. From the XRD results the size of nanograins was estimated using the Scherrer formula. With increasing annealing temperatures the FeCo grains become larger; their size increases from below 5 nm (at $T_a \approx 470°C$) to about 19 nm (at 570°C) and beyond 20 nm (at 620°C) depending on the Co-content in the nanocrystalline alloy. The size of $Fe_{23}B_6$-type grains is somewhat larger: about 26 nm at $T_a = 570°C$.

In summary, the Mössbauer study of crystallization behaviour of amorphous $Fe_{31}Ni_{50}Zr_7B_{12}$ (Co-free) and $Fe_{31}Ni_{40}Co_{10}Zr_7B_{12}$ (Co_{10}) and $Fe_{31}Ni_{30}Co_{20}Zr_7B_{12}$ (Co_{20}) precursors have shown that the hyperfine field of nanocrystalline grains of FeCo phase is indeed substantially larger than that characteristic of bcc-FeNi phase formed in Co-free alloy. The addition of Co atoms increased the hyperfine field and saturation magnetization of the Co-containing nanocrystalline alloy.

Acknowledgement The financial support from the Grant no. KBN 2 P03B 075 24 from the State Committee of Scientific Research (Poland) is gratefully acknowledged.

References

1. Kopcewicz, M., Idzikowski, B., Kalinowska, J.: J. Appl. Phys. **94**, 638 (2003)
2. Kondoro, J.W., Campbell, S.J.: Hyperfine Interact. **55**, 993 (1990)
3. Itoh, T., Onodera, H., Miura, S., Inoue, A.: Mater. Trans., JIM **41**, 1392 (2000)
4. Johnson, C.E., Ridout, M.S., Cranshaw, T.E.: Proc. Phys. Soc. **81**, 1079 (1963)

Hyperfine Interact (2005) 165: 195–201
DOI 10.1007/s10751-006-9265-x

Mössbauer and XRD study of pulse plated Fe–P and Fe–Ni thin layers

Annamária Mikó · Ernő Kuzmann · Magda Lakatos-Varsányi · Attila Kákay · Ferenc Nagy · Lajos Károly Varga

Published online: 14 November 2006

Abstract ^{57}Fe conversion electron Mössbauer spectroscopy, X-ray diffraction, electrochemical and magnetic measurements were used to study pulse electroplated Fe–P and Ni–Fe coatings. XRD and ^{57}Fe CEMS measurements revealed the amorphous character of the novel pulse plated Fe–P alloys. CEM spectra indicated significant differences in the short range order and in the magnetic anisotropy between the Fe–P deposits pulse plated at medium long deposition time (t_{on}=2 ms), with short relaxation time (t_{off}=9 ms) and low current density (I_p=0.05 Acm^{-2}) or at short deposition time (t_{on}=1 ms) with long relaxation time (t_{off}=250 ms) and high current density (I_p=1.0 Acm^{-2}). The broad peaks centred around the fcc reflections in XRD of the pulse plated Ni-22 wt.% Fe deposit reflected a microcrystalline Ni–Fe alloy with a very fine, 5–8 nm, grain size. The CEM spectrum of the pulse plated Ni-22 wt.% Fe coating corresponded to a highly disordered solid solution alloy containing a minute amount of ferrihydrite. Extreme favourable soft magnetic properties were observed with these Ni–Fe and Fe–P pulse plated thin layers.

Key words Fe–P and Fe–Ni thin layers · electrochemical deposition · pulse plating · Mössbauer spectroscopy · XRD · soft magnetic properties

1 Introduction

Pulse plating technique of electrodeposition [1] offers a powerful synthesis for the production of nano-crystalline and amorphous magnetic thin films and multilayers. This method gives an

A. Mikó · M. Lakatos-Varsányi
Bay Zoltán Institute for Material Science, Fehérvári út 130, Budapest, Hungary

E. Kuzmann (✉) · F. Nagy
Research Group for Nuclear Methods in Structural Chemistry, Hungarian Academy of Sciences,
Department of Nuclear Chemistry, Eötvös Loránd University,
Pázmány P. s. 1/a, Budapest 1117, Hungary
e-mail: kuzmann@para.chem.elte.hu

A. Kákay · L. K. Varga
Research Institute for Solid State Physics and Optics, 1525 Budapest, P.O.B. 49, Hungary

Figure 1 X-ray diffractograms of Fe–P pulse plated samples from group 1 (a) and from group 2 (b).

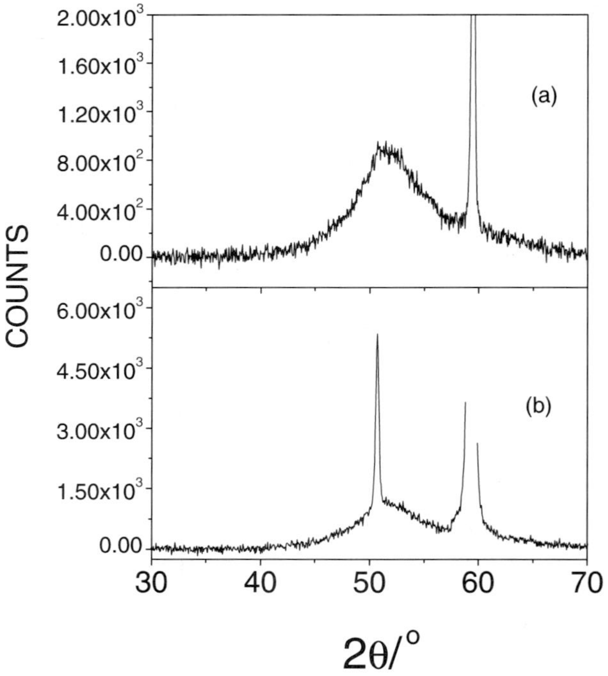

alternative way besides the evaporation and sputtering as well as rapid quenching techniques. The pulse plating electrodeposition could provide a new range of nanostructured composite materials possessing a useful range of similar or even improved properties.

Conversion Electron Mössbauer Spectroscopy (CEMS) measurements have already been successfully applied to investigate the local atomic arrangement of the constituent elements in pulse plated multilayers [2, 3] especially when pulse electrodeposition of nanostructured iron and the subsequent anodic passivation has already been used to prepare Fe/Fe-oxide multilayers [3, 4].

Since the Fe/Fe-oxide multilayers were found to be good inductive elements up to several kHz frequencies [3, 4] one of the goals of our present work was to prepare novel soft magnetic materials exhibiting even more favourable high frequency magnetic properties.

In this work the pulse plating electrodeposition technique was applied to prepare amorphous Fe–P and nanostructured Ni–Fe thin layers. For the characterisation of these films CEMS, XRD, electrochemical and magnetic measurements were used.

2 Experimental

The Fe–P and Ni–Fe samples were prepared by pulse plating electrochemical deposition onto Cu substrate. In case of rectangular waveform current modulation, the pulse-electrochemical method can be characterized by the following pulse parameters: t_{on} is the pulse time when cathodic current is applied; t_{off} is the relaxation time and I_p is the cathodic current density during the pulse. Based on previous studies on the deposition of amorphous and nanostructured Fe–P and Ni–Fe [1, 5–9], two groups of samples were prepared: Group (1) Medium long deposition time (t_{on}), with short relaxation time (t_{off}) and low current

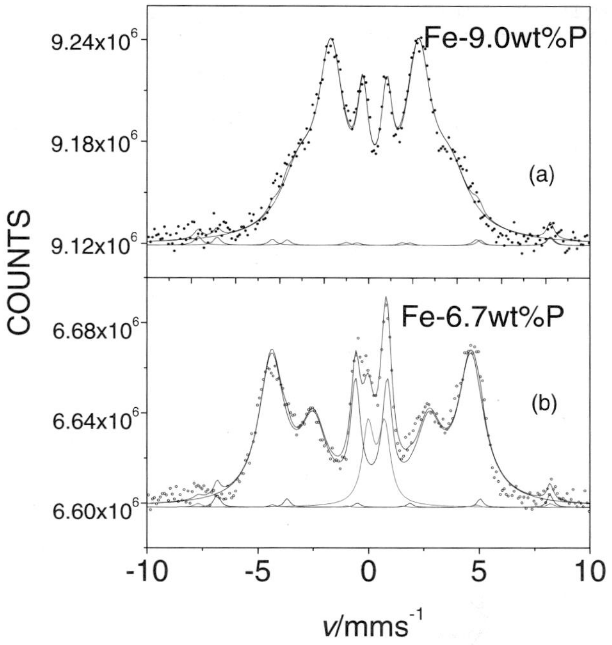

Figure 2 CEM spectra of Fe–P pulse plated samples from group 1 (a) and from group 2 (b).

density; (t_{on}=2 ms, t_{off}=9 ms, I_p=0.05 Acm^{-2}) and Group (2) Short t_{on} time with long relaxation time and high current density; (t_{on}=1 ms, t_{off}=250 ms, I_p=1 Acm^{-2}).

In the case of Fe–P samples the electrolyte consisted of the following components: $FeSO_47H_2O$ 300 g/l, $NaH_2PO_2{\cdot}H_2O$ 21 g/l, H_3BO_3 30 g/l, L-ascorbic acid 5 g/l, NH_4Cl 20 g/l. The electrolyte composition for Ni–Fe samples was: $FeSO_47H_2O$ 5.6 g/l, $NiSO_47H_2O$ 140 g/l, H_3BO_3 25 g/l, L-ascorbic acid 8.8 g/l, saccharin 3 mg/l. The electrolyte solutions have been prepared by reagent grade chemicals and distilled water. The electrolytes were deaerated by bubbling Ar gas for 30 min before and during the pulse plating. In all cases the temperature of the electrolyte was 25°C and the electrolyses were performed at pH=3 using Pt anode electrodes. The plating bath was agitated using a magnetic stirrer at a rate of 500 rpm during each deposition process. The Fe–P and Fe–Ni samples were deposited with a layer thickness between 5–30 μm.

X-ray diffractograms of electrodeposited samples were recorded using a computer controlled powder diffractometer (DRON-2). The measurements were performed with a goniometer speed of 1° cm^{-1} and β-filtered CoK$_\alpha$ radiation (λ=1.7890), in the range of $2\Theta=3-120°$ at temperature 295 K. The powder diffraction patterns were evaluated by "EXRAY" peak searching computer program (Klencsár, 1998).

The magnetic coercivity of the samples have been characterised by 1–100 kHz AC hysteresis loop measurements. The frequency limit of the layers was determined at the layer thickness 10 μm by initial permeability measurements.

^{57}Fe conversion electron Mössbauer spectra of the samples were recorded with a conventional Mössbauer spectrometer (Wissel) using He-4%CH_4 gas flow CEM detector (RANGER) at 300 K. The γ-rays were provided by a 3×10^9 Bq ^{57}Co/Rh source. Isomer shifts are given relative to α-iron. The Mössbauer spectra were analyzed by the least-square fitting of lines and by the evaluation of hyperfine field distribution with the help of MOSSWINN code [10].

Table I Average Mössbauer parameters of pulse plated Fe–P and Fe–Ni samples

Sample	δ_1 (mms^{-1})	B_1 (T)	A_1 (%)	δ_2 (mms^{-1})	Δ_2 (mms^{-1})	A_2 (%)	Calculated ferri-hydrite (%)	δ_3 (mms^{-1})	B_3 (T)	A_3 (%)	δ_4 (mms^{-1})	B_4 (T)	A_4 (%)	Calculated Fe_3O_4	
														%	nm
Fe–P from group 1	0.23	22.8	97	–	–	–	–	0.67	46.2	1.9	0.26	49.3	0.8	2.8	5.8
Fe–P from group 2	0.17	27.9	86	0.35	0.75	11	–	0.68	46.1	1.6	0.27	49.2	1.4	3.0	6.5
Fe–Ni	0.03	29.7	96	0.35	0.82	4	4.8								

3 Results and discussion

Figure 1a shows the X-ray diffractogram of a Fe–P pulse plated sample, from group 1. The sharp peaks at $2\theta=50.8$, 59.4, 88.9 and 110.4° in both Fe–P and Ni–Fe spectra (see Figure 3) are belonging to the Cu substrate, which can be reflected since the penetration depth of the X-ray is larger than thickness of the Fe–P layer (20 μm). The very broad peak centred at $2\theta=51.8°$ evidently reflects an amorphous phase being attributed to the Fe–P phase. The *P* content of this sample was found to be 9.0±0.1 wt.%. At this composition the formation of the amorphous phase can also be obtained both with rapid quenching and also by conventional electrodeposition methods [11, 12].

The diffractogram of the sample from group 2 (Figure 1b) looks like to be similar to that shown in Figure 1a, however/expect, the relative occurrence of the Cu peaks compared to the broad amorphous peak is more intensive. This can be mainly attributed to the smaller thickness of the sample (10 μm) than that from group 1. The *P* content of samples from group 2 was found to be 6.7±0.1 wt.%. As the results of samples from group 2 indicate, the suitable selection of pulse parameters allows the formation of amorphous structure at relative low phosphorous content, too.

Figure 2 shows typical room temperature Mössbauer spectra of pulse plated Fe–P samples from group 1 and 2. The envelope of spectrum (on Figure 2a) of the sample deposited at short relaxation time and low current density (group 1) exhibits six broad lines typical for the magnetically ordered amorphous alloy. This spectrum can be considered as a superposition of a large number of sextets having different hyperfine parameters due to iron atoms being in slightly different surroundings in the amorphous alloy. The average Mössbauer parameters, shown in Table I, roughly correspond to those of Fe–P amorphous alloys prepared by dc electrodeposition or rapid quenching in the corresponding composition range [11, 12]. Consequently, the Mössbauer spectroscopy together with the XRD reveal that it was succeeded to prepare an amorphous Fe–P alloy layer using pulse plating method. The hyperfine field value of pulse plated Fe-9 wt.% P alloy from group 1 is smaller than that observed previously [11, 12], indicating differences in the short range ordering due to the different preparation routes.

The relative area of the second and fifth lines compared to those of first and sixth lines of the sextet, determined by the angle between the direction of the gamma ray and the average direction of the effective magnetic field, can be found to be 4:3 (Figure 2a) which reflects shape anisotropy causing in-plane orientation of the spins, according to the expectation [13].

The Mössbauer spectrum (Figure 2b) of Fe–P sample pulse plated with long relaxation time and high current density (group 2) differs significantly from that of sample from group 1 (Figure 2a). The broad six line envelope indicates the presence of a dominant amorphous

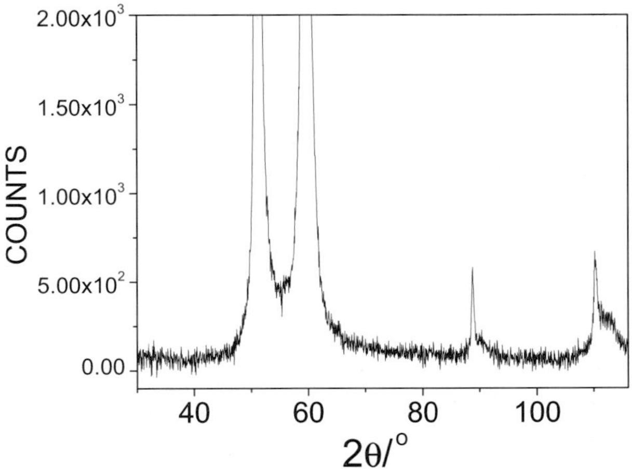

Figure 3 X-ray diffractogram of a Ni–Fe pulse plated sample.

phase in this sample, too. The asymmetry of the third and fourth line requires to involve new components (at least a doublet) in the decomposition of the spectra. The change in the average hyperfine field, belonging to the ferromagnetic amorphous Fe–P phase, compared to that of sample from group 1 can be connected with the change of the *P* content to 6.7 wt.% [11] and with the change in the chemical short range order. The doublet can be assigned to the paramagnetic amorphous Fe–P phase. This can be connected with Fe surroundings, not the same but somewhat similar to those occurring in Fe_2P [14, 15]. The presence of analogous paramagnetic phase has already been shown by the help of Mössbauer spectroscopy in the case of Fe–P amorphous alloys prepared by dc electrodeposition or by rapid quenching techniques [16]. Our present results are consistent with the transmission electron microscopy observations which reveal that these samples contain few nano-crystallites embedded into the amorphous matrix. It was also found that the occurrence and the size of nano-crystallites increased considerably with decreasing the relaxation time at high current density of the pulse deposition. The increase of the deposition time in this case has already resulted well visible crystalline phases (besides the amorphous one) both by TEM and by XRD (to be published [9]).

Visible minor sextets were also present in both Fe–P spectra which sextets can be associated with a magnetite due to the surface oxidation of the sample after the electrolyses.

It can be also well visible in the spectrum on Figure 2b that the relative intensity of the second and fifth lines of the sextets decreased considerably in comparison to that belonging to sample from the group 1 (Figure 2a). This change in the magnetic anisotropy can be connected with the difference in the amorphous structure between the samples prepared with short or long relaxation period, namely it can be attributed to changes in the topological defects, in the nano-crystalline inclusions, in the geometrical arrangement of atoms and in their stress fields which will cause corresponding changes in the orientation of spins [17].

The distribution of the layer components (Table I) was estimated from the CEM spectrum data using the BEATRICE code [18] (a modified Liljequist method [19]), similarly as was done in our previous work [3].

Figure 3 shows the X-ray diffractogram of the pulse plated Fe–Ni layer, deposited with same pulse parameters as used for Fe-P group 1. The relative broad peaks besides the narrow reflections of Cu substrate can be well identified as the pattern of disordered solid

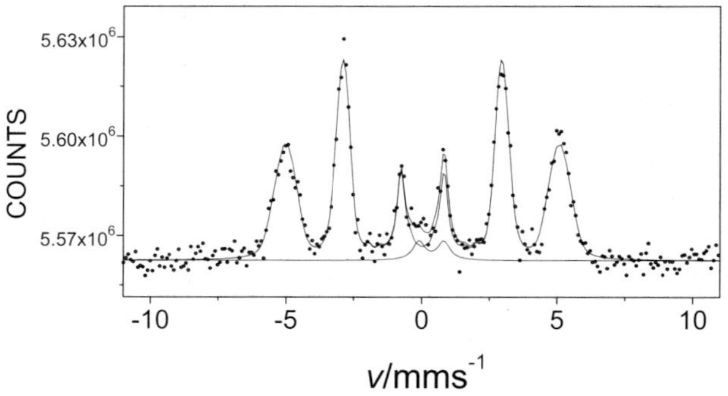

Figure 4 CEM spectrum of a Ni–Fe pulse plated sample.

solution fcc Ni–Fe alloy with a lattice parameter $a=0.35766$ nm. The broad lines can be associated with the microcrystalline character of the Ni–Fe electrodeposit, 5–8 nm very fine grain size was calculated from the profile analyzed XRD lines.

The CEM spectrum of the pulse plated Ni–Fe sample (Figure 4) exhibits a sextet with relatively broad lines. The average parameters of the spectrum (Table I) correspond to those of Ni–Fe alloys [20, 21]. The spectrum can be understood as a Ni-22 wt.% Fe solid solution by decomposition of the spectrum into sextets taking into consideration the polarization effect of the Ni neighbors [20], located in the different coordination spheres of iron, on the iron, in agreement with the 78 wt.% Ni content determined by chemical analysis. Consequently, the Mössbauer spectroscopy together with the XRD reveal that a highly disordered microcrystalline solid solution Ni–Fe layer was successfully prepared by the pulse plating electrochemical route. The small intensity doublet in Figure 4 (Table I) can indicate the presence of the ferrihydrite which can occur in Ni rich Fe–Ni system [22]. Similar component was also observed with fine particle (20 nm) evaporated Ni–Fe alloys [20].

The amorphous Fe–P and microcrystalline Fe–Ni coatings showed unique magnetic and electrochemical properties (to be published in details [9]). The magnetic hysteresis measurements of amorphous Fe–P samples from group 1 and group 2 and of Ni–Fe samples gave 76, 85 and 20 A/m coercitive field, 0.62, 0.82 and 0.97 T saturation magnetisation and 16.3, 16.5 and 6.76 W/kg power loss values at 1 kHz frequency, respectively. The high frequency limit (at 10 μm layer thickness) is extended up to 5 MHz for the pulse plated Ni–Fe and 85 MHz for the pulse plated Fe–P. This latter threshold is 5–85 times better than that of the best ferrite magnets.

Acknowledgements This study was supported by the Hungarian Science Foundation (OTKA T043687). The technical help of Mr S. Stichleutner is also acknowledged.

References

1. Landolt, D., Marlot, A.: Surf. Coat. Technol. **169–170**, 8–13 (2003)
2. Kuzmann, E., Stichleutner, S., El-Sharif, M., Chisholm, C.U., Sziráki, L., Vértes, A.: Hyp. Int. **41**, 425–433 (2002)
3. Kuzmann, E., Lakatos-Varsányi, M., Varga, L.K., Mikó, A., Kálmán, E., Homonnay, Z., Nagy, F., Vértes, A.: ISIAME 04, Madrid, Proc. AIP **765** 211–216 (2005)
4. Lakatos-Varsányi, M., Mikó, A., Varga, L.K., Kálmán, E.: Corros. Sci. **47**, 681–693 (2005)
5. Vitkova, S.T., Kjuchukova, M., Raichevski, G.: J. Appl. Electrochem. **18**, 673–678 (1988)
6. Kamei, K., Maehara, Y.: Mater. Sci. Eng., **A181/A182**, 906–910 (1994)
7. Trudeau, M.L.: Nanostruct. Mater. **12**, 55–60 (1999)

8. Natter, H., Schmelzer, M., Löffler, M.-S., Krill, C.E., Fitch, A., Hempelmann, R.: J. Phys. Chem. B **104**, 2467–2476 (2000)
9. Mikó, A., Hempelmann, R., Lakatos-Varsányi, M., Kálmán, E.: Electrochim. Acta (submitted) (2005)
10. Klencsár, Z., Kuzmann, E., Vértes, A.: J. Radioanal. Nucl. Chem. **210**, 105–112 (1996)
11. Kuzmann, E., Vitkova, S., Láng, Gy., Raichewski, G., Vértes, A.: Hyp. Int. **67**, 671–675 (1991)
12. Cserei, A., Kuzmann, E., Vértes, A., Pöppl, L., Novochatskii, I.A.: J. Radioanal. Nucl. Chem.-Lett. **176**, 429–442 (1993)
13. Johnson, M.T., Lungblut, R., Kelly, P.J., den Broeder, F.J.A.: J. Magn. Magn. Mater. **148**, 118–124 (1995)
14. Wappling, R., Haggstom, L., Ericsson, T., Devarayana, S., Karlsson, E., Karlsson, B., Rundquist, R.: J. Solid State Chem. **13**, 258–271 (1975)
15. Zecevic, S.C., Zotovic, J.B., Gojkovic, S.Lj., Radmilovic, V.: J. Electroanal. Chem. **448**, 245–252 (1998)
16. Kuzmann, E., Vértes, Cs., Vértes, A., Vitkova, S., Novochatskii, I.A., Chisholm, C.U.: Electrochim. Acta **40**, 2013–2015 (1995)
17. Vértes, A., Vankó, Gy., Németh, Z., Klencsár, Z., Kuzmann, E., Homonnay, Z., Kármán, F.H., Szücs, E., Kálmán, E.: Langmuir **18**, 1206–1210 (2002)
18. Nagy, F., Klencsár, Z.: Nucl. Instrum. Methods **245**, 528–538 (2006)
19. Liljequist, D., Ekdahl, T., Bäverstam, U.: Nucl. Instrum. Methods B **155**, 529–538 (1978)
20. Drijver, J.V., Van der Woude, F., Radelaar, S.: Phys. Rev., B **16**, 985–992 (1977)
21. Huang, R.S., Widatallah, H.M., Hsia, Y.F., Lee, X.M., Wang, J.H., Lu, H.X., Tong, G.H.: Nuovo Cím. **50**, 227–230 (1996)
22. Refait, Ph., Drissi, S.H., Abdelmoula, M., Jeannin, M., Reffass, M., Genin, J.-M.R.: Proc. AIP **765**, 79–84 (2005)

Hyperfine Interact (2005) 165: 203–208
DOI 10.1007/s10751-006-9266-9

Mössbauer studies on ultraporous Fe-Oxide/SiO_2 aerogel

A. Lančok · K. Závěta · M. Popovici · C. Savii · M. Gich · A. Roig · E. Molins · K. Barčová

Published online: 10 November 2006

Abstract Magnetic aerogels with very low volume density of ~0.2 g/cm^3 were prepared by sol-gel method and supercritical drying. The resulting materials were monolithic and displayed high surface area. By X-ray diffraction and Mössbauer spectroscopy the crystalline phase formed inside the mesopores of the SiO_2 matrix was identified as a spinel iron oxide. Comparison of the magnetic measurements with Mössbauer spectra at various temperatures contributed to the elucidation of the magnetic state of this nanocomposite system with restricted magnetic interactions, in particular its transition to a superparamagnetic state.

Key words aerogels · Fe_2O_3 · nanoparticles · superparamagnetic state · Mössbauer spectroscopy

1 Introduction

Nanostructured materials have been a subject of intense research during the past 10 years. The nanocomposites that contain a reinforcement component in the form of one or more ultrafine phase with dimensions less than 100 nm, have been used to improve various

A. Lančok (✉) · K. Závěta
Joint Laboratory of Low Temperatures, V Holešovičkách 2, 180 00 Prague, Czech Republic
e-mail: lancok@seznam.cz

M. Popovici · C. Savii
Institute of Chemistry Timisoara of Romanian Academy, 24 Mihai Viteazul Blvd., 300223 Bucharest, Romania

M. Gich · A. Roig · E. Molins
Institut de Ciència de Materials de Barcelona (CSIC), Esfera UAB, 08193 Bellaterra, Catalonia, Spain

K. Barčová
Institute of Physics, Technical University of Ostrava, 17. listopadu 15, 70833 Ostrava, Czech Republic

material characteristics, including mechanical, chemical, structural, optical and electric/ magnetic properties. Porous amorphous matrices play an active role in shaping the physical properties of intergrown nanoparticles in addition to providing a means of their confinement and dispersion [1]. These matrices are easily obtained by sol-gel method which offers the opportunity to produce not only ultra-homogeneous materials but also heterogeneous or nanocomposite materials [2]. Sol-gel process consists of hydrolysis and condensation reactions of an alkoxide resulting in formation of a wet gel, a solid network containing liquid phase entrapped inside the pores. Materials called aerogels are obtained by supercritical drying process of the wet gels and consist of a continuous polymeric inorganic network comprising open pores in the mesoporous range. These materials are of interest due to the combination of high porosity with small pore size [3] that assures physical and chemical properties difficult to obtain by other synthesis methods at low temperatures. Despite the growing interest in magnetic nanoparticles, due to their peculiar properties such as superparamagnetism [4], quantum tunnelling of the magnetization [5], and the broad range of applications, from medicine to magnetic recording [6, 7], relatively few studies on the magnetic properties of nanocomposite aerogels [8, 9] have been published. The iron oxide/silica aerogels obtained by us as monolithic samples by sol-gel process using the impregnation method, but with considerably higher bulk density than in the present work, are described in [8].

The aim of the present work is to prepare and characterize aerogels containing iron oxide nanoparticles with increased spatial separation and thus suppressed magnetic interactions. Their magnetic state, in particular the transition to superparamagnetic behaviour, has been investigated by magnetometry and Mössbauer spectroscopy in the temperature range from 5 to 300 K.

2 Experimental

Silica gel matrices were prepared by sol-gel synthesis in neutral medium via hydrolysis and condensation of tetraethoxysilane (Aldrich 98%) in hydroethanolic solution. All the gels, obtained after six days at room temperature, were transparent monolithic slabs, and were washed several times with pure ethanol to remove residual water; then they were kept in closed vessels. Anhydrous ferrous acetylacetonate (Aldrich 99.95%) was used as iron source for the magnetic nanoparticles. The impregnation of silica gels with a supersaturated ethanolic solution of iron (II) acetylacetonate took place during four days. Finally, ethanol supercritical drying was carried out in an autoclave at 260°C and 131 bar. The whole drying cycle lasted about 24 h.

The iron content of the samples was determined by chemical analysis with flame atomic absorption spectrophotometry. Samples were characterized by X-ray diffraction (XRD) with the crystallite sizes estimated from the width of the diffraction peaks using the Scherrer formula. Using an ASAP 2000 equipment, the pore-size distribution was derived from N_2 adsorption/desorption isotherms by the Barrett–Joyner–Halenda (BJH) method and surface area was determined by the Brunauer–Emmett–Teller (BET) method. Hysteresis loops were measured by an extraction method in a PPMS facility by Quantum Design at 5, 130, and 300 K with applied fields up to 5 T. Mössbauer spectra (MS) were obtained at the same temperatures using a conventional spectrometer with a ^{57}Co/Rh source and constant acceleration and fitted by means of the NORMOS program [10] using Voigt line profiles for the sextets.

Table I Textural parameters of the aerogels

Sample	Surface area BET (m^2/g)	Bulk density (g/cm^3)
A	717±50	0.19±0.02
AFe	774±50	0.21±0.02

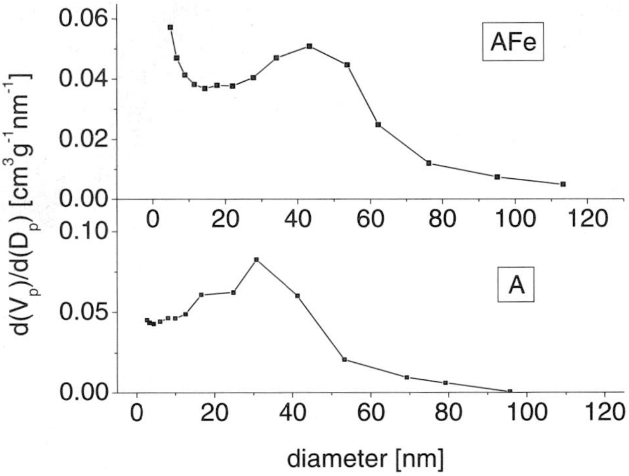

Figure 1 Pore-size distribution from BJH method.

3 Results and discussion

The textural parameters for the pure silica matrix A, and iron oxide-impregnated silica AFe are given in Table I. There are no large differences between the density and surface area of the A and AFe samples with rather large BET surface area of 774 m^2/g for the latter case. The iron content was estimated at 6 wt.% by chemical analysis.

The wide pore-size distribution determined from the adsorption/desorption isotherms at 77 K is shown in Figure 1. It indicates a high volume porosity in both samples which is also evident from the nitrogen adsorption mainly occurring at high relative pressure.

From the XRD pattern of the AFe nanocomposite, see Figure 2, the main diffraction lines of cubic iron oxide spinel phase are clearly seen though it is not possible to decide whether they belong to maghemite or magnetite phases. The Scherrer diameter of the particles was estimated to 5.7±2 nm taking into consideration the instrumental broadening.

The magnetization curves of the AFe sample are displayed in Figure 3. The saturated magnetizations $m_S(T)$ were estimated from the extrapolation of the Langevin-type function to high fields. The reduced magnetizations m/m_S are plotted against H/T in the inset. The curves for 300 and 130 K practically merge, which, together with negligible coercivity, indicates mainly superparamagnetic (SPM) behaviour. The rather different curve with a coercivity of ~14 mT points to dominating blocked moments at 5 K.

The Mössbauer spectra of the AFe sample acquired at 300, 130, and 5 K are shown in Figure 4 and the parameters of their fits are given in Table II. The 300 and 130 K spectra were fitted by one doublet and a narrow (S1) and "broad" (S2) sextets. The widths of the S1 sextets were ~4 T while those of the S2 sextets were comparable with the hyperfine field itself. From the relative areas of the narrow and broad sextets, the ratio of particles with blocked and fluctuating moments, in the time window relevant to Mössbauer effect, was roughly estimated [11]. The 5 K spectrum could be fitted by two doublets and two relatively narrow sextets S1 (the widths ~4 T) that were definitely different from the

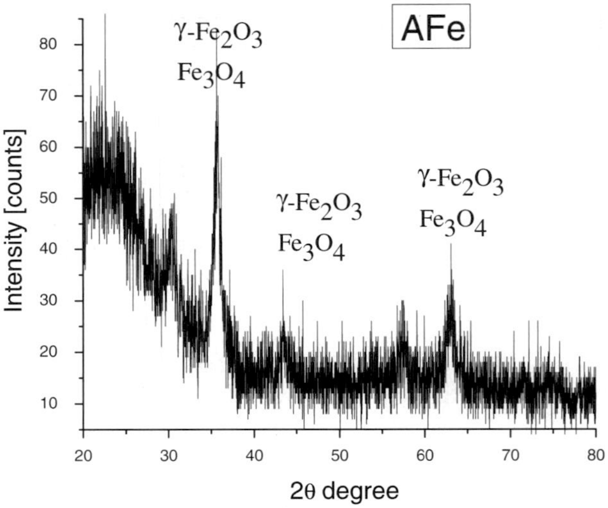

Figure 2 X-ray diffraction pattern.

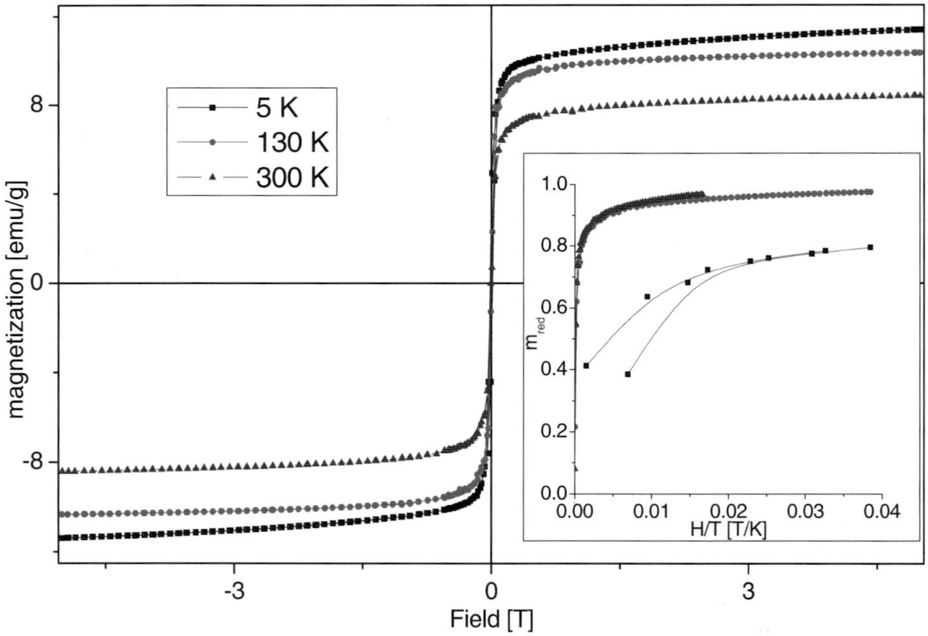

Figure 3 Magnetization curves of aerogel.

magnetite spectrum. Taking into account the distribution of the particle sizes, the effect of SPM relaxation, and the large relative volume of the nanoparticle surfaces, all of them decreasing the mean hyperfine field [11, 12], we may most probably ascribe these sextets to spinel-like oxide of three-valent Fe, i.e., maghemite. It is worth noting that the relative areas of the doublets in our tentative fit do not essentially change with temperature and only the relative areas of the resolved and broad sextets are affected. This is in contrast to [13], where the SPM relaxation of slightly larger γ-Fe_2O_3 nanoparticles lead to a higher share of the doublets at the expense of the resolved sextets.

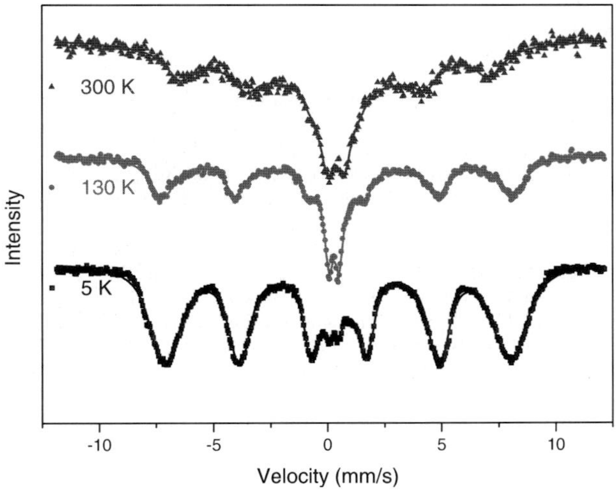

Figure 4 Mössbauer spectra at various temperatures.

Table II Parameters of Mössbauer spectra

	300 K				130 K				5 K			
	B_{hf} (T)	QUA (mm/s)	ISO (mm/s)	r.a. (%)	B_{hf} (T)	QUA (mm/s)	ISO (mm/s)	r.a. (%)	B_{hf} (T)	QUA (mm/s)	ISO (mm/s)	r.a. (%)
Sextet 1	42.0	−0.05	0.34	24.2	47.5	0.03	0.41	51.4	47.0	0.19	0.52	67.3
									46.5	−0.78	0.35	17.6
Sextet 2	24.6	−0.54	0.03	60.6	19.9	−0.49	0.53	31.1				
Doublet 1		−0.68	0.37	15.2		−0.45	0.29	17.5		0.08	0.46	12.6
Doublet 2										0.41	0.27	2.4

4 Conclusions

Dispersed nanoparticles of iron oxide embedded in silica aerogels have been obtained by impregnation of the wet silica gel with an anhydrous iron(II) acetylacetonate salt and subsequent supercritical drying without post annealing. The nanocomposite has a rather low bulk density of ~0.2 g/cm^3, large surface area and the pore distributions of the matrix exhibit a maximum around 40 nm. The size of the iron oxide particles, most probably γ-Fe_2O_3, is in the nanometer range (~6 nm) and we may assume that their interactions are effectively avoided. From the magnetization curves follows that at 130 K most of the particles are already in the SPM state, while in the time window seen by the Mössbauer effect the transition to the SPM behaviour is not complete even at 300 K.

Acknowledgements The discussions and help of J. Nogués are gratefully acknowledged. M. Popovici thanks the Marie Curie Program for fellowship No. HPMT-CT-2000-0006. This work was supported by Ministerio de Educación y Ciencia of Spain (MAT2003-01052) and by the Grant Agency of the Czech Republic under grants Nos. 202/05/2111 and 202/03/P158; the participation of K.B. on this conference was enabled by IGS-VSB-2005-516/1.

Our respectful thanks go to the referee, whose remarks considerably helped to improve our text.

References

1. Shull, R.D., Ritter, J.J., Swartzendruber, L.J.: J. Appl. Phys. **69**, 5144 (1991)
2. Sternitzke, M.: J. Eur. Ceram. Soc. **17**, 1061 (1997)
3. Scherer, G.W.: Adv. Colloid Interface Sci. **76–77**, 321 (1998)
4. Martinez, B., Roig, A., Obradors, X., Molins, E., et al.: J. Appl. Phys. **79**, 2580 (1996)
5. Chudnovsky, E.M., Gunther, L.: Phys. Rev. B **37**, 9455 (1988)
6. Klem, M.T., Young, M., Douglas, T.: Materials Today **8**, 28 (2005)
7. Martín, J.I., Nogués, J., Liu, K., et al.: J. Magn. Magn. Mater. **256**, 449 (2003)
8. Popovici, M., Gich, M., Roig, A., et al.: Langmuir **20**(4), 1425 (2004)
9. Racka, K., Gich, M., Slawska-Waniewska, A., et al.: J. Magn. Magn. Mater. **290–291**, 127 (2005)
10. Brand, R.A.: Normos Programs. Duisburg University (1989)
11. Kündig, W., Bömmel, H., et al.: Phys. Rev. **142**, 327 (1966)
12. Prene, P., Tronc, E., Jolivet, J.-P., et al.: IEEE Trans. Magn. **29**, 2658 (1993)
13. Shafi, K.V.P.M., Ulman, A., Dyal, A., et al.: Chem. Mater. **14**, 1778 (2002)

Hyperfine Interact (2005) 165: 209–213
DOI 10.1007/s10751-006-9267-8

A novel investigation of tin-doped ferrihydrite nanoparticles

D. R. Mabe · A. M. Khasanov · J. G. Stevens

Published online: 7 November 2006

Abstract Tin-doped ferrihydrite was synthesized and characterized with powder X-ray diffraction and ^{57}Fe and ^{119}Sn Mössbauer spectroscopy to obtain diagnostic and structural information. Samples were prepared with doping concentrations (molar percentages) of 0, 10, 25, 40, 50, 60, 75, and 90% tin. As Sn concentration increases, the 110 reflection plane of ferrihydrite shifts to higher *d*-values, while the 300 plane *d*-values increase, reach a maximum and then sharply drop. The δ (isomer shift) and Δ (quadrupole splitting) values for ^{57}Fe Mössbauer spectra increase, while the Γ (line width) values stay constant. In ^{119}Sn Mössbauer spectra, δ values decrease with increasing tin substitution while Δ and Γ values increase. After 50% Sn has been substituted, the ferrihydrite becomes oversaturated with surface tin and a secondary tin structure is proposed to develop.

Key words ^{57}Fe and ^{119}Sn Mössbauer spectroscopy · powder X-ray diffraction · 2-line ferrihydrite · tin

1 Introduction

Ferrihydrite is an iron oxyhydroxide given the molecular formula $Fe_5HO_8{\cdot}4H_2O$ [1], although $5Fe_2O_3{\cdot}9H_2O$ [2] and others [3, 4] have been accepted. Ferrihydrite can show from two to six X-ray diffraction lines and nanoparticles are typically 2–10 nm in size [2, 5]. Adsorption of electrolytes to 2-line ferrihydrite has been shown to reduce its particle sizes [6]. Structurally, ferrihydrite is believed to be based on simple chains of iron octahedra, although tetrahedrally coordinated iron has also been proposed [3, 7]. Ferrihydrite's large surface area to volume ratio gives it high sorptive capabilities, especially towards the heavy and transition metals, which are of environmental and industrial importance [8–13].

D. R. Mabe · A. M. Khasanov · J. G. Stevens (✉)
University of North Carolina at Asheville, One University Heights, Asheville, NC 28804, USA
e-mail: stevens@unca.edu

Table I X-ray diffraction data

Sample	% Sn	d [110] ±0.001 nm	d [300] ±0.001 nm
F1	0	0.257	0.148
F2	10	0.257	0.149
F3	25	0.258	0.149
F4	40	0.259	0.151
F5	50	0.26	0.153
F6	60	0.261	0.15
F7	75	0.263	0.146
F8	90	0.264	0.145

Table II Mössbauer spectral data

		^{57}Fe				^{119}Sn		
	% Sn	δ	Δ	Γ	% area	δ	Δ	Γ
F1	0	0.34	0.57	0.40	64			
	0	0.34	1.00	0.40	36			
F2	10	0.34	0.57	0.41	60	0.08	0.52	0.87
	10	0.34	1.01	0.40	40			
F3	25	0.35	0.57	0.46	60	0.06	0.58	1.04
	25	0.35	1.08	0.47	40			
F4	40	0.35	0.57	0.38	61	0.05	0.60	1.06
	40	0.36	1.10	0.38	39			
F5	50	0.36	0.56	0.40	60	0.05	0.61	1.10
	50	0.36	1.16	0.41	40			
F6	60	0.36	0.54	0.39	58	0.03	0.71	1.26
	60	0.36	1.20	0.43	42			
F7	75	0.37	0.54	0.32	54	0.02	0.85	1.68
	75	0.37	1.22	0.47	46			
F8	90					0.01	0.91	1.74

δ, Δ and Γ values are ±0.01 mm/s. Area values are ±5.0%.

2 Experimental

2-line ferrihydrite was precipitated from a $FeCl_3{\cdot}6H_2O$ solution with NaOH. Sn(IV) was added in molar ratios $((Sn/(Sn+Fe))^{*}100)$ ranging from 0 to 90% by substituting the necessary amount of $SnCl_4{\cdot}5H_2O$. Sn(IV) was used instead of Sn(II) because Fe(III) reacts with Sn(II) to produce Fe(II) and Sn(IV). All samples were repeatedly washed with distilled water and centrifuged until tests for chloride ion with silver nitrate were negative. Powder X-ray diffraction was performed using a Philips X'pert diffractometer with a Cu source and only $\alpha 1$ data is reported. A step size of 0.02° (2θ) and a 10 s step time was used. ^{57}Fe and ^{119}Sn Mössbauer measurements were performed at 297 K with a constant acceleration spectrometer and at least 1×10^6 per channel counts were collected. ^{57}Fe spectra were obtained with a ^{57}Co source in a Rh matrix, calibrated with α-Fe and fit using single and two doublet procedures. ^{119}Sn spectra were collected using a ^{119}Sn source calibrated with

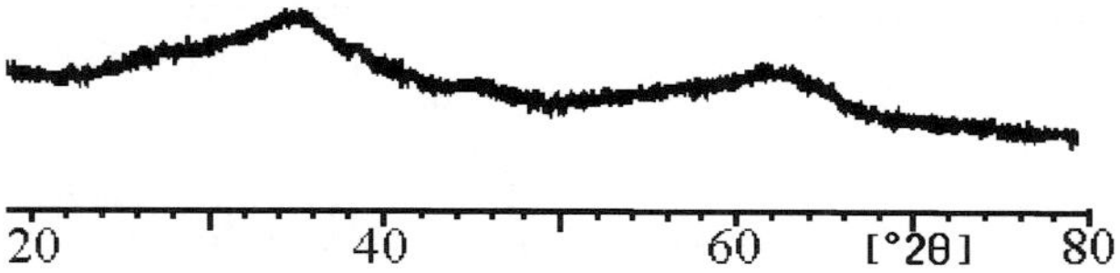

Figure 1 X-ray diffractogram of F2.

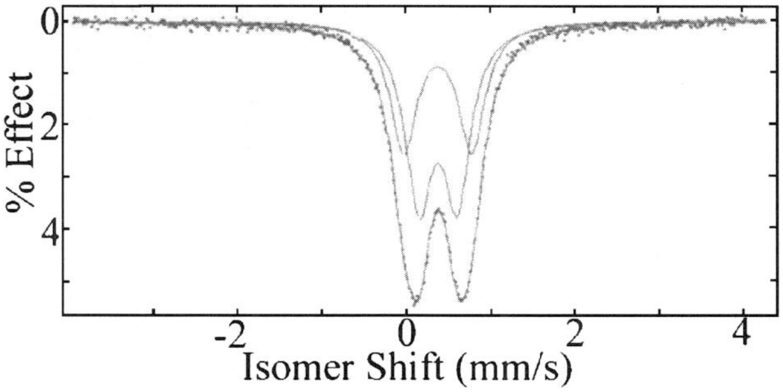

Figure 2 57Fe Mössbauer spectrum of F3.

αFe, referenced to barium stannate, and fit with one doublet component. X-ray diffraction data is reported in Table I and Mössbauer spectral data in Table II.

3 Results

Samples F1–F5 are consistent with published color observations for ferrihydrite (reddish-brown) [2]. After F5 the samples become increasingly tan, then white in F8. X-ray diffraction data indicates the main phase present is two line ferrihydrite, showing two very broad peaks at 0.257 nm and 0.148 nm. The diffractogram in Figure 1 is representative of all samples. The 0.257 nm peak (110 plane) shifts to 0.264 nm in F8. The 0.148 nm peak (300 plane) data was not able to be fit with a simple linear regression, but instead shows an increasing trend followed by a decreasing tendency. No significant change is noticed in peak widths as Sn substitution increases. The ^{57}Fe Mössbauer spectrum presented in Figure 2 is representative of all spectra collected. Two doublet fitting is shown because it typically reduced χ^2 by a factor of two and has a logical connection to the ferrihydrite structure [14–16]. An increasing trend was noted for δ values. For Δ values, the outer doublet increases while the inner doublet does not change (Figure 3). Changes among Γ and percent area values fell well within error limits. The ^{119}Sn Mössbauer spectrum of F3 is shown in Figure 4, containing a broad doublet with low δ, diagnostic of Sn(IV). As Sn substitution increases, δ decreases from 0.08 to 0.01 mm/s. Δ increases sharply from 0.52 to 0.91 mm/s and Γ increases as well, rising from 0.87 to 1.74 mm/s.

4 Discussion

Initially, as Fe octahedra are condensing in solution, Sn octahedra are absorbed onto ferrihydrite's surface. Sn in octahedral coordination with O has very diagnostic ^{119}Sn Mössbauer parameters; δ values very close to 0.00 mm/s and Γ values larger than 1.00 mm/s.

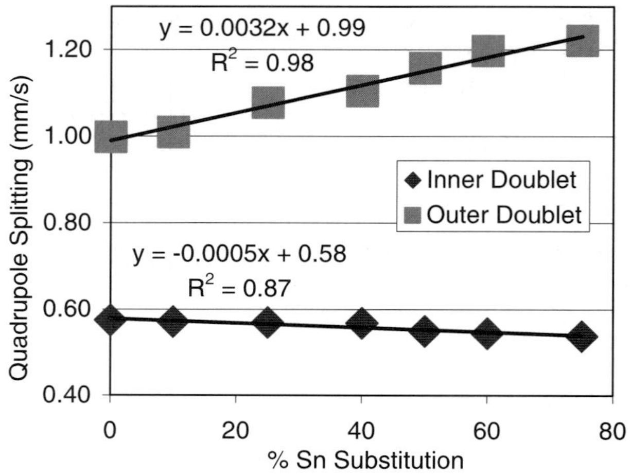

Figure 3 57Fe QS values.

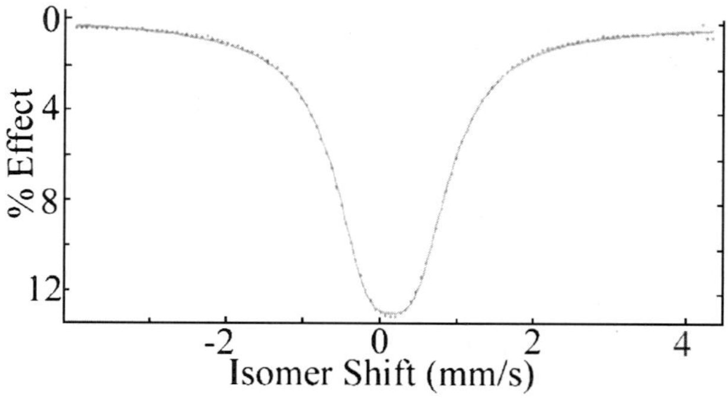

Figure 4 119Sn Mössbauer spectrum of F3.

After 50% Sn concentration there is an excess of Sn in solution, allowing Sn octahedra to begin their own packing pattern on the ferrihydrite surface. The δ values for ^{119}Sn Mössbauer spectra support increased connectivity of Sn octahedra since they approach zero as Sn substitution increases. As seen in previous research with silicon [17] and borate [18], absorption of cationic or anionic species onto the surface of ferrihydrite produces greater *d*-values. For Sn substitutions performed in this study, the *d*[110] values follow the predicted increasing trend; however, the *d*[300] values do not. Since the *d*[300] values begin decreasing in F6, the excess Sn present is the most logical reason for the decrease, possibly compacting the nanoparticles along the 300 plane. ^{57}Fe Mössbauer spectral data shows increasing δ values, which are accompanied by the decreasing trend of ^{119}Sn's δ values, providing evidence that Sn is the source of Fe's added electron density. As expected, the Δ values for Fe and Sn increased significantly since their respective electronic environments become increasingly distorted as Sn octahedra add to ferrihydrite. If some Sn is forming a secondary surface structure after 50%, there would be two distinct Sn octahedral sites, much like there is for Fe in ferrihydrite. The initial Δ value of 0.52 mm/s for F2 slightly increases up to 0.61 mm/s for F5; however, there is a larger increase to 0.71 mm/s for F6. The same trend is noted for Γ values. Logically, it seems after 50% Sn concentration there is the added Sn site that is linked to other Sn octahedra, which produces much higher quadrupole splitting and line width values. X-ray diffraction indicates ferrihydrite as the only phase present in all samples, so ferrihydrite samples with high Sn concentration are simply overwhelmed, and Sn begins to organize into its own oxyhydroxide structure on ferrihydrite's surface, producing the lighter white color characteristic of Sn oxyhydroxides.

Acknowledgements Appreciation is expressed to Sara Boshamer for some of the initial work conducted on the project. The UNC-Asheville Undergraduate Research Program is acknowledged for providing a grant.

References

1. Towe, K.M., Bradley, W.F.: J. Colloid Interface Sci. **24**, 384 (1967)
2. Jambor, J.L., Dutriziac, J.E.: Chem. Rev. **98**, 2549 (1998)
3. Eggleton, R.A., Fitzpatrick, R.W.: Clays Clay Miner. **36**, 111 (1988)
4. Chukhrov, F.V., Zvyagin, B.B., Gorshkov, A.I., Yermilova, L.P., Balashova, V.V.: Int. Geol. Rev. **16**, 1131 (1973)
5. Schwertmann, U., Friedel, J., Kyek, A.: Clays Clay Miner. **52**, 221 (2004)
6. Hofmann, A., Pelletier, M., Michot, L., Stradner, A., Schurtenberger, P., Kretzschmar, R.: J. Colloid Interface Sci. **271**, 163 (2004)
7. Cardile, C.M.: Clay Miner. **36**, 537 (1988)
8. Carlson, L., Bigham, J.M., Schwertmann, U., Kyek, A., Wagner, F.: Environ. Sci. Technol. **36**, 1712 (2002)
9. Simeoni, M.A., Batts, B.D., McRae, C.: Appl. Geochem. **18**, 1507 (2003)
10. Badruzzaman, M., Westerhoff, P., Knappe, D.R.: Water Res. **38**, 4002 (2004)
11. Loyaux-Lawniczak, S., Refait, P., Ehrhardt, J.J., Lecomte, P., Genin, J.R.: Environ. Sci. Technol. **34**, 438 (2000)
12. Gustafsson, J.P.: Chem. Geol. **200**, 105 (2003)
13. Neri, G., Rizzo, G., Galvagno, S., Loiacono, G., Donato, A., Musolino, M.G., Pietropaolo, R., Rombi, E.: Appl. Catal., A. **274**, 243 (2004)
14. Murad, E., Bowen, L.H., Long, G.J., Quin, T.G.: Clay Miner. **23**, 161 (1988)
15. Johnston, J.H., Logan, N.E.: J. Chem. Soc. Dalton Trans. **13**, 13 (1979)
16. van der Kraan, A.M.: Phys. Status Solidi **18**, 215 (1973)
17. Seehra, M.S., Roy, P., Raman, A., Manivannan, A.: Sol. St. Comm. **130**, 597 (2004)
18. Stevens, J.G., Khasanov, A.M., Mabe, D.R.: Hyp. Int. **161**, 83 (2005)

Hyperfine Interact (2005) 165: 215–220
DOI 10.1007/s10751-006-9268-7

Mössbauer study of nanodimensional nickel ferrite – mechanochemical synthesis and catalytic properties

E. Manova · C. Estournès · D. Paneva · J.-L. Rehspringer · T. Tsoncheva · B. Kunev · I. Mitov

Published online: 10 November 2006

Abstract Iron–nickel spinel oxide $NiFe_2O_4$ nanoparticles have been prepared by the combination of chemical precipitation and subsequent mechanical milling. For comparison, their analogue obtained by thermal synthesis is also studied. Phase composition and structural properties of iron–nickel oxides are investigated by X-ray diffraction and Mössbauer spectroscopy. Their catalytic behavior in methanol decomposition to CO and methane is tested. An influence of the preparation method on the reduction and catalytic properties of iron–nickel samples is established.

Key words X-ray diffraction · Mössbauer spectroscopy · mechanochemical synthesis · methanol decomposition, ferrite nanoparticles

1 Introduction

The synthesis of spinel ferrite nanoparticles has been intensively studied in the recent years, because of their large-scale application. The principal role of the preparation conditions on the morphological and structural features of the ferrites is discussed [1–6]. In the last two

E. Manova · D. Paneva · B. Kunev · I. Mitov
Institute of Catalysis, Bulgarian Academy of Sciences,
Acad. G. Bonchev Str., bl. 11, 1113 Sofia, Bulgaria

T. Tsoncheva
Institute of Organic Chemistry, Bulgarian Academy of Sciences,
Acad. G. Bonchev Str., bl. 9, 1113 Sofia, Bulgaria

J.-L. Rehspringer
I.P.C.M.S. Groupe des Matériaux Inorganiques, UMR7504 CNRS, ULP, ECPM.,
23 rue du Loess, BP 43, 67034 Strasbourg Cedex 2, France

C. Estournès (✉)
CIRIMAT, UMR5085 CNRS, UPS,
118, route de Narbonne, 31062 Toulouse Cedex 4, France
e-mail: estourne@chimie.ups-tlse.fr

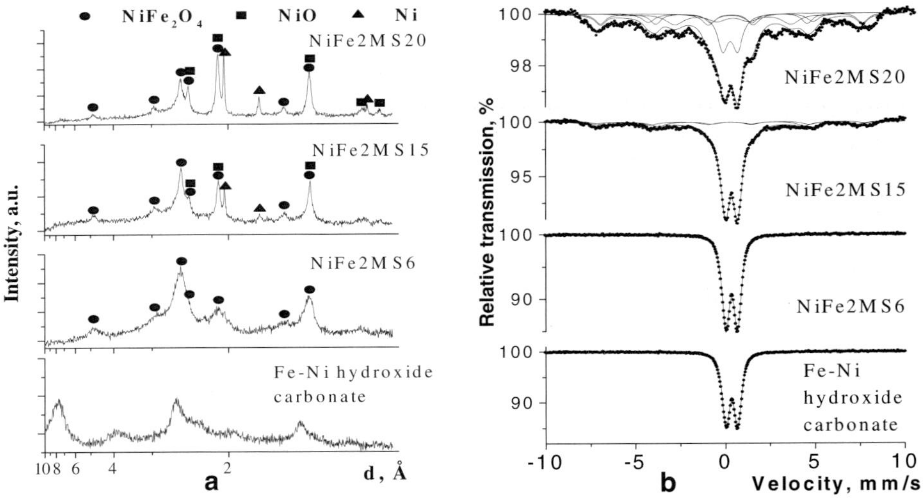

Figure 1 XRD patterns (a, abscissa is d-spacing) and Mössbauer spectra (b) of the samples before and after different milling times.

decades, the decomposition of methanol to CO and H_2 has received growing attention because of its possible use as an alternative, ecological fuel. The reaction is also applicable to the recovery of waste heat from industries, but significant improvement of the catalysts must be achieved. In our previous studies [7] it was shown that nanosized cobalt ferrites present a good activity in methanol decomposition, and both the preparation method and Fe/Co ratio influence the phase transformations and catalytic behavior of the samples. The present paper aims at studying the effects of the preparation method on the catalytic properties in methanol decomposition of nanosized nickel ferrites obtained by mechanochemical and thermal methods.

2 Experimental

The starting materials used were $Fe(NO_3)_3.9H_2O$ (purity 99%), $Ni(NO_3)_2.6H_2O$ (purity 96%) and Na_2CO_3. Nickel ferrite powders ($NiFe_2O_4$) were prepared in two steps: (1) co-precipitation to obtain the precursor, and (2) mechanical milling of the precursor. For the co-precipitation the procedure described in [7] was followed. The obtained precursor powders were milled using a Fritsch Planetary miller in a hardened steel vial together with fifteen grinding balls having different diameters (ranging from 3 to 10 mm). The ball-to-powder mass charge ratio was 10:1. The powders were milled for 6, 15 and 20 h (samples denoted as NiFe2MS6, NiFe2MS15 and NiFe2MS20, respectively).

The thermal synthesis (NiFe2TS) was performed as described in [7].

Phase composition and structural properties of the obtained samples are investigated by X-ray diffraction (XRD) and Mössbauer spectroscopy. The powder XRD patterns were recorded by use of a TUR M62 diffractometer with Co Kα radiation. The observed patterns were cross-matched with those in the JCPDS database. Mössbauer spectra of nanosize nickel ferrite particles were recorded at 295 K on a Wissel electromechanical Mössbauer spectrometer (Wissenschaftliche Elektronik GmbH, Germany) working in a constant acceleration mode. A $^{57}Co/Cr$ (activity≅10 mCi) source and an α-Fe standard were used.

Table I Mössbauer parametra of mecanochemically synthesized samples

Sample	Components	IS (mm/s)	QS (mm/s)	H_{eff} (kOe)	FWHM (mm/s)	G (%)
NiFe2MS6	SPM $NiFe_2O_4 - Fe^{3+}_{octa}$	0.33	0.67	–	0.50	100
NiFe2MS6	$Fe_3O_4 - Fe^{3+}_{tetra}$	0.29	0.03	485	0.48	19
after catalytic test	$Fe_3O_4 - Fe^{2.5+}_{octa}$	0.71	0.01	457	0.63	17
	Fe–Ni	0.01	0.04	307	0.68	44
	Fe_3C	0.21	0.03	197	0.44	20
NiFe2MS15	$NiFe_2O_4 - Fe^{3+}_{tetra}$	0.26	0.00	435	1.00	15
	$NiFe_2O_4 - Fe^{3+}_{octa}$	0.36	0.00	476	1.00	12
	SPM $NiFe_2O_4$–Fe^{3+}	0.33	0.70	–	0.63	73
NiFe2MS20	$NiFe_2O_4 - Fe^{3+}_{tetra}$	0.25	0.00	435	1.26	13
	$NiFe_2O_4 - Fe^{3+}_{octa}$	0.35	0.00	476	0.50	42
	Fe–Ni	0.12	0.10	289	0.50	14
	SPM $NiFe_2O_4$–Fe^{3+}	0.33	0.65	–	0.50	28
	Fe^{2+}	0.86	2.33	–	0.40	3
NiFe2MS20	$Fe_3O_4 - Fe^{3+}_{tetra}$	0.28	0.00	478	0.52	18
after catalytic test	$Fe_3O_4 - Fe^{2.5+}_{octa}$	0.55	0.00	441	0.74	13
	Fe–Ni	0.02	0.00	306	0.70	60
	Fe_3C	0.18	0.04	189	0.46	9
NiFe2TS	$NiFe_2O_4 - Fe^{3+}_{tetra}$	0.26	0.00	464	0.78	57
	$NiFe_2O_4 - Fe^{3+}_{octa}$	0.36	0.00	500	0.71	43
NiFe2TS	$NiFe_2O_4 - Fe^{3+}_{tetra}$	0.26	0.00	492	0.28	15
after catalytic test	$NiFe_2O_4 - Fe^{3+}_{octa}$	0.36	0.00	510	0.45	24
	$Fe_3O_4 - Fe^{3+}_{tetra}$	0.28	0.00	483	0.26	10
	$Fe_3O_4 - Fe^{2.5+}_{octa}$	0.61	0.00	457	0.42	12
	Fe–Ni	0.15	0.08	299	0.50	27
	Fe_3C	0.18	0.02	205	0.40	12

The experimental spectra were treated using the least squares method. The parameters of hyperfine interaction such as isomer shift (IS), quadrupole splitting (QS) and effective internal magnetic field (H_{eff}) as well as the line widths (FWHM) and the relative weight (G) of the partial components of the spectra were determined.

The catalytic experiments were performed in a fixed-bed reactor (0.1 g catalyst) at a methanol partial pressure of 1.57 kPa and a WHSV of 1.5 h^{-1}. The temperature was raised with a rate of 1 K/min in the range of 350–723 K. The on-line gas chromatographic analysis was made on a Porapak Q and a molecular sieve column using an absolute calibration method [8].

3 Results and discussion

XRD patterns of the precursor and those after different milling times are shown in Figure 1. The pattern of the precursor is characteristic of layered double hydroxides as found for pyroaurite (PDF 25-0521) and hydrotalcite (PDF 41-1428). After 6 h of milling the diffraction pattern of the sample consists of broad peaks. Their position and intensities suggest the formation of a spinel phase with cubic structure and lattice constant a_0= 8.341 Å, corresponding to $NiFe_2O_4$.

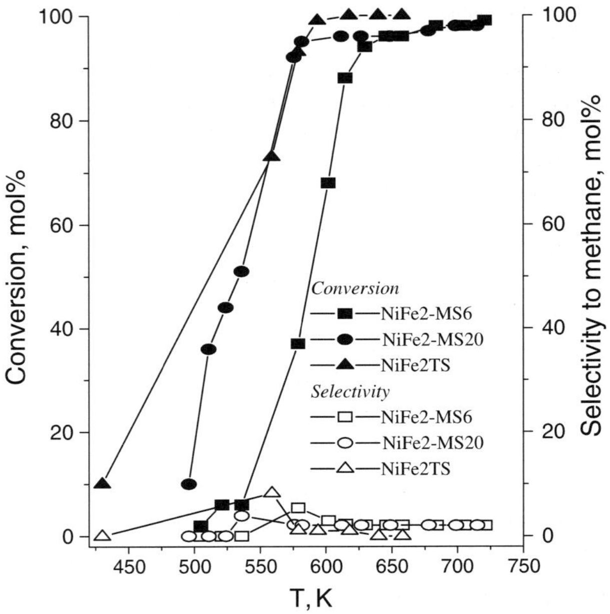

Figure 2 Temperature dependencies of methanol conversion and methane selectivity for mechanochemically and thermally synthesized nickel ferrites.

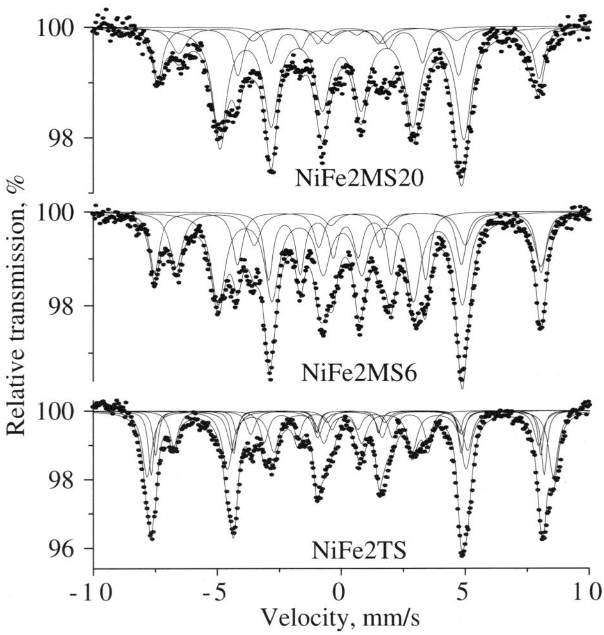

Figure 3 Mössbauer spectra of the samples after catalytic test.

The diffraction patterns show that with the increase of milling time some decomposition of the spinel phase into Ni and FeNi alloy occur. The average particle size calculated by the Scherrer equation is 3.5, 5.2 and 8.1 nm for NiFe2MS6, NiFe2MS15 и NiFe2MS20, respectively. It is interesting to note that the process of crystallization continues as a function of treatment time and at later stages is accompanied by the reduction to nickel. In our previous study we evidenced that high-energy milling of cobalt ferrite spinels in a steel vial and using steel balls induces the reduction of cobalt ions to metallic cobalt [9].

Mössbauer spectroscopy was applied to obtain more information about the phase composition, cationic occupations and/or different state distribution of iron ions in the studied ferrite materials. Figure 1b shows the RT Mössbauer spectra of samples taken from different steps of the processing route. The corresponding parameters are listed in Table I. The co-precipitated precursor exhibits a quadrupole doublet with IS=0.33 mm/s, QS= 0.65 mm/s, indicating that the hydroxide carbonate precursor is paramagnetic. As shown in Figure 1b the spectrum of NiFe2MS6 presents a doublet. The spectra of NiFe2MS15 and NiFe2MS20 consist of a central doublet superposed on a sextet pattern. A considerable diminution of the doublet component for the spectrum of NiFe2MS20 is observed. The doublets observed in all milled samples arise from Fe(III)-ions in ultradispersed ferrite particles exhibiting superparamagnetic (SPM) behavior [10, 11]. This particle size effect is in good agreement with the XRD data.

In Figure 2 the temperature dependencies of methanol conversion and methane selectivity for mechanochemically and thermally synthesized nickel ferrites are shown. Total (100%) methanol conversion is registered just above 575 K for NiFe2TS and NiFe2MS20, and at about 625 K for NiFe2MS6. CO and H_2 are the main products in all cases. Different degrees of phase transformations in mechanochemically and thermally synthesised samples during the catalytic test are found. The presence of Ni-doped Fe_3O_4, NiFe alloy and Fe_3C is observed for the all studied samples (Figure 3, Table I), but in the case of NiFe2TS a significant amount of unchanged ferrite phase is also found. The formation of iron–cobalt alloys and iron carbides after catalytic test is found for the mechanosynthesized $CoFe_2O_4$ too and is correlated with the observed low selectivity to methane [7].

4 Conclusion

The high-energy ball milling of the layered nickel–iron hydroxide carbonates results in the formation of nanocrystalline nickel ferrite, where the particle size can be controlled by the treatment time. Differences in the phase composition and catalytic behavior in methanol decomposition of nickel ferrites are observed depending on the preparation method used. Both, the variations in the phase composition and the particles dispersion affect the catalytic properties of the samples.

Acknowledgments The authors thank the National Science Fund of the Bulgarian Ministry of Education and Science for the financial support of Project X-1504/05.

References

1. Shaft, K.V.P.M., Gedanken, A.: Chem. Mater. **10**, 3445 (1998)
2. Kim, C.S., Yi, Y.S.: J. Appl. Phys. **85**, 5223 (1999)
3. Kim, Y.J., Kim, D., Lee, C.S.: Physica B **337**, 42 (2003)
4. Shi, Y., Ding, J., Liu, X., Wang, J.: J. Magn. Magn. Mater. **205**, 249 (1999)
5. Lee, S., John, V.T., O'Connor, C., Harris, V., Carpenter, E.: J. Appl. Phys. **87**, 6223 (2000)
6. DeGuide, M.R., O'Handley, R.C., Kaionji, G.: J. Appl. Phys. **65**, 3167 (1989)
7. Manova, E., Tsoncheva, T., Paneva, D., Mitov, I., Tenchev, K., Petrov, L.: Appl. Catal., A General **277**, 119 (2004)

8. Tsoncheva, T., Dimitrov, M., Paneva, D., Mitov, I., Köhn, R., Fröba, M., Minchev, Ch.: React. Kinet. Catal. Lett. **74**, 385 (2001)
9. Manova, E., Kunev, B., Paneva, D., Mitov, I., Petrov, L., Estournès, C., d'Orléans, C., Rehspringer, J.L., Kurmoo, M.: Chem. Mater. **16**, 5689 (2004)
10. Shi, Y., Ding, J., Yin, H.: J. Alloys Compd. **308**, 290 (2000)
11. Ahn, Y., Choi, E.J., Kim, S., Ok, H.N.: Mater. Lett. **50**, 47 (2001)

Hyperfine Interact (2005) 165: 221–225
DOI 10.1007/s10751-006-9269-6

Mössbauer study of iron oxide modified montmorillonite

H. Bartonkova · M. Mashlan · R. Zboril · J. Pechousek · O. Schneeweiss · P. Martinec

Published online: 7 November 2006

Abstract Montmorillonite particles were modified by iron oxides using the precipitation process with the aim to monitor the differences in the structural and magnetic properties of intercalated and adsorbed Fe^{3+}. The Mössbauer spectra recorded at 5 K in zero and 6 T external fields, IR spectra and TG curves measured in zero and 32 mT fields identified the ferrihydrite pillars in an interlayer space of the montmorillonite structure and γ-Fe_2O_3 nanoparticles adsorbed on the mineral surface. The temperature dependent Mössbauer spectra (25–300 K) reflect the superparamagnetic behaviour of maghemite nanoparticles and ferrihydrite pillars with the blocking temperatures of about 80 and 25 K, respectively.

Key words Mössbauer effect · iron oxide · montmorillonite

1 Introduction

Montmorillonite (MMT) is a fine-grained (<2 μm) sheet silicate mineral and belongs under the smectite group of clays. Smectites are subject of interest for their very good swelling and sorption properties. Each MMT particle is designed for layers with a net negative charge; the volume between layers is occupied by exchangeable cations, Na^+ or Ca^{2+}. In the environment with a high enough concentration of other cations, the original sodium and calcium ions can be replaced by these new cations, maintaining a charge balance [1]. Generally, the clay can be modified by external ions in two ways: on the surface due to an adsorption process and in the interlayer space owing to the swelling properties and cation exchange.

H. Bartonkova · M. Mashlan (✉) · R. Zboril · J. Pechousek
Research Centre of Nanomaterials, Palacky University, Olomouc, Czech Republic
e-mail: miroslav.maslan@upol.cz

O. Schneeweiss
Institute of Physics of Materials, Academy of Sciences of Czech Republic, Brno, Czech Republic

P. Martinec
Institute of Geonics, Academy of Sciences of Czech Republic, Ostrava, Czech Republic

Table I Hyperfine parameters of RT Mössbauer spectra

Sample	Dried at RT		Heated at 150°C		Heated at 300°C	
	A	*B*	*A*	*B*	*A*	*B*
δ_{Fe} (mm/s)	0.34	0.34	0.34	0.35	0.33	0.34
ΔE_Q (mm/s)	0.73	0.69	0.77	0.71	0.79	0.71

Clays modified by metal oxides in their interlayer space are known as pillared clays (PILC) because the oxide particles act as pillars, keeping silicate layers apart and forming two dimensional molecular sieves with a higher surface area compared to the original clay. PILC have been studied for their superior sorption [2] and catalytic properties [3, 4]. A surface modified clay combines properties of both metal oxides and the clay, useful, for example, for solving environmental problems with water contamination [5].

In this comparative study, the structural and magnetic properties of MMT modified by pillared and adsorbed iron oxides are investigated using low temperature and external magnetic field Mössbauer spectroscopy, IR spectroscopy, TG and BET surface area measurements.

2 Experimental methods

For the synthesis of iron oxide modified MMTs, the known amount of MMT was well dispersed in distilled water by a vigorous stirring to obtain 1% (w/w) suspension. Concentrated solutions of iron (III) chloride and iron (II) sulphate heptahydrate, with a Fe^{2+}/Fe^{3+} molar ratio of 1:2 and MMT/Fe weight ratio of 2:1, were added to the suspension. The suspension was stirred for 1 h and then divided into two parts. The one part of the initial suspension was released from the rest of iron salts by a filtration, few times washed and then newly suspended in distilled water, resulting in MMT with Fe^{3+} and Fe^{2+} in the interlayer space (intercalated ions) and with utmost lowered amount of adsorbed iron ions (sample *A*). The other part of the original suspension was not filtrate (sample *B*), leading to MMT with intercalated and adsorbed iron ions, in which the amount of adsorbed ions is severalfold higher then that of intercalated ions. The excess of NaOH solution (1 mol/l) was added to both stirred suspensions. Brown precipitate was made and left to age for 0.5 h. Two samples (*A* and *B*) of iron modified MMT were separated out by the filtration and the parts of both samples were consequently thermally treated at 150 and 300°C for 3 h and a part was dried at room temperature. Thus, three and three samples of iron oxide pillared MMT (samples *A*) and MMT with iron oxide bonded on the surface (samples *B*) were prepared, respectively.

The transmission ^{57}Fe Mössbauer spectra were collected in constant acceleration mode with a $^{57}Co(Rh)$ source at temperatures ranging from 25 to 300 K in zero external magnetic fields. In-field spectra were recorded at 5 K in 6 T field, applied parallel to the gamma-ray direction. TG curves (THASS GmbH) were measured in temperature range of 20–600°C with the rate of 5°C/min in zero and 32 mT magnetic fields. Fourier transform infrared spectra were recorded in the wave number range 4,000–400 cm^{-1} using a Nicolet Nexus 670 spectrometer equipped with DTGS detector and XT-KBr beamsplitter. The specific surface areas were determined using a Coulter SA 3100 instrument with N_2 as an adsorption gas.

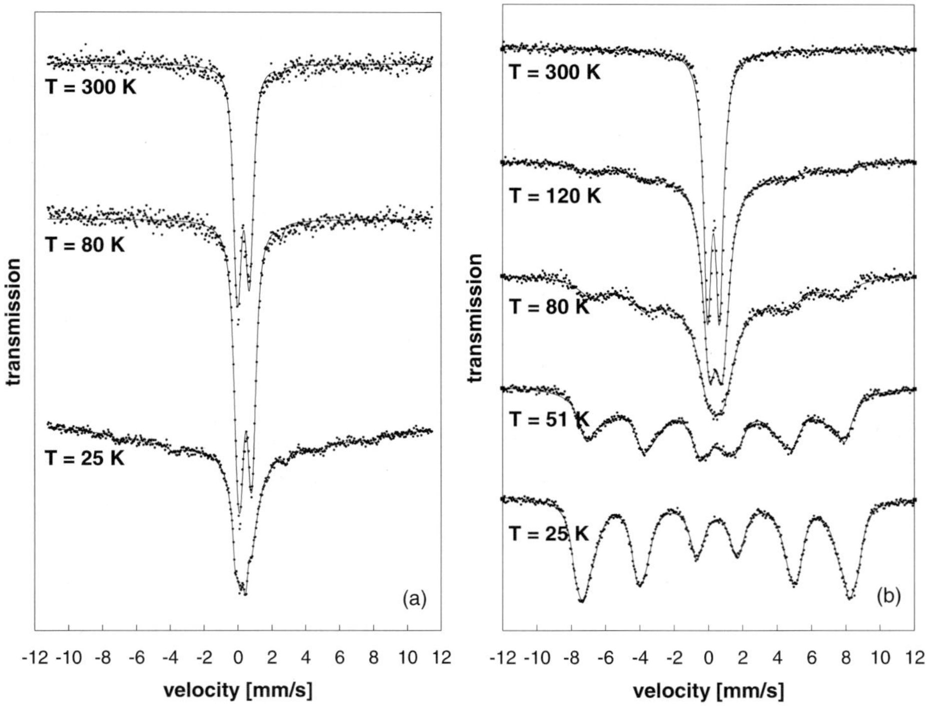

Figure 1 Mössbauer spectra of samples *A* (a) and *B* (b) heated at 150°C.

3 Results and discussion

Although the amount of iron in original MMT (KUNIPIA, Japan) is very low, its RT Mössbauer spectrum consists of Fe^{2+} doublet (δ_{Fe}=1.12 mm/s, ΔE_Q=2.83 mm/s) and Fe^{3+} doublet (δ_{Fe}=0.30 mm/s, ΔE_Q=0.73 mm/s).

RT Mössbauer spectra of all iron oxide modified samples (Table I) show the broadened superparamagnetic doublet with isomer shift parameter ranging between 0.33 and 0.35 mm/s as typical for the high spin Fe^{3+}. The subspectra of the original MMT are invisible because of a low amount of iron in the mineral. The higher values of QS for sample *A* can be connected with lower particle size of iron oxide pillars due to the restriction of their growth in the MMT interlayer space. Nevertheless, the lower symmetry of iron environments in pillared MMT can relate also with a difference in the structures of intercalated and adsorbed iron oxides.

To compare the structure and magnetic properties of intercalated and adsorbed iron oxides, Mössbauer spectra of the *A* and *B* samples heated at 150°C were collected at various temperatures between 25 and 300 K. Following such temperature dependent spectra (Figure 1a), the sample *A* remains superparamagnetic even at very low temperatures (25 K) as characteristic for ferrihydrite ($Fe_2O_3{\cdot}nH_2O$) nanoparticles [6]. The IR spectrum of sample *A* confirms such structural assignment as typical bands (1,468 cm^{-1} and 3,800 cm^{-1}) of ferrihydrite [7] were identified.

The temperature evolution of Mössbauer spectra in sample *B* (Figure 1b) reflects completely other magnetic character of iron oxide particles adsorbed on the MMT surface compared to those intercalated in the MMT interlayers. With lowering the temperature, the

Figure 2 Mössbauer spectra of sample *B* recorded at 5 K in zero and 6 T external magnetic fields.

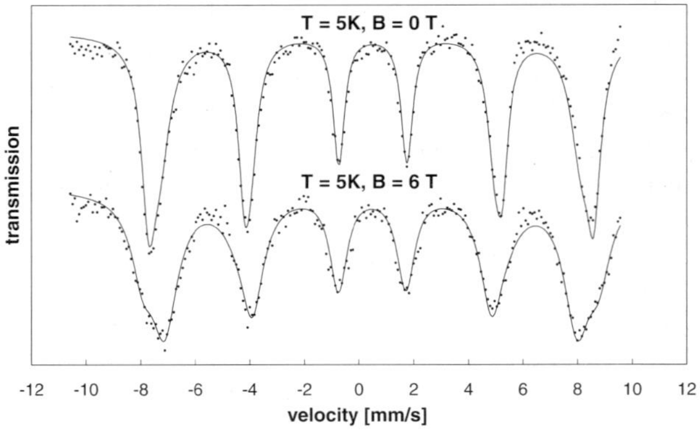

Figure 3 TG curves of the sample *A* (a) and *B* (b) in a zero and 32 mT magnetic fields.

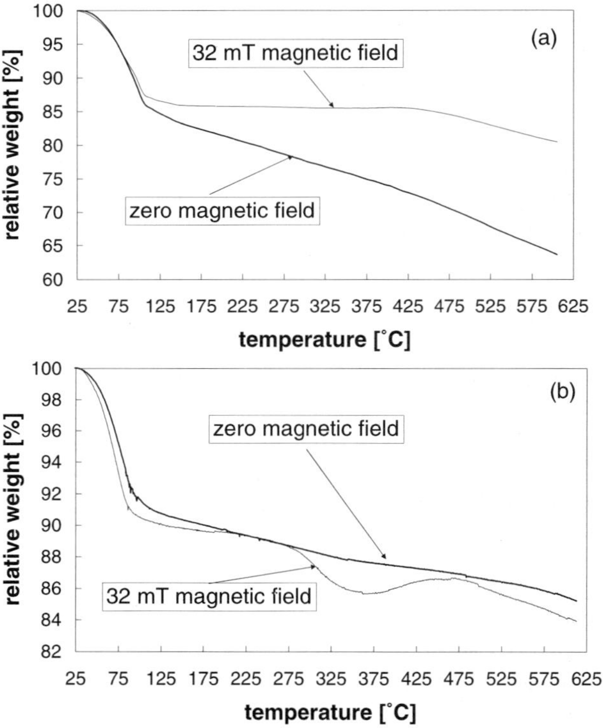

content of superparamagnetic doublet gradually decreases and broad sextet, corresponding to the distribution of magnetic hyperfine fields, prevails. The range of the coexistence of the superparamagnetic doublet and magnetic sextet components is very broad and lies between 50 and 120 K. The blocking temperature was found to be about 80 K. Generally, the thermal evolution of Mössbauer spectra of sample *B* represents the system of strongly interacting iron oxide nanoparticles with a broad size distribution [8]. The Mössbauer spectrum of sample *B* recorded at a 5 K in zero external magnetic field (Figure 2 up) shows the broadened sextet which can be fitted by two magnetic hyperfine field distributions with the average isomer shift of 0.46 mm/s and quadrupole shift closed to zero. This value of quadrupole shift is typical for nanocrystalline γ-Fe_2O_3 particles [8, 9].

The in-field spectrum (Figure 2 down) exhibits the non-zero intensities of the second and fifths spectral lines although they should be suppressed for ferrimagnets in parallel applied field. Such behaviour is usually observed in poorly crystalline maghemite due to the high degree of the spin frustration [8, 9].

To monitor the thermal stability of the MMT matrix as well as the magnetic behaviour of modifying iron oxides under heating in air, TG curves in zero and 32 mT magnetic fields were measured. TG curve of the original non-modified MMT (not shown) demonstrated the destruction of the clay structure above 475°C. The rapid falls in weight below 100°C is obvious in the zero filed TG curves of MMT modified by ferrihydrite pillars (Figure 3a) and MMT with the surface modified by maghemite (Figure 3b), clearly due to an evaporation of the adsorbed water. Further weight decrease up to 475°C registered in a zero magnetic field is caused by the consecutive evolution of water from the interlayer space of the MMT structure.

In the external field TG curve of pillared MMT (Figure 3a) between 100 and 475°C, the almost negligible weight decrease is recorded compared to the zero field curve, evidently due to the gradual decrease of the magnetic moment of iron oxide pillars with increasing temperature. Moreover, no peaks indicating the change of the magnetic regime induced either by the crystallization process or the structural rearrangements were detected. This is an indirect evidence that ferrihydrate particles act as nanopillars which are well-separated and, thus, prevented against the thermally induced crystallization and structural transformation. An increase of the surface area of the sample *A* (63 m^2/g) compared to the original MMT (18 m^2/g) also confirms the pillared character of the modifying iron oxide resulting in the expansion of the interlayer MMT space. On the other hand, the external field TG curve of the surface modified MMT (Figure 3b) exhibits the rapid fall in the weight above 300°C as a result of the increase of the magnetic moment of maghemite particles. Such effect is evidently due to the crystallization and growth of γ-Fe_2O_3 particles, which induces their transition from the superparamagnetic to the ferrimagnetic state.

Acknowledgement Financial support from The Ministry of Education, Youth and Sports of the Czech Republic under research project MSM 6198959201 is gratefully acknowledged.

References

1. Nesse, W.D.: Introduction to Mineralogy. Oxford University Press, Oxford, (2000)
2. Heylen, I., Vanhoof, C., Vansant, E.F.: Microporous Mater. **5**, 53 (1995)
3. Sathyamoorthy, A., Raj, P., Gupta N.M., Samant, S.D.: Catal. Letters **81**, 97 (2002)
4. Feng, J.Y., Hu, X.J., Zue, P.L., Zhu, H.Y., Lu, G.Q.: Ind. Eng. Chem. Res. **42**, 2058 (2003)
5. Oliveira, L.C.A., Rios, R.V.R.A., Fabris, J.D., Sapag, K., Garg, V.K., Lago, R.M.: Appl. Clay Sci. **22**, 167 (2003)
6. Murad, E., Cashion, J.: Mössbauer Spectroscopy of Environmental Materials and their Industrial Utilization, p. 162. Kluwer, Dordrecht (2004)
7. Cornell, R.M., Schwertmann, U.: The Iron Oxides, p. 130. VCH, Weinheim (1996)
8. Zboril, R., Mashlan, M., Petridis, D.: Chem. Mater. **14**, 969 (2002)
9. de Baker, P.M.A., de Grave, E., Vandenberghe, R.E., Bowen, L.H.: Hyp. Int. **54**, 493 (1990)

Hyperfine Interact (2005) 165: 227–230
DOI 10.1007/s10751-006-9270-0

Magnetic properties of metallic glasses of the type $Fe_{80-x}Pd_xB_{20}$ and $Fe_{80-x}Pt_xB_{20}$

Janusz Oleniacz · Przemysław Duda · Włodzimierz Zych · Jan Grabski

Published online: 26 October 2006

Abstract Amorphous alloys of the type $Fe_{80-x}Pd_xB_{20}$ and type $Fe_{80-x}Pt_xB_{20}$ for $0 \leq x \leq 50$ have been investigated by means of ^{57}Fe Mössbauer spectroscopy and magnetisation measurements in temperatures from 4.2 up to 300 K. Curie temperatures and crystallisation temperatures are found by DTMG-DTA method. Mössbauer spectroscopy magnetic field is observed to visible increase for $x=1$ and 1.5% at room temperature for Pd, while a decrease is observed for higher x values. Curie temperature for Pd alloys has a maximum at $x=4$ with $T_C=753$ K, which supports enforcing influence of Pd at low concentrations of Pd for magnetic interactions. We discuss different explanations for these measurements and compare with other findings for high Pd concentrations and alloys with Pt instead of Pd.

Key words Mössbauer spectroscopy · amorphous alloys · magnetic alloys · Fe alloys

1 Introduction

The magnetic properties of metallic glasses – amorphous alloys obtained usually by melt spinning technology are posing lots of interesting phenomena. One of the first metals obtained as a metallic glass was Pd–Si alloy because of easy creation. Both crystalline and amorphous Pd–Fe alloys (see [4] for Pd–Fe–Si system) are nonmagnetic when Pd is high, and then start to be ferromagnetic when Fe concentration is big enough. There is also a spin glass state reported for $Fe_xPd_{80-x}Si_{20}$ at x in the range $1<x<7$ [4, 5]. ^{57}Fe Mössbauer spectroscopy was applied already to study $Fe_xPd_{80-x}Si_{20}$ alloys with low Fe concentrations in [3]. In this work we study amorphous alloys of the type $Fe_{80-x}Pd_xB_{20}$ and type $Fe_{80-x}Pt_xB_{20}$ for $0 \leq x \leq 50$, from low to high additive concentration (Pd or Pt).

2 Experimental

Metallic glasses of the type $Fe_{80-x}Pd_xB_{20}$ (where x has been equal to 0, 1, 1.5, 2, 3, 6) and type $Fe_{80-x}Pt_xB_{20}$ (with x taking values 0.5, 1, 3, 5, 6, 7, 10, 20) where prepared by simple

J. Oleniacz (✉) · P. Duda · W. Zych · J. Grabski
Faculty of Physics, Warsaw University of Technology, PL-00-662 Warsaw, Poland
e-mail: oleniacz@if.pw.edu.pl

Table I Experimental values for $Fe_{80-x}Pd_xB_{20}$ type amorphous alloys

x (at%)	T_{Curie} (K)	$T_{crystallisation}$ (K)	μ (μ_B) 4.2 K	H_{max} (kOe), 80 K
0-Fe	665	703	2.01	290
1-Pd	673	713	2.11	285
1.5-Pd	653	713	2.56	280
2-Pd	683	718	2.03	285
2.5-Pd	678	723	2.52	–
3-Pd	713	723	2.14	285
4-Pd	753	758	2.06	
6-Pd	703	713	2.07	
10-Pd	730	723	2.23	
20-Pd			2.36	
27-Pd–$Si_{12}B_8$	658	768	2.18	285
50-Pd–$Si_{12}B_8$	323	784	3.37	255
62-Pd–Si_{18}	192 [5]		2.8 [5]	
67-Pd–Si_{18}	158 [5]			
72-Pd–Si_{18}	50 [5]		2.5 [5]	
73-Pd–Si_{18}	40 [5]			

roller quenching technique in argon atmosphere. Because of quick crystallisation for higher Pd content with x=27 and 50, the mixture $Fe_{80-x}Pd_xSi_{12}B_8$ has been used, where Si causes easier quenching. Partially crystalline samples were excluded by X-ray diffraction checking for amorphous atomic structure. Magnetisation was measured by using a vibrating sample magnetometer (from Oxford Instruments) in magnetic fields 1 and 2 kOe. Simultaneous DTMG-DTA method (Derivative Thermo-Magneto-Gravitometry and Differential Thermal Analysis described in [6]) was used to measure Curie as well as crystallization temperature. ^{57}Fe Mössbauer spectroscopy was used in transmission geometry and in temperatures from liquid helium to room temperature.

3 Results and discussion

We continue our previous studies of theses materials [1, 2] which are also interesting for others and have been measured with Si instead of B in [3, 4] and with low Fe concentration in [3] in which there is expected an effect of "giant magnetic moment" (like 12 μ_B per Fe atom for about 0.28 at.% Fe in Pd). Our samples are on the other end – high Fe concentration, but still the influence of Pd or Pt is worth investigation.

Table I presents measured values for Curie temperature, crystallisation temperature as well as magnetic moment per Fe atom deduced from magnetisation measurements. H_{max} value is hyperfine field of maximal probability deduced from Mössbauer spectra measured for samples at liquid helium temperature. Spectra were analysed by Hesse-Rubartsch method and MossWin commercial computer program.

Mössbauer spectroscopy magnetic field is observed to visible increase for x=1 and 1.5% at room temperature for Pd (analysed by Hesse-Rubartsch method and MossWin commercial program), while a decrease is observed for higher x values. Curie temperature for Pd alloys has a maximum at x=4 with T_C=753 K, which supports enforcing influence of Pd at low concentrations of Pd for magnetic interactions. On the other hand calculated average magnetic moment per Fe atom in $Fe_{80-x}Pd_xB_{20}$ (see Table I) has an maximum at

Figure 1 Magnetic moment per Fe atom versus Pd at %.

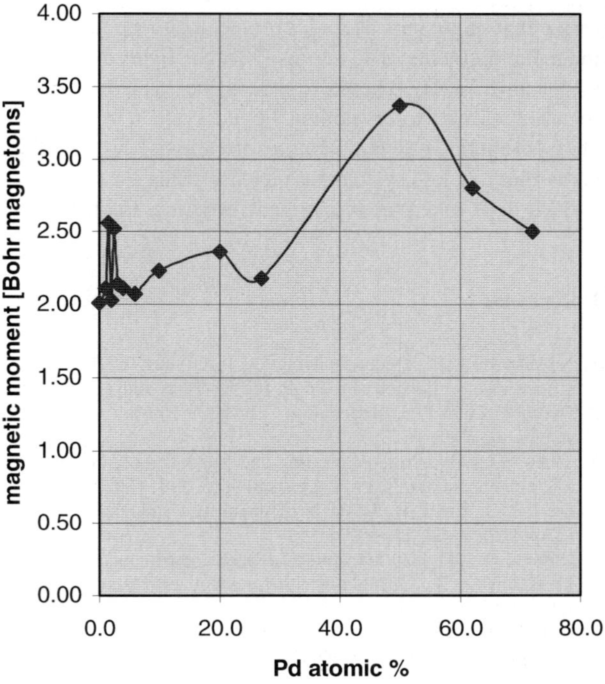

Table II Experimental values for $Fe_{80-x}Pt_xB_{20}$ type amorphous alloys

x (at %)	T_{Curie} (K)	$T_{crystallisation}$ (K)	μ (μ_B) 4.2 K	H_{max} (kOe), 80 K
0-Fe	665	703	2.01	290
0.5-Pt	553	713	2.56	290
6-Pt	728	778	2.53	290
10-Pt	753	796	2.57	290
20-Pt–$Si_{12}B_8$	782	815	2.83	307

x=1.5 with μ=2.56 μ_B. For low Pd content high values about μ=2.56 μ_B were obtained at present, while older ones close to 2.1 μ_B have been done by us a few years ago [1] (and the difference may stem from fabrication conditions). The increase of this value for higher Pd concentrations is evidently due to Pd magnetic moments of 4 day shells which interact with Fe magnetic moments and enlarge average value. Hyperfine magnetic field does not reflect such Fe magnetic moments increase, so as discussed in [4] this is an effect of Pd atoms (with about 0.4 μ_B per Pd atom as found in [7] for crystalline Fe–Pd system). A good illustration of this is the alloy $Fe_{30}Pd_{50}Si_{12}B_8$ as seen from Table I and Figure 1 where the value of average magnetic moment per Fe atom is 3.37 μ_B with H_{max}=255 kOe only compared to $Fe_{80}B_{20}$ alloy where these values take 2.01 μ_B and 290 kOe, respectively.

4 Conclusions

There is a need to discuss different explanations for these measurements and compare with other findings for high Pd concentrations and alloys with Pt instead of Pd. Current studies show that high increase of average magnetic moment per Fe atom cannot be due to magnetic moments of Fe atoms (because hyperfine fields for them stay more or less equal as for pure Fe–B). Therefore it must be caused by magnetic splitting of Pd 4 day shells, which is similar to crystalline Fe–Pd system, where magnetic moment of 0.4 μ_B per Pd atom is found. Curie temperatures for alloys $Fe_{80-x}Pd_xB_{20}$ and type $Fe_{80-x}Pt_xB_{20}$ (Table II) show more or less systematic increase when Pd or Pt is added. This confirms that both Fe and Pd or Pt take part in magnetic ordering and that it is a positive influence.

References

1. Nowakowska, E., Duda, P., Grabski, J., Gradys, E., Kupczak, R., Oleniacz, J., Zych, W.: Magnetic properties of amorphous Fe–Pd–B and Fe–Pt–B alloys. Phys. Stat. Sol.(a) **185**, 277 (2001)
2. Duda, P.: 2004 PHD dissertation "Zbadanie wplywu Pt i Pd na wlasności magnetyczne amorficznych stopów metalicznych"
3. Chien, C.L.: Hyperfine interactions and magnetic ordering temperatures of amorphous $Fe_xPd_{80-x}Si_{20}$ with low Fe concentrations. Phys. Lett. **68A** (no. 3–4) (1978)
4. Dublon, G., Lin, C.H., Bevk, J.: Magnetisation and magnetoresistance study of amorphous $Fe_xPd_{82-x}Si_{18}$. J. Appl. Phys. **50**, 11 (1979)
5. Dublon, G.: Magnetic transitions in $Fe_{82-x}Pd_xSi_{18}$ metallic glasses. Phys. Stat. Sol. (a) **60**, 287 (1980)
6. Moskalewicz, R.: Quantitative interpretation of TG and DTG curves obtained for some ferromagnetic substances in low intensity AC generated magnetic field. Thermochim. Acta **28**, 229 (1979)
7. Crangle, J.: Ferromagnetism in Pd-rich Palladium–Iron Alloys. Philos. Mag. **5**, 335 (1960)

Hyperfine Interact (2005) 165: 231–237
DOI 10.1007/s10751-006-9271-z

Mössbauer and D.C. magnetization studies of $(CuFe_2O_4)_{1-x}$ $(SnO_2)_x$ (x=0 and 5 wt.%) nanocomposites

R. Kalai Selvan · C. O. Augustin · M. I. Oshtrakh · O. B. Milder · V. A. Semionkin

Published online: 31 October 2006

Abstract $CuFe_2O_4$ nanoparticles and $(CuFe_2O_4)_{0.95}(SnO_2)_{0.05}$ nanocomposites were studied by XRD, Mössbauer spectroscopy and SQUID magnetometer. The XRD pattern shows the well-defined sharp peaks characterized tetragonal structure of $CuFe_2O_4$. The Mössbauer spectrum of $CuFe_2O_4$ showed ferrimagnetic structure while spectrum of $(CuFe_2O_4)_{0.95}(SnO_2)_{0.05}$ composite demonstrated a mixture of ferrimagnetic and paramagnetic components. The D.C. magnetization revealed that SnO_2 addition leads to overcoming the "superparamagnetic limit."

Key words Mössbauer spectroscopy · XRD · SQUID · nanocomposites

1 Introduction

During the past decades, the synthesis and characterization of nanocomposites are of great interest for their end uses in wide range of applications because of their fascinating properties, which are superior to their individual counterparts. Very recently, nanoferrites and their composites have gained much attention due to their indigenous properties for using in many technological fields like information storage devices, magnetic resonance imaging contrast agent, quantum tunneling, ferrofluids, sensors, catalysts and electrodes for electrochemical applications [1–3]. Among the synthesis techniques [4] solution based combustion method is the most suitable for preparing nanomaterials with advantageous

R. K. Selvan (✉) · C. O. Augustin
Central Electrochemical Research Institute, Karaikudi 630006, India
e-mail: selvankram@rediffmail.com

M. I. Oshtrakh (✉)
Faculty of Physical Techniques and Devices for Quality Control,
Ural State Technical University – UPI, Ekaterinburg 620002, Russian Federation

O. B. Milder · V. A. Semionkin
Faculty of Experimental Physics, Ural State Technical University – UPI,
Ekaterinburg 620002, Russian Federation

Figure 1 XRD patterns of nanocrystalline $CuFe_2O_4$ (a) and $(CuFe_2O_4)_{0.95}(SnO_2)_{0.05}$ nano-composites (b).

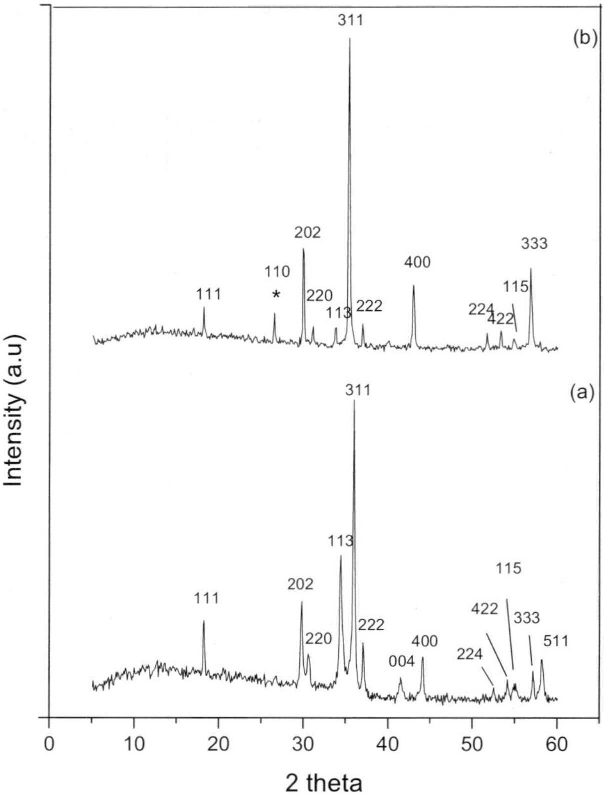

properties. Hence, this study has been undertaken considering the importance of ferrite-based nanocomposites as anode material for the replacement of conventionally used carbon anodes [5]. Structural and magnetic properties of $CuFe_2O_4$ nanoparticles prepared by the different way were studied by Mössbauer spectroscopy and magnetic measurements [6, 7]. In the present work we discuss the results of comparative study of $CuFe_2O_4$ nanoparticles and $(CuFe_2O_4)_{0.95}(SnO_2)_{0.05}$ nanocomposites by Mössbauer spectroscopy and SQUID magnetometer.

2 Materials and methods

The nanocrystalline $CuFe_2O_4$ and $(CuFe_2O_4)_{0.95}(SnO_2)_{0.05}$ nanocomposite were prepared by novel urea-nitrate combustion method. The synthetic procedure was explained elsewhere [8]. The compound formation, phase purity and crystalline nature of the ferrite materials were identified by powder X-ray diffraction technique using an X-ray diffractometer (Rigaku Dmax/2C, Japan, with $Cu-K_\alpha$ radiation; α=1.5405 Å). The particle size was analyzed by TEM (JEOL-JEM 100SX microscope at an accelerating voltage of 200 kV). Mössbauer spectra of the sample powders were measured at room temperature with the constant acceleration computerized high precision, sensitive and stable spectrometer that was a part of multi-dimension parametric Mössbauer spectrometer SM-2201 [9]. Character-

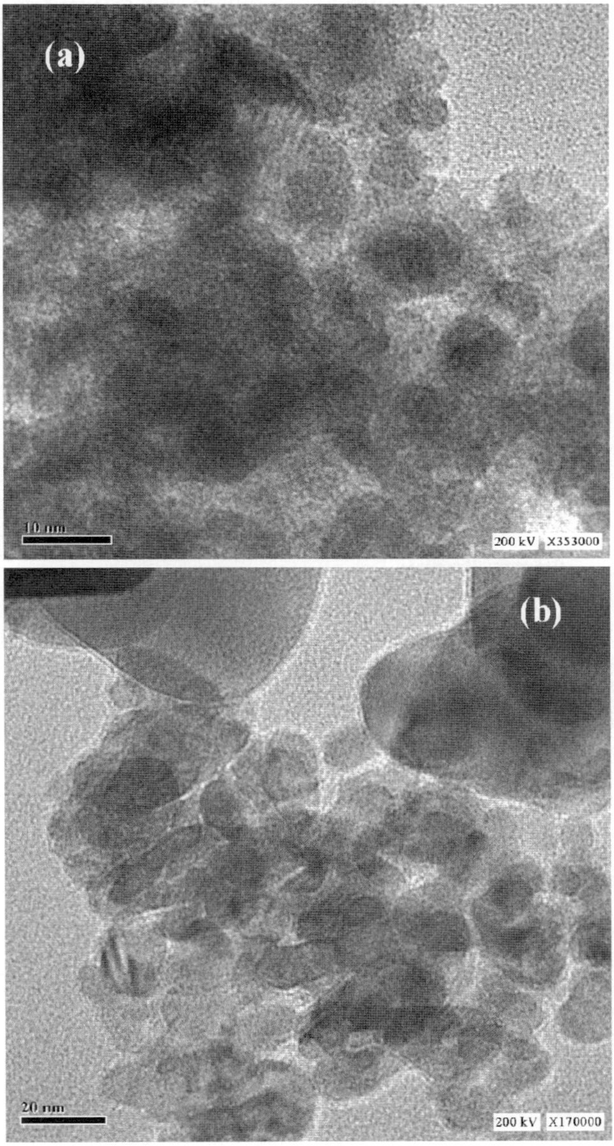

Figure 2 TEM images of nanocrystalline $CuFe_2O_4$ (a) and $(CuFe_2O_4)_{0.95}(SnO_2)_{0.05}$ nanocomposites (b).

istics of Mössbauer spectrometer were given in [10]. Mössbauer spectra were computer fitted with the least squares procedure using Lorentzian line shape. Mössbauer parameters isomer shift δ, quadrupole splitting ΔE_Q, magnetic hyperfine field H_{eff}, line width Γ and subspectrum area S were determined. The values of isomer shift are given relative to α-Fe at 295 K. The effects of temperature and applied magnetic field on magnetization have been studied using superconducting quantum interference device (SQUID). For ZFC magnetization measurements the sample was first cooled down to 2 K without any magnetic field and then the magnetization of the sample was measured from 2 to 350 K.

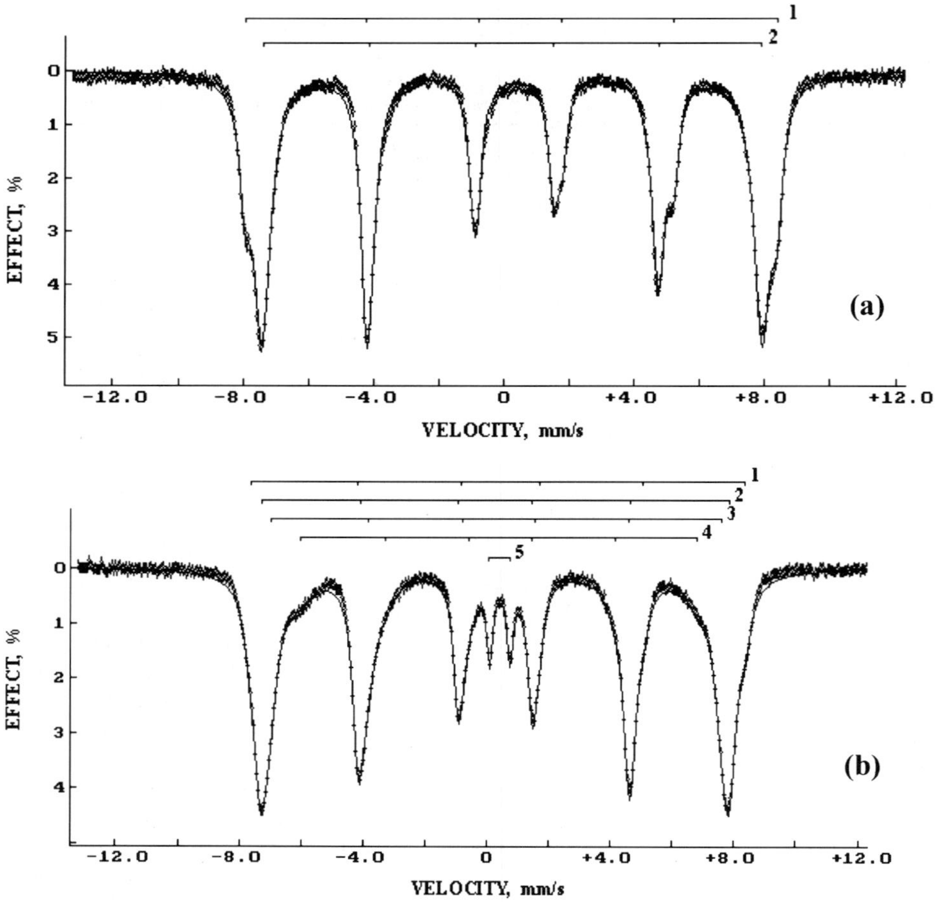

Figure 3 Mössbauer spectra of nanocrystalline $CuFe_2O_4$ (a) and $(CuFe_2O_4)_{0.95}(SnO_2)_{0.05}$ nanocomposites (b). T=295 K.

3 Results and discussion

The X-ray diffraction patterns of nanocrystalline $CuFe_2O_4$ and $(CuFe_2O_4)_{0.95}(SnO_2)_{0.05}$ nanocomposite are shown in Figure 1. Figure 1a demonstrates the sharp well defined peaks corresponding to the planes of (111), (202), (220), (113), (311), (222), (004), (400), (224), (422), (115), (333), (511) indicating spinel structure of $CuFe_2O_4$ without any impure phase. From the lattice constant values it can be confirmed that the synthesized $CuFe_2O_4$ is a tetragonal structure with a=8.2740 Å and c=8.4814 Å (PDF no. 6-545). The XRD pattern of $(CuFe_2O_4)_{0.95}(SnO_2)_{0.05}$ nanocomposite (Figure 1b) shows the extra peak of (110) corresponding SnO_2. The changing of XRD parameters for $(CuFe_2O_4)_{0.95}(SnO_2)_{0.05}$ gives the tetragonal distorted structure with a=8.4125 Å and c=8.3876 Å due to the addition of SnO_2 [8]. The TEM images of $CuFe_2O_4$ and $(CuFe_2O_4)_{0.95}(SnO_2)_{0.05}$ nanocomposite are shown in Figure 2. TEM studies show that both samples exhibited the presence of 20–30 nm-sized particles.

Room temperature Mössbauer spectra of nanocrystalline $CuFe_2O_4$ and $(CuFe_2O_4)_{0.95}$ $(SnO_2)_{0.05}$ nanocomposites are shown in Figure 3. Mössbauer parameters are given in

Table I Mössbauer parameters of nanocrystalline $CuFe_2O_4$ and $(CuFe_2O_4)_{0.95}(SnO_2)_{0.05}$ nanocomposites (T=295 K)

Sample	Components[a]	Γ^b (mm/s)	δ (mm/s)	ΔE_Q (mm/s)	H_{eff} (kOe)	S (%)
$CuFe_2O_4$	1	0.40±0.10	0.36±0.05	−0.26±0.05	504.4±1.5	21
	2	0.63±0.10	0.28±0.05	−0.05±0.05	474.4±1.5	79
$(CuFe_2O_4)_{0.95}+(SnO_2)_{0.05}$	1	0.45±0.10	0.40±0.05	−0.07±0.05	494.1±1.5	9
	2	0.54±0.10	0.28±0.05	0.02±0.05	468.9±1.5	42
	3	0.63±0.10	0.34±0.05	−0.08±0.05	451.6±1.5	34
	4	0.81±0.10	0.40±0.05	−0.03±0.05	398.0±1.5	10
	5	0.25±0.10	0.40±0.05	0.66±0.05	–	5

[a] Numbers of components are corresponded to the component numbers in Mössbauer spectra in Figure 3.

[b] Line width of the first and sixth lines of sextets or the both lines of doublet.

Table I. Mössbauer spectrum of $CuFe_2O_4$ nanoparticles consists of two magnetic sextets which were attributed to two different sublattices in which Fe ions are occupied tetrahedral and octahedral sites. High value of H_{eff} (504.4 kOe) was assigned to Fe^{3+} in the octahedral (B) sites while low value of H_{eff} (474.4 kOe) was related to Fe^{3+} in the tetrahedral (A) sites. Hence it can be concluded that the synthesized $CuFe_2O_4$ is an inverse spinel structure.

Mössbauer spectrum of $(CuFe_2O_4)_{0.95}(SnO_2)_{0.05}$ nanocomposite consists of at least four magnetic sextets and one paramagnetic doublet. Addition of SnO_2 led to decrease of the values of H_{eff} for A and B sites. The value of H_{eff} in $CuFe_2O_4$ was higher for B-sites than that for A-sites due to hyperfine interaction of A–O–B which is stronger at B-sites as well as due to covalent nature of Fe^{3+}–O^{2-} bond which is lower in octahedral sites compared to tetrahedral sites.

The variation of H_{eff} depends on the redistribution of Fe ions on the A and B sites due to the doping effect. The decreasing of H_{eff} is due to the diamagnetic Sn^{4+} ions, which substitute Fe^{3+} ions on B-sites; therefore, the Fe^{3+} ions are shifted from B-site to A-site. The values of H_{eff} are decreased due to the lesser concentration of Fe^{3+} ions on the B-site. The similar observations have also been reported on $CoFe_2O_4$ substituted with Sn^{4+} ions [11]. Observation of new additional magnetic sextets with lower values of H_{eff} may be related to the tetrahedral lattice site distortion resulting from the incorporation of Sn^{4+} ions. This compares well with the results already observed for milled $CuFe_2O_4$ [7]. Appearance of the paramagnetic doublet may be related to the Sn^{4+} ions substitution of Fe^{3+} ion neighbors leading to paramagnetic state of iron. Similar paramagnetic doublets were observed in Mössbauer spectra of milled $(NiFe_2O_4)_{0.3}(SnO_2)_{0.7}$ [12] and $SnFe_2O_4$ nanoparticles (internal region) [13].

The ZFC and FC curves of nanocrystalline $CuFe_2O_4$ and $(CuFe_2O_4)_{0.95}(SnO_2)_{0.05}$ nanocomposites are shown in Figure 4. The ZFC curve of nanocrystalline $CuFe_2O_4$ indicates that magnetization reaches a maximum value of 26.45 emu/g at 83 K and starts to decrease with temperature increase. The FC magnetization shows the maximum magnetization value of 26.9 emu/g at 2 K and then steadily decreases with temperature increase. The separation was observed below the blocking temperature (225 K) may be due to the freezing of disordered surface spins that leads to the glassy state. The ZFC and FC magnetization curves of $(CuFe_2O_4)_{0.95}(SnO_2)_{0.05}$ nanocomposites are separated up to 350 K which shows that blocking temperature of nanocomposite (350 K) is higher than that of nanocrystalline $CuFe_2O_4$ (225 K). It indicates a great improvement in the thermal stability of the magnetic moments as well as in overcoming the "superparamagnetic limit" of the ferrite-based nanocomposites [14]. The saturation magnetization values are decreased as

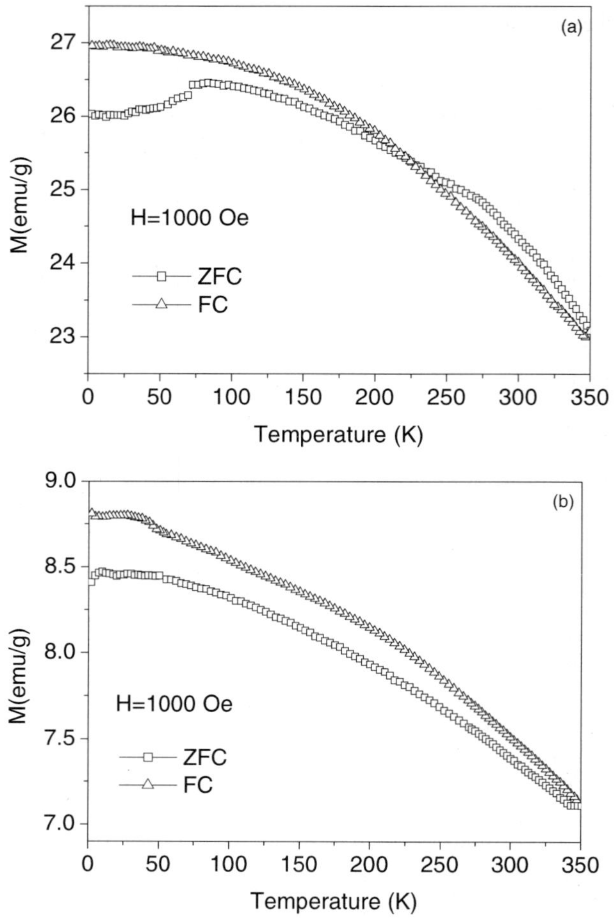

Figure 4 D.C. magnetization of nanocrystalline $CuFe_2O_4$ (a) and $(CuFe_2O_4)_{0.95}(SnO_2)_{0.05}$ nanocomposites (b).

compared to the parent $CuFe_2O_4$ due to the magnetic dilution effect and the reduction of ferrite particle–particle interaction.

4 Conclusion

Nanocrystalline $CuFe_2O_4$ and $(CuFe_2O_4)_{0.95}(SnO_2)_{0.05}$ nanocomposites demonstrated different Mössbauer spectra as a result of Sn^{4+} ions dilution in $CuFe_2O_4$ and formation at least four magnetic phases and one paramagnetic phase. The ZFC and FC magnetization curves show the ferrimagnetic behaviour of nanocrystalline $CuFe_2O_4$ and $(CuFe_2O_4)_{0.95}(SnO_2)_{0.05}$ nanocomposites as well as demonstrate that SnO_2 addition to $CuFe_2O_4$ overcomes the "superparamagnetic limit." It can be concluded that SnO_2 addition to $CuFe_2O_4$ has favourable effect on its end uses.

Acknowledgments The authors express their gratitude to Prof A. Gedanken, Bar-Ilan University, Israel, for extending the TEM facility and Prof A.K. Shukla, Director of CECRI for his keen interest.

References

1. Vestal, C.R., Zhang, Z.J.: Nano Lett. **125**, 9828 (2003)
2. Chen, N.-S., Yang, X.-J., Liu, E.S., Huang, J.-L.: Sens. Actuators **66B**, 178 (2000)
3. Berchmans, L.J., Kalai Selvan, R., Augustin, C.O.: Mater. Lett. **58**, 1928 (2004)
4. Augustin, C.O., Kalai Selvan, R.: Bull. Electrochem. **19**, 319 (2003)
5. Kalai Selvan, R., Kalaiselvi, N., Augustin, C.O., Doh, C.H., Sanjeeviraja, C.: J. Power Sources **157**, 522 (2006)
6. Goya, G.F., Rechenberg, H.R., Jiang, J.Z.: J. Appl. Phys. **84**, 1101 (1998)
7. Jiang, J.Z., Goya, G.F., Rechenberg, H.R.: J. Phys. Condens. Matter **11**, 4063 (1999)
8. Kalai Selvan, R., Augustin, C.O., Sanjeeviraja, C., Pol, V.G., Gedanken, A.: Mater Chem Phys **99**, 109 (2006)
9. Irkaev, S.M., Kupriyanov, V.V., Semionkin, V.A., Sokolov, M.M.: British Patent no. 10745 (7 May, 1987)
10. Oshtrakh, M.I., Milder, O.B., Semionkin, V.A.: Anal. Chim. Acta **506**, 155 (2004)
11. Chhaya, U.V., Kulkarni, R.G.: Hyperfine Interact. **128**, 511 (2000)
12. Albuquerque, A.S., Ardisson, J.D., Macedo, W.A.A., Plivelic, T.S., Torriani, I.L., Larrea, J.J., Saitovitch, E.B.: J. Magn. Magn. Mater. **272–276**, 2211 (2004)
13. Liu, F., Li, T., Zheng, H.: Phys. Lett., A **323**, 305 (2004)
14. Li, J.J., Xu, W., Chen, J.S.: Solid State Commun. **131**, 519 (2004)

Hyperfine Interact (2005) 165: 239–245
DOI 10.1007/s10751-006-9272-y

Hybrid magnetic nanoparticles derived from wüstite disproportionation reactions at the nanoscale

G. C. Papaefthymiou · F. X. Redl · C. T. Black · R. L. Sandstrom · M. Yin · C. B. Murray · S. P. O'Brien

Published online: 26 October 2006

Abstract We present Mössbauer studies of hybrid iron-oxide nanoparticles obtained by the disproportionation of oleic acid stabilized wüstite (Fe_xO) nanoparticles produced by selective oxidation of iron pentacarbonyl in high boiling temperature organic solvents. The results support X-ray diffraction and Transmission Electron Microscope studies of the presence of mixed Fe_xO and Fe_3O_4 phases within the nanoparticles whose relative content can be altered through heat treatment in nitrogen atmosphere. Furthermore, the Mössbauer study gives evidence of the presence of an amorphous, spin-glass like phase due to spin frustration at the Fe_xO/Fe_3O_4 interface.

Key words Mössbauer · wüstite · magnetite · mixed iron oxide nanoparticles

1 Introduction

Our interest in wüstite, Fe_xO, derives from its metastability to disproportionate into metallic iron, α-Fe, and magnetite, Fe_3O_4; and the possibility of utilizing this metastability to generate mixed-phase nanoparticles with adjustable composition and tunable magnetic properties. The large surface energies, surface capping ligands and encapsulating surfactants involved in nanoscale synthesis stabilize phases otherwise thermodynamically unstable in the bulk [1–5]. Many novel iron-oxide based magnetic nanophases have been reported, including supramolecular iron-oxo complexes [6–8], maghemite [9] and magnetite nanoparticles [10, 11]. These systems exhibit enhanced coercivities [12], affording tunable magnetic behavior through surface chemistry modification [13]. Multi-component nanostructures would have magnetic properties controlled additionally by the relative amounts of the various iron

G. C. Papaefthymiou (✉)
Department of Physics, Villanova University, Villanova, PA, USA
e-mail: gcp@villanova.edu

F. X. Redl · C. T. Black · R. L. Sandstrom · C. B. Murray
T. J. Watson Research Center, Nanoscale Materials and Devices, IBM, Yorktown Heights, NY, USA

M. Yin · S. P. O'Brien
Department of Applied Physics and Applied Mathematics, Columbia University, New York, NY, USA

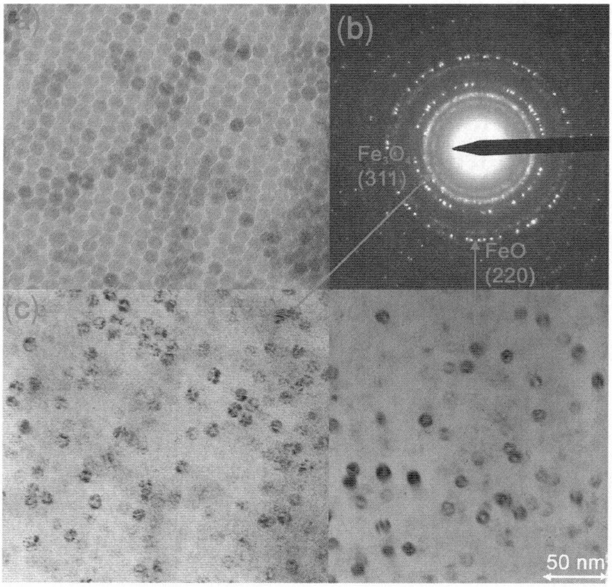

Figure 1 (a) TEM image of wüstite nanocrystals with seeds of magnetite inside. (b) SAED shows a speckled pattern for Fe_xO reflections and diffuse rings for Fe_3O_4 reflections. (c) Dark-field image of the region in (a) (shown as negative); a part of Fe_3O_4 reflections were selected with the objective aperture. (d) Dark-field image of the region in (a) (shown as negative); a part of the Fe_xO reflections were selected with the objective aperture.

oxide phases within the particle and degree of interfacial magnetic exchange coupling between phases, such as antiferromagnetic Fe_xO and ferrimagnetic Fe_3O_4. Assemblies of such mixed-phase nanoparticles could further exhibit spin dependent electronic transport properties due to the involved interfaces and half-metallic nature of magnetite [14].

Bulk Fe_xO is a non-stoichiometric iron oxide phase with a known stability range of $x=0.83$ to 0.96, above 560°C [15]. Fe_xO can be oxidized to magnetite, Fe_3O_4, and finally to maghemite, γ-Fe_2O_3 [15]. For densely packed super-lattices of surfactant stabilized nanosized Fe_xO particles, surface chemistry alteration and surfactant destabilization upon heating lead to the formation of larger mixed-oxide nanoparticles of various size and shape [16]. Herewith, we present Mössbauer spectroscopic investigations of thus derived hybrid nanostructures that confirm the stabilization of Fe_xO/Fe_3O_4 mixed-oxide nanoparticles. The results of X-ray diffraction (XRD), Transmission Electron Microscope (TEM) and Selected Area Electron Diffraction (SAED) studies on these nanostructures are also presented.

2 Materials and methods

With tight control over temperature and reaction time, reproducible generation of Fe_xO nanocrystals was achieved by selective oxidation of iron pentacarbonyl ($Fe(CO)_5$) with pyridine N-oxide in organic solvents with high boiling temperatures, trioctylamine and dioctyl ether [16]. Oleic acid (OA) was used as surfactant to stabilize Fe_xO in the nanoscale. Figure 1 shows typical TEM and SAED micrographs of the derived wüstite nanocrystals containing magnetite inclusions, due to partial oxidation upon handling, taken on a Philips CM12 transmission electron microscope. The narrow particle size distribution (5–10%) facilitated self-assembly of the nanocrystals into densely packed super lattices. Annealing in N_2-atmosphere at different temperatures and lengths of time varied the Fe_xO/Fe_3O_4 content ratios, as monitored by XRD-measurements [16].

The complex and metastable behavior of wüstite at the nanoscale was studied by Mössbauer spectroscopy. Spectra were taken from $4.2<T<300$ K, of the as prepared

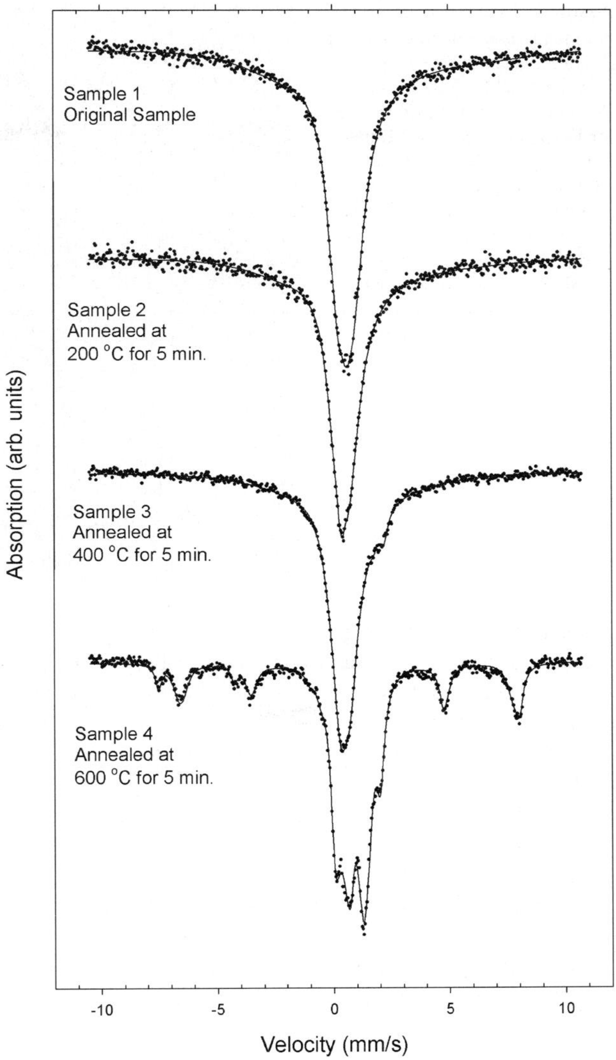

Figure 2 Room temperature Mössbauer spectra of mixed oxide nanoparticles derived from OA stabilized Fe_xO nanocrystals. The *solid lines* are theoretical fits to the data assuming a superposition of spectral signatures corresponding to four distinct phases: *1* paramagnetic Fe_xO, *2* superparamagnetic Fe_3O_4, *3* magnetic Fe_3O_4 and *4* interfacial Fe_xO/ Fe_3O_4 phases. See Table I.

nanoparticles and after annealing in N_2-atmosphere for 5 min at various temperatures. The short annealing time prevented excessive aggregation and particle growth. The source was 50-mCi ^{57}Co in Rh maintained at room temperature. A 7-μm thick metal iron foil enriched in ^{57}Fe was used for calibration. Isomer shifts are reported relative to α-Fe at RT. Spectra were fitted to a superposition of iron subsites assuming Lorentzian absorption lines and the presence of distributions in internal hyperfine magnetic fields.

3 Results and discussion

Figure 2 shows room temperature (RT) Mössbauer spectra of the as prepared (sample # 1) and annealed samples at temperatures 200°C (sample #2), 400°C (sample #3) and 600°C (sample #4), while Figure 3 shows their 4.2 K spectra.

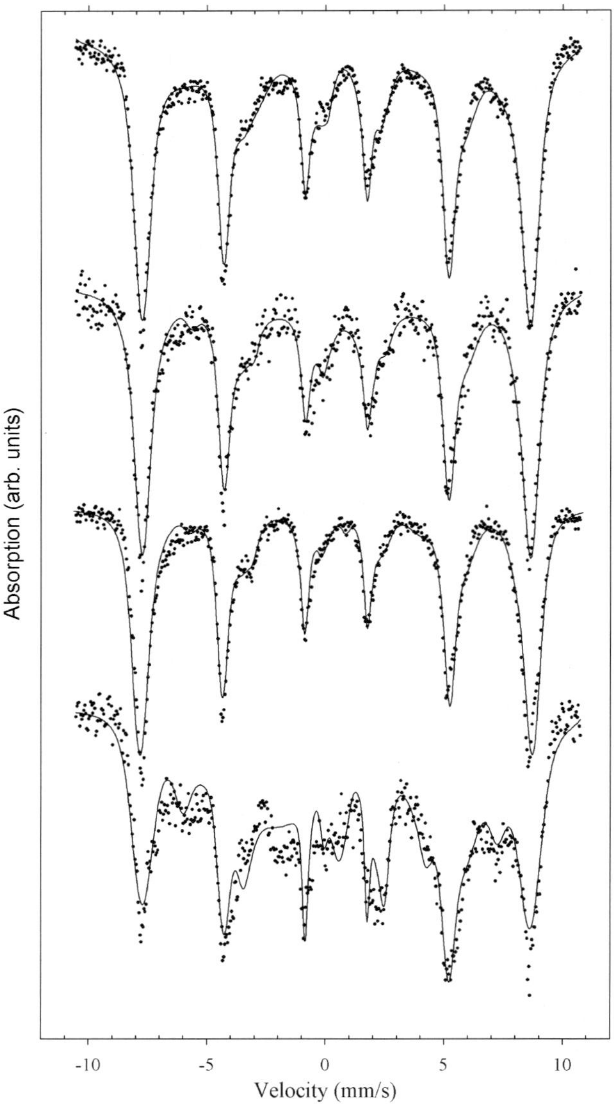

Figure 3 Mössbauer spectra of the mixed-oxide nanoparticles of Figure 2, *samples #1 to #4 from top down*, at 4.2 K. The *solid lines* are least square fits of the experimental data to the superposition of magnetic subsites corresponding to iron oxide phases as indicated in Table II.

Featureless, collapsed spectra were observed at RT consistent with the presence of paramagnetic Fe_xO and superparamagnetic Fe_3O_4, in agreement with SAED and XRD studies which showed both iron-oxide phases within individual nanoparticles, Figure 1 [16]. Up to four spectral components were assumed: (1) paramagnetic Fe_xO, simulated by a broad absorption line with isomers shift δ=0.88 mm/s [17], (2) superparamagnetic Fe_3O_4 [18] and/or (3) interfacial Fe_xO/Fe_3O_4 and (4) magnetic Fe_3O_4. Components (2) and (3) are difficult to distinguish at RT, but the 4.2 K data give clear evidence of the presence of such interfacial component (vide infra). At this expanded velocity range and in the presence of distributions in Mössbauer parameters at the nanoscale, the small quadrupole splittings associated with bulk Fe_xO (~0.50 and 0.78 mm/s [17]) are not resolvable. With increasing annealing temperature, the average particle size and the percentage contribution of magnetite increased, as shown in Table I, where results are summarized.

Table I Relative absorption areas associated with various phases and processes from Mössbauer spectral fits at T=300 K, blocking temperatures, XRD and TEM data

Sample #	Fe_xO (%)	Fe_3O_4 (%)	T_B (K)	XRD, TEM
1 (as prepared)	51 (paramag.)	49 (superparamag/interfacial)	175	7 nm Fe_xO, 3 nm Fe_3O_4
2 (annealed at 200°C)	41 (paramag.)	59 (superparamag/interfacial)	138	7 nm Fe_xO, 4 nm Fe_3O_4
3 (annealed at 400°C)	7 (paramag.)	93 (intermediate relaxation)	180	10–12 nm Fe_3O_4
4 (annealed at 600°C)	6 (paramag.)	67 (superparamag and interfacial) 27 (magnetic)	190	18 nm Fe_3O_4, 6–7 nm Fe_xO

Table II Weighted content of various phases from Mössbauer spectral fits at T=4.2 K and associated average internal magnetic field values

Sample #	Fe_xO (%)	Fe_3O_4 (%)	Spin-glass/Interfacial phase (%)
1	4	65	31
2	3	77	20
3	3	87	6
4	8	35	57
Average H_{int} (kOe)	Fe^{2+}, 410	Fe^{3+}, 507 Fe^{2+}, 495	Fe^{2+}, 319

Wüstite undergoes a first order magnetic phase transition from a paramagnet to antiferromagnet at ~198 K. Mössbauer spectra taken below room temperature ($4.2<T<300$ K, not shown) exhibited complex intermediate-relaxation and/or superparamagnetic behavior, typical of small magnetically ordered particles [18], eventually passing to slow-relaxing, fully magnetically split spectra by 4.2 K, for all samples. Associated blocking temperatures, T_B, are given in Table I. For samples #2, #3 and #4, T_B increases monotonically (138, 180 and 190 K) with increasing annealing temperature from 200, 400 to 600°C, in accord with their increasing particle size indicated by TEM (7, 11, and 18 nm diameter). Only T_B of sample #1 (~7 nm Fe_xO particles with inclusions of 3 nm Fe_3O_4, similar to sample #2) appears to deviate. We attribute this deviation to different surface characteristics for the heat-treated particles. The OA surfactant deteriorates upon heating, leaving a carbon shell on the surface [16], modifying the surface magnetic anisotropy energy density [13] and producing the observed discontinuity in the dependence of T_B on particle size.

At 4.2 K, far below their respective T_B's, spin relaxation processes are frozen, resulting in relatively sharp absorption spectra as shown in Figure 3 (sample #1 to #4, from top to bottom). Nevertheless, fits necessitated the inclusion of a severely broadened component (Table II). This component is most pronounced in the case of sample #4, whose sharp magnetic features are superimposed on a broad, featureless background envelope, as seen more clearly in Figure 4, together with details on the fitting procedure. We interpret the broad spectral component and associated broad distribution in the magnitudes of internal magnetic fields to the presence of an interfacial Fe_xO/Fe_3O_4 amorphous component. Competing ferromagnetic and antiferromagnetic interactions at the interface, produce a frustrated-spin system and a spin-glass like phase. Surface states, where the crystallographic and magnetic order is broken, may also contribute.

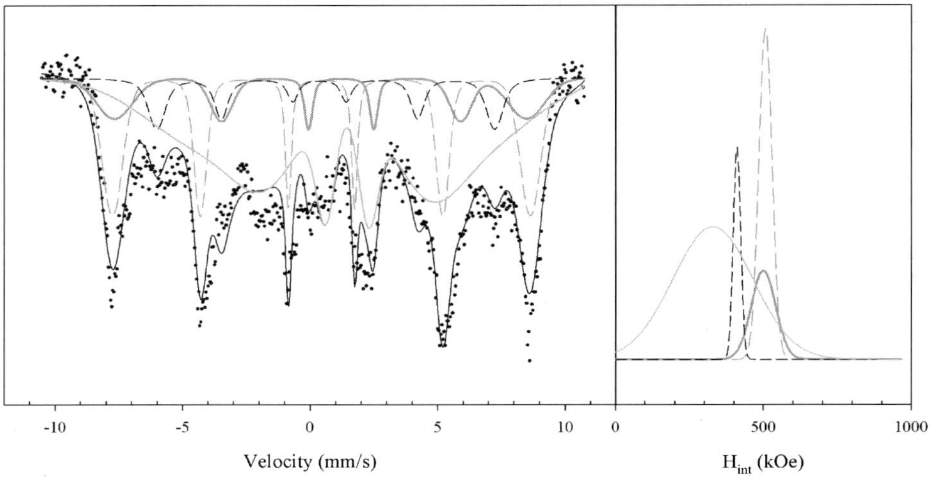

Figure 4 Mössbauer spectra of sample #4 at 4.2 K including individual component spectra that depict contributions by various iron sites. The *solid line* through the experimental points gives the least-squares fit of the experimental data to the resultant, convoluted theoretical spectrum. On the *right*, the internal hyperfine magnetic field distribution for each iron site is given. Code: *broken lines in light gray* depict Fe^{3+} sites of magnetite, *solid lines in dark gray* depict Fe^{2+} sites of magnetite; *broken lines in dark gray* depict Fe^{2+} sites in crystalline, antiferromagnetic Fe_xO; and, the *solid line in light gray* depicts Fe^{2+} at interfacial Fe_xO/Fe_3O_4 sites or surface sites contributing the broadened amorphous, spin-glass like phase. See Table II.

The detailed characteristics of the glassy phase may depend on the length of annealing time, cooling rate and other experimental conditions. For this sample, annealed at 600°C for only 5 min, the glassy component dominates, contributing about half of the total absorption intensity. The annealing temperature of 600°C is higher than 560°C above which Fe_xO is stable. Thus, magnetite is expected to be transformed back to wüstite at this temperature, in agreement with our conclusions in Table II, which indicate a larger sum for the percent contributions of Fe_xO and spin-glass phase in sample # 4, as compared to the other three samples.

Acknowledgments This work was supported by NSF at Villanova University and by NSF and DOE at Columbia University. FXR thanks DAPRA for postdoctoral support.

References

1. Ayyub, P., Palkar, V.R., Chattopadhyay, S., Multani, M.: Phys. Rev. B **51**, 6135 (1995)
2. Herhold, A.B., Chen, C.-C., Johnson, C.S., Tolbert, S.H., Alivisatos, A.P.: Phase Transit. **68**, 1 (1999)
3. Qadri, S.B., Skelton, E.F., Hsu, D., Dinsmore, A.D., Yang, J., Gray, H.F., Ratna, B.R.: Phys. Rev. B **60**, 9191 (1999)
4. Diehl, M.R., Yu, J.-Y., Heath, J.R., Held, G.A., Doyle, H., Sun, S., Murray, C.B.: J. Phys. Chem., B **105**, 7913 (2001)
5. Sun, S., Murray, C.B.: J. Appl. Phys. **85**, 4325 (1999)
6. Papaefthymiou, G.C.: Hyperfine Interact. **113**, 357 (1998)
7. Taft, K.L., Papaefthymiou, G.C., Lippard, S.J.: Science **259**, 1302 (1993)
8. Papaefthymiou, G.C.: Phys. Rev. B **46**, 10366 (1992)
9. Zhang, L., Papaefthymiou, G.C., Ying, J.Y.: J. Phys. Chem., B **105**, 7414 (2001)
10. Sun S., Zeng, H.: J. Am. Chem. Soc. **124**, 8204 (2002)

11. Li, Z., Wei, L., Gao, M., Lei, H.: Adv. Mater. **17**, 1001 (2005)
12. Papaefthymiou, G.C., In: Stranick, S.K., Searson, P., Lyon, L.A., Keating, C.D. (eds.) Anisotropic Nanoparticles-Synthesis, Characterization and Applications. MRS Symp. Proc. (Materials Research Society, Pittsburgh), vol. 635, p. C2.4.1. (2001)
13. Mørup, S., Topsøe, H., Lipka, J.: J. Phys. **37**, C6 (1976)
14. Dedkov, Y.S., Rüdiqer U., Günther, G.: Phys. Rev. B **65**, 064417 (2002)
15. Cornell, R.M., Schwertmann, U.: The Iron Oxides. Wiley, New York (1997)
16. Redl, F.X., Black, C.T., Papaefthymiou, G.C., Standstrom, R.L., Ying, M., Zeng, H., Murray, C.B., O'Brien, S.P.: J. Am. Chem. Soc. **126**, 14583 (2004)
17. Greenwood, N.N., Gibb, T.C.: Mössbauer Spectroscopy. Chapman & Hall, London (1971)
18. Dorman, J.L., Fiorani, D. (eds.): Magnetic Properties of Fine Particles. Elsevier, Amsterdam (1992)

Hyperfine Interact (2005) 165: 247–252
DOI 10.1007/s10751-006-9273-x

Superparamagnetic relaxation Mössbauer spectra by phonon–magnon interaction using the multilevel method

H.-D. Pfannes · D. S. Vieira Junior · R. Paniago · J. H. Dias Filho

Published online: 9 November 2006

Abstract A theory for superparamagnetic relaxation based on phonon-modulated exchange interaction is developed. A one-phonon direct–two-magnon inelastic scattering process is considered and it is assumed that this leads to a spin transition $|S_z\rangle \rightarrow |S_z \pm 1\rangle$ of the total spin S. Corresponding transition probabilities and from these Mössbauer patterns are calculated. The influence of sample temperature, particle size and external magnetic field is demonstrated.

Key words superparamagnetism · Mössbauer spin relaxation spectra · phonon–magnon interaction

The magnetism of ultra-fine magnetic particles is of fundamental interest, e.g., for magnetic data storage where it sets a limit for the storage density (superparamagnetic limit [1]). However, there are only few tentatives to explain superparamagnetism on the basis of the microscopic processes involved [2–4]. Direct spin–phonon interaction has been suggested as basic mechanism [2, 3] as well as magnon–phonon interaction [4]. In [2] a superparamagnetic relaxation rate has been calculated in the pseudo-spin-1/2 model, i.e., relaxation of the magnetization between antiparallel uniaxial anisotropy directions. In [4] the rate of energy transfer between the phonon and magnon systems of the particle was derived but no Mössbauer spectra have been calculated. Known the transition rates between spin states the correct way to calculate Mössbauer spectra would be by the superoperator method [5]. However this is only feasible for paramagnetic impurities with spins of up to 5/2 whereas the total spin of superparamagnetic particle is of the order of several thousand. For (predominant) diagonal fine and hyperfine interaction the multilevel (many-state) method [6]

H.-D. Pfannes (✉) · D. S. Vieira Junior · R. Paniago
Departamento de Física, Universidade Federal de Minas Gerais,
C. P. 702, 30123-970 Belo Horizonte, M.G., Brazil
e-mail: hansdiet@fisica.ufmg.br

J. H. Dias Filho
Departamento de Ciências Exatas, Universidade Estadual de Montes Claros,
C. P. 126, 39401-089 Montes Claros, M.G., Brazil

can be used. In [3] Mössbauer relaxation patterns have been calculated using this method and presuming spin–phonon interaction induced transitions between adjacent states $|S_z>$ of the collective (total) spin S of the particle and coherent (unisson) rotation of the single spins.

As in [2–4] we assume for the anisotropy energy of the superparamagnetic particle a spin-Hamiltonian H_{an} of the form

$$H_{an} = -AS_z^2 - g\mu_B B_z S_z, \tag{1}$$

where $AS^2 = KV$ is the height of the energy barrier (for $B_z=0$) between antiparallel easy directions of the magnetization (K=anisotropy energy/volume V) and the second term represents the Zeeman interaction in a magnetic field B_z parallel to the $+z$-direction. Since $V \alpha S$, AS is independent from S for a given K. Figure 1 shows the anisotropy energy and possible spin transitions between adjacent spin levels accompanied by phonon annihilation (left) or phonon creation (right).

In the present contribution we analyze the inclusion of phonon–magnon interaction induced collective spin transitions in the multilevel scheme in order to simulate Mössbauer superparamagnetic relaxation spectra. The basic idea is that by annihilation or creation of a phonon a magnon is created and another annihilated (one-phonon direct process–two-magnon inelastic scattering) and that the phonon energy is taken up by S resulting in a spin transition $|S_z> \rightarrow |S_z \pm 1>$. As in [6] only transitions between adjacent spin levels are considered. We showed in [3] that this results in more realistic spectra. For the phonon and magnon density of states the Debye model in the long wavelength approximation is used [2, 6]. Interparticle interactions and size distribution of the particles are not considered.

In order to account for the phonon-modulated exchange interaction we consider the dynamic Hamiltonian [6, 7]

$$H = -2\sum_{n>m} J'(R_{mn})\mathrm{d}R_{mn}\vec{s}_m\vec{s}_n \tag{2}$$

where $J' = \mathrm{d}J/\mathrm{d}R_{mn}$ is the dynamic exchange constant (gradient of exchange interaction constant J) and R_{mn} is the distance of a pair of spins $\vec{s}_m$ and $\vec{s}_n$. By introducing Holstein–Primakoff operators [8] in the product $\vec{s}_m\vec{s}_n$ and allowing only for linear and bilinear terms in the magnon variables, expressing $\mathrm{d}R_{mn}$ by phonon operators and taking into account exchange interaction only between nearest neighbors (lattice constant a) this can be written as [4]

$$\begin{gathered} H = -\sum_{q,k} C(q,k) b_{q-k}^+ b_q \left(a_k^+ - a_{-k}\right) \quad \text{with} \\ C(q,k) = 2sJ' a^3 (h/2Nm\omega_k)^{1/2} q(q-k)k, \end{gathered} \tag{3}$$

where q is the magnon wave vector and b_{q-k}^+ and b_q are magnon creation and annihilation operators, respectively. a_k^+ and a_{-k} are phonon operators of phonons with wave vector k and frequency ω_k. N is the total number of magnetic ions in the superparamagnetic particle and m the mass of the ion.

The transition rate $W_{S_z-1}^{S_z}$, given by the golden rule, between adjacent total spin levels S_z-1 and S_z (transitions indicated on the left side of the barrier in Figure 1) due to creation and annihilation of magnons with wave vectors q and $q-k$, respectively, accompanied by the annihilation of a phonon with wave vector k is [4]

$$W_{S_z-1}^{S_z} = \frac{2\pi}{\hbar}\sum_q |C(q,k)|^2 n_{q-k}\left(n_q+1\right)N_k\delta\left(E_q - E_{q-k} - E_k\right), \tag{4}$$

Figure 1 Anisotropy energy vs. eigenvalues with applied magnetic field. Spin transitions from $|-S>$ to $|S>$ with $\Delta m = \Delta S_z = \pm 1$ are indicated. The reduced field h is given by $h = g\mu_B g M_B B_z / 2AS$. The *left* barrier height is zero for $h=1$.

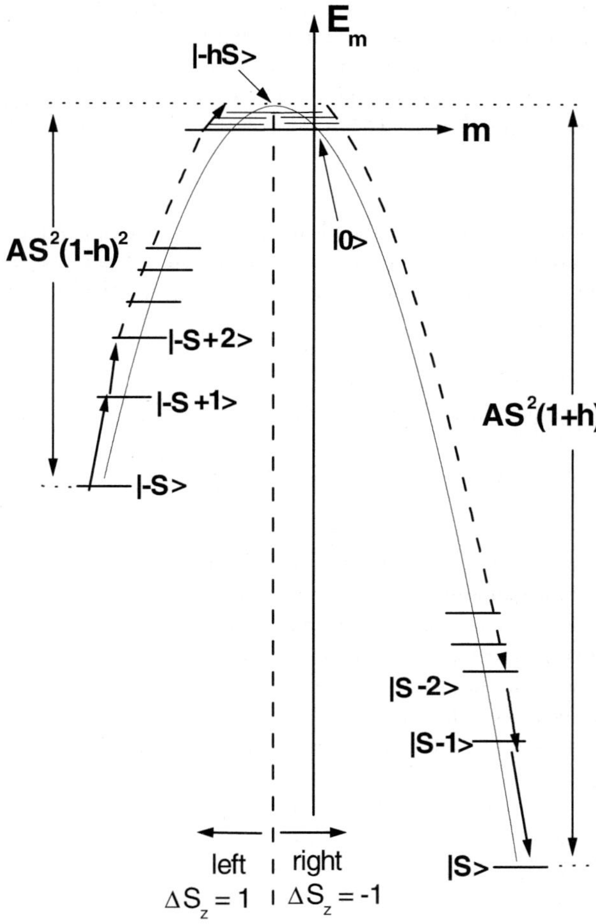

where n_q and n_{q-k} or N_k are magnon or phonon occupation numbers, respectively, given by the Bose–Einstein distribution and the δ-function secures energy conservation. E_q and $E_k = \hbar\omega_k$ are the magnon and phonon energies, respectively. Changing the sum to an integral ($\Sigma \rightarrow \int dq g(q)$), where $g(q)$ is the spectral density of magnons $g(q)dqg = Na^3q^2dq/2\pi^2$, results in

$$W^{S_z}_{S_z-1} = \frac{2s^2J'^2a^9k^2}{\pi m\omega_k} \times \frac{1}{e^{E_k/k_BT}-1} \int dqq^4(q-k)^2 \frac{e^{E_q/k_BT_s}\delta(E_q - E_{q-k} - E_k)}{\left(e^{E_q/k_BT_s}-1\right)\left(e^{E_{q-k}/k_BT_s}-1\right)}, \tag{5}$$

where T_S is the spin temperature, s the single spin (e.g., $s=5/2$ for Fe^{3+} in octahedral environment) and a^3 the volume of the unitary cell. Introducing the dimensionless variable $y=qa$ and the energies $Eq = k_BT_Cy^2$, $E_{q-k} = k_BT_C(y-ka)^2$ and $E_k = k_BT_Dka = \hbar\omega_k$, with $T_C = 3\Theta_C/z(s+1)$ (Θ_C=Curie temperature, z=number of nearest neighbors of the iron atom), and $T_D=\Theta_D/(6\pi^2)^{1/3}$ (Θ_D=Debye temperature), the δ-function can be carried out and the above expression be calculated to

$$W_{S_z-1}^{S_z}=\frac{s^2J'^2\hbar}{2^6\pi mk_B^2T_CT_D}(ka+\alpha)^4(ka-\alpha)^2\frac{e^{E_q/k_BT}}{\left(e^{E_q/k_BT}-1\right)\left(e^{E_{q-k}/k_BT}-1\right)\left(e^{E_k/k_BT}-1\right)},\tag{6}$$

where we neglected the (small) difference between T_S and the (phonon) temperature T and α is given by $\alpha=T_D/T_C$. For phonon creation we obtain an identical expression. In order to account for the magnon–phonon interaction induced spin-transitions of the superparamagnetic spin S we allow for phonons whose energy is equal to the energy difference of adjacent spin states, i.e.,

$$E_k^{\text{left}}=E_{S_z}-E_{S_z-1}=-2A(\mathrm{S}_z-1/2+hS),\mathrm{S}_z=-\mathrm{S}+1,-\mathrm{S}+2,\ldots,-hS-1,-hS$$

(left side in Figure 1) and

$$E_k^{\text{right}}=E_{S_z-1}-E_{S_Z}=+2A(\mathrm{S}_z-1/2+hS),\mathrm{S}_z=\mathrm{S},\mathrm{S}-1,\ldots,-hS+2,-hS+1\tag{7}$$

(right side in Figure 1).

In the case of isotropic magnetic hyperfine interaction and diagonal nuclear Hamiltonian (nuclear quadrupole interaction << magnetic hyperfine interaction) the general expression for Mössbauer relaxation spectra can be simplified to an expression which contains no superoperators and Mössbauer spectra can be calculated by the multilevel method [6]. A Mössbauer spectrum can be simulated by calculating three separate partial spectra consisting of the pairs of lines (1–6), (2–5), (3–4) of the six-lines spectrum. The expression for one pair of lines reads [3, 6]

$$\mathrm{I}(\omega)\propto-2\mathrm{Re}\left(\overrightarrow{\boldsymbol{W}}\overleftrightarrow{M}^{-1}\overrightarrow{1}\right)\quad\text{with}\quad\overleftrightarrow{M}=(i(\omega-\omega_i)+\Gamma)\overleftrightarrow{1}-\overleftrightarrow{\mathrm{R}},\tag{8}$$

where $\overrightarrow{W}$ is a row vector consisting of probabilities (Boltzmann factors) of the ($2S$+1) states, $\overleftrightarrow{M}$ contains the static line positions ω_i, given by the splitting (1–6), which is assumed to be proportional to S_z, $\omega\propto$ source velocity and Γ is the natural line width. The relaxation matrix $\overleftrightarrow{R}$ has as elements transition rates W_i^j between the states $|i>\rightarrow|j>$ ($i\neq j$) and $W_i^i=-\Sigma_j W_i^j(i\neq j)$. The total Mössbauer spectrum is obtained by summing up the calculated two-lines spectra (1–6), (2–5) and (3–4), where in the calculation of the ω_i adequately reduced splittings (2–5) and (3–4), still proportional to S_z, appropriate relative intensities (e.g., 3:2:1) and the above calculated transition rates between adjacent states are used. The Boltzmann factors contained in $\overrightarrow{W}$ are

$$W_z\propto\exp\left[-(\mathrm{A}S/k_\mathrm{B})(S/T)\left\{(1+h)^2-(S_z/S+h)^2\right\}\right].\tag{9}$$

Introducing these expressions in $I(\omega)$ one can simulate Mössbauer spectra. The inverse of $\overleftrightarrow{M}$ is required for each value of ω (256 Mössbauer velocities), involving a order of 256 times N^3 operations for a matrix of rank N. In our case $\overleftrightarrow{M}$ is tridiagonal and thus only of the order of $256N^2$ operations are necessary for inversion. A further significant saving in computer time is obtained by observing that in the expression for $I(\omega)$ in fact not the inverse $\overleftrightarrow{M}$ but only $\overleftrightarrow{M}^{-1}\overrightarrow{1}$ is needed. Writing $\overleftrightarrow{M}^{-1}\overrightarrow{1}=\overrightarrow{V}$ or $\overleftrightarrow{M}\ \overrightarrow{V}=1$ the problem is reduced to solve a set of simultaneous linear equations for $\overrightarrow{V}$ which requires only an order of $256N$ operations. This method allows one to simulate spectra for spins S up to several thousand, corresponding to spins found in real systems, on a modest computer in a few minutes.

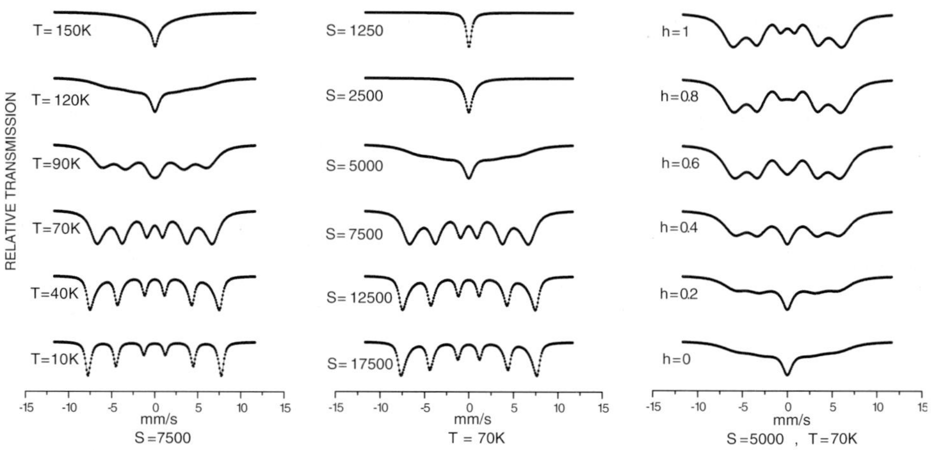

Figure 2 Simulation of Mössbauer patterns using the theory developed and the parameters given in the text. *Left*: Spin S=7,500 (diameter 89.5 Å) at temperatures indicated. *Middle*: T=70 K at spins indicated. *Right*: S =7,500, T=70 K at reduced fields h indicated.

For the parameters contained in Eqs. 7 and 9 we use the same values as in [4] which are similar to the values used by us in [2]. They are: s=5/2, $J'=5\times10^{-10}$ J m^{-1}, $m=57\times1.66\times10^{-27}$ kg (mass of a $^{57}Fe^{3+}$−ion) $T_C \approx \theta_C/7 \approx 900K/7 \approx 128.6K$, $T_D \approx \theta_D/4 \approx 700K/4 = 175K$, ka for $h=0$ is ka $= E_k/k_B T_D \approx 5.6 \times 10^{-8} S_z \ll \alpha = 1.36$ and $E_k/k_B T \ll E_q/k_B T = (KV + 2ASh)/k_B T + 0.25\alpha T_D/T$ (cf. [4]) for S and T in the usual range (i.e., $E_{q-k} \approx E_q$), $K=10^4$ J/m^3, AS$=5\times10^{-25}$ J, $V=Na^3=N(5\ \text{Å})^3$, where $N=S/s$. With this the transition rates (in s^{-1}) for phonon annihilation on the left side are given from Eq. 6 by

$$\begin{aligned} W_{S_z-1}^{S_z} &= \frac{s^2 J'^2 \hbar \alpha^6}{2^6 \pi m k_B^2 T_C T_D} \cdot \frac{e^{E_q/k_B T}}{\left(e^{E_q/k_B T}-1\right)^2 \left(e^{E_{k\text{left}}/k_B T}-1\right)} \\ &\approx \frac{4.41\times10^{13} ST/Kelvin}{-S_z+1/2-hS}\left[\sin h\,\frac{0.0181S+29.8}{T/Kelvin}\right]^{-2} s^{-1},\ S_z \\ &= -S+1, -S+2, \ldots, -hS \end{aligned} \tag{10}$$

(left side of Figure 1, phonon annihilation) and correspondingly for $W_{S_z}^{S_z-1}$ (phonon creating transitions on the right side). The reverse rates (phonon creation on the left and annihilation on the right) are identical for identical energy differences.

The time τ^S_{-S} of a spin-transition from the state $-S$ to S for h=0 can be estimated in the following way

$$\tau_{-S}^{S} = \left\{ 2\left[\frac{1}{4.41\times10^{13} ST/\text{Kelvin}}\right]\cdot\left[\sin h\,\frac{0.0181S+29.8}{T/\text{Kelvin}}\right]^2 \sum_{S_z=-S+1}^{0} (-S_z + 1/2) \right\} s$$

$$\text{or } W_{-S}^{S} \approx \left\{ 4.41 \times 10^{13}\, \frac{T/\text{Kelvin}}{S} \exp\left(-\frac{0.0362S+59.6}{T/\text{Kelvin}}\right) \right\} s^{-1}. \tag{11}$$

These rates lay in the usual range and exhibit the classical Arrhenius-like behavior.

Figure 2 shows simulations of Mössbauer spectra effectuated using Eqs. 8, 9, and 10 for the temperatures T, total spins S (proportional to the volume V) and reduced fields h indicated.

With the above given values $S=7{,}500$ means a diameter of 89.5 Å for a spherical particle. Computation time for a spin of $S=17{,}500$ is typically 2 min on a personal computer.

The simulated spectra in Figure 2 reproduce the patterns of experimental spectra found in Mössbauer spectra of superparamagnetic samples. The dependence on sample temperature, particle volume and external field corresponds to that of real spectra, e.g., of Ni- and Co-ferrite powders [3].

Acknowledgement The support of the Brazilian agencies CNPq and FAPEMIG is gratefully acknowledged.

References

1. Skumryev, V., Stoyanov, S., Zhang, Y., Hadjipanayis, G., Givord, D., Nogués, J.: Nature **423**, 850 (2003)
2. Pfannes, H.-D., Mijovilovich, A., Magalhães-Paniago, R., Paniago, R.: Phys. Rev. B **62**, 33722 (2000)
3. Pfannes, H.-D., Dias Filho, J.H., Paniago, R., Magalhães-Paniago, R.: Hyp. Interactions **148/149**, 243 (2003)
4. Srivastava, K.K.P.: J. Phys., Condens. Matter **15**, 549 (2003)
5. Clauser, M.J., Blume, M.: Phys. Rev. B **3**, 583 (1971)
6. Jones, D.H., Srivastava, K.K.: Phys. Rev. B **34**, 7542 (1986)
7. Sinha, K.P., Upadhyaya, U.N.: Phys. Rev. **127**, 432 (1962)
8. Holstein, T., Primakoff, H.: Phys. Rev. **58**, 1098 (1940)

Hyperfine Interact (2005) 165: 253–259
DOI 10.1007/s10751-006-9274-9

Local structural order in nanostructured hematite

J. M. Florez · J. Mazo-Zuluaga · J. Restrepo

Published online: 26 October 2006

Abstract Nanostructured α-Fe_2O_3 powders were prepared by high-energy ball milling. The milling process spans grinding times from 30 min to 24 h. The as-milled samples were characterized by means of ^{57}Fe Mössbauer spectrometry, Rietveld analysis of X-ray diffraction data and particle size analysis. The obtained results evidence the presence of disordered hematite characterized by a hyperfine field distribution with a well-behaved dependence on the mean crystallite size for which the mean hyperfine field decreases asymptotically as the grain size decreases. A new relationship is proposed in order to describe such behavior. Finally the presence of superparamagnetic grains, the occurrence of a partial topotactic phase transformation into a spinel phase and tool induced contamination are also presented and discussed.

Key words Mössbauer · hematite · Rietveld · ball milling · nanostructure

1 Introduction

During the last decade, great attention has been paid to the magnetic and structural properties of nanostructured alloys [1], fluorides [2] and some iron oxides [3] for which a wide variety of grain boundary effects takes place. However a wider and deeper insight on the structural and magnetic properties of some nanostructured iron oxides like hematite is still remaining. These features are important to understand some mechanisms like those involved in the Morin transition [4–7]. In this work we stress on nanostructured hematite powders obtained through a vacuum grinding route for different milling times in a planetary ball mill. Their structural, microstructural and magnetic properties are investigated by

J. M. Florez · J. Mazo-Zuluaga · J. Restrepo (✉)
Grupo de Estado Sólido, Instituto de Física, Universidad de Antioquia, A.A. 1226 Medellín, Colombia
e-mail: jrestre@fisica.udea.edu.co

J. Mazo-Zuluaga · J. Restrepo
Grupo de Instrumentación Científica y Microelectrónica, Instituto de Física,
Universidad de Antioquia, A.A. 1226 Medellín, Colombia

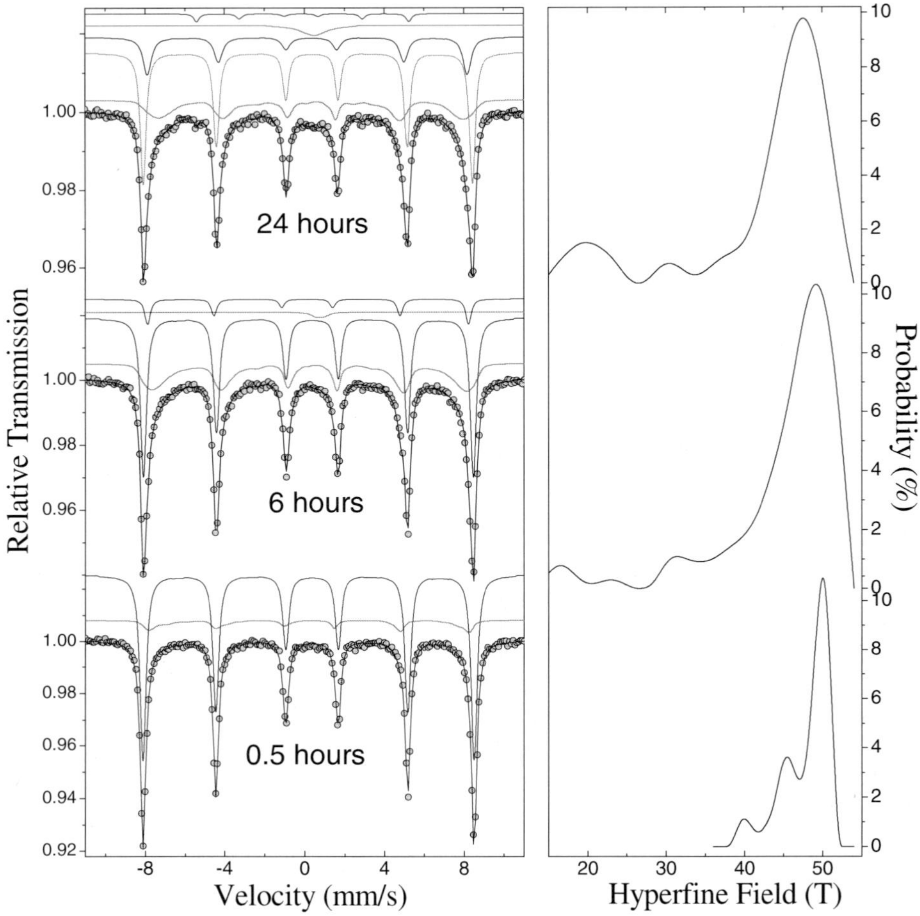

Figure 1 Room temperature Mössbauer spectra for 0.5, 6 and 24 h of milling and their respective hyperfine field distributions. The decomposition into components is shown.

means of transmission ^{57}Fe Mössbauer spectrometry at room temperature, Rietveld analysis of X-ray diffraction patterns and particle size analysis by laser scattering.

2 Experimental

Mechanical milling of 99.9% pure hematite (Merck) powders, with an average particle size of 4 μm, was carried out in a planetary ball mill Fritsch-Pulverisette5 by using Chromium-based stainless steel jars of 250 ml with balls of the same material and a diameter of 13 mm. The employed vessels were previously evacuated at 0.1 atm. Milling times of 0.5, 1, 3, 6, 12, 15, 18 and 24 h were considered using an average sample-balls weight ratio of 1:10. Every sample for every milling time was prepared separately in order to preserve the sample to balls weight ratio and to guarantee the absence of statistical correlation among them. Every sample was milled at 390 rpm for intervals of 1 h and break periods of 30 min up to complete the desired milling time. At the end of every milling interval both the

Table I Hyperfine parameters of the distributed component for the first milling stages

Milling time (h)	$2\varepsilon_Q$ (mm/s)	δ (mm/s)	B_{hf} (T)
0.5	0.010	0.671–0.059	36–54
1	−0.079	0.431–0.258	15–54
3	−0.138	0.508–0.243	15–54
6	−0.154	0.550–0.277	15–54

pressure and temperature were monitored. This last one ranged between 35 and 41°C. The obtained as-milled samples were analyzed by means of ^{57}Fe Mössbauer transmission spectrometry at 296 K, X-ray diffraction and laser scattering for particle size analysis. Mössbauer spectra were fitted using the Normos-Distri program [8]. All fittings involved the presence of at least one hyperfine field distribution (HFD) according to the observed line broadenings. Isomers shifts are quoted relative to α-Fe. X-ray diffraction analysis was performed using a Bruker AXS Advance D8 diffractometer with Co tube and Fe filter in the 2θ range of 20–70° every 0.014° and 4 s per step. Diffraction patterns were fitted using the MAUD program (http://www.ing.unitn.it/~luttero/maud/) based on the Rietveld method from which phase distribution, cell parameters, mean crystallite size, microstrain and volume fractions were obtained.

3 Results and discussion

Figure 1 shows the Mössbauer spectra for three selected milling times. In addition to the HFD, a well defined crystalline component accounting for the observed kinks in the external absorption lines was also considered. The fitted hyperfine parameters of this last component are B_{hf}=51.4(3) T, $2\varepsilon_Q$=−0.20(1) mm/s and δ=0.38(1) mm/s, which clearly correspond to the precursor hematite. The corresponding relative area of this component decreases as milling time increases. Concomitantly, the spectral area of the distributed component increases and the HFD becomes progressively broader in agreement with a higher degree of disorder. Table I shows the hyperfine parameters of the distributed component for the first milling stages whereas the milling time dependence of the relative areas is shown in Figure 2. Figure 3 shows some selected diffractograms including Rietveld fitting. In all cases two hematite components were employed: One of them preserving structural parameters of the precursor hematite. The second one was fitted revealing a progressive volume fraction increase and a marked reduction of the mean crystallite size in correspondence with the HFD via Mössbauer. Several features are particularly interesting like the fractional variations of the lattice parameters a and c of the free hematite component as shown in Figure 4. The observed behavior reveals a mechanical dilation anisotropy where the unit cell volume tends to be more easily deformed along the basal plane instead of the c axis. In fact, the average unit cell volume expansion was as higher as 0.5%. We think the observed deformation anisotropy should play a key role as a magnetovolume effect in the explanation of the Morin transition, since as is established the spin canting observed in hematite involves a preference of the spins to lie along the basal plane [9]. Such a degree of deformation agrees with the observed variation of the root mean square microstrain ranging from 0.0049 up to 0.011 and becoming stable at around this last value beyond 3 h of milling. On the basis of these results and in order to elucidate the influence of the structural parameters upon the hyperfine ones we have carried out several correlations including the dependence of the mean hyperfine field on both the unit cell average volume of hematite and the mean crystallite size. In the former, the mean

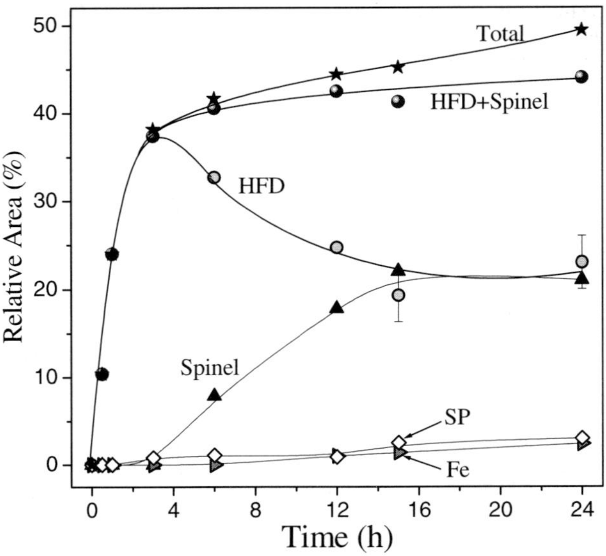

Figure 2 Milling time dependence of the increasing relative areas obtained via Mössbauer. The different employed components are exhibited. Label *SP* stands for superparamagnetic grains. *Lines* are guides to the eye.

hyperfine field decreases smoothly with the unit cell volume up to around 909 $Å^3$ where $\langle B_{hf} \rangle$ exhibits a sharp decrease down to almost 40 T. Such reduction corresponds to 21% relative to the hyperfine field of pure hematite. This remarkable decrease is ascribed to the milling-driven increase of the nearest neighbor distance for which the dipolar contribution to the magnetic hyperfine field becomes smaller. The correlation of $\langle B_{hf} \rangle$ with the mean crystallite size is shown in Figure 5 which can be fitted by the following proposed relationship:

$$\langle B_{hf} \rangle = B_0 - \frac{\alpha}{\langle D \rangle + (\alpha / B_0)}, \tag{1}$$

where $\langle D \rangle$ accounts for the mean crystallite size, B_0 is the hyperfine field of the precursor hematite and α is a fitting parameter ruling out the way as the mean hyperfine field goes to zero as the mean crystallite size decreases. This expression complements that proposed by Murad and Schwertmann [10], and it describes better the behavior of the mean hyperfine field in those limit cases where the mean crystallite size goes to infinity (an extended and homogeneous single-crystalline hematite) or when it goes to zero, for which $\langle B_{hf} \rangle$ vanishes, corresponding to the superparamagnetic limit. Such decrease is attributed to a milling-induced higher degree of disorder. Moreover, the decrease of the mean hyperfine field must be linked to those grain boundaries and interfacial regions where the degree of disorder is greater. In correspondence, the employed hematite crystalline component can be attributed to the core of the grains with a progressive decrease of the relative area as we have already mentioned.

On the other hand, as milling time goes beyond 3 h, a doublet (Γ=0.9 mm/s, δ= 0.55 mm/s, ΔE_Q=0.31 mm/s) is necessary to get a good fit of the central region in Mössbauer spectra (see Figure 1). Such a signal is attributed to the occurrence of superparamagnetic grains having sizes smaller than 8 nm for which a superparamagnetic relaxation is expected [12]. The presence of those grains is endorsed by the obtained grain size distribution centered at around 11 nm in agreement with that fraction of hematite corresponding to the distributed component, i.e. grain boundaries and interfacial regions.

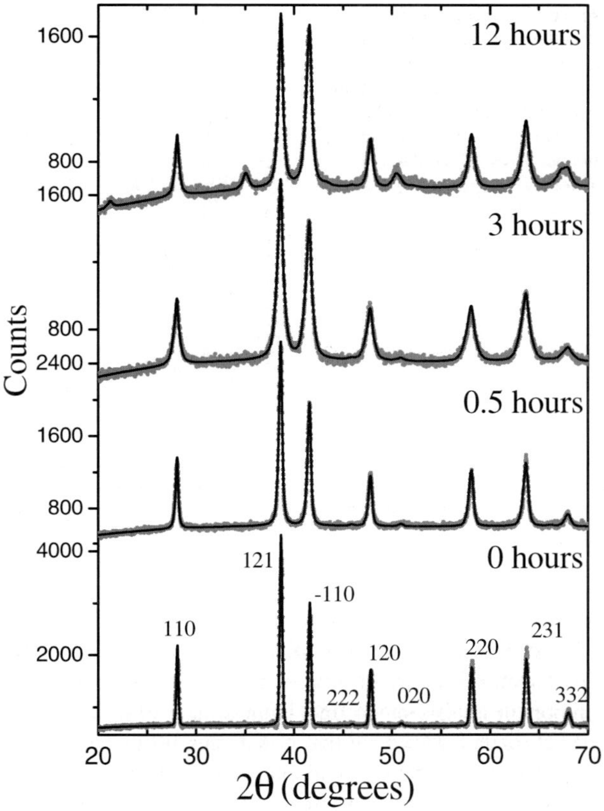

Figure 3 X-ray diffractograms, including Rietveld fitting for the precursor material, 0.5, 3 and 12 h of milling. Two hematite components were employed, one of them for the core of the grains and the other one accounting for grain boundaries and interfacial regions.

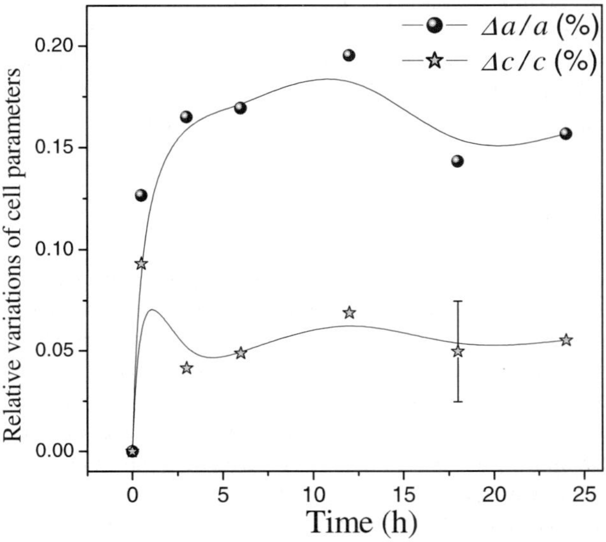

Figure 4 Fractional variations of the lattice parameters a and c for the distributed hematite component obtained via Mössbauer. *Lines* are guides to the eye.

Figure 5 Mean crystallite size dependence of the mean hyperfine field $\langle B_{hf} \rangle$. Inset shows the fitting relationship describing the observed behavior, where B_0 corresponds to the hyperfine field of pure hematite (51.6 T) and $\langle D \rangle$ in Å accounts for the mean crystallite size. The parameter α giving the best fit was 900 TÅ.

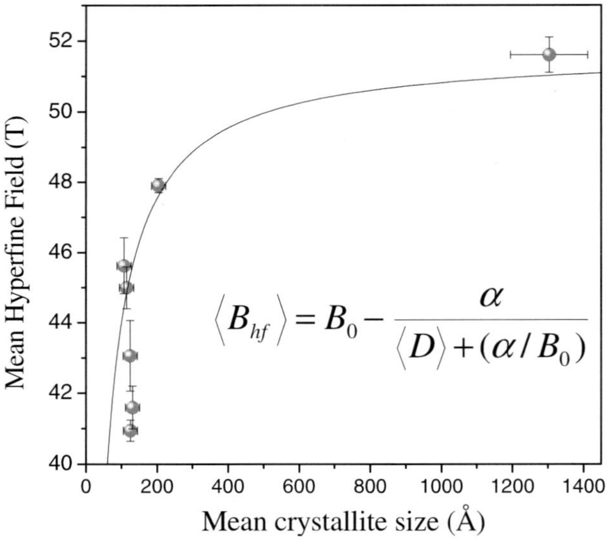

For milling times above 6 h, an additional component must be included in the fitting process corresponding to a spinel phase having B_{hf}=49.8 T, $2\varepsilon_Q$=−0.08 mm/s, and δ= 0.30 mm/s, which could be ascribed to a maghemite phase [10]. The relative area of this component is also shown in Figure 2. The onset of this component is endorsed by X-ray diffraction data for which two new Bragg peaks become evident at 35.0 and 50.6° and can easily be indexed by Rietveld analysis to a magnetite or maghemite phase. However, since magnetite involves the presence of Fe^{2+} ions in octahedral crystallographic sites, for which a reduction from hematite is needed, it makes more sense to assign this new phase to maghemite, which involves only Fe^{3+} like in the precursor material, preserving the molecular formula. From this point of view, this spinel component can therefore be ascribed to a topotactic isochemical transformation from hematite due to the imparted energy during the milling process. For milling times above 12 h a sextet corresponding to α-iron was detected with relative area below 2.5%. The presence of this phase, which is also observable through diffractograms, is attributed to contamination arising from milling tools [13]. Finally, measurements on particle size through laser scattering, evidenced particle sizes in the range 0.1–500 μm endorsing a scenario of micrometric nanostructured particles.

4 Conclusions

The obtained results reveal a partial transformation (less than 23%) of the precursor hematite into a spinel phase attributed to maghemite. However, despite the presence of this spinel phase, the employed milling conditions delays the complete transformation of hematite contrary to what has been found in wet milling conditions by using the same type of mill [14]. The observed doublets in Mössbauer spectra (less than 2.5%), for milling times above 3 h, are attributed to superparamagnetic grains. On the other hand, a worthy relationship between the mean crystallite size and the mean hyperfine field of the distributed hematite component has been proposed through the correlation between Mössbauer results and Rietveld analysis. Our results endorse some of the findings reported by Stewart et al. [5] and Bruzzone and Ingalls [15] related to the effect of milling-induced disorder and the effect of pressure upon the magnetic properties of hematite.

Finally, the observed results allow proposing a scenario of nanometric crystalline grains with interfacial regions characterized by a high degree of atomic disorder. Such a state is ruled out by an anisotropic mechanical dilation for which a higher fractional variation of the *a* parameter was observed. Correspondingly the mean hyperfine field decreases with the progressive evolution of this state as the milling time increases.

Acknowledgements This work was supported by Universidad de Antioquia through the projects SIU24-1-28, CODI-GICM 2005–2007 and by COLCIENCIAS-CENM contract no. 043-2005. We are also very grateful to Prof. Herley Casanova from the Chemistry Institute for performing the particle size analysis.

References

1. Miglierini, M., Greneche, J.M., Idzikowski B.: Mater. Sci. Eng. **A304-306**, 937 (2001)
2. Guérault, H., Bureau, B., Silly, G., Buzaré, J.Y., Greneche J.M.: J. Non-Crryst Solids **287**, 65 (2001)
3. Randrianantoandro, N, Mercier, A.M., Hervieu, M., Greneche, J.M.: Mater. Lett. **47**, 150 (2001)
4. Kletetschka, Günther, Wasilewski, Peter J.: Phys. Earth Planet. Inter. **129**, 173 (2002)
5. Stewart, S.J., Borzi, R.A., Cabanillas, E.D., Punte, G., Mercader, R.C.: J. Magn. Magn. Mater. **260**, 447 (2003)
6. Amin, N. Arajs, Sigurds: Phys. Rev., B. **35**, 4810 (1987)
7. Zysler, R.D., Fiorani, D., Testa, A.M., Suber, L., Agostinelli, E., Godinho, M.: Phys. Rev., B. **68**, 212408–212411 (2003)
8. Brand, R.A.: Nucl. Instrum. Methods Phys. Res., B. **28**, 417 (1987)
9. Dang, M.Z., Rancourt, D.G., Dutrizac, J.E., Lamarche, G., Provencher, R.: Hyperfine Interact. **117**, 217 (1998)
10. Murad, E., Schwertmann, U.: Clays Clay Miner. **34**, 1 (1986)
11. Cornell R.M., Schwertmann U.: The Iron Oxides. VCH, Weinheim (1996)
12. Rancourt, D.G., Julian, S.R., Daniels, J.M.: J. Magn. Magn. Mater. **49**, 305 (1985)
13. Sánchez, F.H., Rodríguez Torres, C.E., Fernández van Raap, M.B., Mendoza Zélis, L.: Hyperfine Interact. **113**, 269 (1998)
14. Betancur, J.D., Restrepo, J., Palacio, C.A., Morales, A.L., Mazo-Zuluaga, J., Fernández, J.J., Pérez, O., Valderruten, J.F., Bohórquez, A.: Hyperfine Interact. **148/149**, 163 (2003)
15. Bruzzone, C.L., Ingalls, R.: Phys. Rev. B. **28**, 2430 (1983)

Hyperfine Interact (2005) 165: 261–266
DOI 10.1007/s10751-006-9275-8

Mössbauer studies of Cr substituted bulk and nano Co–Zn ferrites

Ram Kripal Sharma · Omprakash Suwalka · N. Lakshmi · K. Venugopalan

Published online: 26 October 2006

Abstract X-ray and Mössbauer studies have been carried out for bulk and nano sized $Cr_xCo_{(0.5-x)}Zn_{0.5}Fe_2O_4$ (x=0, 0.1, 0.2, 0.25, 0.3, 0.4, 0.5) prepared through wet chemical co-precipitation method. The hyperfine fields in the bulk samples at tetrahedral A site is independent of the substitution while it decreases sharply at octahedral B site due to the quenching of A–B super-exchange interaction due to the addition of Cr ions which preferentially occupy the octahedral B sites. The average particle size for the nano samples was seen to decrease exponentially with the addition of Cr. The RT Mössbauer spectra for all the nano samples show only two paramagnetic doublets showing the superparamagnetic relaxation phenomena in these samples due to small particle sizes involved.

Key words Mössbauer spectra · spinel ferrites · nano Co–Zn ferrites

1 Introduction

Ferrites are widely used in microdevices due to their high resistivity, low eddy current losses and high saturation magnetization. They are also used for high frequency applications as in radio frequency coils, transformer cores etc. [1, 2].

Ferrimagnetic spinels have the general formula XY_2O_4. In ferrites, Y is Fe and X is some transition metal like Fe, Co, Mn etc. Their magnetic properties originate mainly from the magnetic interactions between cations with magnetic moments that are situated in the tetrahedral (A) and the octahedral (B) sites [3]. $ZnFe_2O_4$ is a normal spinel with all the Fe^{3+} ions in the B sites and all the Zn^{2+} ions in the A sites, whereas $CoFe_2O_4$ has an inverse spinel structure with the Co^{2+} ions mainly in the B sites and Fe^{3+} ions distributed almost equally between the A and the B sites [4]. The nano sized ferrites have received a growing and renewed interest in last few years due to the high dependence of the magnetic properties on particle sizes involved.

R. K. Sharma (✉) · O. Suwalka · N. Lakshmi · K. Venugopalan
Department of Physics, Mohanlal Sukhadia University, Udaipur 313001, India
e-mail: ramkripal_sharma@yahoo.com

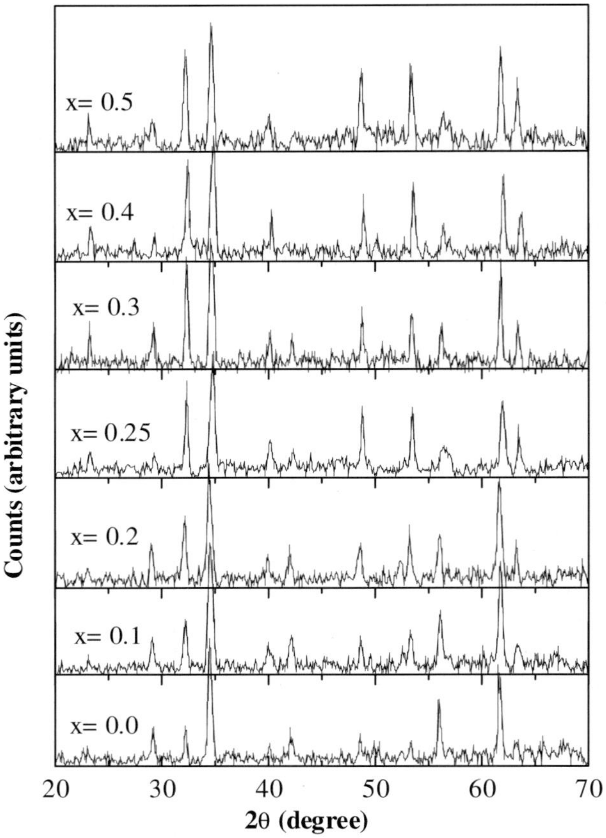

Figure 1 X-ray diffractograms of $Cr_xCo_{(0.5-x)}Zn_{0.5}Fe_2O_4$ (x=0, 0.1, 0.2, 0.25, 0.3, 0.4, 0.5) for Bulk samples.

The spinel structure also allows introduction of different metallic ions, which can change the magnetic and electric properties considerably [5]. Thus magnetic parameters like remanence and coercivity in technologically important materials can be modified and controlled by the simple substitution of M ions in MFe_2O_4 by other ions like Cr^{3+} [6]. In this paper we report the results of room temperature Fe-57 Mössbauer studies made on the Cr substituted series $Cr_xCo_{(0.5-x)}Zn_{0.5}Fe_2O_4$ (x=0, 0.1, 0.2, 0.25, 0.3, 0.4, 0.5), for both bulk and nano samples, to investigate the effect of Cr substitution in $Co_{0.5}Zn_{0.5}Fe_2O_4$, which has a mixed type of spinel structure.

2 Experimental

Bulk and nano particles of $Cr_xCo_{(0.5-x)}Zn_{0.5}Fe_2O_4$ for (x=0, 0.1, 0.2, 0.25, 0.3, 0.4, 0.5) were prepared by a chemical co-precipitation method using analytical grade Fe $(NO_3)_3.9H_2O$, $CoCl_2.6H_2O$, anhydrous $ZnCl_2$ and $CrCl_3.6H_2O$ as starting materials and maintaining an optimum pH value with 30% ammonia solution. The precipitate was dried by heating at 100°C for 24 h and then annealed at 300°C for 24 h to obtain the nano samples. Bulk samples were obtained by heating at 900°C for 24 h after initial drying.

X-ray diffractograms of samples were obtained on a Rigaku Miniflex diffractometer at room temperature using Cu Kα-1 radiation. Mössbauer studies were done using a standard

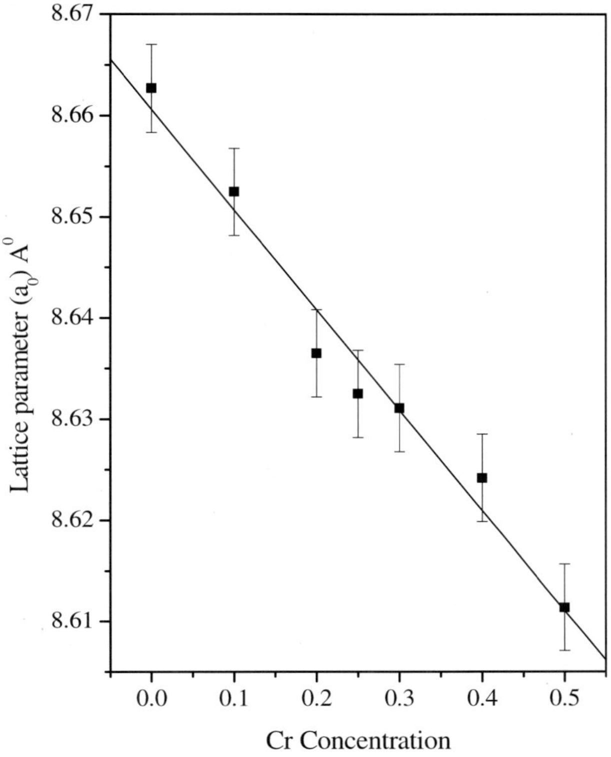

Figure 2 Linear least squares fit of variation of lattice parameter a_0 with Cr concentration for bulk samples.

Austin drive and controller assembly in the constant acceleration mode. The source used was a 5-mCi Co^{57} in Rhodium matrix for nano samples and 25 mCi (Rh) for bulk samples. Velocity calibration was done using metallic iron foil. The Mössbauer spectra were fitted using the Meerwaal programme [7].

3 Results and discussion

X-ray diffractograms of the bulk samples (Figure 1) show all of them to have a single phased spinel structure. They also show the presence of 2–3% α-Fe_2O_3 peaks. The presence of such peaks has also been reported in other ferrite samples prepared by the chemical co-precipitation method [8]. The variation of the lattice parameter vs. Cr concentration is shown in Figure 2 which shows that the lattice parameter a_0 decreases linearly with the increase in Cr concentration. This is due to the replacement of the bigger Co ions by the smaller Cr ions in the spinel structure. This result is well in agreement with Vegard's law and has been reported earlier for other ferrites [9, 10]. It can be seen from Figure 3 that average particle size for the nano series decreases exponentially with the addition of Cr [13].

Figure 4a shows the room temperature Mössbauer spectra for the bulk samples. All the samples have been fitted for two doublets and two sextets that can be attributed to Fe at two inequivalent crystallographic sites A (tetrahedral) and B (octahedral). The line widths of sextets corresponding to the B site are large enough to justify fitting with a subspectrum. In many studies, fitting of at least two sextets at the B site have been reported [11, 12].

Figure 3 Exponentially fitted variation of average particle size with Cr concentration for nano samples.

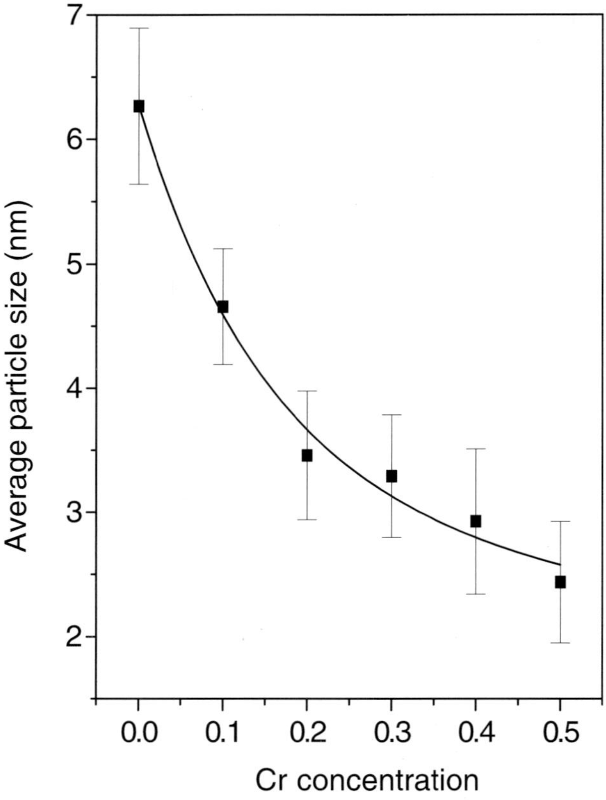

However, in these samples, the line intensities of the B site sextet decreases sharply with the addition of Cr. We have, therefore fitted for only a single sextet at the B site, which has line widths in the range quoted for ferrite samples in general.

The value of the hyperfine fields at the two sites is strongly dependent on cation distribution between the A and B sites, which in turn depends on the heat treatment given during sample preparation. Petitt and Forester [3] obtained a value of 0.19 for the inversion parameter for $CoFe_2O_4$ prepared by them, while Sawatzky et. al. [14] report an inversion parameter equal to 0.24 for their sample of $CoFe_2O_4$. The difference in the two values has been attributed [3] to the difference in the heat treatment at the time of sample formation. In our case, it can be seen from Table I that 84% of the Fe is present at site A displacing almost 68% of Zn to octahedral B site, since the temperature maintained during the formation was only 900°C. This is nearer to the value assumed by W.A. Kaczmarek et al. [15] for $Co_{0.5}Zn_{0.5}Fe_2O_4$, prepared through aerosol route, in which the temperature given during formation was 1,027°C. The higher amount of nonmagnetic zinc at the B sites instead of magnetic Fe leads to a decrease in the B–B interaction, reducing the hyperfine field at B site. This may be the reason for the observed smaller value of the hyperfine field at B site in our case, as compared to values obtained by earlier workers for $Co_{0.5}Zn_{0.5}Fe_2O_4$.

The fitted Mössbauer parameters for bulk samples are shown in Table I. From this table, it is evident that the area occupied by the sextet at A site rises sharply as the Cr concentration increases. The value of isomer shifts at A site is almost constant while there is

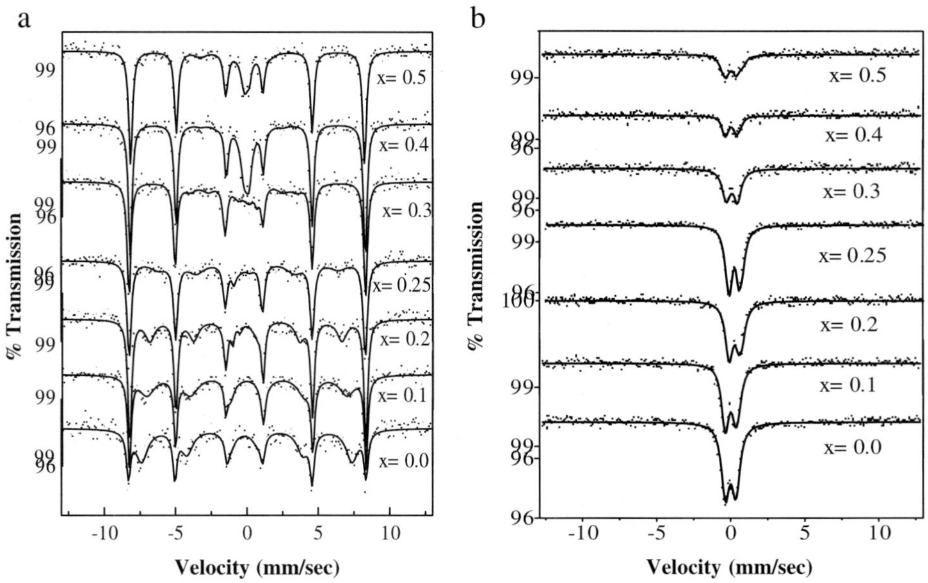

Figure 4 Room temperature Mössbauer spectra of $Cr_xCo_{(0.5-x)}Zn_{0.5}Fe_2O_4$ (x=0, 0.1, 0.2, 0.25, 0.3, 0.4, 0.5) for (a) Bulk and (b) Nano samples (*Solid lines* give the fit to experimental data).

a continuous decrease in its value at the B site. This implies that the Cr ions preferentially occupy the octahedral B site, displacing Fe^{+3} to tetrahedral A site, hence increasing the area of sextet A. These trends are similar to those seen for the corresponding nano samples [13].

The kinds of interactions possible between the cations situated at tetrahedral A site and octahedral B site are A–A, B–B and super-exchange A–O–B interaction. As is evident from Table I, the magnetic hyperfine field at A site is almost unaffected by the addition of Cr, while the field at B site decreases sharply. Hence we can conclude that the super-exchange A–B type of interaction is quenched or shielded by the addition of the Cr ions at B sites, leaving only A–A or B–B type of interactions to give the observed values of fields. Since the addition of Cr does not very much affect the A–A interaction, the hyperfine field at A site is almost constant, whereas, since Cr moments order antiferromagnetically with respect to iron magnetization, induction of Cr at the B site leads to a sharp fall in the value of hyperfine field at B site.

In comparison, the room temperature Mössbauer spectra of the corresponding nano samples (Figure 4b) consist of only two quadrupole doublets. This shows the nano samples to be in the superparamagnetic state due to the very small particle sizes involved [13].

4 Conclusions

The cation distribution in spinel ferrites depends strongly on the temperature treatment given during their formation. RT Mössbauer spectra show that Cr preferentially occupies octahedral B sites in both bulk and nano samples of $Cr_xCo_{(0.5-x)}Zn_{0.5}Fe_2O_4$. The addition of Cr leads to quenching or shielding of A–O–B super-exchange interaction between cations distributed at tetrahedral A and octahedral B sites, leaving A–A and B–B type of interaction to play the deciding role in determining the hyperfine fields at A and B sites.

Table I Fitted Mössbauer parameters for $Cr_xCo_{(0.5-x)}Zn_{0.5}Fe_2O_4$ bulk samples for (x=0, 0.1, 0.2, 0.25, 0.3, 0.4, 0.5)

x	Hyperfine fields (kOe)		Percent area of sextets		I. S.[a](mm/sec)		Γ (mm/sec)	
	A site	B site	A site	B site	A site	B site	A site	B site
0	514.62	456.33	41.89	62.04	0.163	0.279	0.338	1.016
0.1	513.87	434.94	44.49	48.71	0.147	0.258	0.286	1.296
0.2	512.36	417.15	63.76	29.51	0.149	0.178	0.312	0.781
0.25	511.17	389.92	66.0	27.29	0.157	0.116	0.354	1.414
0.3	511.7	231.04	69.04	17.45	0.157	0.062	0.316	1.048
0.4	510.48	178.68	74.19	1.09	0.157	−0.086	0.311	0.622
0.5	508.28	160.39	78.24	4.72	0.149	−0.227	0.324	0.649

[a] I.S. values are quoted with respect to metallic iron

Mössbauer spectra of all the nano samples show only two paramagnetic doublets while their corresponding bulk samples show sextets as well as doublets clearly showing the superparamagnetic relaxation phenomena in nano samples.

Acknowledgments This work has been supported by the UGC-DRS and COSIST schemes of the Department of Physics, M.L. Sukhadia University, Udaipur. One of us (Ram Kripal Sharma) acknowledges financial support by CSIR. We also thank the UGC-DAE CSR, Indore, India for the use of Mössbauer facilities for the bulk samples.

References

1. Smit, J., Wijn, H.P.J.: Ferrites. Philips Technical Library (1959)
2. Snoek, J.J.: New Development in Ferrimagnetic Material. Elsevier, New York (1974)
3. Petitt, G.A., Forester, D.W.: Phys. Rev. B. **4**, 3912 (1971)
4. Vander Woude, F., Swatzky, G.A.: Phys. Rev. B. **4**, 3159 (1971)
5. Ahmed, T.T., Rahman, I.Z., Rahman, M.A.: J. Mater. Process. Technol. **153–154**, 797 (2004)
6. Kuznetsov, M.V., Barquín, L.F., Pankhurst, Q.A., Parkin, I.P.: J. Phys. D. Appl. Phys. **32**, 2590 (1999)
7. von Meerwall, E.: Comput. Phys. Commun. **9**, 117 (1975)
8. Rath, C., Sahu, K.K., Anand, S., Date, S.K., Mishra, N.C., Das, R.P.: J. Magn. Magn. Mater. **202**, 77 (1999)
9. Ladgaonkar, B.P., Vasambekar, P.N., Vaingakar, A.S.: J. Magn. Magn. Mater. **210**, 289 (2000)
10. Chandrasekaran, G., Nimy Sebastian, P.,: Mater. Lett. **37**, 17 (1998)
11. Mahmoud, M.H., Hamdeh, H.H., Ho, J.C., O'Shea, M.J., Walker, J.C.: J. Magn. **220**, 139 (2000)
12. Amer, M.A., Hiti, M.E.: J. Magn. Magn. Mater. **234**, 118 (2001)
13. Sharma, R.K., Suwalka, O., Lakshmi, N., Venugopalan, K., Banerjee, A., Joy, P.A.: Mater. Lett. **59**, 3402 (2005)
14. Sawatzky, G.A., Van Der Woude, F., Morrish A.H.: Phys. Rev. **187**, 747 (1969)
15. Kaczmarek, W.A., Calka, A., Ninham, B.W.: IEEE Trans. Magn. **29**(6), 2649 (1993)

Hyperfine Interact (2005) 165: 267–271
DOI 10.1007/s10751-006-9276-7

Surface oxidation control of nanosized magnetite and Mössbauer measurements

R. E. Vandenberghe · I. Nedkov · T. Merodiiska · L. Slavov

Published online: 10 November 2006

Abstract Selected highly homogeneous powders of Fe_3O_4 with different particle size on the nanometer scale (10 ± 2 and 3 ± 2 nm) obtained by soft-chemical methods were studied by Mössbauer spectroscopy. The study shows clearly the powerful possibilities of Mössbauer spectroscopy to analyze the surface oxidation of nanostructured powders of magnetite. On the other hand, it is shown that for very small superparamagnetic particles the spectrum of magnetite might be quite similar to that of maghemite, making it difficult to distinguish between both phases.

Key words surface oxidation control · nanosized magnetite · Mössbauer measurements

1 Introduction

Mössbauer spectroscopy (MöS) is a widely used tool for the characterization of nanosized magnetite, the latter being one of the most interesting magnetic ferrospinels, for instance as carrier in medical applications. The structural formula of magnetite can be written as $\left(Fe^{3+}\right)_A\left[Fe_2^{2.5+}\right]_BO_4$, where the octahedral (B) sites contain iron with valence $Fe^{2.5+}$ due to the fast electron hopping above the Verwey transition (T>130 K). Typical values for the hyperfine parameters for Fe^{3+} and $Fe^{2.5+}$ in Mössbauer spectra for the respective tetrahedral and octahedral sublattices of well-crystallized magnetite at room temperature are $H_{hf,A}$= 49 T, δ_A=0.28 mm/s, $H_{hf,B}$=46 T, δ_B=0.66 mm/s with ε_Q=0 mm/s for both sites [1]. So, for stoichiometric magnetite the spectral area ratio is about 1:1.9 at RT. The atoms on the surface have fewer neighbors as compared with those inside, which disturbs the

R. E. Vandenberghe (✉)
Department Subatomic and Radiation Physics, Ghent University, 9000 Ghent, Belgium
e-mail: robert.vandenberghe@ugent.be

I. Nedkov · T. Merodiiska · L. Slavov
Institute of Electronics, BAS, 72 Tzarigradsko Chaussee, 1784 Sofia, Bulgaria

equilibrium and the symmetry in the distribution of the forces of attraction, which, in combination with the large surface forces, create conditions for fast oxidation. If the vacancies are assumed to be only located at octahedral sites in the spinel structure, the oxidized state should be described by the formula: $(Fe^{3+})_A[Fe^{2+}_{1-3x}Fe^{3+}_{1+2x}\square_x]_BO_4$ where □ represents the vacancies. The fast electron hopping, which is known to be a pair-localized phenomenon in magnetite, results in $Fe^{2.5+}$ to be present in an equal amount of octahedral Fe^{2+} and Fe^{3+}, yielding the following formula: $(Fe^{3+})_A\left[Fe^{2.5+}_{2(1-3x)}Fe^{3+}_{5x}\square_x\right]_BO_4$. The room temperature (RT) Mössbauer spectrum contains in this case three sextets: one for $Fe^{2.5+}$ and two for Fe^{3+}, the latter with slightly different hyperfine parameters. This results in practice in a similar spectrum as for stoichiometric magnetite, however, with a different area ratio for the subspectra [1]. Maghemite is the end-member of non-stoichiometric magnetite (x=0.33) and has a general formula: $(Fe^{3+})_A\left[Fe^{3+}_{5/3}\square_{1/3}\right]_BO_4$. For small particles down to 10 nm the RT Mössbauer spectra exhibit asymmetric lines complicating to some extent the analysis of the spectra. For very small particles of magnetite a superparamagnetic behavior at RT is observed. Mørup et al. [2] have investigated 6 nm particles of magnetite and showed for the first time the presence of a superparamagnetic singlet-like spectrum at 227 K by using an in situ technique. On the other hand, most of the superparamagnetic magnetite reported in literature [3] is represented by a doublet with $\Delta E_Q \approx 0.6$ mm/s, pointing to a more maghemite-like phase. At first sight, the nanofabrication techniques are bringing about considerable problems with the oxidation of the small particles.

The object of this investigation are single-domain spherical and quasi-spherical nanoparticles of magnetite (Fe_3O_4) which in the extreme case can be considered as a state of matter on the borderline between atomic clusters and the crystalline state. As X-ray diffraction (XRD) is far from conclusive, the aim of this study is to investigate by MöS the deviations in the spectra related with surface oxidation and particle size of powders on the nanometer scale.

2 Experimental

The here reported samples with average particle sizes 10±2 nm (Mag 1) and 3±2 nm (Mag 2) were prepared in laboratory conditions [4] by two different technological approaches. Sample Mag 1 was obtained by co-precipitation in NaOH (0.3 M) at pH=13 from the main solution $FeCl_2$ $4H_2O$ (0.03 M) and $NaNO_2$ (0.02 M) as oxidizing solution. Sample Mag 2 was prepared by co-precipitation at pH=13 from strongly diluted solutions of $FeCl_2$ $4H_2O$ (0.03 M) and $FeCl_3$ $6H_2O$ (0.03 M). Spinel arrangement occurs instantaneously in both approaches. Intermediate oxyhydroxides do not form at different main oxidizing solution ratios. The last sample (Mag 3) has also been prepared in inert atmosphere (N_2) in which the particles were further covered with β-cyclodextrin [5].

X-ray diffraction (XRD) analyses were carried out using Co$-K_\alpha$ radiation on a TUR-M62 diffractometer with Bragg–Brentano geometry and a computer controlled goniometer HZG-3. High-resolution transmission electron microscopy (HRTEM) investigations were performed with a JOEL JEM-4000 EX microscope. Infrared (IR) spectra were recorded on a Nicolet 320 FTIR spectrometer, using the KBr pellet technique. Mössbauer spectra were collected on a conventional time–mode spectrometer using a $^{57}Co/Rh$ source. Integrated low-energy Mössbauer spectroscopy (ILEEMS) was performed on a channeltron-based spectrometer, described elsewhere [6]. The spectra were analyzed taking hyperfine field distributions into account where necessary.

3 Results and discussion

The HRTEM images of the Mag 1 sample showed that the particle shape is spherical and the size distribution is very homogeneous, both depending on the co-precipitation regime. The particle size was calculated using the average particle size statistics from the TEM photographs and by applying Sherrer's formula from the line widths in the XRD diagrams. Both methods yield 10 ± 2 nm.

The analysis of the Mössbauer spectrum of the Mag 1 sample (Figure 1a) is not straightforward due to the presence of asymmetric line shapes evoked by small-particle effects. However, the presence of about 5% of a doublet at RT is clearly observed which can be attributed to the presence of superparamagnetic maghemite. The low temperature (80 K) spectrum (not shown) is comparable to the one of maghemite and does not allow distinguishing between oxidized Fe_3O_4 and γ-Fe_2O_3. Whether maghemite is present as separate particles or as an oxidation layer on the magnetite particles cannot be directly concluded from the transmission spectra. To elucidate this and to estimate the degree of oxidation at the surface, the sample has been measured by the ILEEMS technique. In contrast to the conventional MöS, this technique allows to study the surface of the nano-sized particles. The electron emission spectrum of sample Mag 1 is displayed in Figure 1b. One can clearly observe that the outer Fe^{3+} sextet lines are much broader and that the intensity of the $Fe^{2.5+}$ spectrum has fairly decreased in comparison with the transmission spectrum, pointing to the presence of a more maghemite-like phase. The relatively large Fe^{3+} hyperfine field implies that this phase follows the relaxation behavior of the whole particle, inferring to be a surface layer. The observation of an equal amount of superparamagnetic doublet in the ILEEMS spectrum can only be attributed to the presence of some smaller particles which are fully oxidized. As the depth sensitivity for ILEEMS is estimated to be up to about 3 nm one might suggest that the surface evolves from a quasi-maghemite phase to a magnetite phase within a few nanometers. From the ILEEMS spectrum the average formula for the iron oxide layer is estimated to be $\left(Fe^{3+}\right)_A\left[Fe^{2.5+}_{0.8}Fe^{3+}\square_{0.2}\right]_BO_4$.

For the Mag 2 sample the average size was estimated to be 3 ± 2 nm from the HRTEM photographs as well as from XRD line broadening. The Mössbauer spectrum (Figure 1c) exhibits mainly a superparamagnetic doublet with ΔE_Q=0.64 mm/s in addition to a relaxation spectrum being the onset to a magnetic sextet. The Mössbauer spectrum of the sample Mag 3, which has been prepared in N_2 and afterwards encapsulated in β-cyclodextrin shows a similar behavior, although the relaxation part in the spectrum is larger (Figure 1d), most probably pointing to a slightly larger particle size. The latter might be due to a very small difference in the preparation conditions, being very critical at this size range. The quadrupole splitting of the doublet in the Mag 3 sample amounts to 0.50 mm/s and seems to turn out somewhat smaller than in the case of the non-covered Mag 2 sample. On the other hand, encapsulated smaller particles prepared in a similar way as Mag 3 but in polyvinyl alcohol during the synthesis, appear as a pure doublet at RT with ΔE_Q=0.69 mm/s. At 80 K, both samples reveal maghemite-like spectra (not shown), similar to the one at 80 K of the Mag 1 sample, although the lines are somewhat more asymmetrically broadened.

Apparently, all 3 nm samples seem to produce maghemite-like spectra at RT. However, the spectral behavior can be explained as follows. The Mag 2 sample is expected to oxidize directly in a similar way as the Mag 1 sample. Because of the very small dimensions (3 nm) the whole particle will consist of the quasi-maghemite phase, which after some time will be completely oxidized to maghemite. This is also corroborated by the brownish color of the

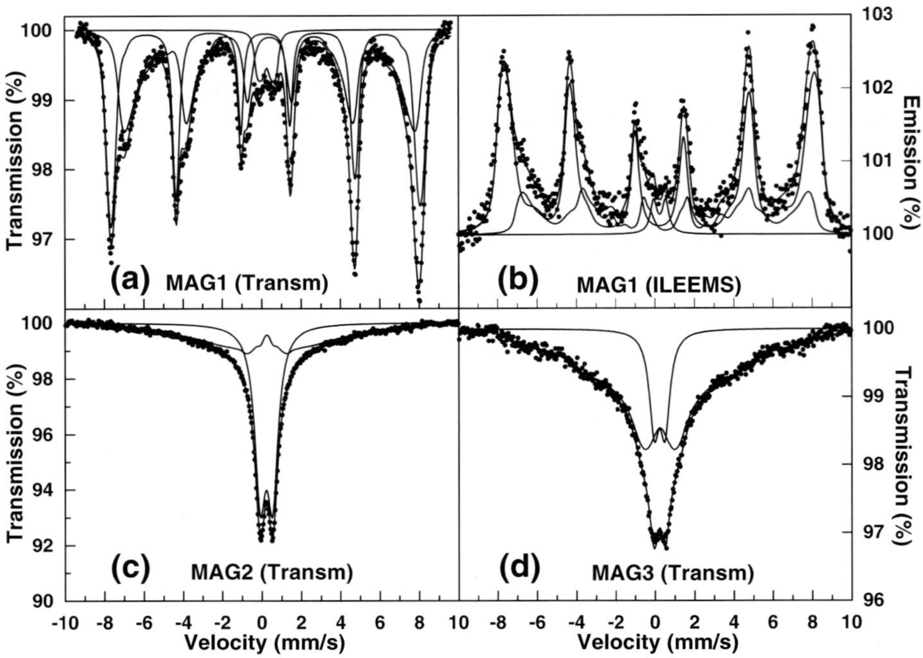

Figure 1 (a) Mössbauer spectrum at RT of sample Mag 1; (b) ILEEMS spectrum at RT of sample Mag 1 (c) Mössbauer spectrum at RT of sample Mag 2; (d) Mössbauer spectrum at RT of sample Mag 3.

powder. On the other hand the color of the organically covered Mag 3 sample was and remained black, showing a more stable magnetite-like phase. The phase characterization of the different samples has also been confirmed by IR measurements. The typical split band of maghemite (580 and 630 cm^{-1}) [7] is well observed in the IR spectrum of the Mag 2 sample, whereas the spectra of the samples Mag 1 and Mag 3 show predominantly the magnetite band at 580 cm^{-1}, with a small side band for the Mag 1 sample.

The reason why the superparamagnetic spectrum of very small-particle magnetite appears as a doublet and not as a singlet might be explained by the larger local distortions in the lattice. It can indeed be expected that the strong reduction of the lattice parameter in very small particles will enhance the distortion (oxygen parameter) in the spinel structure. Moreover, the energetically more favorable spherical shape in such small particles may also introduce strong deviations from the ideal lattice positions. All this will lead to a substantial contribution to the quadrupole splitting in comparison with bulk magnetite. This might equally explain the superparamagnetic doublet observed for starch-coated magnetite nanoparticles, reported by Kim et al. [3]. The singlet-like spectrum obtained by Mørup et al. [2] can then be explained by a larger effective particle size.

4 Conclusions

In conclusion, our data showed that transmission Mössbauer spectroscopy in combination with ILEEMS enables to describe qualitatively well the oxidation behavior of nanosized magnetite particles. However, for very small superparamagnetic particles for which the size commensurates with the depth of oxidation, the interpretation becomes more complicated. In spite of having a stoichiometry close to magnetite in the coated samples, as confirmed by the IR analyses, MöS indicates a quasi-maghemite structure. Although the color of the samples is already quite indicative, additional measurements such as IR spectroscopy are needed to confirm the results.

Acknowledgments The work was supported in part by Bulgarian National Council for Scientific Research under contract TH-1/01 and by collaboration between the Bulgarian Academy of Sciences and the FWO – Flanders, Belgium.

References

1. Vandenberghe, R.E., Barrero, C.A., da Costa, G.M., Van San, E., De Grave, E.: Hyperfine Interact. **126**, 247 (2000)
2. Mørup, S., Topsøe, H., Lipka, J.: J. de Physique, Colloque C6, Suppl. 12, C6–287 (1976)
3. Kim, D.K., Mikhaylova, M., Ahyang, Y., Muhammed, M.: Chem. Mater. **15**, 1617 (2003)
4. Patent BG No 107837/22.05.2003/
5. Bocanegra-Diaz, A., Mahallem, N.D.S., Sinistera, R.D.: J. Braz. Chem. Soc. **14**, 936–941 (2003)
6. De Grave, E., Vandenberghe, R.E., Dauwe, C.: Hyperfine Interact. **161**, 147 (2005)
7. Poling, G.W.: Infrared reflection studies of the oxidation of copper and iron. J. Electrochem. Soc. **116**, 958 (1969)

Hyperfine Interact (2005) 165: 273–278
DOI 10.1007/s10751-006-9277-6

Mössbauer study of nano-TiO_2 doped with Fe

Sanyuan Zhu · Wenhan Liu · Chongzheng Fan · Yuzhi Li

Published online: 2 November 2006

Abstract A Mössbauer study of nano-TiO_2 doped with Fe is presented. The samples are prepared by sol-gel method, doping Fe by 5, 10 and 15 wt.%, respectively, which are measured with XRD, TEM and Raman spectra. Especially, Mössbauer spectra are emphasized in this study. The anatase phase is major in both doped and no-doped sample. The α-Fe_2O_3 phase is also in the doped samples. The grain size of doped sample is in 5–20 nm range, the major grains are about 13 nm. And the grain size of no-doped sample is about 8 nm. Studying Mössbauer spectra and Raman spectra, we concluded that in the doping process the Fe^{3+} ions entered anatase lattice and substituted Ti^{4+} ions. However, the amount of Fe ions in the site is limited to about 1.5 wt.%. It does not increase as the doping Fe increase. The more Fe doped, the more α-Fe_2O_3 formed. For comparing conveniently, it also can be described as $(Ti_{0.98}Fe_{0.02})$ O_2 by atomic percent.

Key words Mossbauer spectroscopy · sol-gel materials

1 Introduction

There are many studies which focus on the nano-TiO_2 because of its value in industrial application and scientific interesting [1].

Karvinen [2] reported that doping Fe by mM% level accelerates anatase-to-rutile (A–R) phase transformation. Zhang et al. [3] reported that the TiO_2 film by pulse dc reactive magnetron sputtering with low concentration 0.58 at.% of Fe performs better photocatalytic activity.

There are several authors study the structure of Fe-doped TiO_2 and report different results.

S. Zhu · Y. Li (✉)
Structure Research Laboratory, University of Science and Technology of China, Hefei 230026, China
e-mail: liyz@ustc.edu.cn

S. Zhu · W. Liu
Department of Physics, University of Science and Technology of China, Hefei 230026, China

C. Fan
Department of Chemical Physics, University of Science and Technology of China, Hefei 230026, China

D. Cordishi et al. [4] reported the poly-crystalline TiO_2–Fe_2O_3 prepared by co-precipitation and by impregnation is always in the rutile phase, in which the Fe^{3+} has a rather low solubility (~0.1 at.%) as the samples are treated at 800 and 1000°C. However, it is in anatase phase by co-precipitation at 500°C, in which the Fe^{3+} has a higher solubility (≥1%). J. A. Wang et al. [5] studied the samples prepared by sol-gel method, there are majority of anatase phase along with some brookite when sample is calcined at below 400°C. The authors assume that Fe ions are inserted into the lattice of titanium, located at interstice or occupied some of the titanium lattice site, formed an iron–titania solid solution. When temperature reaches 800°C, the Fe_2TiO_5 is produced in 5 to 10 wt.% Fe samples. Zhang et al. [6], who prepared their sample by sol-gel, found that the temperature of A–R phase transformation remained approximately 600–650°C for iron concentration above 0.2%. They suggested that the iron atoms are incorporated into the lattice of the anatase structure for concentration up to 30%. In contrast, only up to 0.2% Fe can be incorporated into the rutile TiO_2.

Amolli et al. [7] reported the dopant ions (Fe^{3+}) migrated from the surface into bulk of the brookite by EPR of Fe Impregnated Brookite TiO_2 heated to 650°C.

Kim et al. [8] indicated that there is a pure rutile phase and no fine Fe particles in the sample ($(Ti_{0.99}{}^{57}Fe_{0.01})O_2$ with ^{57}Fe enriched and heated at 900°C.

The site and quantity of doping Fe are important to study structure and property of TiO_2 doped with Fe, which will be studied in this paper.

2 Preparation of samples

According to quantity needed for doping iron, suitable amount of $Ti(SO_4)_2$ and $Fe(NO_3)_3$ were dissolved in the water. Small amount of solution of low concentration HNO_3 was added to keep the pH value of 2 for this solution. The solution had been stirred for 2 h at room temperature, then Na_2CO_3 solution of 0.05 mol/l was added as hydrolysis catalyst; the solution was stirred again for two more hours until the value of pH of solution was 6.5. After that it was stirred for 4 h and heated at 80°C, then it was put into oven at 120°C for 10 h. Finally, it was calcined for 10 h at 650°C in a muffle furnace. Thus the samples were formed.

According to the quantity of doping iron, the samples were named as TF5, TF10 and TF15 for 5, 10 and 15 wt.% doping, respectively; the no-doped sample was referred to TF0.

3 Experiment

The samples are measured with several instruments. Rotating Anode X-Ray Diffractometer (Cu Kα; λ=0.154 nm, Rigaku Rotaflex, D/MAX-RA); Transmission Electron Microscopy (TEM) (HITACHI, H-800); and Confocal Laser MicroRaman Spectrometer (λ=514.5 nm excitation light of an Ar^+ laser, JY, LABRAM-HR). The Mossbauer spectra are recorded with Mossbauer Spectrometer, (MS-500, OXFORD), with a ^{57}Co/(Pa) (25 mCi) source, at room temperature. The experiment data are fitted by the Gauss-Newton method, the ISs are referred to α-Fe (25 μm film).

4 Results

The X-ray diffraction (XRD) patterns of samples (TF0, TF5, TF10 and TF15) are represented in Figure 1. All samples show diffraction peaks of anatase phase, especially those at 2θ=25.6, 38.3, 47.8 and 62.5° are obvious. The very weak peak, at 2θ=27.6°, is from rutile phase. For doped sample TF15, there are several peaks from α-Fe_2O_3. The

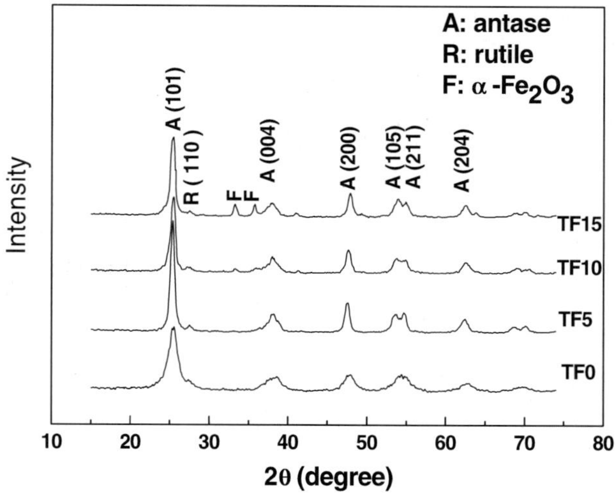

Figure 1 X-ray diffraction patterns of TF0, TF5, TF10 and TF15, respectively.

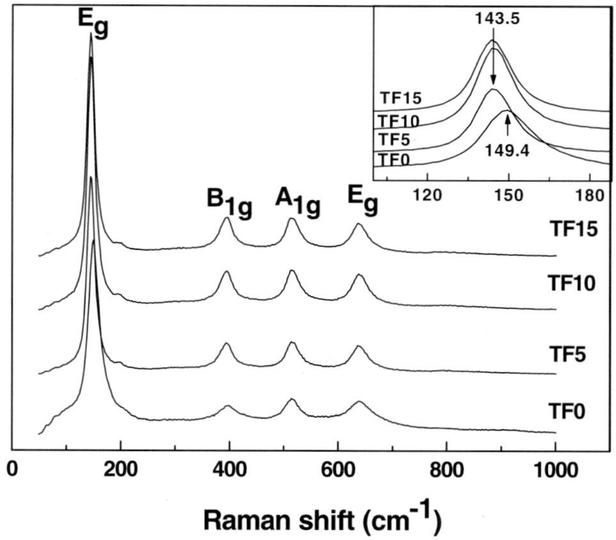

Figure 2 Raman spectra of TF0, TF5, TF10 and TF15, respectively.

peaks, at 2θ=33.5 and 36.1° are obvious. But these peaks are very weak and unconfirmed in TF10. As well as they are not found in TF5.

The size of grains is calculated from Figure 1 by Scherrer's formula: $D=0.89\lambda/(w\cos\theta)$. Where λ=0.154 nm is the wavelength of the X-ray, and w is the full width at half height of the XRD peak at the diffraction angle θ.

The grain size of anatase phase is about 8 nm in TF0, and 13 nm in doped samples. The grain size of α-Fe_2O_3 is also about 13 nm in TF15.

The TEM micrograph shows that most particles are in 5–20 nm range, this is in agreement with the results of XRD.

Figure 2 shows the Raman spectra of the samples. Four peaks, E_g(1st and 4th), B_{1g} and A_{1g}, are in all samples. The first E_g is at 143.5 cm^{-1} for all doped samples. However, it is at 149.4 cm^{-1} for no-doped sample. Obviously, a Raman shift of 5.9 cm^{-1} is occurred in doped samples.

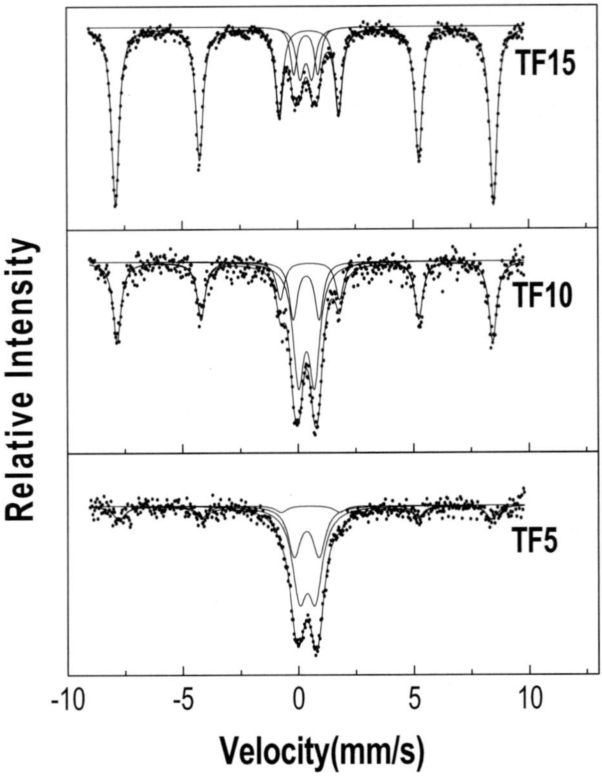

Figure 3 Mossbauer spectra of TF5, TF10 and TF15, respectively.

Figure 3 shows Mössbauer spectra at room temperature. Both doublet and sextet are present in the samples. Obviously, the area of sextet spectra is increased and the area of doublet is decreased as increasing iron content.

5 Analysis and discussion

The XRD patterns show that there are two phases of titania, anatase and rutile, in all samples. Since the content of rutile is small, less than 5%, it is negligible. Diffraction peaks of α-Fe_2O_3 are only present in TF15, but not in TF5. Comparing with TF15, the peaks of α-Fe_2O_3 are very weak in TF10 and seem unconfirmed. However, the Mössbauer sextet is obviously in all doped samples for Mössbauer effect is more sensitive than that of the XRD for fine particles and local structure.

Raman shift describes the variance of the lattice structure. Comparing with no-doped TiO_2, the shift of all three doped sample is 5.9 cm^{-1}, it is small. That indicates the structural distortion of doped TiO_2 samples is minor and implies a small amount of Fe substitute Ti in TiO_2 lattice.

It is not possible for Fe atom entering the lattice of TiO_2 to be interstitial as their big size. Usually the interstitial atoms are the smaller ones, for example, carbon atoms [9]. If the interstitial atom is large, it will destroy the equilibrium, result in much more distortion of crystal lattice, result in a larger shift of Raman spectrum.

Table I Mössbauer parameters of samples of doping iron

Sample	Site (state)	IS (mm/s)	QS (mm/s)	B (T)	Area (%)		HWH (mm/s)		Fe (wt.%) in each state
TF5	A	0.36	1.08		28.4		0.30		1.4
	B	0.39	0.69		53.9		0.37		2.7
	C	0.40	−0.22	50.2	17.7	0.32	0.27	0.31	0.9
TF10	A	0.38	1.11		16.6		0.22		1.7
	B	0.38	0.68		38.1		0.28		3.8
	C	0.41	−0.22	50.6	45.3	0.21	0.19	0.21	4.5
TF15	A	0.38	1.04		8.5		0.17		1.3
	B	0.39	0.51		10.2		0.20		1.5
	C	0.40	−0.21	51.0	81.3	0.19	0.18	0.21	12.1

(The isomer shifts are referred to α-Fe)

There are both sextet and doublet in the doped samples (Figure 3). Assuredly, the sextet is corresponding to α-Fe_2O_3. How to fit the doublet? There are two kinds of particles. One is the Fe particle entered lattice of TiO_2, which locates in asymmetric site, as the structure of anatase is distorted octahedral. Another one is the particle of α-Fe_2O_3, which is superparamagnetic as small size at room temperature, for example, 13 nm [10]. So the doublet is fitted by two sub-spectra. The fitting data are in Table I.

In Table I, the site C is corresponding to sextet obviously. The superfine magnetic fields are 50.2, 50.6 and 51.0 T for TF5, TF10 and TF15, respectively. The values are close to the 50.3 T, which is for α-Fe_2O_3 with grain size of 18 nm at room temperature (296 K) [10]. The ISs are 0.40, 0.41 and 0.40 mm/s; the QSs are −0.22, −0.22 and −0.21 mm/s, respectively. Both values of IS and QS are in agreement with ones of typical α-Fe_2O_3. The area is 17.7, 45.3 and 81.3%, respectively. It increases with quantity of doping Fe. For the site B, the values of QS are 0.69, 0.68 and 0.51 mm/s, respectively. Where, the values of 0.69 and 0.68 mm/s (for TF5 and TF10) are close to the value of 0.68 mm/s for 10 nm grains of α-Fe_2O_3 [10]. So this site is from superparamagnetic phase of α-Fe_2O_3.

Finally, for the site A, it should be from Fe particles that entered lattice of anatase. In Table I, the QSs are 1.08, 1.11 and 1.04 mm/s, respectively. As we know, the anatase has a structure of distorted octahedral. If iron ions substitute and locate the site of titanium ions, they will locate in asymmetric site. According to Mössbauer theory, these values of QS should be larger. Similarly, Cordishi et al. assigned that the value of QS is 1.24 mm/s for Fe^{3+} ions in the anatase phase [4].

On the analysis, the site A is corresponding to the iron ions which substituted titanium ions. The site B is corresponding to superparamagnetic grains of α-Fe_2O_3 (less critical size), the site C is corresponding to α-Fe_2O_3.

Also, all values of IS in each state (A, S and C) are corresponding to the trivalent of iron.

In Table I site A, the Fe wt.% is 1.4, 1.7 and 1.3 for TF5, TF10 and TF15, respectively, which does not increase with the quantity of doping Fe. It implies the amount of iron ion entering anatase lattice is limited or saturated. This result is in agreement with Raman spectra, which shows all doped samples have the same Raman peaks, in spite of different quantity of doping iron ion. The reason is that: When doping irons are more, the iron ions go together to form α-Fe_2O_3 easily, instead of entering TiO_2 lattice.

6 Conclusions

Studying Mössbauer spectra and Raman spectra, we concluded that in the doping process the Fe^{3+} ions entered anatase lattice and substituted Ti^{4+} ions. However, the amount of Fe ions in the site is limited to about 1.5 wt.%, which does not increase as increasing Fe-dopant. The more Fe doped, the more α-Fe_2O_3 formed.

For comparing conveniently, it is also described as $(Ti_{0.98}Fe_{0.02})O_2$ by atomic percent.

Acknowledgements This work has been supported by National Natural Science Funds under projects No.10275061.

References

1. Linsewbigler, A.L., Lu, G., Yates, J.T. Jr: Chem. Rev. **95**, 735 (1995)
2. Karvinen, S.: Solid State Sci. **5**, 811–819 (2003)
3. Zhang, W.J., Li, Y., Zhu, S.L., Wang, W.H.: Chem. Phys. Lett. **373**, 333 (2003)
4. Cordishi, D., Burriesci, N. D'Alba, F., Petrera, M., Polizzotti, G., Schiavello, M.: J. Solid State Chem. **56**, 182 (1985)
5. Wang, J.A., Limax-Ballesteros, R.: J. Phys. Chem. B **105**, 9692 (2001)
6. Zhang, Y., Ebbinghaus, S.G., Weidenkaff, A., Kurz, T., von Nidda, H.-A.K., Klar, P.J., Güngerich, M., Reller, A.: Chem. Mater. **15**, 4028 (2003)
7. Amorelli, A., Evans, J.C., Rowlans, C.C.: J. Chem. Soc., Faraday Trans. I **85**, 4031 (1989)
8. Kim, E.C., Moon, S.H., Woo, S. II, Cho, J.H., Joh, Y.G., Kim, D.H.: Solid State Commun. **132**, 477 (2004)
9. Wert, C.A., Thomson, R.M.: Physics of solids, Chapter 3, Imperfection in crystals, p. 49. McGraw-Hill, New York (1964)
10. Kündig, W., Bommel, H., Constabaris, G., Lindquist, R.H.: Phys. Rev. **142**, 327 (1966)

Hyperfine Interact (2005) 165: 279–283
DOI 10.1007/s10751-006-9286-5

Quadrupole splitting temperature dependence of high and low affinity deoxyhemoglobin encapsulated in wet silica gel

S. Croci · O. J. Babalola · S. Bettati · C. Valenti · I. Ortalli · F. Parak

Published online: 16 November 2006

Abstract Hemoglobin (Hb) is still one of the most studied proteins due to its physiological function and because of its allosteric properties. It is actually well known that hemoglobin has at least two quaternary structures, one when the protein is fully oxygenated (*R* structure) and the other one when it is fully deoxygenated (*T* structure). In the last years, the encapsulation of hemoglobin in wet silica gel, preventing a change in the quaternary structure, gave the opportunity to show the existence of two *T* states: high affinity (HA) and low affinity (LA). Hb *T* state, in presence (LA) and in absence (HA) of allosteric effectors when it is encapsulated in wet silica gel, binds oxygen in non-cooperative way. Small differences between the quadrupole splitting values of HA and LA samples can be pointed out at low temperature regime. However, at present, these are too small differences to deduce that HA and LA Hbdeoxy states are different within the iron surrounding.

Key words deoxyhemoglobin · silica gel · quadrupole splitting

1 Introduction

Hemoglobin (Hb) is a tetrameric protein that has cooperative binding behaviours. This positive cooperative behaviour allows hemoglobin to efficiently bind or release $4O_2$ molecules in response to relatively small changes in their concentrations throughout our bodies. Moreover this sigmoidal binding curve can be changed by many factors, the so-called allosteric effectors. Particularly relevant in this respect is the action of pH and of 2,3-bisphosphoglycerate (2,3-BPG). The pH lowering, as it happens in the tissues, decreases

S. Croci (✉) · S. Bettati · C. Valenti · I. Ortalli
Public Health Department – Physics Division, University of Parma, Via Volturno 39, 43100 Parma, Italy
e-mail: simo@unipr.it

O. J. Babalola
Department of Chemistry, University of Ibadan, Ibadan, Nigeria

F. Parak
Physik-Department E17, TUM, 85747 Garching, Germany

the O_2 Hb affinity: this is called Bohr effect. The 2,3-bisphosphoglycerate (2,3-BPG), derived from the glycolytic intermediate 1,3-bisphosphoglycerate, stabilizes the tense (*T*) structure, thus an increased 2,3-BPG concentration lowers the O_2 affinity. In the *T* structure, the two β-subunits form a binding site for the 2,3-bisphosphoglycerate. In the *R* structure, the gap between the two β chains becomes too narrow to accommodate the BPG. However many models tried to explain the cooperative behaviour in particular the *T* to *R* transition in presence of allosteric effectors [1, 2]. It has been shown that Hb in *T* state encapsulated in wet silica gel binds oxygen in non-cooperative way both with allosteric effectors and in absence of allosteric effectors. This clearly shows that the cooperative behaviour is coming from the change of the quaternary structure. In addition, two distinct affinity conformations [3] in *T* state have been found: a low affinity (LA) and a high affinity (HA) [4]. Separation of tertiary from quaternary conformational changes seems the way to understand this allosteric behaviour. The encapsulation of hemoglobin in wet silica gels gave the opportunity to study the tertiary dynamics slowing down dramatically the quaternary one, which can be considered blocked. The encapsulation does not alter the Hb optical spectral and the functional properties, in fact no restriction was observed on tertiary conformational changes triggered by homeotropic and heterotropic effectors but the kinetics of ligand-linked quaternary conformational changes are significantly slowed down. Since the quaternary structure is practically blocked, there is the possibility to study if there are differences in the tertiary structure of these two *T* quaternary states. In particular using Mössbauer spectroscopy we want to have information about the iron surrounding, to see if there is a connection between the different *T* state affinity and the iron bindings. Therefore, we studied the quadrupole splitting temperature dependence of the high and low affinity Hbdeoxy encapsulated in silica gel. In fact, the temperature dependence of the quadrupole splitting of heme proteins, that was widely studied by Eicher et al. [5–7] gives information on the iron geometry and the low-lying electronic levels of Fe^{2+}. According to this theory, it was possible to establish a different binding strength between the iron and the proximal histidine in case of Mbdeoxy and Hbdeoxy. This is related to a difference of the distance between the iron cation and the heme plane of about 0.1 Å [5]. Therefore, using this theory, it is possible to look at the differences between the bindings that iron has with the proximal histidine and with the nitrogens of the pirrole rings on the two quaternary *T* states. We measured the quadrupole splitting temperature dependence of the ^{57}Fe enriched Hbdeoxy encapsulated in wet silica gel in high and low affinity *T* states.

2 Material and methods

Adult normal human hemoglobin was purified according to the method described by Rivetti et al. [8]. Human hemoglobin was preserved in the methemoglobin (Hbmet) in 4 M ammonium sulphate buffer. This was dialyzed with bi-distilled water before enrichment procedure to remove the buffer. The hemoglobin sample was enriched with iron-57 using the Acid–Butanone-Method [9]. The sample preparation of enriched Fe^{57} Hbmet was checked with optical measurements. The encapsulation of deoxyhemoglobin was carried out using a slightly different method already reported by Bettati and Mozzarelli [10], Bruno et al. [4], and Viappiani et al. [2] in presence and in absence of allosteric effectors. There are many kinds of allosteric effectors, non-physiological but much more efficient. One of the most studied is Inositol Hexa Phosphate (IHP) [11, 12]. BPG and IHP change the Hb's partial oxygen pressure when 50% of heme is bound, called p50, and what is more they

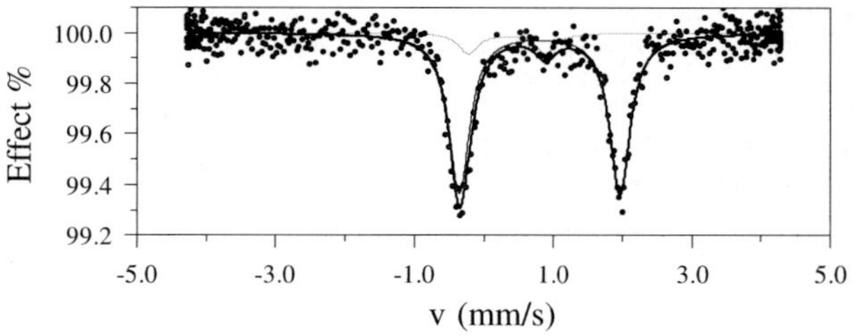

Figure 1 Hbdeoxy HA encapsulated in wet silica gel at 78 K: *solid line* – Hbdeoxy QS= 2.32 mm/s, IS=0.89 mm/s, Γ= 0.36 mm/s; *dotted line* – low spin complex coordinated with the N^{τ}-His E_7: QS=1.1 mm/s, IS=0.42 mm/s Γ=0.35 mm/s.

increase the dissociation constant for *T* state (K_T) and for *R* state (K_R). As reported by Marden [13], in case of BPG and IHP at 1 mM the p50 values are 3 kPa (15 mmHg) and 10.4 kPa (78 mmHg), respectively. Both BPG and IHP bind the same site of hemoglobin, that is in the central cavity of the protein and it is formed by residues of β chains. These positive-charged residues interact with the negative charges of the BPG and IHP phosphoric groups. Another allosteric effector is bezafibrate that was studied as antilipidaemic drugs that might prevent the aggregation in deoxy-hemoglobin *S* in patient with sickle anaemia. This compound and the analogous one stabilize the *T* structure, binding to symmetric sites of the central cavity formed mainly by residues of α chain. The effect of IHP and BPG on allosteric equilibrium is additive [14]. The low affinity Hbdeoxy was prepared with the following procedure. A solution containing 10 mM *N*-(2-hydroxyethyl)piperazine-*N*-(2-ethanesulfonic acid) (HEPES), 2 mM inositol hexaphosphate (IHP) (pretitrated to pH 6), 1 mM ethylenediamine tetraacetic acid (EDTA) at pH 6, was added to an equal volume of tetramethyl orthosilicate (TMOS). This sol was thoroughly mixed and vortexed for a few minutes. The resulting solution was deoxygenated by continuous mixing under nitrogen flow for 40 min at 4°C. This process removes the fear of oxidation and denaturation of ferrous hemoglobin because any methanol produced is eliminated from the solution. Finally, a deoxygenated solution containing 2.7% w/v deoxyhemoglobin, 50 mM HEPES, 30 mM dithionite, 1 mM EDTA, 2 mM IHP was added to the sol anaerobically in the ratio 1.5 to 1. A homogeneous mixture of this solution was distributed on a glass (solid) support vial. Gelation occurred within a few minutes and a solution of 100 mM HEPES, 30 mM dithionite, 1 mM EDTA, 2 mM IHP, 2 mM benzafibrate and 2 mM chloride ion at pH 7 was poured on the gel and was stored anaerobically at 4°C. The encapsulation of the high affinity *T* state hemoglobin is similar to the protocol used for the low affinity *T* state hemoglobin except that the negative heterotropic allosteric effectors were not used. The samples after gelification were immediately put in Mössbauer sample holder and frozen in liquid nitrogen. Measurements of HA and LA *T* Hbdeoxy samples were performed from 5.5 to 200 K, with a conventional Mössbauer spectrometer in transmission mode. Isomer shift is reported respect to α-iron.

3 Results and discussion

Hbdeoxy shows a quadrupole splitting temperature dependent behaviour due to the different electric field gradient (EFG) felt at the nuclear site produced by electrons in the perturbed $3d^6$-configuration and by the second order components of the crystalline electric field weighted by an appropriate Boltzmann factor. According to the theory developed by Eicher et al. [5–7] it is possible to calculate the eigenvalues and eigenstates of the

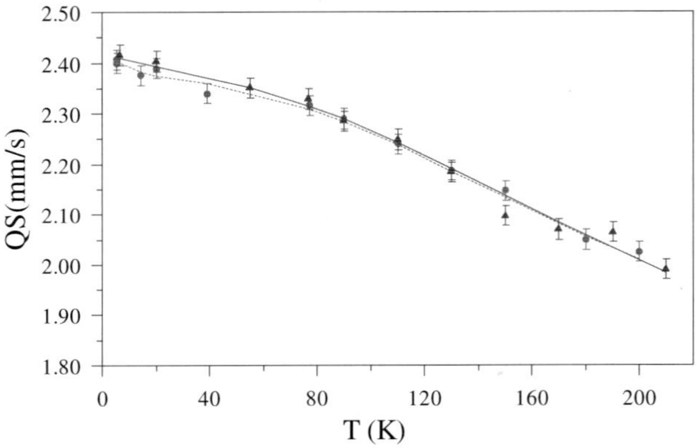

Figure 2 Quadrupole Splitting (*QS*) as function of temperature *T* (K) of Hbdeoxy samples encapsulated in silica gel. *Circle*: low affinity (*LA*) Hbdeoxy sample prepared with allosteric effectors. *Triangle*: high affinity (*HA*) Hbdeoxy sample prepared in absence of allosteric effectors. *Solid* and *dashed lines* are the least squares fitting curves according to Eicher's theory.

hamiltonian of the $3d^6$ electrons. The hamiltonian in [5] describes in first approximation a system with a nearly tetragonal point symmetry of the iron cation and the Coulomb repulsion of the 3d-electrons, afterwards the spin orbit coupling of the 3d electrons and the effect of a small rhombic perturbation of the C_{v4} point symmetry [6] where taken into account. As reported in [6] it was found that the many-electron levels $^1A_1$, ^3E, $^5B_2$, and ^5E, lay lowest in energy. The eigenvectors and eigenfuntions of these 22 states depend on five adjustable parameters: ε_1, ε_2, ε_3, λ, D. In particular λ is the spin orbit constant and D is proportional to the splitting amount caused by rhombic perturbation of the orbital doublets ^3E and ^5E. The ε_1, ε_2, ε_3 parameters determine the energy gaps of the antibonding single 3d electron orbitals, giving important information about the iron displacement respect to the heme plane and the proximal histidine. Therefore we measured the quadrupole splitting temperature dependence of Hbdeoxy HA and LA encapsulated in wet silica gel to get information about the iron position. A representative spectrum of Hbdeoxy HA at 78 K is reported in Figure 1.

The quadrupole splitting temperature dependence of Hbdeoxy High Affinity (HA) and Hbdeoxy Low Affinity (LA) samples has been fitted with the theory developed by Eicher et al. [5] using the Levenberg–Marquardt algorithm to minimize the sum of the squares of nonlinear function. We also used the same parameter notations. In Figure 2, the values of the quadrupule splinting as function of temperature are reported together with the least square fits in case of Hbdeoxy LA and Hbdeoxy LA.

The error associated with QS was estimated to be 0.02 mm/s for each measurement. In this way it is also possible to calculate the error associated with the fit parameters. This calculation was done using the covariance matrix. The data points of the two samples are very similar as one can see from Figure 2. In particular the slope of the curve is practically the same. In the Bade's thesis [15] the temperature dependence of the quadrupole splitting have been simulated varying the parameters ^3E, ^5E, and $^1A_1$, D. Their variation is connected with the shape of the curves used to fit the quadrupole splitting values. In particular the slope can be influenced by the variation of ^5E term. Differences between the data values of HA and LA Hbdeoxy samples are only visible at T lower then 50 K. The shape of the fit curve in this temperature range is connected to the variation of ^3E and also $^1A_1$. During the fit procedures, the spin-orbit coupling constant λ and the covalency factor α^2 are kept constant. In particular, for the Hbdeoxy sample the values reported in [5] are

Table I ε_1, ε_2, ε_3, values obtained by the fit procedure of high affinity and low affinity Hbdeoxy samples

Hbdeoxy	ε_1 (cm^{-1})	ε_2 (cm^{-1})	ε_3 (cm^{-1})
HA	496.74 (±0.04)	9,420.1 (±0.3)	16,746.1 (±0.2)
LA	619.3 (±0.2)	10,146.9 (±0.1)	16,332.89 (±0.02)

used: λ=69 cm^{-1} and α^2=0.88 – for both high and low affinity Hbdeoxy samples. In Table I are reported the fitting values.

From the fitting of QS data, it comes out firstly that the parameters are in agreement with those reported by Eicher et al. [5]. The energy ε_2 of the $a_{1g}\left(d_z^2\right)$ orbital depends on the binding strength with the imidazole nitrogen of the proximal histidine. Consequently, the binding seems to be stronger in the LA with respect to HA sample. Similarly, the ε_3 value is rising when the distance of iron with respect to the heme plane decreases. In this case ε_3 is higher in the HA sample than in the LA one. In this way one could deduce a correlation between the different affinities of the two *T* states, with the different distance of iron with respect to the heme plane and to the proximal histidine. Due to the small differences in the QS values and because of the quality of the data it is not possible to conclude that these two sets of measurements are different. A more accurate set of measurements especially in the low temperature regime is necessary. At present the HA and LA samples behave in the same way supporting the thesis that the change in the affinity is due to the ligand binding and not to a different tertiary structure already present when the hemoglobin has no ligands bound.

References

1. Eaton, W.A., Henry, E.R, Hofriechter, J., Mozzarelli, A.: Nat. Struct. Biol. **6**(4), 351 (1999)
2. Viappiani, C., Bettati, S., Bruno, S., Ronda, B., Abbruzetti, S., Mozzarelli, A., Eaton, W.A.: PNAS **101**, 14414 (2004)
3. Shibayama, N., Saigo, S.: FEBS Lett. **492**, 50 (2001)
4. Bruno, S., Bonaccio, M., Bettati, S., Rivetti, C., Viappiani, C., Abruzzetti, S., Mozzarelli, A.: Protein Sci. **10**, 2401 (2001)
5. Eicher, H., Bade, D., Parak, F.: J. Chem. Phys. **64**(4), 1446 (1976)
6. Eicher, H.,Trautwein, A.: J. Chem. Phys. **50**(6), 2540 (1969)
7. Eicher, H., Trautwein, A.: J. Chem. Phys. **52**(2), 932–934 (1970)
8. Rivetti, C., Mozzarelli, A., Rossi, G.L., Henry, E.R., Eaton, W.A.: Biochem. **32**, 2888 (1993)
9. Teale, F.W.J.: Biochim. Biophys. Acta **35**, 543 (1959)
10. Bettati, S., Mozzarelli, A.: J. Biol. Chem. **272**, 32050 (1997)
11. Imai, K.: Cambridge University Press (1982)
12. Arnone, A., Perutz, M.F.: Nature **249**, 36 (1974)
13. Marden, M.C., Bohn, B., Kister, J., Poyart, C.: Biophys. J. **57**(3), 397 (1990)
14. Perutz, M.F., Heidner, E.J., Ladner, J.E., Beetlestone, J.G., Ho, C., Slade, E.F.: J. Am. Chem. Soc. **108**, 1064 (1986)
15. D. Bade Diplomarbeit – Technischen Universität München (1975)

Hyperfine Interact (2005) 165: 285–288
DOI 10.1007/s10751-006-9287-4

Human nigral and liver iron – comparison by Mössbauer spectroscopy, electron microscopy and ELISA

J. Galazka-Friedman · E. R. Bauminger · A. Friedman · D. Koziorowski · K. Szlachta

Published online: 26 October 2006

Abstract The properties of iron present in human liver and human substantia nigra (SN) were compared. Mössbauer measurements have shown that iron is present in similar concentrations in SN and in the liver, mainly as ferritin-like iron. The size of the iron cores of ferritin, as obtained from electron microscopy, is much smaller in SN (3.6±0.4 nm) than in liver (5.7±0.5 nm). The small size in SN is in agreement with the low blocking temperature (about 10 K), determined by temperature dependent Mössbauer studies on whole tissues. ELISA studies have shown differences between SN and liver in the structures of the protein shells of ferritin. The H/L ratio (concentration of heavy to light chains) in liver is 0.40±0.02, while in SN it is 4.3±0.3. Another possible iron binding compound in SN is neuromelanin (NM). Mössbauer studies of neuromelanin, isolated from 22 SNs, demonstrated that this iron is superparamagnetic. The blocking temperature found for this neuromelanin iron is however much higher than that for ferritin.

Key words iron · ferritin · human liver · substantia nigra · Mössbauer spectroscopy · electron microscopy · ELISA

1 Introduction

Iron in the liver is homogenously distributed and its role is well known. This is not the case in the human brain, where high concentrations of iron are present in specific areas of various physiological functions [1]. We compared the properties of iron present in human liver and human substantia nigra (SN) – a small structure in the brain stem, whose nervous

J. Galazka-Friedman (✉) · K. Szlachta
Faculty of Physics, Warsaw University of Technology, Warsaw, Poland
e-mail: friedman@amwaw.edu.pl

E. R. Bauminger
Racah Institute of Physics, The Hebrew University, Jerusalem, Israel

A. Friedman · D. Koziorowski
Department of Neurology, Medical University in Warsaw, Warsaw, Poland

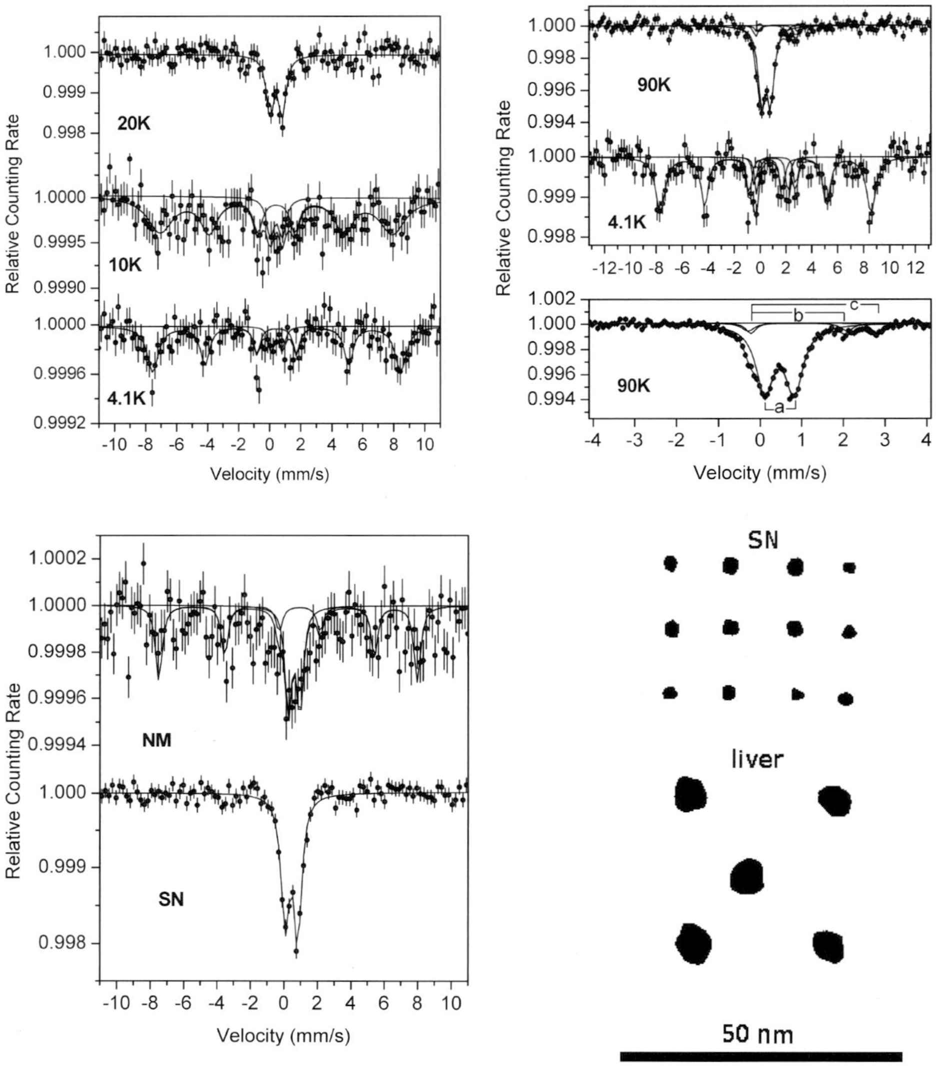

Figure 1 *Top left*: Mössbauer spectra obtained at 20, 10 and 4.1 K from fresh substantia nigra tissue; *Top right*: Mössbauer spectra obtained at 90, 4.1, 90 K narrow scale from human liver tissue; *Bottom left*: Mössbauer spectra obtained at 90 K from neuromelanin isolated from 22 SNs and from fresh substantia nigra tissue; *Bottom right*: Electron microscopy of iron-cores of ferritin isolated from human SN and liver.

cells produce dopamine, a neurotransmitter essential for controlling voluntary movements. A destruction of SN, in which iron may play a crucial role, is the cause of Parkinson's disease (PD). Our previous studies have shown some differences in the Mössbauer spectra of SN between PD and control, suggesting more non-ferritin iron in PD [2]. Iron in the liver is present mainly in ferritin and hemosiderin. Iron in SN is stored in ferritin, hemosiderin and, as suggested by several authors, also in neuromelanin [3]. The comparison of the properties of iron and ferritin in human SN and liver may allow a better understanding of the function of iron in SN.

Table I Mössbauer parameters of SN, liver and NM obtained at 90 and 4.1 K

Sample	90 K			4.1 K		
	IS (mm/s)	QS (mm/s)	H (T)	IS (mm/s)	QS (mm/s)	H (T)
SN	0.47 (1)	0.67 (4)	–	0.48 (4)	–	49.2 (5)
Liver – a	0.46 (2)	0.71 (2)	–	0.46 (2)	–	50.3 (5)
Liver – b	0.90 (4)	2.37 (5)	–	0.97 (4)	2.41 (6)	–
Liver – c	1.28 (5)	3.00 (1)	–	1.18 (5)	3.20 (1)	–
NM	0.60 (4)	0.68 (7)	–			
	0.58 (4)	−0.69 (8)	47.9 (3)			

Table II Concentrations of H and L ferritins, and H/L ratio obtained by ELISA study in human SN and liver

	H ferritin ng/mg wet tissue	H ferritin ng/μg protein	L ferritin ng/mg wet tissue	L ferritin ng/μg protein	H/L ratio
SN	375±38	19.3±1.5	98±12	5.0±0.5	4.3±0.3
Liver	179±25	2.8±0.5	550±100	8.5±1.7	0.40±0.02

2 Material and methods

The samples for measurements were separated from human liver and substantia nigra at autopsies, using plastic blades. Autopsies were performed within 48 h after death. The samples were obtained from patients who died without any symptoms of neurological diseases. Samples were frozen immediately after autopsy in dry ice and were kept in liquid nitrogen until assayed. Neuromelanin was obtained from 22 SNs by an isolation procedure described by Double et al. [3]. Ferritin for electron microscopy studies was isolated according to Williams et al. [4]. Twenty samples of SN and 20 samples of liver were used for ELISA (enzyme-linked immunoabsorbent assay) studies. The experimental procedures used for Mossbauer and ELISA studies are described in reference [2]. Electron microscopy (EM) studies were performed using a JEM-1200 EX II microscope operating at a voltage of 80 kV. The sizes of the iron cores were assessed from EM pictures obtained at 100 K magnification by a computerized procedure.

3 Results

Mössbauer spectra obtained from fresh tissues of SN at different temperatures (20, 10, and 4.1 K) are shown in Figure 1 top left. At 20 K a pure doublet, with ferritin-like parameters, is observed. At 10 K a superposition of a doublet and a broadened sextet, with more than 50% of the spectral area in the sextet, is obtained. At 4.1 K the sextet shows much narrower lines and only a very small fraction of the spectral area is in a doublet. Mössbauer spectra obtained from fresh samples of human liver are shown in Figure 1 top right. At 90 K the spectrum appears as three doublets (a – ferritin, b – deoxy Hb, c – probably a Hb derivative). At 4.1 K the spectrum appears as one sextet and two doublets. In Figure 1 bottom left Mössbauer spectra obtained at 90 K from fresh tissues of SN and from neuromelanin isolated from 22 SNs, are shown. The Mössbauer spectrum of SN shows a pure ferritin-like doublet, while the Mössbauer spectrum of NM shows a superposition of a doublet and a sextet. The NM spectrum was measured with statistics of 65 million counts per channel. Due to the small signal to noise ratio in this spectrum, the line width in the analysis was kept constant at 0.4 mm/s. All other parameters were free. The Mössbauer parameters of SN, liver and NM obtained at 90 and 4.1 K are summarized in Table I.

Electron microscopy of iron-cores of ferritin isolated from human SN and liver are shown in Figure 1 bottom right. The average diameter of the iron cores of ferritin isolated from human SN was 3.6±0.4 nm and the average diameter of the iron cores of ferritin isolated from human liver tissue was 5.7±0.5 nm.

The results of ELISA studies performed on 20 samples of SN and 20 samples of human liver are presented in Table II.

4 Discussion

ELISA results demonstrated important differences in the structure of the protein shell of ferritin in human liver and SN. In the liver L ferritin is dominant, while in SN H ferritin is dominant. H chains contain ferroxidase centers and rapidly convert divalent to trivalent iron, which is then stored within the ferritin molecule. L chains lack the ferroxidase center and are slow in sequestering iron, yet promote the building of the iron core for long-term storage. This difference in H/L ratio may lead to the difference in the size of the iron cores of ferritin between SN and liver: the iron cores of liver ferritin are much larger than those of SN ferritin.

Temperature dependent measurements of fresh SN tissue show that the iron storage compounds in SN have a blocking temperature between 10 and 20 K, much lower than the blocking temperature of ferritin isolated from human spleen (40 K) [5] or rat liver (35 K) [4]. This finding fits well with the small size of the iron cores of ferritin found in SN by EM. The Mossbauer parameters obtained from SN and liver point to ferritin-like iron clusters.

Mössbauer spectra obtained from NM isolated from 22 SNs, show at 90 K a superposition of a doublet and a sextet, suggesting that the blocking temperature of superparamagnetic particles stored in NM is around 90 K, much higher than the blocking temperature of ferritin. This may be due to clusters different from ferrihydrite, the iron core of ferritin, with a different anisotropy constant, or due to larger clusters than the iron-cores of ferritin. The Mossbauer parameters obtained from neuromelanin are similar, yet not identical, to those of ferritin (e.g., QS=0 in the sextet of ferritin, but QS≠0 in the sextet of NM). No large iron clusters were observed by electron microscopy studies performed on whole SN tissues. It may be that the iron clusters found in isolated NM are formed by aggregation of iron during the isolation procedure of NM [6].

References

1. Hallgren, B., Sourander, P.: J. Neurochem. **3**, 41 (1958)
2. Galazka-Friedman, J., Bauminger, E.R., Koziorowski, D., Friedman, A.: Biochim. Biophys. Acta. **1688**, 130 (2004)
3. Double, K.L., Gerlach, M., Schuenemann, V., Trautwein, A.X., Zecca, L., Gallorini, M., Youdim, M.B.H., Riederer, P., Ben-Shachar, D.: Biochem. Pharmacol. **66**, 489 (2003)
4. Williams, J.M., Andrew, S.C., Treffry, A., Harrison, P.M.: Hyperfine Interact. **29**, 1447 (1986)
5. St. Pierre, T.G., Bell, S.H., Dickson, D.P.E., Mann, S., Webb, J., Moore G.R., Williams, R.J.P.: Biochim. Biophys. Acta. **870**, 127 (1986)
6. Zecca, L., Stroppolo, A., Gatti, A., Tampellini, D., Toscani, M., Gallorini, M., Giaveri, G., Arosio, P., Santambrogio, P., Fariello, R.G., Karatekin, E., Kleinman, M.H., Turro, N., Hornykiewicz, O., Zucca, F.A.: Proc. Natl. Acad. Sci. U.S.A. **101**, 9843 (2004)

Hyperfine Interact (2005) 165: 289–294
DOI 10.1007/s10751-006-9288-3

Mössbauer study of iron uptake in cucumber root

K. Kovács · E. Kuzmann · F. Fodor · A. Vértes · A. A. Kamnev

Published online: 14 November 2006

Abstract ^{57}Fe Mössbauer spectroscopy was used to study the uptake and distribution of iron in the root of cucumber plants grown in iron-deficient modified Hoagland nutrient solution and put into iron-containing solution with 10 μM Fe citrate enriched with ^{57}Fe (90%) only before harvesting. The Mössbauer spectra of the frozen roots exhibited two Fe^{3+} components with typical average Mössbauer parameters of δ=0.5 mm s^{-1}, Δ=0.46 mm s^{-1} and δ=0.5 mm s^{-1}, Δ=1.2 mm s^{-1} at 78 K and the presence of an Fe^{2+} doublet, assigned to the ferrous hexaaqua complex. This finding gives a direct evidence for the existence of Fe^{2+} ions produced via root-associated reduction according to the mechanism proposed for iron uptake for dicotyledonous plants. Monotonous changes in the relative content of the components were found with the time period of iron supply. The Mössbauer results are interpreted in terms of iron uptake and transport through the cell wall and membranes.

Key words iron uptake · iron(III) reduction · plant nutrition · cucumber (*Cucumis sativus* L.) root · Mössbauer spectroscopy

1 Introduction

Nowadays, the transport and storage of metal ions (including iron) in plants present a considerable scientific interest, since plants can accumulate trace elements and, in

K. Kovács · E. Kuzmann (✉) · A. Vértes
Research Group for Nuclear Methods in Structural Chemistry,
Hungarian Academy of Sciences, Department of Nuclear Chemistry,
Eötvös Loránd University,
Pázmány P. s. 1/a, Budapest 1117, Hungary
e-mail: kuzmann@para.chem.elte.hu

F. Fodor
Department of Plant Physiology and Molecular Plant Biology,
Eötvös Loránd University,
Pázmány P. s. 1/c, Budapest 1117, Hungary

A. A. Kamnev
Institute of Biochemistry and Physiology of Plants and Microorganisms,
Russian Academy of Sciences, Prosp. Entuziastov 13, Saratov 410049, Russia

Figure 1 Mössbauer spectra, recorded at 78 K, of roots of cucumbers grown in iron-deficient nutrient solution and which were supplied with iron only before harvesting for times indicated in the figure (*on the left*).

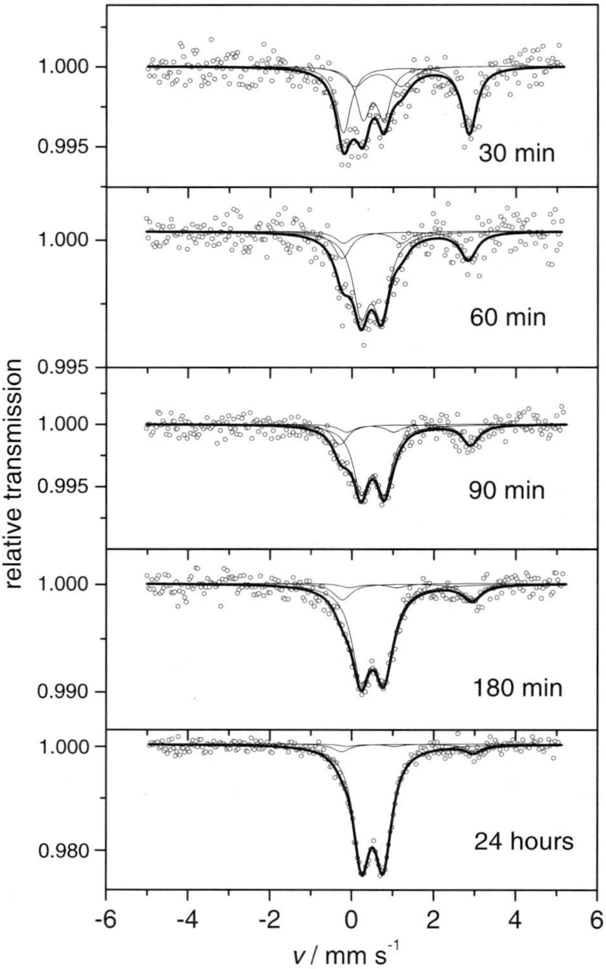

particular, heavy metals. Thus plants are a potential intermediate "reservoir" for trace elements originated from the lithosphere, hydrosphere and atmosphere [1–3]. This is of importance for plant nutrition studies and phytoremediation-based biotechnology.

Some recent studies [2, 3] of metal transport in cucumber indicated significant differences for different metals. Iron is considered as an important plant micronutrient; it is also a key metal in energy transformations needed for synthesis and other vital processes in cells.

Plants use two distinct strategies [4] to solubilize and absorb iron from soil. Plants either reduce Fe(III) complexes at the root surface and absorb the resulting Fe^{2+} ions produced via root-associated reduction (strategy group I), or excrete specific Fe(III)-binding, low-molecular-weight organic polydentate ligands known as phytosiderophores, which solubilize Fe^{3+} ions and make them available for absorption (strategy group II). These two strategies are used by two distinctly separate taxonomic groups of plants, with the phytosiderophore-based mechanism restricted to the grass family and the Fe(III) reduction mechanism restricted to the dicots and monocots not included in the grass family. The cucumber belongs to the latter group.

Table I Mössbauer parameters, recorded at around 78 K, of roots of cucumbers grown in iron-deficient nutrient solution which were supplied with iron only before harvesting for times indicated in the table

Sample characterisation		Fe(II) component D			Fe(III) A component DA			Fe(III) B component DB		
Time interval of iron supply	W mm s^{-1}	δ_1 mm s^{-1}	Δ_1 mm s^{-1}	A_1 %	δ_2 mm s^{-1}	Δ_2 mm s^{-1}	A_2 %	δ_3 mm s^{-1}	Δ_3 mm s^{-1}	A_3 %
30 min	0.42	1.35	3.19	48.0	0.49	0.46	32.0	0.52	1.11	20.0
60 min	0.46	1.31	3.05	22.9	0.46	0.44	60.0	0.45	1.15	17.1
90 min	0.47	1.30	3.19	22.6	0.50	0.50	68.7	0.44	1.14	8.7
180 min	0.49	1.35	3.17	15.6	0.50	0.48	81.7	0.50	1.14	3.9
24 h	0.48	1.37	3.23	6.4	0.51	0.52	91.0	0.47	1.12	2.5

^{57}Fe Mössbauer spectroscopy is a promising and useful tool to study the chemistry of Fe in plants. However, the iron content of plant tissues under normal nutrition conditions is insufficient for Mössbauer measurements. Previous Mössbauer investigations, performed with plants grown either at a high iron content of isotopically enriched nutrient solutions [5–7] or in an extremely acidic soil environment with a very high iron content [8], indicated only Fe(III) species in the roots of duckweed, stocks, pea [5, 6], and a perennial grass [8].

The aim of this work was to get information about the microenvironments and valence state of iron taken up by cucumber roots using Mössbauer spectroscopy.

2 Experimental

Cucumber (*Cucumis sativus* L.) seeds were germinated in complete darkness on wet filter paper at 26°C for 30 h. The seedlings were then kept on 0.5 mM $CaSO_4$ solution for 24 h before transferring to light and nutrient solution. The nutrient solution was a modified quarter strength Hoagland solution of the following composition: 1.25 mM KNO_3, 1.25 mM $Ca(NO_3)_2{\cdot}4H_2O$, 0.5 mM $MgSO_4{\cdot}7H_2O$, 0.25 mM KH_2PO_4, 11.56 μM H_3BO_3, 4.6 μM $MnCl_2{\cdot}4H_2O$, 0.19 μM $ZnSO_4{\cdot}7H_2O$, 0.12 μM $Na_2MoO_4{\cdot}H_2O$, 0.08 μM $CuSO_4{\cdot}5H_2O$. Iron was supplied as Fe(III) citrate at 10 μM concentration containing 90% enriched ^{57}Fe. Culture solutions were renewed three times a week and the pH was kept below 5.5. The plants were illuminated with 100 μmolm^{-2}s^{-1} ppfd for 12/12 h light/dark photoperiod. The temperature was 20/26°C night/day. Cucumbers were grown in 400-ml plastic pots. The plants were put into iron-containing solution only before harvesting for 0.5 to 24 h. After three weeks of growth the plants were harvested and the roots were immediately frozen to 78 K.

^{57}Fe Mössbauer spectra were recorded in the transmission geometry with a conventional Mössbauer spectrometer (Wissel). γ-rays were provided by a 3×10^9 Bq ^{57}Co/Rh source. The measurements were performed at between 74 and 250 K in a temperature-controlled cryostat (Leybold, Germany). Isomer shifts are given relative to α-iron. The Mössbauer spectra were analysed by the least-squares fitting using Lorentzian lines using the MOSSWINN code [9].

3 Results and discussion

Figure 1 shows the Mössbauer spectra, recorded at around the temperature of liquid nitrogen, of roots of cucumbers grown in iron-deficient nutrition solution and then kept in iron-containing solution for times between 30 min and 24 h immediately before harvesting.

Figure 2 Time dependence of the relative content of components, Fe(II) (a), $Fe(III)_A$ (b) and $Fe(III)_B$ (c) (*on the right*).

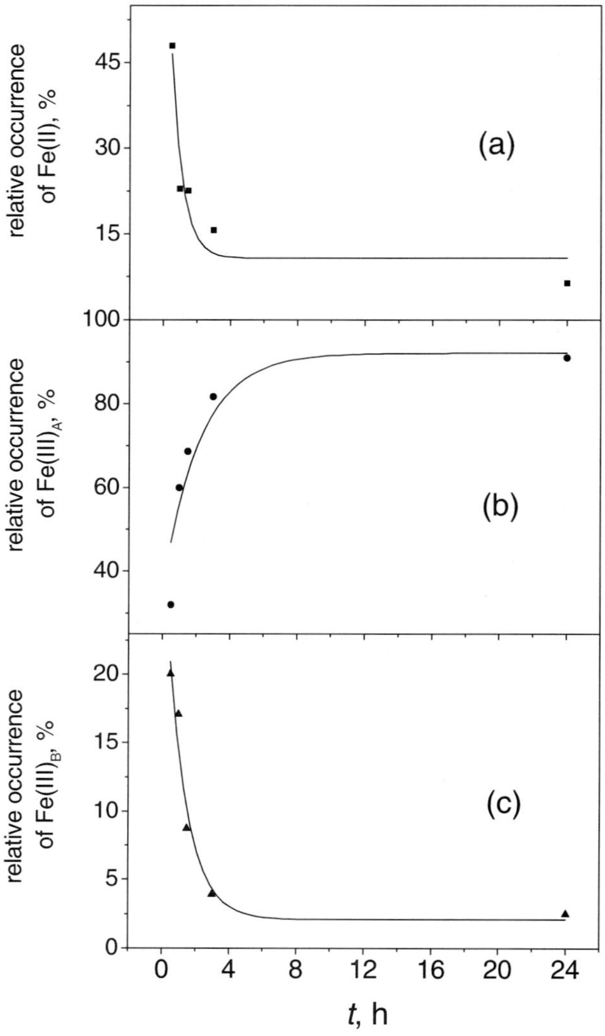

The spectra were decomposed into three symmetrical doublets, D, DA and DB. The Mössbauer parameters are presented in Table I. The line width for all components remained around or below 0.5 mm/s even in the case of simultaneous fitting of the spectra. The doublets can be associated with three different iron microenvironments in those cucumber roots. Doublet D represents an Fe(II) species while doublets DA and DB are attributed to two different Fe(III) microenvironments denoted as $Fe(III)_A$ and $Fe(III)_B$. We have found monotonous changes in the relative content of the components with the time period of iron supply, as can be seen in Figures 1 and 2, as well as in Table I. The Fe(II) species, which is the dominant component when cucumber was kept for 30 min in the iron-containing solution, decreases exponentially to a very minor component after 24 h (Figure 2a). At the same time, the $Fe(III)_A$ component increases exponentially and reaches 91% after 24 h (Figure 2b). On the other hand, an exponential decrease of the $Fe(III)_B$ component occurs from 20 to 2.5% within the investigated time interval (Figure 2c).

A further strong confirmation of the correctness of the above spectrum decomposition was given by evaluating the spectra of those cucumber roots which had been kept in iron-sufficient solution when components DA and DB were also present but no ferrous iron could be identified. A detailed description and discussion of these results will be published elsewhere because of the limited size of this paper.

The Mössbauer results presented here can be interpreted in terms of iron uptake and transport through the cell wall and membranes. The component D is attributed to ferrous hexaaqua complex based on its characteristic Mössbauer parameters [10]. This species is expected to be formed from Fe(III) compounds outside the membranes of roots growing in the nutrient solution (or soil), in the case of plants belonging to strategy group I, via the mechanisms of active electron and proton transport through the membrane. The transported protons can enhance the solubility of iron at the cell wall, while the transported electrons reduce the Fe(III) to Fe(II) which can already enter into the cytoplasm. The transport mechanisms are accelerated in iron-deficient plants.

We have succeeded in showing by Mössbauer spectroscopy the presence of divalent iron in the plant root when the nutrient solution contained only Fe(III). This gives a direct evidence for the existence of Fe^{2+} ions produced via root-associated reduction according to the mechanism proposed for iron uptake in plants belonging to strategy group I. The changes in the relative content of the components shown in Figure 2 can indicate the transformation of Fe(II) hexaaqua complex into the $Fe(III)_A$ species. This agrees with the accepted viewpoint that the translocation and storage of iron inside the root cells take place in the form of Fe(III) compounds. The reaction rate of this Fe(II)-to-Fe(III) transformation is much higher than that of the Fe(III) reduction outside the membrane, when sufficient amounts of iron are taken up by the root, possible due to significantly lower Fe(II) reduction rate in these plants. Consequently, the Fe(II) species cannot be detected on the background of the Fe(III) components using Mössbauer spectroscopy in the case of iron-sufficient roots. The $Fe(III)_A$ component, which is the dominant and characteristic iron species in all cucumber roots, can be associated with an Fe(III) compound being in the cytoplasm (symplast) inside the cell. A spectral component similar to DA has already been found but not assigned in the case of rice root [7], too. The $Fe(III)_B$ species can be attributed to jarosite, $KFe_3(OH)_6(SO_4)_2$, or its analogous compound, since the Mössbauer parameters of DB agree with those of jarosite [11], and the nutrient solution contained suitable amount of sulfates. Moreover, jarosite has already been shown to be a major component in the rhizome of a perennial grass [8]. Jarosite is assumed to be loosely connected with the cell wall, which is supported by our observation that this component disappeared when the root was washed with EDTA which is supposed to mobilize the cell-wall-bound iron [3].

Acknowledgements This study was supported by the Hungarian Science Foundation (OTKA T043687 and TS044742), NATO (Grants LST.EV.980141, CBP.NR.NREV.981748) and under the Agreements on Scientific Cooperation between the Russian and Hungarian Academies of Sciences for 2002–2004 and 2005–2007.

References

1. Yang, X., Feng, Y., He, Z., Stoffella, P.: J. Trace Elem. Med. Biol. **18**, 318–353 (2005)
2. Sárvári, E., Fodor, F., Cseh, E., Varga, A., Záray, G., Zolla, L.: Z. Naturforsch. **54c**, 746–753 (1999)
3. Varga, A., Záray, Gy., Fodor, F.: J. Inorg. Biochem. **89**, 149–154 (2002)
4. Römheld, V., Marschner, A.: In: Trinker, B., Lauchli, A. (eds.) Advances of Plant Nutrition, vol. 2, pp. 155–204. Preager Scientific, New York (1986)

5. Goodman, B.A., DeKock, P.C.: J. Plant Nutr. **5**, 345–353 (1982)
6. Goodman, B.A., DeKock, P.C., Rush, J.D.: J. Plant Nutr. **5**, 355–362 (1982)
7. Kilcoyne, S.H., Bentley, P.M., Thongbai, P., Gordonm D.C., Goodman, B.A.: Nucl. Instrum. Methods Phys. Res. B. Beam Interact. Mater. Atoms **160**, 157–166 (2000)
8. Rodríguez, N., Menéndez, N., Tornero, J., Amils, R., de la Fuente, V.: New Phytol. **165**, 781–789 (2005)
9. Klencsár, Z., Kuzmann, E., Vértes, A.: J. Radioanal. Nucl. Chem. **210**, 105–112 (1996)
10. Vértes, A., Korecz, L., Burger, K. (eds.): Mössbauer Spectroscopy. Elsevier, Amsterdam (1979)
11. Stevens, J.G. (ed.): Mössbauer Effect Reference and Data Index (MERDI). Interscience, New York (1958–2002)

Hyperfine Interact (2005) 165: 295–298
DOI 10.1007/s10751-006-9289-2

Nuclear resonance vibrational spectroscopy of iron–sulfur proteins

A. X. Trautwein · P. Wegner · H. Winkler · H. Paulsen · V. Schünemann · C. Schmidt · A. I. Chumakov · R. Rüffer

Published online: 26 October 2006

Abstract Nuclear inelastic scattering in conjunction with density functional theory (DFT) calculations has been applied for the identification of vibrational modes of the high-spin ferric and the high-spin ferrous iron–sulfur center of a rubredoxin-type protein from the thermophylic bacterium *Pyrococcus abysii.*

Key words nuclear inelastic scattering · DFT calculations · rubredoxin

Abbreviations
NIS nuclear inelastic scattering
DFT density functional theory

1 Introduction

Iron–sulfur proteins function as electron-transfer proteins in all living cells. They are involved in photosynthesis, cell respiration as well as in nitrogen fixation. Most iron–sulfur proteins have either single-, two-, three- or four-iron centers. Mössbauer spectroscopy has

A. X. Trautwein · P. Wegner · H. Winkler · H. Paulsen
Institut für Physik, Universität zu Lübeck, 23538 Lübeck, Germany

V. Schünemann (✉)
Fachbereich Physik, Technische Universität Kaiserslautern, 67663 Kaiserslautern, Germany
e-mail: schuene@physik.uni-kl.de

C. Schmidt
Institut für Biochemie, Universität zu Lübeck, 23538 Lübeck, Germany

A. I. Chumakov · R. Rüffer
ESRF, 38043 Grenoble Cedex 9, France

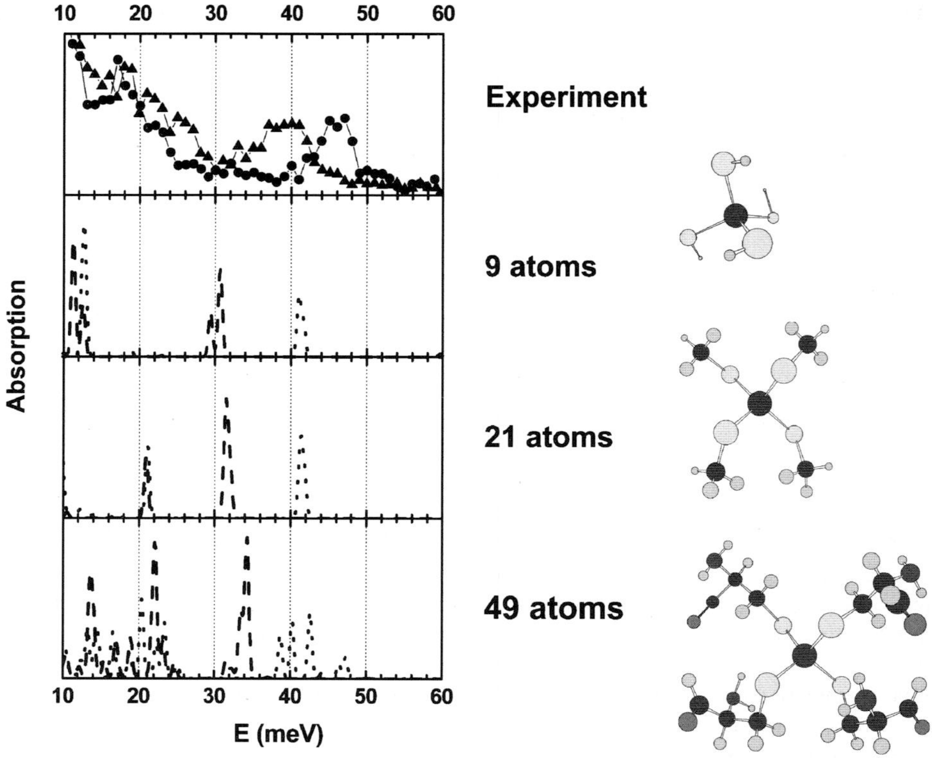

Figure 1 NIS-spectra of oxidized (*filled circle*) and reduced rubredoxin mutant Rm 2-4 (*filled triangle*) from *Pyrococcus abyssi* obtained at 25 K. Theoretically calculated NIS spectra based on DFT calculations (B3LYP/CEP-31G) of 9, 21 and 49 atoms are shown below. The *dashed lines* represent calculated NIS spectra for the oxidized Fe^{III}–S_4 center and the *dotted lines* for the reduced Fe^{II}–S_4 center.

contributed a lot to the understanding of the electronic ground states of these clusters [1]. The vibrational properties of the iron centers, however, which are related to their biological function, are much less studied. This is partly due to the fact that the vibrational states of the iron centers are masked by the vibrational states of the protein backbone and thus conventional spectroscopy like Resonance Raman or infrared spectroscopy does not provide a clear picture of the vibrational properties of the iron centers. Nuclear inelastic scattering of synchroton radiation (NIS) measurements allow to identify individual iron modes and therefore are sensitive to iron-ligand bonds which are in turn related to the formal oxidation state of the iron.

2 Materials and methods

NIS measurements have been performed on the rubredoxin type iron–sulfur protein mutant Rm 2-4 from *Pyrococcus abysii* [2], which has a single Fe–S_4 center, at the beamline ID 18 of ESRF in Grenoble, France at T=25 K. Protein samples have been prepared with ^{57}Fe concentrations of *up to* 10 mM. DFT calculations were performed for three different models

Table I Assignment of vibrational iron-ligand modes based on a DFT calculation on the model with 21 atoms (Figure 1)

Character of vibrational modes	Energy (meV)	
	Fe^{III}–S_4	Fe^{II}–S_4
4 torsional modes	1–6	2–5
2 S–Fe–S bending modes (NIS vis.)	8–10	6–8
7 torsional modes	10–14	9–13
4 Fe–S–C bending modes (NIS vis.)	17–21	16–21
4 Fe–S stretching modes (NIS vis.)	35–42	28–32

(Figure 1) in vacuo with the B3LYP hybrid functional [3, 4] together with the CEP-31G basis [5–8] as included in the Gaussian 03 program package [9].

3 Results and discussion

Figure 1 shows NIS measurements on the oxidized Fe^{III}–S_4 protein which reveal a broad symmetric band around 15–25 meV (121–202 cm^{-1}) and a broad asymmetric band around 42–48 meV (339–387 cm^{-1}) consistent with the results of Bergmann et al. on rubredoxin from *Pyrococcus furiosus* [10]. These results confirm Resonance Raman studies on oxidized rubredoxin from *Desulfovibrio gigas* which indicates that there are three bands (43.15, 45.01 and 46.62 meV; 348, 363 and 376 cm^{-1}) in the region where asymmetric Fe^{III}–S stretch modes are expected [11, 12]. Resonance Raman data of the reduced rubredoxin with its iron in the Fe^{II}–S_4 (S=2) state could not be obtained up to now [12]. Thus NIS is the method of choice to study the dynamical properties of Fe^{II}–S_4 centers. Upon reduction the asymmetric stretch modes observed by NIS shift to lower energies (36–42 meV; 291–339 cm^{-1}). This can be rationalized by the fact that upon reduction the Fe^{II}–S bond lengths increase, which in turn is accompanied by a decrease of Fe^{II}–S binding energy and therefore by a decrease of the force constant of the Fe^{II}–S bond. This correlation between iron-oxidation state and Fe–S stretch-mode position is in line with DFT calculations for three different models (Figure 1), which have been performed on the basis of the obtained crystal structure of the protein. The results for the model with 21 atoms are shown in Table I. NIS visible S–Fe–S bending modes are expected in the region from 6–10 meV, Fe–S–C bending modes from 16–21 meV and Fe–S streching modes around 28–42 meV.

However, although the shift of the asymmetric stretch modes is well reflected by the DFT calculation, the line positions are less well reproduced. In fact the line positions may depend on the choice of the B3LYP hybrid functional. Another reason could be that obviously more than 21 and even more than 49 atoms (see Figure 1) have to be regarded in the DFT calculations.

In conclusion this NIS study has provided information about the vibrational properties of a reduced iron–sulfur center in a protein. These studies demonstrate that it is profitable to apply nuclear inelastic scattering as a complementary method to IR and Raman spectroscopy for investigating the dynamical properties of iron sites in iron sulfur proteins.

Acknowledgement This work has been supported financially by ESRF (experiment SC-1217).

References

1. Schünemann, V., Winkler, H.: Rep. Prog. Phys. **63**, 263–353 (2000)
2. Wegner, P., Bever, M., Schünemann, V., Trautwein, A.X., Schmidt, C., Bönisch, H., Gnida, M., Meyer-Klaucke, W.: Hyperfine Interact. **156/157**, 293–298 (2004)
3. Lee, C., Yang, W., Parr, R.G.: Phys. Rev., B **37**, 785–789 (1988)
4. Miehlich, B., Savin, A., Stoll, H., Preuss, H.: Chem. Phys. Lett. **157**, 200–206 (1989)
5. Watchers, A.J.H.: J. Chem. Phys. **52**, 1033 (1970)
6. Hay, P.J.: J. Chem. Phys. **66**, 4377–4384 (1977)
7. Stevens, W.J., Krauss, M., Basch, H., Jasien, P.G.: Can. J. Chem. **70**, 612–630 (1992)
8. Cundari, T.R., Stevens, W.J.: J. Chem. Phys. **98**, 5555–5565 (1993)
9. Frisch, M.J., et al.: Gaussian 03, Revision C.02. Gaussian, Wallingford, CT (2004)
10. Bergmann, U., Stuhrhahn, W., Linn, D.E. Jr., Jenney, F.E. Jr., Adams, M.W.W., Rupnik, K., Hales, B.J., Alp, E.E., Mayse, A., Cramer, S.P.: J. Am. Chem. Soc. **125**, 4016 (2003)
11. Czernuszewicz, R.S., LeGall, J., Moura, I., Spiro, T.G.: Inorg. Chem. **25**, 696 (1986)
12. Spiro, T.G., Czernuszewicz, R.S., Han S.: In: Spiro, T.G. (ed.) Biological Applications of Raman Spectroscopy, p. 523.Wiley, New York (1988)

Hyperfine Interact (2005) 165: 299–302
DOI 10.1007/s10751-006-9278-5

Tin–DNA complexes investigated by nuclear inelastic scattering of synchrotron radiation

G. Barone · L. H. Böttger · J. A. Wolny · H. Paulsen · A. X. Trautwein · A. Silvestri · I. Sergueev · G. LaManna

Published online: 2 November 2006

Abstract Nuclear inelastic scattering (NIS) of synchrotron radiation has been used to investigate the dynamics of tin ions chelated by DNA. Theoretical NIS spectra have been simulated with the help of density functional theory (DFT) calculations using 12 models for different binding sites of the tin ion in $(CH_3)Sn(DNA_{Phosphate})_2$. The simulated spectra are compared with the measured spectrum of the tin–DNA complex.

Key words nuclear inelastic scattering · tin–DNA

Abbreviations

NIS	nuclear inelastic scattering
DFT	density functional theory
PVDOS	partial vibrational density of states

1 Introduction

Organotin(IV) compounds are generally toxic and due to chemical applications of their derivatives, widely spread in the environment. A possible explanation of their toxicity seems to be their ability to interact with DNA. The interactions between organotin cations and DNA have been studied by ^{119}Sn-Mössbauer spectroscopy and by electronic structure calculations [1]. The Mössbauer studies suggest a direct octahedral coordination of the

G. Barone · A. Silvestri
Dipartimento di Chimica Inorganica e Analitica, Universitá degli Studi di Palermo, 90128 Palermo, Italy

L. H. Böttger · J. A. Wolny · H. Paulsen · A. X. Trautwein
Institut für Physik, Universität zu Lübeck, 23538 Lübeck, Germany

I. Sergueev (✉)
ESRF, F-38043, Grenoble Cedex 9, France
e-mail: sergueev@esrf.fr

G. LaManna
Dipartimento di Chimica Fisica, Universitá degli Studi di Palermo, 90128 Palermo, Italy

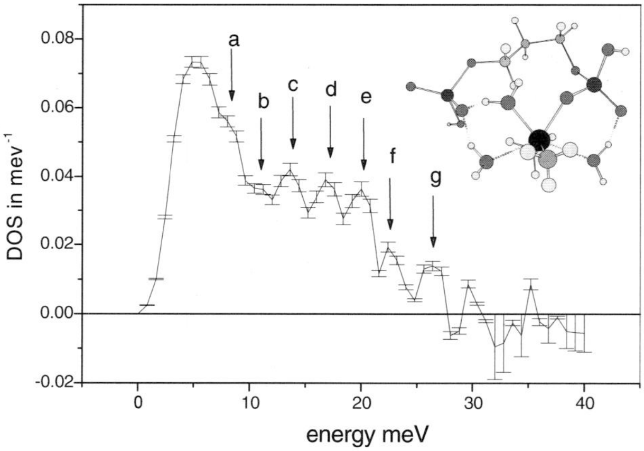

Figure 1 PVDOS extracted from the measured NIS spectrum. The inset shows the structure model used for calculation of model DD1.

organotin cation to the DNA, most probably to the negatively charged phosphate groups. Electronic structure calculations at the level of the semiempirical method PM3 [2, 3], have been performed for the determination of geometries. PM3 is especially suitable to predict intermolecular hydrogen bondings. The geometries obtained are generally in good agreement with ^{119}Sn Mössbauer and X-ray data. Calculations have also been performed for a model complex of dinucleoside triphosphate duplex (pApCp)–(pTpGp), which mimics DNA, and the cationic dimethyltin(IV) moieties in different binding positions. Among the possible DNA coordination sites for metal ions only the two most important are mentioned here i.e., the oxygen atoms of phosphate groups and the heteroatoms of nitrogen bases. Four structures have been obtained (DD1 to DD4) by calculations for the case of interaction with phosphate groups and for the case of interaction with nitrogen bases, respectively [4]. Electronic structure calculations also provide information about the dynamical behaviour of the ^{119}Sn nucleus. The dynamics are strongly connected with the strength of the bonds between tin and its organic ligands. Experimental access to the dynamics via conventional vibrational spectroscopy (IR, Raman) is hampered by the complexity of the spectra which are dominated by the vibrations of the large organic ligands. Therefore, our aim was to focus on those vibrational modes connected with a significant contribution of the tin nucleus. For this reason we have applied nuclear inelastic scattering of synchrotron radiation (NIS) to detect tin-centred molecular vibrations. The measured NIS spectra allow for the first time to determine the frequency of bond stretching vibrations of tin atoms bound to DNA fragments. The measured spectra are compared with simulated NIS spectra obtained from electronic structure calculations [5]. This comparison corroborates our hypothesis about the binding sites of tin [6]. To our knowledge, NIS measurements on biological samples (e.g., iron porphyrins or rubredoxines) have been performed up to now only in cases where the structure of the binding site is known by other methods like X-ray diffraction. This study thus demonstrates the usefulness of the NIS method in an extended field of biological samples.

2 Experimental

A detailed description of the sample preparation is given in ref. [6]. Measurements were performed at beamline ID22 [7, 8] ESRF Grenoble. The beamline setup was optimized for

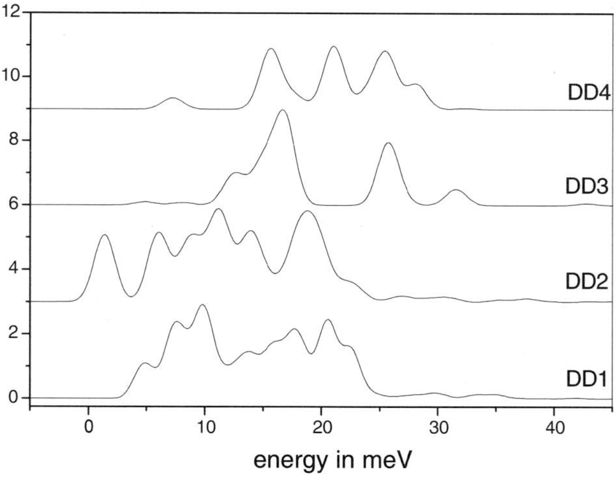

Figure 2 Calculated PVDOS based on DFT calculations for models DD1 to DD4.

measurements at the Sn-resonance energy of 23.875 keV. The sample was placed in a slit sample holder (10mm×3mm×2mm) and attached to a constant flow cryostat. The temperature was set to 50 K and had a temperature stability of about 1 K. The 10 mm slit of the sample holder was orientated parallel to the synchrotron beam and the detector for the delayed photons was placed directly beneath (~5 mm) the sample and perpendicular to the synchrotron beam, with the detector area covering the whole length of the slit. The energy resolution of the beamline setup was about 0.8 meV. The NIS spectra were collected in multiple scans, each having an energy step width of 0.2 meV.

3 Results and discussion

The NIS spectra were processed by the method of Kohn [9] to calculate the partial vibrational density of states (PVDOS) [10]. After the subtraction of the elastic scattering contribution [10] the NIS spectra show at least five shoulders on the positive energy side, which also appear in the phonon density of states as five clear peaks (c–g) in the energy range of 11 to 27 meV and as two clear shoulders (a–b) between eight and 11 meV (Figure 1). The spectral region above 27 meV is not discussed here because of a very low signal to noise ratio.

Barone et al. [4, 6] have considered 12 possible configurations of organotin–DNA complexes in their electronic structure calculation at the level of density functional theory and PM3, respectively. Two models (DD1 and DD2) describe the Sn–O–P interaction with two symmetrically placed water molecules around the tin atom and with two water molecules placed on the same side of the tin atom, respectively. Two other models (DD3 and DD4) describe the interaction of the tin atom via different nitrogen-molecules of the DNA. Detailed description of the models can be found in [4, 6].

Models DD3 and DD4 yields in our calculations 3 almost equally strong peaks in the range of 15 meV to 25 meV, but had different position and energy spacing compared to experiment. Model DD2 yields peaks, which are in between observed peak positions. The obtained peak positions for model DD1 resemble closed those of the observed spectra. This model exhibits an additional peak (resolved as shoulder) at about 16 meV in reasonable agreement with peak (d) of the experimental spectrum (Figure 1). The calculated normal modes that we assigned to the observed peaks are strongly mixed. The main contributions are in all cases Sn–OH_2 bond stretching and H_2O–Sn–OH_2 bending modes (Figure 2).

4 Conclusion

From comparing calculated PVDOS with PVDOS extracted from a measured NIS spectrum for different tin-ligand configurations we conclude that the tin atom is bound through an oxygen bridge towards the phosphor of the DNA fragment and has symmetrically arranged water molecules around the tin atom. Thus nuclear inelastic scattering together with electronic structure calculations is very suitable to study the metal-ligand coordination geometry and its vibrational behaviour of Mössbauer-active metal centres in even large biomolecules such as DNA.

Acknowledgements One of the authors (J.A.W.) gratefully acknowledges financial support by the Deutsche Forschungsgemeinschaft (DFG, Tr 97/32).

References

1. Barbieri, R., Ruisi, G., Silvestri, A., Giuliani, A. M., Barbieri, A., Spina, G., Pieralli, F., Del Giallo, F.: J. Chem. Soc., Dalton Trans.; Inorg. Chem. p. 467
2. Stewart, J.J.P.: J. Comput. Chem. **10**:209 (1989)
3. Stewart, J.J.P.: J. Comput. Chem. **12**:320 (1991)
4. Barone, G., Ramusino, M.C., Barbieri, R., La Manna, G.: J. Mol. Struct., Theochem **469**:143 (1999)
5. Paulsen, H., Benda, R., Herta, C., Schünemann, V., Chumakov, A.I., Duelund, L., Winkler, H., Toftlund, H., Trautwein, A.X.: Phys. Rev. Lett. **86**:1351 (2001)
6. Barone, G., Böttger, L.H., Wolny, J.A., Paulsen, H., Trautwein, A.X., Silvestri, A., Sergueev, I., Rüffer, R., LaManna, G.: Hyperfine Interact. (submitted)
7. Chumakov, A., Rüffer R.: Hyperfine Interact. **113**:59 (1998)
8. Rüffer, R., Chumakov, A.I.: Hyperfine Interact. **97–98**:589 (1996)
9. Kohn, V.G., Chumakov, A.I., Rüffer, R.: Phys. Rev., B **58**:8437 (1998)
10. Sturhahn, W., Toellner, T.S., Alp, E.E., Zhang, X., Ando, M., Yoda, Y., Kikuta, S., Seto, M., Kimball, C.W., Dabrowski, B.: Phys. Rev. Lett. **74**:3832 (1995)

Hyperfine Interact (2005) 165: 303–308
DOI 10.1007/s10751-006-9282-9

Application of ^{57}Co emission Mössbauer spectroscopy to studying biocomplexes in frozen solutions

A. A. Kamnev · L. A. Kulikov · Yu. D. Perfiliev ·
L. P. Antonyuk · E. Kuzmann · A. Vértes

Published online: 26 October 2006

Abstract Emission Mössbauer spectroscopy with the ^{57}Co isotope was used to study very dilute rapidly frozen aqueous solutions of cobalt(II) complexes with low-molecular-weight biomolecules (aromatic amino acids – anthranilic acid and L-tryptophan) and within a sophisticated biopolymer, bacterial glutamine synthetase, a key enzyme of nitrogen metabolism. The appearance of after-effects of the ^{57}Co→^{57}Fe nuclear transformation as well as the coordination properties of the cation and the ligands in the complexes are discussed on the basis of their Mössbauer parameters.

Key words cobalt(II) complexes · frozen aqueous solutions · anthranilic acid · L-tryptophan · metalloenzymes · after-effects · ^{57}Co emission Mössbauer spectroscopy

1 Introduction

In emission Mössbauer spectroscopy (EMS), ^{57}Co is the most widely used radionuclide. The ^{57}Co EMS technique is several orders of magnitude more sensitive than its ^{57}Fe absorption counterpart [1] that, nevertheless, has been widely applied in biology and biomedicine (for recent reports and reviews see, e.g., [2–6]). As for EMS, the obvious methodological difficulties, arising due to using the radionuclide in the samples under study, restrain EMS applications in bioscience [7]. Cobalt has a broad range of physiological and biochemical

A. A. Kamnev (✉) · L. P. Antonyuk
Laboratory of Biochemistry of Plant–Bacterial Symbioses, Institute of Biochemistry and Physiology of Plants and Microorganisms, Russian Academy of Sciences, Saratov, 410049, Russia
e-mail: aakamnev@ibppm.sgu.ru

L. A. Kulikov · Y. D. Perfiliev
Laboratory of Nuclear Chemistry Techniques, Department of Radiochemistry, Faculty of Chemistry, M.V. Lomonosov Moscow State University, Moscow, 119992, Russia

E. Kuzmann · A. Vértes
Research Group for Nuclear Techniques in Structural Chemistry, Hungarian Academy of Sciences, Department of Nuclear Chemistry, Eötvös Loránd University, Budapest 112, 1518, Hungary

functions; e.g., in many organisms it is involved in diverse enzymatic activities. Therefore its biochemical speciation is of importance but presents a challenge, as the participation of cobalt in physiological processes is limited by its trace concentrations.

In this report, some examples are discussed illustrating the applicability of EMS in studying various biocomplexes in rapidly frozen aqueous solutions, as well as the unique information which can thus be obtained non-destructively in situ. As the Mössbauer effect is observed in a solid matrix only, solutions or liquid samples are commonly studied rapidly frozen [1], so that the overall structure of the resulting frozen (glassy) matrix would adequately reflect that of the initial liquid. The appearance of after-effects of the $^{57}Co \rightarrow ^{57}Fe$ nuclear transformation, though complicating the EMS spectra, yet can provide additional information related to the electron acceptor properties of the ^{57}Co microenvironment [7, 8].

2 Materials and methods

Complexes of [^{57}Co]-cobalt(II) with anthranilic (*o*-aminobenzoic) acid and L-tryptophan ("Serva") were prepared using carrier-free $^{57}CoCl_2$ (i.e., without natural Co^{2+} salt). An aliquot of its aqueous solution (1 mCi) was put into a PTFE sample holder, completely dried in air at about 40°C, and 1 ml of 0.01 M aqueous solution of an amino acid was added to the resulting solid remainder (thus giving over three orders of magnitude in molar excess to $^{57}Co^{2+}$). The solution was stirred, closed to prevent evaporation, after 1 h rapidly frozen in liquid nitrogen and used for EMS measurements.

Preparation of Co^{2+}-containing samples of glutamine synthetase (GS), a ubiquitous key enzyme of nitrogen metabolism [9] isolated from the nitrogen-fixing plant-associated rhizobacterium *Azospirillum brasilense* (strain Sp245), was described in detail elsewhere [10, 11]. Treatment of GS with 5 mM EDTA for 30 min with further dialysis led to the removal of native cations from the enzyme (with subsequent loss of its activity in the absence of divalent cations in the medium). Further incubation of the resulting cation-free GS with Co^{2+} salt restored its activity, thus indicating that Co^{2+} cations enter the GS active centres (total 12 active centres per GS molecule, with *two* cation-binding sites each [9]), which is necessary for the activity of the enzyme to be expressed [9, 10]. Basing on this, the incubation of cation-free GS with a pre-calculated amount of $^{57}Co^{2+}$ salt gave the $^{57}Co^{II}$-doped biocomplex used for EMS measurements.

Emission Mössbauer spectra were collected using a conventional constant-acceleration spectrometer combined with a PC-operated multi-channel analyser and calibrated using α-Fe foil. Samples, which in EMS are sources of γ-radiation, were kept in a specially designed cryostat (with a window for γ-rays) filled with liquid nitrogen (at ca. 80 K). The absorber ($K_4[^{57}Fe(CN)_6]\ 3H_2O$) was attached to a vibrator in the γ-ray beam path (sample–absorber–detector). Other details of EMS measurements and statistical treatment of emission spectra were described elsewhere [7, 12]. All isomer shift values are presented relative to α-Fe at ambient temperature (converted into a form compatible with that of conventional absorption ^{57}Fe Mössbauer measurements, positive with regard to α-Fe).

3 Results and discussion

3.1 Amino acid complexes

Anthranilic (*o*-aminobenzoic) acid (AA) and L-tryptophan (Trp) are biologically important amino acids. They also occur in soil as a result of microbial or plant-root activity, or among

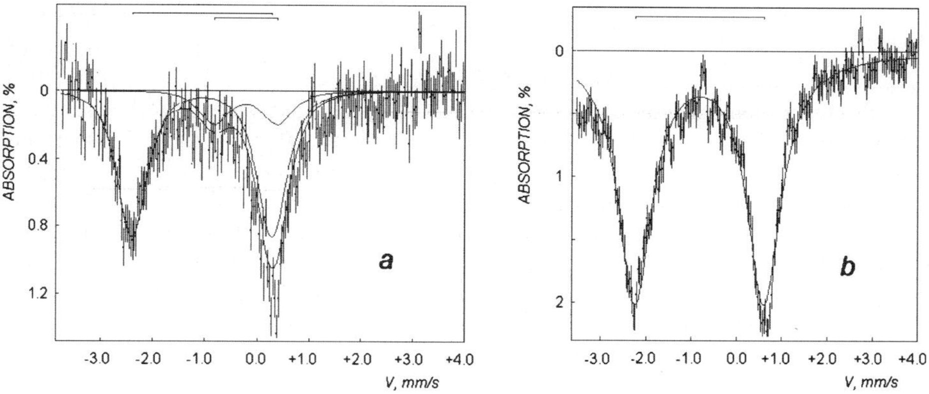

Figure 1 Emission Mössbauer spectra of [^{57}Co]-cobalt(II) complexes with anthranilic acid (a) and L-tryptophan (b) in frozen aqueous solutions (T=80 K).

Table I Mössbauer parameters of emission spectra of ^{57}Co biocomplexes in frozen aqueous solutions (T= 80 K)

Biomolecule	Oxidation state[a]	δ, mm/s	Δ, mm/s	Γ_{exp}, mm/s	S_r, %
AA	+2	1.10(2)	2.71(5)	0.62(4)	75(1)
	+3	0.32(14)	1.1(2)	0.8(2)	25(1)
Trp	+2	0.88(6)	2.84(13)	0.92(1)	100(1)
GS	+2	1.08(2)	3.08(8)	0.48(5)	18(1)
	+2	1.05(2)	2.39(6)	0.75(8)	60(1)
	+3	0.34(10)	1.12(20)	1.25(30)	22(1)

δ, isomer shift (vs. α-Fe at room temperature; converted to the absorption convention); Δ, quadrupole splitting; Γ_{exp}, experimentally observed line width at half maximum; S_r, the relative area of a subspectrum (representing the relative content of the related form, assuming a common recoilless fraction for all forms in a sample).

Errors (in the last digits) are indicated in parentheses.

AA, anthranilic acid; Trp, L-tryptophan; GS, glutamine synthetase (from *A. brasilense*).

[a] For the stabilised nucleogenic ^{57}Fe form.

decay products of proteins (Trp). Thus their interactions with metal ions, including complexation and possible redox transformations [13, 14], are of interest.

Emission Mössbauer spectra of AA and Trp complexes with [^{57}Co]-cobalt(II) (Figure 1a, b) show the presence (Figure 1a) and absence (Figure 1b) of a second doublet of a (+3)-form with a smaller isomer shift and quadrupole splitting (see also Table I). Note that under the conditions applied, no oxidation of Co^{2+} to the (+3) state can take place, so that the (+3) component in AA complex is ascribed to the aliovalent form resulting from after-effects.

The absence of any signs of after-effects, which for $^{57}Co^{2+}$-containing species usually show up as a (+3) component, in the $^{57}Co^{II}$–Trp system is relatively unusual. Considering the possibility for both AA and Trp to chelate Co^{2+} in a similar way (involving the carboxylate and the amino group), the difference in exhibiting after-effects (i.e., in electron-acceptor properties of the metal microenvironment) is probably related to the nature and/or degree of "isolation" of the aromatic moiety. In AA, the benzene ring (which could act as

Figure 2 Emission Mössbauer spectrum of $^{57}Co^{2+}$-doped glutamine synthetase from *A. brasilense* in frozen aqueous solution (T=80 K).

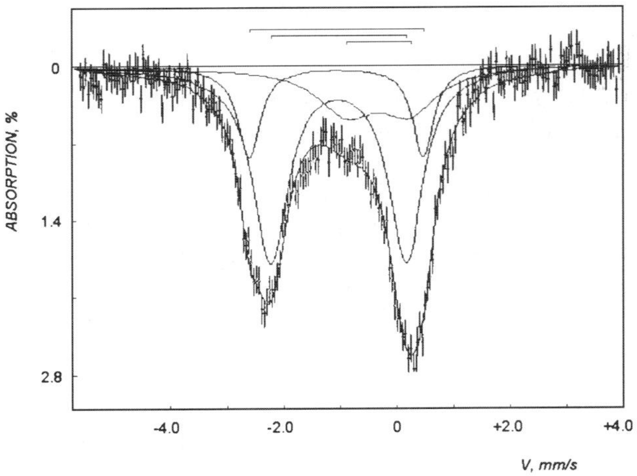

an electron acceptor for Auger electrons emitted in the Auger cascade [8]) is closer to the central ion, whereas in Trp it is isolated by an additional >CH–CH_2– moiety.

Furthermore, for the ^{57}Co(II) anthranilate complex, the δ value at T=80 K (δ=1.10 mm/s; see Table I) is noticeably lower than for solid Fe(II) anthranilate [$Fe(C_6H_4(NH_2)CO_2)_2$] (δ= 1.25 mm/s at T=80 K [14]), for which an octahedral coordination has been assumed based on FTIR and ^{57}Fe transmission Mössbauer data (with a bidentate coordination of each of the two COO^- groups, besides the two coordinated amino groups) [14]. Note that the $^{57}Co \rightarrow ^{57}Fe$ nuclear transformation, proceeding via electron capture by the ^{57}Co nucleus (which thereby turns into ^{57}Fe) with a subsequent Auger cascade developing within 10^{-14} to 10^{-15} s, is followed by emission of a 14.4 keV γ-quantum (ca. 10^{-7} s after the electron capture) [1, 7, 8]. Thus, the resulting substance under investigation using the EMS technique may be described as an ^{57}Fe complex substituted for the 'parent' ^{57}Co binding site (retaining its geometry). Therefore the above mentioned lower δ value found for the $^{57}Co^{II}$–AA complex may indicate that Co^{2+} is coordinated tetrahedrally in complex with AA, in contrast to Fe^{II}–AA complex. The same holds for $^{57}Co^{II}$ complex with Trp, where the δ value at T=80 K is even lower (δ=0.88 mm/s; see Table I). Note that tetrahedral coordination (T_d) is relatively common for cobalt(II) complexes, including biomolecules [15], along with the octahedral symmetry (O_h).

3.2 Metalloenzyme

The structural organisation of enzymic cation-binding sites (bearing Co^{2+} alone or together with other cations, as in cases of double-site active centres related to two-metal-ion catalysis [16]) can also be probed by EMS using $^{57}Co^{2+}$-doped enzyme samples. Thus the coordination of Co and its distribution between the sites can be analysed. This is illustrated by EMS data on $^{57}Co^{2+}$-doped glutamine synthetase isolated from *A. brasilense* [10, 11] (Figure 2; see also Table I).

Note that bacterial GS molecules are dodecamers formed from two face-to-face hexameric rings of subunits, with 12 active centres between monomers, each of the active centres having two divalent cation-binding sites separated by 6 Å, with different coordination and affinity for cations [9]. Earlier we have shown that, along with Mg^{2+}

and Mn^{2+}, cobalt(II) is one of the main activating cations for the *A. brasilense* GS, also playing a role in maintaining its secondary structure [10, 11].

According to Figure 2 and Table I, there are two forms of cobalt(II) in the sample represented two doublets corresponding to high-spin nucleogenic Fe^{2+} (the Fe^{+3} form results from after-effects). The two different Δ values reflect a higher symmetry (the lower Δ) and a lower symmetry (the higher Δ) at each of the two microenvironments of the cation. This is in good agreement with the existence of two separate cation-binding sites at each of the 12 active centres of the GS molecule, viz. a more symmetric site (with three Glu residues as ligands) and a less symmetric site with a significantly lower affinity to the cation (with one His and two Glu residues as ligands) [9–11]. In line with this conclusion, the former site (with a higher affinity for cations and $\Delta\approx2.4$ mm/s) may be expected to be primarily saturated with $^{57}Co^{2+}$ (featured by a higher S_r value), whereas the latter site (with $\Delta\approx3.1$ mm/s) binds the remaining $^{57}Co^{2+}$ characterised by a lower S_r value (see Table I).

As in the case with AA and Trp, the relatively low δ values for both the (+2) forms in GS (δ=1.05 and 1.08 mm/s at T=80 K) may indicate a tetrahedral symmetry of Co^{2+} (in this case, the coordination of all the Glu residues has to be monodentate, which is often observed for cation-binding sites in metalloproteins [15]; note that there is also at least one water molecule as a ligand, according to Eads et al. [17]). As mentioned above, tetrahedral coordination (T_d) of cobalt(II) is possible [15, 18, 19], though in many proteins cobalt was found to have preference for higher coordination numbers, i.e., 5 and 6 (see [19] and references cited therein).

4 Conclusions

Highly sensitive ^{57}Co emission Mössbauer spectroscopy is a valuable tool for structural investigations of biocomplexes with cobalt cations, including their aqueous solutions. Rapid freezing allows one to obtain emission spectra which represent the coordination state of the cation in solution at very low concentrations, as illustrated by the data on amino acid complexes and a metalloenzyme (bacterial glutamine synthetase) doped with $^{57}Co^{2+}$.

Acknowledgements Parts of this work were supported by INTAS (Grant 96-1015), NATO (Collaborative Linkage Grants LST.CLG.977664, LST.NR.CLG.981092; Expert Visit Grants LST.EV.980141, CBP.NR. NREV.981748) and under the Agreements on Scientific Cooperation between the Russian and Hungarian Academies of Sciences for 2002–2004 and 2005–2007.

References

1. Vértes, A., Nagy, D. (eds.): Mössbauer Spectroscopy of Frozen Solutions. Akadémiai Kiadó, Budapest (1990)
2. Matzanke, B.F., Anemüller, S., Schünemann, V., Trautwein, A.X., Hantke, K.: Biochemistry **43**, 1386–1392 (2004)
3. Oshtrakh, M.I.: Faraday Discuss. **126**, 119–140 (2004)
4. Oshtrakh, M.I.: Spectrochim. Acta **A60**, 217–234 (2004)
5. Paulsen, H., Schünemann, V., Trautwein, A.X., Winkler, H.: Coord. Chem. Rev. **249**, 255–272 (2005)
6. Jørgensen, A.M., Parak, F., Christensen, H.E.M.: Phys. Chem. Chem. Phys. **7**, 3472–3477 (2005)
7. Kamnev, A.A.: J. Mol. Struct. **744–747**, 161–167 (2005)
8. Perfiliev, Yu.D., Rusakov, V.S., Kulikov, L.A., Kamnev, A.A., Alkhatib, K.: J. Radioanal. Nucl. Chem. **266**, 557–560 (2005)

9. Eisenberg, D., Gill, H.S., Pfluegl, G.M.U., Rotstein, S.H.: Biochim. Biophys. Acta **1477**, 122–145 (2000)
10. Antonyuk, L.P., Smirnova, V.E., Kamnev, A.A., Serebrennikova, O.B., Vanoni, M.A., Zanetti, G., Kudelina, I.A., Sokolov, O.I., Ignatov, V.V.: BioMetals **14**, 13–22 (2001)
11. Kamnev, A.A., Antonyuk, L.P., Smirnova, V.E., Kulikov, L.A., Perfiliev, Yu.D., Kudelina, I.A., Kuzmann, E., Vértes, A.: Biopolymers **74**, 64–68 (2004)
12. Kamnev, A.A., Antonyuk, L.P., Kulikov, L.A., Perfiliev, Yu.D.: BioMetals **17**, 457–466 (2004)
13. Kamnev, A.A., Kuzmann, E.: In: Carmona, P., Navarro, R., Hernanz A. (eds.) Spectroscopy of Biological Molecules: Modern Trends. Annex, pp. 85–86. Annex, UNED, Madrid (1997)
14. Kamnev, A.A., Kuzmann, E., Perfiliev, Yu.D., Vankó, Gy., Vértes, A.: J. Mol. Struct. **482–483**, 703–711 (1999)
15. Holm, R.H., Kennepohl, P., Solomon, E.I.: Chem. Rev. **96**, 2239–2314 (1996)
16. Kamnev, A.A., Antonyuk, L.P., Smirnova, V.E., Kulikov, L.A., Perfiliev, Yu.D., Kuzmann, E., Vértes, A.: FEBS J. **272**(Suppl. 1), 10 (2005)
17. Eads, C.D., LoBrutto, R., Kumar, A., Villafranca, J.J.: Biochemistry **27**, 165–170 (1988)
18. Schneider, G., Eklund, H., Cedergren-Zeppezauer, E., Zeppezauer, M.: Proc. Natl. Acad. Sci. USA **80**, 5289–5293 (1983)
19. Innocenti, A., Zimmerman, S., Ferry, J.G., Scozzafava, A., Supuran, C.T.: Bioorg. Med. Chem. Lett. **14**, 3327–3331 (2004)

Hyperfine Interact (2005) 165: 309–312
DOI 10.1007/s10751-006-9283-8

Is the hematin derived from dodecaphenylporphyrin a μ-oxo dimer?

T. Ohya · J. Takeda · M. Sato

Published online: 26 October 2006

Abstract We have prepared DPP-hematin (DPP=2, 3, 5, 7, 8, 10, 12, 13, 15, 17, 18, 20-dodecaphenylporphyrin), and characterized it by Mössbauer and magnetic measurements. It is concluded that DPP-hematin is most likely a monomer, as is expected from the distorted nonplanar structure of the porphyrin, in contrast to usual hematins: hematins derived from usual porphyrins are μ-oxo dimers.

Key words hematin · dodecaphenyl porphyrin · iron porphyrin complexes · spin admixture · Mössbauer spectra

1 Introduction

Hematins are formed by treating hemins (five-coordinate ferric porphyrin complexes) with aqueous base, and usually are μ-oxo dimers. They contain two high-spin (S=5/2) ferric ions antiferromagnetically coupled, and have different features [1, 2] from those of mononuclear species in Mössbauer spectra and magnetic susceptibilities. 2, 3, 5, 7, 8, 10, 12, 13, 15, 17, 18, 20-Dodecaphenylporphyrin (H_2(DPP)) has 12 phenyl groups at the porphyrin periphery, and adopts a very distorted nonplanar structure [3, 4] in the crystalline state to relieve the steric crowding of the substituents. The hematin derived from dodecaphenyl porphyrin is expected to show steric restraint against μ-oxo dimer formation (Figure 1). We have prepared DPP-hematin, and characterized it by Mössbauer and magnetic measurements.

2 Materials and methods

H_2(DPP) and the iron bromide complex Fe(DPP)Br were prepared as previously reported [4]. The iron complex was treated with an aqueous solution of potassium hydroxide to yield

T. Ohya (✉) · J. Takeda · M. Sato
Laboratory of Biophysics, Faculty of Pharmaceutical Sciences, Teikyo University,
Sagamiko, Kanagawa 199-0195, Japan
e-mail: ohyat@pharm.teikyo-u.ac.jp

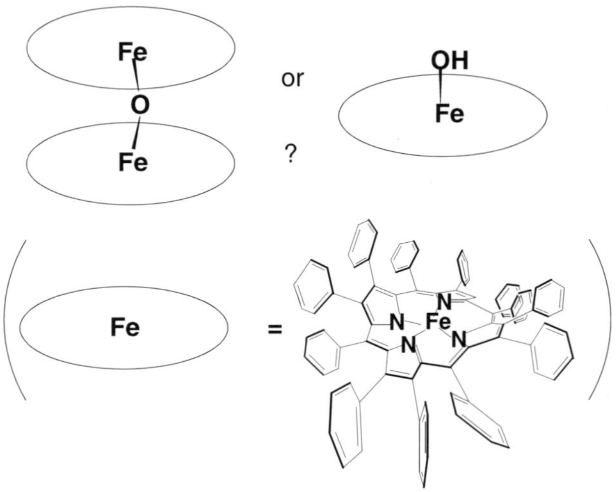

Figure 1 Two possible structures of DPP-hematin (DPP=2, 3, 5, 7, 8, 10, 12, 13, 15, 17, 18, 20-dodecaphenylporphyrin).

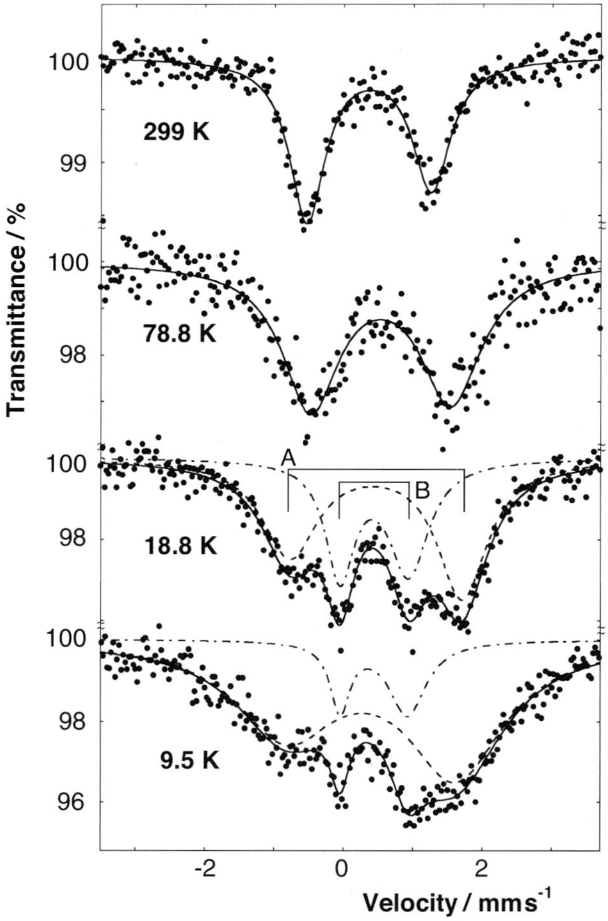

Figure 2 Mössbauer spectra of DPP-hematin. The component with larger quadrupole splitting is denoted as A, and the one with smaller value as B.

Table I Mössbauer data for DPP-hematin and Fe(DPP)X [4] (X = Cl, Br or I)

T [K]	Component	δ [mm s^{-1}]	ΔE_Q [mm s^{-1}]
DPP-hematin			
299		0.36(1)	1.79(1)
78.8		0.53(2)	2.02(3)
18.8	A	0.46(2)	2.48(5)
	B	0.44(2)	0.98(3)
9.5	A	0.43(5)	2.43(6)
	B	0.45(3)	0.97(5)
Fe(DPP)Cl			
300		0.35(1)	0.96(1)
7.6	A	0.37(1)	1.88(11)
	B	0.44(2)	0.85(3)
Fe(DPP)Br			
300	A	0.36(2)	2.18(4)
	B	0.38(2)	1.38(4)
6.9	A	0.38(3)	2.85(6)
	B	0.46(2)	0.96(2)
Fe(DPP)I			
300		0.35(1)	2.41(1)
5.3	A	0.40(2)	3.03(3)
	B	0.49(10)	1.23(14)

The isomer shifts δ are given relative to α-iron at room temperature.

DPP-hematin. A powder sample has been used for Mössbauer and magnetic measurements. ^{57}Fe Mössbauer spectra were measured with a Topologic Systems MA-100 Mössbauer spectrometer in transmission geometry using a ^{57}Co/Rh source. The spectra were fitted with appropriate combinations of Lorenzian lines. The isomer shift values are given relative to α-iron at room temperature. The molar magnetic susceptibilities were measured at room temperature with a modified Gouy balance (Sherwood MSB-MK I).

3 Results and discussion

Figure 2 shows the zero-field Mössbauer spectra of DPP-hematin at various temperatures. The spectra at 9.5 and 18.8 K have been fitted as two doublets A and B, the quadrupole splitting ΔE_Q of A being larger than that of B, whereas the spectra at 78.8 and 298 K have not satisfactorily been resolved into components (Table I). The general characteristics of the spectra are similar to those for Fe(DPP)X (X = Cl, Br or I) [4]. The average value of ΔE_Q for component A of DPP hematin lies between those of Fe(DPP)Cl and Fe(DPP)Br, and corresponds to admixed S=3/2, 5/2 ferric state. The second component B has ΔE_Q again between those of Fe(DPP)Cl and Fe(DPP)Br, and is assigned to a predominantly high-spin S=5/2 state with slight mixing in of the S=3/2 state.

The Mössbauer spectra of oxo-bridged iron(III) dimers are characterized by ΔE_Q values within a very narrow range (0.5 to 0.8 mm s^{-1}) [2]. The quadrupole doublets remain sharp and nearly symmetric even to room temperature because of the antiferromagnetic coupling between the two iron(III) ions. The view that DPP-hematin is a μ-oxo dimer or that the present sample contains both monomeric and dimeric species fails to explain the observed Mössbauer spectra: neither component A nor B has features compatible with a μ-oxo dimer, and it is impossible to give any proper reason why the two components merge into one at higher temperatures. The observed features for DPP-hematin strongly suggest the existence of two distinct mononuclear states with different extent of spin admixture. It is likely that the two components in Figure 2 result from slow relaxation between the two states. The relaxation probably becomes fast at higher temperatures making it impossible for the spectra to be resolved into components. Mergers of the two components of the Mössbauer spectra at higher temperatures have also been observed for Fe(DPP)Cl and Fe(DPP)I [4].

The effective magnetic moment has been calculated to be $\mu_{eff} = 4.64\ \mu_B$ per iron(III) ion from the magnetic susceptibility measured at 300 K after corrections of diamagnetic contributions. μ-Oxo dimers are diamagnetic at lower temperatures and μ_{eff} remains small ($<1.8\ \mu_B$) [5] even at room temperature. The observed μ_{eff} lies between spin-only values for $S=5/2$ ($5.92\ \mu_B$) and $S=3/2$ ($3.87\ \mu_B$) states, and strongly support the view described above.

4 Conclusion

The Mössbauer spectra of DPP-hematin contain two components A and B at lower temperatures, the former corresponding to strongly spin-admixed iron(III). The effective magnetic moment of the iron(III) ion in the complex is much higher than the value expected for a μ-oxo dimer, and lies between the spin-only values for $S=5/2$ and $S=3/2$ states. In conclusion, DPP-hematin is most likely a monomer, as is expected from the distorted nonplanar structure of the porphyrin.

References

1. Miller, J.R., Taies, J.A., Silver, J.: Inorg. Chim. Acta **138**, 205 (1987)
2. Sams, J.R., Tsin, T.B.: In: Dolphin, D.H. (ed.) The Porphyrins, vol. 4, p. 425. Academic, New York (1979)
3. Medforth, C.J., Senge, M.O., Smith, K.M., Sparks, L.P., Shelnutt, J.A.: J. Am. Chem. Soc. **114**, 9859 (1992)
4. Ohya, T., Takeda, J., Sato, M.: Hyp. Interact. **156/157**, 265 (2004)
5. Cohen, I.A.: J. Am. Chem. Soc. **91**, 1980 (1969)

Hyperfine Interact (2005) 165: 313–320
DOI 10.1007/s10751-006-9284-7

Mössbauer spectroscopy: Application in biomedical research

M. I. Oshtrakh

Published online: 2 November 2006

Abstract The results of the application of Mössbauer spectroscopy in biomedical research were briefly reviewed. The main directions of biomedical application and possibilities of Mössbauer spectroscopy to study the molecular nature of diseases were considered.

Key words Mössbauer spectroscopy · biomedical application

1 Introduction

Progress in basic and applied biomedical research is supported in part by the application of modern physical techniques. In the framework of this review any biological, biophysical and biochemical research related to medical problems (normal and pathological physiology, environment factors and health, medicine and pharmaceutics, etc.) were considered as biomedical. The presence of iron and some other Mössbauer elements in biological subjects as well as in species which are used in medicine or may affect the living systems is an incentive to use Mössbauer spectroscopy in biomedical research (see [1, 2]). The results of various biomedical applications of Mössbauer spectroscopy were grouped as quantitative studies (studies of changes of the content of biological molecules during molecular diseases), qualitative studies (studies of any structural changes of biological molecules during molecular diseases), studies of the effect of environmental factors on biological molecules, studies of metabolic processes with Mössbauer elements, studies of dynamic properties of biological subjects, studies of pharmaceuticals containing Mössbauer elements [3, 4]. Some results of the biomedical applications were also considered in terms of "drastic" and "small" variations of the iron electronic structure [5–7]. "Drastic" changes imply a formation of new iron compounds within a tissue or a change of the valence and/or spin state of iron resulting from the transformation or destruction of proteins. In these cases, Mössbauer spectra or subspectra may be well defined. "Small" changes imply a change of

M. I. Oshtrakh (✉)
Division of Applied Biophysics, Faculty of Physical Techniques and Devices for Quality Control, Ural State Technical University-UPI, Ekaterinburg 620002, Russian Federation
e-mail: oshtrakh@mail.utnet.ru

the electronic structure of iron without any change of the valence and spin state resulting from structural modifications of proteins. The study of "small" changes is more complicated and requires a high precision, sensitive and stable Mössbauer spectrometer. The structural modifications may be related to protein heterogeneity and functional variety in human and different animals and in different organs due to its specific structure-function relationship as well as related to pathological changes in the body. About 40 years experience of biomedical applications of Mössbauer spectroscopy is briefly reviewed in this work.

2 Study of quantitative changes of biomolecules

The iron status disorder, which may be caused by hereditary diseases, related to an increase of iron uptake or to the synthesis of unstable iron containing proteins, has grave consequences. Iron overload or other changes of iron containing species could be detected by Mössbauer spectroscopy. Study of red blood cells from patients with β-thalassemia, sickle-cell anemia and hemoglobin Hammersmith demonstrated the appearance of a new component in these cells, which was identified as ferritin-like iron [8, 9]. This study of thalassemic blood samples showed that reticulocytes contained higher amounts of ferritin-like iron than old red blood cells. Ferritin-like iron was also found in serum of thalassemic blood while in patient with idiopathic hemochromatosis ferritin-like iron was found in serum only [10]. The studies of heart, spleen and liver tissues from patients with β-thalassemia showed that Mössbauer spectra of patient's tissues contained higher quantities of ferritin and hemosiderin in comparison to normal tissues [11–15]. The toxicity of a large amount of iron in tissues is a serious problem in iron overload diseases. Mössbauer studies of blood samples from thalassemic patients treated with the iron chelating drug desferrioxamine (Desferal) showed that ferritin-like iron content in serum decreased strongly, following chelation therapy, while that in red blood cells was not changed [10]. A further study showed that Desferal removed a major part of iron aggregates from the iron overload rat myocardial cells, however, this drug did not remove ferritin iron [16]. An analysis of additional components in serum and red blood cells from patients with thalassemia formed during the course of treatment was also made using Mössbauer spectroscopy [17, 18]. An additional component was found also in spectra of red blood cells in some cases of erythremia and was identified as ferritin-like iron [19]. Incorporation of iron into ferritin and hemoglobin in erythroleukemia cell lines was studied by Mössbauer spectroscopy for analyzing the variation of the quantity of these proteins during induction of cell differentiation, as well as during intracellular hemoglobin denaturation [20–22]. Study of red blood cells from patients with high altitude polycythemia revealed new additional components, besides oxy- and deoxyhemoglobins, which were considered as denatured hemoglobin [23].

The possible role of iron in Parkinson's disease was also studied by Mössbauer spectroscopy. Contrary to what had been suggested previously by other methods, Mössbauer studies found no divalent iron in the brain, and did not detect a substantial increase of iron in parkinsonian *Substantia nigra* – the part of the brain damaged in Parkinson's disease [24]. As other methods suggested that iron increase may be associated with Parkinson's disease and that neuromelanin maybe the main iron storage compound in *Substantia nigra* neurons, neuromelanin isolated from *Substantia nigra* was also studied by

Mössbauer spectroscopy [25]. In another study, the presence of up to 8% of non-ferritin-like iron in parkinsonian *Substantia nigra* or possible slight differences in the composition of the ferritin-like iron cores in Parkinsonian *Substantia nigra* compared to control, were suggested [26].

3 Study of qualitative changes of biomolecules

Synthesis of anomalous iron containing biomolecules during molecular diseases may lead to some variations of the iron electronic structure and environment. Structural and functional changes of abnormal biomolecules are of interest for studying the molecular nature of these diseases. Comparative studies of normal adult hemoglobin and hemoglobin Zürich in which the amino acid substitution *His* E7(63)β→*Arg* is in the distal heme region showed slight differences of temperature dependences of quadrupole splittings (ΔE_Q) for both oxy-hemoglobins [27]. This result could be explained by a change of the Fe(II)-O_2 bond due to the influence of the substituted amino acid on oxygen molecule. Slight differences of ΔE_Q were also observed for several abnormal deoxy- and oxyhemoglobins [28, 29]. Further changes in the Fe(II) electronic terms scheme between normal and abnormal deoxyhemoglobins were revealed based on small differences of ΔE_Q temperature dependences [30, 31]. The study of hemoglobin from patients with leukemia and erythremia revealed a slight increase of ΔE_Q and isomer shift (δ) for oxyhemoglobins of patients compared to normal adults [19, 32, 33]. These results were analyzed in terms of the changes of the Fe(II) electronic structure due to modifications of molecular structure of the active site in patient's oxyhemoglobin. A further study confirmed earlier results for small variation of Mössbauer hyperfine parameters for normal adult and patient's oxyhemoglobins as well as showed that normal adult and patient's deoxyhemoglobins have the same Mössbauer parameters [34]. This fact was compared with the above mentioned results for anomalous hemoglobin Zürich and small structural variations in the distal heme region of patient's hemoglobins which may affect the Fe(II)-O_2 bond were suggested. Mössbauer spectroscopy revealed some differences in the magnetic and iron core structure in liver hemosiderins isolated from patients with primary idiopathic hemochromatosis and with secondary (transfusional) hemochromatosis [35–37]. The values of ΔE_Q for liver hemosiderins from patients with secondary and primary hemochromatosis at 77 K were slightly different: 0.71±0.01 and 0.67±0.01 mm/s, respectively. Low temperature Mössbauer spectra of patient's hemosiderin demonstrated different magnetic behavior and different blocking temperatures for hemosiderin iron cores from different patients. These results indicated that the iron cores in hemosiderin from patients with secondary hemochromatosis were more crystalline than those from patients with primary hemochromatosis [36, 37]. Mössbauer spectroscopy of crude hemosiderin from spleen, liver, pancreas and heart tissues from patients with β-thalassemia/hemoglobin E demonstrated that spleen and heart contained hemosiderin with iron cores in the non-crystalline form of ferric oxyhydroxide which remained paramagnetic below 4.2 K while pancreas and liver contained hemosiderin with mostly ferrihydrite cores with some defect-goethite-like cores [38]. Study of ferritin from heart and spleen tissues from patients with β-thalassemia/hemoglobin E also showed some differences: the cores of heart ferritin were found larger and more crystalline [39]. Comparative studies of the iron oxide deposits in liver, spleen and pancreas from Thai patients with β-thalassemia/hemoglobin E and Australian patients with β-thalassemia revealed some differences of mineral iron core structures shown by small variations of ΔE_Q and δ [40]. Preliminary studies of liver and spleen tissues from normal chicken and chicken with lymphoid leukemia revealed small variations of ΔE_Q and δ between normal and leukemia tissues [41, 42]. These works may be also considered as studies of "small" variations of the iron electronic structure.

4 Metabolic processes with Mössbauer elements

Mössbauer spectroscopy can be used for analysis of metabolic processes by means of determination of the pathways of compounds containing Mössbauer isotopes. The study of the chemical state and metabolism of ^{153}Sm(III) citrate in mice by emission spectroscopy showed the formation of the $Sm(H_2O)_6^{3+}$ complex within the serum protein and higher accumulation of the ^{153}Sm in tumor tissues than in other tissues [43]. Mössbauer emission studies of the metabolism of ^{57}Co-bleomycin, containing both Co(II) and Co(III) compounds, showed preferential accumulation of Co(III) species in tumor [44]. The possibilities of the trivalent ^{161}Tb metabolic uptake in bone crystals were also shown by emission Mössbauer spectroscopy [45, 46]. The study of ^{57}Fe (Fe^{2+} and Fe^{3+}) uptake into the bones and teeth demonstrated that iron was presented in bones and teeth as a Fe^{3+} compound mainly (similar to FeOOH) and that iron release from teeth was much lower than from bones [47]. The study of sump beaver's dental enamel showed the presence of iron in the form similar to ferritin iron core or ferrihydrite [48]. Mössbauer spectroscopy was also used to study the ^{57}Fe metabolism in animals to analyze the iron distribution in mitochondria, cells and tissues [49–51].

5 Study of the effect of environmental factors

Various diseases are caused by environmental factors. Mössbauer spectroscopy can be used to investigate some molecular mechanisms of biological effects of exogenous agents. The study of deoxy- and oxyhemoglobin irradiated by X-rays identified iron containing compounds resulting from radiolysis of proteins and showed differences of both hemoglobin forms destruction [52, 53]. Another study of oxyhemoglobin radiolysis demonstrated the formation of different products of protein damage by γ-rays of various energies, by electrons and by microwave heating [54–58]. ^{119}Sn Mössbauer spectroscopy was used to study the tin state in triethyltin bound to cat hemoglobin as well as triethyltin and triethyltin chloride binding sites on rat liver mitochondria [59–61]. The effect of hydrazine and its derivatives on oxyhemoglobin was studied to analyze of the pathways of protein destruction [62, 63]. Further Mössbauer spectroscopy was applied to study the different oxidative stresses caused by phenylhydrazine and acethylphenylhydrazine on hemoglobin in normal adult and neonatal red blood cells, as well as in normal and cancer patients' red blood cells [64–66]. Hemoglobin oxidation in solution with vanadyl sulphate and in red blood cells from cerium chloride feeding rats was detected by Mössbauer spectroscopy [67, 68]. Mössbauer spectroscopy was also used to follow hemoglobin degradation caused by malaria parasites. The malarial pigment iron was investigated and the influence of antimalarial drugs on the pigment size in red blood cells infected by chloroquine sensitive malaria parasite strains was shown [69, 70]. Further Mössbauer studies were made for analyzing malarial pigment formation and structure as well as the effect of antimalarial drugs [71–75]. Human lung tissues from various occupational groups were studied by

Mössbauer spectroscopy and different iron containing compounds were observed in the lung tissues from coal workers, foundry workers, pottery workers, hematite workers and asbestos workers [76–79]. It was shown that iron containing dusts were deposited in the lung and stimulated an increase of iron storage proteins in lung tissues. Further Mössbauer studies were applied to analyze bioavailable iron in asbestos fibers and coal fly ash and to investigate the effect of iron chelators [80–83]. Some of these works were related to the study of "drastic" changes of the iron electronic structure resulting from protein destruction.

6 Study of dynamic properties

Dynamic properties of biological materials from molecules up to organs were used to compare normal and pathological subjects. Physiological studies of basilar membrane motion were made using vibration measurements by the Mössbauer technique [84]. The possibilities of ocular pulse measurements for comparative estimation of patients' ocular and cerebral circulations were also shown [85]. Mössbauer spectroscopy was applied to study the effective viscosity of chick embryo fibroblast membranes [86]. It was found that the effective viscosity of the membrane of Rous sarcoma virus infected chick embryo fibroblasts was about two times lower than that of normal ones.

7 Study of pharmaceutical compounds and blood substitutes

Mössbauer spectroscopy was used to study pharmaceutical compounds containing Mössbauer elements. Iron containing pharmaceuticals for treatment of iron deficiency anemia were studied for characterization of the iron state and for the detection of iron containing impurities. An analysis of injectable iron dextran and iron polysaccharide complexes, which were models of ferritin and contained ferric hydrous oxides, showed small variations of Mössbauer parameters and the presence of ferrous impurities [41, 87–95]. Vitamins and dietary supplements containing various ferrous compounds were also characterized and the presence of ferrous and ferric impurities was detected in several vitamins and dietary supplements [96–99]. Mössbauer spectroscopy was used for chemical and structural characterization of various antitumor drugs and antibiotics containing iron, tin and gold compounds [100–104] and for analyzing iron containing magnetic particles, which could be used or developed for drug delivery and contrast agents [105–107]. Mössbauer spectroscopy could also be used to study blood substitutes (artificial oxygen carriers), produced on the basis of iron porphyrin complexes and modified hemoglobins. Investigation of hemoglobin cross-linked by glutaraldehyde and pyridoxylated hemoglobin cross-linked by glutaraldehyde detected small variations of ΔE_{Q} and heme iron electronic structure as a result of protein modification [108–111].

8 Conclusion

More than 290 articles which may be related to biomedical applications of Mössbauer spectroscopy were published since 1967. The results obtained demonstrate wide possibilities of biomedical applications of Mössbauer spectroscopy for quantitative and qualitative characterization and for obtaining physical parameters of studied subjects for further understanding the molecular nature of diseases and pathological processes.

Acknowledgement This work was supported in part by the grant of the Russian Foundation for Basic Research # 06-08-00677-a.

References

1. Dickson, D.P.E., Johnson, C.E.: In: Cohen, R.L. (ed.) Applications of Mössbauer Spectroscopy, vol. 2, p. 209. Academic, New York (1980)
2. Oshtrakh, M.I.: J. Mol. Struct. **480–481**, 109 (1999)
3. Mints, R.I., Oshtrakh, M.I.: Archive Pathol. (Moscow). **48**, 82 (1986)
4. Oshtrakh, M.I.: Hyperfine Interact. **66**, 127 (1991)
5. Oshtrakh, M.I.: Z. Naturforsch. **51a**, 381 (1996)
6. Oshtrakh, M.I.: Spectrochim. Acta. Part A. **60**, 217 (2004)
7. Oshtrakh, M.I.: Hyperfine Interact. **159**, 337 (2004)
8. Bauminger, E.R., Cohen, S.G., Ofer, S., Rachmilewitz, E.A.: Proc. Natl. Acad. Sci. U.S.A. **76**, 939 (1979)
9. Bauminger, E.R., Ofer S.: In: Int. Conf. Appl. Mössbauer Effect. Proc. Indian Natl. Sci. Academy, p. 61. Indian National Science Academy, New Delhi, (1982)
10. Rachmilewitz, E.A., Bauminger, E.R., Eikelboom, J., Ofer, S., Peters, S.W., Jacobs, A.: In: Saltman, P., Hegenauer, J. (eds.) The Biochemistry and Physiology of Iron, p. 373. Elsevier, New York (1982)
11. Kaufman, K.S., Papaefthymiou, G.C., Frankel, R.B., Rosenthal, A.: Biochim. Biophys. Acta. **629**, 522 (1980)
12. Bell, S.H., Weir, M.P., Dickson, D.P.E., Gibson, J.F., Sharp, G.A., Peters, T.J.: Biochim. Biophys. Acta. **787**, 227 (1984)
13. Rimbert, J.N., Dumas, F., Kellershohn, C., Girot, R., Brissot, P.: Biochimie. **67**, 663 (1985)
14. St. Pierre, T.G., Tran, K.C., Webb, J., Macey, D.J., Pootrakul, P.: In: Suga, S., Nakahara, H. (eds.) Mechanisms and Phylogeny of Mineralization in Biological Systems, p. 291. Springer, Berlin Heidelberg New York (1991)
15. St. Pierre, T.G., Chua-anusorn, W., Webb, J., Macey, D.J., Pootrakul, P.: Biochim Biophys. Acta. **1407**, 51 (1998)
16. Bauminger, E.R., Iancu, T.C., Link, G., Pinson, A., Hershko, C.: Hyperfine Interact. **33**, 249 (1987)
17. Zhang, X., Shen, L., Bao, H., Din, X., Wang, R., Liu, Y., Gao, N., Huang, Y.: Hyperfine Interact. **71**, 1251 (1992)
18. Jiang, K., Ma, W.Y., Ortalli, I., Pedrazzi, G., Zhang, X., Izzi, G.C.: Hyperfine Interact. **91**, 859 (1994)
19. Oshtrakh, M.I., Semionkin, V.A.: FEBS Lett. **257**, 41 (1989)
20. Ofer, S., Fibach, E., Kessel, M., Bauminger, E.R., Cohen, S.G., Eikelboom, J., Rachmilewitz, E.A.: Blood. **58**, 255 (1981)
21. Fibach, E., Bauminger, E.R., Konijn, A.M., Ofer, S., Rachmilewitz, E.A.: Blood **62**, 928 (1983)
22. Fibach, E., Konijn, A.M., Bauminger, R.E., Ofer, S., Rachmilewitz, E.A.: J. Cell. Physiol. **130**, 460 (1987)
23. Zhang, X., Shen, L., Chen, S., Liu, Y., Gao, N., Zheng, Y., Ao, Z., Shong, L.: Hyperfine Interact. **58**, 2413 (1990)
24. Galazka-Friedman, J., Bauminger, E.R., Friedman, A., Barcikowska, M., Hechel, D., Nowik, I.: Mov. Disord. **11**, 8 (1996)
25. Double, K.L., Gerlach, M., Schünemann, V., Trautwein, A.X., Zecca, L., Gallorini, M., Youdim, M.B. H., Riederer, P., Ben-Shachar, D.: Biochem. Pharmacol. **66**, 489 (2003)
26. Galazka-Friedman, J., Bauminger, E.R., Koziorowski, D., Friedman, A.: Biochim. Biophys. Acta. **1688**, 130 (2004)
27. Bill, E., Di Iorio, E., Trautwein, A., Winterhalter, K.: In: Proc. Ind. Natl. Sci. Academy, Proc. Int. Conf. Appl. Mössbauer Effect, p. 648. Indian National Science Academy, New Delhi (1982)
28. Ding, X.Q., Hsia, Y.F., Shi, S.Y., Liu, R.C., Wang, S.X., Wang, H.B., Chen, Y., Chen, P.Z., Liu, Y.T.: Acta. Biochim. Biophys. Sin. **18**, 313 (1986)
29. Zeng, X.S., Ren, B.Z.: Jinan Liyi Xuebao **1**, 36 (1987)
30. Zeng, X.S., Lian, Y., Ren, B.Z.: Jinan Liyi Xuebao **3**, 27 (1988)
31. Zeng, X.S., Lian, Y.: Hyperfine Interact. **71**, 1327 (1992)
32. Oshtrakh, M.I., Semionkin, V.A.: FEBS Lett. **208**, 331 (1986)
33. Oshtrakh, M.I., Semionkin, V.A.: Biofizika (Mosk.) **32**, 197 (1987)
34. Oshtrakh, M.I., Semionkin, V.A.: Stud. Biophys. **139**, 157 (1991)

35. Mann, S., Wade, V.J., Dickson, D.P.E., Reid, N.M.K., Ward, R.J., O'Connell, M., Peters, T.J.: FEBS Lett. **234**, 69 (1988)
36. Dickson, D.P.E., Reid, N.M.K., Mann, S., Wade, V.J., Ward, R.J., Peters, T.J.: Biochim. Biophys. Acta. **957**, 81 (1988)
37. Dickson, D.P.E., Reid, N.M.K., Mann, S., Wade, V.J., Ward, R.J., Peters, T.J.: Hyperfine Interact. **45**, 225 (1989)
38. St. Pierre, T.G., Tran, K.C., Webb, J., Macey, D.J., Pootrakul, P., Dickson, D.P.E.: Hyperfine Interact. **71**, 1279 (1992)
39. Webb, J., St. Pierre, T.G., Macey, D.J., Tran, K.C., Pootrakul, P.: In: Skinner, H.C.W., Fitzpatrick, R.W. (eds.) Biomineralization Processes, Iron, Manganese, p. 169. Catena, Cremlingen, (1992)
40. St. Pierre, T.G., Chua-anusorn, W., Webb, J., Macey, D.J., Pootrakul, P.: Biochim. Biophys. Acta. **1407**, 51 (1998)
41. Oshtrakh, M.I., Milder, O.B., Semionkin, V.A., Prokopenko, P.G., Malakheeva, L.I.: Izv. Russ. Acad. Sci. Ser. Phys. (Bull. Russ. Acad. Sci. Phys.) **69**, 1547 (2005)
42. Oshtrakh, M.I., Milder, O.B., Semionkin, V.A., Malakheeva, L.I., Prokopenko, P.G.: J. Radioanal. Nucl. Chem. **269**, 671 (2006)
43. Friedman, A.M., Sullivan, J.C., Ruby, S.L., Lindenbaum, A., Russell, J.J., Zabransky, B.J., Rayudu, G. V.S.: Int. J. Nucl. Med. Biol. **3**, 37 (1976)
44. Zalutsky, M.R., Friedman, A.M., Sullivan, J.C., Ruby, S.L., Rayudu, G.V.S.: Int. J. Nucl. Med. Biol. **4**, 216 (1977)
45. Rimbert, J.N., Kellershohn, C., Dumas, F., Fortier, D., Maziere, M., Hubert, C.: Biochimie **63**, 931 (1981)
46. Rimbert, J.N., Dumas, F., Kellershohn, C.: Hyperfine Interact. **14**, 221 (1983)
47. Bauminger, E.R., Ofer, S., Gedalia, I., Horowitz, G., Mayer, I.: Calcif. Tissue Int. **37**, 386 (1985)
48. Madsen, M.B., Mørup, S., Koch, C.J.W., Lindemann, G.: Hyperfine Interact. **29**, 1431 (1986)
49. Aleksandrov, A.Y., Druzhenkov, V.V., Ivanov, I.I., Kuzmin, R.N., Novakova, A.A., Rubin, A.B., Semin, B.K.: Biofizika (Mosk.) **24**, 663 (1979)
50. Semin, B.K., Novakova, A.A., Aleksandrov, A.Y., Ivanov, I.I., Rubin, A.B., Kuzmin, R.N.: Biochim. Biophys. Acta. **715**, 52 (1982)
51. Vanin, A.F., Kovalenko, O.A., Kubrina, L.N., Prusakov, V.E., Stukan, R.A., Tarasova, N.I.: Biofizika (Mosk.) **27**, 804 (1982)
52. Kellershohn, C., Rimbert, J.N., Chevalier, A., Hubert, C.: J. Phys. **37**(Suppl. 12), C6–C185 (1976)
53. Chevalier, A., Kellershohn, C., Rimbert, J.N.: Radiat. Res. **94**, 51 (1983)
54. Oshtrakh, M.I., Semionkin, V.A.: Radiat. Environ. Biophys. **30**, 33 (1991)
55. Oshtrakh, M.I., Kopelyan, E.A., Semionkin, V.A.: Nucl. Instrum. Methods in Phys. Res. **B131**, 226 (1997)
56. Oshtrakh, M.I., Semionkin, V.A., Kopelyan, E.A., Milder, O.B.: Rad. Phys. Chem. **55**, 549 (1999)
57. Oshtrakh, M.I., Semionkin, V.A., Milder, O.B.: Nucl. Instrum. Methods Phys. Res. **B151**, 227 (1999)
58. Oshtrakh, M.I.: Nucl. Instrum. Methods Phys. Res. **B185**, 129 (2001)
59. Elliott, B.M., Aldridge, W.N., Bridges, J.W.: Biochem. J. **177**, 461 (1979)
60. Farrow, B.G., Dawson, A.P.: Eur. J. Biochem. **86**, 85 (1978)
61. Dawson, A.P., Farrow, B.G., Selwyn, M.J.: Biochem. J. **202**, 163 (1982)
62. Garibov, R.E., Russkikh, E.V., Khrapov, V.V.: Dokl. Akad. Nauk SSSR **245**, 238 (1979)
63. Garibov, R.E., Khrapov, V.V., Russkikh, E.V.: Dokl. Akad. Nauk SSSR **252**, 739 (1980)
64. Shahal, Y., Bauminger, E.R., Zmora, E., Katz, M., Mazor, D., Horn, S., Meyerstein, N.: Pediatr. Res. **29**, 119 (1991)
65. Passeri, G., Piccolo, P., Croci, S., Ortalli, I., Pedrazzi, G.: Hyperfine Interact. **126**, 47 (2000)
66. Croci, S., Pedrazzi, G., Passeri, G., Piccolo, P., Ortalli, I.: Biochim. Biophys. Acta. **1568**, 99 (2001)
67. Xue, D., Lu, J., Wang, K.: J. Inorg. Biochem. **71**, 79 (1998)
68. Cheng, Y., Li, Y., Li, R., Lu, J., Wang, K.: Chem. Biol. Interact. **125**, 191 (2000)
69. Bauminger, E.R., Ginsburg, H., Ofer, S., Yayon, A.: Hyperfine Interact. **15/16**, 885 (1983)
70. Yayon, A., Bauminger, E.R., Ofer, S., Ginsburg, H.: J. Biol. Chem. **259**, 8163 (1984)
71. Blauer, G., Akkawi, M., Bauminger, E.R.: Biochem. Pharmacol. **46**, 1573 (1993)
72. Adams, P.A., Berman, P.A.M., Egan, T.J., Marsh, P.J., Silver, J.: J. Inorg. Biochem. **63**, 69 (1996)
73. Bohle, D.S., Debrunner, P., Jordan, P.A., Madsen, S.K., Schulz, C.E.: J. Am. Chem. Soc. **120**, 8255 (1998)
74. Bauminger, E.R., Akkawi, M., Blauer, G.: Inorg. Chim. Acta. **286**, 229 (1999)
75. Egan, T.J., Combrinck, J.M., Egane, J., Hearne, G.R., Marques, H.M., Ntenteni, S., Sewell, B.T., Smith, P.J., Taylor, D., van Schalkwyk, D.A., Walden, J.C.: Biochem. J. **365**, 343 (2002)
76. Guest, L.: Ann. Occup. Hyg. **19**, 49 (1976)

77. Guest, L.: Ann. Occup. Hyg. **21**, 151 (1978)
78. Guest, L.: Analyst **106**, 663 (1981)
79. Stroink, G., Lim, D., Dunlap, R.A.: Phys. Med. Biol. **32**, 203 (1987)
80. Gold, J., Amandusson, H., Krozer, A., Kasemo, B., Ericsson, T., Zanetti, G., Fubini, B.: Environ. Health Perspect. **105**(Suppl. 5), 1021 (1997)
81. Gulumian, M., Nkosibomvu, Z.L., Channa, K., Pollak, H.: Environ. Health Perspect. **105**(Suppl. 5), 1041 (1997)
82. Veranth, J.M., Smith, K.R., Huggins, F., Hu, A.A., Lighty, J.S., Aust, A.E.: Chem. Res. Toxicol. **13**, 161 (2000)
83. Ball, B.R., Smith, K.R., Veranth, J.M., Aust, A.E.: Inhal. Toxicol. **12**(Suppl. 4), 209 (2000)
84. Sellick, P.M., Patuzzi, R., Johnstone, B.M.: J. Acoust. Soc. Am. **72**, 131 (1982)
85. Rogozev, B.I., Silin, M.Y.: Hyperfine Interact. **112**(Pt. 2), 279 (1998)
86. Bauminger, E.R., Cohen, S.G., Ofer, S., Bachrach, U.: Biochim. Biophys. Acta **720**, 133 (1982)
87. Berg, K.A., Bowen, L.H., Hedges, S.W., Bereman, R.D., Vance, C.T.: J. Inorg. Biochem. **22**, 125 (1984)
88. Yang, C.-Y., Bryan, A.M., Theil, E.C., Sayers, D.E., Bowen, L.H.: J. Inorg. Biochem. **28**, 393 (1986)
89. Oshtrakh, M.I., Kopelyan, E.A., Semionkin, V.A., Livshits, A.B., Krylova, V.E., Prostakova, T.M., Kozlov, A.A.: J. Inorg. Biochem. **54**, 285 (1994)
90. Coe, E.M., Bowen, L.H., Speer, J.A., Wang, Z., Sayers, D.E., Bereman, R.D.: J. Inorg. Biochem. **58**, 269 (1995)
91. Kilcoyne, S.H., Gorisek, A.: J. Magn. Magn. Mater. **177–181**, 1457 (1998)
92. Knight, B., Bowen, L.H., Bereman, R.D., Huang, S., De Grave, E.: J. Inorg. Biochem. **73**, 227 (1999)
93. Oshtrakh, M.I., Semionkin, V.A., Milder, O.B., Livshits, A.B., Kozlov, A.A.: Z. Naturforsch. **55a**, 186 (2000)
94. Oshtrakh, M.I., Semionkin, V.A., Prokopenko, P.G., Milder, O.B., Livshits, A.B., Kozlov, A.A.: Int. J. Biol. Macromol. **29**, 303 (2001)
95. Funk, F., Long, G.J., Hautot, D., Büchi, R., Christl, I., Weidler, P.G.: Hyperfine Interact. **136**, 73 (2001)
96. Sud, M., Bala, N., Liveleen, Vishwamittar, Puri, S.P.: Ind. J. Pure Appl. Phys. **26**, 701 (1988)
97. Coe, E.M., Bowen, L.H., Bereman, R.D.: J. Inorg. Biochem. **58**, 291 (1995)
98. Oshtrakh, M.I., Milder, O.B., Semionkin, V.A.: Anal. Chim. Acta **506**, 155 (2004)
99. Oshtrakh, M.I., Milder, O.B., Semionkin, V.A.: Hyperfine Interact. **156/157**, 273 (2004)
100. Capolongo, F., Giomini, M., Giuliani, A.M., Matzanke, B.E., Russo, U., Silvestri, A., Trautwein, A.X., Barbieri, R.: J. Inorg. Biochem. **65**, 115 (1997)
101. Pellerito, A., Fiore, T., Pellerito, C., Fontana, A., Di Stefano, R., Pellerito, L., Cambria, M.T., Mansueto, C.: J. Inorg. Biochem. **72**, 115 (1998)
102. Canumalla, A., Shaw III, C.F., Wagner, F.E.: Inorg. Chem. **38**, 3268 (1999)
103. Di Stefano, R., Scopelliti, M., Pellerito, C., Fiore, T., Vitturi, R., Colomba, M.S., Gianguzza, P., Stocco, G.C., Consiglio, M., Pellerito, L.: J. Inorg. Biochem. **89**, 279 (2002)
104. Natha, M., Pokhariaa, S., Engb, G., Songb, X., Kumar, A.: Eur. J. Med. Chem. **40**, 289 (2005)
105. Jung, C.W., Jacobs, P.: Mag. Res. Imaging. **13**, 661 (1995)
106. Rudge, S.R., Kurtz, T.L., Vessely, C.R., Catterall, L.G., Williamson, D.L.: Biomaterials **21**, 1411 (2000)
107. Avivi (Levi), S., Felner, I., Novik, I., Gedanken, A.: Biochim. Biophys. Acta **25198**, 1 (2001)
108. Guillochon, D., Vijayalakshmi, M.W., Thiam-Sow, A., Thomas D., Chevalier, A.: Biochem. Cell Biol. **64**, 29 (1986)
109. Chevalier, A., Guillochon, D., Nedjar, N., Piot, J.M., Vijayalakshmi, M.W., Thomas, D.: Biochem. Cell Biol. **68**, 813 (1990)
110. Oshtrakh, M.I., Milder, O.B., Semionkin, V.A., Berkovsky, A.L., Azhigirova, M.A., Vyazova, E.P.: Z. Naturforsch. **55a**, 193 (2000)
111. Oshtrakh, M.I., Milder, O.B., Semionkin, V.A., Berkovsky, A.L., Azhigirova, M.A., Vyazova, E.P.: Int. J. Biol. Macromol. **28**, 5 (2000)

Hyperfine Interact (2005) 165: 321–326
DOI 10.1007/s10751-006-9285-6

Study of chicken liver and spleen by Mössbauer spectroscopy

M. I. Oshtrakh · O. B. Milder · V. A. Semionkin · L. I. Malakheeva · P. G. Prokopenko

Published online: 2 November 2006

Abstract A preliminary study of purified normal human liver ferritin, normal chicken liver and spleen tissues in lyophilized form showed differences in room temperature Mössbauer hyperfine parameters. An additional study of liver and spleen tissues with lower iron content from chicken with lymphoid leukemia indicated small differences between the quadrupole splittings in these samples compared with those in normal tissues.

Key words Mössbauer spectroscopy · human liver ferritin · chicken liver and spleen · lymphoid leukemia

1 Introduction

Iron storage protein ferritin is deposited in various tissues. Liver and spleen tissues contain a large part of iron storage proteins. It is well known that ferritin molecules vary in different animals as well as in different tissues [1]. Comparative Mössbauer study of normal human

M. I. Oshtrakh (✉)
Division of Applied Biophysics, Faculty of Physical Techniques and Devices for Quality Control, Ural State Technical University—UPI,
Ekaterinburg 620002, Russian Federation
e-mail: oshtrakh@mail.utnet.ru

O. B. Milder · V. A. Semionkin
Faculty of Experimental Physics, Ural State Technical University—UPI,
Ekaterinburg 620002, Russian Federation

L. I. Malakheeva
Science Consultation Department, Simbio Holding,
Ekaterinburg 620044, Russian Federation

P. G. Prokopenko
Faculty of Biochemistry, Russian State Medical University,
Moscow 117869, Russian Federation

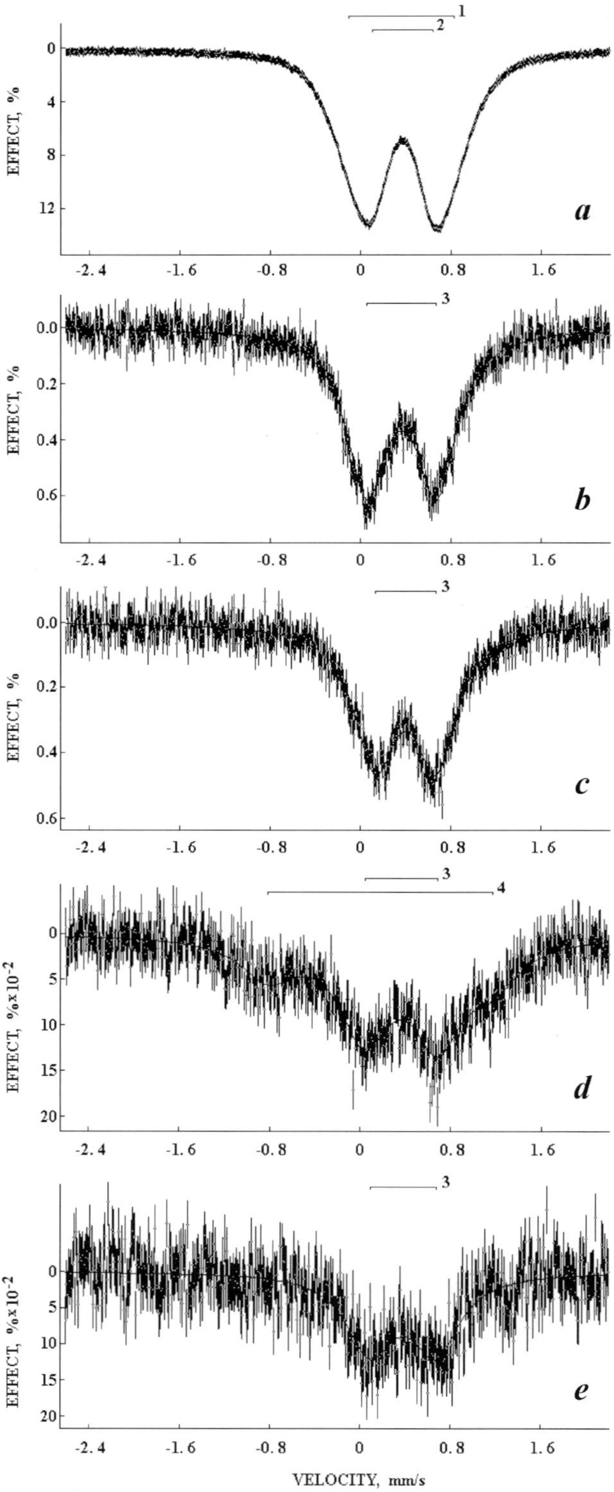

Figure 1 Mössbauer spectra of normal human liver ferritin (**a**), normal chicken liver (**b**) and spleen (**c**) tissues, liver (**d**) and spleen (**e**) tissues from chicken with lymphoid leukemia. *1* and *2* are the results of two quadrupole doublets fit of ferritin, *3* is the result of one quadrupole doublet fit of ferritin-like iron, 4-methemoglobin. T=295 K.

Table I Mössbauer parameters of normal human liver ferritin and chicken liver and spleen tissues obtained by one or superposition of two quadrupole doublets fit

Sample	Γ (mm/s)	δ (mm/s)	ΔE_Q (mm/s)	S (%)	χ^2
Normal human liver ferritin	0.472±0.020	0.368±0.010	0.685±0.010	100	17.24
	0.347±0.020	0.370±0.010	0.545±0.010	44	3.40
	0.426±0.020	0.363±0.010	0.930±0.010	56	
Normal chicken liver tissue	0.436±0.020	0.361±0.010	0.618±0.010	100	0.98
Normal chicken spleen tissue	0.446±0.020	0.399±0.010	0.537±0.010	100	1.00
Liver tissue from chicken with lymphoid leukemia	0.536±0.020	0.363±0.010	0.639±0.010	69	0.97
	0.599±0.020	0.176±0.010	1.978±0.010	31	
Spleen tissue from chicken with lymphoid leukemia	0.506±0.020	0.382±0.010	0.587±0.010	100	1.12

liver and spleen tissues at 78 K demonstrated small differences in quadrupole splittings and isomer shifts [2]. Small differences were observed between Mössbauer parameters obtained at room temperature for normal human liver ferritin and normal chicken liver tissue [3]. Some differences in quadrupole splittings were found at 78 K for spleen tissues from normal human and patients with β-thalassemia [4]. These variations of Mössbauer parameters indicate small variations of the iron electronic structure and the iron core structure in iron storage proteins. To continue these investigations we present further results of a Mössbauer comparative study of purified human liver ferritin, normal chicken liver and spleen tissues and liver and spleen tissues from chicken with lymphoid leukemia by Mössbauer spectroscopy.

2 Materials and methods

Preparation of lyophilized normal human liver ferritin was described earlier [5]. Liver and spleen tissues from one normal chicken and from one chicken with lymphoid leukemia were washed from the blood and lyophilized. The sample of ferritin contained about 200 mg of protein with effective thickness of about 10 mg Fe/cm^2 while chicken tissue samples contained 1,200–1,600 mg of lyophilized tissue powders. Mössbauer spectra of the sample powders were measured at room temperature with the constant acceleration computerized high precision, sensitive and stable spectrometer that was a part of multi-dimension parametric Mössbauer spectrometer SM-2201 [6]. The noise of velocity signal of spectrometer was 1.5×10^{-3} mm/s, the drift of zero velocity point was $\pm2.6\times10^{-3}$ mm/s, the nonlinearity of velocity signal was 0.01%, the harmonic distortion factor was 0.005% for the frequency band in the range of 0–1120 Hz. The 0.3×10^9 Bq ^{57}Co(Cr) source was used at room temperature. Spectra of ferritin were measured several days with statistics of 1.8×10^6 counts per channel while liver and spleen samples were measured up to 20 days with statistics from 10×10^6 counts per channel for normal tissues to 29×10^6 counts per channel for leukemia tissues. Mössbauer spectra were measured in transmission geometry with moving absorber to exclude parabolic distortion of the spectrum and contribution of the ^{57}Fe in the beryllium window of the scintillator detector to the measured spectrum for samples with low iron content [7]. Mössbauer spectra were computer fitted with the least squares procedure using Lorentzian line shapes. Mössbauer parameters such as isomer shift δ, quadrupole splitting ΔE_Q, line width Γ and subspectrum area S were determined. The values of isomer shift are given relative to α-Fe at 295 K.

Figure 2 The plot of Mössbauer hyperfine parameters for normal human liver ferritin (◇), normal chicken liver (□) and spleen (△) tissues, liver (■) and spleen (▲) tissues from chicken with lymphoid leukemia. Data are the results of one quadrupole doublet fit. T=295 K.

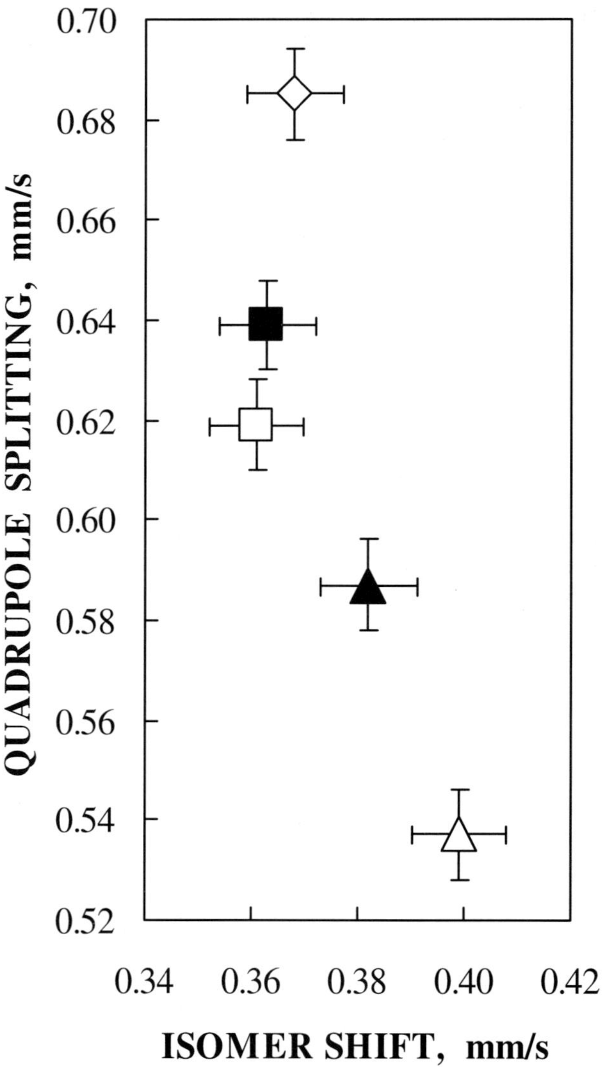

3 Results and discussion

Mössbauer spectra of ferritin and chicken tissues are shown in Figure 1. Spectra of normal chicken liver and spleen demonstrate the presence of ferritin-like iron in tissues. Mössbauer spectra of liver and spleen tissues from chicken with lymphoid leukemia show significantly lower iron content and decreasing of the absorption effect. The spectrum of liver from leukemia chicken contains additional component which can be related to residue quantity of hemoglobin in the sample. It is well known that paramagnetic Mössbauer spectra of ferric hydrous oxides have non-Lorentzian line shape and, therefore, were better fitted using more than one quadrupole doublet or using a distribution of quadrupole splitting [8, 9].

Therefore, Mössbauer spectra of human liver ferritin and chicken liver and spleen tissues were fitted using one or superposition of two quadrupole doublets.

The results of the fits to the Mössbauer spectra of normal human liver ferritin and chicken liver and spleen tissues are given in Table I. An additional component 4 in the spectrum of leukemia chicken liver tissue (Figure 1d) has parameters close to those of methemoglobin [10]. Small differences between the Mössbauer hyperfine parameters (one doublet fit) of normal human liver ferritin and ferritin-like iron from chicken tissues are clearly seen in the plot of ΔE_Q and δ, shown in Figure 2. The fit of the Mössbauer spectrum of normal human ferritin using superposition of two doublets showed that χ^2 value decreased significantly in comparison with that for one doublet fit. These doublets may be related to various structural regions (for instance, external and internal regions) in the iron core in ferritin molecules from normal human liver. On the other hand, χ^2 values for both fits of the Mössbauer spectra of normal chicken liver and spleen tissues were in fact the same and, therefore, the results of two doublets fit were omitted from consideration. Moreover, the fit of the Mössbauer spectra of ferritin-like iron in leukemia tissues using two quadrupole doublets did not reach convergence. It is possible that these facts may be a result of low absorption effects in the Mössbauer spectra of chicken tissues and need further measurements. Thus, the results of one doublet fit for normal chicken spleen tissue are different from those of normal human liver ferritin and normal chicken liver tissue, while there were only small differences between normal and leukemia chicken tissues.

4 Conclusion

Mössbauer spectra were obtained from normal human liver ferritin, normal and leukemia chicken liver and spleen tissues. The obtained results showed some differences in the values of ΔE_Q and δ in case of one quadrupole doublet fit. These differences may be a result of small variations of the iron core structure in normal human liver ferritin and ferritin-like iron in normal chicken liver and spleen tissues. It was interesting to observe small variations of the values of ΔE_Q and δ for ferritin-like iron in liver and spleen tissues from chicken with lymphoid leukemia in comparison with normal ones. These variations may reflect additional structural differences in the iron cores related to pathology, however, further studies for such samples are required.

Acknowledgement This work was supported in part by the grant of the Russian Foundation for Basic Research #06-08-00677-a.

References

1. Theil, E.C.: In: Theil, E.C., Eichhorn, G.L., Marzilli, L.G. (eds.) Iron Binding Proteins without Cofactors or Sulfur Clusters. Advances in Inorganic Biochemistry, vol. 5, p. 1. Elsevier, New York (1983)
2. Chua-anusorn, W., St. Pierre, T.G., Webb, J., Macey, D.J., Yansukon, P., Pootrakul, P.: Hyperfine Interact. **91**, 905 (1994)
3. St. Pierre, T.G., Chua-anusorn, W., Webb, J., Macey, D.J., Pootrakul, P.: Biochim. Biophys. Acta **1407**, 51 (1998)
4. Oshtrakh, M.I., Milder, O.B., Semionkin, V.A., Prokopenko, P.G., Malakheeva, L.I.: Hyperfine Interact. **156/157**, 279 (2004)

5. Oshtrakh, M.I., Semionkin, V.A., Prokopenko, P.G., Milder, O.B., Livshits, A.B., Kozlov, A.A.: Int. J. Biol. Macromol. **29**, 303 (2001)
6. Irkaev, S.M., Kupriyanov, V.V., Semionkin, V.A.: British Patent no. 10745 (7 May, 1987)
7. Oshtrakh, M.I.: Faraday Discuss. **126**, 119 (2004)
8. Murad, E., Johnston, J.H.: In: Long, G.J. (ed.) Mössbauer Spectroscopy Applied to Inorganic Chemistry, vol. 2, p. 507. Plenum, New York (1987)
9. Murad, E., Bowen, L.H., Long, G.J., Quin, T.G.: Clay Miner. **23**, 161 (1988)
10. Thomanek, U.F., Parak, F., Wintergerst, B.: Z. Naturforsch. **32c**, 11 (1977)

Hyperfine Interact (2005) 165: 327–331
DOI 10.1007/s10751-006-9279-4

Ab initio study of the ^{57}Fe quadrupole splitting in the heme models of α- and β-subunits in tetrameric deoxyhemoglobin

E. I. Yuryeva · M. I. Oshtrakh

Published online: 27 October 2006

Abstract Ab initio Xα discrete variation method was used for calculation of quadrupole splitting for the rough heme models in α- and β-subunits of tetrameric deoxyhemoglobin accounting small stereochemical variations. The differences of theoretical values of quadrupole splitting for these heme models were obtained.

Key words quadrupole splitting · deoxyhemoglobin · heme models · Xα-DVM calculation

1 Introduction

Hemoglobin, a vitally important oxygen transport protein, consists of two pairs of non-identical α- and β-subunits. X-ray structural analysis of human deoxyhemoglobin revealed small stereochemical differences for the heme iron in α- and β-subunits of tetrameric protein [1]. Basing on this data and some features of Mössbauer spectra of deoxyhemoglobin the supposition about superimposed character of deoxyhemoglobin Mössbauer spectra was made [2–5]. To evidence this supposition semi-empirical iterative extended Hückel calculations for deoxyheme models accounting iron stereochemical differences for α- and β-subunits were made [3, 6, 7]. These results showed differences of quadrupole splitting for deoxyheme models of these subunits. However, further quantum chemical studies using better heme models and ab initio calculations are needed. One of ab initio calculations based on Xα discrete variation method (Xα-DVM) [8] was recently used for theoretical study of deoxyheme [9]. Therefore, in this work we present the results of the first study of rough heme models for α- and β-subunits in tetrameric deoxyhemoglobin using ab initio Xα-DVM.

E. I. Yuryeva
Institute of Solid State Chemistry, Ural Branch of the Russian Academy of Sciences, Ekaterinburg 620219, Russian Federation

M. I. Oshtrakh (✉)
Division of Applied Biophysics, Faculty of Physical Techniques and Devices for Quality Control, Ural State Technical University – UPI, Ekaterinburg 620002, Russian Federation
e-mail: oshtrakh@mail.utnet.ru

Table I Geometry parameters for the heme models

Geometrical parameters[a]	Heme model structure	
	α-subunits	β-subunits
Distance	Å	Å
$N_{porph}(1)$-Fe	0.2117	0.2076
$N_{porph}(2)$-Fe	0.2166	0.2026
$N_{porph}(3)$-Fe	0.2155	0.2017
$N_{porph}(4)$-Fe	0.2130	0.2047
Average N_{porph}-Fe	0.2142	0.2042
$N_{his}(5)$-Fe	0.2199	0.2095
Angles	Grad. (°)	Grad. (°)
$N_{porph}(1)$-Fe-$N_{porph}(3)$	158.13	159.66
$N_{porph}(2)$-Fe-$N_{porph}(4)$	157.79	159.55
$N_{porph}(1)$-Fe-$N_{his}(5)$	96.50	94.16
$N_{porph}(2)$-Fe-$N_{his}(5)$	96.96	99.49
$N_{porph}(3)$-Fe-$N_{his}(5)$	105.36	106.15
$N_{porph}(4)$-Fe-$N_{his}(5)$	106.58	100.96
$N_{porph}(1)$-Fe-$N_{porph}(4)$	89.61	90.03
$N_{porph}(1)$-Fe-$N_{porph}(2)$	85.17	88.50
$N_{porph}(2)$-Fe-$N_{porph}(3)$	91.53	87.10
$N_{porph}(3)$-Fe-$N_{porph}(4)$	85.32	87.22

[a] N_{porph} – nitrogen atoms numbered 1–4 in the porphyrin plane, N_{his} – nitrogen atom numbered 5 in the distal histidine which binds Fe.

2 Materials and methods

Initial heme models were made using X-ray structural data [1] obtained from the Protein Data Bank (4HHB) and consisted of 53 atoms. We used averaged coordinates for hemes in α_1- and α_2-subunits and in β_1- and β_2-subunits, respectively. For simplification the rough heme models including Fe atom and its nearest neighbors (four N atoms of porphyrin and one N atom of distal histidine) were used for calculations of the electronic structure (bond distances and angles for Fe and nearest neighbors in the heme models are given in Table I). In this case dangling nitrogen bonds in chosen clusters were saturated with additional electrons to form completed N2p atomic orbitals. Electronic structure and parameters of atomic interactions were calculated by ab initio self-consistent spin-polarized Xα-DVM [8] in the framework of the electronic density functional theory with exchange-correlation potential in the Gunnarsson–Lundqvist form [10]. The calculations were carried out in a completely asymmetrical basis of atomic orbitals (AO) according to the magnetic orbital quantum number distribution. The frozen-core approximation was used; the rest of computation parameters corresponded to those described in [11]. Atomic orbital populations (P), effective charges (Q_{eff}) and magnetic moments of atoms (μ) as well as model densities of electronic states (DOS) were calculated. Electric field gradient (EFG) and contributions to EFG from electrons (q^{el}) and lattice (q^{lat}) as well as quadrupole splitting (ΔE_Q) were calculated as following [12]:

$$\Delta E_Q = \frac{e^2 q_{zz} Q}{2}\left(1+\frac{\eta^2}{3}\right)^{1/2},$$

Table II Calculated parameters of the iron electronic structure for heme models

Parameters of electronic structure	Heme model for α-subunits	Heme model for β-subunits
P_{3d}	6.0300	6.0375
P_{4s}	0.0724	0.0894
P_{4p}	0.0217	0.0303
Q_{eff}	1.8759	1.8428
μ	−3.9623	−3.9547

Mulliken population P (e), effective atomic charge Q_{eff} (e) and electron-spin magnetic moments μ (μ_B).

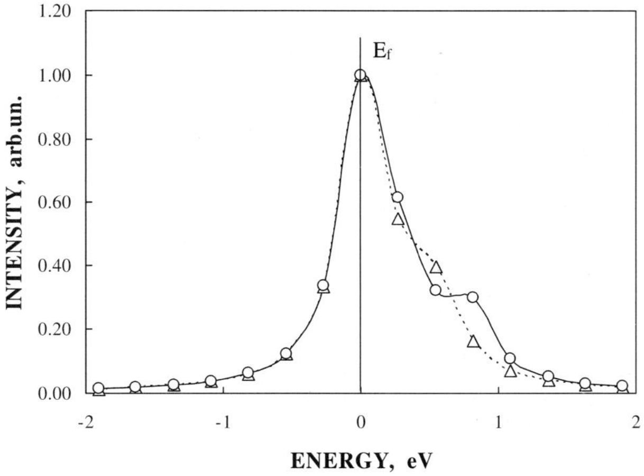

Figure 1 Densities of electronic states for atomic orbitals for the heme models of α- and β-subunits (Δ and ○, respectively) in tetrameric deoxyhemoglobin.

where Q is nuclear quadruple moment, Q=0.16 barn for ^{57}Fe [13], q_{zz} is the maximum value of diagonalized EFG, $\eta = \frac{|q_{xx}-q_{yy}|}{q_{zz}}$ is the asymmetry parameter. EFG tensor was determined by equation: $q_{ij} = q_{ij}^{el}(1-R) + q_{ij}^{lat}(1-\gamma_\infty)$, where $q_{ij}^{el}(A) = -\left\langle \psi \left| \sum_k q_{ij}^{el}(A_k) \right| \psi \right\rangle$ denotes the contribution of electrons of Fe atoms (with MO LCAO approximation $q_{ij}^{el}(A) = -\sum_p n_p \sum_{q,q'} C_{pq} C_{pq'} \left\langle \chi_q \left| \sum_k q_{ij}^{el}(A_k) \right| \chi_{q'} \right\rangle$), q_{ij}^{lat} denotes the contribution of nearest neighbors cluster calculated with the point charges approximation, R and γ_∞ are the Sternheimer screening and antiscreening factors, respectively (R=0, $\gamma_\infty = -9.64$). Temperature dependencies of E_Q for ^{57}Fe were calculated as a result of temperature dependent redistribution of Fe3d$^\uparrow$ atomic orbitals population near Fermi level at T>0. This is realized by smearing of electron density near Fermi level described by Fermi–Dirac distribution for degenerated electron gas.

3 Results and discussion

The results of the iron electronic structure calculations are given in Table II. Small differences of the iron 3d, 4s and 4p orbital populations for the heme models of α- and β-subunits were obtained. These populations were higher for Fe in the heme model for β-subunits while effective charge and magnetic moment were higher for Fe in the heme model for α-subunits. The difference of the DOS for Fe3d$^\uparrow$ atomic orbitals for both models is shown in Figure 1. The Fe3d-AO splitting by crystal field was different for the heme

Figure 2 Calculated temperature dependences of the ^{57}Fe quadrupole splitting for the heme models of α- and β-subunits (Δ and ○, respectively) in tetrameric deoxyhemoglobin: (a) with accounting of electronic contribution only; (b) with accounting of both electronic and nearest neighbors contribution.

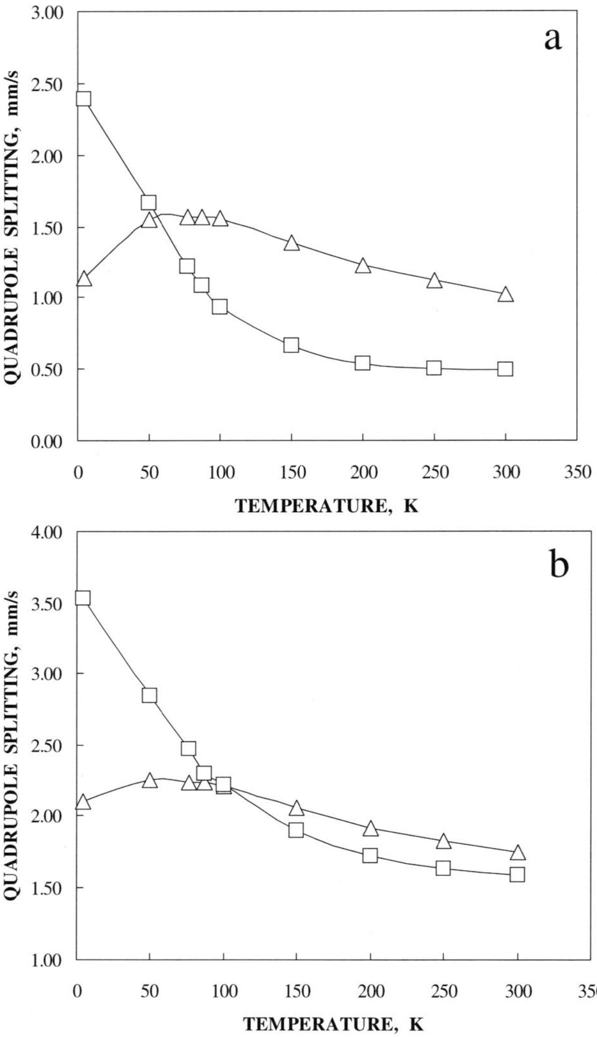

models of α- and β-subunits due to small differences of nearest neighbors' geometry. The crystal field was stronger for the heme model of β-subunit than for that of α-subunit. Therefore, temperature effect on electron density smearing near Fermi level for the heme model in α-subunit was more significant.

Further calculations of the temperature dependences of ^{57}Fe quadrupole splitting for the heme models of α- and β-subunits were made with accounting of electronic and sum of electronic and nearest neighbors' contributions. These results were shown in Figure 2 and demonstrated some differences also. Calculated values of ΔE_Q (Figure 2b) were slightly higher than experimental data as a result of rough approximation. Earlier fitting of experimental Mössbauer spectra of deoxyhemoglobin measured at 87 K using superposition of two quadrupole doublets with almost the same areas (see Figure 3) [3–5] showed 0.17 mm/s difference for ΔE_Q values related to α- and β-subunits, respectively. In the present calculations theoretical difference of ΔE_Q values at 87 K was 0.06 mm/s. Thus, qualitative differences of ΔE_Q values for the heme models of α- and β-subunits obtained in

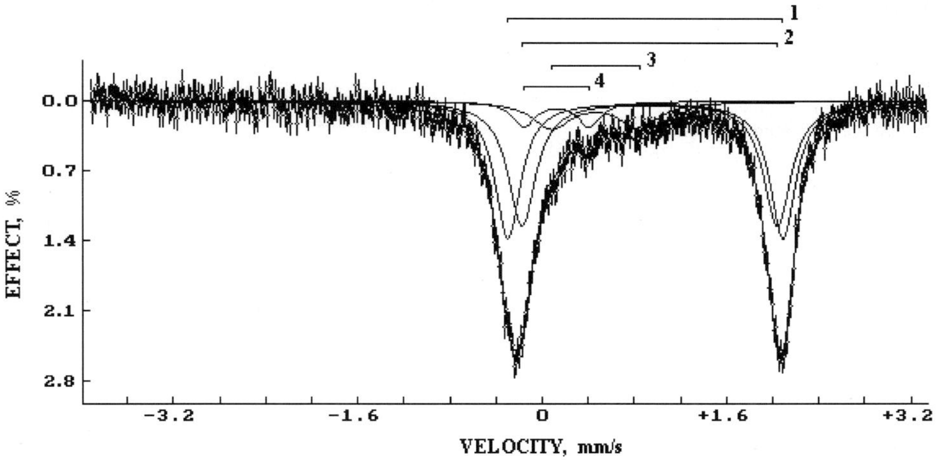

Figure 3 Mössbauer spectrum of normal human deoxyhemoglobin in solution [5]. 1 – α-subunits in tetramer, 2 – β-subunits in tetramer, 3 – unknown Fe^{3+} compound and 4 – Be(^{57}Fe). T=87 K.

this rough approximation were in agreement with results of previous semi-empirical calculations (the difference of ΔE_Q values was 0.13 mm/s at 87 K) [3, 6, 7] and supported approach to fit experimental Mössbauer spectra of deoxyhemoglobin using superposition of two quadrupole doublets with almost the same areas considered in [3–5].

4 Conclusion

Ab initio Xα-DVM calculations of the iron electronic structure and quadrupole splitting for the rough heme models accounting stereochemical differences of the nearest neighbors in α- and β-subunits of tetrameric deoxyhemoglobin demonstrated small differences which may be an additional evidence of the Fe(II) electronic structure non-equivalence in α- and β-subunits of tetrameric deoxyhemoglobin and the necessity of accounting this fact during the fitting of Mössbauer spectra of deoxyhemoglobin.

References

1. Fermi, G., Perutz., M.F., Shaanan, B., Fourme, R.: J. Mol. Biol. **175**, 159 (1984)
2. Oshtrakh, M.I., Semionkin, V.A.: Mol. Biol. (Moscow) **19**, 1310 (1985)
3. Oshtrakh, M.I., Semionkin, V.A., Burykin, B.N., Khleskov, V.I.: Mol. Phys. **66**, 531 (1989)
4. Oshtrakh, M.I.: Z. Naturforsch. **53a**, 608 (1998)
5. Oshtrakh, M.I., Semionkin, V.A.: Hyperfine Interact., **159**, 345 (2004)
6. Burykin, B.N., Khleskov, V.I., Oshtrakh, M.I., Semionkin, V.A.: Mol. Biol. (Moscow) **21**, 1677 (1987)
7. Khleskov, V.I., Burykin, B.N., Smirnov, A.B., Oshtrakh, M.I.: Biochem. Biophys. Res. Communs. **155**, 1255 (1988)
8. Averil, F.V., Ellis, D.E.J.: Chem. Phys. **59**, 6412 (1973)
9. Zhi, Z., Guenzburger, D., Ellis, D.E.: J. Mol. Struct., Theochem **678**, 145 (2004)
10. Gunnarsson, O., Lundqvist, B.I., Ros, P.: Phys. Rev. B. **13**, 4274 (1976)
11. Yuryeva, E.I., Pletnev, R.N., Skripov, A.V.: Int. J. Quant. Chem. **104**, 143 (2005)
12. Yuryeva, E.I., Zhukov, V.P., Medvedeva, N.I., Gubanov, V.A.: Phys. Status Solidi **146**, 557 (1994)
13. Dufek, P., Blaha, P., Schwarz, K.: Phys. Rev. Lett. **75**, 3545 (1995)

Hyperfine Interact (2005) 165: 333–338
DOI 10.1007/s10751-006-9280-y

Deuterium isotope effects on iron core formation in ferritin

G. C. Papaefthymiou · A. J. Viescas · R. Horn · E. Carney · G. Zhao · N. D. Chasteen · J. Lee · S. M. Gorun

Published online: 9 November 2006

Abstract We present comparative Mössbauer investigations of nanosized FeOOH and FeOOD biomineral phases nucleated within the 7-nm diameter cavity of horse-spleen apoferritin in order to assess deuterium isotopic effects on nanoscale, bioinorganic lattice structures with extended hydrogen bond networks. Differences in magnetic anisotropy energy, packing density and degree of crystallinity in the resulting iron oxo-hydroxide nanophases obtained via D_2O (heavy water) vs. H_2O (light water) solution chemistry are noted. These observations point to the possibility of stabilizing new thermodynamic states in the solid-state by utilizing isotope effects, with important implications for new synthetic pathways to novel nano materials.

Key words Mössbauer · ferritin · deuteration · heavy water · superparamagnetism

1 Introduction

Deuterium isotope substitution is routinely used for structural determinations by magnetic resonance and neutron diffraction studies. Such substitution has minimal effect on molecular structure and thermodynamic stability due to similar O–H and O–D bond

R. Horn and E. Carney are REU Summer Students.

G. C. Papaefthymiou (✉) · A. J. Viescas · R. Horn · E. Carney
Department of Physics, Villanova University, Villanova, PA 19085, USA
e-mail: gcp@villanova.edu

G. Zhao · N. D. Chasteen
Department of Chemistry, University of New Hampshire, Durham, NH 03824, USA

J. Lee
Department of Chemistry, Brown University, Providence, RI 02912, USA

S. M. Gorun
Department of Chemistry and Environmental Science, New Jersey Institute of Technology, Newark, NJ 07102, USA

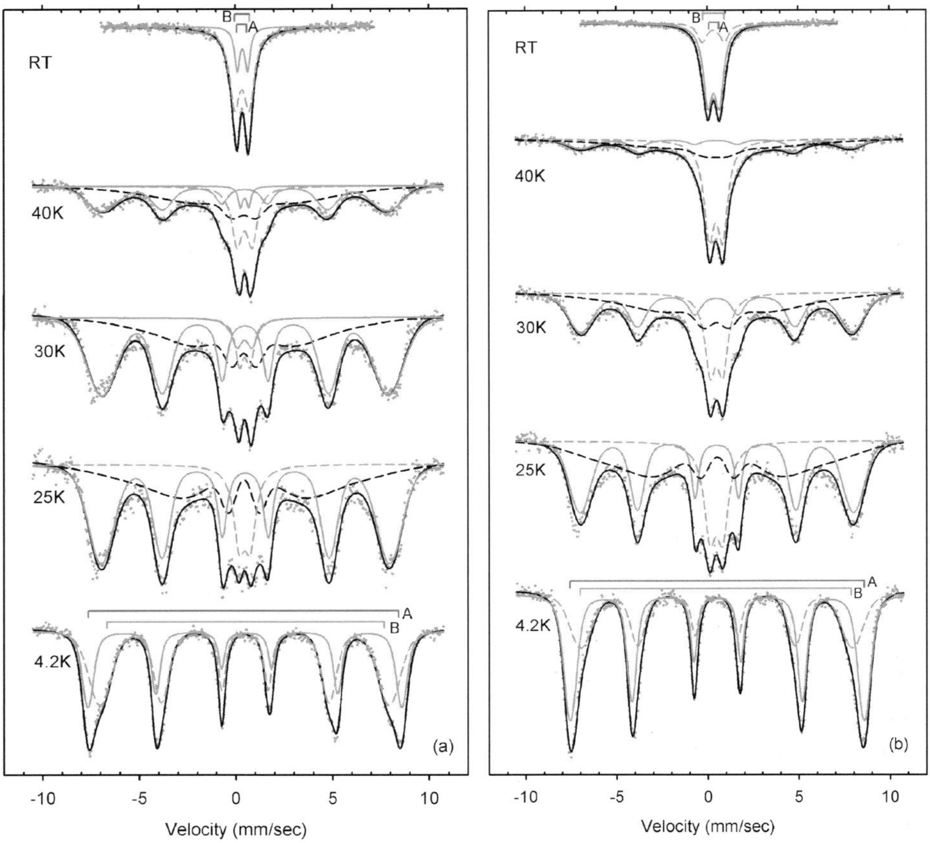

Figure 1 (a) Mössbauer spectra of H_2O reconstituted HoSF at various temperatures. (b) Mössbauer spectra of D_2O reconstituted HoSF at various temperatures. *Solid lines* are theoretical fits to the experimental data assuming superposition of poorly resolved fast, slow and intermediate relaxation components (see text, Table I).

energies, resulting in hydrogen bonded O···O distances that differ by less than 0.03 Å [1, 2]. Accumulated effects of such small structural differences in hydrogen bonded networks in the solid-state, however, may result in the stabilization of different structures [3]. Well-known effects of deuteration include the higher melting/boiling temperatures of D_2O, 3.82 and 101.42°C, respectively, compared to H_2O; the smaller degree of dissociation of D_2O compared to H_2O [4]; the formation of Ice-IV, a well-defined phase of D_2O, but absent in H_2O [5]; the altered Jahn–Teller distortions of copper Tutton salts [6] and; the shifts in ferro- and antiferro-electric transition temperatures of KH_2PO_4 upon deuteration [7].

Large, polynuclear transition metal aggregates with extensive oxo/hydroxo bridging groups, such as the ferritin iron core [8], afford experimental model systems with large hydrogen-bond networks for the investigation of isotopic effects on nanoscale lattice structures. Newly engineered, ferritin-inspired, bio-compatible nanosize systems are of great current interest for various biotechnological applications [9–11]. Herewith, we present comparative Mössbauer studies of horse spleen ferritin (HoSF) reconstituted in vitro from apoferritin and ^{57}Fe-enriched ferrous sulfate in H_2O and D_2O solutions. Ferritin reconstitution is achieved by iron oxidation followed by site-directed iron nucleation and

Table I Selective Mössbauer parameters derived from least square fits of the experimental data presented in Figure 1a and b

Sample	T (K)	δ^a(mm/s) (±0.02)	ΔE_Q (mm/s) (±0.03)	Γ (mm/s) (±0.02)	<H_{ave}> (kOe) (±5)	ΔH^b (kOe) (±1)	Area (%) (±0.02)	Sub-site identification
H_2O HoSF	300	0.37	0.54	0.32		–	0.22	A(core)
		0.36	0.75	0.66		–	0.78	B(surface)
	4.2	0.50	−0.10	0.39	502	29	0.37	A(core)
		0.46	−0.02	0.39	463	73	0.63	B(surface)
D_2O HoSF	300	0.36	0.59	0.57		–	0.49	A(core)
		0.35	0.91	0.78		–	0.51	B(surface)
	4.2	0.50	−0.07	0.41	499	38	0.60	A(core)
		0.46	−0.02	0.41	462	76	0.40	B(surface)

[a] Isomer shifts are relative to α-Fe at room temperature.

[b] FWHM of the magnetic field distribution.

hydrolytic polymerization resulting in the growth of a FeOOH/FeOOD mineral core within the 7-nm inner-diameter apoferritin shell [12].

2 Materials and methods

Deuterated (>99.99%) horse spleen ferritin was prepared by repeated ultra filtration of 15 μM apo-protein against deuterated 0.2 M MOPS buffer, pD=7.5, over a period of two days. The protein was diluted to 2 μM concentration in deuterated pD=7.5 buffer to a volume of 6 ml. To this solution was added 15 increments of 24 μl 0.049 M $^{57}FeSO_4$ in D_2O, pD=2, at intervals of at least 0.5 h over a period of two days to a total loading of 1,500 ^{57}Fe/protein. The MOPS buffer was then replaced by 99.99% D_2O using ultra filtration. In the final step the protein was lyophilized. The protein sample in H_2O was prepared similarly at pH=7.5 using normal buffers.

Mössbauer spectra were obtained using a conventional transmission spectrometer, with temperature variation $4.2<T<300$ K. The source was 50-mCi ^{57}Co in Rh maintained at room temperature. The spectrometer was calibrated with a 7 μm thick metal iron foil enriched in ^{57}Fe. Isomer shifts are referenced to α-Fe at RT. WMOSS (Web Research Co.) was used for spectral fitting to superpositions of iron sites, assuming Lorentzian lines and distributions in hyperfine magnetic fields.

3 Results and discussion

Mössbauer spectra for $4.2<T<300$ K are shown in Figure 1a and b. The intensity of the absorption varied from 4% effect at 4.2 K to 8% effect at 300 K, for either sample. Thus, some thickness broadening may be present, which, however, will not affect comparative studies of FeOOH and FeOOD presented here. At 4.2 K the spectra show the superposition of two slightly inequivalent magnetic subsites, indicated by components A (solid, gray line) and B (dashed, gray line). For $25<T<60$ K the spectra exhibit a superposition of slow (solid, gray line), intermediate (dashed, black line) and fast (dashed, gray line) spin-relaxation components. In fitting these spectra a distribution of hyperfine magnetic fields is

Figure 2 Magnetic hyperfine field distributions obtained from least square fits of the experimental data at 4.2 K; (a) H_2O reconstituted HoSF and (b) D_2O reconstituted HoSF, for subsites A and B.

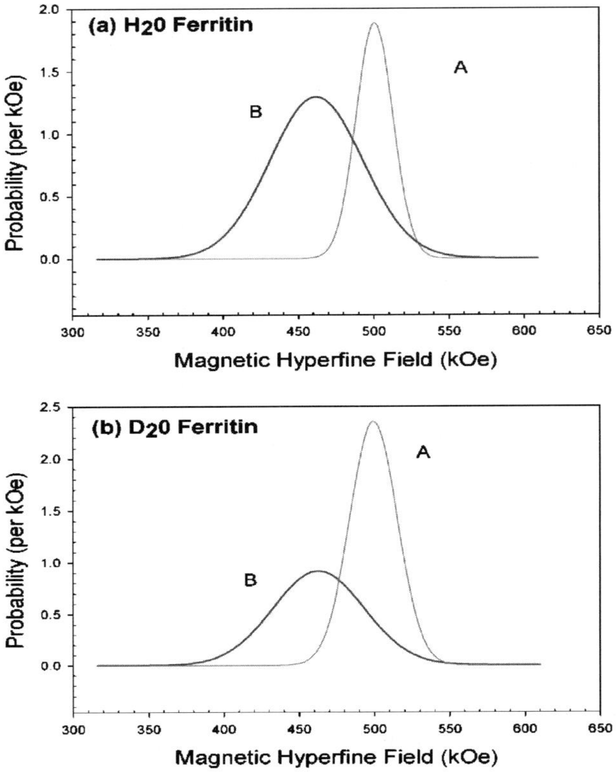

used to simulate relaxation effects. At 300 K fast-relaxing components A (solid, gray line) and B (dashed, gray line) produce a slightly asymmetric doublet. Mössbauer parameters are tabulated in Table I.

This is typical of superparamagnetism where competition between magnetic anisotropy, KV, and thermal, k_BT, energy determines the spin relaxation time, $\tau_s = \tau_0 \exp(KV/k_BT)$ [13] and thus, the blocking temperature, T_B, of the sample. Analysis of the Mössbauer data gave T_B=60 K for H_2O ferritin and T_B=40 K for D_2O ferritin [3]. The broad spectra, even at 4.2 K, necessitated fits to distributions of magnetic fields (Figure 2a and b; Table I). $\langle H_{ave} \rangle \sim 500$ kOe, for sub-spectrum A, and ~460 kOe for B. The main difference between the FeOOH/FeOOD spectra is in the breath of the absorption lines. FeOOD has broader line-width, Γ, and broader magnetic field distribution, ΔH, compared to FeOOH (Table I and Figure 2a, b). This and the higher T_B for H_2O ferritin indicate that FeOOH-cores possess higher crystallinity and denser iron packing than FeOOD.

We have associated subsites A and B with iron atoms within the interior vs. the surface of the FeOOH (FeOOD) biomineral core, respectively. This is according to a ferritin-core model proposed by Brooks et al. [14], which depicts a highly crystalline interior surrounded by an amorphous surface layer (Figure 3). Component B exhibits reduced $\langle H_{ave} \rangle$ due to the magnetic order discontinuity at the surface, and increased quadrupole splitting due to surface strain [15].

The 4.2 K data imply that for the H_2O-cores surface atoms contribute ~65% of the spectral absorption area, and ~40% for D_2O-cores. The surface/volume ratio can become

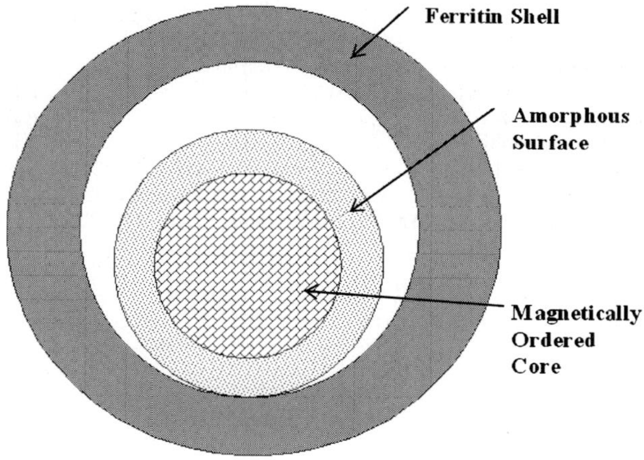

Figure 3 Schematic representation of the ferritin core formation within the protein cavity, identifying the core and surface phases discussed in the text.

very high for small particles, with significant implications for magnetic nano-lattices [16], owing to large contributions of surface and strain anisotropies [15]. The larger surface/volume ratio for the H_2O-derived cores points to a somewhat smaller core size compared to FeOOD. Thus, the higher T_B=60 K for FeOOH compared to T_B=40 K for FeOOD, cannot be attributed to particle-size, but rather, to structural differences. Our observations and the recent use of genetically engineered apoferritin templates for the growth of non-native metal cores, semiconducting nanoparticles as well as iron arsenate, vanadate and molybdate minerals [17] make ferritin nanotemplating a promising route in the search for new materials at the nanoscale.

Acknowledgments This research was supported in part by Brown University and Cambridge Isotopes Laboratories (S.M.G.), by NIH grant R37 GM20194 (N.D.C.) and NSF grant DMR-0074537 (G.C.P).

References

1. Olovsson, I., Jönsson, P.-G., In: Schuster, P., Zundel G., Sandorfy C. (eds.) The Hydrogen Bond. North-Holland, Amsterdam (1976)
2. Gallagher, K.J., In: Hazi, D. (eds.) Hydrogen Bonding. Pergamon, New York (1959)
3. Lee, J., Chasteen, N.D., Zhao, G., Papaefthymiou G.C., Gorun, S.M.: J. Am. Chem. Soc. **124**, 3042 (2002)
4. Salomaa, P.: Acta Chem. Scand. **25**, 367 (1971)
5. Whalley, E., In: Schuster, P., Zundel G., Sandorfy, C. (eds.) The Hydrogen Bond. North-Holland, Amsterdam (1976)
6. Henning, R.W., Schultz, A.J., Hitchman, M.A., Kelly, G., Astley, T.: Inorg. Chem. **39**, 765 (2000)
7. Ichikawa, M.: Ferroelectrics **168**, 177 (1995)
8. Chasteen, N.D., Harrison, P.M.: J. Struct. Biol. **126**, 182 (1999) (and references there in)
9. Häfeli, U. et al. (ed.): Scientific and Clinical Applications of Magnetic Carriers. Plenum, New York (1997)
10. Perez, J.M., Simeone, F.J., Tsourkas, A., Josephson, L., Weissleder, R.: Nano Lett. **4**, 119 (2004)
11. Perez, J.M., O'Loughin, T., Simeone, F.J., Weissleder R., Josephson, L.: J. Am. Chem. Soc. **124**, 2856 (2002)

12. Zhao, G., Bou-Abdallah, F., Arosio, P., Levi, S., Janus-Chandler C., Chasteen, N.D.: Biochemistry **42**, 3142 (2003) (and references there in)
13. Brown W.F. Jr.: Phys. Rev. **130**, 1677 (1963)
14. Brooks, R.A., Vymalaz, J., Goldfarb, R.B., Butle J.W.M., Aisen, P.: Magn. Reson. Med. **40**, 227 (1998)
15. Papaefthymiou, G.C.: Strain anisotropies in core/shell magnetic nanostructures. In: Stanick. S.J., Searson, P., Lyons L.A., Keating C.D. (eds.) Anisotropic Nanoparticles-synthesis, Characterization and Applications. Mater. Res. Soc. Symp. Proc. **635**, C2.4 (2000)
16. Dorman J., Fiorani D. (eds.): Magnetic Properties of Fine Particles. North-Holland, Amsterdam (1991)
17. Polanams, J., Ray A.D., Watt, R.K.: Inorg. Chem. **44**, 3203 (2005) (and references there in)

Hyperfine Interact (2005) 165: 339–344
DOI 10.1007/s10751-006-9281-x

A ^{57}Fe Mössbauer characterization of Fe-biopolymer complexes and their relevance to biological molecules

Subhash C. Bhatia · Beatriz H. Cardelino · Natarajan Ravi

Published online: 26 October 2006

Abstract ^{57}Fe Mössbauer spectroscopy is used to study the interactions, geometry, and the coordination characteristics of the Fe-complexes of biopolymers such as chitosan, glucosamine, and chondritin sulfate. In addition, a computational effort is undertaken for predicting the geometries and energies of the metal complexes by the Density Functional Theory (DFT) methods as implemented in the Gaussian 2003 quantum mechanical program. Both experimental and computational results suggest that the structure of the metal complexes resemble closely the structure of the active sites of metalloenzymes in 2+ or 3+ oxidation states and is at least tetracoordinated and can possibly have six ligands.

Key words biopolymers · metal complexes · Mössbauer spectroscopy

1 Introduction

There are several useful biopolymers such as cellulose, chitin, chitosan, and chondritin sulfate. Cellulose is a polymer of glucose with $\beta(1\rightarrow4)$ linkage resulting in the minimum energy conformation. It is the ideal conformation for making fibers, because parallel polysaccharide chains can readily be linked by *H*-bonds (Figure 1). Chitin, the second most abundant polysaccharide after cellulose, is the fibrous component of the exoskeleton of arthropods: crabs, spiders, and insects. This polymer, which resembles cellulose structurally and chemically, offers diverse applications due to the presence of an acetylamide group at one of the carbons. Chitosan, the *N*-deacetylated chitin, is similar to chitin but easier to

S. C. Bhatia
Department of Chemistry, Morehouse College, Atlanta, GA 30314, USA

B. H. Cardelino
Department of Chemistry, Spelman College, Atlanta, GA 30314, USA

N. Ravi (✉)
Physics Department, Spelman College, Atlanta, GA 30314, USA
e-mail: nravi@spelman.edu

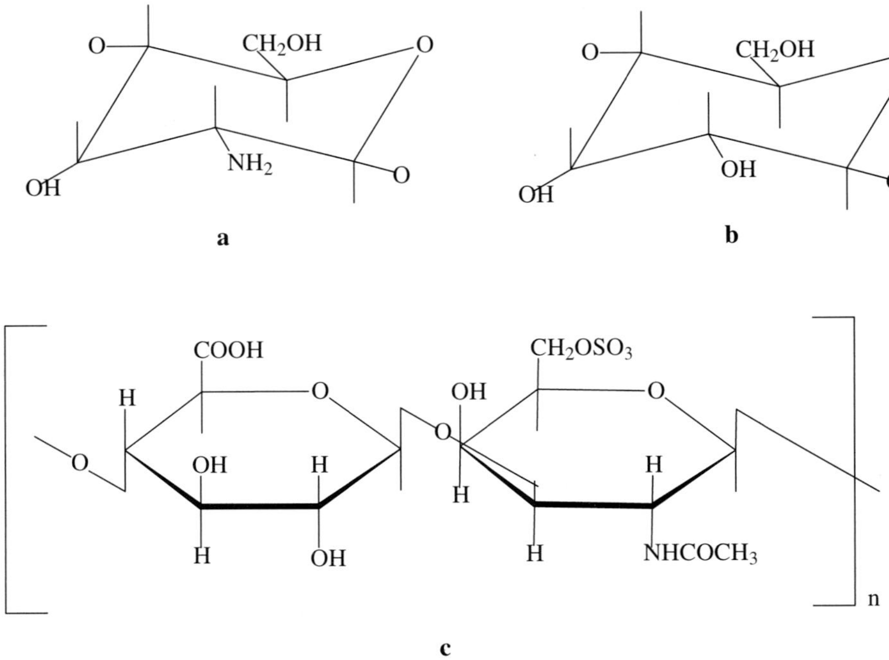

Figure 1 Chemical structures of (a) chitosan, (b) cellulose, and (c) chondritin sulfate.

work with due to its increased solubility in different solvents [1]. Chondritin sulfate, on the other hand, is an alternating copolymer of β(1→4)-D glucuronic acid and β(1→3)-*N* acetyl-D-galactosamine which can be sulfated at C_4 and C_6 positions. The presence of –COOH and SO_4^{2-} groups makes this polymer highly soluble in water. Chondritin is involved in forming network with collagen in the connective tissues and allows the transfer of globular proteins, stimulates the growth and differentiation of neurons and has also been implicated in spinal cord injury [2, 3].

The polar nature of the carbohydrates of these biopolymers, gives them the ability to chelate with metal ions for the environmental remediation. Such metal complexes structurally resemble the active sites of metalloenzymes due to the presence of common ligands [4]. A theoretical approach for predicting the geometries and energies of the metal complexes is to adopt the Density Functional Theory (DFT) methods as implemented in the Gaussian 2003 quantum mechanical program [5, 6]. We report in this manuscript the ^{57}Fe Mössbauer results and preliminary computational results of metal complexes of these biopolymers.

2 Experimental

Preparation of the Fe-complexes with the polymer, water-soluble polymer, and the monomeric forms of chitin and chitosan were undertaken in acetic acid medium for the polymers and in aqueous medium for the monomer and chondritin sulfate. The chemicals glucosamine, *N*-acetyl glucosamine, gluconic acid, sulfonated *N*-acetyl glucosamine, chitin, chitosan, and chondritin sulfate were purchased from WAKO Bioproducts. In one set of

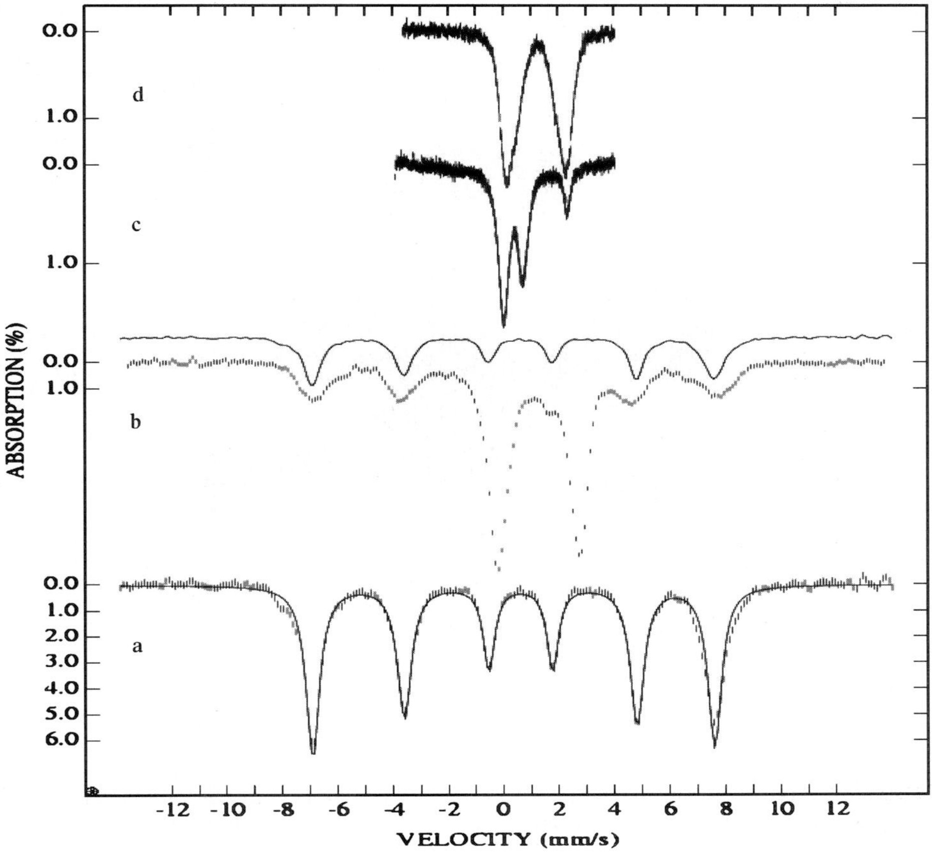

Figure 2 Mössbauer spectra of crystalline powder of (a) Fe-chitosan sample measured at 4.2 K. (b) Fe-chitosan (water soluble) at 4.2 K. The *solid line* plotted above the spectrum *B* is the experimental data of Fe-chitosan polymer complex scaled to 30% of absorption. (c) Fe-glucosamine (monomer) at pH 9.2 measured at 293 K, and (d) Fe-chondritin sulfate complex at pH 4.0 at 293 K.

experiments, a known amount of a given biopolymer was dissolved in acetic acid. The pH of the solution was adjusted to the desired level and an appropriate amount of chloride salt of metal ion was added to the solution. The mixture was stirred for approximately 2 h, filtered, and was dried at 50°C for approximately 12 h. If no precipitate was formed at room temperature then the water was removed by warming the sample at approximately 50°C until the solid was formed. A Mössbauer sample was prepared in the solid state corresponding to a thickness of ~2 mg/cm^2 of ^{57}Fe in a shallow derlin sample cup with an inner cup sealed and frozen with liquid nitrogen. The Mössbauer spectrometer (Ranger model MS 1200, Ranger drive VA900) was used operating in a constant acceleration mode in a transmission geometry. For low temperature measurements, a Cryo Industries 8CN variable-temperature cryostat model was used with liquid He and the temperatures were measured by the calibrated silicon diode, sensitive to different temperature regions. The zero velocity of the Mössbauer spectra referred to the centroid of the room-temperature spectrum of a metallic Fe foil. The spectra were analyzed using the software WMOSS (WEB Research Co., Edina, MN).

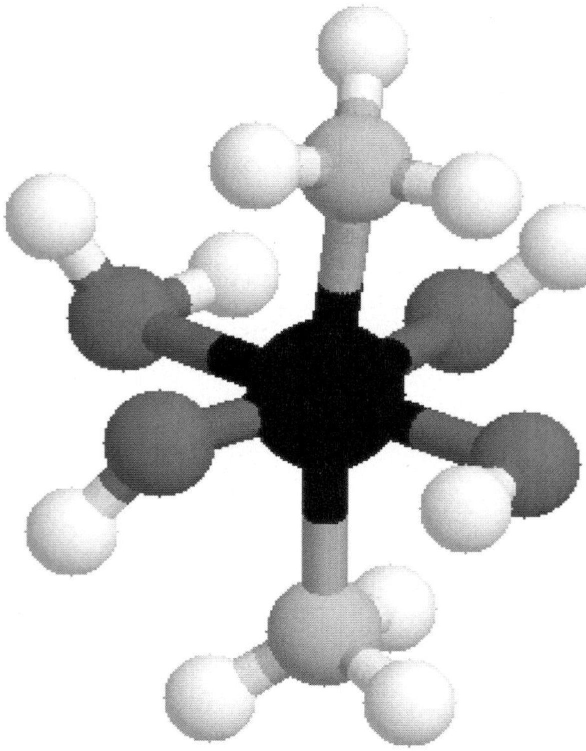

Scheme 1 $Fe(III)(NH_3)_2(H_2O)(OH)_3$. The *horizontal plane* contains the water molecule and the three-hydroxide groups. The two ammonia molecules are *above* and *below* the horizontal plane.

3 Results and discussion

Mössbauer data analysis of Fe complexes of chitin, and chitosan, and chondritin sulfate clearly indicates that the metal ion in the 3+ state is stabilized in the case of polymer, while the 2+ state is partially stabilized in the case of water-soluble chitosan and complete stabilization is achieved with D-glucosamine which is the monomer of chitosan. The pertinent information about the oxidation state of the metal ion was obtained by using the isomer-shift deduced from the Mössbauer spectra (see Figure 2). A working structural model for the Fe-chitosan complex has been proposed based on the results obtained from these investigations. While the data suggested that there is at least five coordination around Fe in the +3 state, a four coordinated Fe^{2+} state cannot be completely ruled out by a comparison with Mössbauer data of some metalloenzymes [7, 8]. Involvement of the amide nitrogen in the bonding with the metal ion is also indicated. A plausible explanation for the stabilization of the respective oxidation states is offered using the simple carbohydrate chemistry where the presence of β(1–4) glycosidic linkage has been invoked as a possible mechanism. In addition, the effect of pH of these metal complexes (Fe:D-glucosamine and Fe:chondritin sulfate) was studied, and the isomer-shift obtained from the Mössbauer data unambiguously revealed the stabilization of the iron ion in 2+ state at low pH values and 3+ state at higher pH. Mössbauer spectra of these complexes are magnetic for the polymer chitosan and quadrupole doublets for the monomer D-glucosamine and chondritin sulfate and the isomer-shift is measured to be around 0.45(04) and 1.20(04) mm/s for 3+ and 2+ states at 293 K, respectively [9, 10].

4 Computational studies

In the calculations, Becke's three-parameter functionals with Lee-Yang-Parr's correlation correction were mainly used, together with the largest suitable basis sets available in the Gaussian program, such as the 6-311G(3df, 3pd) for atoms with less than 36 electrons, and augmented 3–21 g(d,p) basis sets for larger atoms [11, 12]. We have obtained preliminary results on two model compounds, consisting of octahedral complexes of Fe(III) and Fe(II). For the Fe(III) complex, we selected as ligands two ammonia groups, one water molecule and three hydroxide ions, resulting in a neutral complex with one unpaired electron. The selection of ligands was based on the expected complexation of iron with ammonia and hydroxide groups of chitosan. For the Fe(II) complex, we chose two ammonia groups, two water molecules and two hydroxide ions as ligands resulting in a neutral complex with no unpaired electrons. Scheme 1 displays the $Fe(III)(NH_3)_2(H_2O)(OH)_3$ complex.

The results of these calculations suggest that there is an increase in bond length of 1, 3, and 7% corresponding to Fe-N, Fe-O(H), and Fe-O(H_2), respectively, in going from Fe(III) to Fe(II). In order to compare the relative stability of the two model compounds, we calculated the energy for the following reaction:

$$\mathrm{Fe(III)(NH_2)_2(H_2O)(OH)_3 + H_2O \rightarrow Fe(II)(NH_3)_2(H_2O)_2(OH)_2 + OH}$$

In this reaction, the radical is transferred from the Fe(III) compound to the OH molecule. The energy for the reaction was calculated to be 280 kJ/mol. A crude population analysis of the Fe(III) compound showed that the open molecular orbital in $Fe(III)(NH_2)_2(H_2O)(OH)_3$ was one of the t_{2g} orbitals, and that the separation between the t_{2g} and the e_g sets were about 0.03 eV. These results seem to be consistent with the experimental observations and other calculations involving other coordination schemes are currently underway.

The experimental and computational results suggest that the structure of the metal complexes resemble closely the structure of the active sites of metalloenzymes in either oxidation state and is at least tetracoordinated and can possibly have six ligands.

Acknowledgements Natarajan Ravi acknowledges the support by a Research Infrastructure in Minority Institutions Award from National Center on Minority Health and Health Disparities (Grant no. MD 00215).

References

1. Brine, C.J.: In: Zikakis, J.P. (ed.) Chitin, Chitosan and Related Enzymes. Academic, New York (1984)
2. Tully, S.E., Mabon, R., Gama, G.I., Tsai, S., Liu, X., Hseieh-Eilson, L.C.: J. Am. Chem. Soc. **126**, 7736 (2004)
3. Bradbury, E.J., Moon, L.S.F., Popat, R.J., King, V.R., Bennett, G.S., Patel, P.N., Fawcett, J.W., McMohan, S.B.: Nature **416**, 636 (2002)
4. Holm, R.H., Solomon, E. (guest eds.): Bioinorganic enzymology. Chem. Rev. 96 (1996)
5. Gaussian 03, Frisch, M.J., et.al.: Gaussian Inc., (2003)
6. Hohenberg, P., Kohn, W.: Phys. Rev. **136**, B864 (1964)

7. Lynch, J.B., Juarez-Garcia, C., Münck, E., Que, L.: J. Biol. Chem. **264**, 8091 (1989)
8. Nordlund, P., Eklund, H.: Curr. Opin. Struc. Biol. **5**, 758 (1955)
9. Bhatia, S.C., Ravi, N.: Biomacromolecules **1**, 413 (2000)
10. Bhatia, S.C., Ravi, N.: Biomacromolecules **4**, 723 (2003)
11. Becke, A.D.: Phys. Rev. A **38**(6), 3098 (1988)
12. Lee, C., Yang, W., Parr, R.G.: Phys. Rev. B **37**, 785 (1988)

Hyperfine Interact (2005) 166:345–349
DOI 10.1007/s10751-006-9291-8

Search for the stannous ion in a chloride/fluoride matrix: cases of $Ba_{1-x}Sn_xCl_{1+y}F_{1-y}$ and of Ba_2SnCl_6

Georges Dénès · Abdualhafeed Muntasar · Sihem Boufas · Hocine Merazig

Published online: 16 November 2006

Abstract With the exception of anhydrous $SnCl_2$, in divalent tin fluorides and chlorides, tin(II) is always covalent bonded, i.e. its valence orbitals are hybridized and the tin lone pair is located in one of the hybrid orbitals. This lone pair is highly stereoactive and generates a large efg, resulting in a large quadrupole splitting. A doubly disordered $Ba_{1-x}Sn_xCl_{1+y}F_{1-y}$ solid solution has been prepared and found to contain either ionic tin(II) (Sn^{2+} ions) or a mixture of ionic and covalent tin(II), depending on x, y and the method of preparation. The ionic tin(II) spectrum in $Ba_{1-x}Sn_xCl_{1+y}F_{1-y}$ gives a Mössbauer single line that is broadened by the lattice efg, like in $SnCl_2$. Now, Sn^{2+} has been found to be present in an undistorted octahedral coordination in a newly isolated compound, Ba_2SnCl_6. It should be the first example of Sn^{2+} that is fully ionic and has a perfectly spherical lone pair.

Key words barium tin chloride fluoride · stannous ion · Mössbauer spectroscopy · solid solution · lone pair

1 Introduction

Tin belongs to group 14, and therefore it has four valence electrons (electronic structure *[Kr]* $5s^2 5p^2$) and in the +2 suboxidation state, it has a lone pair (non-bonded electron pair) that can be in either of the two following situations: (a) in the Sn^{2+} stannous ion, the tin valence orbitals are not hybridized and the lone pair is

G. Dénès (✉) · A. Muntasar
Laboratory of Solid State Chemistry and Mössbauer Spectroscopy, Department of Chemistry and Biochemistry, Concordia University, Montréal, Québec, Canada
e-mail: gdenes@alcor.concordia.ca

S. Boufas · H. Merazig
Laboratoire de Chimie Moléculaire, du Contrôle de l'Environnement et de Mesures Physico–Chimiques, Département de Chimie, Faculté des Sciences, Université Mentouri, Constantine, Algeria

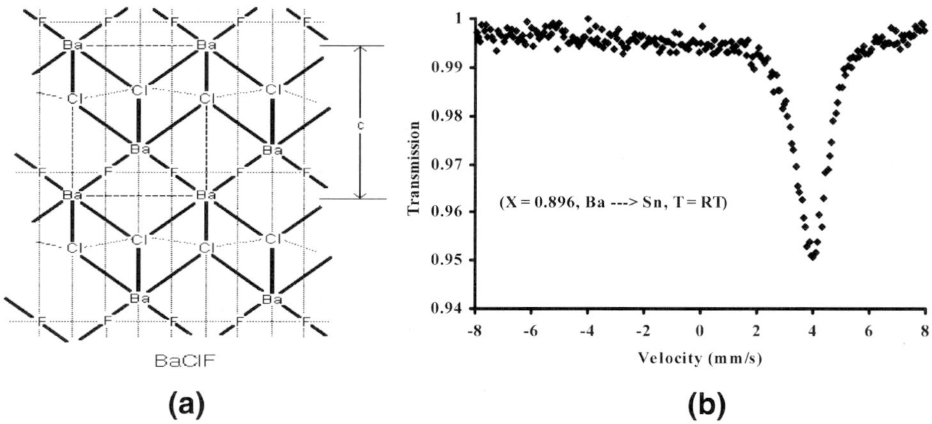

Figure 1 **a** Crystal structure of BaClF, **b** ^{119}Sn Mössbauer spectrum of precipitated $Ba_{1-x}Sn_xCl_{1+y}F_{1-y}$.

located on the spherical native $5s^2$ orbital and is said to be *non-stereoactive* since it does not distort the tin coordination, (b) when tin(II) is covalently bonded, its valence orbitals are hybridized, the lone pair is located on one of the hybrid orbitals and it is said to be *stereoactive* since hybrid orbitals are highly axial and the lone pair reduces the coordination number and strongly distorts the polyhedron of coordination [1]. In fluorides and chlorides, tin(II) is always found covalently bonded, with the exception of anhydrous tin(II) chloride [2].

The structure of BaClF belongs to the PbClF type. It is a tetragonal distortion of the fluorite type (CaF_2 type), whereby each second layer of fluoride ions is replaced by chloride ions to give alternating layers of fluoride ions and chloride ions parallel to the *(a,b)* plane of the unit–cell [3]. Their large size forces the chloride ions to move up or down parallel to the *c* axis, while a shift parallel to *(a,b)* brings them in axial position below or above barium, to give a total shift of (1/4, 1/4, ±z) compared to the fluoride ions they replace. It results in that the cubic coordination of Ba in fluorite type BaF_2 changes to a $BaF_4ClCl'_4$ distorted monocapped square antiprism in BaClF (Figure 1a).

In the present work, a doubly disordered $Ba_{1-x}Sn_xCl_{1+y}F_{1-y}$ solid solution was prepared ($0 \leq x \leq 0.25$, $-0.15 \leq y \leq 0.25$) and the tin electron configuration was studied by ^{119}Sn Mössbauer spectroscopy. In addition, the search for purely ionic tin(II) has resulted in the preparation of Ba_2SnCl_6, which contains Sn^{2+} in a regular octahedron.

2 Experimental

The preparation of the $Ba_{1-x}Sn_xCl_{1+y}F_{1-y}$ was carried out by both the dry method and the wet method as described elsewhere [4]. Characterization and Mössbauer studies were performed using equipment and methods that have already been reported [4].

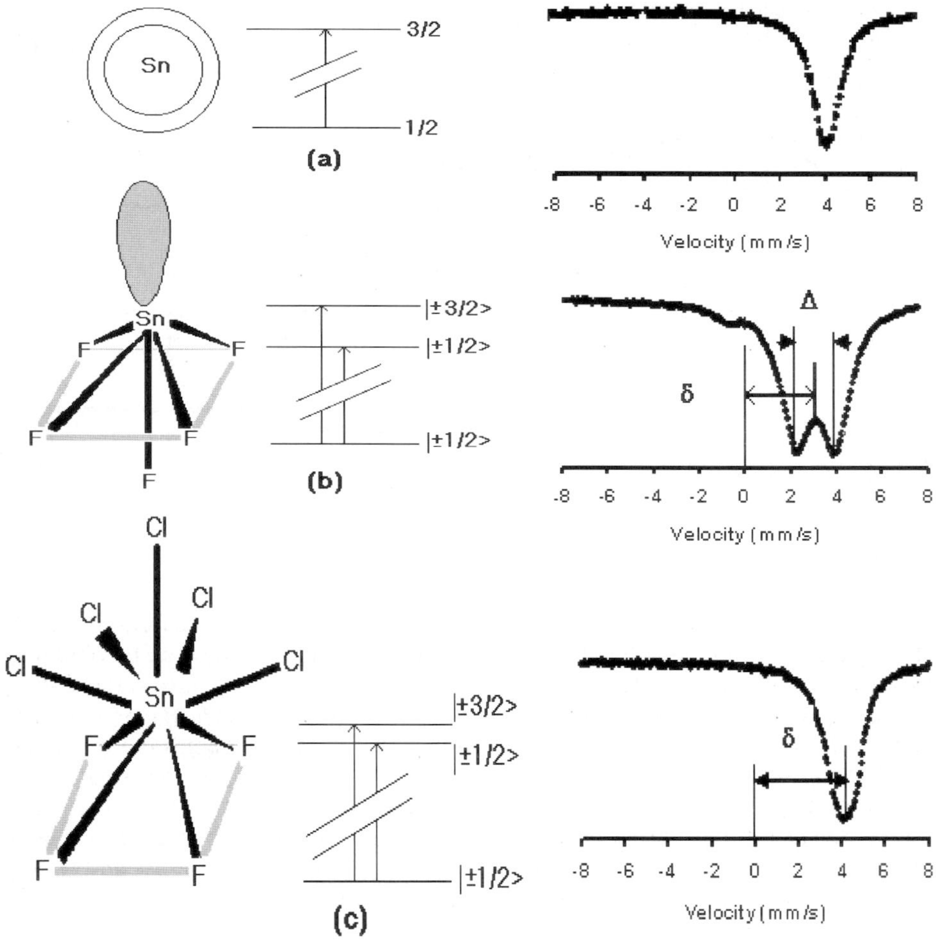

Figure 2 Electrical monopolar and electrical quadrupolar interactions generated by different kinds of divalent tin, and their effect on the Mössbauer spectrum: **a** unhybridized lone pair (e.g. semiconducting $CsSnBr_3$), **b** hybridized lone pair (e.g. $BaSnF_4$) [5], **c** unhybridized lone pair in the case of $Ba_{1-x}Sn_xCl_{1+y}F_{1-y}$ discovered in the present work.

3 Results and discussions

When aqueous solutions of SnF_2 and $BaCl_2.2H_2O$ are mixed, precipitation of a white solid takes place, which was found by chemical analysis to be a $Ba_{1-x}Sn_xCl_{1+y}F_{1-y}$ (x=0.15, y=0.11) solid solution at high X barium molar ratio ($X \geq 0.87$). Its X-ray diffraction pattern gives the same set of lines as that of BaClF. However, a close examination of their diffraction patterns shows some small but significant line shift, consistent with a 0.91% decrease of the a parameter and a 0.97% increase of c. The change of unit–cell parameters is the result of the non-stoichiometry. Preparations carried out by the dry method also yielded $Ba_{1-x}Sn_xCl_{1+y}F_{1-y}$ ($0 \leq x \leq 0.25$, $-0.15 \leq y \leq 0.25$). The difference of unit–cell parameters of the solid solution prepared by the two methods shows that something

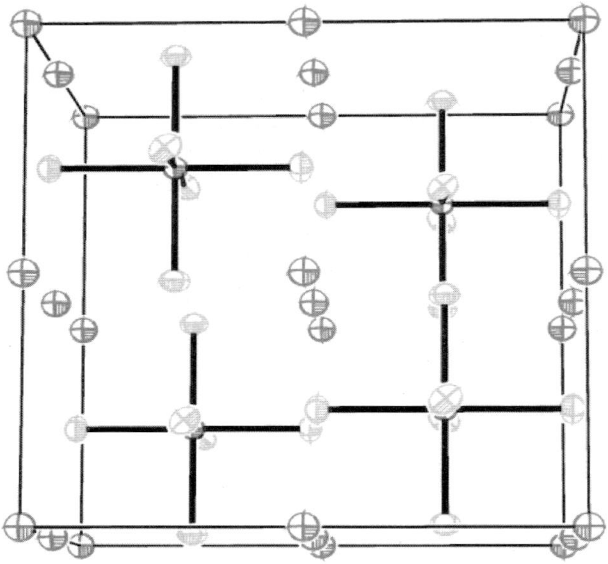

Figure 3 Octahedral coordination of the Sn^{2+} stannous ions in Ba_2SnCl_6.

differentiates them from one another, although both are doubly disordered BaClF structures. For these disordered phases, only average Ba/Sn values can be obtained by diffraction alone. Could the difference be due to the way tin binds to the BaClF network? A selective probing of the tin sites, such as by Mössbauer spectroscopy could help. Figure 2 shows the expected Mössbauer spectrum versus the modes of bonding of divalent tin. Figure 2a describes the case of ionic bonding in an undistorted coordination, the *5s* orbital is perfectly spherical and there is no electric field gradient (efg) at the Sn^{2+} site. It results no splitting of the nuclear levels, and therefore only one nuclear transition (1/2→3/2), and a single Mössbauer absorption line is observed. This contrasts with the case of covalent bonding, where the tin lone pair is located on one of the hybrid orbitals (*stereoactive* lone pair). This lone pair is highly axial and therefore generates a large efg, $(V_{zz})_{val}$. In addition, the stereoactive lone pair distorts very strongly the tin polyhedron of coordination, which generates another term of the efg, $(V_{zz})_{latt}$. The overall effect is the splitting of the first excited state which results in two nuclear transitions, and the Mössbauer spectrum is a doublet, with a large *quadrupole splitting* Δ (Figure 2b) [5]. In addition, the higher the valence *s* electron density acting at tin, the higher the isomer shift δ. It results that the isomer shift of Sn^{2+}, 4.0–4.1 mm/s (*5s* electron density concentrated around the tin atom), is higher than that for $Sn(II)_{cov}$, 2.5–3.5 mm/s (*5s* electron density pulled away in hybrid orbitals). The ambient temperature spectrum of precipitated $Ba_{1-x}Sn_xCl_{1+y}F_{1-y}$ is a single line at 4.07 mm/s (Figures 1b and 2c) which shows that tin is present in the form of Sn^{2+} stannous ions. For the $Ba_{1-x}Sn_xCl_{1+y}F_{1-y}$ solid solution prepared in dry conditions, the Mössbauer spectrum is a Sn^{2+} single line, pure or mixed with a quadrupole doublet, depending on the *x* and *y* composition parameters, and it will be published elsewhere. The theoretical spectrum for purely ionic tin(II), i.e. the stannous Sn^{2+} ion, is a single line at high velocity (≥4.1 mm/s) shown on Figure 2a. However, $SnCl_2$ has an orthorhombic structure, and therefore the tin site is distorted, which results in a small efg due to possible polarization of the lone pair and site distortion, therefore one can expect a moderate broadening of the line. The same applies to ionic tin in $Ba_{1-x}Sn_xCl_{1+y}F_{1-y}$, where the presence of a mixture of fluorine and chlorine around tin also contributes to the efg (Figure 2c). Therefore, no strictly purely ionic tin(II) had been obtained before in a chloride or in a fluoride. Now, this has been done: crystals of Ba_2SnCl_6 were obtained and its crystal structure has been solved. The unit–cell is cubic, and tin is located in an undistorted octahedral site (Figure 3). This is the first case of a perfectly spherical, i.e. completely non-stereoactive, lone pair, in a tin(II) chloride. Our efforts are now focused on the preparation of a sufficient quantity of Ba_2SnCl_6 in order to collect its Mössbauer spectrum. It should have the narrowest possible line for Sn^{2+}, and give, for the first time, the true isomer shift of purely ionic tin in a chloride matrix. The full crystal structure of Ba_2SnCl_6 will be published elsewhere.

Acknowledgements This work was made possible by the support of Concordia University, the Natural Science and Engineering Research Council of Canada, and the Université Mentouri. Grateful thanks are also due to the Procter and Gamble Co. (Mason, Ohio) for supporting our Mössbauer laboratory.

References

1. Dénès, G.: In: Wynter, C.I., Alp, E.E. (eds.) Proceedings of the Second Nassau Mössbauer Conference, p. 109. W. C. Brown, Dubuque, IA (1994)
2. Donaldson, J.D., Silver, J.: J. Chem. Soc. Dalton Trans. 666 (1973)
3. Flahaut, J.: J. Solid State Chem. **9**, 124 (1974)
4. Dénès, G., Muntasar, A.: Hyperfine Interact. **153**, 91 (2004)
5. Dénès, G., Yu, Y.H., Tyliszczak, T., Hitchcock, A.P.: J. Solid State Chem. **91**, 1 (1991)

Hyperfine Interact (2005) 166:351–355
DOI 10.1007/s10751-006-9292-7

Exploring metal atom dynamics in organometallics: methodology and results for C_{60} ferrocenes

Rolfe H. Herber · Israel Nowik

Published online: 14 November 2006

Abstract Two new bucky-ferrocene compounds, incorporating two η^5-CpFe ligands in the C_{60} framework, have been examined by temperature-dependent ^{57}Fe Mössbauer spectroscopy over the temperature range $86<T<350$ K. The root-mean-square-amplitude-of-vibration of the iron atom have been determined and compared to those observed for related mono-iron bucky ferrocenes, as well as for the parent ferrocene complex.

Key words metal atom dynamics · Mössbauer temperature · vibrational amplitude · ferrocene derivatives

Temperature-dependent ^{57}Fe Mössbauer effect (ME) measurements have recently been reported from these laboratories [1] for a number of C_{60} compounds incorporating an η^5-CpFe fragment (bucky-ferrocenes) in their structure. In that study, which also included the determination of the usual isomer shift (I.S.) and quadrupole splitting (Q.S.) hyperfine parameters, the root-mean-square-amplitudes-of-vibration (rmsav) of the iron atom were also reported. It was noted that these amplitudes were largest for the parent ferrocene and smallest when the C_{60} ligand was substituted with five phenyl (Φ) groups.

This study has now been extended to two compounds in which each C_{60} group carries two ferrocenoid ligands, located at 180° with respect to each other and the center of the buckyball sphere. The objective here was to elucidate the effect (if any) of the introduction of the second η^5-CpFe group on the dynamics of the iron atom ligated to the C_{60} framework.

The two new compounds, $CpFe(CH_3)_5C_{60}(CH_3)_5FeCp$, **1**, and $CpFe(CH_3)_5C_{60}$-(Φ_5)FeCp, **2**, were generously made available [2] as air-stable powders, and transferred to plastic sample holders before introduction into the variable temperature

R. H. Herber (✉) · I. Nowik
Racah Institute of Physics, The Hebrew University of Jerusalem, 91904 Jerusalem, Israel
e-mail: herber@vms.huji.ac.il

Table I Summary of Mössbauer parameters for Compounds **1** and **2**

	1		**2**		Units
I.S.(90)	0.572(4)		0.858(4)		mm s^{-1}
Q.S.(90)	0.246(4)		2.473(4)		mm s^{-1}
−d IS/d T	4.63(2)	110–350 K	4.68(4)	130–350 K	$\times10^{-4}$ mm s^{-1} K^{-1}
−d ln A/d T	5.45(12)	86–350 K	6.03(6)	90–350 K	$\times10^{-3}$ K^{-1}
M_{eff}	83.5(3)		89(1)		Da
Θ_M	131(2)		121(4)		K

The parenthetical values are the experimental errors in the last significant figure(s).

cryostat, as described earlier [3, 4]. ME measurements were carried out in standard transmission geometry, making use of a 50 mCi ^{57}Co(Rh) source and a fast proportional counter. Temperature control was effected by a high-gain feedback circuit and monitored with a digital multiparameter program (Glaberson and Brettschneider, personal communication; available at http://www.Phys.HUJI.AC.IL/~glabersn) which confirmed a temperature stability better than ±0.2 K over the time interval of the ME measurement at a given temperature (normally 8–24 h per point). Spectrometer calibration was effected by use of a 10.25 mg m/cm^2 α-Fe absorber at room temperature. The isomer shifts are reported with respect to the centroid of such calibration spectra. The observe line widths (fwhm) at 300 K were 0.228(3) and 0.243(4) mm s^{-1} for **1** and **2**, respectively.

As is true of all other neutral (diamagnetic) ferrocene-related compounds so far examined, the ME spectra of **1** and **2** consist of a single QS doublet with two well-resolved components. The hyperfine parameters at 90 K are summarized in Table I, together with the temperature dependent derived parameters to be discussed below. For **1**, both the IS and QS parameters at 90 K are identical to those reported earlier [1] for $CpFe(CH_3)_5C_{60}$, and suggest an identical bonding environment for the Fe atom in both cases. The temperature dependence of the IS parameter in the high temperature regime (170<T<350 K) is well fitted by a linear regression with −d IS/d T=(4.98±0.20)$\times10^{-4}$ mm s^{-1} K^{-1}, (cc: 0.998 for 13 data points). From this dependence, M_{eff} is calculated [5] to be 84±1 Da. The QS parameter is strikingly temperature independent over the range 110<T<350 K with an average value of 2.460± 0.002 mm s^{-1}, in contrast to the negative slope commonly observed due to thermal expansion effects in related covalent solids [6, 7]. The area ratio, defined as $R=A_\pi/A_\sigma$ (where π and σ refer to the ±3/2→±1/2 and ±1/2→±1/2 transitions, respectively) shows a significant temperature dependence, reflecting the anisotropy of the metal atom motion with respect to the local symmetry axis passing through the center of the Cp ring and the metal atom. This anisotropy has been previously reported [1] for other bucky ferrocenes and appears to be a general phenomenon associated with the local environment around the metal atom.

The temperature dependence of ln A (where A is the area under the resonance curve) is well fitted by a linear regression over the total temperature range (cc: 0.996 for 20 data points) and is summarized graphically in Figure 1 (filled data points) together with the corresponding data for the mononuclear complex (open data points). For an optically “thin” absorber, ln A scales directly with the recoil-free fraction, $f=\exp(-k^2<x_{ave}^2>)$, where k is the wave number of the 14.4 keV γ-ray of ^{57}Fe and $<x_{ave}^2>$ is the average value of the mean-square-amplitude-of-vibration of

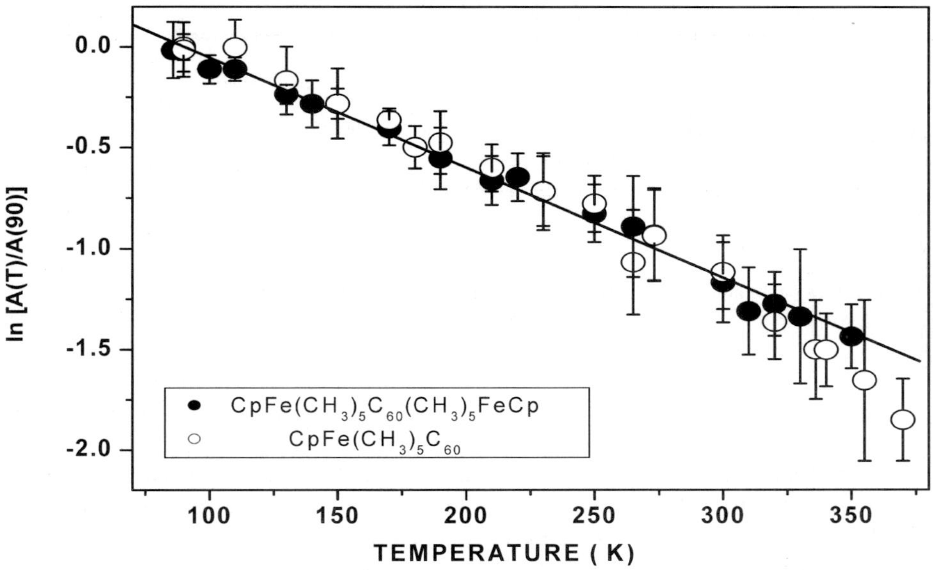

Figure 1 The temperature dependence of ln A (*filled data points*) together with the corresponding data for the mononuclear complex (*open data points*).

the metal atom at the given temperature. From these data it is possible to extract the rmsav of the iron atom at any temperature over the indicated range. An underlying assumption in this analysis is that when d ln A/d T evidences a linear relationship and ln A passes through zero at T=0°, there is good agreement between the $U_{i,j}$ values extracted from single crystal X-ray diffraction data, and the $k^2\langle x_{ave}^2\rangle$ values derived from Mössbauer effect measurements. This assumption has been well validated for a number of iron-organometallic compounds, including pure and intercalation solids [8]. Finally, in this context it should be noted that the Mössbauer lattice temperature, Θ_M, calculated [5] from the temperature dependencies of IS and A, is found to be ~131 K, so that the "high temperature approximation" that $T>\Theta_M/2$ is well justified in the present case.

The ME parameters for **2** are included in Table I and again it is noted that both the IS and QS parameters at 90 K are very similar to those of $CpFe(\Phi)_5C_{60}$ reported earlier [1]. The temperature-dependence of the IS parameter is well fitted by a linear regression and has a value of $-(4.69\pm0.17)\times10^{-4}$ mm s^{-1} K^{-1} (cc: 0.998 for 12 data points) over the range 130<T<350 K. The M_{eff} value calculated from this slope is 89 Da. As in the case of **1** the QS parameter is essentially temperature independent and has a mean value of 2.466±0.005 mm s^{-1} over the above temperature range. The temperature dependence of $k^2\langle x_{ave}^2\rangle$ is linear from 90 to 350 K and as before yields Θ_M=121 K.

The rmsav of the iron atom in **1** and **2**, as well as for several related bucky ferrocenes and ferrocene itself, are visualized graphically in Figure 2 and summarized in Table II.

The vibrational anisotropy noted for **1** is also observed for **2**, and, in fact, is very similar for the two iron atom environments over the same temperature range. That

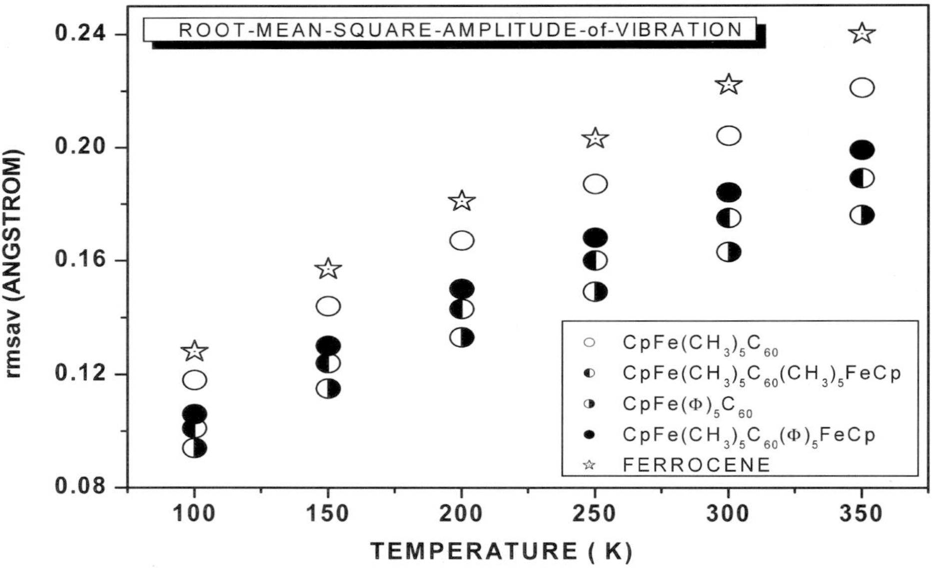

Figure 2 The rmsav of the iron atom in the compounds discussed in the text.

Table II Root-mean-square-amplitudes-of-vibration of the compounds discussed in the text

Temp. K/Compd.	$CpFe(CH_3)_5C_{60}$ (Å)	**1** (Å)	$CpFe(\Phi)_5C_{60}$ (Å)	**2** (Å)	Ferrocene (Å)
100	0.12	0.10	0.094	0.11	0.13
150	0.14	0.12	0.12	0.13	0.16
200	0.17	0.14	0.13	0.15	0.18
250	0.19	0.16	0.15	0.17	0.20
300	0.20	0.18	0.16	0.18	0.22
350	0.22	0.19	0.18	0.20	0.24

is, both the methyl and phenyl groups influence the dynamics of the metal atom to the same extent. Whether this is also true when the FeCp ligands are not located at 180° with respect to each other is the subject of a current study.

From these results it is seen that the rmsav of the metal atom is smaller for the bucky ferrocenes than for the parent ferrocene, and that the data for the two bi-nuclear complexes are intermediate between those of the $(CH_3)_5$ and $(\Phi)_5$ mono-iron molecular species.

Acknowledgements The authors are indebted to Profs. E. Nakamura and Y. Matsuo for generously making compounds **1** and **2** available for study, and to W. Glaberson and M. Brettschneider for the use of their DASWIN program prior to publication.

References

1. Herber, R.H., Nowik, I., Matsuo, Y., Toganoh, M., Kuninobu, Y., Nakamura, E.: Inorg. Chem. **44**, 5629–5635 (2005)
2. Toganoh, M., Matsuo, Y., Nakamura, E.: J. Am. Chem. Soc. **125**, 13974–13975 (2003); see also references in [1]
3. Herber, R.H., Nowik, I.: Hyperfine Interact. **136/7**, 699–703 (2001)
4. Herber, R.H., Nowik, I.: Solid State Sci. **4**, 691–694 (2002)
5. Herber, R.H.: In: Herber, R.H. (ed.) Chemical Mössbauer Spectroscopy, Ch. VII. Plenum, New York (1984)
6. Herber, R.H.: Inorg. Chim. Acta **291**, 74–81 (1999)
7. Gütlich, P., Link, R., Trautwein, A.X.: In: Mössbauer Spectroscopy and Transition Metal Chemistry, pp. 73–75. Springer, Berlin Heidelberg New York (1978)
8. Herber, R.H., Nowik, I.: Hyperfine Interact. **144/5**, 249–253 (2002)

Hyperfine Interact (2005) 166:357–361
DOI 10.1007/s10751-006-9293-6

Reactions of ^{57}Mn implanted into solid oxygen

Y. Kobayashi · H. Nonaka · J. Miyazaki · M. K. Kubo · H. Ueno · A. Yoshimi · H. Miyoshi · D. Kameda · K. Shimada · D. Nagae · K. Asahi · Y. Yamada

Published online: 27 October 2006

Abstract The in-beam Mössbauer spectroscopy using a short-lived ^{57}Mn ($T_{1/2}$=1.5 min) beam was carried out to study the production of the exotic chemical species of ^{57}Fe atoms arising from ^{57}Mn implanted into solid oxygen. The obtained spectra can be analyzed by four components of doublets at least, which are assigned to be novel Fe species of FeO, $Fe(O_2)$, $(O_2)FeO_2$, and $Fe(O_2)_2$, on the basis of ab initio molecular orbital calculations.

Key words Mössbauer spectroscopy · ^{57}Mn · solid oxygen

1 Introduction

We have investigated the in-beam Mössbauer spectroscopy using a short-lived radioactive beam, ^{57}Mn ($T_{1/2}$=1.45 min), produced by the nuclear projectile-fragment reaction of a primary stable beam. Implantation of these RI particles directly into an appropriate substance should provide us with the unique information about chemical structures and bonds. The RI atoms can react with molecules and atoms in the substance. It is possible to produce a variety of novel chemical species and exotic

Y. Kobayashi (✉) · H. Ueno · A. Yoshimi · K. Asahi
RIKEN, Hirosawa, Wako, Saitama 351-1098, Japan
e-mail: kyoshio@riken.jp

H. Nonaka · J. Miyazaki · Y. Yamada
Department of Chemistry, Science University of Tokyo, Kagurazaka, Shinjuku, Tokyo 162-8601, Japan

M. K. Kubo
College of Liberal Arts, International Christian University, Osawa, Mitaka, Tokyo 181-8585, Japan

H. Miyoshi · D. Kameda · K. Shimada · D. Nagae · K. Asahi
Department of Physics, Tokyo Institute of Technology, Oh-okayama, Meguro, Tokyo 152-8551, Japan

valence states unavailable under normal conditions. In our previously obtained in-beam Mössbauer spectra of $^{57}Fe/^{57}Mn$ implanted into $KMnO_4$ at low temperatures, the ^{57}Fe atoms occupying Mn sites in a $[MnO_4]^-$ tetrahedron were found to be in a high valence state of Fe(VIII) [1]. In this system, however, since a *substitutional* Fe atom was surrounded by tetrahedral symmetry of oxygen atoms, it was thought that Fe(VIII) species might not be formed by a basic reaction process between an isolated Fe atom and oxygen in $KMnO_4$.

In this study, we report the in situ characterization of the reaction products between Fe and oxygen by $^{57}Fe/^{57}Mn$ implantation Mössbauer spectroscopy and ab initio molecular orbital calculations. It is expected that the high electron affinity of oxygen favors the production of relatively high valence states of Fe. On the other hand, the oxidation of Fe is a fundamentally and chemically interesting subject from the viewpoint of the corrosion of materials and catalytic reactions. The simplest reaction between a localized Fe atom and reactive O_2 molecules provides important information on the chemistry under non-equilibrium conditions and on the production of exotic species.

2 Experimental

^{57}Mn particles were produced as a secondary RI beam following a nuclear projectile-fragmentation of $^{58}Fe^{21+}$ beams (E=63 MeV/nucleon, I=40 pnA) accelerated by the LINAC and the RING Cyclotron with a Be production target (the thickness of which was 0.5 mm). In the present experiment, an ^{58}Fe isotope was used as the primary beam, which could produce orders of magnitude higher intensities of ^{57}Mn than the ^{59}Co beam used in the former experiments [1]. The ^{57}Mn was separated and focused by the in-flight isotope separator using the RIPS (*Ri*ken *P*rojectile-fragment *S*eparator) in the RIKEN Accelerator Research Facility. Subsequently, the ^{57}Mn was implanted into a solid O_2 sample after passing through an Al plate with an appropriate thickness to diminish the beam energy down to about 26 MeV/nucleon. The stopping range of ^{57}Mn in the solid O_2 was calculated to be 180 μm from the surface with the straggling range of ±100 μm. The beam intensity of ^{57}Mn was typically 5×10^6 particle/s.

The solid O_2 sample was prepared at the position just above the beam line by an aggregation of a pure O_2 gas on an Al plate cooled at 18 K under 10^{-6} mbar for 5 h. The thickness of the sample was estimated to be approximately 1 mm, and the sample was transparent with light blue color. After the aggregation, the solid O_2 sample was moved down to the ^{57}Mn beam position.

The 14.4 keV γ-rays emitted from ^{57}Mn were detected by a parallel-plate avalanche counter (PPAC) with an ^{57}Fe-enriched stainless-steel foil. The PPAC was connected with the gas-flow controller, which maintains the optimized pressure and flow rate of C_3F_8 as a counter gas.

3 Results and discussion

Solid O_2 occurs in α-($0\leq T\leq24$ K), β-($24\leq T\leq44$ K), and γ-phases below 54.4 K of the melting point. It is known that α-O_2 forms the only molecular antiferromagnet,

Figure 1 In-beam Mössbauer spectra of $^{57}Fe/^{57}Mn$ atoms implanted into solid oxygen at **a** 18 and **b** 32 K.

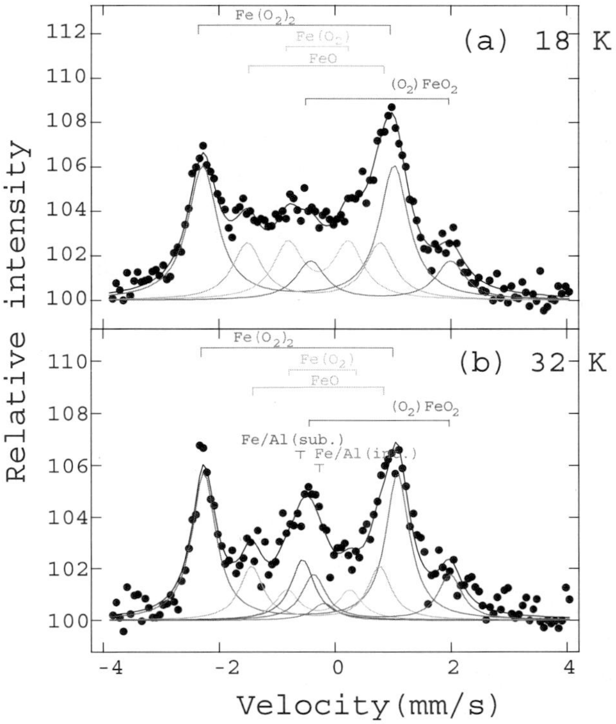

because two electrons in the outer shell of O_2 form a triplet spin state. The $^{57}Fe/^{57}Mn$ in-beam Mössbauer spectra were measured at 18, 32 and 40 K during the implantation of ^{57}Mn. The counting times were 18, 12, and 12 h, respectively. The total implantation dose of ^{57}Mn was 5×10^{11} $^{57}Mn/cm^2$ typically for one spectrum. The spectra obtained at 18 and 32 K are shown in Figure 1a and b, respectively. The spectra obtained at 32 and 40 K exhibited two components of Fe in the Al substance, which were a substitutional Fe atom (I.S.=−0.56 mm/s at 30 K) and an interstitial one (I.S.=−0.19 mm/s at 30 K). The reason was that part of the ^{57}Mn particles reached the Al substance, since the thickness of solid oxygen became thinner as the O_2 molecules evaporated due to the higher vapor pressure at increasing temperature.

As two dominant lines appeared at the positions of approximate −2.2 and 1.0 mm/s at 18 K, one tends to consider the existence of a magnetic hyperfine field at ^{57}Fe atoms arising from ^{57}Mn because of the magnetic property of α–O_2. It was found, however, that these two lines did not vanish up to 40 K above the Néel temperature (~30 K) of the α–phase. Therefore the in-beam Mössbauer spectra obtained in this study were analyzed with no magnetic interaction but a combination of paramagnetic doublets, referred to the previous experiment of the Ar matrix isolation of laser-ablated Fe atoms and the reaction products with O_2 gas at low temperatures by means of Mössbauer and infrared spectroscopic studies [2].

Our present results could be analyzed by mainly 4 components of Lorentzian doublets. These components were determined to be FeO (*S*=2), $Fe(O_2)$ (*S*=1),

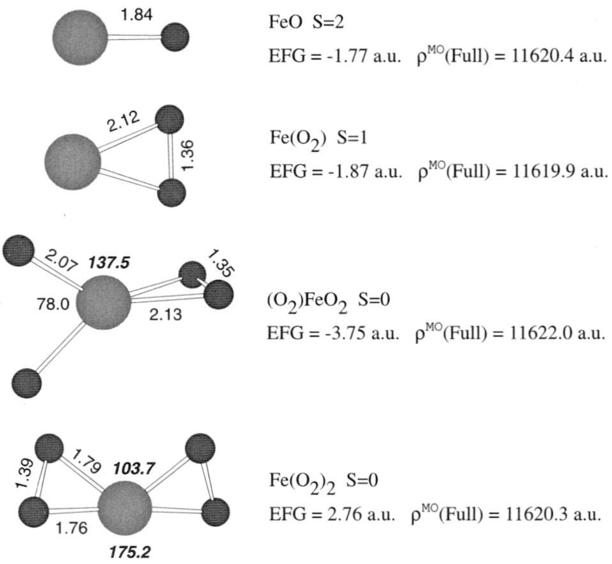

Figure 2 Optimized structures for FeO, $Fe(O_2)$, $(O_2)FeO_2$, $Fe(O_2)_2$ using Gaussian 98 and 03 programs. Bond lengths are given in angstroms and angles (*italics*) in degrees.

$(O_2)FeO_2$ (S=0), and $Fe(O_2)_2$ (S=0) from the Mössbauer parameters (I.S. and Q.S.). There showed up no clusters of Fe atoms in all spectra. The assignment of the Fe species was confirmed by ab initio molecular orbital calculations using Gaussian 98 and 03 programs. The optimized chemical structures for the reaction products after implantation of ^{57}Mn into solid O_2 are shown in Figure 2.

Comparing the reaction products of this ^{57}Mn implantation study with those of the laser-ablated Fe atoms with O_2 gas isolated in the Ar matrix, the species of FeO, $Fe(O_2)$, and $(O_2)FeO_2$ are identified in both methods. The species of OFeO and FeO_3, which are detected in the Ar matrix isolation experiment, are not determined in the ^{57}Mn implantation study. The reaction between the laser-evaporated Fe atoms and O_2 gas occurs in the gas phase. The Fe atoms having enough energy to break the O–O bonds collide with O_2 molecules, and the dissociated O atoms can hardly recombine to O_2 molecules. Thus the yields of reaction products containing Fe–O bonds may increase. On the other hand, it is found that $Fe(O_2)_2$ is newly formed in the ^{57}Mn implantation experiment, and also major components of resonance areas (49% at 18 K, 45% at 32 K) in all the spectra are measured. An energetic ^{57}Mn atom certainly creates defects and decomposes the O_2 molecules until it stops in the suitable final position in solid O_2. But in the final position the total number of ^{57}Fe atom arising from ^{57}Mn is extraordinarily smaller than that of O_2 molecules. The ^{57}Fe atom is almost surrounded by a lot of reactant O_2 molecules in the solid state. It is thought that they primarily react without decomposition of O_2, so that the products containing O–O bonds become dominant compared to the matrix isolation.

Acknowledgements The authors wish to thank Dr. N. Suzuki for preparing the ^{58}Fe-enriched $Fe(C_5H_5)_2$ for an ECR ion source of the LINAC. This study was partially supported financially by Grant-in-Aid for Scientific Research No. 15550058.

References

1. Kobayashi, Y., Kubo, M.K., Yamada, Y., Saito, T., Ueno, H., Ogawa, H., Sato, W., Yoneda, K., Watanabe, H., Imai, N., Miyoshi, H., Asahi, K.: J. Radioanal. Nucl. Chem. **255**, 403 (2003)
2. Yamada, Y., Sumino, H., Okamura, Y., Shimasaki, H., Tominaga, T.: Appl. Radiat. Isotopes **52**, 157 (2000)

Hyperfine Interact (2005) 166:363–366
DOI 10.1007/s10751-006-9316-3

Iron-57 Mössbauer spectroscopy study of phases in the $CaZrTi_{2-2x}Nb_xFe_xO_7$ zirconolite system

Frank J. Berry · Gregory R. Lumpkin · Gordon Oates · Karl R. Whittle

Published online: 26 October 2006

Abstract Materials of composition $CaZrTi_{2-2x}Nb_xFe_xO_7$ with the fluorite-related zirconolite structure have been prepared. The ^{57}Fe Mössbauer spectra show that iron is initially located in the five co-ordinate cation sites. As the iron content increases the iron enters the octahedral sites until, at a composition $CaZrTi_{0.4}Nb_{0.8}Fe_{0.8}O_7$, ca. 50% of the iron is five co-ordinate and the remainder is located in the octahedral sites.

Key words zirconolite · ^{57}Fe Mössbauer spectra

1 Introduction

The material known as zirconolite of composition $CaZrTi_2O_7$ was initially described as having a structure closely related to that of fluorite [1]. The structure was subsequently solved and refined by both powder [2] and single crystal [3] X-ray diffraction. The close relationship to the fluorite structure was confirmed and whilst the zirconium and one titanium site were identified as being octahedrally coordinated by six oxygen atoms the other titanium site was found to be displaced thereby decreasing the coordination from six to five in a trigonal bipyramidal array of oxygen atoms. The material has more recently attracted interest as a host phase for the immobilisation of radioactive lanthanide and actinide elements arising from nuclear waste [4, 5].

We are currently involved in an examination of the structural properties of the $CaZrTi_{2-2x}Nb_xFe_xO_7$ system where niobium and iron can substitute on either the 6-

F. J. Berry (✉) · G. Oates
Department of Chemistry, The Open University, Walton Hall, Milton Keynes MK7 6AA, UK
e-mail: f.j.berry@open.ac.uk

G. R. Lumpkin · K. R. Whittle
Department of Earth Sciences, University of Cambridge, Cambridge CB2 3EQ, UK

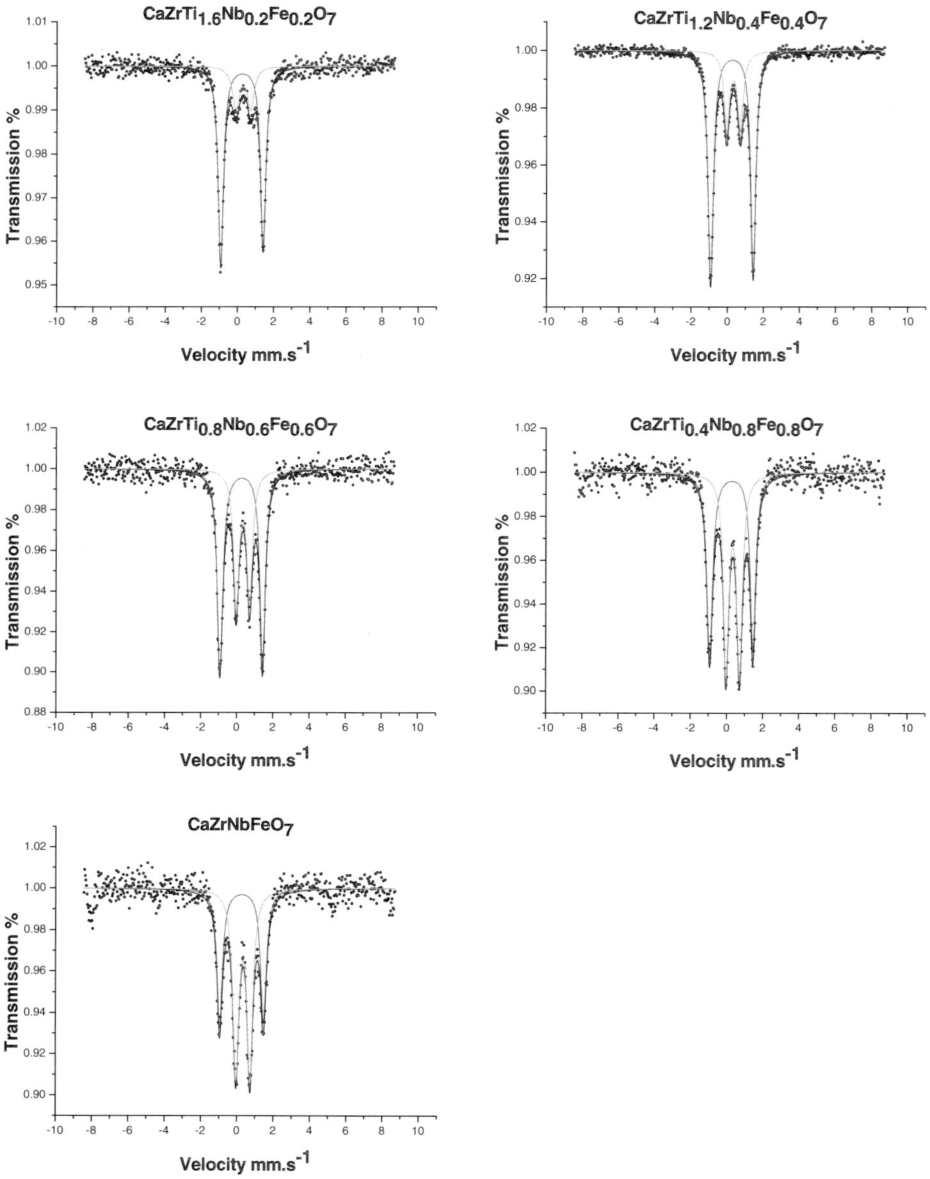

Figure 1 ^{57}Fe Mössbauer spectra recorded at 298 K from materials of composition $CaZrTi_{2-2x}Nb_xFe_xO_7$.

or 5- co-ordinate titanium sites. As a part of planned investigations by synchrotron X-ray- and neutron-diffraction we have initially examined the materials by ^{57}Fe Mössbauer spectroscopy and report here on our interpretation of the results in terms of site substitution by iron and which will, in due course, inform the analysis of our investigations by diffraction techniques.

Table I ^{57}Fe Mössbauer parameters recorded from materials of composition $CaZrTi_{2-2x}Nb_xFe_xO_7$

Compound	Component	$\delta \pm 0.01$/mms^{-1}	$\Delta \pm 0.01$/mms^{-1}	Area (%)
$CaZrTi_{1.6}Nb_{0.2}Fe_{0.2}O_7$	A	0.26	2.34	78
	B	0.33	0.81	22
$CaZrTi_{1.2}Nb_{0.4}Fe_{0.4}O_7$	A	0.26	2.35	73
	B	0.34	0.76	27
$CaZrTi_{0.8}Nb_{0.6}Fe_{0.6}O_7$	A	0.26	2.35	60
	B	0.34	0.74	40
$CaZrTi_{0.4}Nb_{0.8}Fe_{0.8}O_7$	A	0.26	2.39	49
	B	0.34	0.74	51
$CaZrNbFeO_7$	A	0.26	2.42	43
	B	0.35	0.78	57

2 Experimental

Materials of composition $CaZrTi_{2-2x}Nb_xFe_xO_7$ were prepared by adding aqueous solutions of calcium- and iron-nitrates and niobium(V) oxide to a mixture of zirconium- and titanium-alkoxides. The components were mixed to smooth homogeneous slurries, dried on a hot plate, ground, and calcined in air at 750°C for 1 h. The ground powders were then milled in water using 5 mm diameter yttrium-stabilized zirconia milling media for 16 h, pressed into pellets and sintered at 1,400°C for 20 h.

^{57}Fe Mössbauer spectra were recorded at 298 K with a microprocessor controlled spectrometer using a ca. 25 m Ci ^{57}Co/Rh source. All spectra were computer fitted and the chemical isomer shift data are quoted relative to that of metallic iron at 298 K.

3 Results and discussion

The ^{57}Fe Mössbauer spectra recorded from materials of composition $CaZrTi_{2-2x}Nb_xFe_xO_7$ are shown in Figure 1. The ^{57}Fe Mössbauer parameters are collected in Table I.

All the spectra were best fitted to two quadruple split absorptions. We associate the doublet characterised by the larger quadruple splitting with the presence of Fe^{3+} in the more distorted environment of the five co-ordinate trigonal bipyramidal site [6] and that characterised by the smaller quadruple splitting with the presence of Fe^{3+} in the more symmetrical six-fold co-ordination of the octahedral site. The spectrum recorded from $CaZrTi_{1.6}Nb_{0.2}Fe_{0.2}O_7$ indicates that ca. 78% of the iron substitutes on the trigonal pipyramidal titanium site. The result implies that this five co-ordinate site is that initially preferred for substitution.

The spectra recorded from the other materials with increasing iron content show the progressive increase in intensity of the doublet with smaller quadruple splitting charactering Fe^{3+} in octahedral co-ordination and the concomitant decrease in intensity of the doublet characteristic of the five coordinated species. The results suggest that, after the initial preferential substitution by iron on the trigonal bipyramidal site, subsequent substitution by increasing amounts of iron favors occupation by iron of the octahedral site. Interestingly, the spectrum recorded from the compound $CaZrTi_{0.4}Nb_{0.8}Fe_{0.8}O_7$ shows an approximately equal amount of iron on the trigonal bipyramidal- and octahedral-sites.

These results are currently being used to inform the analysis of our investigations of the structural properties of this series of materials and the occurrence of ordering by niobium and iron cations over the different cation sites.

References

1. Pyatenko, Y.A., Pudovkina, Z.: Kristallografiya **9**, 98 (1964)
2. Rossell, H.J.: Nature **283**, 282 (1980)
3. Gatehouse, B.M., Grey, I.E., Hill, R.J., Rossell, H.J.: Acta Cryst. **B37**, 306 (1981)
4. Ringwood, A.E., Vesson, S.E., Ware, N.G., Hibberson, W., Major A.: Nature **278**, 219 (1979)
5. Lumpkin, G.R., Hart, K.P., McGlinn, P.J., Payne, T.E.: Radiochim. Acta **66–67**, 469 (1994)
6. Evans, B.J., Grandjean, F., Lilst, A.P., Vogel, R.H., Gerard, A.: J. Magn. Mater. **67**, 123 (1987)

Hyperfine Interact (2005) 166:367–371
DOI 10.1007/s10751-006-9294-5

X-ray and Mössbauer study of structural changes in $K_3Na(FeO_4)_2$

S. K. Dedushenko · M. G. Zhizhin · Yu. D. Perfiliev

Published online: 2 November 2006

Abstract Mixed potassium–sodium ferrate(VI), $K_3Na(FeO_4)_2$, has been synthesized by precipitation from alkaline solution. At room temperature it decomposes spontaneously giving Fe(III) compounds and ferrate(VI) with a structure similar to that of K_2FeO_4, which is confirmed by X-ray diffraction and Mössbauer spectroscopy.

Key words Mössbauer spectroscopy · X-ray diffraction · iron in high oxidation state · ferrates

1 Introduction

There is rapidly growing interest in iron in oxidation state +6, which exists as the ferrate(VI) ion, FeO_4^{2-}, mainly due to its high potential in wastewater treatment [1]. The high efficiency of ferrates in water purification from a number of toxic pollutants has already been undoubtedly proved. The ferrate(VI) ion is a double-action reactant, comprising the properties of a very powerful oxidant with a highest redox potential and an effective coagulant due to the formation of colloidal iron hydroxide upon its decomposition. Ferrate(VI) can neutralise extremely toxic substances, whereas common oxidising agents used in water treatment, such as chlorine and ozone, often results in much more toxic products. For example, treatment of cyanide solutions with chlorine gives very toxic cyanogen chloride [2]. Moreover, ferrates(VI) per se and their decomposition products have low toxicity, which clears the way to the development of environmentally friendly technologies for water and wastewater treatment. Nowadays, the relatively higher cost of ferrates(VI) seems to be the only problem for their application.

Most of publications on ferrates(VI) within the last years deal with various aspects of their practical application. At the same time, papers on the study of their

S. K. Dedushenko (✉) · M. G. Zhizhin · Y. D. Perfiliev
Department of Chemistry, Moscow State University, Moscow, Russia
e-mail: dedushenko@mail.ru

properties or on the development of new synthetic methods are virtually absent. It can be partly explained by the fact that there are few well-characterised crystalline ferrates(VI) [3]. All these salts contain the tetrahedral complex anion FeO_4^{2-} with the high-spin d^2 electronic configuration of the iron valent shell, the same ion being present in aqueous solutions of ferrates(VI). The stability of solid compounds and their aqueous solutions, as well as some other properties of ferrates(VI) have been studied in more detail within the second half of the twentieth century. This creates the illusion that all the most important properties of ferrates(VI) are already well known.

However, this conclusion is wrong. For example, the possibility of precipitation of the mixed salt $K_3Na(FeO_4)_2$, which has recently been revealed [4], was not even taken into consideration in a number of studies. Note that the determinations of the solubility of K_2FeO_4 in aqueous alkaline solutions (including NaOH) did not attempt analysing the precipitates for Na and K [5].

This paper presents the results of Mössbauer and X-ray study of $K_3Na(FeO_4)_2$ samples kept in inert atmosphere and at room temperature, which demonstrate an unusual mechanism of decomposition of this compound.

2 Experimental

We used formerly prepared samples of $K_3Na(FeO_4)_2$ obtained by adding solid potassium hydroxide to 40% NaOH solution containing ca. 0.1 M FeO_4^{2-} following by the precipitation of $K_3Na(FeO_4)_2$ crystals.

X-ray diffraction patterns were obtained at room temperature in Bragg–Brentano geometry using a SIEMENS D500 powder diffractometer (CuK_α-radiation, λ= 1.5418 Å). The data were collected over the range of 5–90° (2θ) with a step of 0.02°.

Mössbauer absorption spectra were measured on a Perseus spectrometer working at constant velocities, the control and the adjustment of the spectrometer-vibrator rate being performed by a laser interferometer. A standard γ-source of ^{57}Co in metallic rhodium matrix with the activity of 0.9 GBq (a product of Cyclotron, Co., Ltd., Obninsk, Russia) was employed. Isomer shifts in this paper are presented relative to α-Fe.

3 Results and discussion

In this work, the samples of potassium–sodium ferrate(VI) were used that had been obtained earlier and studied using several techniques [4]. The samples were kept at 16÷26°C in hermetically sealed containers.

Mössbauer spectra of these samples were found to change with time. The asymmetric doublet of the initial salt ($\delta=-0.89\pm0.01$ mm·s^{-1}, $\Delta=0.21\pm0.01$ mm·s^{-1}, $\Gamma=0.22\pm0.01$ mm·s^{-1}) transformed into a single line with the following parameters: $\delta=-0.90\pm0.01$ mm·s^{-1} and $\Gamma=0.28\pm0.01$ mm·s^{-1} (Figure 1). Moreover, the relative content of trivalent iron increased.

At first, the occurrence of a phase transition of the ferrate was supposed. Such transitions are well known for the compounds of the glaserite-like structure, but occur at somewhat lower temperatures. For instance, the most close structural analog of $K_3Na(FeO_4)_2$, mixed potassium–sodium chromate(VI) $K_3Na(CrO_4)_2$,

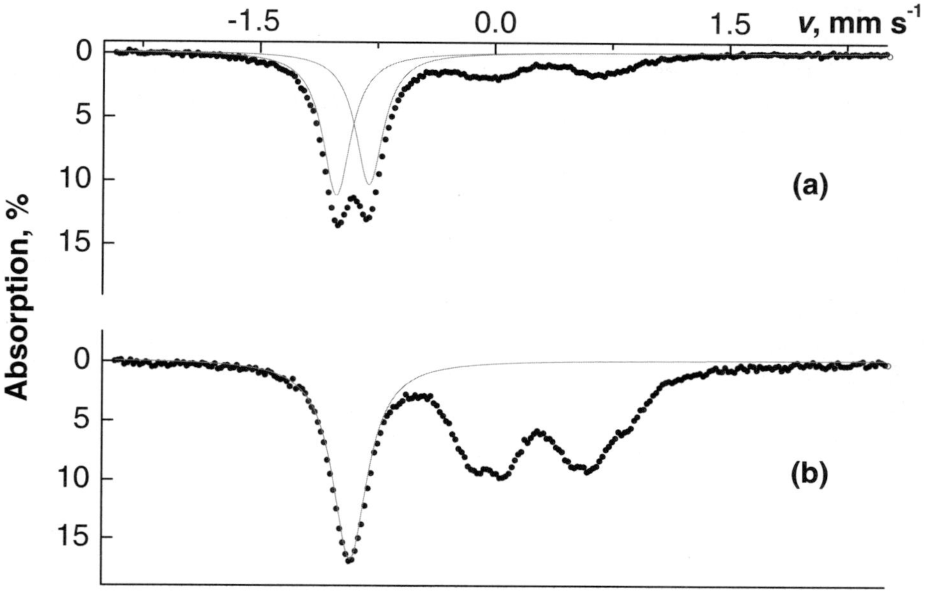

Figure 1 Experimental room-temperature Mössbauer spectra of $K_3Na(FeO_4)_2$ before (**a**) and after (**b**) decomposition; *solid lines* show the refined curves of Fe^{6+}.

transforms into a monoclinic phase at 239 K [6]. However, an X-ray study of the decomposition product showed that the main crystalline phase was orthorhombic and its parameters were very close to those known for K_2FeO_4 [7]. In the X-ray pattern, there were also a few lines of low intensity, which were not prescribed to any known phase. The lines of initial salt were not detected (Figure 2).

Mössbauer measurements at the liquid helium temperature did not show the presence of any magnetic interaction in the substance. Only regular positive shift of iron(VI) line was observed ($\delta = -0.81 \pm 0.01$ mm·s^{-1}), the line width remained changeless. Thus, no signs of transition into the antiferromagnetic state, which is known for K_2FeO_4 ($T_N = 3.6 \div 4.2$ K [8, 9]), were found. It could be explained by small sizes of crystallites, which is also in agreement with the increased line width of the iron(VI) Mössbauer line (Figure 1b).

Two possible interpretations of the observed transformation can be proposed. A first variant is the transformation of the hexagonal lattice into the orthorhombic lattice of K_2FeO_4 with partial random substitution of sodium for potassium. (An ordered substitution would have been detected by X-ray measurements due to the formation of a sublattice.) The second possibility is the crystallisation of K_2FeO_4 and formation of a separate sodium- and iron(III)-containing phase(s). Note the change of relative intensities of tri- and hexavalent iron, which is of particular interest. The decrease of the iron(VI) line intensity was very close to 50%. It may correspond to the following decomposition scheme:

$$K_3Na(FeO_4)_2 \rightarrow K_2FeO_4 + [KNaFe^{3+}O_{2.5}] + 0.75\ O_2 \uparrow .$$

We were unable to propose a univocal model for the part of Mössbauer spectra related to trivalent iron due to low resolution of Fe^{3+} lines. Together with

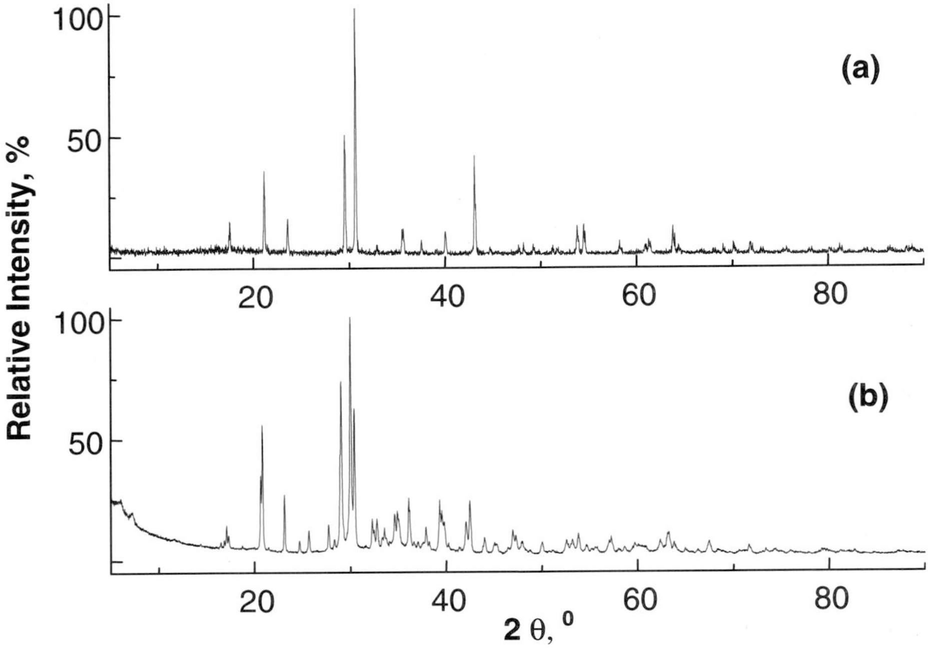

Figure 2 Experimental room-temperature X-ray diffraction patterns of $K_3Na(FeO_4)_2$ [4] (**a**) and the product of its decomposition (**b**), CuK_α-radiation.

[$KNaFe^{3+}O_{2.5}$], the constituent of trivalent iron in the sample should include by-products of the formation of $K_3Na(FeO_4)_2$ as well as the product(s) of Fe(VI) decomposition by reduction mechanism.

It is remarkable that the motive force for the decomposition process of $K_3Na(FeO_4)_2$ is its tendency to transform into a more stable crystal structure; so that decomposition is not caused by the instability of Fe^{6+}. The same holds for decomposition of other ferrates, e.g. for sodium ferrate(IV), Na_4FeO_4. Its decomposition proceeds via the mechanisms of disproportionation, i.e. with the formation of both lower and higher oxidation states of iron, which was clearly observed in Mössbauer spectra [10].

High oxidation states of iron represent an exotic subject of modern inorganic chemistry with a limited number of examples. Due to this fact, classification of such iron compounds is usually based on their oxidation state. Note, however, that some compounds have been reported that contain a complex ion FeO_4^{n-} with iron oxidation states +6, +5, +4 and lower (n=2, 3, 4, etc.). Moreover, some data are available about iron(VIII) oxide, FeO_4 [11]. Therefore, these compounds can be grouped in a separate class of coordination oxoferrates. Other compounds of iron in high oxidation states, such as $LiLa_2Fe^{+5}O_6$ and $CaFe^{+4}O_3$, have another type of structure and thus differ in their chemical properties [12, 13].

It is reasonable to suppose that similar decomposition mechanisms occur in related compounds of other transition metals, e.g. in manganates, MnO_4^{n-}. However, due to the lack of efficient techniques for the analysis of the electronic state of manganese ions, such mechanisms have so far not been observed.

4 Conclusion

Compounds of iron in high oxidation states, which contain complex ion FeO_4^{n-}, show chemical transformations in solid state. These transformations caused by tendency to transform into a more stable crystal structure, but not caused by the instability of Fe^{x+}. We believe these compounds should be considered to be the member of separate class of coordination oxoferrates.

Solid coordination oxoferrates possess many unusual properties. These properties are of theoretical interest now, but they become important in case of ferrate application in industry, e.g. for water cleaning.

Mössbauer spectroscopy is the most efficient technique for analysis of ferrates. It allows one to get maximum information value vs. other analytical techniques. Therefore, it becomes the main method for quality control of ferrates.

Acknowledgement This work was supported by the Russian Foundation for Basic Research (project # 06-03-32990).

References

1. Sharma, V.K., Jiang, J.-Q., Bouzek, K. (eds.): Innovative Ferrate (VI) Technology in Water and Wastewater Treatment. ICT, Prague (2004)
2. Sharma, V.K., Rivera, W., Smith, J.O., O'Brien, B.: Environ. Sci. Technol. **32**, 2608 (1998)
3. Dedushenko, S.K., Perfiliev, Yu.D., Goldfeld, M.G., Tsapin, A.I.: Hyperfine Interact. **136/137**, 373 (2001)
4. Dedushenko, S.K., Perfiliev, Yu.D., Golubev, A.M., Melnikov, P., Corbi, P.P., Saprykin, A.A.: Hyperfine Interact. (C) **5**, 277 (2002)
5. Bailie, A.G., Bouzek, K., Lukasek, P., Rousar, I., Wragg, A.A.: J. Chem. Technol. Biotechnol. **66**, 35 (1996)
6. Fabry, J., Braczewsky, T., Madariaga, G.: Acta Crystallogr. **B50**, 13 (1994)
7. Audette, R.J., Quail, J.W., Black, W.H., Robertson, B.E.: J. Solid State Chem. **8**, 43 (1973)
8. Herber, R.H., Johnson, D.: Inorg. Chem. **18**, 2786 (1979)
9. Corson, M.R., Hoy, G.R.: Phys. Rev. B **29**, 3982 (1984)
10. Dedushenko, S.K., Kholodkovskaya, L.N., Perfiliev, Yu.D., Kiselev, Yu.M., Saprykin, A.A., Komozin, P.N., Lemesheva, D.G.: J. Alloys Compd. **262–263**, 78 (1997)
11. Kiselev, Yu.M., Kopelev, N.S., Spitsyn, V.I., Martynenko, L.I.: Dokl. Acad. Sci. (USSR) **292**, 628 (1987)
12. Buffat, B., Demazeau, G., Pouchard, M., Fournes, L., Dance, J.-M., Fabritchnyi, P., Hagenmuller, P.: C. R. Acad. Sci., Ser. II. **292**, 509 (1981)
13. Takeda, Y., Naka, S., Takano, M., Shinjo, T., Takada, T., Shimada, M.: Mater. Res. Bull. **13**, 61 (1978)

Hyperfine Interact (2005) 166:373–378
DOI 10.1007/s10751-006-9295-4

$BaSnF_4$ fast ion conductor: Variations versus the method of preparation and anomalous temperature variation of the quadrupole splitting

Jamil Hantash · Alan Bartlett · Georges Dénès · Abdualhafeed Muntasar · Philip Oldfield

Published online: 16 November 2006

Abstract A new method of preparation of high performance fluoride ion conductor, $BaSnF_4$, by water leaching of newly discovered barium tin(II) chloride fluorides, has been designed, and the materials have been studied and compared to the solid prepared by the usual dry method. The unit-cell parameters and crystallite dimensions were found to vary with the method of preparation. In addition, the crystallite dimensions were found to be highly anisotropic for the samples obtained by the wet method. The Mössbauer spectrum is made of a large tin(II) quadrupole doublet, and a broad tin(IV) oxide peak due to surface oxidation. The tin(II) spectrum is in agreement with covalently bonded tin(II) having a strongly stereo-active lone pair. An unusually high dependence of the quadrupole splitting at low temperatures was observed (5.8 times larger than for α-SnF_2).

Key words barium tin fluoride · ionic conductor · Mössbauer spectroscopy · synthetic methods

1 Introduction

The highest performance fluoride ion conductors are found in the SnF_2/MF_2 systems, where M is an alkaline earth metal or lead, and their crystal structure is derived from the fluorite-type structure (CaF_2 type), with M and Sn being either ordered or disordered, depending on the compound. The $MSnF_4$ materials (M=Sr,

J. Hantash · G. Dénès (✉) · A. Muntasar
Laboratory of Solid State Chemistry and Mössbauer Spectroscopy, Department of Chemistry and Biochemistry, Concordia University, Montréal, Québec, Canada
e-mail: gdenes@alcor.concordia.ca

J. Hantash · A. Bartlett · P. Oldfield
CTBR, A Charles River Laboratories Company, Senneville, Quebec H9X 3R3, Canada

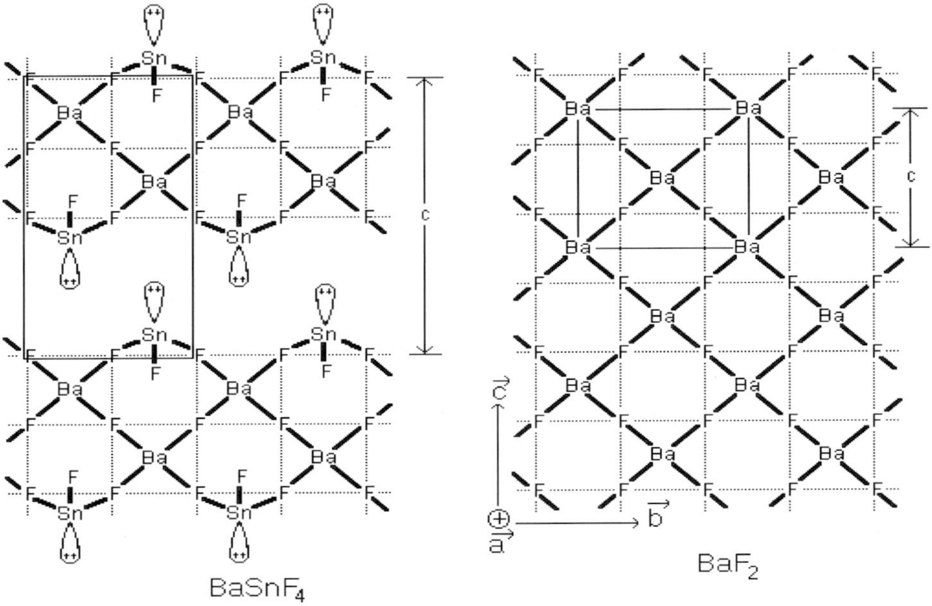

Figure 1 Projection of the structures of $BaSnF_4$ and BaF_2 onto the (*b*,*c*) plane.

Ba and Pb) have a tetragonal distortion relative to fluorite, with a ... M M Sn Sn ... order parallel to the *c* axis of the unit-cell (Figure 1) [1]. The barium ions are in a 12-fold $BaF_4F'_4F''_4$ coordination (4F at 2.626 Å, 4F′ at 2.955 Å, 4 F″ at 3.148 Å) which is a distortion of the BaF_8 cube (8F at 2.685 Å) present in BaF_2, with four longer interactions. Tin is in a $SnF''F'_4$ square pyramidal coordination (F″ at 2.282 Å), which is completed by the tin stereoactive lone pair E, located in trans-position relative to the axial Sn–F″ bond, that make a pseudo-octahedral $SnF''F'_4E$ geometry [1]. Each F″ makes a short bond to Sn and four long interactions (3.148 Å) to neighboring barium ions. In the fluorite-type structure, BaF_8 cubes alternate with $[]F_8$ cubes ([]=metal ion vacancy) parallel to the three axes of the unit-cell. It results that half of the cubes have a vacant position in their center, that can be used for lodging a fluoride ion that has left its regular site. This combination of fluoride ion vacancies and interstitials makes a larger number of Frenkel defects that have been held responsible for the anionic conductivity of the compounds that have the fluorite type structure [2]. In addition, it has been shown that the presence of a metal that contains a lone pair (e.g. Pb(II) in β-PbF_2) enhances the ionic conductivity by a large amount [3]. The conductivity of $BaSnF_4$ was found to be three orders of magnitude larger than that of BaF_2, i.e. is a superionic conductor, and transport number measurements showed that the long range motion of fluoride ions is responsible for more than 99% of its total electrical conductivity [4]. Before the present work, $BaSnF_4$ could be prepared only by direct reactions between BaF_2 and SnF_2 in dry conditions at high temperatures [5]. In this work, a new method of preparation of $BaSnF_4$ by leaching barium tin chloride fluorides in aqueous solutions is described, and the material obtained is studied by X-ray powder diffraction and Mössbauer spectroscopy.

2 Experimental

Sample preparation was carried out either by the dry or the wet method. In the dry method, 1 g of a stoichiometric mixture of intimately mixed SnF_2 and BaF_2 was loaded and sealed in a copper tube according to a procedure designed in our laboratory [6]. After preheating at 200°C for 2 h, the temperature was raised to 500°C and maintained at that temperature for 4 h. Then, the copper tube reactor was taken out of the oven and allowed to cool in air. In the first step of wet preparations, barium tin(II) chloride fluorides were prepared by addition of an aqueous solution of SnF_2 to a solution of $BaCl_2{\cdot}2H_2O$ (Sn→Ba) or vice-versa (Sn→Ba) upon stirring. The precipitate obtained was filtered, washed with 10 ml of cold water, and allowed to dry in air at ambient conditions. All samples were characterized by X-ray diffraction, on a Philips PW-1050 focusing powder diffractometer, using the K_α line of copper (λ=1.54178 Å). Mössbauer spectra were recorded on selected samples, using a nominally 20 mCi $Ca^{119m}SnO_3$ γ-ray source from Ritverc. The detector was a (Tl)NaI scintillation counter. The Doppler velocity (±10 mm/s) was generated by an Elscint driving system. The amplifier and the single channel analyzer are built in the Tracor Northern TN7200 multichannel analyzer, where the data were accumulated. Low temperature spectra were obtained in an APD Cryogenics helium closed cycle refrigerator equipped with a Displex® Mössbauer cold head. The GMFP software was used for fitting the spectra [7]. Isomer shits are referenced relative to the line of a $CaSnO_3$ standard absorber at ambient temperature.

3 Results and discussions

3.1 X-ray diffraction

The mixing of aqueous solutions of SnF_2 and $BaCl_2.2H_2O$, in the X molar fraction of barium chloride, with $X{=}n_{Ba}/(n_{Ba}{+}n_{Sn})$, where n_{Ba} and n_{Sn} are the number of moles of barium chloride and of tin(II) fluoride, respectively, results in the immediate precipitation of a white solid, which was found to be $BaSn_2F_6$ ($X{\leq}0.34$), $BaSn_2Cl_2F_4$ ($0.10{\leq}X{\leq}0.52$), $BaSnClF_3.0.8H_2O$ ($0.40{\leq}X{\leq}0.86$) or a $Ba_{1-x}Sn_xCl_{1+y}F_{1-y}$ solid solution ($X{\geq}0.80$), and mixtures in the X ranges of overlap. If the amount of water used for washing was no higher than 10 ml, no other phase was obtained. For larger volumes of washing water, the chloride fluoride samples were found to contain some $BaSn_2F_6$ and/or $BaSnF_4$. A detailed study of the reaction of the barium tin(II) chloride fluorides with water was carried out by stirring, in water, samples of the chloride fluorides that had been previously filtered, washed, dried and characterized. A 40–60% weight loss was measured in all cases when no chloride fluoride remained, in agreement with a 46 and 67% loss for the full decomposition of $BaSnClF_3.0.8H_2O$ to give $BaSnF_4$ and $BaSn_2F_6$, respectively. The residue was either pure $BaSnF_4$ or a mixture of $BaSnF_4$ and $BaSn_2F_6$ when the chloride fluoride used was $BaSnClF_3.0.8H_2O$. When $Ba_{1-x}Sn_xCl_{1+y}F_{1-y}$ was stirred in water, the transformation was not complete, and a mixture of $BaSnF_4$ and untransformed $Ba_{1-x}Sn_xCl_{1+y}F_{1-y}$ was obtained. Work currently under progress shows that $BaSnF_4$ and/or $BaSn_2F_6$ are also obtained by stirring $BaSn_2Cl_2F_4$ in

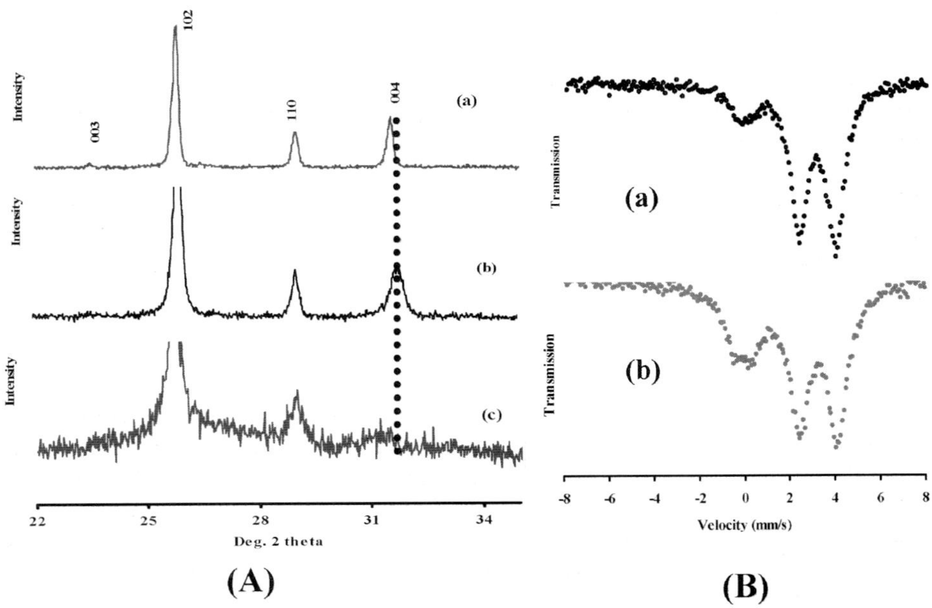

Figure 2 **A** Region of the X-ray diffraction pattern of $BaSnF_4$, (**a**) prepared by direct reaction at 500°C for 2 h, (**b**) AM-1100, (X=0.845, Ba→Sn, stirred for 22 h), (**c**) AM-1206, (X=0.745, Ba→Sn, stirred for 24 h; **B** Ambient temperature ^{119}Sn Mössbauer spectrum of $BaSnF_4$ (**a**) prepared by direct reaction at 500°C for 2 h, (**b**) prepared by precipitation (X=0.845, Ba→Sn) followed by leaching in water.

water. This works has shown that the barium tin(II) chloride fluorides are not stable in water: leaching of all the chloride ions and of the excess Ba and/or Sn. The X-ray diffraction pattern of $BaSnF_4$ has the same set of lines as that of the material prepared by direct reaction in dry conditions. However, a close examination of their diffraction patterns show some subtle differences (Figure 2A). The most obvious difference is the line width.

The line width at half height was used to calculate the average crystallite dimension, using Scherrer formula, after applying Warren correction to account for instrumental broadening [8]. Bragg peak broadening is known to originate from either small crystallite dimensions and from internal strain [8]. Line broadening for both reasons have been observed in precipitated α-$PbSnF_4$ [9]: the very large broadening of (110) is due to an enormous amount of strain parallel to the (a,b) plane of the tetragonal unit-cell, and the milder broadening of (004) results from the small crystallite dimension parallel to the c axis. Internal strain in α-$PbSnF_4$ is understood by the onset of the $\alpha \rightarrow o$ (tetragonal to orthorhombic) paraelastic to ferroealstic transition, however, no such transition has been observed in $BaSnF_4$. The small dimension parallel to c results from the layered structure (Figure 1) that creates cleaving parallel to (a,b). Figure 1 shows typical examples of a section of the diffraction pattern of $BaSnF_4$, encompassing the (110) and (004) peaks, thereby probing the directions of (a,b) and of c. The (110) and (004) peaks are broader in the samples prepared by the wet method, and therefore the average crystallite dimensions are smaller in both directions than in the sample prepared under dry

Table I Unit-cell parameters and average crystallite dimension of $BaSnF_4$ versus the method of preparation

AM number	Prep.	*a* (Å)	*c* (Å)	*V* (Å^3)	*c/2a*	$D_{\parallel}$ (Å)	$d_{\perp}$ (Å)
AM-456	D	4.287	11.336	208.3	1.32	465	462
AM-1100	S	4.344	11.232	212.0	1.29	136	462
AM-1206	S	4.360	11.409	216.9	1.31	72	209

D: High temperature dry method, *S* chloride fluoride stirred in filtrate,
c/2a: tetragonal distortion,
$d_{\parallel}$ (Å): average crystallite dimension parallel to the *c* axis of the unit-cell,
$d_{\perp}$ (Å): average crystallite dimension perpendicularly to the *c* axis of the unit-cell

conditions. In addition, Table I shows a significant crystallite dimension anisotropy for samples prepared by the wet method, with $d_{\parallel}$ (Å), the average crystallite dimension parallel to the *c* axis of the unit-cell being three times smaller than $d_{\perp}$ (Å), the dimension perpendicularly to the *c* axis, i.e. parallel to (*a,b*). A similar behavior was observed in stress-free samples of α-$PbSnF_4$, obtained by ball-milling followed by either stirring in water or annealing [10]. Crystal dimension anisotropy is explained by the layered structure (Figure 1). In contrast with the samples prepared by the wet method, the samples prepared under dry conditions, shows only moderate line broadening, and more particularly there no anisotropy of crystallite dimensions. Obviously the conditions of crystal growth have a significant influence, not only on the crystallite dimensions, but also on the speed of growth versus the direction in the crystal. The unit-cell parameters seem also to vary with the method of preparation (Table I, Figure 2A), particularly *c*, i.e. the interlayer spacing. This might be due to intercalation of fluoride ions between the layers, in a Frenkel defect mechanism that contributes to the conductivity. Why it is related to the method of preparation is unclear at the present time.

3.2 Mössbauer spectroscopy

The ^{119}Sn Mössbauer spectrum (Figure 2B) shows a minor broad tin(IV) signal at ca. 0 mm/s, and a large tin(II) quadrupole doublet in the 2.5–4.1 mm/s. The tin(II) doublet shows that a large efg (electric field gradient) is present at tin, and it is so large that the main contribution can only originate from a stereoactive lone pair, the clustering of which generates cleavage planes responsible for the layered structure (Figure 1). The stereoactivity of the lone pair shows that bonding to tin is covalent and its valence orbitals are hybridized, with the lone pair occupying one of the hybrid orbitals. The pseudo-octahedral $SnF''F'_4E$ tin coordination can be interpreted as using a sp^3d^2 hybridization, or alternatively, as a sp hybridization, with one of the hybrid orbitals being occupied by the lone pair, the other by the fluorine atom (F″) opposite to the lone pair, forming a $[E–Sn–F'']^+$ ion that binds ionically to a layer of fluoride ions with insertion of F″ inside an otherwise empty F_8 cube (Figure 1). The doublet asymmetry could be due to either the Goldanskii–Karyagin effect (anisotropy of thermal vibrations, due to bonding anisotropy), or to preferred orientation, both of which being expected due to the layered structure with tin forming no bond in the lone pair direction. The highly broadened line at about 0 mm/s is in the region of tin(IV) bonded to oxygen and can be attributed to surface

Table II Mössbauer parameters of $BaSnF_4$ versus the method of preparation

AM sample number	Preparation method	δ (mm/s)	Δ (mm/s)	Sn(IV) signal (%)	T (K)
456	D	3.289	1.530	14.24	298
10991321	S	3.289	1.530	30.280	298
108918H	S	3.359	1.530	28.740	298
$BaSnF_4$	L	3.450	1.953	4.500	76
α-$PbSnF_4$		3.24	1.560	–	298
α-SnF_2		3.430	1.532	1–2	298

D: High temperature dry method,
S: chloride fluoride stirred in filtrate,
L: chloride fluoride leached by stirring in water

oxidation of the particles [11]. It is found in all polycrystalline tin(II) compounds that have been exposed to air, and provided bonding to tin(II) is covalent, it does not increase on aging [11]. The isomer shift at ambient temperature changes somewhat for different samples prepared by the wet method, while the quadrupole splitting is remarkably constant at 1.530 mm/s, which is very similar to that of isostructural α-$PbSnF_4$ and to α-SnF_2 (Table II). An unusually large temperature dependence of the quadrupole splitting takes place in $BaSnF_4$ prepared by the dry method (1.9×10^{-3} mm. s^{-1}. K^{-1}), compared to a 5.8 times smaller value for α-SnF_2 (3.3×10^{-4} mm. s^{-1}. K^{-1}). Explaining this phenomenon will require a detailed low temperature study, including by X-ray or neutron diffraction, in order to investigate the existence of a possible anomalous thermal contraction, or a phase transition at cryogenic temperatures.

Acknowledgements This work was made possible by the support of Concordia University, the Natural Science and Engineering Research Council of Canada, and CTBR. Grateful thanks are also due to the Procter and Gamble Co. (Mason, Ohio) for supporting our Mössbauer laboratory.

References

1. Dénès, G., Yu, Y.H., Tyliszczak, T., Hitchcock, A.P.: J. Solid State Chem. **91**, 1 (1995)
2. Schoonman, J.: Solid State Ionics **1**, 121 (1980)
3. Dénès, G.: In: Solid State Ionics IV, Mater. Res. Soc. Symp. Proc., Boston 1994, vol. 369, p. 295 (1995)
4. Dénès, G., Birchall, T., Sayer, M., Bell, M.F.: Solid State Ionics **13**, 213 (1984)
5. Dénès, G., Pannetier, J., Lucas, J.: C. R. Acad. Sci. Paris **280C**, 831 (1975)
6. Dénès, G.: J. Solid State Chem. **77**, 54 (1986)
7. Ruebenbauer, K., Birchall, T.: Hyperfine Interact. **7**, 125 (1979)
8. Cullity, B.D.: In: Elements of X-ray Diffraction, 2nd ed., p. 263 and 444. Addisson-Wesley, Reading, MA (1955)
9. Dénès, G., Madamba, M.C., Parris, J.M.: In: Solid State Ionics IV, Mater. Res. Soc. Symp. Proc., Boston 1994, vol. 369, p. 463 (1995)
10. Dénès, G., Madamba, M.C.: Mater. Struct. **3**, 227 (1996)
11. Dénès, G., Muntasar, A.: Hyperfine Interact. **153**, 121 (2004)

Hyperfine Interact (2005) 166:379–384
DOI 10.1007/s10751-006-9296-3

Phase transition of the doubly disordered $Ba_{1-x}Sn_xCl_{1+y}F_{1-y}$ solid solution at high temperature

Julie Kochuparampil · Georges Dénès · Abdualhafed Muntasar

Published online: 16 November 2006

Abstract The $Ba_{1-x}Sn_xCl_{1+y}F_{1-y}$ solid solution has the BaClF structure, with full disorder of the cations and partial disorder of the anions. It can be prepared either by the wet method or the dry method. Most samples prepared by the aqueous route give a tin(II) single line at ca. 4.1 mm/s characteristic of the Sn^{2+} stannous ion, and therefore of ionic bonding. In some rare cases, a very weak quadrupole doublet indicates the presence of a small amount of covalently bonded tin(II). The solid solution prepared by the dry method has different unit-cell parameters, and it contains a larger amount of covalently bonded tin, except at very low x values or highly positive y values, and the Sn^{2+} recoil-free fraction is much smaller. In the present work, a study of the phase transition that takes place when precipitated $Ba_{1-x}Sn_xCl_{1+y}F_{1-y}$ is heated has been undertaken. A dramatic increase of the amount of covalently bonded tin(II) occurred on heating.

Key words barium tin chloride fluoride · phase transition · Mössbauer spectroscopy · solid solution · lone pair

1 Introduction

The crystal structures of $BaSnF_4$ and BaClF are tetragonal distortions of fluorite-type BaF_2 (Figure 1). In BaF_2, Ba is in a BaF_8 cubic coordination while in $BaSnF_4$, the BaF_8 unit is distorted and four longer Ba- - -F″ interactions give a $BaF_4F'_4F''_4$ 12-fold coordination, and the coordination of tin is a $SnF''F'_4$ square pyramid, where the short Sn−F″ bond is strongly covalent and is located in trans-position relative to the lone pair E, resulting in an octahedral distribution of valence electron pairs, with a sp^3d^2 hybridization. The lone pair (nonbonded electron pair) is located on one of

J. Kochuparampil · G. Dénès (✉) · A. Muntasar
Laboratory of Solid State Chemistry and Mössbauer Spectroscopy,
Department of Chemistry and Biochemistry, Concordia University,
Montréal, Québec, Canada
e-mail: gdenes@alcor.concordia.ca

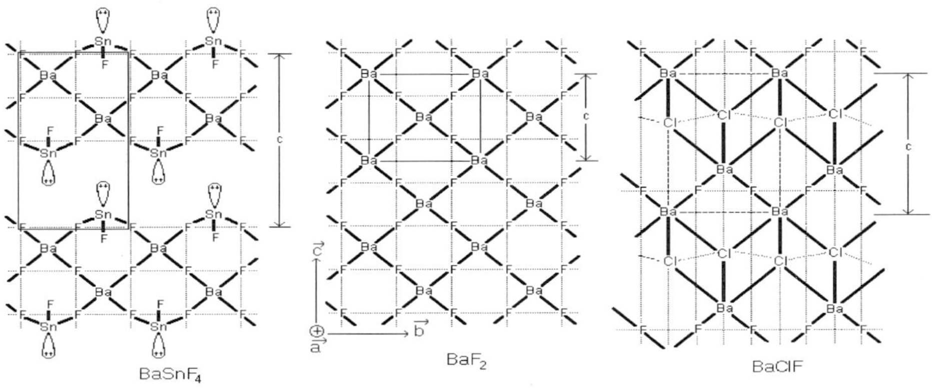

Figure 1 Crystal structure of $BaSnF_4$, BaF_2 and BaClF.

the hybrid orbitals and is therefore highly axial, and it reduces the tin coordination number and distort very strongly the polyhedron of coordination, hence the name *stereoactive* lone pair [1]. In the structure of BaClF (PbClF type), the coordination of Ba is a $BaF_4ClCl'_4$ distorted monocapped square antiprism [2].

A $Ba_{1-x}Sn_xCl_{1+y}F_{1-y}$ solid solution has been prepared by wet and dry methods. The materials obtained by both methods show it is a doubly disordered solid solution, where Ba and Sn are fully disordered while the order between the F and Cl sites, characteristic of the PbClF type structure, is preserved, with however, full disorder of yCl and $(1 - y)$F on the F site for $y > 0$, or of $-y$F and $(1 + y)$Cl on the Cl site for $y < 0$ [3]. X-ray diffraction showed a peak shift that indicates that the unit-cell parameters of the solid solutions precipitated or from dry method are not quite the same. In addition, while most, and often all, of the tin is in the form of Sn^{2+} stannous ions when obtained by the wet method, a mixture of significant amounts of covalently bonded tin and Sn^{2+} ions is observed in samples prepared by the dry method, except at low tin content (small x) or when there is a large excess of chlorine (large y). These differences prompted the present study of the phase transition taking place when the precipitated solid solution is heated.

2 Experimental

Preparation and characterization were carried out according to methods that have already been described elsewhere [3–5].

3 Results and discussions

X-ray powder diffraction shows that all the samples, within the domain of the pure solid solution, give the same set of peaks as BaClF, and no additional lines at low angles and no peak splitting is observed, that would indicate the presence of long range order or lattice distortion, respectively. Therefore, full Ba/Sn^{2+}/Sn(II) covalent disorder and Cl/F disorder, either on the F site ($y > 0$) or on the Cl site ($y < 0$), are present in all $Ba_{1-x}Sn_xCl_{1+y}F_{1-y}$ samples, while the order between the

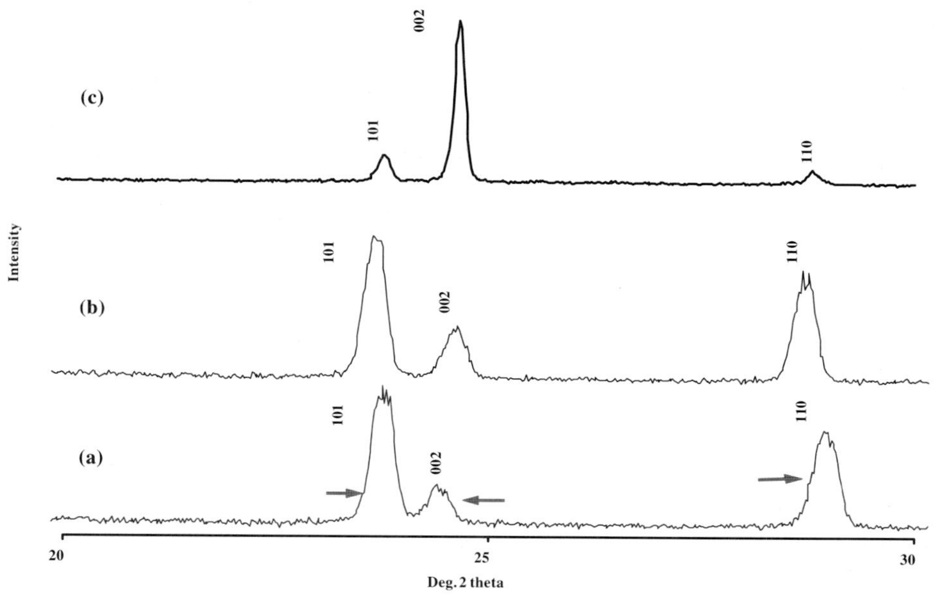

Figure 2 Enlargement of a portion of the X-ray powder diffraction pattern of: **a** precipitated $Ba_{1-x}Sn_xCl_{1+y}F_{1-y}$ (X=0.923), **b** BaClF precipitated by addition of $HF_{(aq)}$ to a solution of $BaCl_2{\cdot}2H_2O$, **c** $Ba_{1-x}Sn_xCl_{1+y}F_{1-y}$ prepared by the dry method (x=0.150, y=0.050, 43 h at 500°C).

Cl and the F sites, characteristic of the PbClF structural type, is preserved. A close examination shows a small, although significant, shift of the peaks of precipitated $Ba_{1-x}Sn_xCl_{1+y}F_{1-y}$ (Figure 2a) relative to unsubstituted BaClF (Figure 2b). The shift of (110) to higher angles (see arrows on Figure 2a) results in a decrease of the *a* unit-cell parameter by 0.91% relative to unsubstituted BaClF (Figure 2b), while the shift of (002) to lower angles gives an increase of *c* by 0.97%. These changes can be accounted for by (a) the substitution of xBa^{2+} by xSn^{2+}, which are smaller (decrease of *a*), and (b) the substitution of *y*F by *y*Cl, which is larger, and therefore shifts out of the fluorine planes by $\pm z$ (increase of *c*).

No change is observed on heating at 100 °C, and the transformation is very slow at 200 °C. At 300 °C and above, it is very rapid. After heating ≥300 °C, the same set of peaks is observed, with however two significant changes. Firstly, the peaks have shifted from the positions of Figure 2a to those of Figure 2c, which are very similar to the positions observed in unsubstituted BaClF (Figure 2b), to give unit-cell parameters very similar to those of BaClF [3]. Secondly, a drastic change of relative peak intensities of (002) and (110) occurs on heating (Figure 2a and c). This can be explained by an increase of preferred orientation. The barium coordination in BaClF is a $BaF_4ClCl'_4$ distorted monocapped square antiprism. While the axial Ba–Cl distance is about equal to the sum of the ionic radii of the Ba^{2+} and Cl^- ions, the four Ba–Cl′ distances are significantly longer. They are the weak links of the structure, making cleavage parallel to the (*a*,*b*) plane favored relative to other directions, hence an increase of the (00l) peaks and a weakening of all (hk0), although it is not as drastic as in $BaSnF_4$ where the sheets of lone pairs parallel to (*a*,*b*) create very efficient cleavage planes (Figure 1). The reason for the drastic enhancement of preferred orientation after heating can be attributed to the large

Table I Relevant Mössbauer parameters for $Ba_{1-x}Sn_xCl_{1+y}F_{1-y}$ sample JK-4, X=0.882) before and after heating

Sn type	Mössbauer parameters	Before heating		After heating	
		RT	16 K	RT	16 K
Sn^{2+}	δ	3.92	4.08	4.08	4.24
	Γ	1.20	1.84	1.20	1.42
	%Abs.	3.6	15.7	1.6	6.9
Sn(II) Covalent	δ	3.04	3.25	3.43	3.58
	Δ	1.76	1.69	1.29	1.32
	%Abs.	0.4	2.5	0.4	3.7
	%Abs. of total Sn(II)	14.3	13.7	20.0	34.9
Sn(IV)	δ	0.05	0.00	−0.14	0.04
	Γ	1.44	1.45	1.44	1.42
	%Abs.	0.8	3.3	2.4	2.9
	%Abs. of total Sn(IV)	16.7	15.3	54.5	21.5

Notes: δ = isomer shift (in mm/s, ±0.01) relative to $CaSnO_3$ at RT, Δ = Quadrupole splitting (in mm/s, ±0.01), Γ = Line width at half height (in mm/s, ±0.01), % Abs. = Maximum γ-ray absorption (at peak maximum), RT=Ambient temperature (not regulated)

increase of the amount of covalently bonded tin(II), all of which can be assumed to have their stereoactive lone pair parallel to the *c* axis of the unit-cell, thereby favoring cleavage parallel to (*a*,*b*).

Table I and Figure 3 show that before heating, the ambient temperature Mössbauer spectrum (Figure 3a top) is mainly made of a single tin(II) line at 3.92 mm/s characteristic of the Sn^{2+} stannous ion and a weak tin(IV) line at 0.05 mm/s that can be attributed to tin(IV) oxide generated by surface oxidation of the particles on contact with air [5]. In addition, a small shoulder on the low velocity site of the Sn^{2+} line indicates the presence of a weak doublet (δ = 3.04 mm/s, Δ = 1.76 mm/s) similar to that of α-SnF_2 (δ = 3.43 mm/s, Δ = 1.53 mm/s) and $BaSnF_4$ (δ = 3.29 mm/s, Δ = 1.53 mm/s), showing that the sample contains a minor amount of covalently bonded tin. This is not surprising since this is the common method of bonding of tin(II) in fluorides and chlorides. The surprising observation is that the major part of tin(II) is in the ionic form. This had been observed before only once, in $SnCl_2$ (δ = 4.07 mm/s, Δ = 0 mm/s at 76 K) [6].

When cooled to 16 K (Figure 3a bottom), the isomer shift of both forms of tin(II) increased, while no significant change was observed for the tin(IV) line. This is due to the second order Doppler shift, and it is therefore a function of $<v^2>$, the mean value of the square of the frequency of the thermal vibrations. The average shift between RT (room temperature) and 16 K is 7.84×10^{-4} mm s^{-1} K^{-1} for covalent Sn(II), 5.97×10^{-4} mm s^{-1} K^{-1} for Sn^{2+}, and about 0 for SnO_2, i.e. the shift decreases with increasing recoil-free fraction, the latter being also a function of thermal vibrations [3]. The observed shifts for tin(II) in $Ba_{1-x}Sn_xCl_{1+y}F_{1-y}$ are about five times as high as the value observed in α-SnF_2 (1.37×10^{-4} $mms^{-1}K^{-1}$), in accordance with the higher recoil-free fraction of α-SnF_2. Freezing the thermal vibrations at low temperature resulted in a dramatic increase of resonant absorption, for example from 3.6 to 15.7% for the Sn^{2+} line.

After heating, significant changes of the RT spectrum (Figure 3b top) have taken place: (a) the intensity of the covalent tin(II) quadrupole doublet has increased, at

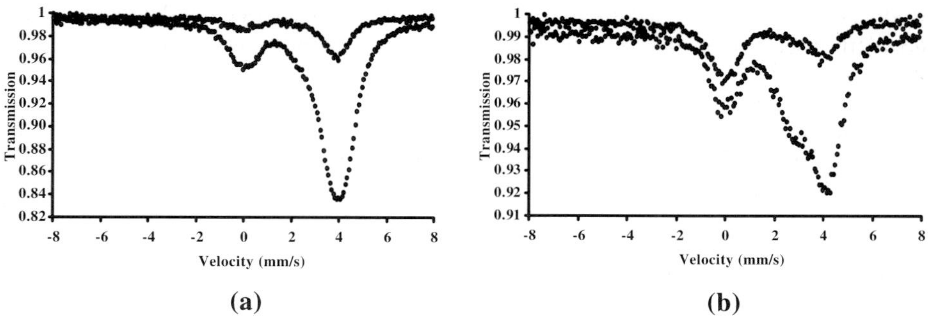

Figure 3 ^{119}Sn Mössbauer spectrum of precipitated $Ba_{1-x}Sn_xCl_{1+y}F_{1-y}$ (X=0.882) at ambient temperature (*top*) and at 16 K (*bottom*): **a** before heating, **b** after heating 96 h at 500°C.

the expenses of the Sn^{2+} line, from 14.3 to 20.0% of the total tin(II) signal; (b) the isomer shift of covalent tin(II) has increased from 3.04 to 3.43 mms^{-1}, while the quadrupole splitting decreased from 1.76 to 1.29 mms^{-1}. The inverse relationship between isomer shift and quadrupole splitting in tin(II) compounds is quite common and shows that the lone pair stereoactivity is the main contributor to the total efg (electric field gradient): the tin hybridization is sp^3d^n (n = 0−2) and an increase of 5*p* and/or 5*d* contribution to the lone pair increases its stereoactivity (more axial lone pair) while increasing the 5*s* contribution that goes in bonding pairs, which are further away from the tin nucleus than the lone pair, thereby decreasing the isomer shift; (c) the Sn^{2+} line has moved from 3.92 to 4.08 mms^{-1} showing a minor increase of ionicity; (d) while the percent absorption of the covalent tin(II) signal remained the same, its contribution to the total Sn(II) signal increased from 14.3 to 20.0%. Therefore, a partial Sn^{2+}→ covalent Sn(II) transition has taken place on heating, and the disturbance it brings to bonding and packing in the solid lattice is responsible for the change of unit-cell parameters observed on heating; (e) the contribution of the Sn(IV) signal to the total Mössbauer spectrum increased from 16.7 to 54.5%, giving the false impression that considerable oxidation took place on heating, even though it was carried out under nitrogen. This large increase is relative to the total Mössbauer spectrum, and it is due to the drastic decrease of the tin(II) recoil-free fraction. The low temperature spectrum (Figure 3b bottom) confirms that there was no significant oxidation of tin(II) on heating, with only a minor increase of Sn(IV) signal from 15.5 to 21.5%. The increase of the covalent Sn(II) contribution to the total Sn(II) signal, from 13.7 to 34.9%, shows that the amount of covalently bonded tin doubled at the phase transition. Tin(II) line shift on cooling is due again to the second order Doppler shift.

4 Conclusion

When the $Ba_{1-x}Sn_xCl_{1+y}F_{1-y}$ solid solution prepared by precipitation, for a barium chloride molar ratio, in the reaction mixture, of X = 0.882, is heated, a phase transition takes place, resulting in a small change of unit-cell parameters (a increases and c decreases) while a partial Sn^{2+}→ covalent Sn(II) transition takes place, that doubles the amount of covalent Sn(II). In addition, a significant decrease of the covalent Sn(II) lone pair stereoactivity takes place, while a minor increase of the ionicity of Sn^{2+} is observed.

Acknowledgements This work was made possible by the support of Concordia University and the Natural Science and Engineering Research Council of Canada. Grateful thanks are also due to the Procter and Gamble Co. (Mason, Ohio) for supporting our Mössbauer laboratory.

References

1. Dénès, G.: In: Wynter, C.I., Alp, E.E. (eds.) Proceedings of the Second Nassau Mössbauer Conference, p. 109. W. C. Brown, Dubuque, IA (1994)
2. Flahaut, J.: J. Solid State Chem. **9**, 124 (1974)
3. Dénès, G., Muntasar, A.: Hyperfine Interact. **153**, 91 (2004)
4. Ruebenbauer, K., Birchall, T.: Hyperfine Interact. **7**, 125 (1979)
5. Dénès, G., Muntasar, A.: Hyperfine Interact **153**, 121 (2004)
6. Donaldson, J.D., Silver, J.: J. Chem. Soc. Dalton Trans. 666 (1973)

Hyperfine Interact (2005) 166:385–390
DOI 10.1007/s10751-006-9297-2

Multifunctional materials exhibiting spin crossover and liquid-crystalline properties

Interplay between spin crossover and liquid-crystal properties in iron(II) coordination complexes

M. Seredyuk · Ana B. Gaspar · V. Ksenofontov · S. Reiman · Y. Galyametdinov · W. Haase · E. Rentschler · P. Gütlich

Published online: 5 December 2006

Abstract The physical characterization of a new class of Fe(II) multifunctional SCO materials exhibiting spin crossover and liquid crystalline properties in the room temperatures region is reported. Mössbauer spectroscopy, magnetic, differential scanning calorimetry (DSC), X-ray powder diffraction (XRD) and optical polarizing microscopy studies have been performed on such materials.

Key words Fe (II) coordination compounds · spin crossover · liquid crystal · multifunctional materials

1 Introduction

Spin crossover (SCO) materials reversibly switch between a nearly diamagnetic low-spin state (LS) and a paramagnetic high-spin state (HS) by varying the temperature, application of pressure or light [1]. A sharp variation of the structure, magnetism, colour and dielectric constant of the system in response to these stimuli may occur in the solid state. Due to their switchable properties the SCO materials may potentially

M. Seredyuk · V. Ksenofontov (✉) · S. Reiman · E. Rentschler · P. Gütlich
Institut für Anorganische und Analystiche Chemie, Johannes-Gutenberg-Universität, Staudinger-Weg 9, D-55099 Mainz, Germany
e-mail: v.ksenofontov@uni-mainz.de

A. B. Gaspar
Institut de Ciència Molecular/Departament de Química Inorgànica, Universitat de València, Edifici de Instituts de Paterna, Apartat de Correus 22085, 46071 València, Spain

Y. Galyametdinov
Kazan Physical technical Institute, Russian Academy of Science, Sibirsky Tract 10/7, 420029 Kazan, Russia

W. Haase
Institute of Physical Chemistry, Darmstad University of Technology, Petersenstrasse 20, 64287 Darmstad, Germany

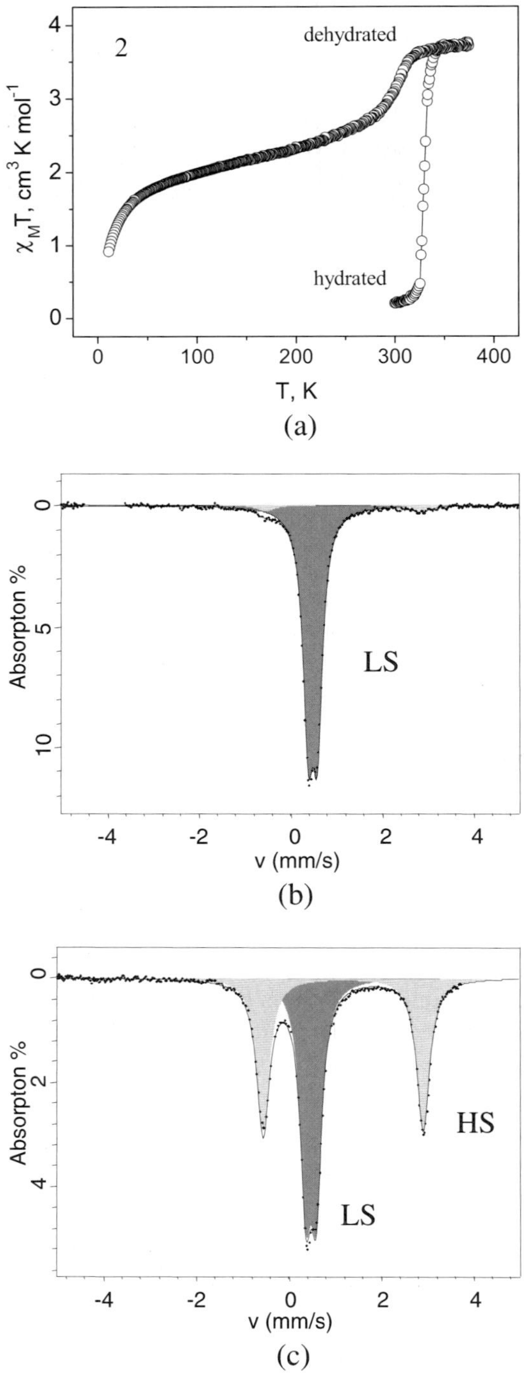

Figure 1 **a** Magnetic properties of the complex 2 in the form of $\chi_M T$ vs. T recorded at a rate of 1 K/min. **b** Mössbauer spectrum of complex 2 acquired at 4.2 K, *dark grey doublet* (*LS*) and *light grey doublet* (*HS*). **c** Mössbauer spectrum of dehydrated 2 at 4.2 K, *dark grey doublet* (*LS*) and *light grey doublet* (*HS*).

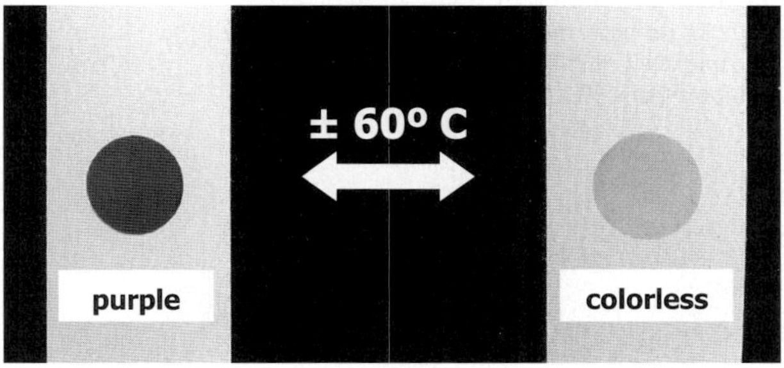

Figure 2 Change of color in the SCO-LC films around 60°C.

be useful for rewritable optical, thermal or pressure memories at a nanometric scale [1–4].

We have recently been investigating the tailoring of multifunctional materials combining spin crossover and liquid crystalline behaviour. The specific peculiarities of SCO and LC materials have stimulated the idea of combining both properties in a single material. Synergic systems with two or more phase transitions of different physical nature may lead to a number of advantages in practical applications, e.g. enhancement of spin transition signals, switching and sensing in different temperature regimes [3, 5]. A first step in this direction was done by Galyametdinov and co-workers, who reported an iron(III) complex combining both LC and SCO properties [6]. Both properties are not synchronous in that material and appear in different temperature regimes. Hayami and co-workers [7] have recently observed a spin transition around 230 K in the mononuclear iron(II) complex [Fe(3C16-L)$_2$(NCS)$_2$] (3C16-L:3,4,5-tris(hexadecyloxy)-N-((pyridin-2-yl)methylene)benzenamine), which subsequently revealed the smetic phase above 350 K.

We have been investigating the possibility to synchronize both transitions, spin transition and crystal–liquid crystal transition, in Fe(II) complexes, where spin crossover takes place between $S = 0$ ($^1A_{1g}$) and $S = 2$ ($^5T_{2g}$) states. Self-assembly of 3,5-dihydroxy-N-4H-1,2,4-triazol-4-ylbenzamide and Fe(p-$CH_3(C_6H_4)SO_3)_2$ have afforded a family of complexes with general formula [Fe(C_n-trz)$_3$](p-$CH_3(C_6H_4)SO_3)_2{\cdot}H_2O$, n being the number of carbon atoms of the alkyl chain (8(1), 10(2), 12(3)) [5, 8, 9], which show spin transition and liquid crystalline properties in the room temperature region. Here we report on the physical characterization of such materials by means of Mössbauer spectroscopy, magnetic, differential scanning calorimetry (DSC), X-ray powder diffraction (XRD) and optical polarizing microscopy studies.

2 Experimental results and discussion

The magnetic properties of the typical complex 2 in the form of $\chi_M T$ vs. T, χ_M being the molar susceptibility and T the temperature, are depicted in Figure 1a. At 300 K the values of 0.03 (1) cm^{3}K mol^{-1}, 0.20 (2) cm^{3}K mol^{-1} and 0.10 (3) cm^{3}K mol^{-1} indicates that all complexes are in the LS state. Mössbauer spectra recorded at 4.2 K confirm that the residual HS fraction does not exceed more than 10% of iron(II) ions in all materials (Figure 1b). For complexes 1–3 the Mössbauer parameters at 4.2 K are: HS population (%): 9.29 (1), 4.76 (2) and 0 (3); IS$_{HS}$: 1.18(1) mm/s (1), $\Delta\Delta E_{QHS}$: 3.262(2) mm/s (1) and IS$_{HS}$: 1.15(3) mm/s (2), ΔE_{QHS}: 3.38(6) mm/s (2); LS

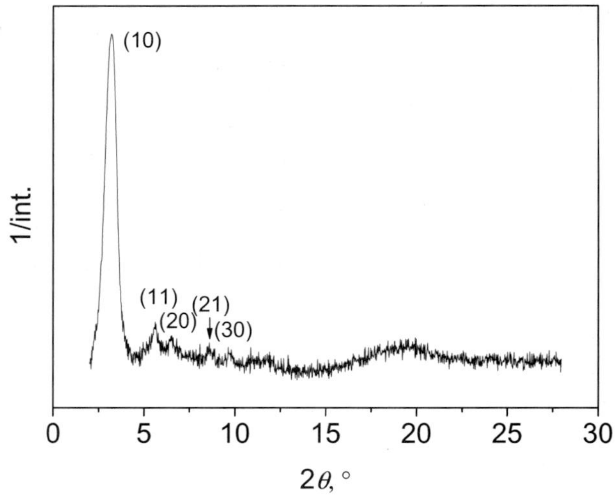

Figure 3 XRD powder diffraction patterns of complexes 2 at different temperatures (K).

population (%): 90.71 (1), 95.24 (2) and 100 (3), IS_{LS}: 0.47(1) mm/s (1), IS_{LS}: 0.48(1) mm/s (2) and 0.47(3) mm/s (3), ΔE_{QLS}: 0.20(1) mm/s (1), ΔE_{QLS}: 0.20(1) mm/s (2) and 0.20(2) mm/s (3).

Upon heating $\chi_M T$ increases abruptly within a few Kelvin, attaining the values of 3.53 (1) $cm^3K\ mol^{-1}$, 3.74 (2) $cm^3K\ mol^{-1}$ and 3.47 (3) $cm^3K\ mol^{-1}$ at 346 K, 342 K, 359 K, respectively. This clearly indicates the occurrence of spin crossover behaviour, which additionally is accompanied by a pronounced change of color [from purple (LS state) to colorless (HS state)] (Figure 2). The characteristic temperatures $T_{1/2}$ at which 50% of the molecules are in the LS and HS states, respectively, are: 330 K (1) and 340 K for 2 and 3. TGA experiments have shown that dehydration takes place in the same temperature region where spin crossover occurs. Dehydration of the materials is accomplished at around 360 K for 1, 350 K for 2 and 370 K in the case of 3. Synergy between dehydration and spin crossover as observed in these materials has been reported in several cases [1, 10].

The magnetic susceptibility of dehydrated samples was recorded down to 4 K. The complexes reveal an incomplete and abrupt spin crossover behaviour in the temperature region of 250–300 K. Around 50% of Fe(II) ions have changed the electronic configuration as can be inferred from the value of $\chi_M T$ at 200 K. A deconvolution analysis of the Mössbauer spectrum at 4.2 K demonstrated that around a half of the iron(II) ions is in the LS state (Figure 2). HS population (%): 46.8, IS_{HS}: 1.18(1) mm/s, ΔE_{QHS}: 3.45(3) mm/s; LS population (%): 53.2, IS_{LS}: 0.48(1) mm/s, ΔE_{QLS}: 0.23(2) mm/s. Decrease of the $\chi_M T$ value below 100 K is attributed to the zero field splitting (ZFS) of the iron atoms remaining in the HS state.

XRD spectra of the complexes 1–3 have been measured in the 300–375 K temperature region and are quite similar. For the typical substance 2 the XRD profile at RT is shown in Figure 3. Diffraction peaks with a *d* spacing of 25.90 Å (1), 27.11 Å (2), 29.03 Å (3) at 298 K are observed in the small angle region. Three minor peaks with *d* spacing values of (1): 14.87 Å, 12.92 Å, 8.42 Å; (2): 15.77 Å, 13.82 Å, 9.12 Å; (3) are also detected, which have been indexed as (11), (20) and

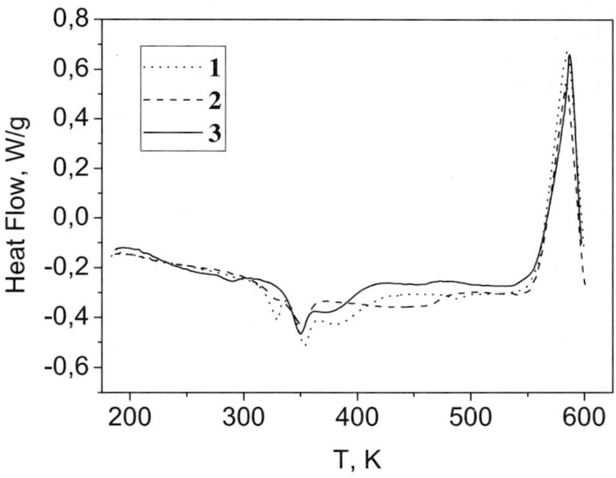

Figure 4 DSC data of complexes 1–3 in the warming mode recorded at the rate of 7 K/min.

(30), respectively. To identify the thermotropic liquid-crystalline mesomophism of 1–3 a hot-stage polarizing optical microscopy apparatus has been used. For all complexes, a discotic columnar mesophase has been identified. On warming, isotropisation occurs near to 500 K, where the clearing points are T_c = 490 K(1), 488 K(2) and 480 K(3). DSC experiments performed at the rate of 7 K/min, in the warming mode, are shown in Figure 4. For compound 1, broad peaks are observed at temperatures of T_1 = 329 K, T_2 = 353 K, T_3 = 381 K, T_4 = 489 K and T_5 = 589 K. The thermal dependence of the heat flow for 2 and 3 is similar to that of 1 despite in these compounds two peaks are observed in the 300–385 K temperature interval. The anomalies are observed at T_1 = 327 K, T_2 = 351 K, T_4 = 486 K, and T_5 = 583 K in 2 and at T_1 = 349 K, T_2 = 377 K, T_4 = 489 K and T_5 = 589 K in the case of 3. The peaks located at temperatures T_1, T_2, T_3 correspond to the SCO and dehydration processes, which match exactly with the reported magnetic and TGA data. T_4 and T_5 are assigned to the isotropisation and decomposition temperatures, respectively, as has been well confirmed by optical polarizing microscopy and TGA analysis. DSC measurements performed at a rate of 5 K/min between 175 K and 400 K, in the cooling and warming modes have proved the reversibility of the processes.

An interesting and important feature is the possibility to obtain these materials in the form of thin films by dissolving ca. 10 mg of complexes in $CHCl_3$ (1 ml) and subsequent layering of the solution onto a glass plate and allowing the solvent to be evaporated (Figure 2). Reversible change between purple (LS state) and colorless (HS state) coloration of the films by heating or cooling around 60°C have been observed without fatigue under sealed conditions.

3 Conclusion

In summary, thermochromic liquid crystals operating in the room temperature region have been successfully synthesized. Mössbauer spectroscopic studies confirmed that SCO behaviour occurs in the temperature range at which the materials show a discotic columnar mesophase. Due to the ability of liquid crystals to form thin layers it is possible to obtain SCO films. Understanding of possible interplay/synergy between the different phase transitions occurring in these materials is still a perspective study, but certainly is exciting and attractive to chemists as well as physicists. We expect that the electrical field should be a suitable physical variable influencing the spin state in these multifunctional materials. The study by means of Mössbauer spectroscopy of influence of an electrical field on the SCO process is currently pursued in our group.

Acknowledgements The authors wish to express their gratitude to Dr. H. Ehrenberg of the Darmstadt University of Technology, Darmstadt, and to Dr. Z. Tomkowicz for their help in carrying out XRD powder diffraction measurements. We also thank Ms. E. Muth and Ms. P. Räder of the Max-Planck-Institute for Polymer Research, Mainz, for their help in performing TGA and DSC measurements. A. B. G. thanks the Alexander von Humboldt Foundation for work visiting fellowships and the MEC for a Ramón y Cajal contract and for the project CTQ2004-03456. Y. G. acknowledges the finantial support from DLR/BMBF project RUS 05/003 and RFBR grant no. 03-03-32571. We also acknowledge the financial help from the Deutsche Forschungsgemeinschaft (Priority Program 1137 "Molecular Magnetism"), the Fonds der Chemischen Industrie and the Materialwissenschaftliches Forschungszentrum der Universität Mainz.

References

1. Gütlich, P., Goodwin, H.A. (eds.): Spin crossover in transition metal compounds. Top. Curr. Chem. **233**, **234**, **235** (2004)
2. Real, J.A., Gaspar, A.B., Niel, V., Muñoz, M.C.: Coord. Chem. Rev. **236**, 121 (2003)
3. Real, J.A., Gaspar, A.B., Muñoz, M.C.: J. Chem. Soc., Dalton Trans. **12**, 2062–2079 (2005)
4. Gütlich, P., Hauser, A., Spiering, H.: Angew. Chem., Int. Ed. Engl. **33**, 2024 (1994)
5. Gaspar, A.B., Ksenofontov, V., Seredyuk, M., Gütlich, P.: Coord. Chem. Rev. **249**, 2661 (2005)
6. Galyametdinov, Y., Ksenofontov, V., Prosvirin, A., Ovchinnikov, I., Ivanova, G., Gütlich, P., Haase, W.: Angew. Chem., Int. Ed. Engl. **40**, 4269 (2001)
7. Hayami, S., Danjobara, K., Inoue, K., Ogawa, Y., Matsumoto, N., Maeda, Y.: Adv. Mater. **11**, 16 (2004)
8. Seredyuk, M., Gaspar, A.B., Ksenofontov, V., Reiman, S., Galyatmedinov, Y., Haase, W., Rentschler, E., Gütlich, P.: Chem. Mater. **18**, 2513–2519 (2006)
9. Gaspar, A.B., Seredyuk, M., Ksenofontov, V., Reiman, S., Rentschler, E., Gütlich, P.: In: Proceedings of ICMM2004, Tsukuba, Japan
10. Kahn, O., Kröber, J., Jay, C.: Adv. Mater. **4**, 718 (1992)

Hyperfine Interact (2005) 166:391–396
DOI 10.1007/s10751-006-9298-1

Ordering in $Fe^{II–III}$ Hydroxysalt green rusts from XRD and Mössbauer analysis (chloride, carbonate, sulphate, oxalate...); about the structure of hydrotalcite-like compounds

J.-M. R. Génin · M. Abdelmoula · R. Aïssa · C. Ruby

Published online: 3 November 2006

Abstract The domain of existence of $Fe^{II–III}$ hydroxysalt green rusts (GR) is characterised by the ratio $x = [Fe^{III}]/[Fe_{total}]$ lying in [1/4, 1/3] as it is for any layered double hydroxide (LDH). Since Fe^{III} cations balance the charges of intercalated anions, cation and anion distributions are correlated. However, the distribution of anions depends on their charge, shape and size. For Cl^- and CO_3^{2-}, x belongs to the complete range whereas for $C_2O_4^{2-}$ and SO_4^{2-} it is limited to $x = 1/4$ and 1/3, respectively. Mössbauer spectra present in all cases two ferrous and one ferric doublets, which characterise each GR by their relative abundance. Intermediate values of x are in fact obtained by intrincate domains of two types, one where the cation periodicity is $2\times a_0$ and the other one $\sqrt{3}\times a_0$ if a_0 is the lattice parameter.

Key words green rust · Mössbauer · ordering · chloride · carbonate · sulphate · oxalate · layered double hydroxide

1 Introduction

Main crystallographic features of $Fe^{II–III}$ hydroxysalts named green rusts are known from X-ray diffraction (XRD). They belong to the general family of layered double hydroxides (LDH) among which divalent and trivalent cations come from the same species, iron. The structure is classified in green rust one, GR1, and green rust two, GR2, according to the sequence of stacking of $Fe(OH)_2$ brucite-like layers that depends on the shape and type of inserted anions. Thus a general formula can be written depending on the anion A^{n-} as $[Fe^{II}_{(1-x)}Fe^{III}_x(OH)_2]^{x+} \cdot [(x/n)\ A^{n-} \cdot (mx/n)\ H_2O]^{x-}$. The sequences are $A, c, B, i, B, a, C, j, C, b, A, k, A$... and A, c, B, i, j, A...

J.-M. R. Génin (✉) · M. Abdelmoula · R. Aïssa · C. Ruby
Laboratoire de Chimie Physique et Microbiologie pour l'Environment,
UMR 7564 CNRS-Université Henri Poincaré-Nancy 1, Equipe Microbiologie et Physique,
Département Matériaux et Structures, ESSTIN, 405 rue de Vandoeuvre,
F-54600 Villers-lès-Nancy, France
e-mail: jmg@lcpe.cnrs-nancy.fr

where *A*, *B* and *C* designate OH^- ions layers, *a*, *b*, and *c* Fe cation layers and *i*, *j* and *k* interlayers constituted of anions and water molecules leading to $R\bar{3}m$ and $P\bar{3}m1$ space group, respectively [1, 2]. Long range order was determined recently by Rietveld analysis for hydroxysulphate GR2(SO_4^{2-}), corresponding to the distribution of cations and anions as well [2]. In contrast, XRD did not allowed us to detect any long range order in GR1 and the question stayed open to discussion as it was for other LDHs. In this paper, Mössbauer spectroscopy is used to demonstrate that long range order also exists in GR1 even though it cannot be detected by XRD and that two types of periodicities for Fe^{III} ions, one corresponding to $2 \times a_0$ and the other one to $\sqrt{3} \times a_0$ if a_0 is the lattice parameter can explain any intermediate composition. These types of long range ordering are likely existing also in other LDHs which are isomorphous to GRs as for example hydrotalcite, pyroaurite or reeversite and where Mössbauer spectroscopy is of no use.

2 Experimental

Controlled preparations of green rusts have been fully documented. There are essentially two methods: (1) the aerial oxidation of a precipitate of $Fe(OH)_2$ in the presence of an excess of foreign anions within the solution, (2) the coprecipitation of ferrous and ferric salts aiming at a given value of the ratio $x = [Fe^{III}]/[Fe_{total}]$ in the presence of foreign anions. In the first method, the electrode potential of the solution *E* and its pH are recorded versus time while stirring in the air the solution contained in a flask with a magnetic rod at constant velocity. The curves display three stages and the green rust is obtained at the first inflection point, i.e. at the end of the first stage when all ferrous hydroxide gets transformed into GR. Hydroxychloride GR1(Cl^-) was prepared this way by adjusting the excess of Cl^- with respect to the stoichiometry of $Fe(OH)_2$ when mixing $FeCl_2$ and NaOH. A large range of composition was obtained from $\{[Cl^-]/[OH^-]\} = 1.14$ to 1.20, which led to compounds $Fe^{II}_3\,Fe^{III}\,(OH)_8\,Cl$ [3] and $Fe^{II}_2\,Fe^{III}\,(OH)_6\,Cl$ [4], respectively. XRD patterns were those of a GR1 with parameters $a_0 = 0.319$ nm and $c = 2.385$ nm. Mössbauer spectra were measured by use of a constant-acceleration spectrometer with a 50 mCi source of ^{57}Co in a Rh matrix that was calibrated with a 25 μm foil of α-Fe at room temperature. Since GRs are very sensitive to oxidation, the spectra were measured at 78 or 20 K (Figure 1a–b). Once achieved the formation of the GR, an event which was accurately detected on the *E* vs time curve, the precipitate was filtered under inert atmosphere, set in the sample holder and introduced in the cryostat for measurements, where the oxidation was definitively stopped.

Coprecipitation was used to prepare GR1(CO_3^{2-}) as explained elsewhere [5]. Some phosphate was necessary to stabilise the GR. Measured parameters were $a_0 =$ 0.317 nm and $c = 2.23$ nm. As previously explained, values of ratio $x = [Fe^{III}]/[Fe_{total}] < 0.25$ led to a mixture of GR1(CO_3^{2-}) and $Fe(OH)_2$ whereas values of $x >$ 0.33 led to a mixture of GR1(CO_3^{2-}) and magnetite [6]. Only values of *x* in the range [0.25, 0.33] led uniquely to GR1(CO_3^{2-}) phase as checked by XRD and Mössbauer spectroscopy [6].

A Fe^{II-III} hydroxyoxalate was also recently synthesised by using the first method [6]. However only one value for *x* at 0.25 was able to obtain a unique GR phase, which is a GR1($C_2O_4^{2-}$) of formula $Fe^{II}_6\,Fe^{III}_2\,(OH)_{16}\,C_2O_4$ according to the XRD

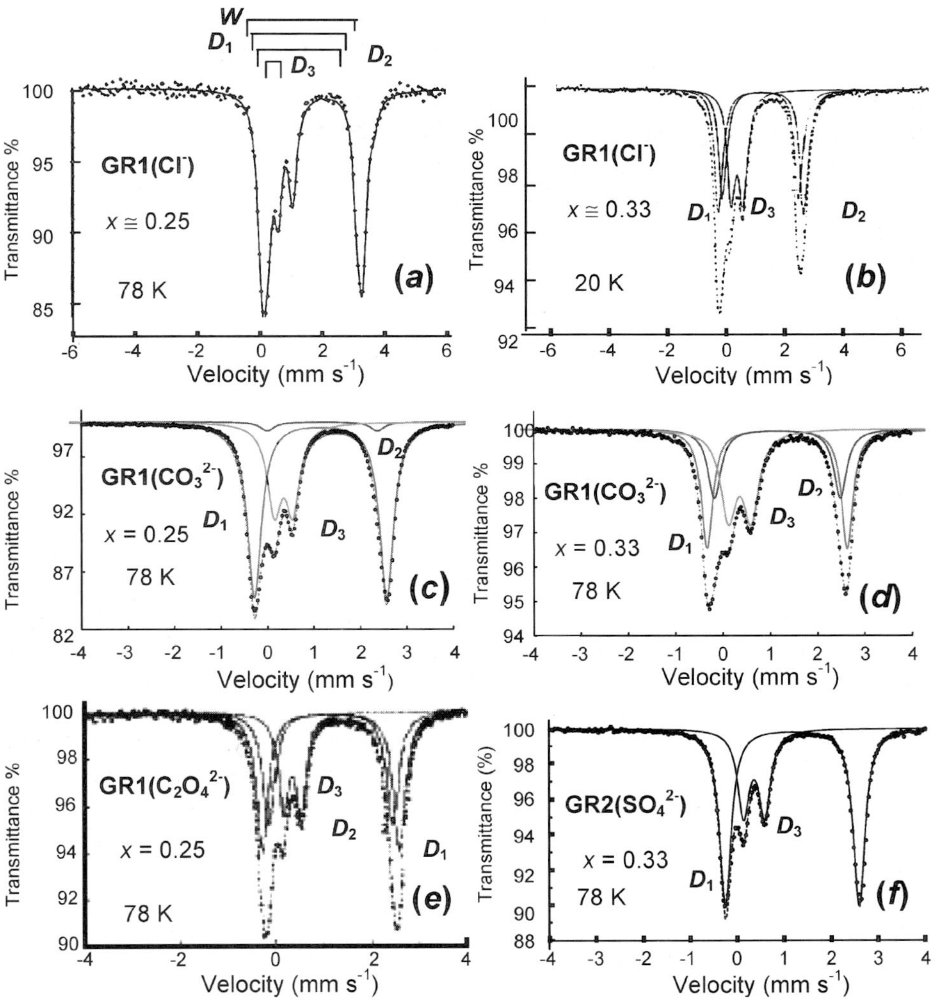

Figure 1 Mössbauer spectra measured at 20 or 78 K of **a** and **b** Fe^{II-III} hydroxychloride green rust for $x = 0.25$ and 0.33, **c** and **d** Fe^{II-III} hydroxycarbonate green rust for $x = 0.25$ and 0.33, **e** Fe^{II-III} hydroxyoxalate green rust for $x = 0.25$, the only possible value, **f** Fe^{II-III} hydroxysulphate green rust for $x = 0.33$, the only possible value.

pattern that gives $a = 0.3195$ nm and $c = 2.34$ nm. In contrast studies concerning the other type of stacking for GR, i.e. GR2 as obtained for SO_4^{2-} or SeO_4^{2-}, led to one unique value of x at 0.33 [2, 8].

3 Results and discussion

Mössbauer spectra for all GRs are presented in Figure 1 and hyperfine parameters obtained by fitting Lorentzian shape lines are gathered in Table I. Spectra have strong similarities and consist of two doublets for GR2 and three for GR1s. The spectrum for $GR2(SO_4^{2-})$ (Figure 1f) comprises one ferrous doublet D_1 with isomer

Table I Hyperfine parameters of Mössbauer spectra of GR1s measured at various temperatures from Figure 1

x	GR1(Cl^-) 0.25			GR1(Cl^-) 0.33			GR1(CO_3^{2-}) 0.25			GR1(CO_3^{2-}) 0.33			GR1($C_2O_4^{2-}$) 0.25		
Ref T (K)	[3] 78			[4] 20			[6] 78			[6] 78			[7] 78		
	δ	Δ	RA	δ	Δ	RA	δ	Δ	RA	δ	Δ	RA	δ	Δ	RA
D_1	1.26	2.88	48	1.31	2.89	37	1.28	2.97	62	1.28	2.97	48	1.28	2.84	42.7
D_2	1.25	2.60	24	1.29	2.57	32	1.28	2.55	12	1.28	2.55	18	1.28	2.58	31.7
D_3	0.47	0.41	24	0.40	0.40	31	0.47	0.43	34	0.47	0.43	34	0.48	0.37	25.6
W^*	1.37	3.36	4												

δ (mm s^{-1}): isomer shift (α-iron as reference at ambient); Δ (mm s^{-1}): quadrupole splitting; *RA* (%): relative abundance. The temperature of measurements is noted with the corresponding reference.

* *W*: Fe^{II} in ice.

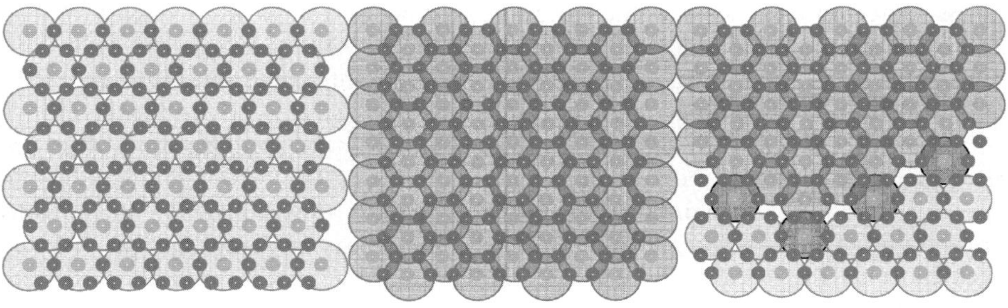

Figure 2 Domains with long-range order of Fe^{III} cations in the hexagonal pavement of brucite-like layers found in Fe^{II-III} hydroxysalt green rusts. **a** Long range order at x = 0.25 with periodicity $a = 2\times a_0$. **b** Long range order at x = 0.33 with periodicity $a = \sqrt{3}\times a_0$. Ordered domains that can adjust any value of x in the [0.25, 0.33] range. These domains can be extended to any LDH.

shift δ = 1.27 mm s^{-1} and quadrupole splitting Δ = 2.88 mm s^{-1} and one ferric doublet D_3 with δ = 0.47 mm s^{-1} and Δ = 0.44 mm s^{-1} at 78 K [2]. The relative abundance of $D_1 : D_3$ is 2:1 that is consistent with the observed value of x at 0.33 and that leads to the formula $Fe^{II}_4\ Fe^{III}_2\ (OH)_{12}\ SO_4$. As obtained from the Rietveld analysis, there is a long range order and lattice parameters are a = 0.5524 nm and c = 1.101 nm. Note that $a = \sqrt{3}\times a_0$ and thus only one Fe^{II} site exists giving one unique D_1 ferrous doublet [2, 6].

In contrast, all spectra of GR1s have two ferrous doublets D_1 and D_2 and one ferric doublet D_3. Hyperfine parameters are practically the same for a given doublet whatever the intercalated anion in the GR1 but the relative abundance is quite different (Table I). Firstly, the intensity of D_3 gives directly the ratio x. Secondly, there is only one ferric doublet with small line-width. This means that there exists only one Fe^{III} site and that the environments of Fe^{III} ions are all the same. This observation is most important since it demonstrates that cations are ordered and that each Fe^{III} ion is surrounded by six Fe^{II} ions within the first nearest neighbour shell. A random distribution among Fe^{III} cations would yield to binomial distributions of various environments as found in alloys. Thus electrostatic repulsion between Fe^{III} cations explains why the value of x that is observed is limited to 0.33. This upper limit for x gives rise to a long range order where the lattice parameter is in fact $a = \sqrt{3}\times a_0$ but it cannot be detected by XRD since it is the cation sublattice.

Two ferrous doublets D_1 and D_2 are observed. A quadrupole splitting value Δ of 2.9 mm s^{-1} for D_1 is attributed to Fe^{III} cations that do not feel the influence of intercalated anion since it is the same for all GRs and the electric field gradient is similar whatever the anion. The corresponding value Δ for D_2 varies slightly depending on the type of anion around 2.6 mm s^{-1} and is attributed to Fe^{III} in the vicinity of an anion in the interlayer above or below. The intensity ratio $D_1 : D_2$ depends strongly on the size and shape of the concerned anion and its value can only be explained according to models of distribution of the anions as proposed recently for Cl^- and CO_3^{2-} [6]. For GR1(Cl^-), the relative abundance for D_1:D_2:D_3 is $(1-2x)$: x: x as can be checked for 0.25 [3] and 0.33 [4] (Table I). Each Fe^{III} is associated to one Cl^- ion and each Cl^- ion has a Fe^{II} in register. For GR1(CO_3^{2-}) the relative abundance of $D_1 : D_2 : D_3$ is $[1-(3/2)x]$: $(x/2)$: x [6]. Each CO_3^{2-} anion is associated to three sites in the cation hexagonal pavement and to two Fe^{III} and one Fe^{II} cations. Thus, the Fe^{II} ions associated to CO_3^{2-} are half the Fe^{III} ions, leading to $D_2 : D_3$ equal to $(x/2) : x$. For $x = 0.25$ the intensity ratio becomes 0.625 : 0.125 : 0.25 [6] and for $x = 0.33$, it becomes 0.50 : 0.167 : 0.333 [5, 6] (Table I).

Experimentally, it is found that the order obtained with oxalate anions does not exceed $x = 0.25$, which corresponds to $Fe^{II}_6\ Fe^{III}_2\ (OH)_{16}\ C_2O_4$ [7]. Firstly, the existence of a long range order where each Fe^{III} cation is surrounded by six Fe^{II} ions can be met also by a periodicity $a = 2 \times a_0$ as illustrated in Figure 2a that must be compared to the periodicity $a = \surd 3 \times a_0$ (Figure 2b). Therefore, it is the room occupied by the anion, which determines the value of x that can be reached. When comparing CO_3^{2-} and $C_2O_4^{2-}$ anions, carbonate occupies three sites whereas oxalate occupies four for the same negative charge. Consequently $C_2O_4^{2-}$ occupies in the hexagonal pavement (4/3) of the area occupied by CO_3^{2-} and the upper limit of x at 0.33 for $GR1(CO_3^{2-})$ becomes $0.33 \times (3/4) = 0.25$ for $GR1(C_2O_4^{2-})$. Incidentally, by the same reasoning one suggests that GR1 intercalating NO_3^- anions has no chance to exist since for an occupancy identical to that of CO_3^{2-} the charge would be half of it. Obtaining a GR1 needs a balance between the size and the charge of the anion.

Any intermediate composition given by a value of x between the limits 0.25 and 0.33 may be accommodated by intricate domains, each one displaying one of the two types of long range order as illustrated in Figure 2c consistently with the hexagonal symmetry. For Mössbauer spectroscopy, there exist three sites, one Fe^{III} and three Fe^{II} in the right proportions. All arguments, which have been used concerning the Fe^{II-III} hydroxysalts here and based on the Mössbauer data are still valid for other LDHs often designated as hydrotalcite-like compounds among which pyroaurite (Mg, Fe), reeversite (Ni, Al) and hydrotalcite (Mg, Al) itself are the most famous.

References

1. Bernal, J.D., Dasgupta, D.R., Mackay, A.L.: Clay Miner. Bull. **4**, 15 (1959)
2. Simon, L., François, M., Refait, Ph., Renaudin, G., Lelaurain, M., Génin, J.-M.R.: Solid State Sci. **5**, 327 (2003)
3. Génin, J.-M.R., Bauer, Ph., Olowe, A., Rézel, D.: Hyperfine Interact. **29**, 1355 (1986)
4. Refait, Ph., Abdelmoula, M., Génin, J.-M.R.: Corros. Sci. **40**, 1547 (1998)
5. Bocher, F., Géhin, A., Ruby, C., Ghanbaja, J., Abdelmoula, M., Génin, J.-M.R.: Solid State Sci. **6**, 117 (2004)
6. Génin, J.-M.R., Aïssa, R., Géhin, A., Abdelmoula, M., Benali, O., Ernstsen, V., Ona-Nguema, G., Upadhyay, C., Ruby, C.: Solid State Sci. **7**, 545 (2005)
7. Refait, Ph., Charton, A., Génin, J.-M.R.: Eur. J. Solid State Inorg. Chem. **35**, 655 (1998)
8. Refait, Ph., Simon, L., Génin, J.-M.R.: Environ. Sci. Technol. **34**, 819 (2000)

Hyperfine Interact (2005) 166:397–402
DOI 10.1007/s10751-006-9299-0

Mössbauer studies of mixed-valence spin-crossover iron complexes with a hexadentate tripod ligand

S. Iijima · F. Mizutani · O. Niwa · N. Matsumoto · Y. Sunatsuki · M. Kojima

Published online: 17 November 2006

Abstract The spin-crossover behaviors of mixed-valence iron compounds $[Fe^{II}H_3L][Fe^{III}L](NO_3)_2$ (1) and $[Fe^{II}H_3L^{Me}][Fe^{III}L^{Me}](NO_3)_2$ (2) have been investigated by ^{57}Fe Mössbauer spectroscopy, where H_3L is a hexadentate N_6 tripod ligand containing three imidazole groups and H_3L^{Me} is its 2-methylimidazole derivative. Deconvolution analyses of the Mössbauer spectra revealed that a two-step SCO (LS Fe^{II}–LS Fe^{III}→HS Fe^{II}–LS Fe^{III}→HS Fe^{II}–HS Fe^{III}) proceeds in each compound on elevating the temperature. Compound 2 exhibited lower spin-transition temperatures than 1. "Frozen-in effect" was observed below 120 and 50 K for 1 and 2, respectively.

Key words spin-crossover · mixed-valence · Mössbauer spectroscopy · iron complex · tripod ligand · two-dimensional structure · hydrogen bonds

1 Introduction

Spin-crossover (SCO) and mixed-valence phenomena are important research topics in metal coordination chemistry. It is expected that the combination of both phenomena would produce new multifunctional materials. We have reported the syntheses and magnetic properties of mixed-valence SCO complexes, $[Fe^{II}H_3L][Fe^{III}L]$

S. Iijima (✉) · F. Mizutani · O. Niwa
National Institute of Advanced Industrial Science and Technology,
Higashi 1-1-1, Tsukuba, Ibaraki 305-8566, Japan
e-mail: s.iijima@aist.go.jp

N. Matsumoto
Department of Chemistry, Faculty of Science, Kumamoto University,
Kurokami 2-39-1, Kumamoto 860-8555, Japan

Y. Sunatsuki · M. Kojima
Department of Chemistry, Faculty of Science, Okayama University,
Tsushima-naka 3-1-1, Okayama 700-8555, Japan

$(NO_3)_2$ (1) and $[Fe^{II}H_3L^{Me}][Fe^{III}L^{Me}](NO_3)_2$ (2) with preliminary Mössbauer results [1,2], where H_3L = tris[2-(((imidazole-4-yl)methylidene)amino)ethyl]amine and H_3L^{Me} = tris[2-(((2-methylimidazole-4-yl)methylidene)amino) ethyl]amine. These compounds assume a 2D sheet structure constructed by imidazole–imidazolate hydrogen bonds that yield extended interactions between SCO sites. In the present work, we have investigated in detail the temperature-dependences of the Mössbauer spectra of 1 and 2 to get information about the individual low-spin (LS)/high-spin (HS) equilibriums at the Fe^{II} and Fe^{III} centers.

2 Experimental

The procedures for preparing compounds 1 and 2 were reported previously [1,2].

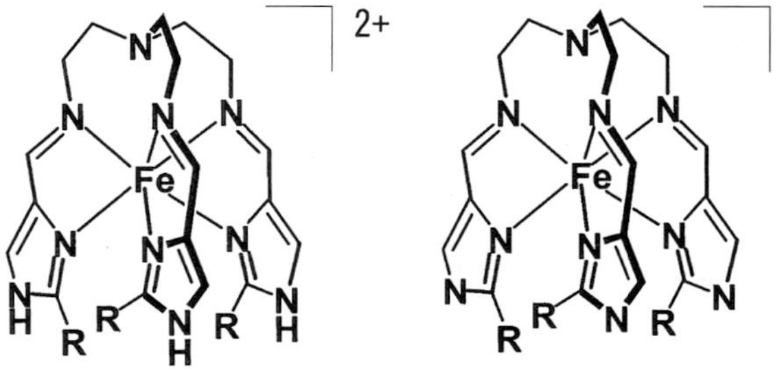

R = H; $[Fe^{II}H_3L]^{2+}$ R = H; $[Fe^{III}L]^0$

R = Me; $[Fe^{II}H_3L^{Me}]^{2+}$ R = Me; $[Fe^{III}L^{Me}]^0$

The ^{57}Fe Mössbauer spectra were recorded using a Wissel 1200 spectrometer and a proportional counter. ^{57}Co(Rh) moving in a constant acceleration mode was used as radioactive source. The calculated lines were obtained by least-squares fitting to Lorentzian peaks. The isomer shifts are reported relative to metallic iron foil. The sample temperature was controlled by a Heli-tran liquid transfer refrigerator (Air Products and Chemicals, Inc.) with an accuracy of ± 0.5 K.

3 Results and discussion

The temperature dependence of the ^{57}Fe Mössbauer spectra of 1 was investigated in the 78–298 K range. The sample was initially rapidly cooled to 78 K (within 3 min from room temperature), and the spectra were measured at selected temperatures in the warming mode, and then in the cooling mode. The representative spectra of 1 in the warming mode are shown in Figure 1.

The spectrum at 78 K measured after the rapid cooling from room temperature (Figure 1e) consisted of two main quadrupole doublets corresponding to LS Fe^{II} (isomer shift δ = 0.43 mm/s, quadrupole splitting ΔE_Q = 0.32 mm/s) and LS Fe^{III}

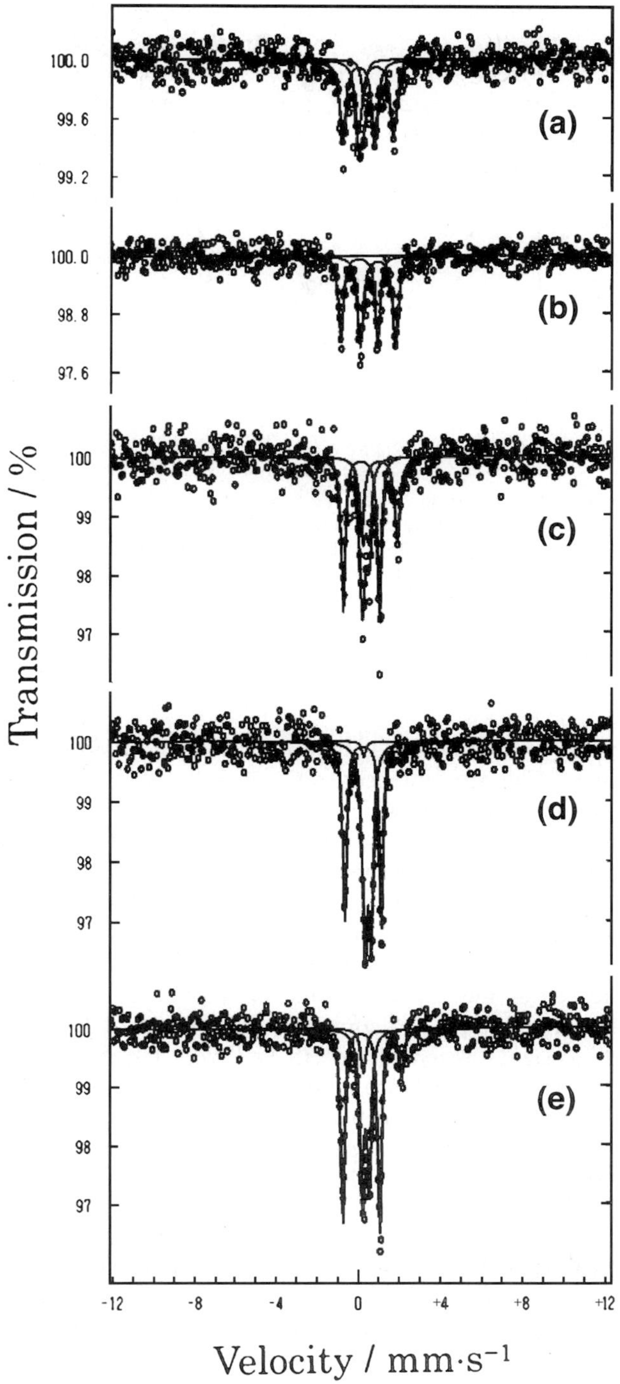

Figure 1 ^{57}Fe Mössbauer spectra of 1 at (**a**) 298 K, (**b**) 200 K, (**c**) 150 K, (**d**) 120 K and (**e**) 78 K. The spectra (**a–d**) were recorded upon warming the sample after rapid cooling to 78 K.

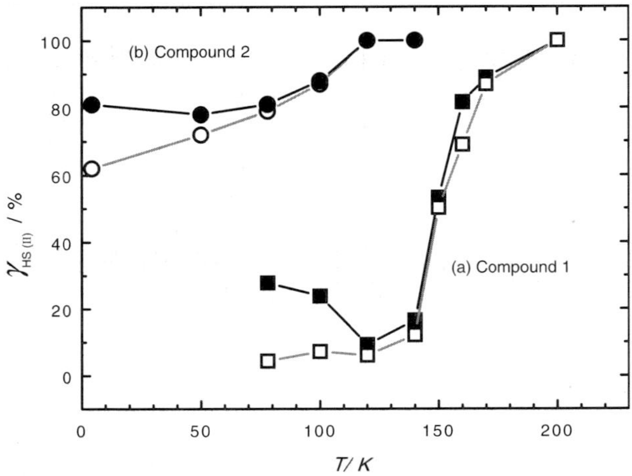

Figure 2 High-spin molar fraction of the Fe^{II} sites ($\gamma_{HS(II)}$) vs. T plots obtained by the deconvolution analyses of the Mössbauer spectra of 1 and 2. The ■ values were obtained from the spectra of 1 upon warming the sample after rapid cooling to 78 K, and the □ values were obtained from the spectra on subsequent cooling. The ● values were obtained from the spectra of 2 upon warming the sample after rapid cooling to 4.2 K, and the ○ values were obtained from the spectra on subsequent cooling.

(δ = 0.20 mm/s, ΔE_Q = 1.83 mm/s), with a weak doublet of HS Fe^{II} (δ = 1.25 mm/s, ΔE_Q = 1.87 mm/s). The area intensity ratio of Fe^{II} (LS Fe^{II} + HS Fe^{II}) to Fe^{III} (LS Fe^{III}) was found to be 54:46, which is close to the 50:50 ratio expected from the chemical formula of 1. When the temperature was raised to 120 K, the relative intensity of the HS Fe^{II} component decreased to reach a minimum value (Figure 1d), suggesting the presence of "frozen-in effect" below this temperature. On further elevation of the temperature, the onset of SCO from LS Fe^{II}–LS Fe^{III} to HS Fe^{II}–LS Fe^{III} state was observed in the Mössbauer spectra. That is, the HS-Fe^{II} absorption increased at the expense of the LS Fe^{II} absorption in the 120–200 K range (Figure 1c), and, the LS Fe^{II} component disappears at 200 K (Figure 1b). On the other hand, no signal assignable to HS Fe^{III} was detected in this temperature range. The temperature dependence of the molar fraction of HS Fe^{II} versus total Fe^{II} ($\gamma_{HS(II)}$), derived from the Mössbauer peak areas, is shown in Figure 2a. From the $\gamma_{HS(II)}$-T plots, the transition temperature ($T_{1/2}$) was estimated to be ca. 150 K both in the warming and in the cooling modes. The "frozen-in effect" in the Fe^{II} sites is clearly shown by the higher $\gamma_{HS(II)}$ value in the warming mode than that in the cooling mode at each temperature below 120 K. The temperature dependence of γ_{HS} mentioned here is essentially consistent with the magnetic susceptibility results [1]. The thermal variation curve of $\chi_M T$ for 1 (χ_M=molar magnetic susceptibility) was reported to exhibit further but gradual increase over ca. 200 K, which suggests the onset of the next step of spin transition, HS Fe^{II}–LS Fe^{III}→HS Fe^{II}–HS Fe^{III}. In the Mössbauer spectrum, the HS-Fe^{III} species became detectable over ca. 250 K, which proves the LS→HS transition at the Fe^{III} sites. Figure 1a shows the spectrum at 298 K, where a small singlet (δ = 0.37 mm/s) of HS Fe^{III} is overlapped with the HS-Fe^{II} (δ = 0.97 mm/s, ΔE_Q = 1.71 mm/s) and LS-Fe^{III} (δ = 0.13 mm/s,

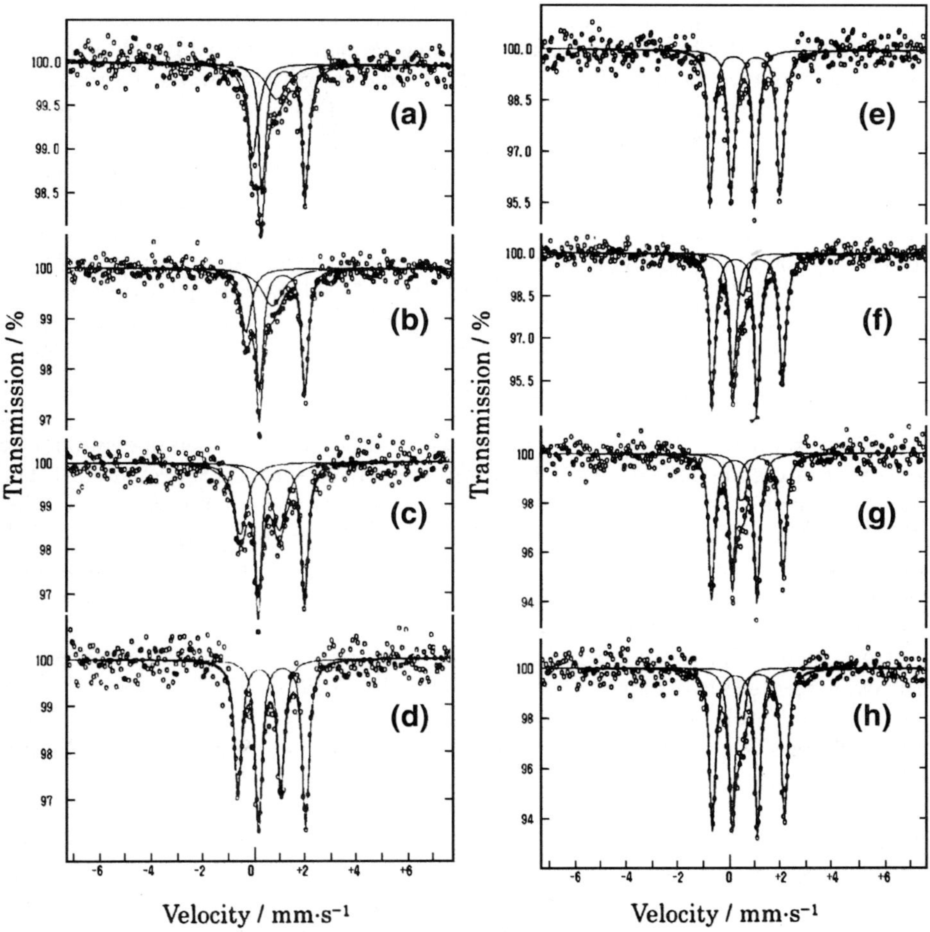

Figure 3 ^{57}Fe Mössbauer spectra of *2* at (**a**) 298 K, (**b**) 250 K, (**c**) 200 K, (**d**) 170 K, (**e**) 120 K, (**f**) 78 K, (**g**) 50 K and (**h**) 4.2 K. The spectra (**a–g**) were recorded upon warming the sample after rapid cooling to 4.2 K.

E_Q = 1.57 mm/s) absorptions; the molar fraction of HS Fe^{III} versus total Fe^{III}, $\gamma_{HS(III)}$, was estimated to be around 20% at 298 K.

The variable-temperature Mössbauer measurements for 2 were carried out in the temperature range of 4.2–309 K. The sample was initially rapidly cooled to 4.2 K (within 3 min from room temperature), and the spectra were recorded at selected temperatures in the warming mode, and then in the cooling mode. The Mössbauer spectra in the two modes were nearly the same except for those below 50 K, due to "frozen-in effect," similarly to the case of 1. The representative spectra of 2 in the warming mode are shown in Figure 3, and $\gamma_{HS(II)}$ values of 2 in the warming and cooling modes are plotted in Figure 2b.

The spectrum at 4.2 K measured after the rapid cooling from room temperature (Figure 3h) consisted of two main quadrupole doublets corresponding to HS Fe^{II} (isomer shift $\delta = 1.13$ mm/s, quadrupole splitting $\Delta E_Q = 2.04$ mm/s) and LS Fe^{III} ($\delta =$ 0.22 mm/s, $\Delta E_Q = 1.76$ mm/s), with a small doublet of LS Fe^{II} ($\delta = 0.47$ mm/s, $\Delta E_Q =$

0.20 mm/s). The $\gamma_{HS(II)}$ values at 4.2 K were estimated from the area intensities to be 0.82 and 0.61 in the warming and cooling modes, respectively, showing the incomplete nature of the HS Fe^{II}→LS Fe^{II} transition even at the extreme low temperature. On elevating the temperature, the LS Fe^{II} absorption disappeared around at 120 K (Figure 3e) to indicate a pure HS Fe^{II}–LS Fe^{III} state. The above Mössbauer results in the 4.2–120 K range agree well with the results of the magnetic susceptibility measurements [2]. On the elevation of the temperature over 120 K, it was reported that the $\chi_M T$ versus T curve shows a narrow plateau between ca. 120–140 K, and then, the $\chi_M T$ begins to increase further, suggesting the one set of SCO from HS Fe^{II}–LS Fe^{III} to HS Fe^{II}–HS Fe^{III} state. In the Mössbauer spectrum of 2, on the other hand, a broadening of the Fe^{III} doublet and a decrease of its ΔE_Q (ΔE_Q = 1.67 mm/s at 170 K and 0.90 mm/s at 298 K) became apparent over 170 K (Figure 3a–d). This behavior indicates the presence of LS/HS equilibrium at the Fe^{III} sites in the higher temperature region, with a spin interconversion rate comparable to the hyperfine frequency (10^7 to 10^8 s^{-1}) [3,4], while the broadened spectra could not be reliably analyzed by a simple model with the conversion rate and the $\gamma_{HS(III)}$ [3]. The high-energy peak of the Fe^{III} doublet is broader than its low-energy side peak (200–298 K); this peak-width asymmetry implies that the line-shape of the Fe^{III} absorption is influenced also by the slow paramagnetic relaxation (spin flop) process of the HS ($^6A_{1g}$) state with positive V_{zz} (the principal component of the electric field gradient tensor) [4,5].

Combined with the cryomagnetic data [1,2], the present Mössbauer investigation has evidenced the two-step SCO process, LS Fe^{II}–LS Fe^{III}→HS Fe^{II}–LS Fe^{III}→HS Fe^{II}–HS Fe^{III}, for the mixed-valence complexes, 1 and 2. It is probable that the LS→HS transition of the Fe^{II} unit triggers the SCO of the otherwise LS Fe^{III} unit.

References

1. Sunatsuki, Y., Ikuta, Y., Matsumoto, N., Ohta, H., Kojima, M., Iijima, S., Hayami, S., Maeda, Y., Kaizaki, S., Dahan, F., Tuchagues, J.-P.: Angew. Chem., Int. Ed. Engl. **42**, 1614 (2003)
2. Ikuta, Y., Ooidemizu, M., Yamahata, Y., Yamada, M., Osa, S., Matsumoto, N., Iijima, S., Sunatsuki, Y., Kojima, M., Dahan, F., Tuchagues, J.-P.: Inorg. Chem. **42**, 7001 (2003)
3. Maeda, Y., Oshio, H., Takashima, Y., Mikuriya, M., Hidaka, M.: Inorg. Chem. **25**, 2958 (1986)
4. Timken, M.D., Abdel-Maugoud, A.M., Hendrickson, D.N.: Inorg. Chem. **25**, 160 (1986)
5. Blume, M., Tjon, J.A.: Phys. Rev. **165**, 446 (1968)

Hyperfine Interact (2005) 166:403–407
DOI 10.1007/s10751-006-9300-y

Study on charge transfer interaction and valence state of layered perovskite-type mixed valence complex $[NH_3(CH_2)_nNH_3]_2[(Au^{I}I_2)(Au^{III}I_4)(I_3)_2]$ (n = 7, 8), by means of ^{197}Au Mössbauer spectroscopy

Kazuhiro Ikeda · Norimichi Kojima · Yasuhiro Kobayashi · Makoto Seto

Published online: 16 November 2006

Abstract $Cs_2[Au^{I}X_2][Au^{III}X_4]$($X$ = Cl, Br, and I) is well known for the three-dimensional perovskite-type gold mixed valence system. Recently, layered perovskite-type gold mixed valence complexes, $[NH_3(CH_2)_nNH_3]_2[(Au^{I}I_2)(Au^{III}I_4)(I_3)_2]$ (n = 7 and 8), have been synthesized. We have investigated the relationship between the structural dimensionality and the Au^{I}–Au^{III} charge transfer interaction for $Cs_2[Au^{I}I_2][Au^{III}I_4]$ and $[NH_3(CH_2)_nNH_3]_2[(Au^{I}I_2)(Au^{III}I_4)(I_3)_2]$ (n = 7 and 8) by means of ^{197}Au Mössbauer spectroscopy.

Key words ^{197}Au Mössbauer spectroscopy · mixed valence complex · layered perovskite

1 Introduction

$Cs_2[Au^{I}X_2][Au^{III}X_4]$($X$ = Cl, Br, and I) is well known for the perovskite-type gold mixed valence system [1–4]. Their characteristic properties of crystal structure and mixed valency are quite similar to those of $BaBiO_3$, which is the parent compound of the superconductors, $Ba_{1-x}KBiO_3$ and $BaBi_{1-x}PbO_3$ [5]. The difference between the Au^{I} and Au^{III} valence states decreases in the order of increasing charge-transfer interaction through bridging halogens, i.e. I > Br > Cl. This system undergoes the pressure-induced and photo-induced Au valence transition from the mixed valence state of $Au^{I,III}$ to the single valence state of Au^{II} [4, 6]. From these interests, we have systematically investigated the crystal structure and the charge transfer interaction and electronic states of hetero-halogen bridged gold mixed valence complexes $Cs_2[Au^{I}X_2][Au^{III}Y_4]$($X$, Y = Cl, Br, I) and found that the charge transfer

K. Ikeda (✉) · N. Kojima
Graduate School of Arts and Sciences, The University of Tokyo, Tokyo 153-8902, Japan
e-mail: kazuhiro@activemail.jp

Y. Kobayashi · M. Seto
Kyoto University Research Reactor Institute, Osaka 590-0494, Japan

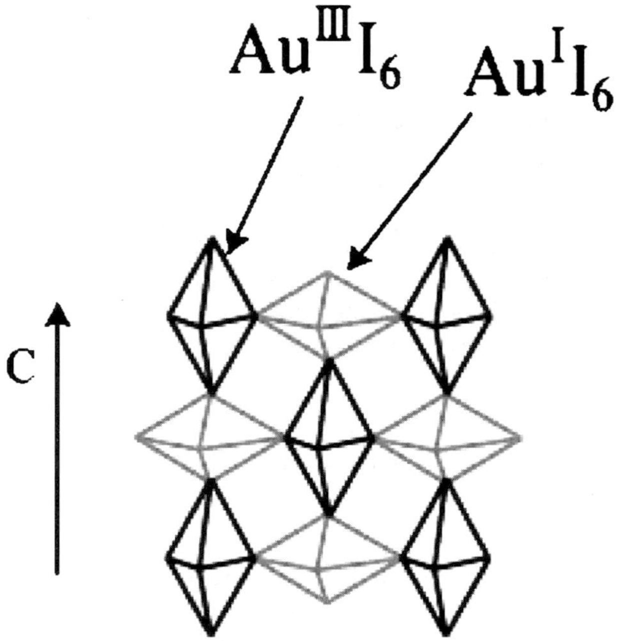

Figure 1 Three-dimensional metal-halogen framework of $Cs_2[Au^{I}I_2][Au^{III}I_4]$.

interaction between Au^{I} and Au^{III} governs the pressure induced valence transition in $Cs_2[Au^{I}X_2][Au^{III}Y_4](X, Y = Cl, Br, I)$ [7]. Recently, Castro and Guloy synthesized layered perovskite-type gold mixed valence complexes $[NH_3(CH_2)_nNH_3]_2[(Au^{I}I_2)(Au^{III}I_4)(I_3)_2]$ (n = 7 and 8) [8]. Being stimulated by the ideal two-dimensional crystal structure of these complexes, we have measured ^{197}Au Mössbauer spectra of the title complexes to investigate the effect of structural dimensionality on the charge transfer interaction between Au^{I} and Au^{III} in perovskite type gold mixed valence complexes. In this paper, we discuss the relationship between the structural dimensionality and the Au^{I}–Au^{III} charge transfer interaction for $Cs_2[Au^{I}I_2][Au^{III}I_4]$ and $[NH_3(CH_2)_nNH_3]_2[(Au^{I}I_2)(Au^{III}I_4)(I_3)_2]$ (n = 7 and 8) by ^{197}Au Mössbauer spectroscopy.

2 Materials and measurement

2.1 Three-dimensional perovskite-type $Cs_2[Au^{I}I_2][Au^{III}I_4]$

The crystal structure of $Cs_2[Au^{I}I_2][Au^{III}I_4]$ is a distorted perovskite-type structure with I4/mmm. In the crystal, all the halogens are distorted from the midpoint between neighboring Au ions in the chains, and the linear $[Au^{I}I_2]^-$ and square-planar $[Au^{III}I_4]^-$ complexes are stacking alternately. Namely, there are two relative orientations between $[Au^{I}I_2]^-$ and $[Au^{III}I_4]^-$. Consequently, its structure consists of three-dimensional metal–halogen frame-works with elongated Au^{III} octahedra and compressed Au^{I} octahedra sharing their corners (9,10; Figure 1).

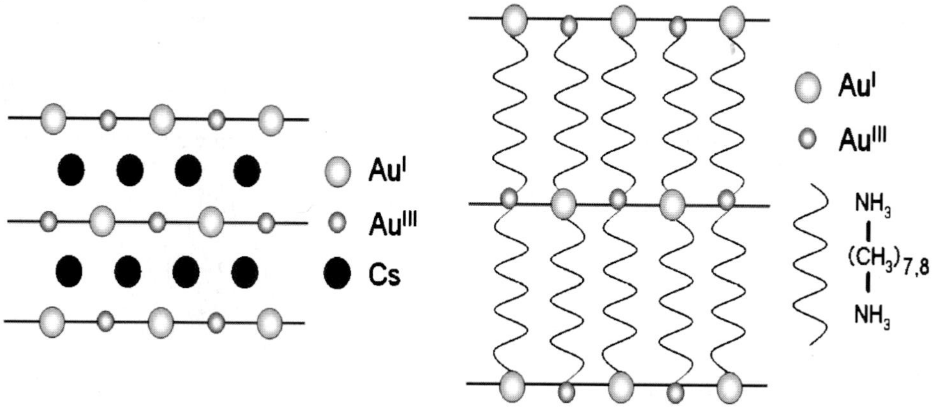

Figure 2 Schematic structures of Cs_2 $[Au^{I}I_2][Au^{III}I_4]$(*left*) and $[NH_3(CH_2)_nNH_3]_2[(Au^{I}I_2)(Au^{III}I_4)(I_3)_2]$ (n = 7, 8) (*right*). For simplicity, iodines are neglected and $Au^{I}\cdots I\text{-}Au^{III}\text{-}I\cdots Au^{I}$ planes are guided by *solid lines*.

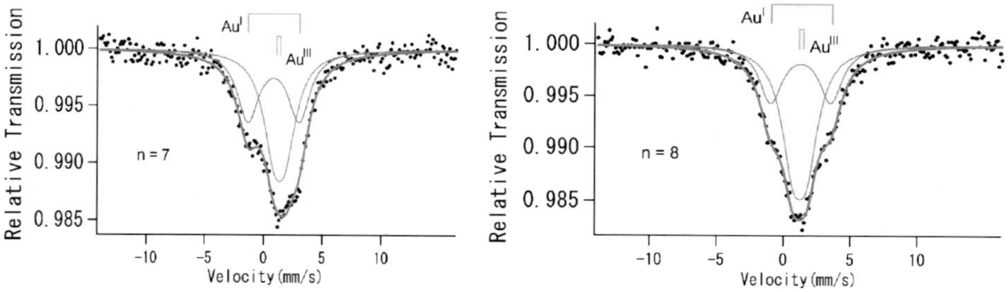

Figure 3 ^{197}Au Mössbauer spectra of $[NH_3(CH_2)_nNH_3]_2[(Au^{I}I_2)(Au^{III}I_4)(I_3)_2]$($n$ = 7, 8) at 13 K.

2.2 Layered perovskite-type $[NH_3(CH_2)_nNH_3]_2[(Au^{I}I_2)(Au^{III}I_4)(I_3)_2]$($n$ =7 and 8)

$[NH_3(CH_2)_nNH_3]_2[(Au^{I}I_2)(Au^{III}I_4)(I_3)_2]$ were synthesized in the similar way as reported in ref. [8]. Iodine was dripped into conc. HI. To this, aqueous solution of $NH_2(CH_2)_nNH_2$ and $HAuCl_4$ was added at r.t. Then a black precipitate was obtained. In the case of $[NH_3(CH_2)_nNH_3]_2[(Au^{I}I_2)(Au^{III}I_4)(I_3)_2]$, since the iodine-bridged $Au^{I,III}$ mixed valence network layers of $Au^{I}\cdots I\text{-}Au^{III}\text{-}I\cdots Au^{I}$ are separated by the $[NH_3(CH_2)_nNH_3]$ cation [8], $[NH_3(CH_2)_nNH_3]_2[(Au^{I}I_2)(Au^{III}I_4)(I_3)_2]$ can be regarded as a two-dimensional $Au^{I,III}$ mixed valence system (Figure 2).

2.3 ^{197}Au Mössbauer spectroscopy

^{197}Au Mössbauer spectroscopy of the 77.34 keV transition in ^{197}Au was employed with both source and absorber cooled down to 13 K in a transmission geometry. The γ-ray source (^{197}Pt) was obtained by neutron irradiation for 98% enriched ^{196}Pt metal with the nuclear reaction $^{196}Pt(n, \gamma)^{197}Pt$ in the Kyoto University Reactor. All isomer shift values were referenced to metallic gold foil.

Table I Isomer shifts (δ), quadrupole splittings (Δ), full width of half maximum (Γ) and relative intensities of $[NH_3(CH_2)_nNH_3]_2[(Au^{I}I_2)(Au^{III}I_4)(I_3)_2]$($n$ = 7, 8) at 13 K

Specimen	Site	δ (mm/s)	Δ (mm/s)	Γ (mm/s)	Relative Intensity $f(Au^{I})$ / $f(Au^{III})$
$Cs_2[Au^{I}I_2][Au^{III}I_4]$[10]	Au^{I}	1.19	4.58	2.0	0.88
	Au^{III}	1.71	1.68	1.89	
n = 7	Au^{I}	0.83	4.35	2.23	0.88
	Au^{III}	1.37	1.00	2.23	
n = 8	Au^{I}	1.29	4.49	2.07	0.58
	Au^{III}	1.40	1.06	2.07	

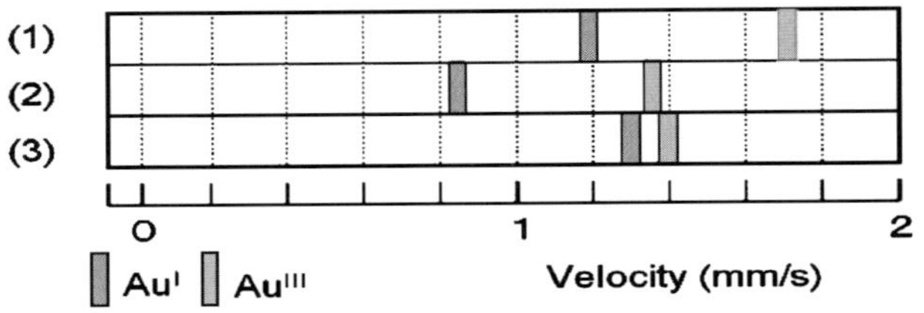

Figure 4 Isomer shifts of Au^{I} and Au^{III} in (*1*) $Cs_2[Au^{I}I_2][Au^{III}I_4]$, (*2*) $[NH_3(CH_2)_7NH_3]_2$ $[(Au^{I}I_2)(Au^{III}I_4)(I_3)_2]$, (*3*) $[NH_3(CH_2)_8NH_3]_2$ $[(Au^{I}I_2)(Au^{III}I_4)(I_3)_2]$.

3 Results and discussion

^{197}Au Mössbauer spectra of the title complexes are shown in Figure 3. There are two doublets in each spectrum. From an analog with $Cs_2[Au^{I}X_2][Au^{III}Y_4]$($X$, Y = Cl, Br, I), we assigned these spectra. ^{197}Au Mössbauer parameters of $Cs_2[Au^{I}I_2]$ $[Au^{III}I_4]$ and $[NH_3(CH_2)_nNH_3]_2[(Au^{I}I_2)(Au^{III}I_4)(I_3)_2]$($n$ = 7 and 8) are listed in Table I, in which these complexes are denoted as $CsAuI_3$, n = 7 and n = 8, respectively.

Figure 4 shows the isomer shifts of Au^{I} and Au^{III} for $Cs_2[Au^{I}I_2][Au^{III}I_4]$ and $[NH_3(CH_2)_7NH_3]_2[(Au^{I}I_2)(Au^{III}I_4)(I_3)_2]$($n$ = 7 and 8). Despite the gravity of isomer shifts of Au^{I} and Au^{III} sites in n = 7 is lower than that of $Cs_2[Au^{I}I_2][Au^{III}I_4]$, the difference in the isomer shifts between Au^{I} and Au^{III} of $Cs_2[Au^{I}I_2][Au^{III}I_4]$ resembles that of $[NH_3(CH_2)_7NH_3]_2[(Au^{I}I_2)(Au^{III}I_4)(I_3)_2]$. In the case of n = 8, however, the isomer shift of Au^{I} is quite similar to that of Au^{III}, which implies that the Au valence state is close to the single valence state of Au^{II}. In the case of perovskite-type Au mixed valence system, there are two possible charge transfer interactions between Au^{I} and Au^{III} in the ab plane. (a) the charge transfer interaction between $Au^{I}(5d_{z2})$ and $Au^{III}(5d_{x2-y2})$, and (b) the charge transfer interaction between $Au^{I}(5d_{x2-y2})$ and $Au^{III}(5d_{x2-y2})$. According to the MO calculation for $Cs_2[Au^{I}X_2]$ $[Au^{III}X_4]$ [3], the $5d_{z2}$ orbital consists of 6s component because of ds hybridization. However, $5d_{x2-y2}$ orbital does not consists of 6s one. Therefore, if the charge transfer interaction (a) is pre-dominant, the isomer shift of Au^{I} becomes to be small since the s-electron population at the Au^{I} site decreases due to the charge transfer from $Au^{I}(5d_{z2})$ to $Au^{III}(5d_{x2-y2})$.On the other hand, if the charge transfer interaction (b) is predominant, the isomer shift of Au^{I} becomes to be large because of the decrease in the screening.

Comparing Mössbauer parameters of $n = 7$ with those of $Cs_2[Au^{I}I_2][Au^{III}I_4]$, both the isomer shifts of Au^{I} and Au^{III} in $n = 7$ are smaller by 0.35 than those of Au^{I} and Au^{III} in $Cs_2[Au^{I}I_2][Au^{III}I_4]$, which implies that the charge transfer interaction (a) is stronger than that of Cs_2 $[Au^{I}I_2][Au^{III}I_4]$. In the case of $n = 8$, the isomer shifts of Au^{I} and Au^{III} are larger and smaller than those of Au^{I} and Au^{III} in $Cs_2[Au^{I}I_2]$ $[Au^{III}I_4]$, and the isomer shifts of Au^{I} and Au^{III} are almost identical, therefore the charge transfer interaction (b) is stronger than that in $Cs_2[Au^{I}I_2][Au^{III}I_4]$. Although, $n = 7$ and 8 of title complex have a small difference in geometric parameter, the effective charge transfer interaction between Au^{I} and Au^{III} is quite different from each other and the Au valence state in $n = 8$ is close to the single valence state of Au^{II}. In order to explain this discrepancy, a comprehensive MO calculation including the tilting between the Au^{I} and Au^{III} octahedra is indispensable and in progress.

4 Conclusion

We have investigated the Au valence state for an ideal two-dimensional gold mixed valence complexes, $[NH_3(CH_2)_nNH_3]_2[(Au^{I}I_2)(Au^{III}I_4)(I_3)_2]$($n = 7$ and 8) by means of ^{197}Au Mössbauer spectroscopy. In the case of $n = 8$, the isomer shifts of Au^{I} and Au^{III} sites are close to each other. Therefore, the single valence state of Au^{II} might realize at ambient pressure for $n = 8$ with small perturbation, such as external field. Although, $n = 7$ and 8 of title complex have a small difference in geometric parameter, the effective charge transfer interaction between Au^{I} and Au^{III} is quite different.

Acknowledgements This work has partly been supported by a Grant-in-Aid for Science Research and a Grant-in-Aid from The 21st Century COE (Center of Excellence) program (Research Center for Integrated Science) from the Ministry of Education, Science, Sports and Culture, Japan and one of the authors (K. Ikeda) was partly supported by Inoue Foundation for Science.

References

1. Elliot, N., Pauling, L.: J. Am. Chem. Soc. **60**, 1846 (1938)
2. Kojima, N.: Bull. Chem. Soc. Jpn. **73**, 1445 (2000)
3. Raubenheimer, H.G., Cronje, S., In: Schmidbauer, H. (ed.) Gold, p. 568. Wiley, New York (1999)
4. Kojima, N., Hasegawa, M., Kitagawa, H., Kikegawa, T., Shimomura, O.: J. Am. Chem. Soc. **116**, 11368 (1994)
5. Prassides, K. (ed.): Mixed Valency Systems: Applications in Chemistry, Physics and Biology. Kluwer, Dordrecht (1991)
6. Liu, X.J., Moritomo, Y., Ichida, M., Nakamura, A., Kojima, N.: Phys. Rev. **B61**, 20 (2000)
7. Ikeda, K., Ono, Y., Enomoto, M., Kojima, N., Kobayashi, Y., Seto, M.: Ceram.-Silik. **48**, 159 (2004)
8. Castro-Castro, M., Guloy, A.M.: Angew. Chem. Int. Ed. **42**, 2771 (2003)
9. Matsushita, N., Kitagawa, H., Kojima, N.: Acta Crystallogr. **C53**, 663 (1997)
10. Kitagawa, H.: Doctoral thesis, Kyoto University (1992)

Hyperfine Interact (2005) 166:409–416
DOI 10.1007/s10751-006-9301-x

Variable temperature Mössbauer spectroscopic studies of oxo-centered trinuclear mixed-valence iron succinate, mesaconate and isophthalate complexes

D. Afroj · M. Katada

Published online: 14 November 2006

Abstract Three new oxo-centered trinuclear mixed-valence iron complexes, iron succinate, iron mesaconate and iron isophthalate have been prepared. Temperature dependent valence delocalization processes were observed for all the complexes on variable temperature Mössbauer spectroscopic studies. Two clear quadrupole split doublets attributed to high-spin Fe(III) and high-spin Fe(II) states were existed for the complexes at liquid nitrogen temperature. At room temperature a nearly averaged singlet peak was observed for mesaconate complex and averaged doublet peaks were observed for succinate and isophthalate complexes with IS values 0.68 and 0.68 mms^{-1} and QS values 0.44 and 0.48 mms^{-1} respectively.

Key words iron complexes · mixed-valence state · trinuclear iron dicarboxylates · Mössbauer spectroscopy

1 Introduction

Many oxo-centered trinuclear mixed-valence iron monocarboxylate complexes of general formula $[Fe^{III}{}_2Fe^{II}O(RCOO)_6L_3]\cdot S$, where L is a monodentate ligand and S is a solvate molecule of crystallization have been studied by many workers with various physico-chemical techniques [1–3]. Dziobkowski et al. [4] have done variable temperature Mössbauer spectroscopy for iron acetates where intramolecular electron transfer phenomena has been ascribed by thermal electron hopping model. A detailed investigation for a series of iron acetates have been carried out by Hendrickson et al. [5–8]. It was assumed by them that the dynamics of coordinated

D. Afroj
Department of Chemistry, Bhawal Badre Alam Government College, Gazipur, Bangladesh

M. Katada (✉)
Graduate School of Science, Tokyo Metropolitan University, Hachioji, Tokyo 192-0397, Japan
e-mail: katada-motomi@c.metro-u.ac.jp

ligand, L and solvate molecule, S influence the rate of intramolecular electron transfer process, likely associated with the valence delocalization process. It was also found that the starting of intramolecular electron transfer occurs cooperatively in a phase transition. As like as interstitial solvate molecules long alkyl chains in non-solvated complex $[Fe_3O(RCO_2)_6(py)_3]$ (RCO_2 = myristate, palmitate or stearate) also affect the rate of intramolecular electron transfer which has been reported by Nakamoto et al. [9].

After discovering the first oxo-centered trinuclear mixed-valence iron complex many monocarboxylate complexes have been studied by various techniques where as there was no report on such dicarboxylate complexes. Some trinuclear iron(III) dicarboxylates have been reported by Dziobkowski et al. [10]. They suggested three-dimensional network for the complexes which can be formed through bridging by the carboxylate groups of the dicarboxylic acid ligand. It can be of great interest to investigate the nature of electron transfer phenomena in such three dimensional mixed-valence complexes. Previously we reported the first oxo-centered trinuclear mixed-valence iron complexes, iron malonate $[Fe_3O(O_2CCH_2CO_2)_3(H_2O)_3]$ and iron fumarate $[Fe_3O(O_2CCH{=}CHCO_2)_3(H_2O)_3]{\cdot}19H_2O$ which were studied by variable temperature Mössbauer spectroscopy [11, 12]. There we succeeded to interpret Mössbauer spectra of fumarate complex by relaxation model based on electron hopping model given by Dziobkowski et al. [4]. In a continuation, in this study three more new complexes of this type, iron succinate, iron mesaconate and iron isophthalate have been succeeded to prepare. All the complexes showed temperature dependent valence delocalization process on variable temperature Mössbauer spectroscopic studies. In this study the relaxation model based on electron hopping model has been applied to fit all the spectra as the spectra at higher temperatures were difficult to fit by Lorentzian line shapes due to relaxation effect.

2 Experimental

Iron succinate $[Fe_3O(O_2CC_2H_4CO_2)_3(H_2O)_3](HO_2CC_2H_4CO_2H)_{0.5}{\cdot}2H_2O$. The complex was prepared as follows. To a solution of succinic acid (2.36 g, 0.02 mol) in 40 ml water $NaHCO_3$ (3.32 g, 0.04 mol) was added slowly. Then the solution was heated for about 1 h at 318 K. The pH of the solution was about 7.9. Then the pH was adjusted to 5.4 by adding excess succinic acid. Nitrogen gas was passed through the solution for about 30 min. The solution was added to an aqueous solution of $FeCl_2{\cdot}4H_2O$ (3.97 g, 0.02 mol in 15 ml water), through which nitrogen gas was passed previously. The resultant solution was transferred immediately in a vacuum desiccator. Black crystal was formed slowly on the upper layer within one week, which was taken off directly for the immediate measurement of Mössbauer spectroscopy as the complex was very sensitive to air. Calcd: C, 24.69; H, 3.70. Found: C, 24.34; H, 3.16.

Iron mesaconate $[Fe_3O(OOCCH{=}C(CH_3)COO)_3(H_2O)_3]{\cdot}20H_2O$. Mesaconic acid of 1.305 g (0.01 mol) and 0.792 g (~0.02 mol) of NaOH was dissolved in 30 ml of water. The pH of the solution was about 5.5. The solution was cooled to 273 K and N_2 gas was passed for 30 min through the solution. Then aqueous solution of $FeCl_2 \cdot 4H_2O$ (1.986 g, 0.01 mol) in about 10 ml water was added to that solution under nitrogen. Nitrogen flow was continued for about 4 h. Then the solution was kept in refrigerator. After about 15 days black crystal was formed which was directly taken off from the solution for Mössbauer spectroscopy measurement. Calcd: C, 18.35; H, 5.95; Fe, 17.06. Found: C, 16.96; H, 5.91; Fe, 17.11. The calculated elemental analysis value for C differs with the observed value, which is may be due to the difficulties of analysis as the complex sublimes during the elemental analysis process.

Iron isophthalate $[Fe_3O(OOCC_6H_4COO)_3(H_2O)_3]\cdot 11H_2O$. The complex was prepared according to the following method. Isophthalic acid of 1.662 g (0.01 mol) and 0.772 g (~0.02 mol) of NaOH was dissolved in 60 ml of water. The pH of the solution was about 5. A solution of 1.988 g (0.01 mol) of $FeCl_2 \cdot 4H_2O$ in about 15 ml water was added to the acid solution under nitrogen. Nitrogen flow was continued for 2 h and then the solution kept covered. Within 1 week firstly brown precipitate was formed and then gradually turned into black crystal. Calcd: C, 31.07; H, 4.34. Found: C, 31.05; H, 3.90.

Mössbauer spectra at various temperature were obtained by a Mössbauer spectrometer (Wissel) using a $^{57}Co(Rh)$ source with a constant acceleration mode over a temperature range 78 K to room temperature. The spectra were fitted to Lorentzian lines by a least-squares fitting. Velocity calibration was done by measuring the magnetic splitting of α-iron foil. All isomer shifts were referred to α-Fe at room temperature. The sample was sealed in a hand-made acryl cell to avoid exposure to high vacuum atmosphere and air oxidation. Differential scanning calorimetry (DSC) and thermogravimetric (TG–DTA) measurements were carried out using a Rigaku Thermoplus DSC and Thermoplus TG at a heating rate of 5 $Kmin^{-1}$.

3 Results and discussion

Temperature dependent Mössbauer fitting spectra of iron succinate complex is shown in Figure 1. The observed Mössbauer spectra of this complex could not be fitted accurately by relaxation model probably due to the presence of Fe(III) ion as an impurity in the complex. The room temperature Mössbauer spectrum fitted by Lorentzian line shape, though the fitting procedure is not perfect, showed two quadrupole doublets with isomer shift values 0.36 and 0.68 mms^{-1} assigning high spin Fe(III) state and the averaged valence state of high spin Fe(III) and high spin Fe(II) states, respectively. The Fe(III) ions may contained by the sample from the beginning as an impurity or the sample can be oxidized partially during the Mössbauer spectroscopy measurements. The amount of Fe(III) state is estimated as about 10% considering the area of the doublet. At liquid nitrogen temperature two clear quadrupole doublets were observed corresponding to high spin Fe(II) state, the less intense one and high spin Fe(III) state, the more intense one with isomer shift values 1.39 and 0.48 mms^{-1}, respectively as can be seen from Figure 1. Considering Fe(III) as one species the area ratio of the Fe(II) and Fe(III) doublet is

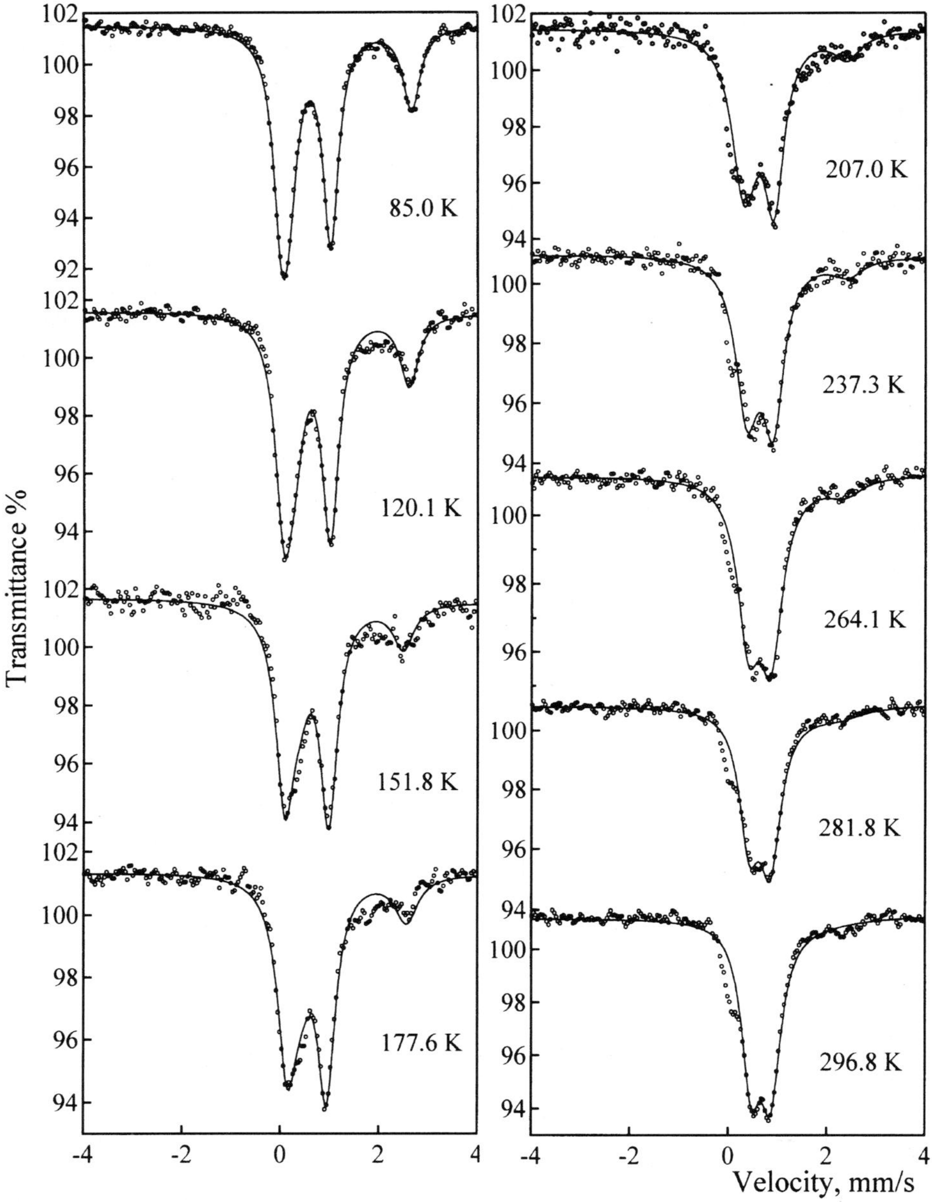

Figure 1 Variable temperature Mössbauer spectra of iron succinate complex.

2.2 which is close to the theoretical area ratio (2.0) of oxo-centered trinuclear mixed valence iron complexes. But the complex contains about 10% Fe(III) ions as discussed before and at liquid nitrogen temperature the area of the observed doublet raised for Fe(III) should be a total of the area of two types of Fe(III) doublets. Consider the same area for Fe(III) state estimated at room temperature is included in the spectrum at liquid nitrogen temperature and deducing this amount from the total area of the observed Fe(III) doublet the calculated area ratio of Fe(II) and Fe(III) doublets in the pure complex is 1.9 which is also very close to the theoretical value. Spectral change was observed from 120 K where spectral line of Fe(II) doublet was broadened with increasing temperature. This change of the spectra can be explained by the change of the rate of intramolecular electron transfer. At low temperature the intramolecular electron transfer rate is slower than the Mössbauer time scale ($\sim 10^{-7}$ s) and the different valence state of iron is distinguishable. The electron transfer rate increases with increasing temperature and in this complex at room temperature become faster than the Mössbauer time scale resulting an averaged doublet of Fe(II) and Fe(III) state i.e. a detrapped valence state of iron. It is clear from the variable temperature Mössbauer spectra that this complex showed a temperature dependent valence delocalization process.

The recoil-free fraction of the total spectral area intensity decrease linearly below 200 K according to the Debye model of solids, after that deviate from linearity without showing any abrupt changes. The solvated water molecules may form weak hydrogen bonds in this complex. At higher temperatures the deviation of recoil-free fraction can be attributed to the increasing thermal motion of these weak hydrogen bonds and also indicates the loose packing arrangement of the components in the complex.

Variable temperature Mössbauer spectra of iron mesaconate and iron isophthalate complexes measured from liquid nitrogen temperature to room temperature are shown in Figures 2 and 3, respectively. At liquid nitrogen temperature three clearly resolved peaks were observed for both the complexes. The more intense doublet corresponds to the high spin Fe(III) state and the less intense one is for high spin Fe(II) state. The presence of two distinct doublets interprets the slower intramolecular electron transfer rate than the Mössbauer time scale at low temperature. Spectral changes were observed at higher temperature where broadening of Fe(II) doublets in both complexes occurred indicating also the change of electron transfer rate, i.e. an increase with increasing temperature. At this temperature the area ratios of Fe(III) and Fe(II) doublets are 2.2 for mesaconate complex and 2.0 for isophthalate complex, which are close to the anticipated value 2.0. The room temperature spectrum of iron mesaconate complex is a nearly singlet peak where the rate of intramolecular electron transfer is greater than the Mössbauer time scale. This is an averaged spectrum for Fe(III) and Fe(II) ions, but the little shoulder indicates that the valences are not completely delocalized at room temperature. Relaxation of the spectra i.e. delocalization of the valence state was started at about 200 K temperature which is very similar to the previously reported fumarate complex [11, 12]. The room temperature spectrum of iron isophthalate complex is a single doublet assigned as the delocalized averaged doublet of Fe(III) and Fe(II) doublets by the isomer shift and quadrupole splitting values. This delocalization can be ascribed to the greater electron transfer rate than the Mössbauer time scale. So it is clear that iron mesaconate and iron isophthalate complexes showed pronounced temperature dependent valence delocalization processes.

The recoil-free fractions of these complexes also decrease linearly below 200 K according to the Debye model of solids, after that deviate largely from linearity without showing any abrupt changes. In these complexes too the increased thermal motion of weak hydrogen bonds formed by the solvated water molecules are

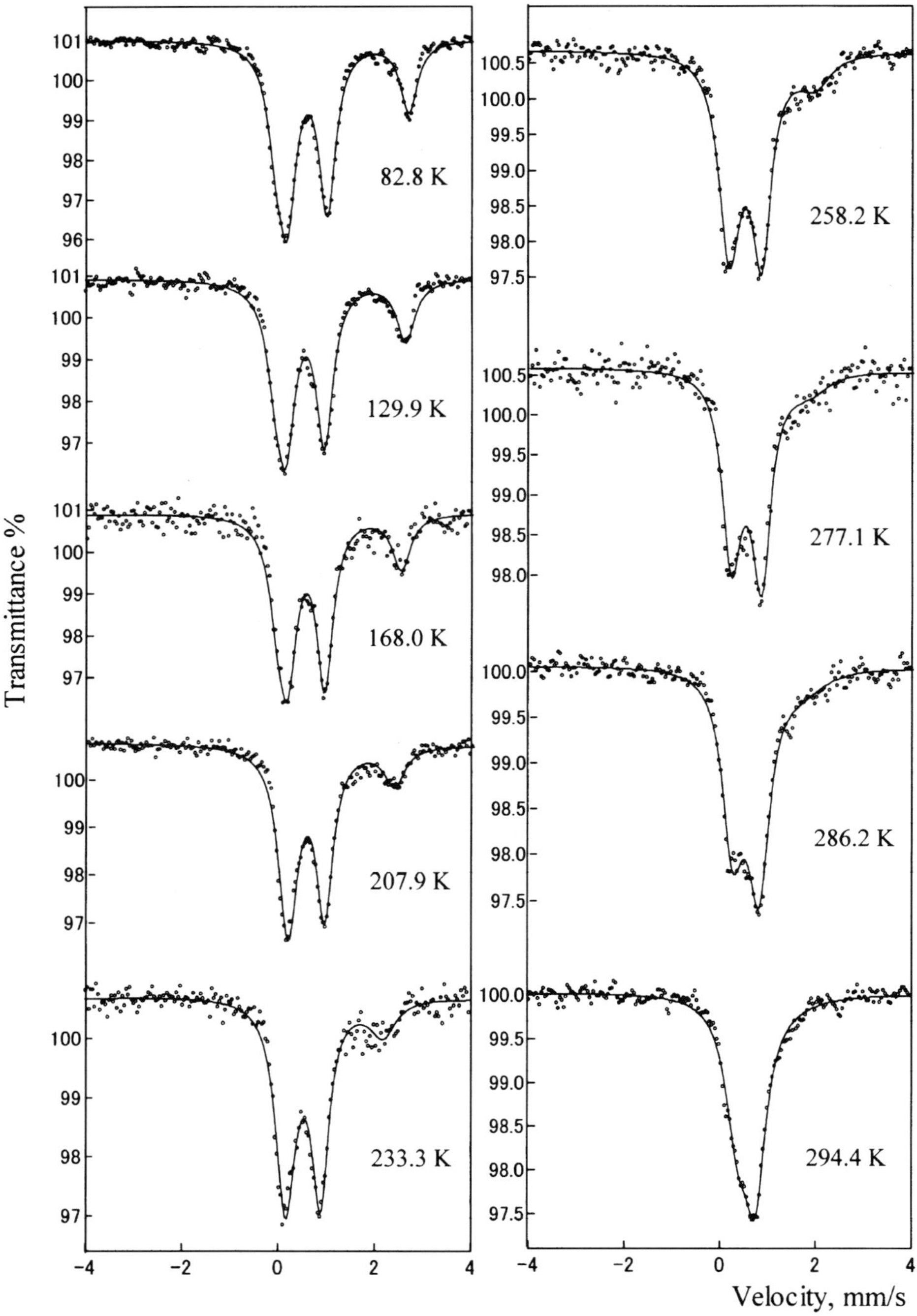

Figure 2 Variable temperature Mössbauer spectra of iron mesaconate complex.

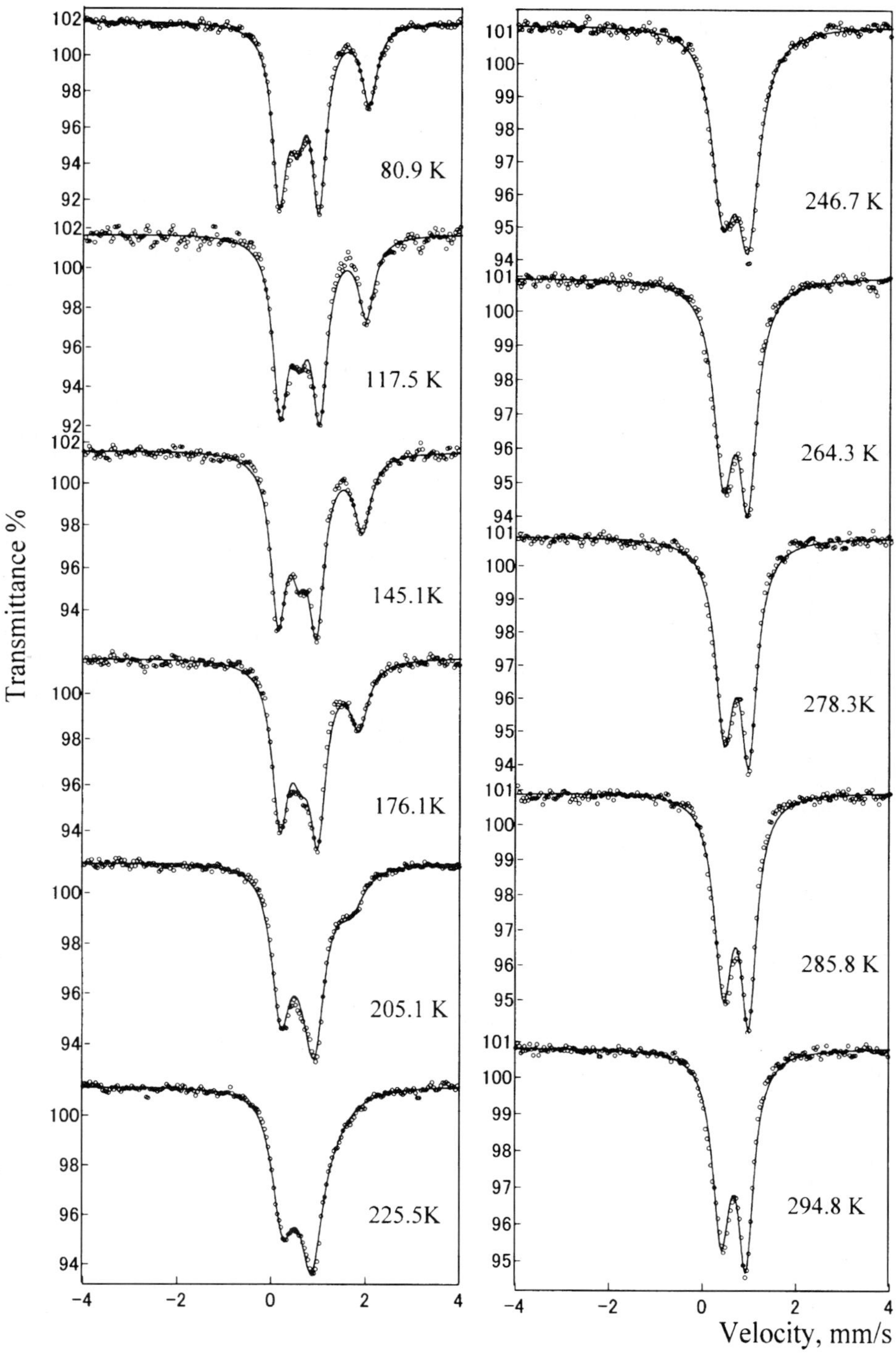

Figure 3 Variable temperature Mössbauer spectra of iron isophthalate complex.

responsible for the deviation at higher temperatures and also indicate the loose packing arrangement of the components. In case of isophthalate complex the deviation of recoil free fraction is smaller than that of mesaconate and fumarate complexes which can be attributed that this complex contains less water molecule than the other two complexes, consequently the thermal motion is weaker in this complex. The less deviation also indicates that the packing arrangement of the component is more rigid in isophthalate complex than that of the other two complexes.

Any endothermic or exothermic absorptions have not been observed on DSC measurements for all of these complexes i.e. phase transition were not observed for the complexes. The complexes were very unstable and readily oxidized in air so detailed structural studies have not been possible to carry out. As like as previously reported fumarate complex mesaconate complex showed almost the same behavior.

In this study it has been seen that these three solvated complexes showed temperature dependent valence delocalization processes on variable temperature Mössbauer spectroscopic studies as like as previously reported solvated fumarate complex. From our studies on dicarboxylates complexes it is found that only non-solvated malonate complex exhibited localized valence state of iron from low temperature to room temperature, where as other three solvated complexes showed delocalized valence state of iron at higher temperatures. At present it can be concluded that in case of dicarboxylate complexes also intermolecular interaction of the solvated water molecules are important for the temperature dependent valence delocalization process which has been reported for many similar monocarboxylate complexes. In dicarboxylate complexes valence delocalization processes are not associated with phase transition which was observed for many monocarboxylates complexes. Though for some oxo-centered trinuclear iron(III) dicarboxylate complexes polymeric structure which is formed through bridging by the carboxylate ligands have been reported [10], it have not been possible to carried out detailed structural studies of these mixed-valence complexes due to their instability.

References

1. Lupu, D., Barb, D., Filoti, G., Morariu, M., Tarina, D.: J. Inorg. Nucl. Chem. **34**, 2803 (1972)
2. Cannon, R.D., White, R.P.: Prog. Inorg. Chem. **36**, 195 (1988)
3. Hendrickson, D.N. In: Prassides, K. (ed.) Mixed Valency Systems: Applications in Chemistry, Physics and Biology, p. 67. Kluwer, Boston (1991)
4. Dziobkowski, C.T., Wrobleski, J.T., Brown, D.B.: Inorg. Chem. **20**, 679 (1981)
5. Oh, S.M., Hendrickson, D.N.: J. Am. Chem. Soc. **106**, 7985 (1984)
6. Oh, S.M., Hendrickson, D.N., Hassett, K.L., Davis, R.E.: J. Am. Chem. Soc. **107**, 8009 (1985)
7. Oh, S.M., Wilson, S.R., Hendrickson, D.N., Woehler, S.E., Wittebort, R.J., Inniss, D., Strouse, C.E.: J. Am. Chem. Soc. **109**, 1073 (1987)
8. Jang, H.G., Geib, S.J., Kaneko, Y., Nakano, M., Sorai, M., Rheingold, A.L., Montez, B., Hendrickson, D.N.: J. Am. Chem. Soc. **111**, 173 (1989)
9. Nakamoto, T., Katada, M., Sano, H.: Chem. Lett. 225 (1990)
10. Dziobkowski, C.T., Wrobleski, J.T., Brown, D.B.: Inorg. Chem. **20**, 671 (1981)
11. Afroj, D., Katada, M.: J. Radioanal. Nucl. Chem. **251**, 493 (2002)
12. Afroj, D., Yamauchi, T., Katada, M.: J. Radioanal. Nucl. Chem. **255**, 407 (2003)

Hyperfine Interact (2005) 166:417–423
DOI 10.1007/s10751-006-9302-9

Relationship between ^{237}Np Mössbauer parameters and bond distances in nitrogen coordinated [$NpO_2(acac)_2py$]

Takeshi Kawasaki · Takafumi Kitazawa · Tatsuru Nishimura · Masami Nakada · Masakatsu Saeki

Published online: 27 October 2006

Abstract We now report the ^{237}Np Mössbauer spectra for [$NpO_2(acac)_2py$] I. The neptunium(VI) ion in I is coordinated by six oxygen atoms and the one nitrogen atom of pyridine. The isomer shift value obtained for I is −40 mm/s, which falls in the range for the Np(VI) oxidation state. The relationship between the isomer shift and Np–O bond distance of the neptunyl group for oxygen coordination compounds is in good agreement with the reported structure of I. It is well known that the mean Np–O bond distance of the neptunium compounds is shorter by 0.01 Å than that of analogous uranium compounds. The good agreement in the ^{237}Np–Mössbauer parameters and the Np–O distance has indicated that the reported analogous [$UO_2(acac)_2py$] structure may have some inaccuracies. Therefore, we have re-determined the crystal structure of [$UO_2(acac)_2py$]. The U–O bond distance we have obtained is reasonable.

Key words ^{237}Np Mössbauer spectra · isomer shift · neptunyl (VI) compounds · Np-O bond distance · crystal data

1 Introduction

Neptunium chemistry has strong relationship with the reprocessing of nuclear fuels and treatment of actinide wastes in the backend chemistry for the nuclear

T. Kawasaki · T. Kitazawa (✉) · T. Nishimura
Department of Chemistry, Faculty of Science, Toho University, 2-2-1 Miyama, Funabashi, Chiba 274-8510, Japan
e-mail: kitazawa@chem.sci.toho-u.ac.jp

T. Kitazawa
Research Center for Materials with Integrated Properties, Toho University, Miyama, Funabashi, Chiba 274-8510, Japan

M. Nakada · M. Saeki
Department of Materials Science, Tokai Research Establishment, Japan Atomic Energy Research Institute (JAERI), 2-4 Shrane, Tokai mura, Naka-gun, Ibaraki 319-1195, Japan

power plants which operate everyday. The fundamental investigation of the bonding and structure of neptunium coordination compounds provides important information on the field of backend chemistry. However, few structural studies of neptunium complexes have been done because of the radioactivity and the laws governing radioactive materials and nuclear fuels. ^{237}Np–Mössbauer spectroscopy is a very powerful tool for structural studies, since the hyperfine parameters are significantly influenced by the electronic structure of the Np site [1]. The sensitivities of the ^{237}Np–Mössbauer spectra are stronger than those of the ^{57}Fe–Mössbauer spectra. It is well known that the ^{237}Np–Mössbauer isomer shifts (δ) range from +40 to -80 mm s^{-1} relative to $NpAl_2$ at 4.2 K and clearly correlated with the oxidation state of neptunium [1, 2]. The correlation between the δ value and coordination number (CN) of the Np atom has also been found for the Np(VI) compounds [1, 2]. The correlation between the δ value and the Np–O bond distance has also been reported for the oxo-coordinated Np(VI) compounds, where the increase in the δ value, corresponding to the decrease in the *s* electron density at the Np nucleus follows an increase in the mean Np–O bond distance/neptunyl Np–O bond distance [1, 2].

We now report ^{237}Np–Mössbauer spectra for [$NpO_2(acac)_2py$] I. I determined by a single crystal X-ray determination [3] exhibits a pentagonal-bipyramidal geometry about the Np(VI) ion. The neptunium(VI) ion in I is coordinated by six oxygen atoms and the one nitrogen atom of pyridine. The δ value of I indicated that the bond distance Np–O reported in [3] is very reasonable since the correlation between the δ value and the Np–O bond distance is high. It is well known that the mean Np–O bond distance of the neptunium compounds is shorter by 0.01 Å than that of analogous uranium compounds. When discussing the structures of Np compounds, the 0.01 Å shorter difference has been used. However, the single crystal data for [$UO_2(acac)_2py$] reported in 1984 [4] may be strange. The U–O(1) distance in [4] is 1.83(1) Å, although the Np–O(1) distance is 1.78(1) Å [3]. Good agreement in the correlation between the δ value and the Np–O bond distance has indicated that the reported analogous [$UO_2(acac)_2py$] structure may have some inaccuracies. Therefore, we have re-determined the crystal structure of [$UO_2(acac)_2py$]. The newly obtained U–O(1) bond distance of 1.77 (1) Å is now reasonable.

2 Experimental

2.1 Synthesis of [$NpO_2(acac)_2py$] I

The $^{237}NpO_2$ powder was dissolved in concentrated HNO_3 by heating to a slight boil. The obtained solution was evaporated to form wet salts. The dry residue was dissolved in distilled water, and the Np(VI) solution was placed in a glass flask. The neptunyl(VI) nitrate used as the starting material, $NpO_2(NO_3)_2{\cdot}H_2O$, was prepared by evaporating the Np(VI) stock solution. All the syntheses and XRD was done in a glove box at room temperature.

To the 3 ml acetonitrile solution containing 250 mg of neptunyl(VI) nitrate were added 0.15 ml of acetylacetone and 0.1 ml of pyridine. The formed brown powder crystals were characterized as [$NpO_2(acac)_2py$] by powder X-ray diffraction patterns using a model RAD-3C (Rigaku Cu Kα radiation) diffractometer.

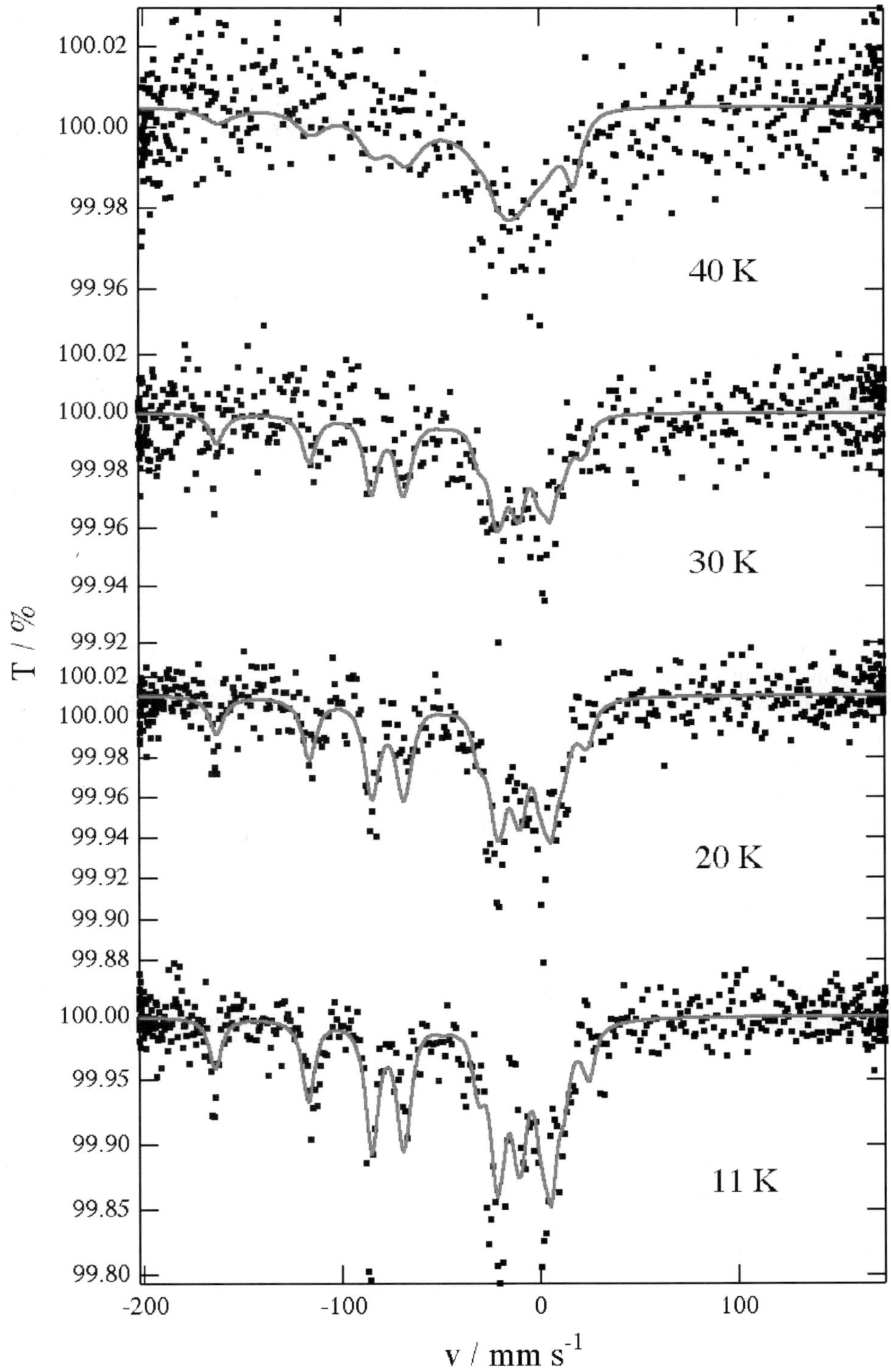

Figure 1 ^{237}Np–Mössbauer spectra of $[NpO_2(acac)_2py]$ I.

Table I ^{237}Np–Mössbauer parameters of I

T (K)	δ ($NpAl_2$) (mm s^{-1})	e^2qQ (mm s^{-1})	H_{eff} (T)	η	τ
40	−39.6	190	220	0	0.23(5)
30	−39.6(5)	189(4)	220	0	0.7(3)
20	−39.5(3)	191(2)	220	0	2(1)
11	−39.9(2)	193(2)	220(1)	0	–

Table II The selected bond lengths and angle of [UO_2(acac)$_2$py]

Bond lengths	Crystal data
U–O(1)	1.77(1) Å
U–O(2)	2.365(2) Å
U–O(3)	2.345(2) Å
U–N(1)	2.602(3) Å
O(1)–U–O(1′)	178.3(2)°

2.2 ^{237}Np–Mössbauer measurement

The Mössbauer spectra were measured using a cryostat equipped with a driving system (WissEl GmbH, System MS II) [5]. An assembled ^{241}Am metal source (~600 MBq) was used. The sample (~100 mg Np/cm^2) was used as an absorber [5].

2.3 Re-determination of crystal structure [UO_2(acac)$_2$py]

To the 10 ml methanol solution containing $UO_2(NO_3)_3.6H_2O$ 1.0 mmol was added 3.0 mmol of acac and 3.0 mmol of py in 5 ml of methanol. After the solvent evaporated slowly at room temperature for a few weeks, suitable single crystals were obtained. An orange crystal of [UO_2(acac)$_2$py] having the approximate dimensions of 0.30 × 0.30 × 0.20 mm was mounted on a glass fiber. All measurements were made using a Rigaku AFC5S diffractometer with monochromated Mo–Kα radiation. The obtained crystallographic data are: Orthorhombic, space group *Fdd2*, a = 11.432(3) Å, b = 29.708(5) Å, c = 10.593(2) Å, U = 3599(1) Å^3, R = 0.049, Rw = 0.073 for 1,089 reflections and 119 variable parameters.

3 Results and discussion

The ^{237}Np–Mössbauer spectra of I are shown in Figure 1. The obtained ^{237}Np–Mössbauer parameters are listed in Table I. The crystal data used for the re-determination of [UO_2(acac)$_2$py] are listed in Table II and shown in Figure 2.

The Mössbauer spectra of [NpO_2(acac)$_2$py] I show absorption peaks due to quadrupole splittings and magnetic splittings at 11 K. As the temperature increases, a magnetic relaxation appears. The Mössbauer spectrum of I at 11 K was computer-fitted by the sum-of-Lorentzian approximation and the 20, 30, and 40 K spectra were computer-fitted by a relaxation procedure based on the Nowik and Wickman

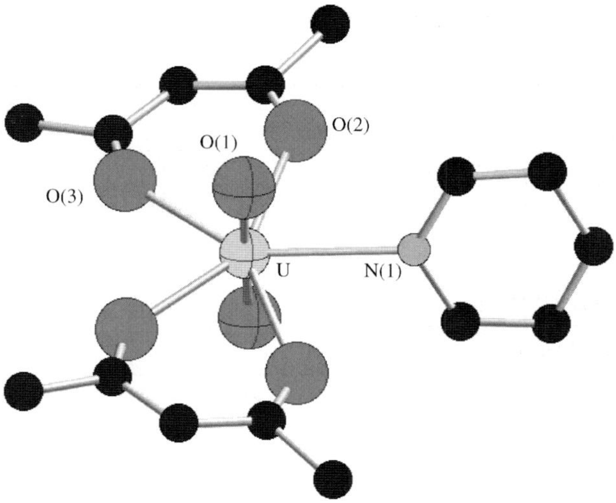

Figure 2 Molecular structure of $[UO_2(acac)_2py]$.

model [6] while fixing values of $\eta = 0$ and $H_{eff} = 220$ T using an "IGOR" (Wave Metrics, Lake Oswego, Oregon) software. The spectrum at 40 K was fitted by fixing the values of the δ and e^2qQ because the absorption peaks were very small and broadened. The δ values were referred to $NpAl_2$ at 4.2 K. The ^{237}Np–Mössbauer parameters, used for the fitting procedure, were $\mu_g = 2.8$ nm and $\mu_e = 1.5$ nm [6]. Here, μ_g and μ_e are the ground and excited state nuclear magnetic moments respectively.

Figure 3a shows a plot of δ versus the mean Np–L bond distance for the Np(V) and Np(VI) compounds including the seven-fold coordinated I. The correlation line between the δ value and the Np–O bond distance has been obtained for various Np(VI) compounds coordinated by oxygen atoms [2, 6–8]. It is well known that the mean Np–O bond distance of neptunium compounds is shorter by 0.01 Å than that of analogous uranium compounds. The correlation of the Np(V) compounds is fitted for the narrower range because the distribution of δ is narrower for the neptunyl(V) compounds than that of the neputunyl(VI) compounds [9, 10]. For the Np(VI) compounds, there are some materials which have a relatively large deviation from linearity.

Figure 3b illustrates the correlation between the δ value and Np=O bond distance of the neptunyl(V) and the neptunyl(VI) compounds. As the neptunyl(VI) compounds coordinated by only oxygen atoms, the correlation in Figure 3b is better than that of Figure 3a. Figure 3a and b also indicate that $[NpO_2(acac)_2py]$ has a 7 CN (coordination number) since the data point of I is located between 8 CN and 6 CN. As for $[NpO_2(acac)_2py]$ I, in which the neptunyl(VI) complex is coordinated by one nitrogen atom of py, the δ values are slightly more positive than the correlation line both in Figure 3a and b. Because the distance between the neptunyl oxygen and the Np(VI) atom is shorter than that of other oxygen atoms of the ligand, the shielding effect of the 5*f* electron associated with the Np=O bonds has a link to the 6*s* electron density which governs the δ value. The electron donation of the nitrogen atom in the py may have a larger effect on the Np 5*f* orbital than that of the oxygen atom in the acac. The δ value of I has a positive trend due to the increase in the shielding effect of the 6*s* electron and the decrease in the total 6*s* electron

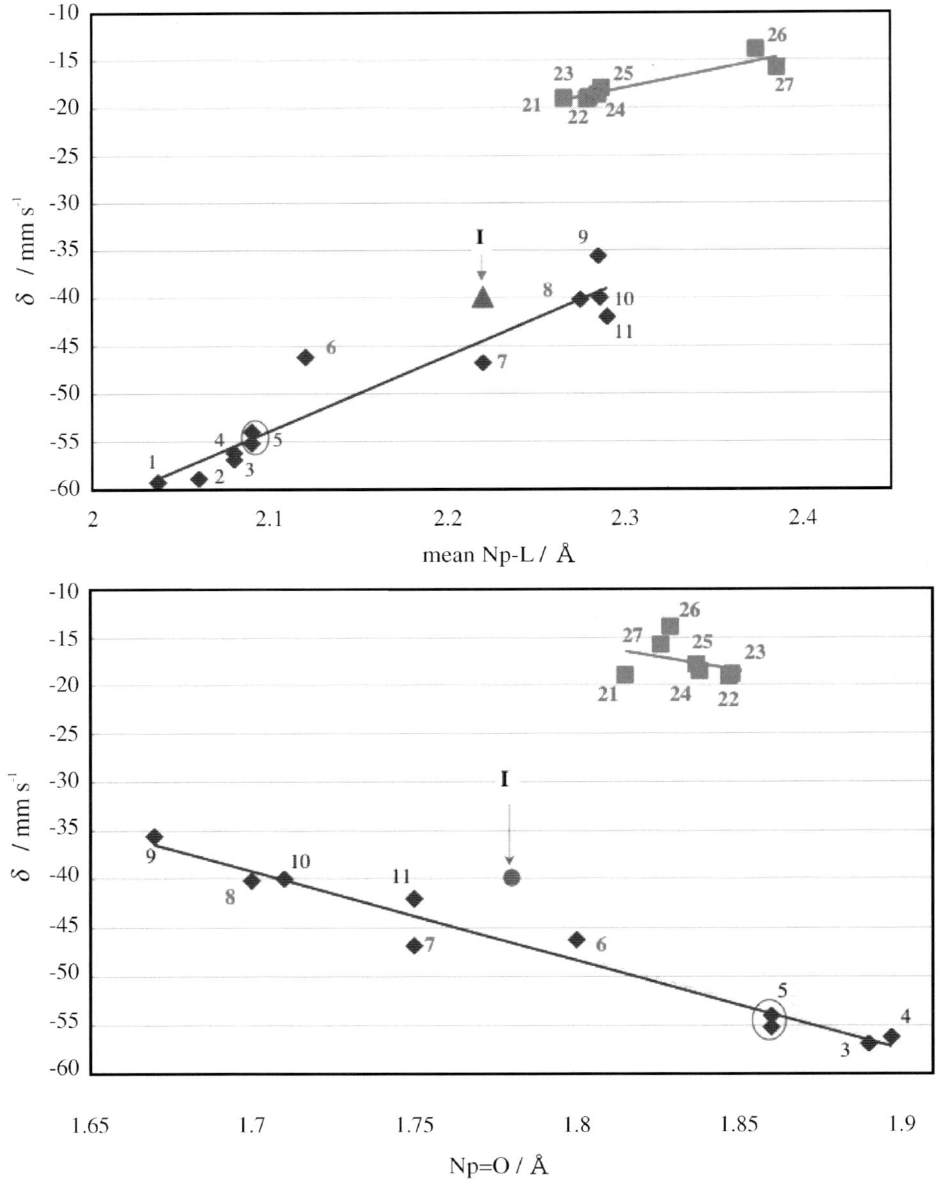

Figure 3 **a** Plot of δ vs mean Np–L bond distance for the Np compounds. **b** Plot of δ vs Np=O bond distance for the neptunyl compounds. Np(VI), CN = 6; 1: Ba_2CoNpO_6, 2: Li_4NpO_5, 3: K_2NpO_4, 4: β–Na_2NpO_4, 5: $BaNpO_4$, 6:β–$NpO_2(OH)_2$Np(VI), CN = 7; 7: $[NpO_2(ox)(OH_2)]{\cdot}2H_2O$Np(VI), CN = 8; 8: α–$NpO_2(OH)_2$, 9: NpO_2CO_3, 10: $NaNpO_2(C_2H_4O_2)_3$, 11: $NpO_2(NO_3){\cdot}6H_2O$Np(V), CN = 7; 21: $NpO_2(CH_2OHCOO){\cdot}H_2O$, 22: $(NpO_2)_2(C_6H_4(COO)_2){\cdot}4H_2O$, 23: $NpO_2(HCOO){\cdot}H_2O$, 24: $(NpO_2)_2(CH_2(COO)_2){\cdot}4H_2O$, 25: $(NpO_2)_2(NO_3){\cdot}5H_2O$ –1Np(V), CN = 8; 26: $(NpO_2)_2(NO_3){\cdot}5H_2O$ –2, 27: $NH_4NpO_2(HCOO)_2{\cdot}H_2O$.

density at the Np nucleus. As for the Np(V) compounds, the correlation line in Figure 3a is much better than that of Figure 3b. The two 5*f* electrons in Np(V) are associated with the equatorial Np–O bonds.

On the other hand, the single crystal data for [UO_2(acac)$_2$py] reported in 1984 [4] may be strange. The U–O(1) distance in [4] is 1.83(1) Å, although the Np–O(1) distance is 1.78(1) Å [3]. The good agreement in the correlation between the δ value and the Np–O bond distance has indicated that the reported analogous [UO_2 (acac)$_2$py] structure may have some inaccuracies. Therefore, we have re-determined the crystal structure of [UO_2(acac)$_2$py]. As listed in Table II, the U–O(1) bond distance of 1.77(1) Å that we obtained now is reasonable. The O–U–O linearity is similar to the O–Np–O in I, although [3] noted that the O–U–O for the large deviation from linearity of the O–U–O bond angle, 173.5(8)°. In this study, the ^{237}Np–Mössbauer data has required our re-determination of [UO_2(acac)$_2$py]. The re-determination also indicated that [UO_2(acac)$_2$py] is analogous to [NpO_2 (acac)$_2$py] regarding the M(VI)–O distance and O–M(VI)–O angle.

Acknowledgements This work was supported by the joint research project between JAERI and University on the Backend Chemistry. We thank Dr K. Minato of JAERI and Prof M. Takeda of Toho University for coordinating the joint research project. This work was also supported by "High-Tech Research Center" Project 2005–2009 from MEXT (Ministry of Education, Culture, Sports, Science and Technology), and the Research Center for Materials with Integrated Properties, Toho University. We also thank Prof M. Takeda and Prof M. Takahashi of Toho University for the helpful discussion and preparation of the uranium compound.

References

1. Dunlap, B.D., Kalvius, G.M.: In: Freeman, A.J., Lander, G.H. (eds) Handbook on the Physics and Chemistry of Actinides, pp. 329–434. Elsevier Science Publishers B.V. (1985)
2. Jové, J., He, L., Proust, J., Pagés, M., Pyykkö, P.: J. Alloys Compd. **177**, 285–310 (1991)
3. Alcock, N.W., Flanders, D.J., Pennington, M.: Acta Crystallogr. **C43**, 1476–1480 (1987)
4. Alcock, N.W., Flanders, D.J.: J. Chem. Soc. Dalton Trans. 679–681 (1984)
5. Saito, T., Wang, J., Kitazawa, T., Takahashi, M., Takeda, M., Nakada, M., Nakamoto, T., Masaki, N.M., Yamashita, T., Saeki, M.: J. Radioanal. Nucl. Chem. **239**(2), 319–323 (1999)
6. Wang, J., Kitazawa, T., Nakada, M., Yamashita, T., Takeda, M.: J. Nucl. Sci. Technol. **Suppl. 3**, 429–432 (2002)
7. Nakada, M., Kitazawa, T., Saito, T., Wang, J., Takeda, M., Yamashita, T., Saeki, M.: Bull. Chem. Soc. Jpn. **76**, 1375–1378 (2003)
8. Wang, J., Kitazawa, T., Nakada, M., Yamashita, T., Takeda, M.: Bull. Chem. Soc. Jpn. **75**, 253–257 (2002)
9. Saeki, M., Nakada, M., Nakamoto, T., Yamashita, T., Masaki, N.M., Krot, N.N.: J. Radioanal. Nucl. Chem. **239**(2), 221–225 (1999)
10. Saeki, M.: J. Crystal Soc. Jpn.(Nippon Kessho Gakkaishi) **46**, 415–420 (2004)

Hyperfine Interact (2005) 166:425–428
DOI 10.1007/s10751-006-9303-8

Neutron in-beam Mössbauer spectroscopic study of iron disulfide at room temperature

M. K. Kubo · Y. Kobayashi · H. Nonaka · Y. Yamada · Y. Sakai · H. Shoji · H. Matsue

Published online: 5 December 2006

Abstract An in-beam emission Mössbauer spectrum of ^{57}Fe arising from the $^{56}Fe(n, \gamma)$ ^{57}Fe reaction in iron disulfide at room temperature was measured with a parallel plate avalanche counter. It was clearly observed that the nuclear reaction and the following process lead to the production of a new chemical species of iron different from the parent compound.

Key words neutron in-beam spectroscopy · iron disulfide

1 Introduction

Physical and chemical changes caused by nuclear events, such as radioactive decay and nuclear reactions have been a major research subject for decades. Study of chemical and physical processes prompted by neutron capture reactions attracted much interest in the previous century and has practical importance in realizing the deterioration of nuclear reactor and fusion plant material. Neutron capture

M. K. Kubo
College of Liberal Arts, International Christian University, Mitaka, Tokyo 181-8585, Japan

Y. Kobayashi (✉)
RIKEN, Wako, Saitama 351-1098, Japan
e-mail: kyoshio@riken.jp

H. Nonaka · Y. Yamada
Department of Chemistry, Science University of Tokyo, Shinjuku, Tokyo 162-8601, Japan

Y. Sakai
Daido Institute of Technology, Minami-ku, Nagoya 457-8530, Japan

H. Shoji
Graduate School of Science, Tokyo Metropolitan University, Hachioji, Tokyo 192-0039, Japan

H. Matsue
Japan Atomic Energy Research Institute, Tokai, Ibaraki 319-1195, Japan

reactions are believed to be unique and different from other nuclear processes because neutron irradiation and neutron capture are free from ionization induced by passage of charged particles or charged particle emission usually leading to molecular decomposition and lattice defects. Only interactions with cascade γ-rays emitted successively during the relaxation process from the excited state formed by the neutron capture, and γ-ray recoil effects cause chemical or physical state change of the target material. In other words, observation of relaxation processes of highly excited atoms in very dilute concentration surrounded by thermally equilibrium material can be carried out. But there are few in situ experimental techniques available for examining the initial processes occurring in materials under neutron irradiation. In-beam Mössbauer spectroscopy is a useful tool to investigate trace amount of chemical species produced by nuclear events, and it has a great advantage of the ability to study solid materials non-destructively. But there have been limited number of research works of in-beam neutron Mössbauer spectroscopy [1–3]. We have started a new ^{57}Fe in-beam Mössbauer spectroscopic study aiming at the detailed in situ investigation on the chemical and physical behaviors of the trace species formed after neutron capture reactions using a parallel plate avalanche counter (PPAC)[4]. In this study we used iron disulfide (FeS_2, pyrite type) as a sample. Pyrite is a binary semiconductor compound, has the rock salt type crystal structure composed of Fe^{2+} cation and $S_2{}^{2-}$ anion. The simple structure and large iron concentration of FeS_2, and relative inertness for γ-ray emission of sulfur were promising to be the first sample of our research work.

2 Experimental

Pyrite type iron disulfide was purchased from Kishida Kagaku Co., and used without further purification. Finely ground powder pressed to form a thin disk of 20 × 20 cm^2 and about 100 mg cm^{-2} in thickness was attached using adhesive tape to a Teflon sample holder frame with thin threads of polyperfluoroethane. The sample was placed at the target position of the large iron shielding box of prompt γ-ray analysis setup [5] at the beam hall of the JRR-3M reactor of the Tokai Establishment of the Japan Atomic Energy Research Institute. The inner Teflon box used in prompt γ-ray analysis experiments for achieving helium atmosphere around the sample was not used since the Mössbauer driving shaft and the PPAC could not be fit in the inner box. The measurement was conducted under ambient atmosphere. The sample disk was oriented both to the beam and the PPAC detector at an angle of 45°. Only the sample, the PPAC with signal and HV cables and gas tubes were placed in the iron target box. The thermal neutron flux was about 1.0×10^8 cm^{-2} s^{-1} from the reactor at a thermal power of 20MW. The PPAC was stably operated in a continuous gas flow mode with perfluoropropane (C_3F_8) as the counter gas at the optimized flow rate and pressure for in-beam Mössbauer spectroscopy [4]. The measurement took 70 h at room temperature.

3 Results and discussion

The absorption spectrum of pyrite obtained by the usual transmission mode showed one doublet of $IS = 0.303$ mms^{-1} and $QS = 0.618$ mms^{-1}, indicating that the Fe^{2+} is

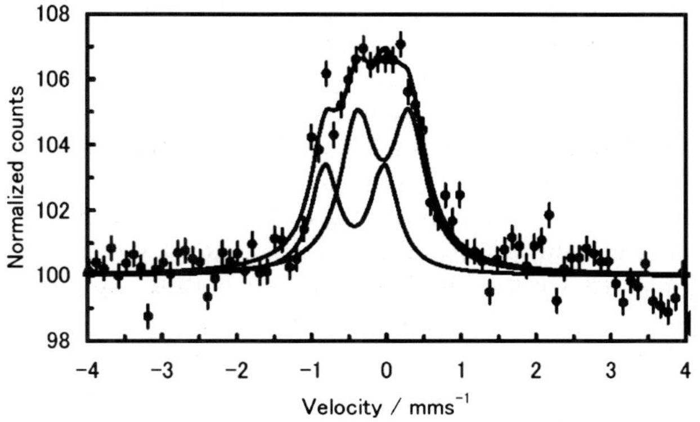

Figure 1 A neutron in-beam Mössbauer spectrum of FeS_2 (pyrite) at room temperature.

Table I Mössbauer parameters of FeS_2 obtained by neutron in-beam and absorption methods

	IS (mms^{-1})	*QS* (mms^{-1})	Intensity
Component I	0.06(3)	0.67(3)	0.66
Component II	0.44(3)	0.79(44)	0.34
Absorption	0.303(1)	0.618(1)	1.00

The sign of IS of in-beam mode is inverted for easy comparison.

in the low spin state. The in-beam spectrum was analyzed as the superposition of two Lorentzian doublets (Figure 1). The minor species, component II with area intensity of 1/3, showed Mössbauer parameters similar to the parent pyrite (Table I, the sign of *IS* is inverted for easy comparison with the absorption spectrum). The slight larger *IS* and *QS* of component II compared with pyrite could indicate that the ^{57}Fe atom formed by the neutron capture reaction experienced a small recoil and disturbed its environment moderately.

On the other hand, component I consisting of 66% of the ^{57}Fe atoms produced by the nuclear reaction considerably changed their states. Since there are only iron and sulfur around the ^{57}Fe atom produced by the neutron capture reaction, we compared the obtained Mössbauer parameters with other compounds made of iron and sulfur. Iron(II) in iron sulfides, FeS (troilite) and Fe_3S_4 (greigite) have *IS* values of around 0.7 − 0.8 mms^{-1} [6–9] larger than that of component II. Mackinawite type FeS has an *IS* of 0.35 mms^{-1} similar to component II, but shows no *QS*. Other stable known Fe(II) compounds of iron and sulfur as far as we know do not have Mössbauer parameters similar to those of component II. Iron(III) in alkali metal iron sulfide compounds of chemical formula $AFeS_2$ (A: alkali metal ion) show *IS* values of around 0.2 mms^{-1} and *QS* around 0.4 − 0.7 mms^{-1} [10] fairly in agree with the parameters of component II. In these compounds Fe(III) is tetrahedrally surrounded by four S^{2-} ions. In pyrite at the interstitial site Fe may be tetrahedrally surrounded by S_2^{2-} anion. Because S^{2-} and S_2^{2-} have the same charge but different size, we cannot directly conclude that component II is interstitial Fe(III), but it may be reasonable that the ^{7}Fe atom was oxidized and relocated from the lattice site by the γ-ray recoil with energies calculated up to 547 eV in FeS_2.

The obtained spectrum was analyzed by only two components. There might be a third species, but its contribution to the spectrum could not be so large. This may be a remarkable result that the relaxation from the 7.65 MeV level after the neutron capture leads to only one new product. There have been no such findings reported in semiconductors for neutron capture reactions. Since there is a possibility that relatively unstable species formed during the relaxation process resulted into the two thermally stable observed species at room temperature within the time window of 140 ns for ^{57}Fe Mössbauer spectroscopy. We plan to carry out experiments on the other polymorph of FeS_2 marcasite and also measure Mössbauer spectra at low temperatures to investigate this system in detail in order to obtain information on the possible short-lived and thermally unstable products.

4 Conclusion

A neutron in-beam emission Mössbauer spectrum of pyrite type iron disulfide showed two components at room temperature. One third of the ^{57}Fe produced via the ^{56}Fe(n, γ)^{57}Fe reaction remained in its original chemical form, while the other two thirds of the iron atoms were considered to be oxidized and relocated to interstitial site.

References

1. Berger, W.G., Fink, J., Obenshain, F.E.: Phys. Lett. **24A**, 466 (1967)
2. Berger, W.G.: Z. Phys. **225**, 139 (1969)
3. Czjzek, G., Berger, W.G.: Phys. Rev. **B1**, 957 (1970)
4. Saito, T., Kobayashi, Y., Kubo, M.K., Yamada, Y.: J. Radioanal. Nucl. Chem. **255**, 519 (2003)
5. Yonezawa, C., Wood, A.K.H., Hoshi, M., Ito, Y., Tachikawa, E.: Nucl. Instrum. Methods Phys. Res. **A329**, 207 (1993)
6. Ono, K., Ito, A., Hirahara, E.: J. Phys. Soc. Jpn. **17**, 1615 (1962)
7. Hafner, S.S., Evans, B.T., Kalvius, G.M.: Solid State Commun. **5**, 17 (1967)
8. Morice, J.A., Rees, L.C., Ricard, D.T.: J. Inorg. Nucl. Chem. **31**, 3797 (1969)
9. Vaughan, D.J., Ridout, M.S.: J. Inorg. Nucl. Chem. **33**, 741 (1971)
10. Taft, C.A., Chuha, S.F., Souza, N.G., Furtado, N.C.: J. Phys. Chem. Solids **41**, 61 (1980)

Hyperfine Interact (2005) 166:429–436
DOI 10.1007/s10751-006-9304-7

Corelationship between local structure and water purifying ability of iron-containing waste glasses

Shiro Kubuki · Norifumi Kawakami · Takamitsu Kamikawa · Masayuki Fukagawa · Toshiharu Nishizumi · Tetsuaki Nishida · Zoltán Homonnay · Ernő Kuzmann

Published online: 14 November 2006

Abstract A relationship between the structure and water purifying ability of waste glass prepared from household garbage and Fe_2O_3 was examined by ^{57}Fe-Mössbauer and induced coupled plasma (ICP) measurements. From the Debye temperature of waste glass, Fe^{II} proved to be loosely bound in the glass network as a network modifier. Dissolution amount of Fe^{III} into artificial drain can be controlled from 0.14 to 0.35 mg/l by changing the Fe_2O_3 content. It proved that chemical oxygen demand (COD) decreases in proportion to the content of Fe^{III}, indicating that iron causes decomposition of organic and phosphorus compounds.

Key words ^{57}Fe-Mössbauer spectra · Debye temperature · iron-containing silicate glass · chemical oxygen demand (COD)

1 Introduction

A compliance rate in a closed water system (e.g. lake or pond) in Japan is reported to be 55.2%, which is relatively lower than that in an open water system like river

S. Kubuki (✉) · N. Kawakami · T. Kamikawa · M. Fukagawa
Department of Chemical and Biological Engineering, Ube National College of Technology, Tokiwadai 2-14-1, Ube 755-8555, Japan
e-mail: kubuki@ube-k.ac.jp

T. Nishizumi · T. Nishida
Department of Biological and Environmental Chemistry, School of Humanity-Oriented Science and Engineering, Kinki University, Kayanomori 11-6, Iizuka 820-8555, Japan

Z. Homonnay
Department of Nuclear Chemistry, Eötvös Loránd University, Pázmány sétány, 1/A 1117 Budapest, Hungary

E. Kuzmann
Research Group for Nuclear Methods in Structural Chemistry, Hungarian Academy of Science, P.O. Box 32, Budapest 1518, Hungary

(87.4%), or ocean and coast (76.2%) [1]. The low compliance rate in closed water systems has not been improved for the decade, showing the difficulty in maintenance, in spite that purification of wastewater in a closed water system is one of the most critical environmental problems. Although some "water purifying minerals" are recently commercially available, but their ability has not been elucidated. Recently, Kubuki et al. developed and evaluated a new water purifying system including porous silicate from waste bottle glass, in which the water quality could be improved to the levels of COD and dissolved oxygen (DO) of 25 and 8.3 mg/l, respectively [2].

The household garbage in Japan is reported to be 90 g/m^2, which is almost comparable to that in Britain (120 g/m^2, being second in the data available in the EU allied nations) and that in Germany (95 g/m^2, third) [3]. Ministry of Environment of Japan reported that the damp yard would become full in the following 4.3 years [4], meaning that the scarcity of damp yard is so serious in Japan. Recently, gasification fusion furnaces have been introduced in several cities in Japan, e.g., Ube (Yamaguchi Pref.) introduced a waste combustion system in 2003. Disadvantage of this system is that the running cost is five times higher than that of earlier techniques. Because of these reasons, it is desired to recycle the wastes into functional materials.

Recently, Nishida et al. revealed that fly ash waste silicate glass shows an electric conductivity from 10^{-8} to 10^{-6} S cm^{-1} due to electron hopping from less distorted $Fe^{II}O_6$ octahedra to distorted $Fe^{III}O_4$ tetrahedra [5]. Precipitation of ferrimagnetic magnetite phase with an internal magnetic field of 46.4–48.2 T was also observed in the Mössbauer spectra when the Fe_2O_3 content of fly ash waste glass was equal to or higher than 14 mass%. This result suggests that waste materials can be recycled successfully by melting them with Fe_2O_3.

It is well known that Fe^{II} plays an important role in the Fenton method, which is effective for water purification [6]. Leupin and Hug reported that Fe^{II} plays an important role in reducing hazardous As^{III} in the ground water [7]. Iron-containing silicate glass can be a good candidate for the effective water purification, because dissolving rate and Fe^{II}/Fe^{III} abundance ratio in the glass are easily regulated by changing the composition. Although the structure of iron containing silicate glass has widely been studied by using Mössbauer spectroscopy [8–10], the relationship between the structure and water purifying ability has not been investigated.

The purpose of this work is to explore the possibility of the recycling of the ash as a water purifying material by investigating the relationship between composition, local structure and properties of ash waste glass. For this purpose, measurements of ^{57}Fe-Mössbauer, X-ray diffraction (XRD), chemical oxygen demand (COD), and induced coupled plasma (ICP) measurements were used for characterizing several ash waste glasses containing different amounts of Fe_2O_3.

2 Experimental

Household waste ash (Ube slag: US) was collected from Ube Waste Treatment Plant (Ube city, Yamaguchi Prefecture, Japan) every week during Dec. 4, 2003 and Feb. 26, 2004. Each rot had a weight of 2.0 kg. They were dried in a dehydrator at 100°C for 5 h. Chemical composition of US was determined by conventional X-ray

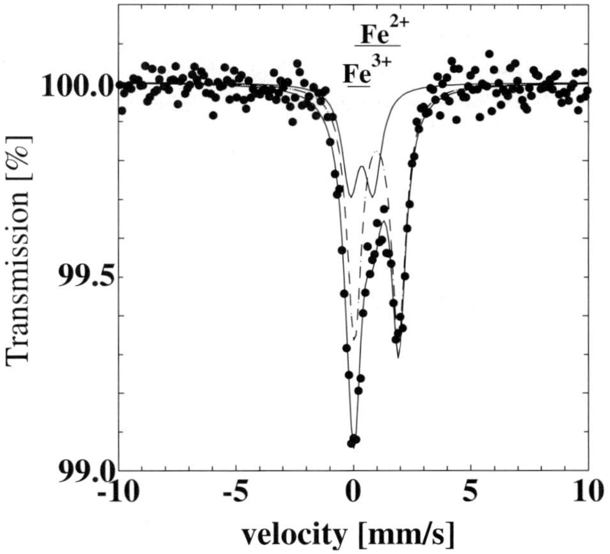

Figure 1 Mössbauer spectrum of an exhausted ash from wastes treatment plant in Ube city (Yamaguchi prefecture, Japan).

fluorescence analysis using ZSX-100e (Rigaku Corporation). A transparent medal-like bead was prepared by melting a pulverized mixture composed of US (0.5 g) and lithium tetraborate (5.0 g) at 1,200°C for 10 min.

Iron-containing US glass was prepared by conventional melt quenching method. One gram of US was finely pulverized in an agate mortar, and mixed with 0.05–0.2 g of Fe_2O_3 of reagent grade (99.99%, Wako Chemicals). Each mixture was melted at 1,400°C for 2 h in an electric muffle furnace (ST-2025D, Motoyama). Dark blown samples were obtained by dipping the bottom of crucible into ice-cold water.

Measurement of ^{57}Fe-Mössbauer spectra was performed by a conventional constant acceleration method under a room temperature. A ^{57}Co(Pd) source of 3.7×10^8 Bq and a reference of metallic iron (α-Fe) were used for the Mössbauer measurement. Temperature dependence of Mössbauer spectra was measured from 77 to 285 K. Data analysis of each spectrum was performed by MossWinn (version 3.0i XP).

XRD pattern was recorded on RINT 2000 (Rigaku Co., Ltd.), using Cu-α as a target, at a scanning rate of 0.02°/s between 2θ of 20 and 90°.

COD was evaluated by redox titration method using $KMnO_4$, which is officially admitted procedure (JIS K 0102 17). Three hundred milliliters of an artificial drain and 10 g of glass sample were installed in an Erlenmeyer flask and were stirred with a digital hot plate stirrer (DP-2L, Iuchi Co., Ltd.) at 20°C at a stirring rate of 300 rpm. The artificial drain is composed of commercially available milk powder (0.59 g) and a mixture of inorganic compounds composed of $(NH_4)_2SO_4$ (0.28 g), KH_2PO_4 (0.026 g), $NaHCO_3$ (0.073 g), and Na_2CO_3 (0.037 g). The mixture was diluted with distilled water until the COD value became 250 mg/l. The value of COD was continuously evaluated by changing the stirring condition from stage 1 to stage 3. In the stage 1, 1 ml of active sewage was added into the artificial drain, and then air was blown into the flask at a rate of 100 ml/min. COD was measured during stirring the

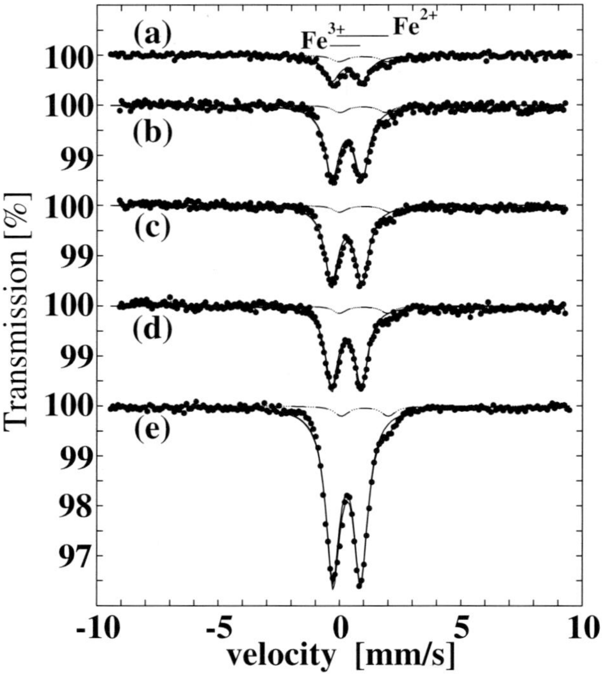

Figure 2 Mössbauer spectra of US (*a*) 0, (*b*) 5, (*c*) 10, (*d*) 15 and (*e*) 20 glasses.

solution in which air was blown in the stage 2, while only stirring was done in the stage 3. Duration of each stage was fixed to be 10 days. Glass samples were slightly washed with distilled water between each period, and the sampling was made twice (10 ml×2). Twenty milliliters of newly prepared artificial drain with COD value of 250 mg/l was added after each sampling.

After COD measurement, determination of iron (III) and that of phosphate ion in the artificial drain were made by an ICP fluorescence analyzer (ICPS-8100, SHIMADZU). Two droplets of nitric acid (14.5 M) was added before each measurement.

3 Results and discussion

3.1 Characterization of US glasses

As a result of chemical analysis, composition of US was determined to be SiO_2 (38.4 mass%), CaO (28.5), Al_2O_3 (14.9), Fe_2O_3 (5.54), P_2O_5 (3.34) and (Na_2O+K_2O+MgO) for the rest. The Mössbauer spectrum of the original US before melting with Fe_2O_3 is depicted in Figure 1. The spectrum can be analyzed into two doublets; one for the $Fe^{II}O_4$ tetrahedra with δ=$0.99_{\pm 0.01}$ mm/s, Δ=$1.88_{\pm 0.02}$ mm/s, Γ=$0.78_{\pm 0.03}$ mm/s and A=$71.9_{\pm 2.2}$%, and $Fe^{III}O_4$ tetrahedra with δ=$0.33_{\pm 0.03}$ mm/s, Δ=$0.97_{\pm 0.07}$ mm/s, Γ=$0.78_{\pm 0.03}$ mm/s and A=$28.1_{\pm 1.6}$%.

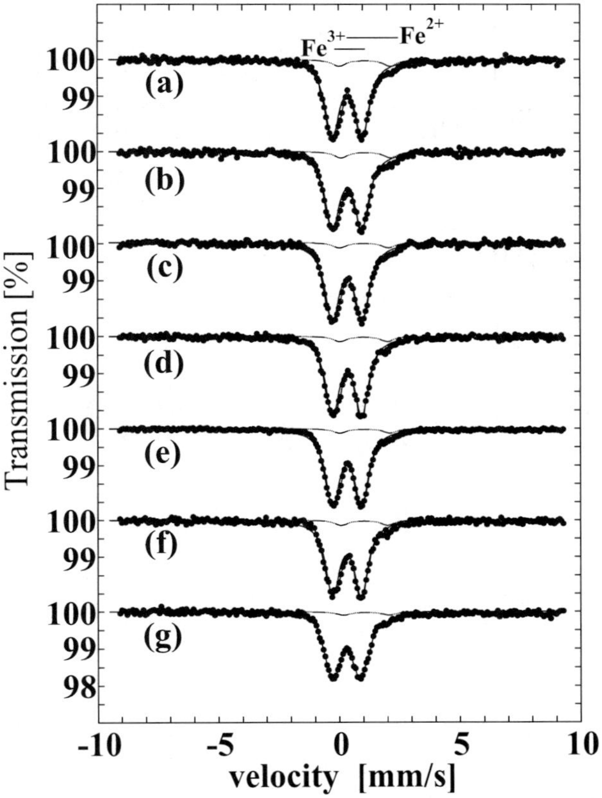

Figure 3 Mössbauer spectra of US20 glass measured at (*a*) 77, (*b*) 100, (*c*) 140, (*d*) 175, (*e*) 210, (*f*) 250 and (*g*) 285 K.

XRD patterns of US glasses showed no peaks for crystal phases, showing that homogeneous glass samples were prepared in the system. Mössbauer spectra of US glasses, US0-US20 are given in Figure 2. Each spectrum can be analyzed into two doublets due to $Fe^{II}O_4$ and $Fe^{III}O_4$ tetrahedra. A gradual increase in δ was observed for Fe^{III} from 0.26 to 0.30 mm/s in the order of US0 to US20 glasses, while identical Δ value of $1.25_{\pm 0.01}$ mm/s was obtained in every sample. For the Fe^{II}, an identical δ value of $1.00_{\pm 0.02}$ mm/s was obtained. A gradual decrease in A for Fe^{II} was observed from $25.0_{\pm 1.3}$ to $5.7_{\pm 0.6}$%. By comparing Figure 2 with Figure 1, we can know a drastic decrease in the absorption area (A) of Fe^{II} in the former, explaining that oxidation of Fe^{II} to Fe^{III} takes place by quenching the melt of US.

In order to estimate Debye temperature (Θ_D), Mössbauer spectra were measured from 77 to 285 K for US0 and US20 glasses. In Figure 3, temperature dependent Mössbauer spectra of US20 glass are depicted. A relationship between recoil-free fraction ($f(T)$) and Θ_D was expressed by the following equation [11];

$$f(T) = exp\left[(-3E_R/2K\Theta_D)\left(1 + 4(T/\Theta_D)^2 \int_0 xdx/(e^x - 1)\right)\right], \qquad (1)$$

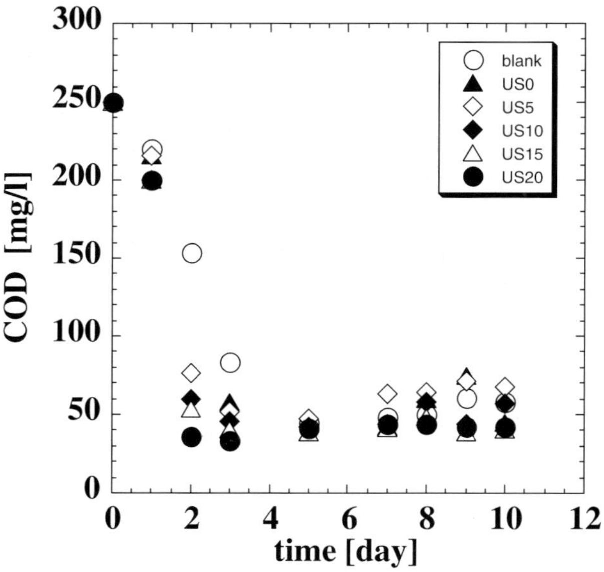

Figure 4 Decrease in COD of artificial drain stirring with US glasses in the 'stage 1.'

where E_R, T and k are recoil energy (=0.002 eV), temperature (K) and Boltzmann constant (=8.62×10^{-5} eV), respectively. The Θ_D values were calculated to be $287_{\pm14.1}$ K for Fe^{III}, $246_{\pm15.1}$ K for Fe^{II} and $275_{\pm10.5}$ K for the total iron (Fe^{II} plus Fe^{III}) in US0 glass, while to be $445_{\pm17.3}$ (Fe^{III}), $224_{\pm19.4}$ (Fe^{II}) and $407_{\pm12.2}$ K (Fe^{II} and Fe^{III}) in US20 glass. These results show that both Fe^{III} and Fe^{II} are more rigidly bound to oxygen atoms in US20 glass than in US0 glass. Johnson and co-workers reported that Θ_D values of $(Fe_2O_3)_{0.05}(Na_2O)_{0.3}(SiO_2)_{0.65}$ glass were 312 K for Fe^{III} and 268 K for Fe^{II} [12]. Nishida reported that Θ_D values of several oxide glasses could be classified into two groups, i.e., one being higher than 280 K and the other less than 270 K [13]. The former was observed when iron played a role of network former (NWF), while the latter when iron played a role of network modifier (NWM). It is concluded that Fe^{II} is located at NWM sites in US0 and US20 glasses in which Θ_D value for Fe^{II} was estimated to be 246 K in the former, while 224 K in the latter.

3.2 Evaluation of water quality

Change of COD for US glasses in the stage 1 is shown in Figure 4. A remarkable difference with the blank value was observed in 2 days. The COD values for the US20 glass under stirring decreased from 250 to 36 mg/l, which is remarkably different from that obtained for the blank; US_{blank} decreased from 250 to 154 mg/l. A consistent decrease in the COD value was also observed in the stages 2 and 3. These results show that iron-containing US glass promotes the decomposition of organic materials.

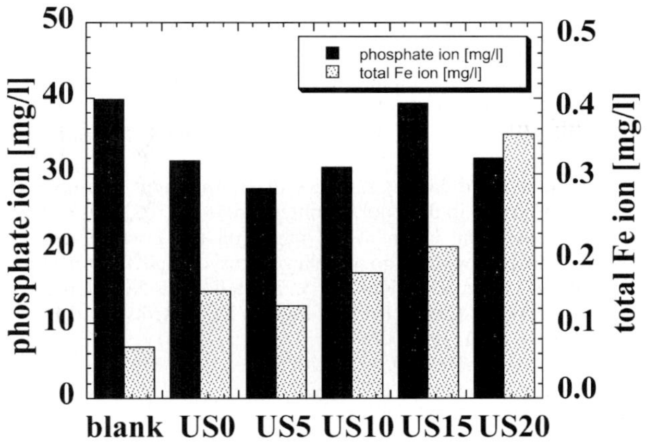

Figure 5 Change of phosphate ion (*left bar*) and total Fe ion (*right bar*) concentrations in the artificial drain after stirring with US glasses in the 'stage 1.'

Concentration of iron and that of phosphate ion obtained from the ICP of the artificial drain at the stage 1 is shown in Figure 5. The iron concentration increased with an increasing amount of Fe_2O_3 from 0.144 to 0.353 mg/l in the stage 1, and from 0.189 to 0.485 mg/l in the stage 2. This result indicates that Fe_2O_3 content in the US glass can regulate the dissolution of the iron. It is reasonable to consider that the iron primarily originates from Fe^{II} occupying NWM sites, because Θ_D value of Fe^{II} site (US0 glass: 246 K, US20 glass: 224 K) is smaller than that of Fe^{III} (US0 glass: 287 K, US20 glass: 445 K). It should also be noted that the concentration of phosphate ion becomes low with an increasing concentration of Fe_2O_3 in the waste glass. This tendency is clear in the US20 glass, in which the phosphate ion is suppressed by adding Fe_2O_3 into the waste glass.

4 Summary

Structure of iron-containing waste glasses, prepared from the household garbage ash was characterized by using ^{57}Fe-Mössbauer spectroscopy. In the iron-containing US glass, an increase in δ was observed for tetrahedral Fe^{III} from 0.26 to 0.30 mm/s, while Δ was identical at 1.25 mm/s. A δ value of 1.00 mm/s and a Δ of 1.96 mm/s were obtained for tetrahedral Fe^{II}. Fe^{II}/Fe^{III} ratio gradually decreased with the Fe_2O_3 content. Debye temperature revealed that Fe^{II} in US glasses occupies the site of NWM, while Fe^{III} occupies the site of NWF. Debye temperature of the US glass became higher along with the introduction of Fe_2O_3, indicating that iron atoms (Fe^{II} and Fe^{III}) are more rigidly bound to oxygen in US20 glass than in US0 glass. It is also considered that Fe^{II} atoms occupying NWM site is loosely bound to oxygen.

Reduction of COD could be associated with the dissolution of the iron in US glass. US20 glass showed a marked effect on the reduction of COD. A step-by-step increase in the dissolution of Fe^{III} was observed in the leaching test of US glass in the artificial drain. It is concluded that iron-containing inorganic glasses prepared from household garbage causes a marked reduction of phosphates ion and hence COD. They can be utilized as a decreasing agent of phosphate ions in wastewater.

References

1. Ministry of the environment of Japan, news release and speeches (homepage), http://www.env.go.jp/en/news/2004/1216a-02.pdf (2004)
2. Kubuki, S., Kawakami, N., Matsumura, H., Fujisato, T., Nishida, T., Fukagawa, M.: J. Technol. Educ. **11**(2), 77 (2004)
3. Ministry of the environment of Japan, statistics of environment (Kankyo Tokei Shiryo shu) (homepage), http://www.env.go.jp/doc/toukei/contents/data/es170831.xls (2004), (in Japanese)
4. Ministry of the environment of Japan, white paper on the environment in Japan (Kankyo Hakusho)(home page), http://www.env.go.jp/policy/hakusyo.php3?kid=219 (2004), (in Japanese)
5. Nishida, T., Tokunaga, M., Sugata, Y., Kubuki, S.: J. Radioanal. Nucl. Chem. **266**(2), 171 (2005)
6. Moura, F.C.C., Araujo, M.H., Costa, R.C.C., Fabris, J.D., Ardisson, J.D., Macedo, W.A.A., Lago, R.M.: Chemosphere **60**, 1118 (2005)
7. Leupin, O.X., Hug, S.J.: Water Res. **39**, 1729 (2005)
8. Kurkjian, C.R.: J. Non-Cryst. Solids **3**, 157 (1970)
9. Coey, J.M.D.: J. Phys. (Paris) **35**, C6 (1974)
10. Frischat, G.H., Tomandl, G.: Glastechn. Ber. **42**, 182 (1969)
11. Homonnay, Z., Music, S., Nishida, T., Kopelev, N.S., Vertés, A.: In: Vértes, A., Homonnay, Z. (eds.) Mössbauer Spectroscopy of Sophisticated Oxides, p. 5. Akademiai Kiado, Budapest (1997)
12. Johnson, J.A., Johnson, C.E., Holland, D., Mekki, A., Appleyard, P., Thomas, M.F.: J. Non-Cryst. Solids **246**, 101 (1999)
13. Fujita, E., Nasu, S., Nishida, T., Yoshida, Y.: In: Fujita, E. (ed.) Introduction to the Mössbauer Spectroscopy – Principle and Applications, p. 204. Agne Gijutu Center, Tokyo (1999) (in Japanese)

Hyperfine Interact (2005) 166:437–442
DOI 10.1007/s10751-006-9305-6

Investigation of primary chemical reactions of electrons by emission Mössbauer and positron spectroscopies

Vsevolod M. Byakov · Leonid A. Kulikov · Yurii D. Perfil'ev · Sergey V. Stepanov

Published online: 14 November 2006

Abstract A set of early chemical reactions induced in frozen medium around decayed ^{57}Co or ^{119m}Sn nuclei by emitted Auger electrons is suggested. The mechanism predicts a correlation between formation probabilities of final products, namely Fe^{2+}-Fe^{3+} or Sn^{2+}-Sn^{4+} ions, positronium atom and molecular hydrogen formed in similar medium after its irradiation by fast positrons and electrons. This correlation indicates a similarity of chemical processes occurring in nanovolumes around Mössbauer ^{57}Fe- and ^{119}Sn-nuclei after radioactive decay as well as in tracks produced by fast positrons and electrons. The important role of quasi-free electrons is revealed.

Key words reaction of electrons · comparative study · emission Mössbauer spectroscopy · positron spectroscopy · radiolysis

For application of the emission Mössbauer spectroscopy, a small amount of special isotopes (^{57}Co, ^{119m}Sn) are added into investigated solid state system. A common property of these isotopes consists in production of Auger ionization of the daughter atoms (^{57}Fe, ^{119}Sn), caused by electron capture or converted isomeric transition. These processes precede emission of the Mössbauer γ-quantum by Mössbauer ^{57}Fe or ^{119}Sn, which carries physico-chemical information about the daughter atoms and their environment. Auger ionization converts the daughter atom into multi-charge ion (electric charge of ^{57}Fe ion could reach +8) and is accompanied by emission of a large number of Auger electrons with total energy about several keV and average energy of the electron above 10^2 eV. Together with the X-rays, Auger electrons

V. M. Byakov · S. V. Stepanov
Institute for Theoretical and Experimental Physics, 117218 Moscow, Russia

L. A. Kulikov · Y. D. Perfil'ev (✉)
Chemical Department, Lomonosov Moscow State University,
Leninskie Gory, 119899 Moscow, Russia
e-mail: perf@radio.chem.msu.ru

produce a spherical grape of ionizations of a nanometer scale around the ion within 10^{-14} s. This grape contains up to 100 ion–electron pairs ($H_2O^{+\cdot}$ and e^-, if the Mössbauer nuclei are dissolved in frozen aqueous medium). Such a grape is similar to numerous ionization grapes that are formed along the track of a fast electron. In radiation–chemical literature they are known as blobs in distinction from the spurs, consisting of several ion–electron pairs [1]. High concentration of the pairs in blobs leads to favorable conditions for reactions between ions and electrons and formation of recombination radiolytic products, such as H_2 in water:

$$e^- + (H_2O, H_2O)^{+\bullet} \rightarrow H_2 + 2O^{\bullet}H. \tag{1}$$

The terminal part of the positron track is also a grape of ionizations, consisting of several tens of ion–electron pairs. In the terminal positron blob, combination of deenergized positron with one of the blob electrons leads to formation of a positronium atom (Ps) [1]:

$$e^+ + e^- \rightarrow Ps. \tag{2}$$

It is quite obvious that in nanovicinity of the multi-charged Mössbauer ions Fe^{n+} or Sn^{n+}, the ion–electron pairs, produced by the Auger process, undergo chemical transformations, recombine with each other and react with Fe^{n+} and Sn^{n+}. The decrease of the ion charge, n+, changes the energy of emitting Mössbauer gamma-quanta. So knowledge of chemical transformations in the environment of the radioactive nuclei is important for understanding and correct treatment of experimental data of the emission Mössbauer spectroscopy and may be also useful for interpretation of some results obtained in radiation and positron chemistry.

The aim of this paper is to reveal early chemical transformations and reactions occurring in nanovolume around the Mössbauer nucleus (we will call this region as the Auger blob). We present a unified interpretation of chemical reactions in molecular media, induced by fast e^+ and e^- irradiation as well as by radioactive nuclear decay accompanied by emission of the Auger electrons and subsequent ionizations. Primarily we shall consider reactions in frozen aqueous solutions (77 K), nevertheless, our findings shed light on intratrack processes at normal conditions.

1 Chemical processes in blobs

Initially electron affinity of $^{57}Fe^{n+}$ significantly exceeds the first ionization potential of surrounding water molecules (~9 eV in liquid water) [1]. So the most earlier chemical reaction in Auger blob appears to be the neutralization of $^{57}Fe^{n+}$ by picking-off-up electrons from the nearest H_2O molecules [2]: $Fe^{n+} + H_2O \rightarrow Fe^{(n-1)+} + H_2O^{+\bullet}$. We assume that decreasing of the electric charge of the Mössbauer ion ends with the formation of the first of the two possible chemically stable states (Fe^{3+} or Fe^{2+}), namely Fe^{3+}. The subsequent possible reduction from Fe^{3+} to Fe^{2+} proceeds via recombination with quasi-free electrons, e^-, of the Auger blob:

$$Fe^{3+} + e^- \rightarrow Fe^{2+}. \tag{3}$$

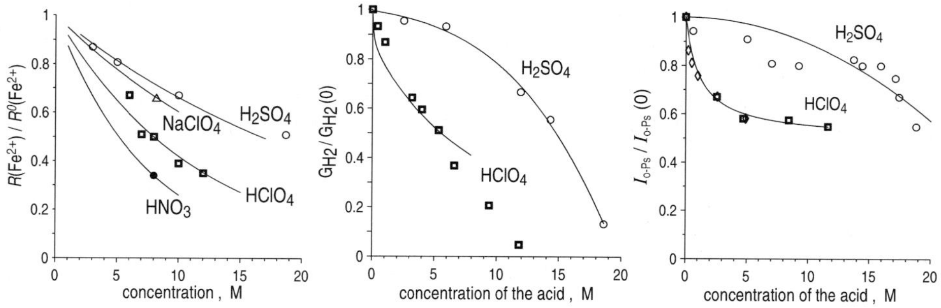

Figure 1 *Left* Relative yields of Fe^{2+} in acid aqueous solutions frozen at 78 K [5]. *Solid lines* represent the first of Eq. 5. *Center* H_2 yields in concentrated solutions of H_2SO_4 (*circle*) [14] and $HClO_4$ (*square*) [15]. *Right o*-Ps yields in concentrated solutions of H_2SO_4 (*circle*) and $HClO_4$ (*square, diamond*) [12, 13].

Physico-chemical processes in the Auger blob are similar to the processes both in the terminal blob of the positron track and in the blobs of fast electrons (e^{-**}, e^{-*}) formed due to ionization losses [1]:

$$e^{-**} + H_2O \to e^{-*} + H_2O^+ + e^{-*}, \qquad e^{-*} \to e^- + H_2O^*, \qquad e^- + trap \to e^-_{tr},$$

$$e^- + (H_2O, H_2O)^{+\bullet} \to H_2 + 2O^\bullet H, \; H_2O^{+\bullet} + H_2O \to H_3O^+ + O^\bullet H, \; e^- + O^\bullet H \to OH^-,$$

$$H_3O^+ + OH^- \to 2H_2O, \qquad e^- + H_3O^+ \to H^\bullet + H_2O, \qquad e^- + S \to S^-,$$

$$H_2O^{+\bullet} + S \to \begin{cases} O^\bullet H + SH^+ \\ H_2O + S^{+\bullet} \end{cases}$$

The probability of the reaction (4) by the time moment of emission of the Mössbauer photon ($\sim 10^{-7}$ s after the birth of ^{57}Fe nucleus) determines the final charge state of ^{57}Fe ion and thus the energy of the photon which depends on the number of electrons, composing electronic shell of $^{57}Fe^{n+}$ ion. It is clear that there is a possibility to govern this probability as well as the formation probabilities of Ps and radiolytic H_2 by introducing, for example, some concentration of solute S which is capable to capture quasi-free electrons. In a concentrated solution of S the ratio $^{57}Fe^{2+}$ to $^{57}Fe^{3+}$ will tend to 0. This effect may serve as a prove for suggested mechanism in Auger blob. The mechanism of radiation chemical transformations in the positron and electron blobs arising in liquid water and aqueous solutions was suggested by us. It gives rather good description of the kinetics of accumulation of the main radiolytic products and their yields vs concentrations of the dissolved electron scavengers.

2 Quantitative consequences of the model

Following relations demonstrating variation of the yields of *ortho*-Ps, $I_{o\text{-Ps}}$, radiolytic hydrogen, G_{H2}, doubly charged Fe ion, $R(Fe^{2+})$, and hydrated electron, $G(e^-_{aq})$ vs

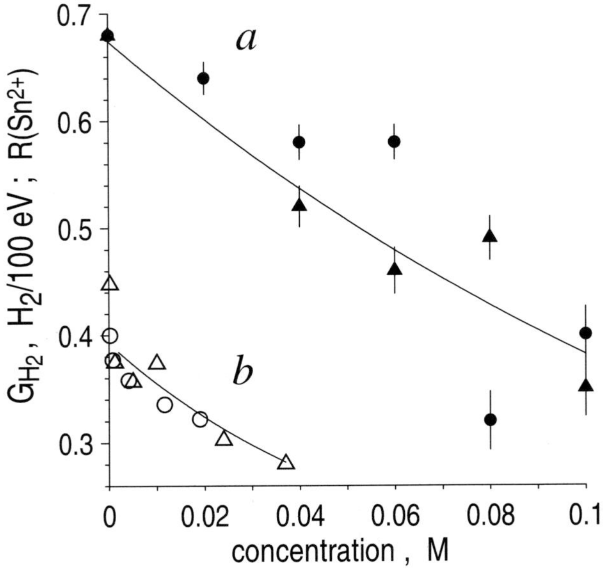

Figure 2 *a* dependence of the Sn^{2+} yield in frozen at 77 K solutions vs concentration of salts $CrCl_2$ (*filled triangle*) and $CuCl_2$ (*filled circle*) [7]. Sn^{2+} is the final product of ion–electron recombination of the ^{119m}Sn nucleus, after its converted isomer transition. *b* dependence of radiolytic hydrogen yields vs concentration of Cu^{2+} (*open circle*) and Cr^{2+} (*open triangle*) ions in aqueous solutions at room temperature [6]. *Solid lines* are calculated from Eq. 5.

concentration c_S of the electron scavenger (S) are obtained from the suggested mechanism [1–4]:

$$I_{o-Ps} = I^0_{o-Ps}/(1 + pc_S), \qquad G(e^-_{aq}) = G^0(e^-_{aq}) \cdot exp(-qc_S), \tag{4}$$

$$R(Fe^{2+}) = R^0(Fe^{2+}) \cdot exp(-fc_S), \qquad G_{H_2} = G^0_{H_2}\Big/(1 + hc_S) \tag{5}$$

Inhibition coefficients of o-Ps, H_2, Fe^{2+} and e^-_{aq} (e^-_{tr}), namely p, h, f and q characterize the reaction ability of S towards thermalized quasi-free electrons (e^-), which we consider as the H_2, o-Ps, Fe^{2+} and e^-_{aq} (e^-_{tr}) precursors. Each coefficient is proportional to the rate constant $k(e^-+S)$ of quasi-free electron with S: $p \propto h \propto f \propto q \propto k(e^- + S)$.

3 Comparison with experiments

There are few radiation-chemical and positron annihilation data, obtained in frozen solutions, which can be directly confronted with data of emission Mössbauer spectroscopy. Nevertheless, correlations between the yields of the above mentioned species even under somewhat different conditions are meaningful and indicate a similarity of the discussed processes in blobs. From the data on Figure 1 in aqueous solutions of $NaClO_4$, H_2SO_4, $HClO_4$ and HNO_3 frozen at 77 K [5] we obtained ratios of f for different electron scavengers: $f(H_3O^+)/f(ClO_4^-)/$

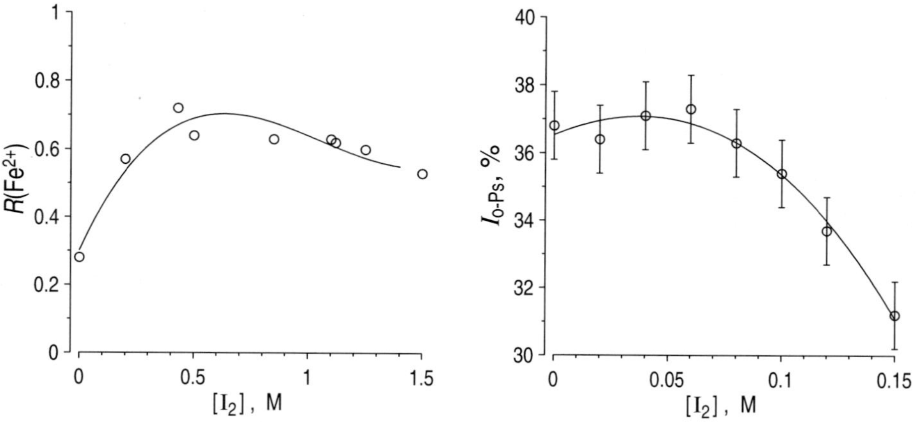

Figure 3 *Left* Fe^{2+} yield in frozen at 77 K ethanol vs I_2 concentration [8]. *Right o*-Ps yield in liquid cyclohexane at room temperature vs I_2 concentration [9]. *Solid lines* approximate the experimental data.

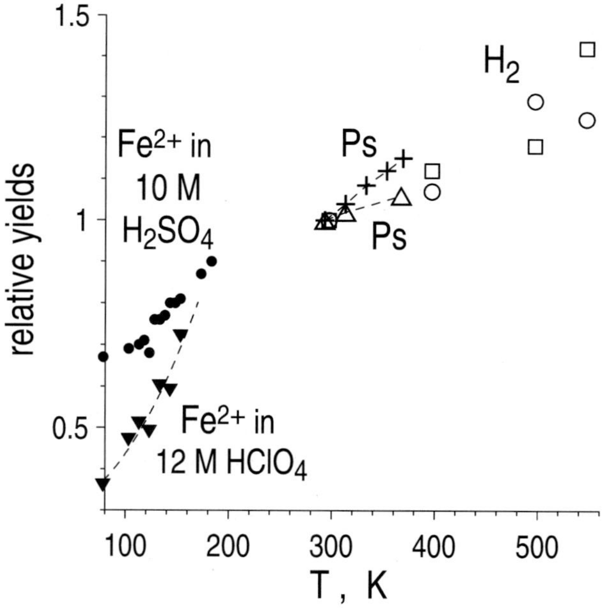

Figure 4 Temperature dependencies of Fe^{2+} yields in frozen glassy solutions of 10 M H_2SO_4 (*filled circle*) and 12 M $HClO_4$ (*filled inverted triangle*) [10], relative yields of Ps (*plus sign, triangle*) in water and relative hydrogen yields under the action of fast neutrons (*open square*) and γ-rays (*open circle*), [11] and references therein.

$f(NO_3^-) \approx 0.04 : 0.05 : 0.1$. These ratios qualitatively agree with the ratios of coefficients q and p, also characterizing reaction ability of the same ions towards quasi-free electron but at room temperature [4, 6]: $q(H_3O^+)/q(ClO_4^-)/q(NO_3^-) \approx (\leq 0.1)/(\leq 0.1)/2.4, p(H_3O^+)/p(ClO_4^-)/p(NO_3^-) \approx (\leq 0.1) : 0.1 : 3.1$. It is noteworthy that all ratios differ significantly from the ratios of the reaction rate constants $k(e_{aq}^{-} + S_i)$ between S_i and hydrated electron [4]: $k(e_{aq}^{-} + H_3O^{+})/k(e_{aq}^{-} + ClO_4^{-})/k(e_{aq}^{-} + NO_3^{-}) = 1 : 0.1 : 1$. The difference of yields of Fe^{2+} in solutions of sulphuric and chloric acids, which are clearly seen at low temperatures, also holds for the yields of radiolytic hydrogen and *o*-Ps in solution of the same acids at room temperature.

Figure 2 demonstrates similar inhibition efficiencies of G_{H2} by Cr^{2+} and Cu^{2+} ions at room temperature, and similar action of these ions on the Sn^{2+} yield in frozen (77 K) aqueous solutions. Figure 3 demonstrates complicate behavior of the Fe^{2+} yield (left) in frozen at 77 K glassy ethanol vs concentration of I_2. In liquid cyclohexane, *o*-Ps formation probability behaves similarly vs concentration of I_2 (right). In more detail this behavior is discussed in [2]. Figure 4 shows correlation in variation of the yields of Fe^{2+}, H_2 and Ps vs temperature of the medium.

4 Conclusion

The above facts tell us about similar state and behavior of electrons in different types of blobs in spite the large difference in temperature and phase state of the investigated samples. Similar character of an influence of concentration of dissolved substances, temperature and phase state on the yields of Mössbauer ions $^{57}Fe^{2+}$, $119Sn^{2+}$, *o*-Ps, radiolytic H_2, and e_s^{-} is revealed. This fact supports our suggestion about similarity of early chemical reactions taking place in positron, electron and Auger blobs. The data demonstrate the leading role of quasi-free electrons in formation of the yields of above products.

Acknowledgment This work was supported by the Russian foundation for Basic Research, project No 05-03-33079 and by the Council of the President of the Russian Federation for support of young scientists and leading scientific schools, project NSh 1907.2003.2.

References

1. Byakov, V.M., Nichiporov, F.G.: Intratrack Chemical Processes. Energo Atomizdat, Moscow (1985)
2. Byakov, V.M., Kulikov, L.A., Perfil'ev, Y.D., Stepanov, S.V.: Phys. At. Nucl. **68**(6), 1016 (2005)
3. Byakov, V.M.: J. Phys. IV **3**, 85 (1993), Colloque C4, Supplement II
4. Lam, K.Y., Hunt, J.W.: Radiat. Phys. Chem. **7**, 317 (1975)
5. Perfil'ev, Y.D., Kulikov, L.A., Bugaenko, L.T., Babeshkin, A.M., Afanasov, M.I.: J. Inorg. Nucl. Chem. **36**(12), 2115 (1976)
6. Faraggi, M.: Int. J. Radiat. Phys. Chem. **5**(2), 197 (1973)
7. Perfil'ev, Y.D., Kulikov, L.A., Byakov, V.M., Stepanov, S.V., Alhatib, H., Bugaenko, L.T.: Khim. Vys. Energ. **37**(5), 390 (2003)
8. Milgrom, Y.M., Perfil'ev, Y.D., Babeshkin, A.M.: Khim. Vys. Energ. **12**(3), 272 (1978)
9. Levay, B., Hautojarvi, P.: J. Phys. Chem. **76**(14), 1951 (1972)
10. Perfil'ev, Y.D., Kulikov, L.A., Babeshkin, A.M.: Zh. Fiz. Khim. **LII**(7), 1631 (1978)
11. Byakov, V.M., Stepanov, S.V.: Nukleonika **42**(1), 45 (1997)
12. Tao, S.J., Green, J.H.: J. Phys. Chem. **73**(4), 882 (1969)
13. Byakov, V.M., Grafutin, V.I., Koldaeva, O.V., Minaichev, E.V., Nichiporov, F.G., Obukhov, Y.V., Stepanova, O.P.: Chem. Phys. **24**, 91 (1977)
14. Boyle, J.W.: Radiat. Res. **17**, 427 (1962)
15. Katakis, D., Allen, A.O.: J. Phys. Chem. **68**, 3107 (1964)

Hyperfine Interact (2005) 166:443–447
DOI 10.1007/s10751-006-9306-5

Mössbauer study on the gamma radiolysis of lithium trisoxalatoferrate (III) pentahydrate

J. Ladriere · V. Senterre

Published online: 14 November 2006

Abstract The final iron product of the 60-Co gamma radiolysis of lithium trisoxalatoferrate(III) pentahydrate has been identified by Mössbauer spectroscopy as $FeC_2O_4.2H_2O$. Quantitative chemical and Mössbauer analyses show that the radiolytic decomposition proceeds as a first order reaction as a function of the absorbed dose. Chemical calibration of the relative peaks areas of the two iron species indicates that the corresponding Lamb–Mössbauer factors ratio is unaffected by the radiolysis.

Key words radiolysis · lithium trisoxalatoferrate(III) · Lamb–Mössbauer factor

1 Introduction

The gamma radiolysis of alkali trisoxalatoferrate(III) has been extensively studied by using Mössbauer spectroscopy for the characterization of the iron radiolytic products. Although these studies generally show a reduction of iron(III) into iron(II), the nature of the iron(II) species has been reported to be dependent on both alkali nature (*M*) and hydration number (*x*) of the initial ferric complex. For heavy alkali (*M*=K,Cs), the ferrous products have been identified as a double oxalate of iron and the corresponding alkali, $M_2Fe(C_2O_4)_2 \cdot y\ H_2O$, with an hydration number *y* depending on the water content of the initial ferric compound [1–4] and, in some cases, on the irradiation conditions [5, 6]. However, as already shown in a previous study [7] on $Li_4Fe(C_2O_4)_3.Cl.9H_2O$, for light alkali, the only ferrous radiolytic product was found to be $FeC_2O_4.2H_2O$.

In this work, we have extended this study to another lithium complex, $Li_3Fe(C_2O_4)_3.5H_2O$, in order to confirm the influence of the nature of the second coordination sphere of the trisoxalatoferrate(III) ion on the nature of the radiolytic

J. Ladriere (✉) · V. Senterre
Laboratoire de Chimie Nucléaire, Université Catholique de Louvain, Chemin du Cyclotron 2, B-1348 Louvain-la-Neuve, Belgium
e-mail: ladriere@chim.ucl.ac.be

products. In addition, quantitative chemical and Mössbauer analyses have been performed to determine the mechanism and the yield of the radiolysis.

2 Experimental

The synthesis, crystal structure as well as Mössbauer and thermogravimetric analyses of lithium trisoxalatoferrate(III) pentahydrate have been described in a previous study [8]. Powdered samples of this compound were irradiated at 298 K in sealed glass tubes, under high purity nitrogen atmosphere. The irradiations have been performed with 60-Co gamma rays at a mean energy of 1.25 MeV, using a dose rate of 0.778 Mrad/h. The absorbed dose was calibrated by Fricke dosimetry and corrected by the ratio (0.9094) of the mass–energy absorption coefficient of the sample to the one of the dosimeter.

After dissolution of the irradiated samples in diluted sulfuric acid, under nitrogen atmosphere, potentiometric titrations of Fe^{2+} and $C_2O_4^{2-}$ were performed with a ceric sulfate standard solution. After reduction of Fe^{3+} into Fe^{2+} by a chromous sulfate solution, the total amount of iron was obtained by back titration with the ceric solution.

The Mössbauer spectra were recorded at 295 and 80 K, using a constant acceleration spectrometer and a 57-Co(Rh) source. The effective thickness of the absorber was adjusted to 1.6 (10 mg Fe/cm^2).The isomer shift (I.S.) values are reported with respect to α-iron. Experimental data were resolved into Lorentzian lines using an iterative least-square program and the goodness of the fits was estimated from the χ^2 values.

3 Results and discussion

3.1 Qualitative Mössbauer analysis

The Mössbauer spectrum of the unirradiated compound (Figure 1a) is characterized by a slightly asymmetric but strongly broadened absorption peak indicating the presence of both paramagnetic relaxation and quadrupolar interaction, as already described in a previous Mössbauer and X-ray diffraction study [8]. The quadrupole splitting (Q.S. = 0.47 ± 0.03 mm/s) and isomer shift (I.S. = 0.38 ± 0.03 mm/s) values at 295 K are consistent with a high spin ferric ion in a non-cubic environment. This result is supported by the crystal structure analysis which shows a deformation of the octahedral coordination sphere of the iron atom from O_h to D_{3h} symmetry, whereas the broadening of the lines results from a relatively fast spin–spin relaxation process between the closest iron ions (6.559 Å)

When the ferric complex is irradiated at increasing doses, the corresponding Mössbauer spectra (Figure 1b to e) show that it is progressively transformed into a high spin ferrous compound, as evidenced by its large isomer shift (I.S. = 1.18 ± 0.02 mm/s at 295 K; I.S. = 1.30 ± 0.02 mm/s at 80 K) and its temperature dependent quadrupole splitting (Q.S. = 1.73 ± 0.02 mm/s at 295 K; Q.S. = 2.01 ± 0.02 mm/s at 80 K). Since these parameters as well as their temperature dependence are identical to those of the dihydrated ferrous oxalate, $FeC_2O_4.2H_2O$, a compound well

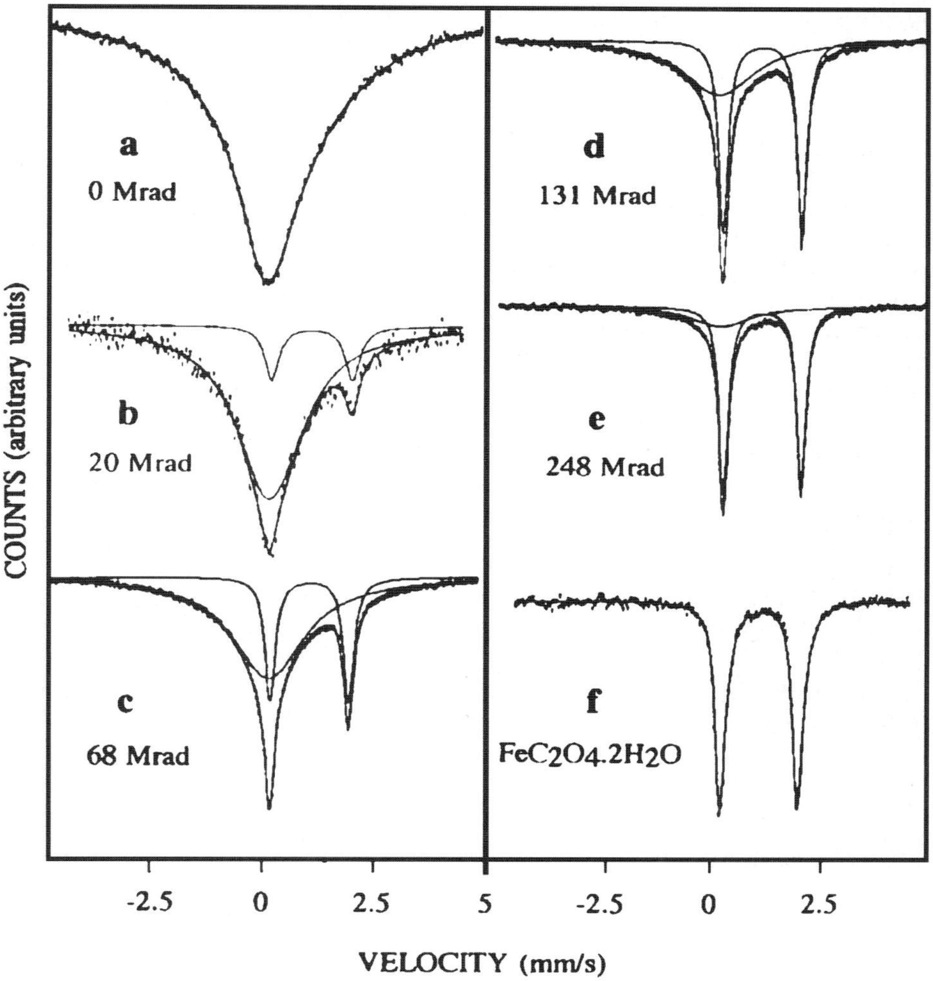

Figure 1 Mössbauer spectra at 295 K of unirradiated (**a**), irradiated $Li_3Fe(C_2O_4)_3.5H_2O$ (**b**–**e**) and $FeC_2O_4.2H_2O$ (**f**).

described in the literature [9], it can be concluded that the latter compound is the iron(II) radiolytic product. For comparison, the Mössbauer spectrum of this compound is shown in Figure 1f.

3.2 Quantitative chemical analysis

The chemical analysis data obtained from potentiometric titration of Fe^{2+} show that the radiolytic reduction of the ferric complex proceeds as a first order process as a function of the absorbed dose, since the chemical radiolytic yield for the ferrous species, Y_C, is related to the absorbed dose D by the classical kinetic law:

$$Y_C = Fe^{2+}/(Fe^{2+} + Fe^{3+}) = 1 - exp(-k_C.D)$$

Figure 2 Chemical yields for the ion(II) formation and the oxalate decomposition as a function of the absorbed dose.

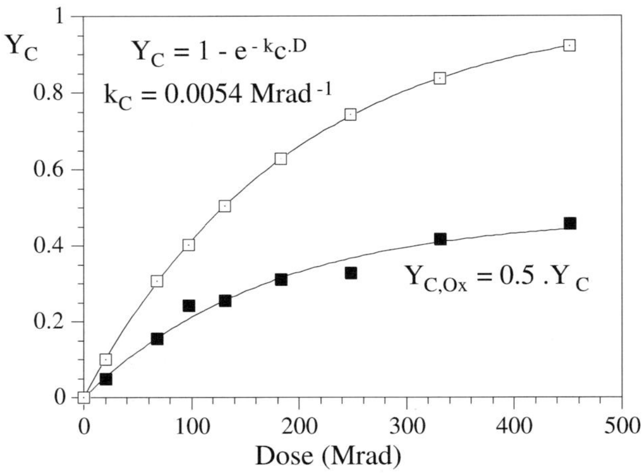

The corresponding least-square calculated function is represented in Figure 2 and is characterized by an apparent rate constant k_C equal to 0.0054 $Mrad^{-1}$. This evolution is due to the original compound depletion and to the radiolytic stability [10] of ferrous oxalate dihydrate.

A similar behaviour is also found for the decomposition yield of the oxalate ligand ($Y_{C,ox}$) which shows a ratio $Y_{C,ox}/Y_C$ equal to 0.5, in agreement with the stoichiometry of the radiolytic reduction process:

$$Fe^{3+} + 0.5(C_2O_4)^{2-} \rightarrow Fe^{2+} + CO_2$$

3.3 Quantitative Mössbauer analysis

From the relative areas of the iron components in the spectra of the irradiated samples, it is possible to quantitatively estimate the relative proportions of Fe^{3+} and Fe^{2+}, provided the corresponding Lamb-Mössbauer *f*-factors of these species are known. Fortunately, in the present case, the irradiated samples are composed of only two well defined iron compounds, $Li_3Fe(C_2O_4)_3.5H_2O$ and $FeC_2O_4.2H_2O$, which can be synthesized separately. It is therefore possible to prepare calibrated absorbers composed of a mixture of these two components with a known molar fractions ($X_2 = Fe^{2+}$; $X_3 = Fe^{3+}$) and to determine the ratio of the corresponding *f*-factors (f_2/f_3) from the Fe^{2+}/Fe^{3+} areas ratio (S_2/S_3), since: $S_2/S_3 = f_2.X_2/f_3.X_3$. The plot of the relative surfaces ratio, S_2/S_3, against the molar fractions ratio, X_2/X_3, is characterized by linear correlations from which calibration constants, f_2/f_3, can be obtained for the two measurement temperatures: 1.11 at 298 K and 1.36 at 80 K.

Since $X_2 = 1-X_3$ and $S_2 = 1-S_3$, the exact proportion of iron(II) ($X_2 = Y_M$) can be calculated from the Mössbauer analysis of the unirradiated mixtures, by using the following expression:

$$Y_M = S_2/[S_2 + (f_2/f_3).S_3]$$

The same procedure can be applied to calculate, from Mössbauer measurements, the iron(II) yields Y_M in the irradiated samples. From these values, represented in

Figure 3 Fe(II) radiolytic yields from Mössbauer (Y_M) and chemical (Y_C) analyses.

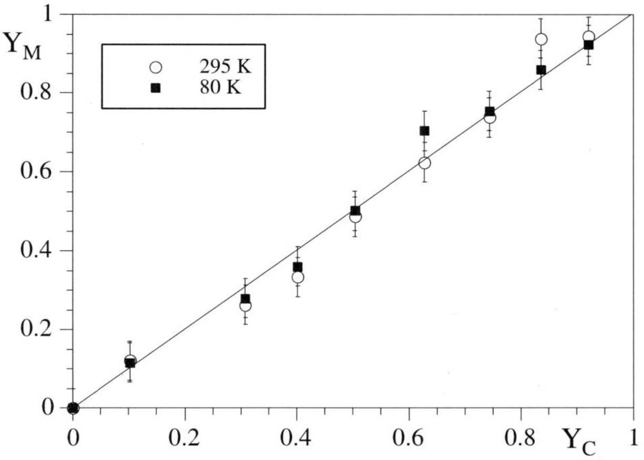

Figure 3, it is seen that there is no significant difference between the chemical (Y_C) and the Mössbauer (Y_M) analyses, within the experimental error of the latter method (±5%). As a consequence, the rate contants calculated from the quantitative Mössbauer analysis are identical (K_M = 0.0054 $Mrad^{-1}$ at 295 K; K_M = 0.0055 $Mrad^{-1}$ at 80 K) to the value found by chemical analysis. This result indicates that the ratio of the Lamb-Mössbauer *f*-factors of the two iron components is unaffected by the radiolysis, presumabely because of fast solid-state annealing reactions during the irradiation time.

Acknowledgment The authors wish to thank the Interuniversitary Institute for Nuclear Sciences (IISN-Belgium) for financial support.

References

1. Temperley, A.A., Pumplin, D.W.: J. Inorg. Nucl. Chem. **31**, 2711 (1969)
2. Sugimori, A.: Bull. Chem. Soc. Jpn. **39**, 2583 (1966)
3. Ladrière, J.: Hypefine Interact. **42**, 993 (1988)
4. Ladrière, J., Senterre, V.: Hyperfine Interact. **70**, 917 (1992)
5. Dharmawardena, K.G., Bancroft, G.M.: J. Chem. Soc., A **10**, 2655 (1968)
6. Bancroft, G.M., Dharmawardena, K.G., Maddock, A.G.: J. Chem. Soc., A **12**, 2914 (1969)
7. Ladrière, J., DeViron, M., Apers, D.: Il Nuovo Cim. **18**, 221 (1996)
8. Declercq, J.P., Feneau-Dupont, J., Ladrière, J.: Polyhedron **14**, 1943 (1995)
9. Devillers, M., Ladrière, J., Apers, D.: Inorg. Chim. Acta **126**, 71 (1987)
10. Gütlich, P., Odar, S., Fitsimmons, B.W., Erickson, N.E.: Radiochim. Acta **10**, 147 (1968)

Hyperfine Interact (2005) 166:449–453
DOI 10.1007/s10751-006-9307-4

Characterisation of the reduction properties of $La_{0.5}Sr_{0.5}Fe_{0.5}Co_{0.5}O_3$

F. J. Berry · J. R. Gancedo · J. F. Marco · X. Ren

Published online: 27 October 2006

Abstract We report here on the characterisation by temperature programmed reduction, ^{57}Fe Mössbauer spectroscopy and X-ray absorption spectroscopy of the phases resulting from treatment of the perovskite-related material $La_{0.5}Sr_{0.5}Fe_{0.5}$-$Co_{0.5}O_3$ in a flowing 90% hydrogen/10% nitrogen atmosphere. The results show that treatment of $La_{0.5}Sr_{0.5}Fe_{0.5}Co_{0.5}O_3$ (which contains approximately 50% Fe^{4+} and 50% Fe^{3+}) in the flowing 90% hydrogen/10% nitrogen atmosphere at 600°C does not result in the reduction of any of the constituent elements of the material and that the perovskite structure is still retained. The Mössbauer spectrum recorded following heating in the gaseous reducing environment at 1,000°C shows the presence of metallic iron, an Fe^{3+}-containing phase with parameters compatible with the presence of $SrLaFeO_4$ which has a K_2NiF_4-type structure, and a paramagnetic Fe^{3+} phase. The X-ray absorption spectroscopy results show the presence of metallic cobalt. The Mössbauer spectrum recorded following heating at 1,200°C continues to show the Fe^{3+}-containing components plus a larger contribution from metallic iron. The X-ray absorption spectroscopy results show the presence of metallic cobalt, $SrLaFeO_4$, La_2O_3 and SrO.

Key words cobalt-substituted ortoferrites · reduced phases
Mössbauer spectroscopy

1 Introduction

Perovskite-related oxygen deficient materials of the type $La_{1-y}Sr_yFe_{1-x}Co_xO_3$ have been the subject of interest for some time. For example, some of these materials

F. J. Berry · X. Ren
Department of Chemistry, The Open University Walton Hal., Milton Keynes MK7 6AA, UK

J. R. Gancedo · J. F. Marco (✉)
Instituto de Química-Física "Rocasolano," CSIC. c/ Serrano, 119, 28006 Madrid, Spain
e-mail: jfmarco@iqfr.csic.es

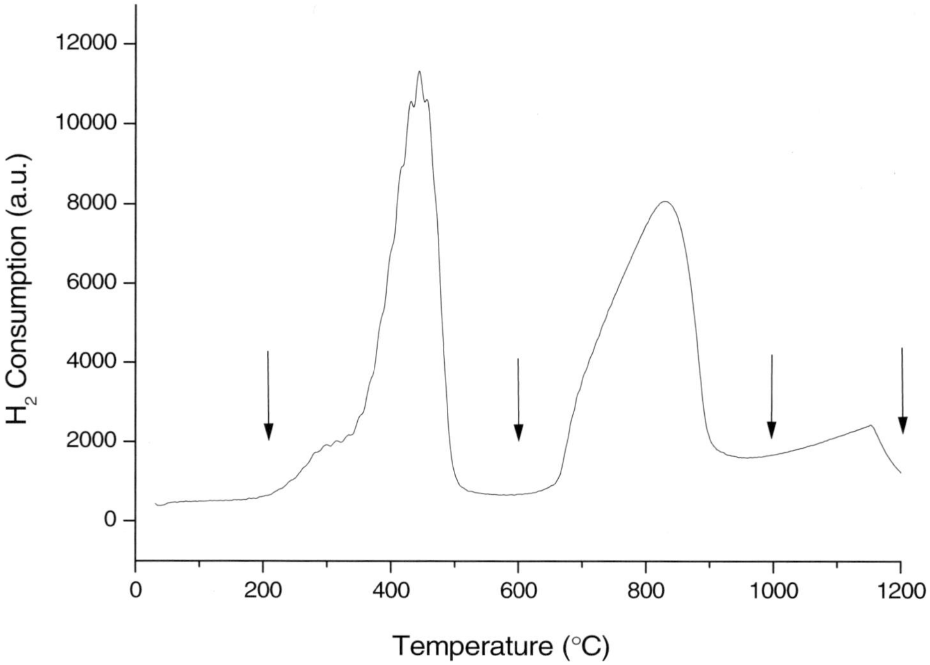

Figure 1 TPR profile recorded from $La_{0.5}Sr_{0.5}Fe_{0.5}Co_{0.5}O_3$.

have potential applications as electrode materials for high temperature fuel cells, as catalysts in automobile exhaust systems, and as gas sensors [1–3]. There is also interest in the unusual magnetic and electronic properties of the materials [4]. Given that some of these applications involve operation under reducing conditions and that the nature of reduced phases in these systems is a matter of sparse attention, we have investigated the nature of materials formed by temperature programmed reduction in hydrogen of the compound of composition $La_{0.5}Sr_{0.5}Fe_{0.5}Co_{0.5}O_3$.

2 Experimental

The compound $La_{0.5}Sr_{0.5}Fe_{0.5}Co_{0.5}O_3$ was prepared by grinding appropriate quantities of lanthanum(III) oxide, strontium(II) carbonate, α-iron(III) oxide and cobalt(II) oxide in an agate pestle and mortar and heating in air at 1,200°C (12 h).

X-ray powder diffraction patterns (XRD) were recorded at 298 K with a Siemens D5000 diffractometer using Cu–Kα radiation. Temperature programmed reduction (TPR) profiles were recorded from *ca.* 150 mg samples in flowing 10% hydrogen/ 90% nitrogen (15–20 ml/min) with the temperature being increased by 5°C/min.

^{57}Fe Mössbauer spectra were recorded at 298 K with a microprocessor-controlled Mössbauer spectrometer using a *ca.* 25 mCi ^{57}Co/Rh source. All the spectra were computer fitted and the isomer shift data are reported relative to that of metallic iron at room temperature.

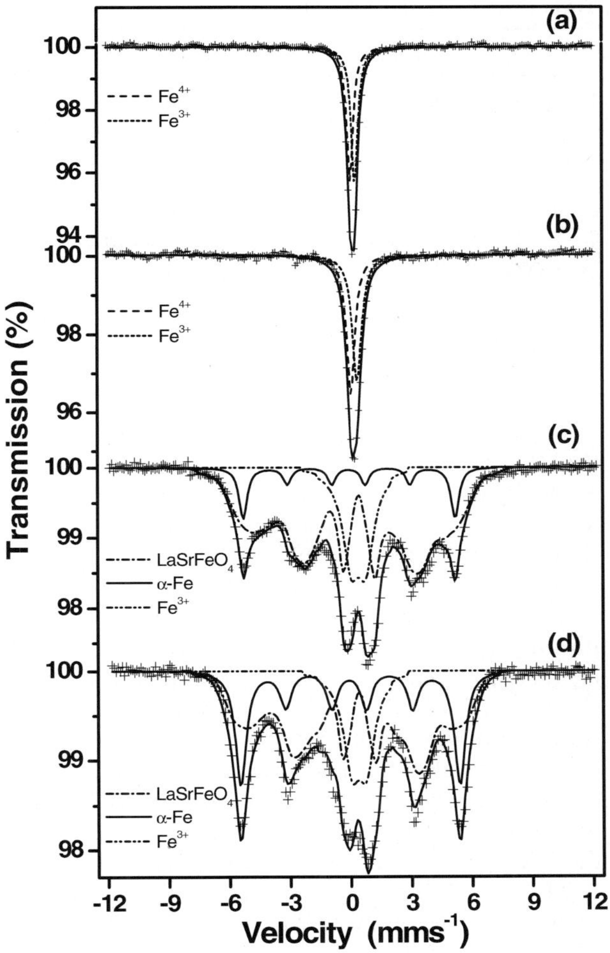

Figure 2 298 K Mössbauer spectra recorded from $La_{0.5}Sr_{0.5}Fe_{0.5}Co_{0.5}O_3$ **a** before tpr treatment and after TPR treatment at **b** 600°C, **c** 1,000°C, and **d** 1,200°C.

X-ray absorption near edge structure (XANES) and extended X-ray absorption fine structure (EXAFS) measurements were performed at Station 8.1 (Fe K-, Co K- and La L_{III}-edge) and at Station 9.2 (Sr K-edge) at the Synchrotron Radiation Source at Daresbury Laboratory operating at an energy of 2.0 GeV and an average current of 200 mA. Data were collected at 298 K and were fitted using the Daresbury Laboratory program EXCURV90.

3 Results and discussion

Figure 1 depicts the TPR profile recorded from $La_{0.5}Sr_{0.5}Fe_{0.5}Co_{0.5}O_3$. It shows two main broad peaks at *ca.* 400 and 800°C, and a smaller peak at around 1,150°C. Samples were taken for examination by ^{57}Fe Mössbauer spectroscopy at the temperatures indicated by arrows in the figure.

Table I Summary of the EXAFS results

	Original	TPR 600°C	TPR 1,000°C	TPR 1,200°C
La L_{III}-edge	Perovskite structure	Oxygen deficient-perovskite structure[a]	$SrLaFeO_4$-type	$SrLaFeO_4$-type + La_2O_3
Sr K-edge	Perovskite structure	Oxygen deficient-perovskite structure[a]	$SrLaFeO_4$-type	$SrLaFeO_4$-type + SrO
Co K-edge	Perovskite structure	Oxygen deficient-perovskite structure[a]	Co metal	Co metal
Fe K-edge	Perovskite structure	Oxygen deficient-perovskite structure[a]	$SrLaFeO_4$ + Fe metal	$SrLaFeO_4$ + Fe metal

[a] Confirmed by the XRD data.

Figure 2a depicts the Mössbauer spectrum recorded at 298 K from $La_{0.5}Sr_{0.5}Fe_{0.5}Co_{0.5}O_3$. It consists of a single absorption which was best-deconvoluted into two singlets with similar relative spectral areas and isomer shifts δ of 0.08 mms^{-1} and δ = 0.28 mm s^{-1}. The lower isomer shift is characteristic of Fe^{4+} while the higher isomer shift is characteristic of Fe^{3+}. The Co K-edge XANES and EXAFS were also compatible with the presence of Co^{3+} and Co^{4+}.

The spectrum recorded from $La_{0.5}Sr_{0.5}Fe_{0.5}Co_{0.5}O_3$ following treatment in the flowing gaseous atmosphere of hydrogen and nitrogen at 600°C (Figure 2b) continued to show a single absorption which was also best fitted to two singlets characteristic of Fe^{4+} and Fe^{3+} but with broader line width than those used to fit the spectrum of the original sample. The Mössbauer spectroscopy results indicate that, after treatment at this temperature, there is no reduction of either the Fe^{4+} or Fe^{3+} to lower oxidation states. The EXAFS and XRD data (not shown) suggest that after the TPR treatment at this temperature the perovskite structure is still retained but the XRD data suggest that the resulting material is deficient in oxygen.

The ^{57}Fe Mössbauer spectrum recorded from $La_{0.5}Sr_{0.5}Fe_{0.5}Co_{0.5}O_3$ following treatment in the flowing gaseous hydrogen and nitrogen atmosphere at 1,000°C (Figure 2c) consisted of a broad magnetic component and a central doublet. The magnetic component was fitted to a single sextet, accounting for the 26% of the total spectral area and with parameters characteristic of α-iron, plus a hyperfine magnetic field distribution accounting for *ca.* 43% of the total spectral area and parameters δ = 0.26 mms^{-1}, 2ε = 0.45 mms^{-1}, average magnetic field H = 27.0 T. The central doublet, accounting for 16% of the total spectral area, had Mössbauer parameters (δ = 0.45 mms^{-1}, Δ = 0.74 mms^{-1}) characteristic of Fe^{3+} in octahedral oxygen coordination. The Mössbauer parameters of the magnetic distribution allow their assignment to a K_2NiF_4-type phase of composition $SrLaFeO_4$ which is probably non-stoichiometric. Stoichiometric $SrLaFeO_4$ gives, at 298 K, a Mössbauer spectrum characterized by a well-defined sextet of parameters δ = 0.31 mm s^{-1}, 2ε = 0.58 mms^{-1}, H = 32.4 mms^{-1} [5,6]. However, it is known that a slight change in La:Sr stoichiometry alters the magnetic and electronic properties of bulk $SrLaFeO_4$ and that in $Sr_{1+x}La_{1-x}FeO_4$ the Néel temperature decreases with increasing values of x and, for some values of x, the oxide can be paramagnetic at room temperature [7]. Therefore, we associate the hyperfine magnetic field distribution to a non-stoichiometric $SrLaFeO_4$ phase with a Néel temperature close to 298 K. The assignment of the paramagnetic doublet is more uncertain, since both the XRD and the Fe K-edge EXAFS are complex and difficult to interpret due to the coexistence of several phases. However, it is plausible to associate the doublet with a phase of the type $Sr_{1+x}La_{1-x}FeO_4$ with a value of x corresponding to the material being paramagnetic at room temperature. The Co K-edge EXAFS data show the total reduction of Co^{4+} and Co^{3+} to metallic cobalt while the Sr K-edge and La L_{III}-edge EXAFS results are compatible with the presence of a phase of the type $SrLaFeO_4$ as shown by Mössbauer spectroscopy. The different phases identified by Fe K-edge, Co K-edge, Sr K-edge and La L_{III}-edge EXAFS are collated in Table I.

The ^{57}Fe Mössbauer spectrum recorded at 298 K from the material following treatment at 1,200°C (Figure 2d), is similar to the spectrum recorded from the material treated at 1,000°C, the main difference being the larger concentration of α-iron found in the material treated at 1,200°C. The Co K-edge EXAFS data show only the presence of metallic cobalt, the Sr K-edge EXAFS indicates the coexistence of SrO and $SrLaFeO_4$ phases and the L_{III}-edge EXAFS results are compatible with the presence of $SrLaFeO_4$ and SrO.

Summarizing, the results suggest that treatment in hydrogen and nitrogen at 600°C does not result in the destruction of the perovskite structure, that treatment at 1,000°C results in the total reduction of Co^{3+} and Co^{4+} to metallic cobalt, the partial reduction of Fe^{4+} and Fe^{3+} to form non-stoichiometric $SrLaFeO_4$ and some segregated metallic iron, and that treatment at 1,200°C results in the partial decomposition of $SrLaFeO_4$ to metallic iron, SrO and La_2O_3.

References

1. Nishihata, Y., Mizuki, J., Akao, T., Tanaka, H., Uenishi, M., Kimura, M., Okamoto, T., Hamada, H.: Nature **418**, 164 (2002)
2. Nakamoro, T., Misona, M., Yoneda, Y.: J. Catal. **83**, 151 (1983)
3. Ohno, J., Nagata, S., Sato, H.: Solid State Ionics **3–4**, 439 (1981)
4. Briceno, G., Xiang, X.-D., Chang, H., Sum, X., Schulz, P.G.: Science **270**, 273 (1995)
5. Soubeyroux, J.L., Courbin, P., Fournès, L., Fruchart, D., Le Flem, G.: J. Solid State Chem. **31**, 313 (1980)
6. Berry, F.J., Marco, J.F., Ren, X.: J. Solid State Chem. **178**, 961 (2005)
7. Mohades-Ardabili, L., Zheng, H., Ogale, S.B., Hannoyer, B., Tian, W., Wang, J., Lofland, S.E., Shinde, S.R., Zhao, T., Jia, Y., Salamanca-Riba, L., Schlom, D.G., Wuttig, M., Ramesh, R.: Nat. Mater. **3**, 533 (2004)

Hyperfine Interact (2005) 166:455–463
DOI 10.1007/s10751-006-9308-3

An investigation of trivalent substituted M-type hexagonal ferrite using X-ray and Mössbauer spectroscopy

T. M. Meaz · C. Bender Koch

Published online: 1 November 2006

Abstract Series of substituted M-type hexagonal ferrite has been synthesized, and studied by means of X-ray diffraction, IR and Mössbauer spectroscopy. The samples have the formula ($BaCr_xFe_{12-x}O_{19}$) (with $x = 0.0$, 1.0, 2.0, 3.0, 4.0, and 5.0). X-ray diffraction studies prove that all samples has a single phase M-type hexagonal structure. The lattice parameters both *a* and *c* were found to be composition dependent. This observation was attributed to the atomic radii of the substituted cations. The Mössbauer spectra change from magnetically ordered ($x = 0$) toward magnetically ordered with strong line broadening. The broadening increases as the Cr content increase. Measurements at low temperature (80 K) restore the magnetic order. The Mössbauer parameters suggest that Cr^{3+} prefers to occupy the $4f_2$ and 2a crystal sites. IR absorption bands were observed between 1,500 and 400 cm^{-1}, and confirm the structure in coincidence with X-ray results.

Key words hexagonal ferrite · Mössbauer spectroscopy · x-ray diffraction · IR · barium ferrite · substitution

1 Introduction

Barium ferrite (BaM) is well established for use as a permanent magnetic material [1]. BaM exhibits a fairly large magnetocrystalline anisotropy and high Curie temperature, and has in addition a relatively large saturation magnetization, chemical stability and corrosion resistance [2]. It is also a suitable material for high-density magnetic recording applications [1–3].

T. M. Meaz (✉)
Physics Department, Faculty of Science, Tanta University, 31527 Tanta, Egypt
e-mail: tmeaz@yahoo.com

C. B. Koch
Department of Natural Sciences, KVL, The Royal Veterinary and Agricultural University, Thorvaldsensvej 40, DK-1871 Frederiksberg C, Denmark

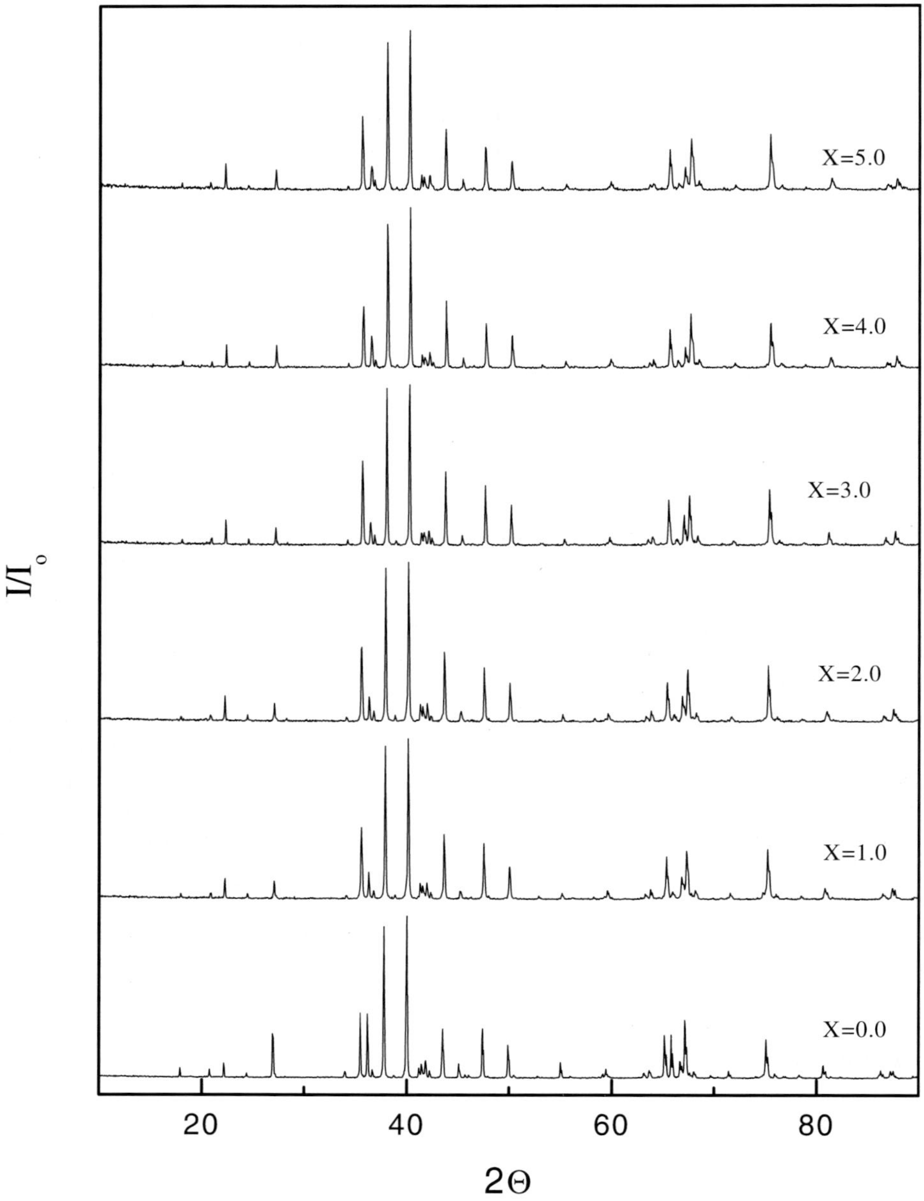

Figure 1 X-ray diffractgrams of the $BaCr_xFe_{12-x}O_{19}$ samples.

Sr^{2+} or Pb^{2+} can replace the Ba^{2+} ion in the M type hexagonal ferrites partly or completely, without changing the crystal structure. Substitutions of Fe^{3+} ion and O^{2-} are also possible. In all cases, substitutions must keep electrical neutrality and have nearly similar ionic radii with the original one. For the various magnetic applications of BaM a high saturation, a suitable coercivity, and low temperature coefficients of coercivity and remanence are desired properties. Therefore, much

Table I Lattice constants (*a* and *c*) of the $BaCr_xFe_{12-x}O_{19}$ samples

x	*a* (Å)	*c* (Å)
0	5.857 ± 0.008	22.981 ± 0.04
1.0	5.847 ± 0.008	22.964 ± 0.04
2.0	5.853 ± 0.008	22.949 ± 0.04
3.0	5.846 ± 0.008	22.911 ± 0.04
4.0	5.845 ± 0.008	22.866 ± 0.04
5.0	5.849 ± 0.008	22.848 ± 0.04

work has been done to modify the magnetic properties through the substitutions of Fe^{3+} ions. Substitution of Fe can be accomplished in two ways: Direct substitution by trivalent ions (Al^{3+}, Ga^{3+}, Mn^{3+}, Cr^{3+}) or by combined substitution of divalent and tetravalent ions (Co^{2+}+Ti^{4+}, Zn^{2+}+Ti^{4+}, Zn^{2+}+Sn^{4+}). Various combinations of substituents (Co–Ti, Co–Sn, Zn–Ti, Co–Zn–Nd, Co–Ti–Sn, Co–Mo and Zn–Zr) have been studied by Mössbauer spectroscopy [4–14].

The magnetic properties of the substituted hexaferrites are strongly dependent upon the electronic configuration of the substituting cations as well as on their site preference. Substitutions by non-magnetic ions are expected to lower the magnetic ordering temperatures and differences primarily in the charge of the substituents may cause different neighbor effects. In previous work coupled substitution of magnetic (Co^{2+}) and non-magnetic ions (Zn^{2+}–Ti^{4+}) for Fe^{3+} in BaM was investigated [15]. The case of the substitution of Fe^{3+} by Cr^{3+} in M-type hexaferrites is of particular interest [12,16–20]. The rapid decrease in the values of the Curie temperature and of the saturation magnetization at 0 K with increasing degree of substitution, cannot be explained on the basis of a long range collinear spin order, and considering that the chromium ions has a magnetic moment of 3 μ_B [16]. In a Mössbauer study of $BaCr_xFe_{12-x}O_{19}$ Wartewing et al. [12] have shown that with an external magnetic field of 6T, applied to samples with x = 0.8 and 1.8 at 4.2 K the Fe^{3+} magnetic moment are aligned along the *c* axis.

A satisfactory explanation for the magnetic structure of the Cr containing ferrites is still a matter of debate.

In this work trivalent substitution of (Cr^{3+}) for Fe^{3+} in BaM was investigated using IR, X-ray diffraction and ^{57}Fe Mössbauer spectroscopy.

2 Experimental

The starting materials used for samples preparation were barium carbonate, chromium oxide and ferric oxide that were mixed together in appropriate ratios to prepare a series of a polycrystalline M-type hexagonal ferrites with the composition $BaCr_xFe_{12-x}O_{19}$ with (x = 0.0, 1.0, 2.0, 3.0, 4.0 and 5.0) using the conventional ceramic technique. The mixed powders were mixed for 5 h using an electrical grinding machine. The final powders were pressed and pre-sintered for 6 h at 950°C. After slowly cooling to room temperature, the reaction mixtures were ground for 5 h and then pressed into a disk shape. The disks were sintered at 1150°C for 6 h, and then the samples were slowly cooled to room temperatures. The last step was repeated, to obtain the final homogeneous samples.

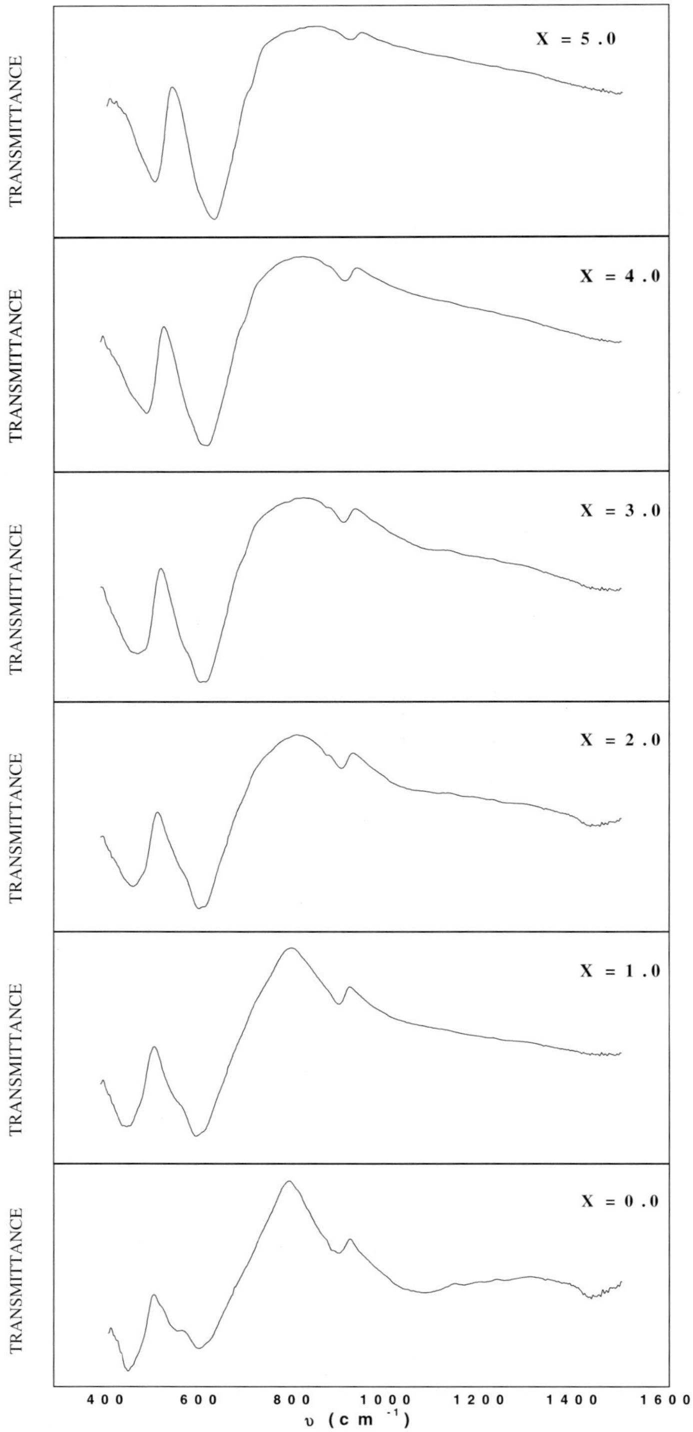

Figure 2 The IR transmission spectra of the $BaCr_xFe_{12-x}O_{19}$ samples.

Table II Compositional-dependence of the position and intensity of the IR transmittance spectra

x	High wave number Bands ν_1 and ν_2 in (cm^{-1})			
	ν_1	T (%)	ν_2	T (%)
0.0	592	87.8	443	86.1
1.0	599	67.1	452	68.6
2.0	605	80.3	469	82.8
3.0	612	77.5	477	81.4
4.0	626	56.1	497	63.7
5	628	76.4	501	76.4

3 Results and discussions

The X-ray diffraction patterns of the $BaCr_xFe_{12-x}O_{19}$ samples are obtained at room temperature (Figure 1). The patterns demonstrate that all the samples of this series have the magnetoplumbite structure, and only a small trace of Fe_2O_3 (at d = 2.51 Å) was detected. On the other hand, sharp lines are observed for all samples, indicating that the samples consist of large crystallites. Although the lattice constant a does not change dramatically with the mole ratio x, a slight decrease in the lattice parameters c are found (Table I). This may be attributed to the small ionic radius of Cr^{3+} compared to Fe^{3+} [15, 20]. This is in accordance with the fact that all hexagonal classes (M-, Y- and W-type) are characterized by constant lattice parameter a (of order 5.88 Å) but the parameter c is varied [1]. The change in the relative intensities may be ascribed to the occupation of the lattice sites by different ions.

Figure 2 shows the IR spectra of the samples. Two main absorption bands have been observed in the displayed wave numbers range (400–1,500 cm^{-1}). The first high-wave number band ν_1 is in the range 595 to 628 cm^{-1}, the lower-wave number band ν_2 is in the range 443 to 501 cm^{-1}. The band positions are given in Table II. It is known that in ferrites four bands are to be expected where the high wave number band ν_1 is assigned to the tetrahedral complexes, the lower wave number band ν_2 is assigned to the octahedral complexes [21]. Therefore in the investigated samples the high wave number band ν_1 represents the vibration of Fe^{3+}–O^{2-} in the sublattice site A while the lower-wave number band ν_2 represents the trivalent metal–oxygen vibrations at the octahedral B-sites. The two weaker bands ν_3 and ν_4 (two low-frequency band) does not appear because they are outside the measurement range (< 400 cm^{-1}). The difference in the ν_1 and ν_2 band positions is expected because of the difference in the Fe^{3+}–O^{2-} distances for the octahedral and tetrahedral complexes [22]. The slight variation in ν_1 and ν_2 bands between different samples my attributed to degree of substitution and the different in radius of Cr^{3-} and Fe^{3+} ions.

The room temperature Mössbauer spectra of the Ba $Fe_{12}O_{19}$ end member and those of Cr^{3+} substituted consists of five magnetically split and overlapping components (Figure 3). All the spectra were fitted with five sextet subpatterns assigned to $4f_1$, $4f_2$, 12k, 2a and 2b sites of the hexagonal crystal structure. The hyperfine parameters of each site deduced from the computer fitting as a function of x are listed in Table III. The hyperfine parameters (of the $BaFe_{12}O_{19}$ end member) are in good agreement with previously published values [6, 8, 16–19]. The

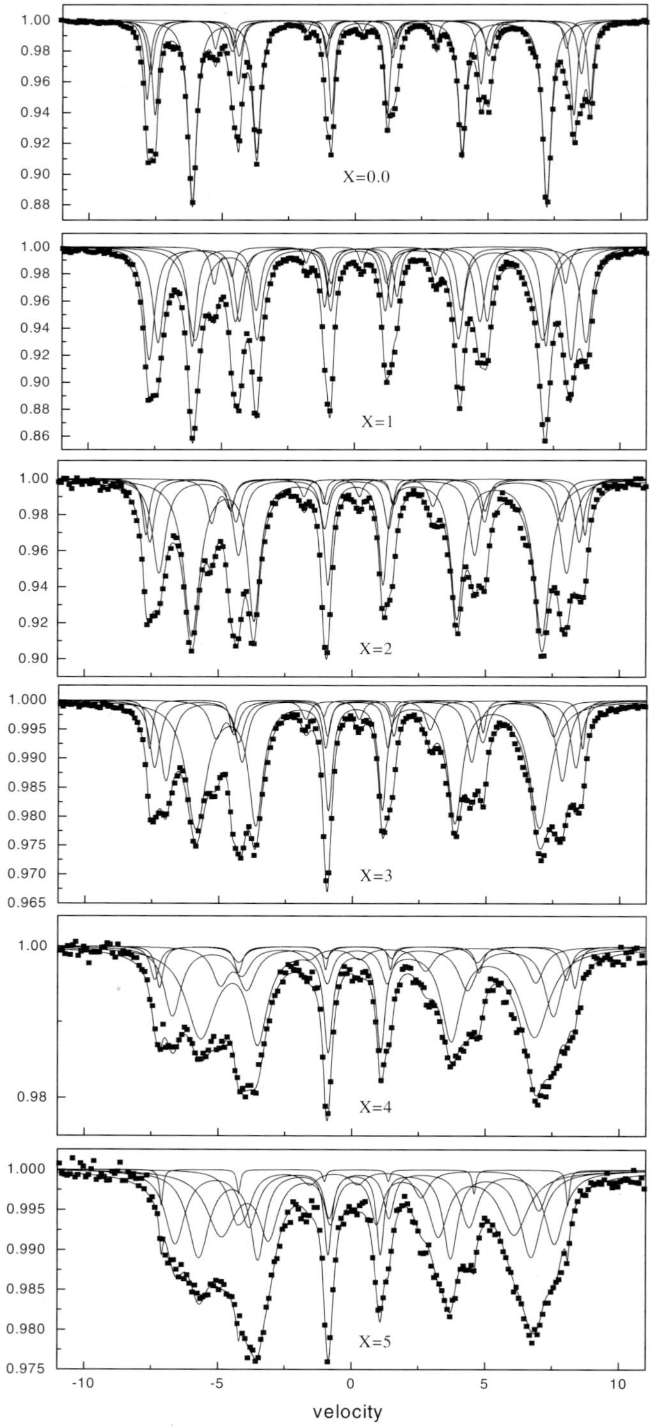

Figure 3 Mössbauer spectra of the $BaCr_xFe_{12-x}O_{19}$ samples obtained at room temperature.

Table III Mössbauer hyperfine parameters of the spectra of the studied system obtained at room temperature

x value	Site	Isomer shift δ (mms^{-1}) ±0.03	Quadrupole shift ε (mms^{-1}) ±0.03	Hyperfine field B_{hf} (Tesla) ±0.2	Width of line 1 Γ (mms^{-1}) ±0.02	Relative area (%) ±2
0.0	12k	0.35	0.19	41.4	0.38	48.5
	$4f_1$	0.26	0.09	48.9	0.29	19.3
	$4f_2$	0.36	0.14	51.7	0.26	14.6
	2a	0.36	0.04	50.2	0.32	12.1
	2b	0.29	1.05	40.9	0.28	5.5
1.0	12k	0.35	0.20	40.9	0.55	48.3
	$4f_1$	0.27	0.11	48.1	0.48	21.8
	$4f_2$	0.36	0.13	51.4	0.31	11.7
	2a	0.37	0.07	50.0	0.43	12.0
	2b	0.29	1.05	41.08	0.32	6.2
2.0	12k	0.34	0.21	40.6	0.72	49
	4f1	0.28	0.11	47.2	0.59	24.4
	4f2	0.36	0.14	50.9	0.30	7.5
	2a	0.39	0.08	49.7	0.42	11.6
	2b	0.26	1.03	40.6	0.40	7.5
3.0	12k	0.34	0.23	39.8	0.89	50.5
	$4f_1$	0.29	0.13	45.9	5.8	21.5
	$4f_2$	0.34	0.15	50.2	0.26	6.3
	2a	0.39	0.08	48.8	0.45	12.5
	2b	0.22	1.00	39.2	0.56	9.2
4.0	12k	0.35	0.24	38.7	1.3	52.0
	4f1	0.31	0.11	44.2	0.77	23.8
	4f2	0.42	0.15	48.16	0.33	6.2
	2a	0.28	0.05	47.7	0.41	6.2
	2b	0.16	0.83	36.6	0.73	11.8
5.0	12k	0.39	0.29	37.8	1.65	55.1
	$4f_1$	0.22	−0.19	37.4	0.93	12.2
	$4f_2$	0.38	0.10	45.8	0.48	10.0
	2a	1.46	−0.99	42.9	0.47	5.3
	2b	0.41	0.83	34.7	0.91	17.4

substitution of the iron ions by other cations leads to line broadening due to perturbation of the magnetic interaction. In particular the spectrum of the sample with $x = 5.0$ exhibits a strong asymmetry in the component and the fit results cannot be trusted. The changes in the relative intensities of the sextet are due to the replacement of iron by chromium.

These results demonstrate strong line broadening and some reduction in the magnetic hyperfine fields. There is no evidence for splitting of the sextet of the predominant k sublattice, as observed in Co substituted M-type hexagonal ferrites [13, 15].

Increasing the Cr^{3+} substitution to $x = 5.0$ causes further line broadening and reduction of magnetic hyperfine field. At these substitution levels fitting becomes meaningless. The results indicate that at high substitution level all sites are affected. Magnetic order in these samples is restored by lowering the temperature to 80 K (spectra not shown). This observation indicates that in the substituted compounds

Fe^{3+} magnetic moment remain axially aligned. However, measurements with applied external magnetic field are needed to confirm this.

For all sites, the observed decrease in the hyperfine field with increase x could be attributed to the less magnetic moment (3 μ_B) for Cr^{3+} replacement (5 μ_B) for Fe^{3+} ions, which cause reduction of the superexchange interaction between the ions due to substitution of ions with smaller magnetic moment.

Assuming that, the relative spectral areas of the five sublattice patterns are corresponding to the number of iron ions on the corresponding sites it is argued that the Cr^{3+} ion substitution does not occur on the 12k or 2b sites. The results suggest that Cr^{3+} occupy $4f_2$ and 2a sites.

It is appears that the substitution for small $x \leq 2.0$ values $4f_2$ sites maybe preferred, while for high value $4 \geq x > 2$ it is plausible that Cr^{3+} occupy the 2a sites. For $x \geq 5$ it is expected other crystal sites rater than $4f_2$ and 2a to host Cr^{3+} ions.

4 Conclusion

The X-ray diffraction analysis confirm that all the studied samples of the series $BaCr_xFe_{12-x}O_{19}$ (with x = 0.0, 1.0, 2.0, 3.0, 4.0, and 5.0) has a single-phase M-type hexagonal structure.

The lattice parameter "a" of the hexagonal cell is almost unchanged, while the c parameter decreases slightly as the substitution increase.

In IR studies, two main absorption bands have been observed and depend on substitution (Figure 2). The high-wave number band ν_1 is in the range 595 to 628 cm^{-1}, while the lower-wave number band ν_2 is in the range 443 to 501 cm^{-1}. The two observed bands confirm the structure of hexagonal ferrites.

The observed decrease of the magnetic hyperfine field of the sites are attributed to the substituted with Cr^{3+} causing a decrease in the interaction between the ions.

We have identified that Cr^{3+} ions enter into the $4f_2$ (antiprism) site of the Ba M-type hexagonal structure and also into the regular 2a octahedral sites.

References

1. Kojima, H.: Fundamental properties of hexagonal ferrites. In: Wohlfahrt, E.P. (ed.) Ferromagnetic Materials, vol. 3, p. 305. North-Holland, Amsterdam (1982)
2. Liu, X., Wang, J., Gan, L.-M., Ng, S.-C., Ding, J.: J. Magn. Magn. Mater. **184**, 344 (1998)
3. Wartewig, P., Krause, M.K., Esquinazi, P., Rösler, S., Sonntag, R.: J. Magn. Magn. Mater. **192**, 93 (1999)
4. Zheng, Y., Yu, Z., Shao, Y.: Hyperfine Interact. **94**, 2035 (1994)
5. Melzer, K., Wartewig, P., Stumm, U., Gütlich, P., Michalk, C.: Hyperfine Interact. **94**, 2045 (1994)
6. Morrish, A.H., Zhou, X.Z., Yang, Z., Zeng, H.-X.: Hyperfine Interact. **90**, 365 (1994)
7. Li, Z.W., Ong, C.K., Yang, Z., Wei, F.L., Zhou, X.Z., Zhao, J.H., Morrish, A.H.: Phys. Rev. B. **V62**(N10), 6530 (2000)
8. Zhou, X.Z., Morrish, A.H., Yang, Z., Zeng, H.-X.: J. Appl. Phys. **75**, 5556 (1994)
9. Batlle, X., Obradors, X., Rodriguez-Carvajal, J., Pernet, M., Cabaǒas, M.V., Vallet, M.: J. Appl. Phys. **70**, 1614 (1991)
10. Hong, Y.K., Paig, Y.J., Agresti, D.G., Shelfer, T.D.: J. Appl. Phys. **61**, 3872 (1987)
11. Rane, M.V., Bahadur, D., Mandal, S.K., Patni, M.J.: J. Magn. Magn. Mater. **153**, L1 (1996)
12. Wartewig, P., Melzer, K., Krause, M.K., Tellgren, R.: J. Magn. Magn. Mater. **140–144**, 2101 (1995)

13. Šimša, Z., Lego, S., Gerber, R., Pollert, E.: J. Magn. Magn. Mater. **140–144**, 2103 (1995)
14. Rane, M.V., Bahadur, D., Nigam, A.K., Srivastava, C.M.: J. Magn. Magn. Mater. **192**, 288 (1999)
15. Meaz, T.M., Koch, C.B.: Hyperfine Interact. **156/157**, 341 (2004)
16. Albanse, G., Watts, B.E., Leccabue, F., Díaz Castañón, S.: J. Magn. Magn. Mater. **184**, 337 (1998)
17. Lanje, N.Y., Kulkarni, D.K., Rewatkar, K.G.: Mater. Lett. **47**, 125 (2001)
18. Pereda, J.A., Isalgué, A., Tejada, J., Litterst F.J., Obradors, X.: Hyperfine Interact. **28**, 569 (1986)
19. Obradors, X., Labarta, A., Tejada, J., Pernet, M., Tholence, J.L., Saint-Paul, M., Barbara, B.: J. Appl. Phys. **63**(8), 4091 (1988)
20. Jei, Y., Oh, I., Shim, B., Jung, H.J., Park, J.Y., lel Park, S., Um, Y.R., Lee, Y.J., Lee, S.W., Kim, C.S.: J. Appl. Phys. **76**(10), 6877 (1994)
21. Mazen, S.A., Elfalaky, A., Mohamed, A.Z., Hashem, H.A.: Mater. Chem. Phys. **44**, 293 (1996)
22. Josyulu, O.S., Sobhanadri, J.: Phys. Status Solidi, A **65**, 479 (1981)

Hyperfine Interact (2005) 166:465–468
DOI 10.1007/s10751-006-9309-2

Metallosupramolecular coordination polyelectrolytes investigated by Mössbauer spectroscopy

A. Lindner · M. Menzel · F. Renz · D. G. Kurth · A. F. Thünemann

Published online: 26 October 2006

Abstract Metallosupramolecular coordination polyelectrolytes (MEPEs) based on rigid and flexible ditopic bis-terpyridine ligands and Fe(II) ions are investigated by Mössbauer spectroscopy. We demonstrate the influence of mechanical stress induced by grinding on the structure of MEPE as well as the ability of MEPE to self-repair through recrystallisation.

Key words metallosupramolecular coordination polyelectrolyte · bis-terpyridine · Mössbauer spectroscopy · supramolecular chemistry

1 Introduction

Inspired by the multitude of natural systems, supramolecular chemistry investigates the assembly of complex chemical structures and systems through weak interactions and is an active area of research in chemistry and material science alike [1, 2]. Assemblies containing transition metal ions open opportunities to fabricate functional materials with novel physicochemical properties for potential applications in electronics, photonics and catalysis [3]. Metal-ion induced self-assembly of ditopic ligands, based on bis-terpyridines, and transition metal ions (i.e. Fe^{2+}, Ir^{3+},

A. Lindner · D. G. Kurth
Max-Planck-Institute of Colloids and Interfaces, Am Mühlenberg 1,
D-14424 Potsdam, Germany

M. Menzel (✉) · A. F. Thünemann
Federal Institute for Materials Research and Testing (BAM), Richard-Willstätter-Str. 11,
D-12489 Berlin, Germany
e-mail: Michael.Menzel@BAM.DE

D. G. Kurth
National Institute for Materials Science, 1-1 Namiki, Tsukuba 305-0044, Japan

F. Renz
Johannes Gutenberg-Universität Mainz, Duesbergweg 10-14, D-55099 Mainz, Germany

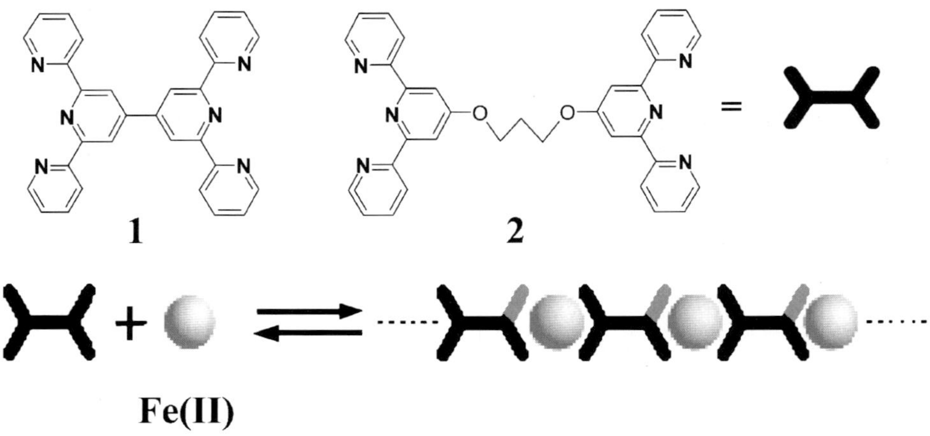

Scheme 1 Metal-ion induced self-assembly of iron acetate and ditopic bis-terpyridines, **1** and **2**, results in metallosupramolecular coordination polyelectrolytes (MEPEs). A chain-typed, positively charged macromolecular assembly is formed as a consequence of the pseudo-octahedral coordination geometry (indicated by the *wedges*).

Ru^{2+}) results in formation of metallosupramolecular coordination polyelectrolytes (MEPEs) [4] (Scheme 1).

Using Fe as central metal ion we can use Mössbauer spectroscopy to study the coordination environment in the corresponding MEPE [5]. The spectra of untreated samples show two quadrupole doublets related to Fe(II) in a pseudo-octahedral environment and in a five-fold coordination geometry, which are considered as end sites in the polymer chain.

From the Mössbauer spectra we conclude, that Fe(III)-species are not present.

Here, we investigate the influence of mechanical stress on the compound induced by grinding. In the following, we present results of MEPEs based on a rigid (**1**) and a flexible ligand (**2**) (see Scheme 1).

2 Experimental

The assembly of MEPEs is shown in Scheme 1.

Reaction of ditopic bis-terpyridines 6′,6″-bis(pyrid-2-yl)-2,2′:4′,4″:2″,2‴-quaterpyridine (**1**), and 1,5-bis(2,2′:6′,2″-terpyridin-4′-yl) 1,5-bioxopentylene (**2**) with $FeOAc_2$ in a 1:1 ratio in glacial acetic acid gave the MEPE 1 and MEPE 2. In order to investigate the influence of mechanical stress induced by grinding on the structure of MEPE the samples were crushed in a mortar at room temperature to create smaller units. To allow self-repair the crushed samples were dissolved in glacial acetic acid and stirred for two days followed by slow evaporation of the solvent and drying in air to recover the solid. The Mössbauer spectra were recorded at room temperature with a constant acceleration Mössbauer spectrometer using a Mössbauer source of ^{57}Co in chromium matrix with an activity of about 0.3 GBq. A proportional counter was used with a voltage of 1,800 V to detect the spectrum. The samples were placed into sealable cylindrical polyethylene sample compartment

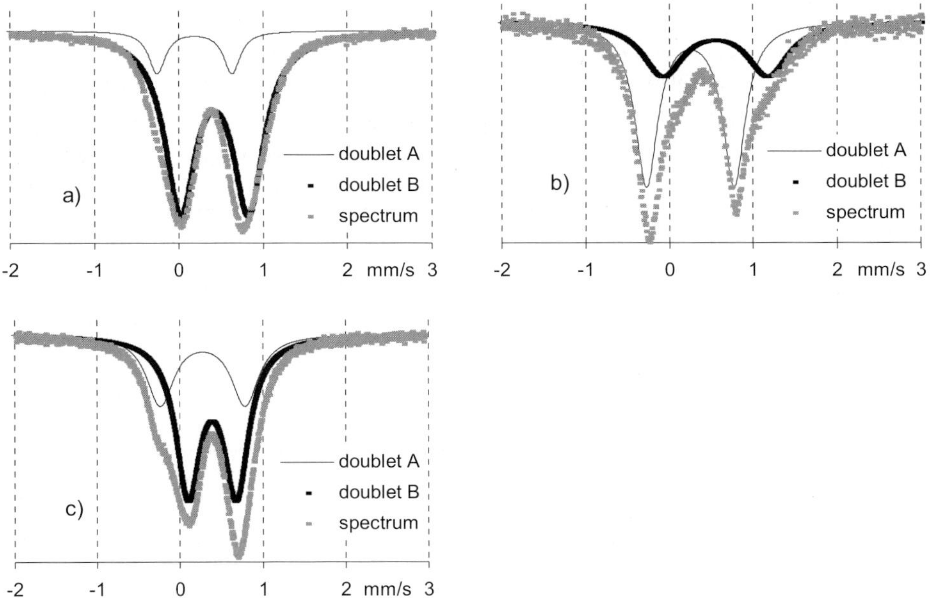

Figure 1 Fe-57 Mössbauer spectrum of MEPE 1. **a** untreated; **b** after grinding; **c** after recrystallisation.

with an inner diameter of *d*=19 mm. The Mössbauer spectra were recorded with a maximum velocity of 3.1 mm/s and referred to a standard α-Fe foil (Goodfellow).

3 Results and discussion

As representative examples, Figure 1 shows the Mössbauer spectra of untreated, ground and recovered MEPE 1.

The spectra are fitted with two quadrupole doublets, named A and B. The values of the Mössbauer parameters of all measurements are summarized in Table I.

The Mössbauer parameters for doublet A are identified as an iron atom in the end site geometry of the MEPE chain, as expected for a metal ion situated in a five-fold coordination geometry given by a terpyridine and two coordinated anions. The Mössbauer parameters for doublet B are significant for a low-spin Fe(II) ion situated in a pseudo-octahedral coordination environment typical for bis-terpyridine Fe(II) complexes [6], as expected for an in-chain position.

After grinding doublet A is enhanced at the expense of doublet B. Therefore, the grinding process raises the number of end sites significantly at the expense of the in-chain positions (see Figure 1b and Table I). This implies a coordination bond breaking on going from coordination number 6, as in the in-chain position, to 5, as in end site position.

In the case of MEPE 1 and MEPE 2, grinding results in an increase of the isomer shift of the doublet A and B. The small changes in all Mössbauer parameters show that grinding causes neither spin crossover nor chemical changes and may be due to disorder of anions and geometrical tensions.

Table I Mössbauer parameters of MEPE 1 and 2 at 295 K (*IS*: isomer shift; *QS*: quadrupole splitting)

Sample		Untreated			Ground			Recrystallised		
		IS (mm/s)	QS (mm/s)	Rel. area (%)	IS (mm/s)	QS (mm/s)	Rel. area (%)	IS (mm/s)	QS (mm/s)	Rel. area (%)
MEPE 1	doublet A	0.18	0.91	11	0.24	1.04	63	0.27	1.02	33
	doublet B	0.40	0.81	89	0.53	1.25	37	0.36	0.58	67
MEPE 2	doublet A	0.16	1.03	25	0.29	1.20	72	0.19	0.99	48
	doublet B	0.44	0.93	75	0.59	1.21	28	0.54	0.83	52

The grinded samples were re-dissolved in glacial acetic acid and dried. The Mössbauer parameters are shown in Table I. The relative areas of doublet A to B changed from 11:89 (untreated), 63:37 (ground) to 33:67 (recrystallised) for MEPE 1. For MEPE 2 the ratios changed in the same line from 25:75, 72:28 to 48:52. Upon grinding the chains show more end sites as if the chain is broken into smaller pieces. Recrystallisation results in the enhancement of the number of in-chain positions at the expense of the number of end sites, which can be interpreted as the partial establishment of the original structure. All Fe ions are still in a divalent state also after grinding and recrystallisation, hence, chemical reaction such as oxidations can be excluded.

References

1. Lehn, J.-M.: Supramolecular Chemistry. VCH, Weinheim (1995)
2. Floriani, C., Merbach, A.E.: In: Williams, A.F. (ed.) Perspectives in Coordination Chemistry. VCH, Weinheim (1992)
3. Balzani, V., Scandola, F.: Supramolecular Photochemistry. Ellis Horwood, New York (1991)
4. Schütte, M., Kurth, D.G., Linford, M.R., Cölfen, H., Möhwald, H.: Angew. Chem. Int. Ed. **37**, 2891 (1998)
5. Lindner, A.: Diploma thesis, University of Applied Sciences Berlin (2004)
6. Gütlich, P., van Koningsbruggen, P., Renz, F.: Struct. Bond. **107**, 27–75 (2004) and references therein

Hyperfine Interact (2005) 166:469–473
DOI 10.1007/s10751-006-9310-9

Spin-crossover phenomena of the assembled iron complexes by enclathrating an organic molecule

S. Nakashima · T. Morita · K. Yamada · K. Inoue

Published online: 16 November 2006

Abstract Assembled complex, $Fe(bpa)_2(NCX)_2$ (bpa=1,2-bis(4-pyridyl)ethane; X=S, Se, BH_3) enclathrated diphenylmethane. The skeletons were interpenetrating structures for NCS and NCSe complexes, while 2D grid structure was obtained for $NCBH_3$ complex. The assembled iron complexes showed spin-crossover phenomena by enclathrating diphenylmethane. The ratio of spin transition and the transition temperature depended on the assembled structure.

Key words assembled iron complexes · 1,2-bis(4-pyridyl)ethane · clathrate · Mössbauer spectroscopy · spin-crossover phenomenon

1 Introduction

Self-assembled coordination polymers containing transition metal ions and organic bridging ligands have attracted intensive interests because of their potential abilities for selective inclusion and transformation of ions and molecules [1]. It is important to synthesize the coordination polymer using 1,2-bis(4-pyridyl)ethane (bpa) as bridging ligand, because the ligand has *anti-gauche* isomer, therefore, there is a possibility to have a variety of supramolecular isomers depending on the isomer of bpa (Scheme 1). Actually we obtained 1D, 2D grid and interpenetrating structures of Fe-bpa-NCS(Se) complex [2]. Spin-crossover phenomenon is one of the interesting topics in the assembled complexes [3, 4]. These assembled complexes have a vacancy and usually the solvent molecule is

S. Nakashima (✉)
Natural Science Center for Basic Research and Development (N-BARD), Hiroshima University, Kagamiyama, Higashi-Hiroshima 739-8526, Japan
e-mail: snaka@hiroshima-u.ac.jp

T. Morita · K. Yamada · K. Inoue
Department of Chemistry, Graduate School of Science, Hiroshima University, Kagamiyama, Higashi-Hiroshima 739-8526, Japan

Scheme 1 *Anti-gauche* isomer of bpa and its schematic assembled structures of the iron complex.

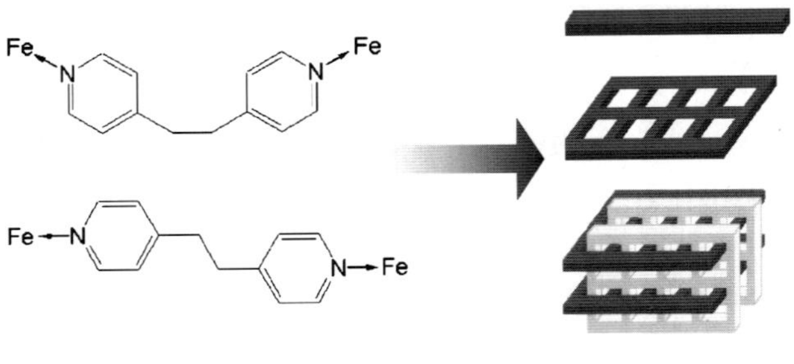

included in the vacancy. In rare cases, the ligand itself is also included [5]. It is interesting to investigate the change in electronic state and spin state of the iron atom located near the inclusion. We have already confirmed that a biphenyl molecule is included to the assembled iron complexes. ^{57}Fe Mössbauer spectroscopy revealed that $Fe_2(bpy)_4(NCS)_4(biphenyl)_5$ (bpy=4,4′-bipyridine), $Fe(bpa)_2(NCS)_2(biphenyl)_2$ and $Fe(bpa)_2(NCSe)_2(biphenyl)$ were in the high-spin Fe^{II} state, while the QS values changed depending on the structure [6].

In the present study, the inclusion of diphenylmethane to $Fe(bpa)_2(NCX)_2$ (X=S, Se, or BH_3) was confirmed using X-ray structural analysis and the spin change was investigated by ^{57}Fe Mössbauer spectroscopy and SQUID measurement.

2 Experimental

Inclusion compounds were obtained by the coexistence of diphenylmethane in the preparation process of the assembled iron complexes with 1,2-bis(4-pyridyl)ethane (bpa) as bridging ligand and NCX^- (X=S, Se, BH_3) as anion.

A ^{57}Co(Rh) source in a constant acceleration mode was used for ^{57}Fe Mössbauer spectroscopic measurements. ^{57}Fe Mössbauer spectra were obtained at 78 and 298 K by using a Wissel Mössbauer spectrometer and a proportional counter. The isomer shift was referred to a metallic iron foil. The Mössbauer parameters were obtained by a least-squares fitting to Lorentzian peaks.

Magnetic susceptibilities were measured at 5,000 G between 5 and 300 K using a SQUID magnetometer.

3 Results and discussion

X-ray structural analysis revealed that the diphenylmethane was included in the host lattice by the coexistence of the guest molecule in the preparation process. The skeletons were interpenetrating structure for NCS and NCSe complexes, while 2D grid structure was obtained for $NCBH_3$ complex. This might be explained by the volume difference between S (or Se) and BH_3. In both structures the bpa ligand had the *anti* form.

Figure 1 shows the ^{57}Fe Mössbauer spectra of $Fe(bpa)_2(NCS)_2$(diphenylmethane) (**1**). The Mössbauer parameters are summarized in Table I. The Mössbauer parameters reveal high-spin state at room temperature. At 78 K, small amount of low-spin species was superimposed on the main Fe^{II} high-spin state. SQUID measurement also revealed the spin-crossover phenomenon and the transition temperature is around 80 K as shown in Figure 2. But the transition from high- to low-spin state is partial even near Liq. He temperature.

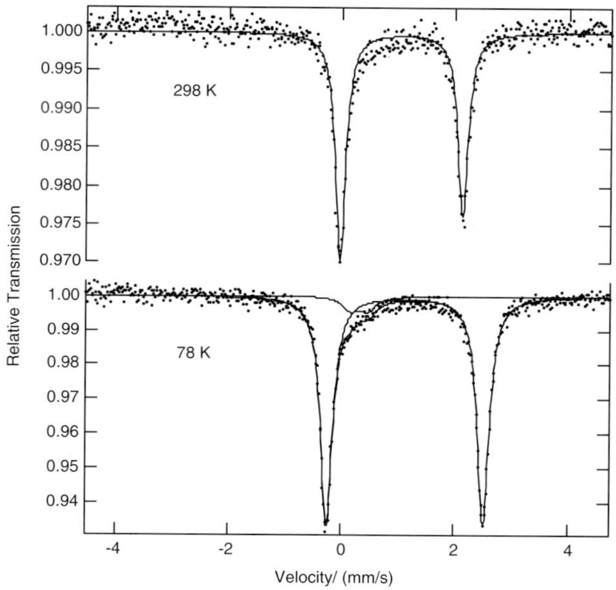

Figure 1 ^{57}Fe Mössbauer spectra of $Fe(bpa)_2(NCS)_2$ (diphenylmethane).

Table I Mössbauer parameters

Sample	*T*/*K*	IS[a] (mm/s)	QS (mm/s)	Relative area (%)
1	298	1.05	2.16	100
	78	1.13	2.77	93
		0.31	0.31	7
2	298	1.05	1.31	100
	78	1.14	1.64	91
		0.36	0.21	9
3	298	1.06	0.36	100
	78	0.33	0.26	100

[a] Relative to iron foil.

$Fe(bpa)_2(NCSe)_2$(diphenylmethane) (**2**) also showed a change similar to that of **1**, except for the change in QS value, whose Mössbauer parameters are summarized in Table I. In this case also the conversion to low-spin state is shown around 80 K and partial even near Liq. He temperature.

Figure 3 shows the ^{57}Fe Mössbauer spectra of $Fe(bpa)_2(NCBH_3)_2(diphenylmethane)_3$ (**3**) and the Mössbauer parameters are summarized in Table I. The Mössbauer parameters reveal high-spin state at room temperature and low-spin state at 78 K. This shows complete spin-crossover phenomenon between the two temperatures. Figure 2 shows the magnetic susceptibility of $Fe(bpa)_2(NCBH_3)_2(diphenylmethane)_3$. This reveals the ca. 180 K of transition temperature. Although the temperature dependence is gradual from 100 to 250 K, the change becomes sharp around the transition temperature. The variation of magnetic susceptibility versus temperature shows some hysteresis. DSC measurement also showed a small thermal anomaly at this temperature.

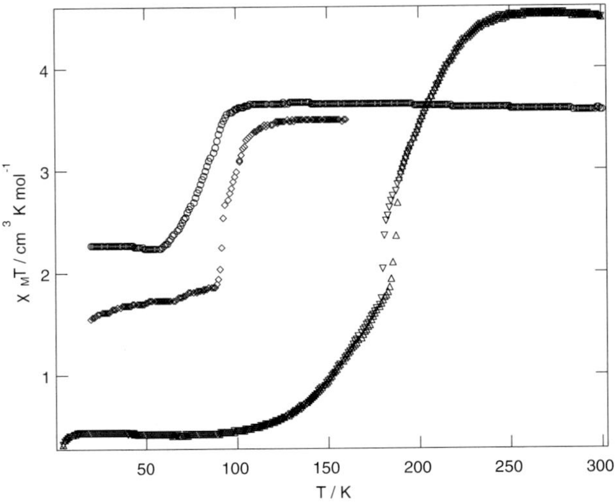

Figure 2 Magnetic susceptibility of **1**(*circle*), **2**(*diamond*) and **3**(*triangle*).

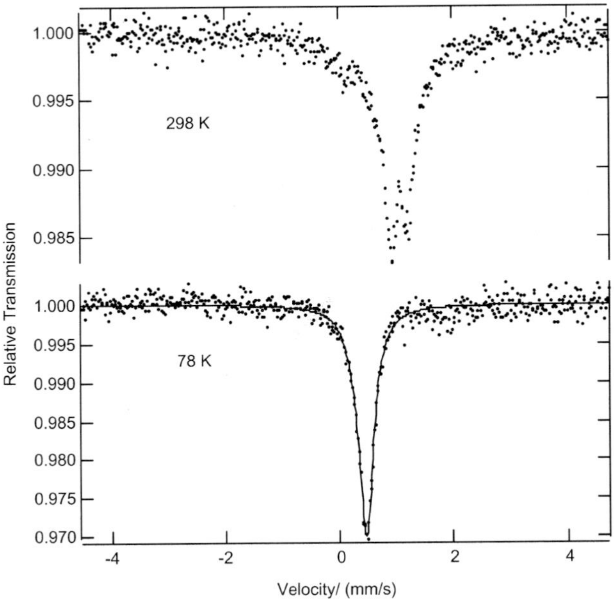

Figure 3 ^{57}Fe Mössbauer spectra of $Fe(bpa)_2(NCBH_3)_2$ (diphenylmethane)$_3$.

By changing the structure from interpenetrating to 2D type, it was achieved to observe a transition that went to completion and to raise the transition temperature. It is known that the $NCBH_3$ complex has higher spin-transition temperature than the corresponding NCS complex [7]. The present difference in transition temperature may reflect the difference in the ligand field strength among the NCX^- (X=S, Se, BH_3). And we may say that the partial transition is due to the interpenetrating structure and the complete one is due to the 2D grid structure.

Three structures (1D, 2D grid and interpenetrating structures) of $Fe(bpa)_2(NCX)_2$ without third organic molecule showed Fe^{II} high-spin state independent of temperature

[2, 8]. Therefore in summary we succeeded to control the spin state of the assembled complexes by including diphenylmethane. And we succeeded to control the transition temperature by changing anion and to control the transition ratio by changing interpenetrating to 2D grid structures. It is thought that the interaction of iron $d\pi$ orbital with the orbital of ligand will slightly change by the inclusion. This may be the reason why the inclusion complexes revealed spin-crossover phenomena.

References

1. Moulton, B., Zaworotko, M.J.: Chem. Rev. **101**, 1629 (2001)
2. Morita, T., Asada, Y., Okuda, T., Nakashima, S.: Bull. Chem. Soc. Jpn. **79**, 738 (2006)
3. Real, J.A., Andrés, E., Muñoz, M.C., Julve, M., Granier, T., Bousseksou, A., Varret, F.: Science **268**, 265 (1995)
4. Halder, G.J., Kepert, C.K., Moubaraki, B., Murray, K.S., Cashion, J.D.: Science **298**, 1762 (2002)
5. Nakashima, S., Yamamoto, A., Asada, Y., Koga, N., Okuda, T.: Inorg. Chim. Acta **358**, 257 (2005)
6. Nakashima, S., Morita, T., Okuda, T.: AIP Conf. Proc. **765**, 269 (2005)
7. Toftlund, H., Pedersen, E., Yde-Andersen, S.: Acta Chem. Scand., Ser. A **38**, 693 (1984)
8. Hernández, M.L., Barandika, M.G., Urtiaga, M.K., Cortés, R., Lezama, L., Arriortua M.I., Rojo, T.: J. Chem. Soc., Dalton Trans. 1401 (1999)

Hyperfine Interact (2005) 166:475–481
DOI 10.1007/s10751-006-9311-8

A temperature and magnetic field dependence Mössbauer study of ε-Fe_2O_3

J.-L. Rehspringer · S. Vilminot · D. Niznansky · K. Zaveta · C. Estournes · M. Kurmoo

Published online: 21 November 2006

Abstract ε-Fe_2O_3 was synthesized as nanoparticles by a pre-vacuum heat treatment of yttrium iron garnet ($Y_3Fe_5O_{12}$) in a silica matrix at 300°C followed by sintering in air at 1,000°C for up to 10 h. It displays complex magnetic properties that are characterized by two transitions, one at 480 K from a paramagnet (P) to canted antiferromagnet (CAF1) and the second at ca. 120 K from the canted antiferromagnet (CAF1) to another canted antiferromagnet (CAF2). CAF2 has a smaller resultant magnetic moment (i.e. smaller canting angle) than CAF1. Analysis of the zero-field Mössbauer spectra at different temperatures shows an associated discontinuity of the hyperfine field around 120 K. In an applied field, the different

J.-L. Rehspringer (✉) · S. Vilminot
Groupe des Matériaux Inorganiques, Institut de Physique et Chimie des Matériaux de Strasbourg, 23 rue du Loess, 67034 Strasbourg Cedex, France
e-mail: rehsp@ipcms.u-strasbg.fr

D. Niznansky
Faculty of Natural Sciences, Department of Inorganic Chemistry, Charles University, Albertov 6, 128 43 Prague 2, Czech Republic

D. Niznansky
Institute of Inorganic Chemistry, ASCR, Rez near Prague, Czech Republic

K. Zaveta
Faculty of Mathematics and Physics, Institute of Physics and Joint Laboratory of Mössbauer Spectroscopy, Charles University, V Holešovičkách 2, Prague 8, Czech Republic

C. Estournes
CIRIMAT, UMR 5085 CNRS – UPS – INP, 118 route de Narbonne, 31062 Toulouse Cedex 04, France

M. Kurmoo
Laboratoire de Chimie de Coordination Organique, CNRS-UMR7140, Tectonique Moléculaire de solide, Institut Le Bel, Université Louis Pasteur, 4 rue Blaise Pascal, 67000 Strasbourg, France

magnetic sublattices were identified and the directions of their moments were assigned. The moments of the two sublattices are antiparallel and collinear at 160 K but are at right angle to each other at 4.2 K.

Key words Mössbauer spectra · nanoparticles · complex magnetic properties

1 Introduction

ε-Fe_2O_3, a rare orthorhombic polar form of iron oxide, was recently prepared as nanoparticles in silica [1–4] and its structure from powder neutron diffraction was successfully modeled to that of the iso-structural $AlFeO_3$ [1]. It was shown to exhibit interesting magnetic properties such as a very high coercive field of ~2 T at 300 K and displays a very complex phase diagram with at least three ground states [1–3].

In our previous study of the magnetic properties and Mössbauer spectroscopy of ε-Fe_2O_3 as a function of temperature [1], we observed two sharp discontinuities; one at 480 K and the other centered at ca. 120 K. It is clear from the magnetization data that the transition at 480 K is from a paramagnet to an ordered state and that at 120 K is accompanied by a severe loss of magnetic hardness and a change in canting angle from a wide angle at high temperature to a narrow one at low temperature. However, the compositions and orientations of the magnetic moments in the respective sublattices remained unresolved. However, due to the similar behavior found for α-Fe_2O_3, we associated this discontinuity to a Morin-like transition. The present work deals principally with the study of the temperature dependence of the Mössbauer spectra in an external applied magnetic field in an attempt to clarify the origin of the change of canting angle at 120 K.

2 Preparation and characterization

The compound was reproducibly prepared as nanoparticles in a silica matrix by the sol-gel technique. The procedure for the production of ε-Fe_2O_3 is quite specific and was described elsewhere [1]. Although the amount of parasitic α-Fe_2O_3 can be minimized by the choice of the sintering temperature between 900 and 1,000°C and the period (5–10 h), no absolute optimization has been reached. Therefore, for all the samples prepared to date Mössbauer spectroscopy suggests the presence of a variable fraction of α-Fe_2O_3.

Powder X-ray diffraction was recorded at room temperature using the Bragg-Brentano geometry on a Siemens D500 employing Co Kα1 radiation (1.789 Å). Variable temperature (2–300 K) neutron powder diffraction was recorded on the D20 High flux multiple detector diffractometer at the Institut Laue Langevin at Grenoble employing an excitation wavelength of 2.41 Å.

Magnetization data were measured using a Quantum Design SQUID magnetometer operating in the temperature range 2 to 400 K and field of up to 50 kOe. All measurements we performed on samples immobilized in polymethylmethacrylate polymer, in their virgin state, to prevent motion of the particles in high applied fields [5]. Susceptibility at high temperatures was measured using a Faraday balance equipped with a variable temperature oven operating between 300 and 1,300 K and a field of up to 10 kOe.

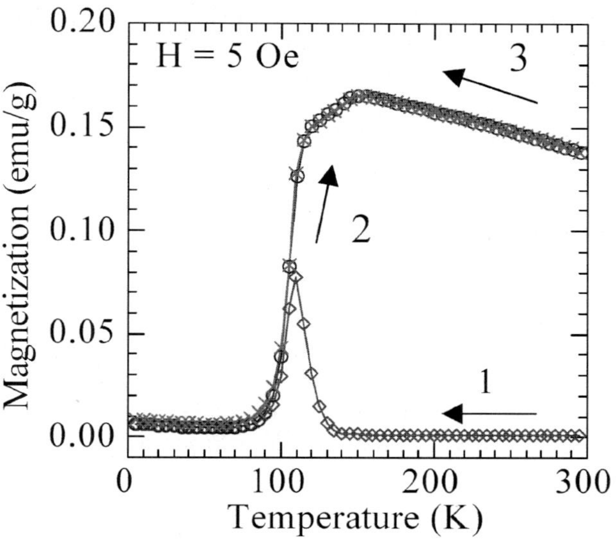

Figure 1 Temperature dependence of the magnetization in 5 Oe on first-cooling (1) and warming (2) and second-cooling (3).

The ^{57}Fe powder absorption Mössbauer spectra in zero-field were recorded at different temperatures and those in 3 and 5 T applied perpendicular to the γ-ray at 4.2, 120, and 160 K by use of a Janis cryosystem. A ^{57}Co/Rh source and a conventional constant acceleration mode were employed. The raw spectra were summed and treated using either DASUM or NORMOSS or MOSFIT [6]. The isomer shift values are relative to that of iron metal at 293 K. The sample contains a small amount of α-Fe_2O_3 (about 10% mol) that was fitted to a sextet with its characteristic parameters. For the evaluation of the angle α_1 between the direction of the applied magnetic field B_{app} and of the effective hyperfine field B_{eff}, we used the equation below:

$$B_{hf}^2 = B_{eff}^2 + B_{app}^2 - 2B_{eff} \times B_{app} \times \cos\alpha,$$

where B_{hf} is the hyperfine field in zero applied field.

3 Results and discussion

X-ray diffraction of the samples (annealed at 1,000°C for 6 h) exhibits very weak Bragg peaks on a high background. In contrast, the variable temperature neutron powder diffraction (λ = 2.41 Å) exhibits a better resolution and a proportionally higher intensity of the sharp peaks compared to the background. Starting from the unit cell parameters of $AlFeO_3$ [7], we successfully performed a pattern matching using FULLPROF [8]. The fitted unit-cell parameters within $Pna2_1$ space group are a = 5.1104(5), b = 8.7362(7), c = 9.466(1) Å [R_p = 13.6 and R_{wp} = 12.2] and these values are in good agreement with those reported in the literature [4, 9–11]. In the structure of ε-Fe_2O_3 iron occupies four different crystallographic positions; Fe1, Fe2 and Fe3 have octahedral coordination and Fe4 is tetrahedral [1]. By valence

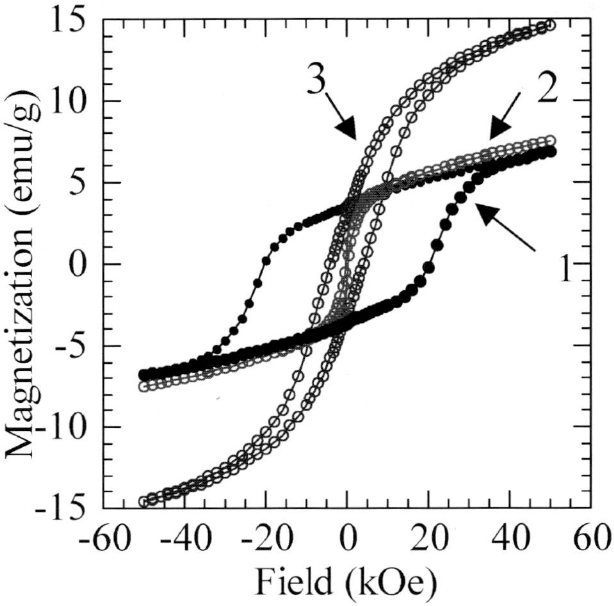

Figure 2 Hysteresis loops at 200 K (1), 100 K (2) and 2 K (3).

consideration all the iron ions are trivalent and are expected to be high spin (5/2) due to the weak oxygen ligand field.

The magnetic properties are quite unusual. Above 480 K the sample is paramagnetic and below it is in an ordered magnetic state. Below 300 K, the evolution of the magnetization as a function of temperature is quite sensitive to the field strength and to the history of the sample as well as the temperature at which the measuring field is applied. On cooling a virgin sample in a small field (<5 Oe) we first observed a sharp peak at 110 K and upon warming in the same field there is a sudden increase around 100 K to two anomalies (120 and 150 K) and then it decreases linearly up to 400 K (Figure 1). Subsequent cooling and warming of the sample in the same field follows the initial warming run.

Measurements of the isothermal magnetization were measured by performing hysteresis loops between +50 kOe and −50 kOe at different temperatures (Figure 2). We note first an unexpectedly wide hysteresis loop at 300 K, similar to that observed by Jin et al. [2, 3]. In addition, on cooling the sample the hysteresis loop disappears at 110 K. In all cases, we notice that the maximum field of 50 kOe of our magnet is not enough to achieve complete saturation of the magnetization of this sample. However, the linear dependence of the magnetization with field is indicative of an antiferromagnetic ground state with a canting [12, 13]. This is also consistent with the low magnetization values in an applied field of 50 kOe. The large change in magnetic hardness is an indication of a change of magneto-crystalline anisotropy that may result from either a change of geometry of the iron sites, or a re-orientation of the moments as found for α-Fe_2O_3. Further analysis of our neutron scattering data should reveal which of the two possibilities is correct.

The Mössbauer spectrum originating from ε-Fe_2O_3 (Figure 3) has contribution from four different Fe sites but two of them (Fe2 and Fe3) have very close hyperfine

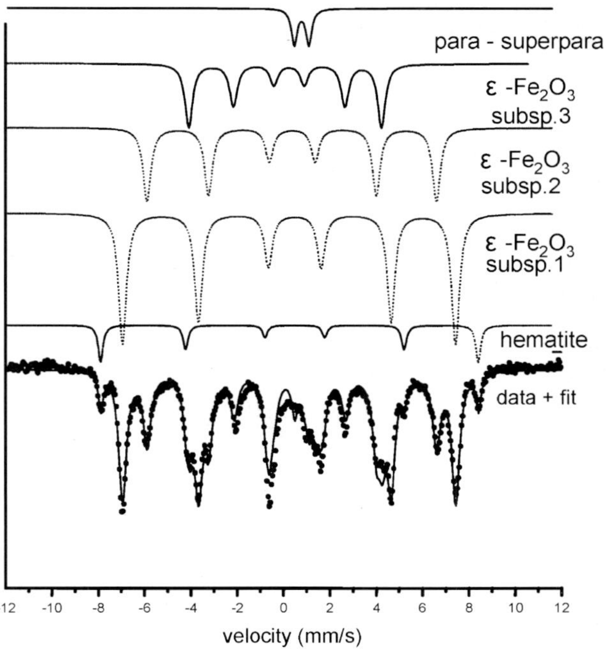

Figure 3 Experimental Mössbauer spectrum and the components as described in the text.

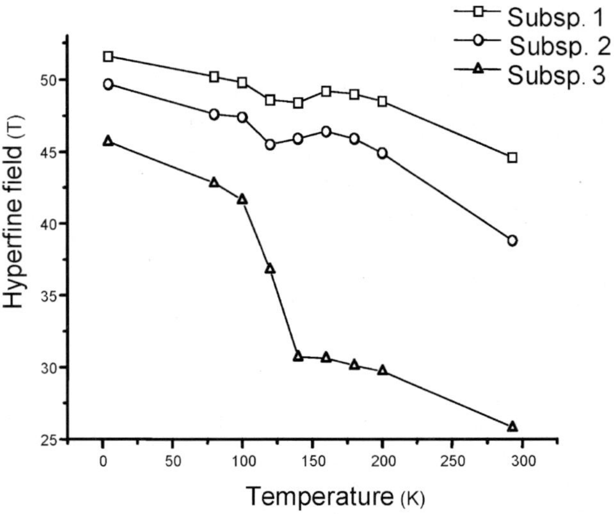

Figure 4 Temperature dependence of the hyperfine field for the three subspectra (*subsp. 1* and *2* for octahedral Fe; *subsp. 3* for tetrahedral Fe).

fields and chemical shifts and cannot be distinguished (labeled here as subsp. 1). Both Fe2 and Fe3 occupy octahedral sites with similar environment and form the edge-sharing double chains. One Fe (Fe1, labeled subsp. 2) is octahedral and is attached on top of the double chains while the fourth Fe is tetrahedral (Fe4, labeled subsp. 3). Each of the spectra was fitted by these three independent sextets (Fe1, Fe2 + Fe3 and Fe4) and one further sextet due to a small amount of residual hematite. The Mössbauer spectra during cooling are presented in [1]. A plot of the

Table I Isomer shift, quadrupole splitting, effective field and magnetic orientation of the different sublattices of ε-Fe_2O_3 at 4.2 and 160 K

	Isomer shift δ (mm/s)	Quadrupole splitting ΔE_Q (mm/s)	Effective field B_{eff} (T)	α_1 (°)
4.2 K; 0 T				
Fe2	0.48	−0.02	51.8	
Fe3	0.51	−0.09	51.7	
Fe1	0.47	−0.03	50.5	
Fe4	0.31	−0.03	45.9	
4.2 K; 5 T				
Fe2	0.49	−0.04	55.8	35
Fe3	0.45	−0.02	48.1	136
Fe1	0.40	−0.02	49.5	99
Fe4	0.43	−0.04	46.4	81
160 K; 0 T				
Fe2	0.47	−0.17	49.3	
Fe3	0.46	−0.35	49.0	
Fe1	0.44	0.01	46.3	
Fe4	0.27	−0.14	30.5	
160 K; 5 T				
Fe2	0.45	−0.09	54.0	19
Fe3	0.45	−0.17	44.7	148
Fe1	0.44	−0.01	42.4	139
Fe4	0.24	−0.18	34.4	36

hyperfine fields of the three sites as a function of temperature (Figure 4) shows a small kink for sextets 1 and 2 leading to a linear extrapolated-zero-temperature value of 51.87 and 50.52 T, respectively. However, variations of the hyperfine field for sextet 3 (tetrahedral Fe4) show a sudden change at the transition near 120 K suggesting a change in the geometry associated with the transition. Subsequently, similar observations were reported by other authors [14].

Therefore, Mössbauer experiments were performed at 160, 120 and 4 K in external applied magnetic field of 3 and 5 T perpendicular to the γ-ray beam propagation. From the fitting procedure the respective hyperfine fields and angles were derived. The fitted parameters of each component are gathered in Table I.

Two magnetic sublattices were identified (Fe2 + Fe3) and (Fe1 + Fe4). The magnetic moments of the two sublattices are antiparallel to one another at 160 K and collinear to the applied magnetic field. At 4.2 K the moments of the (Fe1 + Fe4) sublattice tend to orient perpendicularly to B_{app}. The results can be interpreted as follows: the anisotropy at 160 K is strong, as already experienced by the magnetic measurements while at 4.2 K it is weak. Thus, the moment is easily flopped as also observed by the presence of metamagnetism below 50 K.

4 Conclusion

We have prepared the rare ε-Fe_2O_3 as nanometer size particles by the decomposition of yttrium iron garnet in silica in air. The proposed crystal structure has been confirmed from pattern matching of the powder neutron diffraction data. The magnetic properties show unusual behaviors, viz: temperature dependence magnetic hardness ranging from zero to 22 kOe, three ground states separated by a Néel transition at 480 K and one associated with a change in the canting angle similar to that of a Morin transition below 150 K. The Mössbauer spectra under magnetic field indicate a collinear two-sublattice antiferromagnet above 120 K where one is easily flopped to the perpendicular direction below 120 K.

References

1. Kurmoo, M., Rehspringer, J.-L., Hutlova, A., D'Orleans, C., Vilminot, S., Estournes, C., Niznansky, D.: Chem. Mater. **17**, 1106 (2005)
2. Jin, B., Ohkoshi, S., Hashimoto, K.: Adv. Mater. **16**, 48 (2004)
3. Jin, B., Hashimoto, K., Ohkoshi, S.: J. Mater. Chem. **10**, 1067 (2005)
4. Tronc, E., Chanéac, C., Jolivet, J.P.: J. Solid State Chem. **139**, 93 (1998)
5. Kurmoo, M., Kumagai, H., Hughes, S.M., Kepert, C.J.: Inorg. Chem. **42**, 6709 (2003)
6. Teillet, J., Varret, F.: MOSFIT Program; Université du Maine, France (unpublished)
7. Bouree, F., Baudour, J.L., Elbadraoui, E., Musso, J., Laurent, C., Rousset, A.: Acta Crystallogr., B **52**, 217 (1996)
8. Rodriguez-Carvajal, J.: FULLPROF: Rietveld, Profile Matching and Integrated Intensity Refinement of X-ray and/or Neutron Data, 3.5d version. Léon-Brillouin Laboratory, CEA Saclay, France (1998)
9. Schrader, R., Büttner, G.: Z. Anorg. Allg. Chem. Einzeldarst. **320**, 220 (1963)
10. Powder Diffraction File no. 16-653. JCPDS-International Center for Diffraction Data, Swarthmore, PA (1993)
11. Powder Diffraction File no. 16-895. JCPDS-International Center for Diffraction Data, Swarthmore, PA (1993)
12. Chikazumi, S.: Physics of Ferromagnetism. Oxford University Press, London, UK (1992)
13. Herpin, A.: Théorie du Magnétisme. Presses Universitaires de France, Paris (1968)
14. Tronc, E., Chanéac, C., Jolivet, J.P., Grenèche, J.M.: J. Appl. Phys. **98**, 053901 (2005)

Hyperfine Interact (2005) 166:483–487
DOI 10.1007/s10751-006-9312-7

Structure comparison between Th_2Zn_{17}-type and $TbCu_7$-type Sm–Fe intermetallic compounds and their nitrides by means of ^{57}Fe-Mössbauer spectroscopy

Yoichi Sakai · Ryo Omatsuzawa · Iwao Sakazaki · Sayaka Suzuki · Hayato Hashino · Takanobu Saito · Takahiko Iriyama

Published online: 27 October 2006

Abstract Samarium–iron intermetallic compounds were prepared by a melt spinning method with low and high wheel speeds, which resulted in a Th_2Zn_{17}-type and a $TbCu_7$-type structure, respectively. Structure comparison between these types was investigated for Sm–Fe intermetallic compounds and their nitrides by ^{57}Fe-Mössbauer spectroscopy.

Key words Sm–Fe intermetallic compound · Th_2Zn_{17}-type structure · $TbCu_7$-type structure · permanent magnet material · ^{57}Fe-Mössbauer spectroscopy

1 Introduction

Since the first discovery of rare earth-transition metal (RETM) permanent magnets in 1960s [1, 2], there have been great developments of such RETM materials, especially in this decade. RETM permanent magnets are commercially available for use in electric motors, computer disc drivers, loudspeakers and so on, even in the Mössbauer drive motor used in our work. Among RETM materials, samarium–iron intermetallic compounds and their nitrides have been paid attention as one of the most powerful magnet candidate in the next generation, the high magnetic performance of which was developed in 1989 by Iriyama et al. [3, 4]. The Fe–Sm intermetallic compounds were prepared by slow cooling of mother solutions in a Th_2Zn_{17}-type crystal form, which are known to be stoichiometric and thermodynamically stable. For a few years after the discovery, Mössbauer spectroscopy was applied to characterize such a system by several research groups [5]. It was pointed

Y. Sakai (✉)
Department of Chemistry, Daido Institute of Technology, Nagoya 457-8530, Japan
e-mail: yocsakai@daido-it.ac.jp

R. Omatsuzawa · I. Sakazaki · S. Suzuki · H. Hashino · T. Saito · T. Iriyama
Research and Development Laboratory, Daido Steel Co., Ltd, Nagoya 457-8545, Japan

out by Katter et al. [6] that products prepared by a rapid cooling method existed in a $TbCu_7$-type crystal structure, showing somewhat higher magnet performance. In the late 1990s, Daido Steel Co. group revealed that the $TbCu_7$-type Sm–Fe(–N) prepared at an extremely high cooling rate showed much higher magnet ability than the Th_2Zn_{17} type [7]. It was proved by Buschow and Goot [8] that this type is non-stoichiometric and thermodynamically metastable in contrast with the Th_2Zn_{17} type. The structures have been monitored by a powder X-ray diffraction (XRD) technique, and magnetic performances have been measured as a B-H curve using a conventional magnetometer. In the present work, the structures were investigated by ^{57}Fe-Mössbauer spectroscopy, comparing $TbCu_7$-type Sm–Fe(–N) compounds with Th_2Zn_{17}-ones prepared in the same procedures except for controlled cooling rates.

2 Experimental

Sm–Fe intermetallic compounds were prepared by a melt spinning method with two wheel speeds of 15 m/s and 40 m/s. The mixing atomic ratio (1:x) of Sm and Fe was used, where x was around 9. Nitriding procedures were carried out after thermal annealing at 780°C for 10 min in an argon atmosphere. Details of the procedures are referred to our previous report [7]. Mössbauer measurements were performed in an ordinary mode, using spectrometers fabricated by Topologic Systems, Inc. with a ^{57}Co(Rh) source. The spectra were measured at 80 K for Sm–Fe intermetallic compounds because the much more-resolved magnetic splitting patterns were obtained at a lower temperature. On the other hand, our preliminary investigation showed that the spectral shapes at 80 K for their nitrides were very similar at 80 K to those at 293 K, which let us to select a higher measurement-temperature of 293 K for the nitrides in this work. The isomer shift and Doppler velocity scale were calibrated with respect to metallic iron at 293 K. XRD and B-H curve measurements were carried out by traditional methods.

3 Results and discussion

In Figure 1 are shown the Mössbauer spectra at 80 K of Sm–Fe intermetallic compounds ($SmFe_x$) with x=8.5 prepared at different wheel speeds of (a) 15 m/s and (b) 40 m/s. Our XRD measurements showed that samples (a) and (b) were attributed to the Th_2Zn_{17}- and $TbCu_7$-type crystal, respectively. The Mössbauer line-shape of sample (a) was very similar to that reported previously [5] for the Th_2Zn_{17}-type Sm_2Fe_{17}, and it could be analyzed to four sextets with the intensity ratio of 6:9:18:18, which is consistent to the crystal structure for the Th_2Zn_{17} type with four sites of iron of 6c, 9d, 18f, and 18h, as shown in Figure 2c. On the other hand, it is seen in Figure 1b that the spectrum was also analyzed to be deconvoluted into four sextets with a ratio of 6:9:18:18 at the first approximation. Although the line-shape of sample (b) is apparently seen to be very different from that of sample (a), Mössbauer parameters for samples (a) and (b) were estimated to be close to each other except for the line-widths of each peak ((a)<(b)). This finding could be explained by the comparison of the structures; the Th_2Zn_{17} type is obtained by substitution of a rare

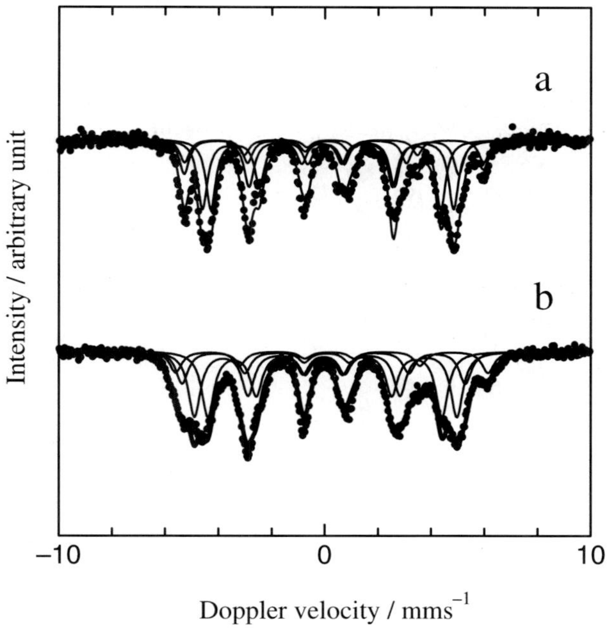

Figure 1 ^{57}Fe-Mössbauer spectra at 80 K of Sm–Fe intermetallic compounds of $SmFe_{8.5}$ prepared by a melt-spinning method with a wheel speed of **a** 15 m/s and **b** 40 m/s.

earth atom of RT_5 by a T–T dumbbell-formed cluster, where R and T is a rare earth atom and a transition metal atom, respectively, with an exact probability of 1/3 exactly and regularly occupying 6c sites. Although not existing actually, the $CaCu_5$-type $FeSm_5$ structure was regarded virtually as an elementary starting structure in our discussion. On the other hand, a rare earth atom is substituted randomly and irregularly by a T–T dumbbell with an identical probability of 1/3 for the $TbCu_7$-type $SmFe_{8.5}$. The above explanation on the structures of Th_2Zn_{17}-and $TbCu_7$-type crystals is illustrated in Figure 2. This should imply that the environments of iron atoms might be not completely same in the two crystal types, but slightly different from microscopic view points around iron atoms, which was for the first time proved from our Mössbauer investigation of Sm–Fe intermetallic compounds synthesized under the same experimental conditions except the cooling rate.

Figure 3 shows the ^{57}Fe-Mössbauer spectra at 293 K of Sm–Fe–N intermetallic nitride compounds prepared by a melt-spin cooling, annealing, and nitriding, where the spectra of (a) and (b) are for the nitrides from the intermetallic compounds shown in Figure 1; a composition of $SmFe_{8.5}$ and a wheel speed of (a) 15 m/s and (b) 40 m/s. There was found little difference in the spectra of the two nitride compounds although the fitting analysis resulted in that the line width of (a) was a little bit smaller than (b), while the intermetallic compounds before the annealing and nitriding show substantially clear difference in the line width. Our XRD measurement showed that the Th_2Zn_{17}-type structure was partly formed as well as the $TbCu_7$-type one for the nitride sample (b), suggesting that a part of the crystalline material should be transformed from the $TbCu_7$ structure to the Th_2Zn_{17} after the annealing and/or nitriding. From the measurements of B-H curves for those two samples, it was revealed that the $(BH)_{max}$ value was higher in the Sm–Fe–N (b) than in the

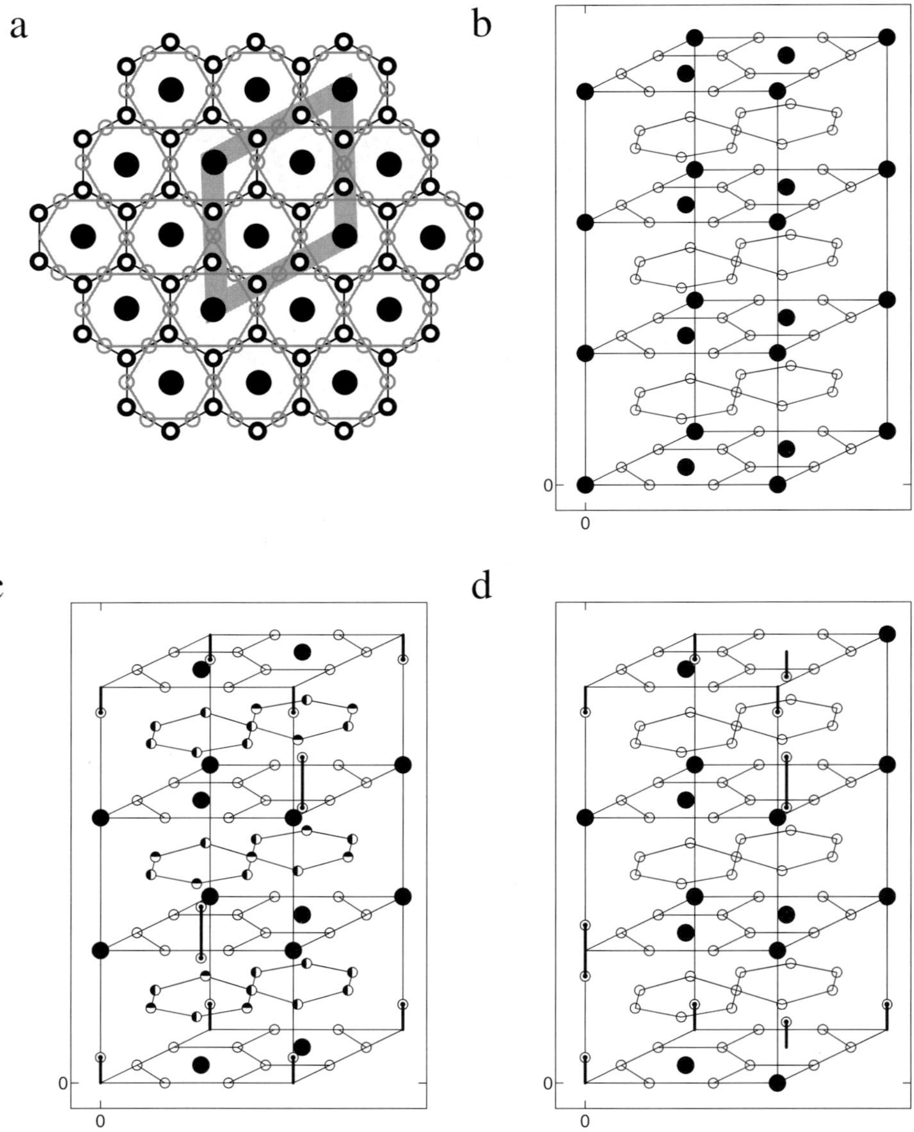

Figure 2 Structures of $CaCu_5$-type, Th_2Zn_{17}-type and $TbCu_7$-type crystals; *larger closed circles* set for Ca, Th, Tb, and Sm (for the present work) atoms, while *smaller circles* for Cu, Zn, or Fe (in this work) atoms. **a.** $CaCu_5$-type structure (top-view), which is regarded as an elementary stating structure. **b.** $CaCu_5$-type structure, which is cut out as a rhombohedron marked by shadow in a. **c.** Th_2Zn_{17}-type structure, which is obtained by the regular replacement of a Sm atom by an Fe–Fe dumbbell in $CaCu_5$-type $SmFe_5$ with an exact probability of 1/3, yielding the following four iron sites;⦿;6c (dumbbell) , ◓;9d, ○;18f, and ◐;18h. **d.** $TbCu_7$-type structure, which is obtained by the irregular replacement of a Sm atom by an Fe–Fe dumbbell in $CaCu_5$-type $SmFe_5$ with a probability of about 1/3.

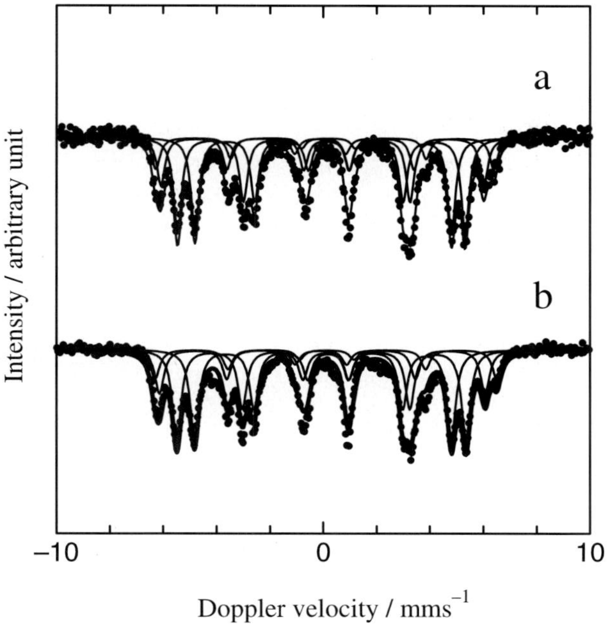

Figure 3 ^{57}Fe-Mössbauer spectra at 293 K of Sm–Fe–N nitride compounds prepared by a annealing and nitriding of the intermetallic compounds of $SmFe_{8.5}$ synthesized by melt-spinning method with a wheel speed of **a** 15 m/s and **b** 40 m/s.

nitride (a). The $(BH)_{max}$ value is well recognized to be an essential measure for magnet performance; the higher the value is, the greater the performance is. Our results might suggest that the determining factor for better permanent RETM magnet should be macroscopic properties such as grain sizes rather than microscopic environment of atoms bearing magnetic moment such as iron. Mössbauer measurements of dependence of iron states on the composition ratio of Sm and Fe were performed in the present work, giving interesting observation preliminarily, which will be published elsewhere after furthermore experiments and considerations.

References

1. Strnat, K.J.: Cobalt **36**, 133 (1967)
2. Tawara, Y., Senno, H.: Japan J. Appl. Phys. **7**, 966 (1968)
3. Iriyama, T., Kobayashi, K., Imai, H.: European Patent Appl. 0-369-097-A1 (1989)
4. Iriyama, T., Kobayashi, K., Imaoka, N., Fukuda, T.: IEEE Trans. Magn. **28**, 2326 (1992)
5. Chen, X., Ryan, D.H., Altounian, Z., Liao, L.X.: J. Appl. Phys. **73**, 6038 (1993)
6. Katter, M., Wecker, J., Schultz, L.: J. Appl. Phys. **70**, 3188 (1991)
7. Omatsuzawa, R., Murashige, K., Iriyama, T.: Denki Seiko **73**, 235 (2002) (in Japanese)
8. Buschow, K.H., van der Goot, A.S.: Acta Crystallogr. **B27**, 1085 (1971)

Hyperfine Interact (2005) 166:489–493
DOI 10.1007/s10751-006-9313-6

Defects induced by electron irradiation in hollandite ceramics, specific radioactive cesium-host wasteforms: a ^{57}Fe Mössbauer study

N. Nguyen · A. Ducouret · F. Studer · V. Aubin · D. Caurant · D. Gourier · J-M. Costantini

Published online: 31 October 2006

Abstract Several single phase hollandite ceramics having $Ba^{2+}{}_{x}Cs^{+}{}_{y}(C = Al,Fe)^{3+}{}_{2x+y}Ti^{4+}{}_{8-2x-y}O_{16}$ composition have been synthesized and irradiated under external electron beams simulating the β-irradiation of radioactive cesium. These samples have been characterized by ^{57}Fe transmission Mössbauer spectroscopy. Mössbauer results show that the irradiation modifies the local arrangements of Ba cations around Fe^{3+} ions by atomic displacement of Ba ions.

Key words Mössbauer spectroscopy · atomic displacement · high level radioactive wastes

1 Introduction

Over the past few years, progress on the enhanced reprocessing of nuclear spent fuel and notably on separating the long-lived fission products from the high level radioactive wastes (HLW) has encouraged investigations on specific host matrices [1]. Particularly, the success of cesium extraction by calixarenes complexes, except for some traces of rubidium, suggests that new materials with higher performance than borosilicate glassy matrices currently used to immobilize unseparated HLW, could be considered.

Cesium appears as one of the most hazardous radionuclide to immobilize because of the high thermal power of ^{134}Cs and ^{137}Cs isotopes, the long lifetime of ^{135}Cs

N. Nguyen · A. Ducouret · F. Studer (✉)
CRISMAT (CNRS-UMR 6508), ENSICAEN, 6, bd du Maréchal Juin,
F-14050 Caen cedex, France
e-mail: francis.studer@ensicaen.fr

V. Aubin · D. Caurant · D. Gourier
LCAES (CNRS UMR 7574), 11, rue Pierre et Marie Curie, 75231 Paris, France

J.-M. Costantini
CEA/DMN/SRMA, 91191 Gif sur Yvette, Paris, France

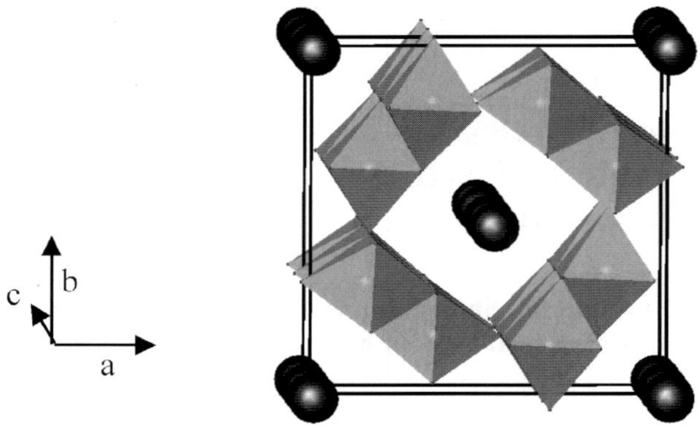

Figure 1 View of the *tetragonal* (I4/m) hollandite-type $A_2B_8O_{16}$ structure approximately along the c-axis: a cations (*black atoms*) and $[TiO_6]$ *octahedra* are shown.

isotope ($T_{1/2}$ = 2.3 10^6 years), its volatility at high temperature during waste forms synthesis and its tendency to form water-soluble compounds. The cesium radioactive isotopes decay into stable barium by emitting β particles (with energy E ranging from 0.09 to 1.17 MeV, mainly 0.5 MeV) and γ rays ($E \approx 0.6$ MeV). Therefore, the choice of new Cs-host matrices rely on their waste (Cs, Rb) loading, their Ba and Sr (decay product of Rb) incorporation ability, their thermal characteristics, their long-term containment properties (chemical durability, β and γ radiation resistance) and the ease of their synthesis.

The hollandite type $A_2B_8O_{16}$ structure, shown in Figure 1, is closely related to the rutile structure. It is tetragonal (I4/m space group) or monoclinic (I2/m) according to the ionic radii of A and B cations. The $[BO_6]$ octahedra are linked by edges and corners to form one-dimensional square tunnels in which are located the large A cations. The tunnel sites are often not full and depending on the hollandite composition, the A cations and vacant sites may take up an ordered arrangement along the tunnels (superlattice ordering).

Hollandite has several advantages. It is able to accommodate chemistry and valence changes due to the cesium transmutation because of its high capacity to incorporate Ba and to the possibility of reduction of a fraction of Ti^{4+} ions to ensure the charge balance [2]. Moreover, its chemical durability is two orders of magnitude higher than the currently used borosilicate glasses, which is required to select Cs-specific matrices [3]. Nevertheless, the synthesis of $[Ba^{2+}{}_xCs^{+}{}_y][(Al^{3+},Ti^{3+})_{2x+y}Ti^{4+}{}_{8-2x-y}]O_{16}$ compositions, named after the Synroc composition, remains actually too complex to be industrialized [4].

The β and γ radiation resistance of hollandite is not yet proved due to the lack of both true natural analogs and experiments on synthetic hollandite samples containing radioactive cesium. Moreover, literature reports only electron external irradiation experiments which overestimate the real effects of β irradiation. Indeed, only experiments under intense electron irradiation in the electron microscope were carried out on single phase hollandite, which generated microtwinnings followed by amorphization [5].

In this work, we report the first Mössbauer spectroscopy measurements of external electron irradiated ceramics of hollandite in the $Ba_xCs_y(Al,Fe)_{2x+y}Ti_{8-2x-y}O_{16}$

Table I Theoretical and EPMA compositions of hollandite ceramics

	Expected compositions	EPMA compositions
1	$Ba_{1.16}Fe_{2.32}Ti_{5.68}O_{16}$	$Ba_{1.13}Fe_{2.32}Ti_{5.70}O_{16}$
2	$Ba_{1.28}Al_{1.64}\ Fe_{0.92}Ti_{5.44}O_{16}$	$Ba_{1.29}Al_{1.71}Fe_{0.93}Ti_{5.38}O_{16}$
3	$Cs_{0.28}Ba_{1}Al_{1.46}Fe_{0.82}Ti_{5.72}O_{16}$	$Cs_{0.25}Ba_{1.05}Al_{1.43}Fe_{0.80}Ti_{5.74}O_{16}$

system. Three compositions were chosen for external electron irradiation in order to investigate the effect of the presence of Al and/or Cs on the damage creation.

2 Experimental

Several ceramics having the $Ba^{2+}{}_{x}Cs^{+}{}_{y}(C{=}Al,Fe)^{3+}{}_{2x+y}Ti^{4+}{}_{8-2x-y}O_{16}$ composition type ($1 \leq x \leq 1.28$; $0 \leq y \leq 0.28$) were prepared by solid-state reaction from Al_2O_3, Fe_2O_3, TiO_2, $BaCO_3$ and $CsNO_3$ in air following a classical solid state chemistry process. Iron is introduced in hollandite because it was reported to facilitate both the synthesis protocol (heat treatment under reducing atmosphere is avoided) and the insertion of cesium in hollandite structure (increase of channels size) [4]. Samples were characterized by X-ray diffraction (XRD) with CoKα wavelength (λ = 1.78897 Å), by Scanning Electron Microscopy (SEM), Energy Dispersive X-ray Analysis (EDX) and by Electron Probe Micro-Analysis (EPMA) (Table I).

(Ba,Cs)-hollandite materials were irradiated by 1, 1.5 or 2.5 MeV electrons generated by a Van de Graaff accelerator to simulate the effect of β particles emitted by radioactive cesium. The thickness of samples (0.5 or 1 mm) was chosen so that electrons go through the sample and radiation effects are quite homogeneous. Different fluences between 8.1 10^{17} and 1.4 10^{19} cm^{-2} were carried out.

The ^{57}Fe transmission Mössbauer spectra were measured at room temperature using a ^{57}Co/Rh source on Fe-rich powders or thin massive samples. The isomer shifts (I.S.) are corrected to the I.S. of metallic iron taken as a reference. The spectra were simulated by a least square fitting routine assuming a Lorentzian line shape (unpublished MOSFIT program).

3 Results and discussion

Sample 1 of composition $Ba_{1.16}Fe_{2.32}Ti_{5.68}O_{16}$ was selected in this study because no ^{57}Fe-enrichment was necessary and its composition is relatively simple with only one trivalent species (Fe^{3+}) and one tunnel cation (Ba^{2+}). Its 300 K Mössbauer spectrum (Figure 2a) exhibits one quadrupolar doublet before irradiation. However, to fit this spectrum, it was necessary to consider two quadrupolar doublets whose parameters are indicated in Table II. Only the quadrupole splitting differentiate these two sites. They are typical of Fe^{3+} ions in octahedral sites with more distortion for site 1. Therefore, no Fe^{2+} ions could be detected before irradiation.

In the tetragonal hollandite structure (I4/m space group), all the octahedral sites are equivalent and Fe^{3+} and Ti^{4+} ions are randomly distributed on them on the evidence of X-ray diffraction [6]. But, if the structure is monoclinic (I2/m space

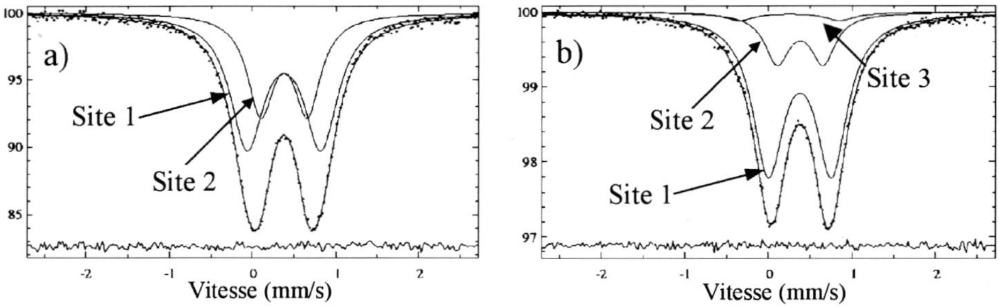

Figure 2 Typical 300 K Mössbauer spectra of sample 1 ($Ba_{1.16}Fe_{2.32}Ti_{5.68}O_{16}$) before (**a**) and after irradiation (**b**) with 1.5 MeV electrons at a fluence Φt = 5.8 10^{18} cm^{-2}. The components used to simulate the experimental spectra are indicated by *solid lines* and the difference between the experimental and simulated spectra *below* each spectrum.

Table II Hyperfine parameters extracted from the least-squares fits of the 300 K ^{57}Fe spectra of the pristine and electron irradiated sample 1 shown in Figure 2: Isomer shift (I.S.), quadrupolar splitting (Q.S.), half width at half maximum (Γ) and fraction (R) of the different sites

	sites	I.S mm/s.	Q.S. mm/s	Γ mm/s	R %		Sites	I.S mm/s.	Q.S. mm/s	Γ mm/s	R %
Pristine	1	0.493	0.884	0.25	63	Irradiated	1	0.490	0.76	0.23	77
sample 1	2	0.494	0.56	0.19	37	sample 1	2	0.492	0.55	0.12	19
							3	0.36	1.21	0.23	4

group), two crystallographic sites occur without preferential occupation [6]. Few studies have been reported on the use of NMR, EPR and Mössbauer spectrocopies of hollandites except two Mössbauer studies on $K_{1.45}Fe_{1.45}Ti_{6.55}O_{16}$ [7] and $Cs_{1.40}Fe_{1.40}Ti_{6.74}O_{16}$ [8] hollandites which reported also one quadrupolar doublet at 300 K whose parameters were close to the ones of site 2. Moreover, few studies have reported a monoclinic structure for a close hollandite composition $Ba_{1.14}Fe_{2.28}Ti_{5.72}O_{16}$ [9].

Therefore, we can wonder if our material does not display a slight monoclinic distortion not clearly detected by XRD and thus these two doublets (Figure 2a) would be related to the two different crystallographic sites. This hypothesis is rejected by comparison with Mössbauer spectra of the samples 2 and 3 (I4/m space group) which also display two quadrupolar doublets, not shown here [10]. The crystal structure of these samples were confirmed by structural determination on single crystals of the same composition [10]. Therefore, only one crystallographic site is assumed for sample 1 and then these two different doublets could be assigned to different local environments of Fe^{3+} ions. First, this difference could be due to the possibility to have Fe^{3+} ions or Ti^{4+} ions as nearest neighbors. This explanation does not seem to be probable given the supposed random distribution of the Fe and Ti ions on the octahedral sites and the presence of only two quadrupolar doublets in samples 2 and 3 containing in addition Al ions [10]. However, these local environment differences could contribute to the full width at half maximum of each doublet. Therefore, the occurrence of these two sites could only be explained by the tunnel site occupancy. Indeed, the tunnel is not full and thus Fe^{3+} ions could have vacancies or Ba cations in their close environment (Figure 1). Moreover, it has been evidenced by E.Fanchon [11] that the C cations (Fe^{3+}) change their position in octahedral site according to the tunnel content, which could explain the different quadrupolar splitting of the two sites. This hypothesis is confirmed by the changes of the site fraction *R* with the hollandite chemical composition [10]. The exact environment of each site is not yet fully determined.

Although 300 K experimental Mössbauer spectrum of the electron irradiated sample 1 revealed only one quadrupolar doublet (Figure 2b), three doublets are necessary to reproduce it properly. The two first sites correspond to those observed before irradiation as shown by their associated parameters in Table II. However, their relative fraction changed in favor of site 1 which becomes also more symmetric. This suggests that electron irradiation modifies the local arrangement of Ba cations around Fe^{3+} ions by atomic displacements of Ba ions.

Finally, a third site was generated in small quantities. Its isomer shift is more typical of fivefold coordinated pyramidal Fe^{3+} ions [12]. Therefore, this electron irradiation generates also some oxygen displacements. Finally, no change of the iron oxidation was observed: no Fe^{2+} ions were detected within the Mössbauer sensitivity scale.

Briefly, Mössbauer results show that electron irradiation ($E = 1.5$ MeV, $\Phi t = 5.8\ 10^{18}\ cm^{-2}$) induced atomic displacements of at least Ba and O ions without reduction of a significant fraction of the Fe^{3+} cations. EFG calculations are currently being performed to correlate the observed quadrupolar splitting to the structural hypothesis.

References

1. Madic, C., Lecomte, M., Baron, P., Boullis, B.: C. R. Physique **3**, 797 (2002)
2. Bursill, L.A.: J. Solid State Chem. **69**, 355 (1987)
3. Carter, M.L., Vance, E.R., Mitchell, D.R.G., Hanna, J.V., Zhang, Z., Loi, E.: J. Mater. Res. **17**, 2578 (2002)
4. Kesson, S.E.: Radioact. Waste Manag. Nucl. Fuel Cycle **4**(1), 53 (1983)
5. Bursill, L.A., Smith, D.J.: J. Solid State Chem. **69**, 343 (1987)
6. Cheary, R.W.: Acta. Crystallogr. **B42**, 229 (1986)
7. Drofenik, M., Hanzel, D.: Mater. Res. Bull. **17**, 1457 (1982)
8. Birchall, T., Greenwood, N.N., Reid, A.F.: J. Chem. Soc. **A**, 2382 (1969)
9. Loezos, J.M., Vanderah, T.A., Drews, A.R.: Powder Diffr. **14**(1), 31 (1999)
10. Aubin, V.: Thesis of the Pierre and Marie Curie University, Paris (2004)
11. Fanchon, E.: Thesis of the Unversité scientifique, technologique et médicale de Grenoble (1987)
12. Studer, F., Houpert, C., Toulemonde, M., Dartyge, E.: J. Solid State Chem. **91**(2), 225 (1991)

Hyperfine Interact (2005) 166:495–498
DOI 10.1007/s10751-006-9314-5

DFT calculations as a tool to analyse quadrupole splittings of spin crossover Fe(II) complexes

J. A. Wolny · H. Paulsen · H. Winkler · A. X. Trautwein · J.-P. Tuchagues

Published online: 7 December 2006

Abstract Density functional methods have been applied to calculate the quadrupole splitting of a series of iron(II) spin crossover complexes. Experimental and calculated values are in reasonable agreement. In one case spin–orbit coupling is necessary to explain the very small quadrupole splitting value of 0.77 mm/s at 293 K for a high-spin isomer.

Key words DFT calculations · iron complexes · Mössbauer spectroscopy · spin crossover

Abbreviations

BTPA	*N,N,N′,N′*-tetrakis(2-pyridylmethyl)-6,6′-bis(aminomethyl)-2,2′-bipyridine
bzTPA	*N,N′*-bis(benzyl)-*N,N′*-bis(2-pyridylmethyl)-6,6′-bis(aminomethyl)-2,2′-bipyridine
DFT	density functional theory
EFG	electric field gradient
HOMO	highest occupied molecular orbital
HS	high spin
LS	low spin
LUMO	lowest unoccupied molecular orbital
phen	1,10-phenanthroline
SCO	spin crossover
tpa	tris(2-pyridylmethyl)amine
tpt	1,1,1-tris{[N(2-pyridylmethyl)-N-methylamino]ethyl}propane
trim	4-(4-imidazolylmethyl)-2-(2-imidazolylmethyl)imidazole

J. A. Wolny (✉) · H. Paulsen · H. Winkler · A. X. Trautwein
Institut für Physik, Universität zu Lübeck, 23538 Lübeck, Germany
e-mail: wolny@physik.uni-luebeck.de

J. A. Wolny
Wydział Chemii, Uniwerystet Wrocławski, 50-383 Wrocław, Poland

J.-P. Tuchagues
Laboratoire de Chimie de Coordination du CNRS, 31077 Toulouse, France

Table I Experimental and calculated (in parentheses) values of ΔE_Q in millimeter per second

Complex	LS	HS	HS[h]
$[Fe(tpa)(NCS)_2]$[a]	0.40 at 77 K (+0.42)	2.5 at 130 K (+2.85)	
$[Fe(tpt)](ClO_4)_2$[b]	0.15 at 77 K (−0.01)	0.77 at RT (+3.4)	
$[Fe(phen)_2(NCS)_2]$[c]	0.34 at 77 K (+0.40)	2.67 at RT (+3.05)	
$[Fe(bzTPA)](PF_6)_2$[d]	0.39 at 77 K (+0.35)	2.70 at 77 K (+3.21)	
$[Fe(BTPA)](PF_6)_2$[d]	0.44 at 77 K (+0.27)	3.21 at 77 K (+3.93)	3.01 at 77 K (+3.41)
$[Fe(trim)_2]X_2$[e]	0.05–0.18[f] at RT (−0.13)	2.7–2.80[f] at 77 K (−3.58)	

[a] [15], [b] [18], [c] [19], [d] [4], [e] [5], [f] depending on anion X, [h] two different HS sites are observed

1 Introduction

Spin crossover (SCO) complexes reveal a crossover from a low-spin (LS) to a high-spin (HS) state that can be induced by a change of temperature or pressure or by irradiation with light [1]. These SCO complexes are not only a fascinating subject of transition metal chemistry, they are also a challenge for electronic structure calculations. Methods based on density functional theory (DFT) turned out to be most suitable for this purpose [2]. A crucial test for the quality of any calculated electronic structure are Mössbauer parameters like the quadrupole splitting ΔE_Q that can be derived from the electric field gradient (EFG) at the ^{57}Fe nucleus. Previously EFGs were calculated for HS and LS isomers of a whole series of differently substituted trispyrazolylomethane ligands with different anions [3]. Here we present calculated and measured ΔE_Q for several SCO compounds, including the recently presented polypyridine and polyimidazole systems [4, 5].

2 Materials and method

In the first step the geometries of the cationic or neutral SCO complexes in vacuo have been fully optimized for the HS and the LS state. These calculations have been performed with the B3LYP hybrid functional [6, 7] together with the CEP-31G basis [8–10] as included in the Gaussian 03 program package [11]. The CEP-31G basis has been shown to give reliable geometry parameters for SCO complexes [12] at modest computational expenses but it does not include core electrons. Therefore, in a second step the geometries optimized with the CEP-31G basis are used to calculate subsequently the EFG with the B3LYP functional together with the 6-311+G(2d,p) basis for H, C, and N and the Wachters-Hay double-ζ basis for Fe [13, 14]. A nuclear quadrupole moment of $Q = 0.15$ barn has been used in order to derive ΔE_Q from the EFG.

3 Results and discussion

The calculated ΔE_Q for LS isomers are in satisfactory agreement with the experimental values (Table I). For the HS isomers the calculated values are approx-

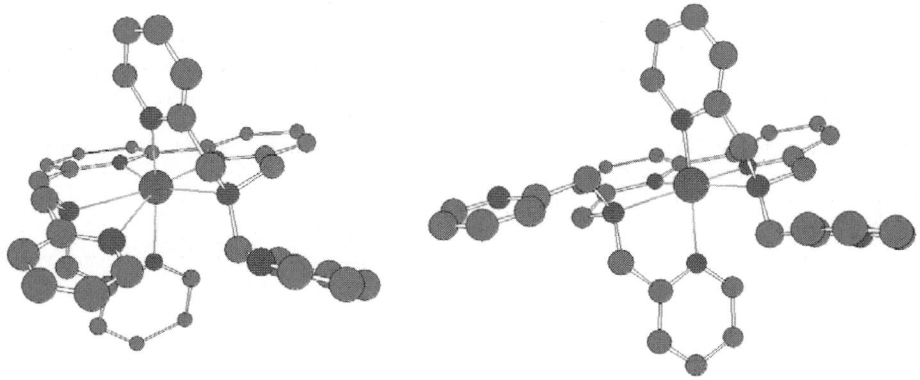

Figure 1 C_1 (*left*) and C_2 (*right*) isomers of HS $Fe(BTPA)^{2+}$.

imately 20–25% higher than the measured values. Two cases of special interest are the complexes $[Fe(tpt)](ClO_4)_2$ [15] and $[Fe(BTPA)](PF_6)_2$ [4].

$[Fe(tpt)](ClO_4)_2$ displays a C_3 symmetry and DFT calculations yield a very small HOMO–LUMO energy gap of about 1,000 cm^{-1} which is of the order of magnitude of the spin–orbit coupling constant (~400 cm^{-1}) for iron(II). Boltzmann population of the LUMO and – in this case even more important – mixing of HOMO and LUMO due to spin–orbit coupling are obvious explanations for a very small ΔE_Q of 0.77 mm/s. The calculated value of 3.4 mm/s is significantly higher since spin–orbit coupling was not included in our calculation.

X-ray and other spectroscopic measurements for $[Fe(BTPA)](PF_6)_2$ reveal the presence of two possible HS stereoisomers with C_1 and approximate C_2 symmetry, that provide quasi-seven- and six-co-ordinated iron centers (Figure 1), respectively [4, 16]. The Mössbauer spectrum of a powder sample exhibits two HS quadrupole doublets. The spectrum of recrystallised $[Fe(BTPA)](PF_6)_2$ as well as the studies of $[Zn(BTPA)](PF_6)_2$ doped with ^{57}Fe [17] indicate that the isomer with C_1 symmetry gives rise to the doublet with the smaller ΔE_Q. This finding is consistent with the calculated values for ΔE_Q. Apart from that, the calculations predict that the C_2 HS isomer yields a higher ΔE_Q value than its isosteric analogue $[Fe(bzTPA)](PF_6)_2$ (Table I).

In conclusion, the quadrupole splittings calculated with DFT methods are in general in agreement with experimental values and corroborate the validity of the calculated electronic structure. In one case, i.e. for the HS isomer of $[Fe(tpt)](ClO_4)_2$, where the experimental quadrupole splitting could not be reproduced by the calculation, the obvious explanation for the very small ΔE_Q value is the mixing of the HOMO and LUMO due to spin–orbit coupling.

Acknowledgment Financial support from the CNRS and the DFG (Tr 97/31 and Tr 97/32) is gratefully acknowledged.

References

1. Gütlich, P., Goodwin, H.A. (eds.): Spin Crossover in Transition Metal Compounds I–III. Springer, Berlin Heidelberg New York (2004)

2. Paulsen, H., Trautwein, A.X.: Top. Curr. Chem. **235**, 197 (2004)
3. Paulsen, H., Duelund, L., Zimmermann, A., Averseng, A.F., Gerdan, M., Trautwein, A.X., Winkler, H., Toftlund, H.: Monatsh. Chem. **134**, 295 (2003)
4. Brady, C., Callaghan, P.L., Ciunik, P.Z., Coates, C.G., Døssing, A., Hazell, A., McGarvey, J.J., Schenker, S., Toftlund, H., Trautwein, A.X., Winkler, H., Wolny, J.A.: Inorg. Chem. **43**, 4289 (2004)
5. Lemercier, G., Bréfuel, N., Shova, S., Wolny, J.A., Dahan, F., Verelst, M., Paulsen, H., Trautwein, A.X., Tuchagues, J.P.: Chem. Eur. J. **12**, 7421–7432 (2006)
6. Lee, C., Yang, W., Parr, R.G.: Phys. Rev. B **37**, 785 (1988)
7. Miehlich, B., Savin, A., Stoll, H., Preuss, H.: Chem. Phys. Lett. **157**, 200 (1989)
8. Stevens, W.J., Krauss, M., Basch, H., Jasien, P.G.: Can. J. Chem. **70**, 612 (1992)
9. Cundari, T.R., Stevens, W.J.: J. Chem. Phys. **98**, 5555 (1993)
10. Wachters, A.J.H.: J. Chem. Phys. **52**, 1033 (1970)
11. Gaussian 03, Revision C.02, M.J. Frisch et al., Gaussian, Inc., Wallingford CT, 2004
12. Paulsen, H., Trautwein, A.X.: J. Phys. Chem. Solids **65**, 793 (2004)
13. Wachters, A.J.H.: J. Chem. Phys. **52**, 1033 (1970)
14. Hay, P.J.: J. Chem. Phys. **66**, 4377 (1977)
15. Paulsen, H., Grünsteudel, H., Meyer-Klaucke, W., Gerdan, M., Winkler, H., Toftlund, H., Trautwein, A.X.: Eur. Phys. J. B **23**, 463 (2001)
16. Schenker, S., Stein, P.C., Wolny, J.A., Brady, C., McGarvey, J.J., Toftlund, H., Hauser, A.: Inorg. Chem. **40**, 134 (2001)
17. Winkler, H., Rusoanov, V., McGarvey, J.J., Toftlund, H., Trautwein, A.X., Wolny, J.A.: Hyperfine Interact. DOI: 10.1007/s10751-006-9456-5
18. Al-Obaidi, A.H.R., Jensen, K.B., McGarvey, J.J., Toftlund, H., Jensen, B., Bell, S.E.J., Carrol, J.G.: Inorg. Chem. **35**, 5005 (1996)
19. König, E., Hüfner, S., Steichele, E., Madeja, K.: Z. Naturforsch. **22A**, 1543 (1967)

Hyperfine Interact (2005) 166:499–503
DOI 10.1007/s10751-006-9315-4

Mössbauer studies of ^{151}Eu in europium oxalate, europium bissalen ammonium and europium benzoate

C. I. Wynter · D. H. Ryan · S. P. Taneja · L. May · F. W. Oliver · D. E. Brown · M. Iwunzie

Published online: 22 November 2006

Abstract Although a number of europium water insoluble chelates have been prepared for several decades, the covalent nature of these compounds has never been established in any quantitative fashion. Shifts in the I.R. bands and conductivity measurements of these salts were hitherto used to qualitatively compare their molecular nature. In this communique we have used temperature coefficients of ^{151}Eu Mössbauer spectra to determine the Debye temperatures (θ_D) of three europium chelates: namely europium oxalate, europium bissalen ammonium (recently reported) and europium benzoate and compared their θ_D with the measured θ_D of the known ionic EuF_3. Additionally, the mean square amplitude ($<x^2>$) of these four compounds was computed and plotted as a function of temperature giving in each case a reasonable linear plot. It was interesting to note that the θ_D of the most ionic EuF_3 was greatest (283 ± 10 K) followed by the oxalate (166 ± 15 K), then the europium bissalen ammonium (133 ± 5 K) and lastly the europium benzoate with a θ_D of (105 ± 5 K).

C. I. Wynter (✉)
Department of Chemistry, Nassau Community College, Garden City, NY 11530, USA
e-mail: wynterc@ncc.edu

D. H. Ryan
Centre for the Physics Materials, Department of Physics, McGill University, Montreal H3A2T8, Canada

S. P. Taneja
Department of Physics, Maharshi Dayanand University, Rohtak 124-001, India

L. May
Department of Chemistry, The Catholic University of America, Washington, DC 20064, USA

F. W. Oliver
Department of Physics, Morgan State University, Baltimore, MD 21251, USA

D. E. Brown
Department of Physics, Northern Illinois University, De Kalb, IL 60115, USA

M. Iwunzie
Department of Chemistry, Morgan State University, Baltimore, MD 21251, USA

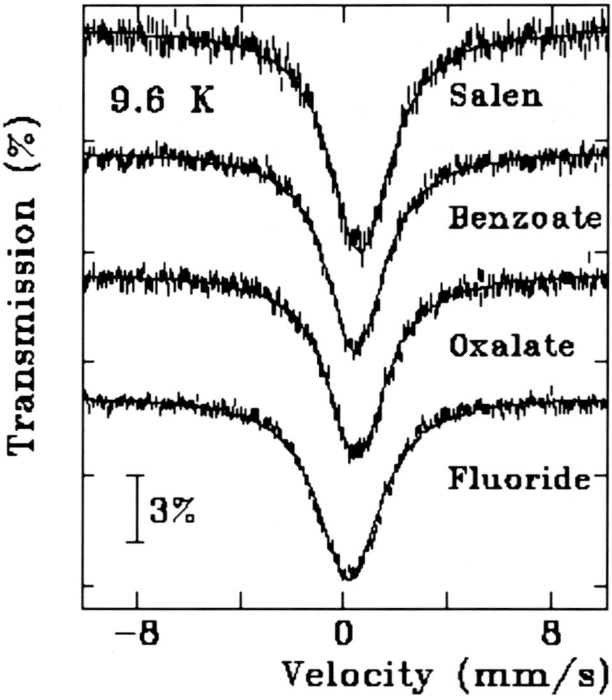

Figure 1 Typical ^{151}Eu Mössbauer spectra of EuF_3 and complexes at 9.6 K.

Key words Mössbauer spectroscopy · europium · Debye temperature · mean square amplitude

PACS 82.80.Eg · 71.15.Fr

1 Introduction

Quite recently, the properties of a new europium salen complex [$NH_4Eu(salen)_2$] were characterized and a Debye temperature value of 133 ± 5 K reported [1]. To determine what effect acid ligands would have, the studies of two more chelates, europium oxalate Eu_2 (C_2 0_4)$_3$ and europium benzoate Eu (C_6H_5C00)$_3$ were made. For comparison, europium fluoride (EuF_3) was also studied as the prototype of the 'truly' ionic europium compound. This paper reports the results of the Mössbauer spectral studies of these compounds at various temperatures.

2 Experimental

The preparation of the Eu (salen)$_2$ NH_4 complex was recently reported [1] while the benzoate and oxalate salts were prepared by standard methods previously reported in the literature [2].

The Mössbauer spectra of each of the four compounds were measured at temperatures ranging from 10 to 295 K in the constant acceleration mode using the

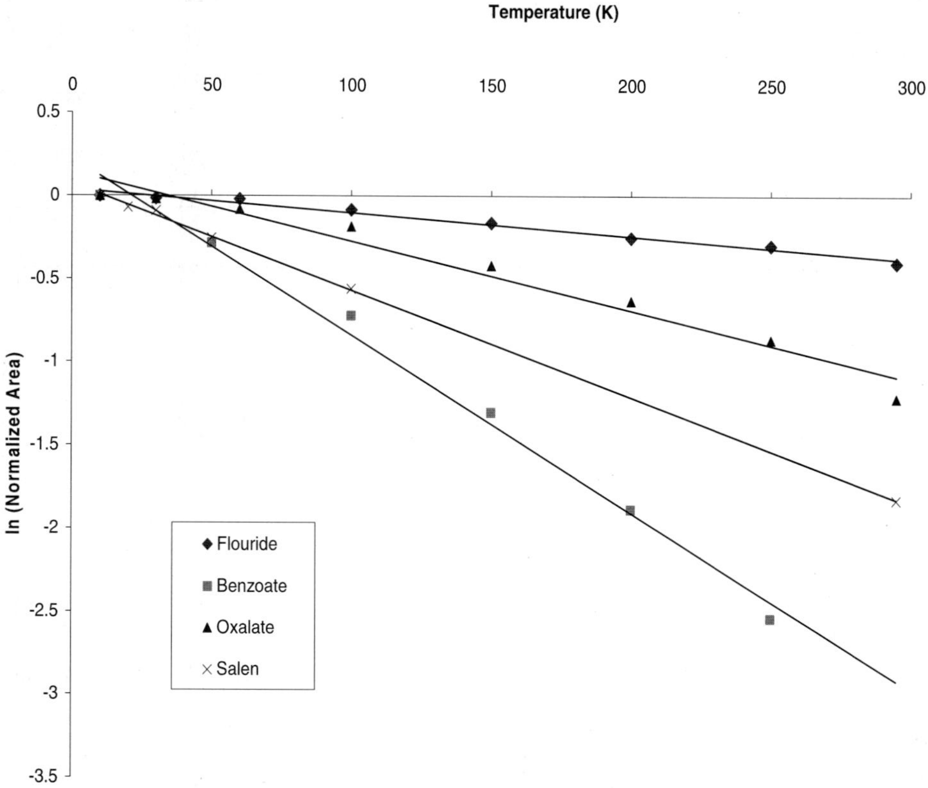

Figure 2 Plot of lnA vs. temperature.

21.6 keV gamma ray from a 100 mCi $^{151}SmF_3$ source. The velocity was calibrated with iron foil at room temperature, and all Eu spectra reported with EuF_3 as the standard at room temperature. A Wissel Mössbauer system was used and spectra fitted using a least square analysis program.

3 Results and discussion

Figure 1 shows the similar gross features of the spectra of the Eu compounds at 9.6 K, the chelates exhibiting among the narrowest of line width shown by Eu^{3+} compounds to date and the positive isomer shifts with respect to EuF_3 confirming trivalent europium [3, 4].

The Debye temperature of the fluoride was 283 ± 10 K whereas the oxalate and the benzoate were 166 ± 15 K and 105 ± 5 K, respectively. Further intermolecular and intramolecular bond strength information was extracted from computation of the mean square amplitude of these compounds.

Of paramount interest was a plot of the natural logarithm of the normalized area versus temperature, Figure 2. The Debye temperature from these plots gave the θ_D's reported in Table I. The obviously most ionic compound, EuF_3, gave the

Table I Debye temperatures (θ_D) of Eu compounds

Substance	θ_D(K)
EuF_3	283 ± 10
$Eu_2\ (C_2O_4)_3$	166 ± 15
Eu NH_4 $(salen)_2$	133 ± 5
Eu $(benzoate)_3$	105 ± 5

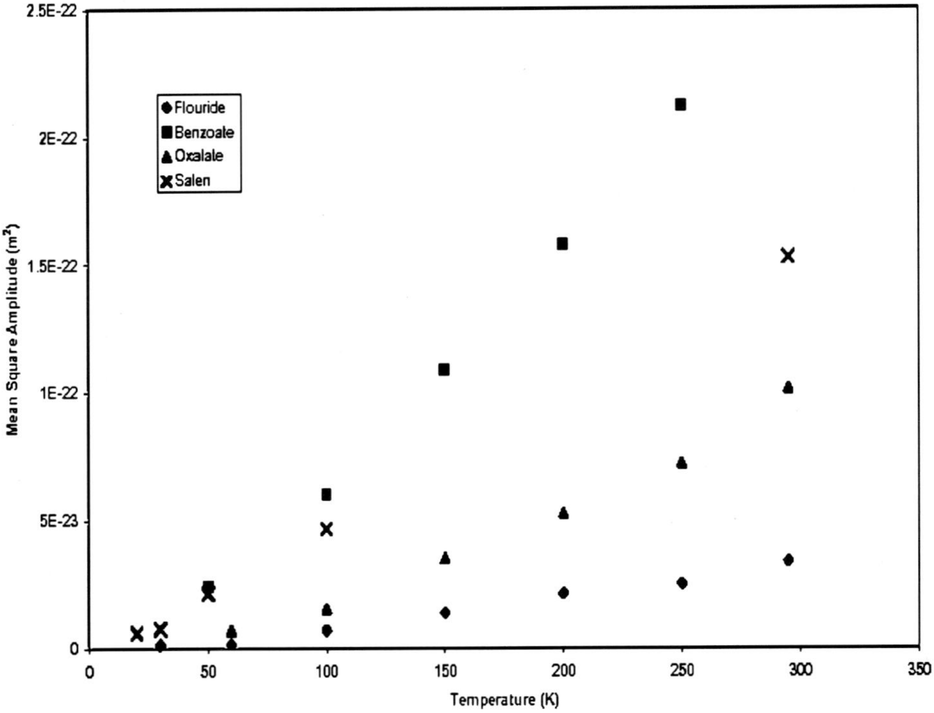

Figure 3 Mean square amplitude vs. temperature.

highest θ_D of 283 K and the least stable or most covalent chelate, the benzoate, recorded 105 K. One can contribute this high covalency as due to the bulk of the benzoate bidentate ligand unabling to bond to the Eu^{3+} ion as firmly as the less bulky oxalate chelate or the more highly symmetrical salen group. The strong ionic Eu-F bonds yielded the smallest mean square amplitude $<x^2>$ at any given temperature, for example 100 K for these four compounds and was greatest for the bulky benzoate, most unstable of these compounds $<x^2>$ was 6.04×10^{-23} m^2.

Recall that the recoil-free fraction *f* is given by the total area under the resonance curve. It has been shown that the recoil-free (f) is related to the mean square amplitude (x^2) by $f = \exp(-k^2<x^2>)$ where k is the wave vector of the gamma radiation.

$$\lambda = hc/E_\gamma$$

and $<x^2>$ is the mean square amplitude of vibration parallel to the gamma ray propagation axis. Figure 3 shows a plot of the mean square amplitude versus temperature for europium fluoride, oxalate, salen adduct and the benzoate. The graph shows a positive slope for $<x^2>$ versus temperature reflecting an increase in the vibration amplitude with increasing temperature. Moreover, for a given temperature $<x^2>$ is least for the fluoride, followed by the oxalate, then the salen complex and the benzoate.

Acknowledgement The authors would like to acknowledge the assistance of William Jackson in the preparation of this manuscript.

References

1. Wynter, C.I., Ryan, D.H., May, L., Oliver, F.W., Brown, E., Hoffman, E.J., Bernstein, D., Gracia, M., Marco, J.F., Plazaola, F. (eds.): Industrial Applications of the Mössbauer Effect, pp. 300–304. American Institute of Physics, New York (2005)
2. Wynter, C.I., Oliver, F.W., Hill, D., Spijkerman, J.J., Bartlett, Y.P.: Radiochim. Acta **55**, 111–112 (1991)
3. Samuel, E.A., Delgass, W.N.: In: Gruverman, I.J., Seidel, C.W. (eds.) Moessbauer Effect Methedology, vol. 10, p. 261. Plenum, New York (1976)
4. Brix, P., Hufner, S., Kienle, P., Quitmann, D.: Phys. Lett. **13**(2), 140 (1964)

Hyperfine Interact (2005) 166:505–509
DOI 10.1007/s10751-006-9317-2

Impact related changes in mineral properties: Ti-magnetite from basaltic rocks at Lonar crater, India

C. Bender Koch · E. A. Ferrow

Published online: 13 December 2006

Abstract Samples from vesicular Lonar impactite have been investigated by Mössbauer spectroscopy and SEM in combination with elemental analysis. Observation reveal the new-formation of oxide minerals have occurred with nucleation in particular at the walls of the vesicles interpreted as having formed from condensation of the gas phase having caused the vesicles to form in the first place. Elemental analysis of Ti-magnetites gives very similar compositions irrespective of the oxides grains being vesicle or matrix associated (Fe/Ti ratios approximately 3). The room temperature spectra of the oxides show magnetically ordered components due to A, B, and C sites in the substituted spinel. The occurrence of these ordered components are not in accordance with expectations from the bulk chemical compositions, suggesting the presence of fine scaled intergrowth of Ti-oxide and Ti-magnetite.

Key words Ti-magnetite · impactite · Lonar crater · vesicles · gas condensation growth

1 Introduction

Impact cratering has over the last decades been established as an important process affecting planetary surfaces [1]. The huge amount of energy potentially being released at the impact certainly holds the potential for both instantaneous changes as well as long-term local hydrothermal changes in the mineralogy.

The stratigraphy of impact craters can be very complex, but often includes an empirical (impact) class description as basis for establishing "fallout" and "throwout"

C. Bender Koch (✉)
Department of Natural Sciences, KVL The Royal Veterinary and Agricultural University, Thorvaldsensvej 40, 1871 Frederiksberg C, Denmark
e-mail: cbk@kvl.dk

E. A. Ferrow
GeoBiosphere Center, Sölvegatan 12, 223 62 Lund, Sweden

units, but essentially unmodified grains may occur adjacent to glass flows pointing to complex energy and matter transfer in the impact induced reactions.

Kieffer et al. [2] conducted a thin section study of samples from cores though the impact units and compared those with experimentally shocked samples and estimated peak pressures around 1,000 kbar. The petrography investigations allowed them to differentiate five different impact classes with glass formation of different minerals, including Ti-oxides, as the most extremely influence.

Lonar crater (and its minor associated crater) remain the only know terrestrial crater in basalt. It consequently servers as an important analogue of craters formed at the surface of other planetary bodies with basaltic composition with long term hydrothermal changes [3].

2 Experimental

The impactite used in this study originates from Lonar Crater in India. It has been collected at the ground surface, but is not labeled with a specific sampling locality. The impactite is approximately 8×5×3 cm in size and distribution and size of the vesicles vary considerably on a cm length scale.

The morphology of Lonar impactite was investigated by SEM using Pt-coated fragments. Quantitative elemental analyses of grains were obtained from microprobe analysis of carbon-coated thin sections. Five fragments (between 80 and 100 mg) were sampled from the interior, powdered and Mössbauer spectra were measured of each at room temperature in transmission geometry, using a constant acceleration spectrometer, that had been calibrated using a thin foil of natural iron at room temperature. The spectra have been fitted using a combination of doublets and sextets having Lorentzian line shape. Attempts to enrich the magnetic minerals using normal magnetic fractionation techniques of powders was not successful. Apparently the silicate and oxides grains remain intimately mixed also after crushing of the rock.

3 Results and discussion

Impact induced processes have transformed the essential non-porous basalt into low-weight vesicular rock with a distribution of vesicle sizes (Figure 1). The vesicles are frequently interconnected by fine holes in the wall of the vesicles. These holes are interpreted as formed from mechanical breakdown of thin parts of the wall due to differential pressure of gas within the neighboring vesicles. Similar features can be observed in vesicles in impactites of granitic composition. However, unlike vesicles from granitic impactites, the wall of the vesicles in Lonar impactite frequently are covered with new formed crystallites. The crystallites have nucleated at multiple sites on the vesicle surface and exhibit preferential growth along the vesicle walls. Fracturing of these new formed minerals can often be observed and are interpreted as release of thermal strain during cooling. In addition to crystals having well developed surfaces, a later period of growth is indicated by the occurrences of delicate structures growing erroneously into the vesicle voids and having nucleated from the surface of the new crystals. These features are tentatively interpreted as an immiscible residue.

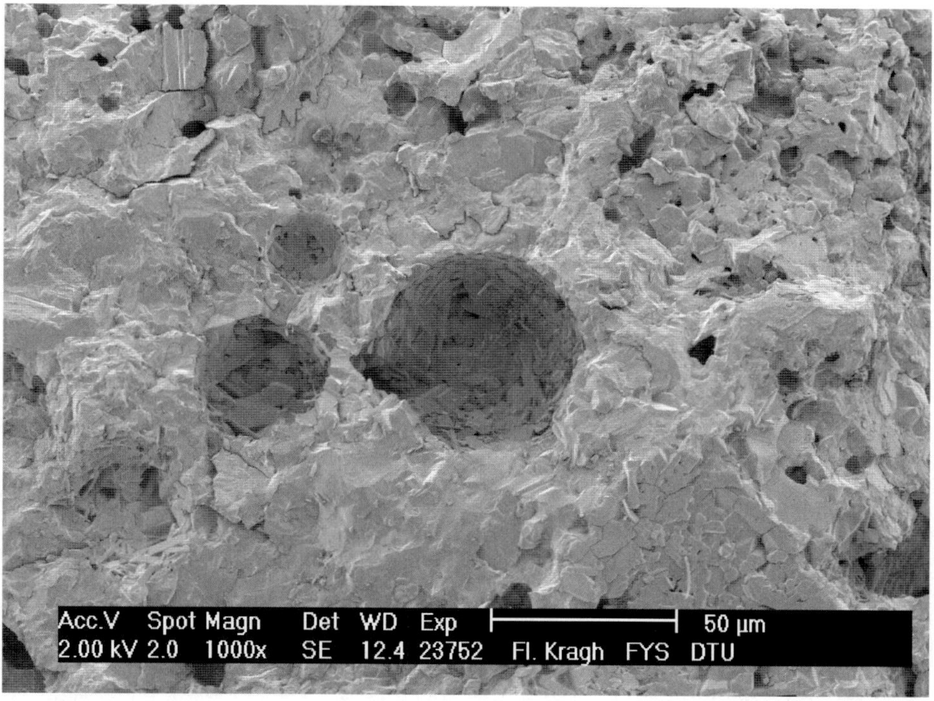

Figure 1 Micrograph of a typical fragment of Lonar impactite. Vesicles of different sizes is a dominant feature in this rock. The inner parts of the vesicles show newly formed crystals (with secondary cooling cracks) and frequently also highly delicate structures are present (immiscible residues?).

Micrographs and analysis of a thin section of the impactite (Figure 2) reveal the presence of silicate and Fe-Ti-oxide minerals. Normal basalts are characterized by a rather even distribution of oxides within the silicate matrix. Such scattered occurrences of oxides are also observed in the Lonar impactite. However, a particularly high concentration of oxides grains are associated with the vesicles. This cannot be explained by a simple redistribution of oxides grains controlled by vesicle formation. Rather, since morphology and elemental analysis reveal that oxides clearly have been growing into the vesicles at least this part of the oxides have been new formed as a result of the impactite. Large scale formation of vesicles in basalts is problematic because of the very low initial porosity and the small amount of volatiles (structural or dissolved) in the minerals. At very high external pressures these can hardly alone account for the observed porosity. However, the new formation of oxides may indicate that the vesicle filling phases after the impact also may have included components of oxides gases. Thus the oxides associated with the vesicle wall have formed by condensation growth in the vacuoles at relative high temperatures and later cooling have induced the cracks in the oxides. Such a redistribution process would explain the observed association of vacuoles and oxides.

Microprobe analysis do not reveal any significant difference in composition among the oxide group distributed in the matrix and associated with the vesicles. Analysis give Fe/Ti ratios between 2.93 and 3.12 corresponding to an ulvöspinel formula with $x = 0.75$.

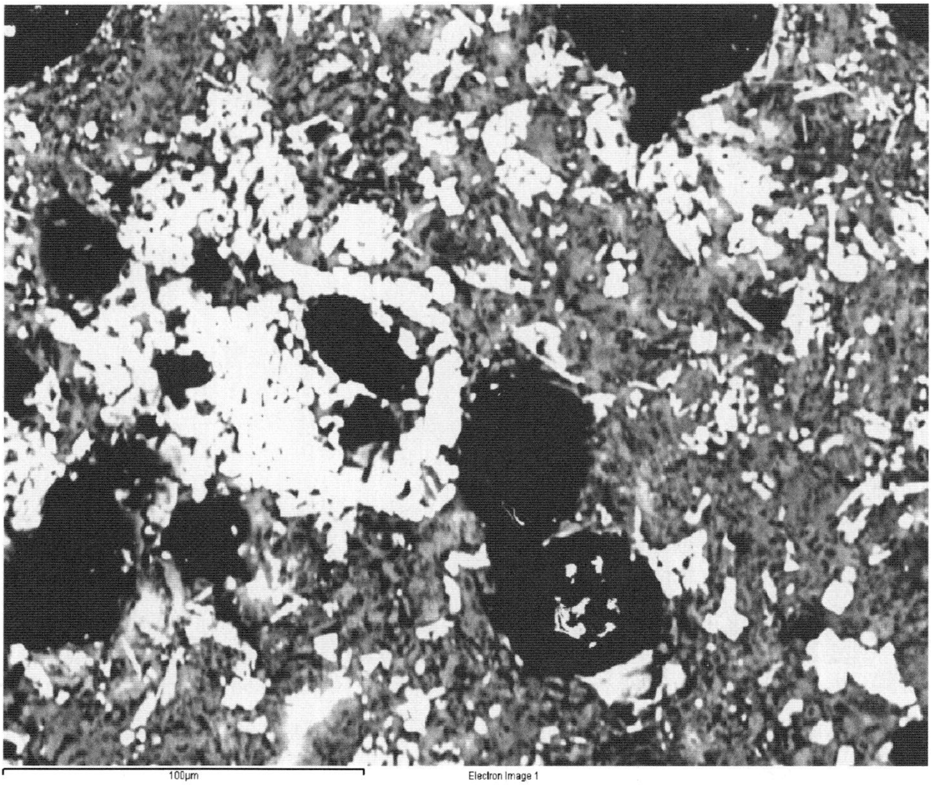

Figure 2 Backscatter electron image of a thin section of Lonar impactite showing Ti-magnetite (*white*) and silicates (*grey colors*). Ti-magnetite is frequently found in aggregates associated with vacuoles implying new formation from the vacuole filling material.

Unlike normal basalts we observed a significant sampling variation (Figure 3). The doublet components are assigned to Fe in silicate minerals and are very similar in all the samples. However, in particular the oxide components in the spectra varied. This may be explained as due to variation in vesicle density among the different samples causing variation in vesicles-associated oxide. The magnetically ordered components can be fitted with two sextets, respectively, assigned to the spinel A site (48.6 T) and the spinel B and C sites of the Ti-substituted structure (44.7 T) that cannot be resolved in the spectra. In comparison to other spinels of similar formal composition the magnetic ordering observed at room temperature is fairly extensive suggesting that the average formulas derived from microprobe analysis may be caused by a very fine scale mixing of Ti-oxide into Fe-Ti-oxide [cp. 4].

The Ti-magnetite in the sample of Lonar impactite have experienced an energy input much higher than samples studied by Kieffer et al. [2]. and establish condensation as an important process for Ti-oxides in high energy impacts.

The new formation of oxides as reported here, caused by the impact process, is in line with the reported occurrence of a magnetic component formed under conditions similar to the present earth magnetic field [5]. Studies of the ultra structure of these oxides are currently being undertaken.

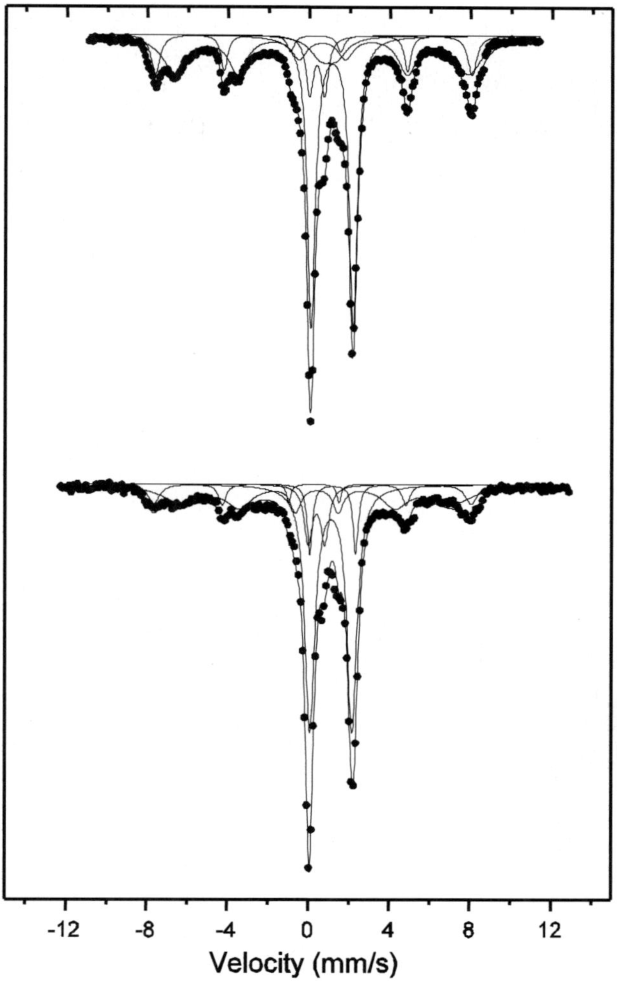

Figure 3 Mössbauer spectra of Lonar impactite samples measured at room temperature. The two spectra show the variation in magnetically ordered components in the spectra of five different samples. The components of the fit and the sum of the components are shown as continuous lines. For explanations: see text.

References

1. Gerasimov, M.V., Ivanov, B.A., Yakovlev, O.I., Dikov, Y.P.: Earth Moon Planets **80**, 209 (1998)
2. Kieffer, S.W., Schaal, R.B., Gibbons, R., Hörz, F., Milton, D.J., Dube, A.: Proc. Lunar Planet Sci. Conf. **7**, 1391 (1976)
3. Hagerty, J.J., Newsom, H.E.: Meteorit. Planet. Sci. **38**, 365 (2003)
4. Hamdeh, H.H., Bargout, K., Ho, J.C., Shand, P.M., Miller, L.L.: J. Magn. Magn. Mater. **191**, 72 (1999)
5. Rao, G.V.S.P., Bhalla, M.S.: Geophys. J. R. Astron. Soc. **77**, 847 (1984)

Hyperfine Interact (2005) 166:511–516
DOI 10.1007/s10751-006-9318-1

On the merits of conversion electron Mössbauer spectroscopy in geosciences

H. P. Gunnlaugsson · P. Bertelsen · C. Butz-Jørgensen · G. Klingelhöfer · I. Kuvvetli · M. B. Madsen · J. P. Merrison · P. Nørnberg · H. Rasmussen · G. Weyer

Published online: 3 November 2006

Abstract Described are some applications of conversion electron Mössbauer spectroscopy (CEMS) in geosciences. It is shown how easily this technique can be applied in existing Mössbauer laboratories to investigate natural samples. Some examples demonstrate the kind of information CEMS can give on the weathering history or mechanical properties of samples. The possible application of CEMS on Mars is discussed.

Key words CEMS · weathering · mechanical properties · Mars

1 Introduction

Conversion Electron Mössbauer Spectroscopy (CEMS) is sensitive to a thin surface layer of samples, in contrast to traditional transmission Mössbauer spectroscopy (TMS) or backscattering Mössbauer spectroscopy (b-MS). Depending upon the composition and density of the sample, CEMS probes a depth of ~0.25 μm from the

H. P. Gunnlaugsson (✉) · J. P. Merrison · G. Weyer
Institute of Physics and Astronomy, Aarhus University, Ny Munkegade, DK-8000 Århus C, Denmark
e-mail: hpg@phys.au.dk

P. Bertelsen · M. B. Madsen
Center for Planetary Science, Juliane Maries Vej 30, 2100 Copenhagen Ø, Denmark

C. Butz-Jørgensen · I. Kuvvetli
The Danish National Space Center, Juliane Maries Vej 30, 2100 Copenhagen Ø, Denmark

G. Klingelhöfer
Institute of Inorganic and Analytical Chemistry, Johannes Gutenberg-University Mainz, Staudinger Weg 9, 55128 Mainz, Germany

P. Nørnberg · H. Rasmussen
Department of Geosciences, Aarhus University, Ny Munkegade, 8000 Århus C, Denmark

surface, while information from ~150 or ~75 μm depths is obtained using, respectively, 14.4 keV γ-rays or 6.4 keV X-rays in b-MS. Typical absorbers for TMS are roughly ~500 μm thick. A comparison between spectra obtained with different methods can thus give information on, e.g., depth dependent oxidation processes that have affected a sample.

While CEMS is conventionally applied in solid state physics, its use in geosciences has not been thoroughly documented, partly due to technical limitations when working with natural materials. Employing gas avalanche detectors of the parallel plate type [1, 2] with small modifications in optimal plate distance and detector gas pressure, these problems can be overcome and CEMS can be applied without difficulty to natural systems. While the detectors operate optimally with plate distances of ~2.5 mm 25 mbar Acetone gas (a few litre in a reservoir) and ~1 kV high voltage, inhomogeneous samples can easily be measured by increasing the plate distance, decreasing the gas pressure (down to 15 mbar) and still maintaining ~1 kV high voltage. For safe operation, the pulse shape has to be monitored, making sure to operate below the Paschen breakdown of the gas mixture.

In this contribution, we illustrate by examples how the technique can give new information about geological systems, which is complementary to current knowledge obtained by other (MS) techniques. This then adds CEMS to the list of promising techniques in the field of geology.

2 Examples of the usage of CEMS

It would be beyond the scope of this paper to give a thorough overview over the possibilities of applying CEMS in geology, and we choose to concentrate on two typical examples that illustrate the usefulness of the method.

2.1 Weathering vs. transport of basaltic sand

Two samples of beach sand derived from basalt were investigated, from Patea, New Zealand and Grotta, Iceland. In the former case, the sand is found roughly 20 km away from the source region due to transportation along the beach. In the latter case, the sand is formed by active breakdown of the rocks at the site. The Mössbauer spectra are shown in Figure 1.

The spectra have been analysed in terms of quadrupole-split doublets originating from paramagnetic Fe(II) in pyroxenes ((Fe, Mg, Ca,...)SiO_3) and olivine ($(Fe,Mg)_2SiO_4$), and mineralogically unspecific Fe(III) compound(s). Additionally, the spectra have been analysed with magnetically split sextets of lines from the iron oxide titanomagnetite ($Fe_{3-x}Ti_xO_4$), using the data from [3] to simulate the effect of titanium on the spectrum and to evaluate the value of x. Finally, a simple sextet, assigned to ferric iron oxide, was applied.

The Patea sample contains unusually high amounts of iron in magnetite, and this alone points to an origin via transport, where less hard silicate minerals have been knocked off from the relatively harder iron oxides. This is also illustrated in the measurement above, as the relative spectral area of the iron oxides increases in the surface layer from 67% in the TMS measurement to 76% in the CEMS

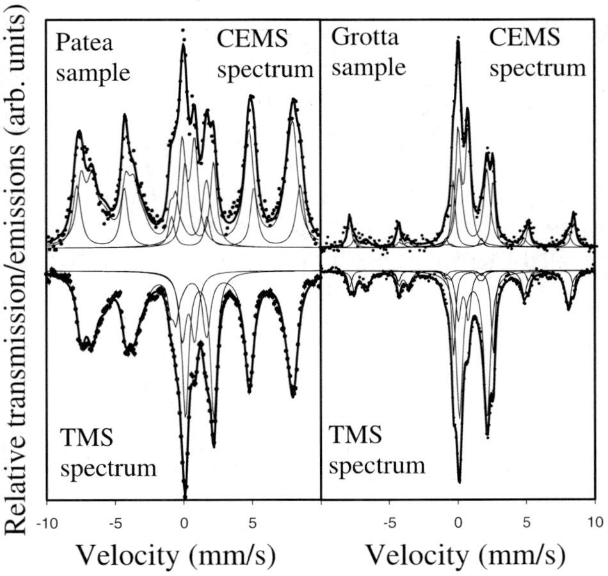

Figure 1 Room temperature CEMS and TMS spectra of beach sand samples from Patea and Grotta.

measurement. The quadrupole shift of the ferric iron oxide component, ΔE_Q (=2ε) = −0.06(3) mm/s, indicates a mixture of hematite (α-Fe_2O_3) and maghemite (γ-Fe_2O_3). The titanomagnetite is found to have a value of x = 0.17(2) in both the surface and the bulk, showing that the sample has not been subjected to a major thermal event (T > 400°C), which would have transformed the titanomagnetite towards more pure magnetite (smaller x) and titanomaghemite [4]. Only a moderate increase of the Fe(III) component is found in the surface, from 10% in the TMS measurement to 12% in the CEMS measurement.

The Grótta sample shows pronouncedly different spectra: Much smaller amounts of iron oxides are found in the CEMS measurement and the Fe(III) component increases by a factor of 3, from 11% in the TMS measurement to 33% in the CEMS measurement, clearly revealing the active oxidation of the sample.

These findings are not a surprise in view of the known origin of the samples, but illustrate nicely the powerfulness of CEMS in particular in combination with other MS techniques. The combination enables to distinguish features of rather complex spectra by the differences in spectral intensities as is also demonstrated in the example.

2.2 Oxidation of olivine basalt

During the solidification of olivine basalt, the mineral olivine is easily oxidized, forming pure magnetite and, depending on the conditions, enstatite (Mg_2SiO_4) [5]. The transformation can take place at both high temperatures (T ~ 1,000°C) and low temperatures (T < 500°C) [6, 7]. The microstructures of samples in these cases are found to differ significantly as illustrated in Figure 2.

The Stardalur sample comes from a drill core from the Stardalur, Iceland, magnetic anomaly [8]. The high remanence magnetisation of samples from the

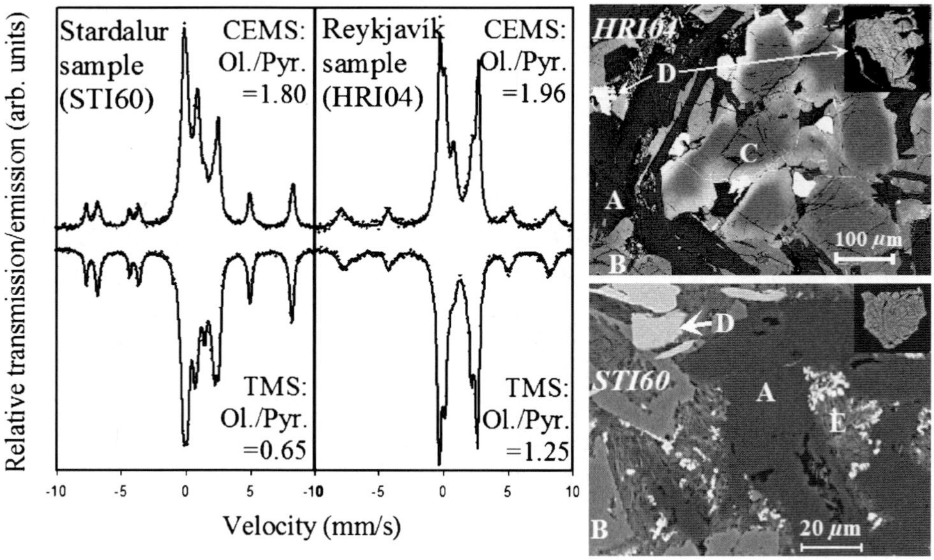

Figure 2 *Left*: Room temperature CEMS and TMS spectra of the samples indicated. *Right*: SEM backscatter images of the same samples. *Capital letters* indicate areas with (*A*) feldspars, (*B*) pyroxene, (*C*) olivine crystals, showing a compositional gradient, (*D*) Fe–Ti oxides, contrast enhanced in the *upper right corners* and (*E*) altered original olivine particles showing voided structures with pure iron oxides embedded.

anomaly (M_r = 61 A/m on average) has been explained by high temperature oxidation of the original olivine [5]. The HRI sample comes from olivine basalt from Reykjavík, Iceland. Based on the magnetic properties of HRI samples [9] the partial oxidation of olivine has most likely taken place at low temperatures. These interpretations are supported by comparison of the morphology of olivine crystals to the simulation studies by Haggerty and Baker [6, 7].

The SEM backscatter image of the Stardalur sample shows voided structures where the primary olivine used to be. Clearly, these represent weaknesses in the material. In case of the HRI sample, the olivine has not lost its strength and iron oxides are found as veins within the olivine host.

For the Mössbauer spectroscopy analysis, the samples were grinded into powder (~50 μm). The main difference between the CEMS and TMS spectra of the Stardalur and HRI samples is in the relative intensity of the olivine line (cf. Figure 2). Characterising this difference by the relative change in the area ratio between olivine and pyroxene by defining:

$$k = \frac{A_{CEMS}(Oliv.)/A_{CEMS}(Pyr.)}{A_{TMS}(Oliv.)/A_{TMS}(Pyr.)}$$

where $A_i(j)$ denotes the spectral area of component j in the type i measurement, we find a clear distinction of the samples. If the material breaks up in a homogeneous way, a value $k = 1$ is expected. In the case of samples where the oxidation of olivine has taken place at high temperatures k is typically much higher than unity, and can be as high as 3 depending on the microstructure (2.77 in case of the Stardalur sample

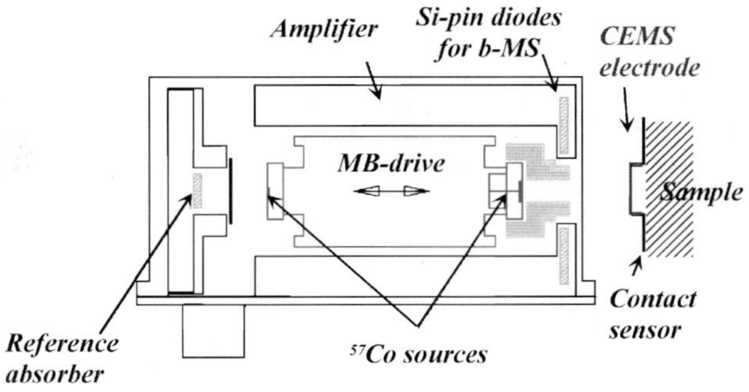

Figure 3 Illustration showing the possible mounting of a CEMS electrode onto the contact sensor of the MER Mössbauer spectrometers. The width of the instrument is roughly 4 cm (adapted from [10]).

above). In the case of low temperature oxidation, the olivine host is not found to have dissolved in the same manner, and only a moderate difference in the spectra is observed (here $k = 1.57$ for the HRI04 sample).

These findings are easily understood with comparison to the SEM images in Figure 2. During the grinding of the samples, these are more likely to break apart where there are weaknesses in the structure. As the weaknesses in the Stardalur sample are where the olivine is located, it will automatically become enriched in the surface on a submicron scale.

2.3 Application to Mars

Given the spectacular success of backscattering Mössbauer spectroscopy on the Mars Exploration Rovers (MER), the inclusion of the method in future robotic missions to Mars seems evident. The addition of a CEMS facility will be possible in a fairly simple way. This is due to the fact that both techniques to a large extent make use of the same hardware and electronics, and that the Martian atmosphere can be used as an electron counting gas.

Our present idea of how to construct a CEMS detector for operation on Mars is illustrated in Figure 3.

The detector uses the parallel plate technique, which is suitable on Mars due to the fact that dust accumulation onto the electrode will not hamper its performance. In the option illustrated in Figure 3, we have placed the CEMS electrode in the opening of the contact sensor. Dust accumulating on the electrode would have negligible effects on the b-MS spectra. We have obtained roughly 30% counting efficiency using simulated Martian atmosphere (7 mbar, CO_2 gas + traces of N_2, Ar, O_2) compared to ideal laboratory conditions. In this design, b-MS spectra and CEMS spectra will be obtained simultaneously.

Apart from the geological interest in the inclusion of the method on a Mars mission as illustrated above, the technique has the potential of extending the usefulness of Mössbauer spectroscopy on Mars by allowing measurements of extremely small samples. In order to get a reasonable measurement by means of CEMS, less than 100 μg of sample is needed. Such miniature samples could be gathered on a robotic mission by magnets or electrodes [11] and the sedimentation of samples onto clean surfaces would allow for a measurement few tenths of Sols (Mars days) into a mission.

3 Conclusions

The examples presented are far from being a exhaustive for the possible benefits of the application of CEMS in geosciences, but can hopefully further stimulate to consider the technique whenever surface information is potentially useful.

References

1. Weyer, G.: In: Gruverman, I.J., Seidel, C.W. (eds.) Mössbauer Effect Methodology, vol. 10, p. 301. Plenum, New York (1976)
2. Campbell, S.J., Klingelhöfer, G., Kaczmarek, W.A., Hofmann, M., Nagel, R., Wang, G.: Hyperfine Interact. **139/140**, 407 (2002)
3. Tanaka, H., Kono, M.: J. Geomagn. Geoelectr. **39**, 463 (1987)
4. Gunnlaugsson, H.P., Weyer, G., Helgason, Ö.: Planet. Space Sci. **50**, 157 (2002)
5. Gunnlaugsson, H.P., Helgason, Ö., Kristjánsson, L., Nørnberg, P., Rasmussen, H., SteinÞórsson, S., Weyer, G.: Phys. Earth Planet. Inter. **154**, 276 (2006)
6. Haggerty, S.E., Baker, I.: Contrib. Mineral. Petrol. **16**, 258 (1967)
7. Haggerty, S.E., Baker, I.: Contrib. Mineral. Petrol. **16**, 233 (1967)
8. Friðleifsson, I.B., Kristjansson, L.: Jökull **22**, 69 (1972)
9. Rasmussen, H., Gunnlaugsson, H.P., Tegner, C., Kristjánsson, L.: Hyperfine Interact. (in press, these proceedings). DOI 10.1007/s10751-006-9341-2 (2006)
10. Klingelhöfer, G., Morris, R.V., Bernhardt, B., Rodionov, D., de Souza Jr., P.A., Squyres, S.W., Foh, J., Kankeleit, E., Bonnes, U., Gellert, R., Schröder, C., Linkin, S., Evlanov, E., Zubkov, B., Prilutski, O.: J. Geophys. Res. **108**, 8067 (2003)
11. Merrison, J., Jensen, J., Kinch, K., Mugford, R., Nornberg, P.: Planet. Space Sci. **52**, 279 (2004)

Hyperfine Interact (2005) 166:517–522
DOI 10.1007/s10751-006-9319-0

Characterization of burned soil profiles by Mössbauer spectroscopy

A. L. Vendelboe · H. P. Gunnlaugsson · Ö. Helgason · P. Nørnberg

Published online: 2 November 2006

Abstract The characteristics of both natural burned and experimentally burned soil profiles as seen by Mössbauer spectroscopy are described. The results help in the identification of profiles where biological and/or chemical processes may have been involved.

Key words soil burning · Mössbauer spectroscopy · hematite · goethite · maghemite

1 Introduction

The rare occurrence of highly magnetic soil ($\sigma_S > 1$ Am^2/kg) in Denmark is normally explained by the hypothesis that the site has been burned. However, at some sites formation by means of biological processes has been suggested [1,2]. It is of considerable interest to determine the characteristics of profiles of burned soil, both in order to recognise these sites, and in order to understand the associated mineralogical alterations and formation of magnetic iron oxides. For this purpose, we have performed a detailed study of properties of burned sites using Mössbauer spectroscopy and X-ray diffraction to determine the Fe(III) mineralogy together with magnetic methods. These have been applied to experimentally burned soil and

A. L. Vendelboe · P. Nørnberg
Department of Earth Sciences, Aarhus University, Ny Munkegade,
DK-8000 Aarhus C, Denmark

H. P. Gunnlaugsson
Institute of Physics and Astronomy, Aarhus University, Ny Munkegade,
DK-8000 Aarhus C, Denmark

Ö. Helgason (✉)
Science Institute, University of Iceland, IS-107 Reykjavik, Iceland
e-mail: ornh@raunvis.hi.is

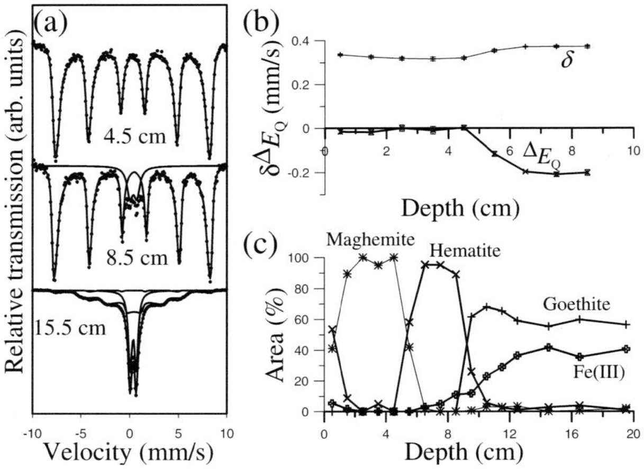

Figure 1 **a** Mössbauer spectra from the experimentally burned profile. The 4.5 cm sample originates from the maghemite layer, the 8.5 cm sample from the hematite layer and the 15.5 cm sample from the unaltered goethite layer. **b** The values of isomer shift (δ) and quadrupole shift, ΔE_Q (=2ε) of the MHFD assigned to the mixture of maghemite and hematite. The quadrupole shift was used to distinguish between hematite and maghemite in **c** assuming $\Delta E_Q = 0.0$ mm/s and $\Delta E_Q = -0.2$ mm/s for maghemite and hematite, respectively.

a natural burned site. In the case of the experimentally burned soil, it was possible to monitor the temperature as a function of depth during the burn, and these data can be used to extract the temperature dependent mineralogical changes. At the naturally burned site, later movement of soil mask some of the properties that are observed [3].

2 Samples and experimental

All analyses were carried out on the <63 μm soil fraction. The natural samples were dry sieved through a 63 μm sieve for 15 min. As a starting material for the experimentally burned profile an iron oxide rich soil (~50 wt.% Fe) from Salten Skov II was used which is dominated by the presence of goethite [1]. The soil was dry sieved at 2 mm and homogenised before the burning experiment. After the experiment the samples were sieved at 63 μm and the <63 μm fraction was used in the analysis. The Mössbauer spectra were recorded in transmission mode at room temperature using 10–50 mCi ^{57}Co:Rh sources. Isomer shifts and velocities are given relative to the centre of the spectrum of α-Fe at room temperature. X-ray diffraction (XRD) patterns were analysed with the Rietveld method. Magnetic hysteresis properties were measured using a vibrating sample magnetometer.

Before the experimental burn, two parallel <2 mm samples from Salten Skov II were placed vertically in the ground in 30 cm long tile drain tubes. The tubes were in line with the soil top. Temperatures were measured at four levels. The maximum temperatures reached were 798 K at 3.75 cm depth, 598 K at 7.5 cm depth, 366 K at 15 cm depth and 321 K at 30 cm depth. The total time of the burn was roughly 24 h.

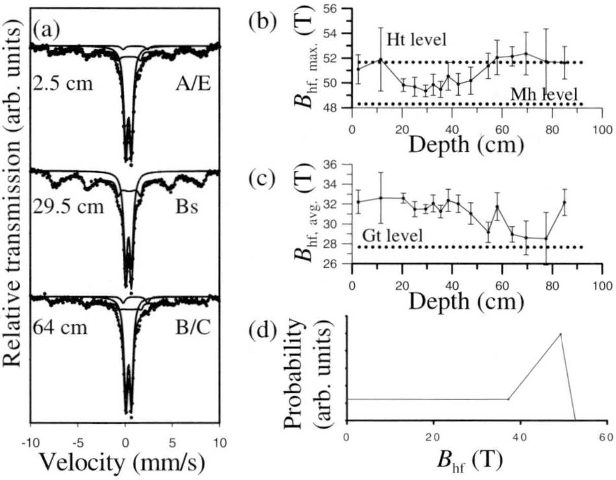

Figure 2 **a** Room temperature Mössbauer spectra from the natural burned site. The *letters* are the horizon designates [9]. **b** The $B_{hf,\ max.}$ value of the natural burned site, with the estimated value of the hematite and maghemite levels shown. **c** The $B_{hf,\ avg.}$ value of the natural burned site, with the estimated value of the goethite level shown. **d** The MHFD obtained from the analysis of the sample from 29.5 cm sample depth.

3 Results and discussion

The Mössbauer parameters, and in particular the magnetic hyperfine field of the pure iron oxides are well known (see e.g. [4]), but both small particle phenomena [5] and cation substitution (generally Al^{3+}, see e.g. [6]) lead to lowering of the magnetic hyperfine field, and an asymmetric line shape, best simulated with a magnetic hyperfine field distribution (MHFD).

The Mössbauer spectra of samples from the experimentally burned profile were analysed in terms of 2–3 spectral components (cf. Figure 1a). Spectra of samples below 10 cm are dominated by a MHFD with an average magnetic hyperfine field close to 28 T assigned to goethite (Gt). In the analysis this component was simulated using a three segment linear distribution in the magnetic hyperfine field [7]. In these samples a small component with magnetic hyperfine field close to ~50 T was analysed with a two linear segment MHFD and assigned to a mixture of maghemite and hematite (Mh/Ht). Additionally, the spectra were analysed with simple Lorentzian doublet (Fe(III)) assigned to paramagnetic and/or superparamagnetic iron oxides. Above 10 cm, the goethite has totally disappeared from the samples due to dehydration, and the spectra are dominated by the Mh/Ht MHFD simulated with a four segment distribution together with a minor contribution from the Fe(III) component. A third component, a simple Lorentzian sextet, was used in the upper most sample of the experimentally burned profile and assigned to hematite.

In the spectra of samples from the natural site, (cf. Figure 2a), the Fe(III) component dominates and goethite could not be identified unambiguously, here a single three segment MHFD was used. In all cases the distribution obtained had a peak around 50 T, and a relatively high intensity in the interval from 0 to ~40 T (cf. Figure 2d). A second Lorentzian doublet with isomer shift ~1.1 mm/s and quadrupole splitting ~2.1 mm/s was assigned to the ferrous iron originating from silicates.

The experimentally burned profile showed four clear layers. Based on XRD results (not presented here), the dominating iron oxides where found to be: (a) (0–1 cm) mixture of maghemite and hematite, (b) (1–5 cm) maghemite, (c) (5–9 cm) hematite and (d) (below 10 cm) none-reacted goethite rich layer. Figure 1a shows typical Mössbauer spectra recorded from three of the four layers. The Mössbauer spectra obtained of the hematite rich and the maghemite rich layer are rather similar, but the values of quadrupole shift (Figure 1b) were used to distinguish between maghemite and hematite above 10 cm. The results are presented in Figure 1c, depicting the relative area fraction of each of the component. It is seen that maghemite is almost the sole magnetic iron oxide in the upper part of the profile, and inversion to hematite is only seen in the top sample. At about 5 cm ΔE_{Q} shifts to a value indicating that hematite is now the sole iron oxide. Below 8.5 cm the errors of δ and ΔE_{Q} are too large and the values of these are not shown in Figure 1b. Figure 1c shows that the parent material has not changed substantially below this level. The spectral area of the doublet component, increases with depth from 6.5 cm indicating that ferrihydrite transforms into better crystalline species or that larger particles are formed from small superparamagnetic particles. These transitions between the different phases are related to the temperature profile, which is not shown here. These findings are in accordance with numerous other studies (see e.g. [8]), that goethite is dehydroxylated to hematite at about 550 K and to maghemite in the upper part of the profile where the temperature has reached about 800 K.

These findings are in good agreement with results from XRD and saturation magnetisation measurements (not presented here). The purity of the iron oxides allows for unambiguous phase identification in XRD and saturation magnetisation measurements are in particular sensitive to the presence of maghemite due to its high saturation magnetisation at room temperature ($\sigma_{\mathrm{S}} \cong 70$ Am^2/kg) compared to hematite ($\sigma_{\mathrm{S}} \cong 0.4$ Am^2/kg) and goethite ($\sigma_{\mathrm{S}} \cong 0$ Am^2/kg).

Three spectra recorded from samples collected at the natural burned site are shown in Figure 2a. Goethite, hematite and maghemite were represented by one MHFD and the ratios between these were estimated comparing the magnetic hyperfine field ($B_{\mathrm{hf,\ max.}}$) where the magnetic hyperfine field distribution peaks at around 50 T, and the average magnetic hyperfine field ($B_{\mathrm{hf,\ avg.}}$). As the hyperfine field of maghemite is slightly lower than that of hematite, the $B_{\mathrm{hf,\ max.}}$ value gives an estimate of the ratio between maghemite and hematite. As goethite has average magnetic hyperfine field well below the peak originating from hematite and maghemite, the $B_{\mathrm{hf,\ avg.}}$ estimates the ratio between (maghemite+hematite) and goethite.

The weight percentages of the iron oxides have been estimated from the XRD analysis. The relative amount of iron oxides are found to coincide with the results from Mössbauer spectroscopy with the levels of average hyperfine splittings of hematite, maghemite and goethite indicated in Figure 2b and c. From the value of $B_{\mathrm{hf,\ max.}}$ it can be seen that hematite dominates the Ht/Mh component in the upper part of the profile, which is consistent with the XRD analysis. From about 25 cm the maghemite content in the profile rises, resulting in a lowering of the maximum hyperfine field. The value of maximum hyperfine field rises again at about 60 cm indicating, although the error has increased due to the low content of iron oxides, that hematite again dominates the Ht/Mh component. This is consistent with the findings from the magnetic hysteresis properties of the samples, where the value of σ_S peaks from about 25 cm to about 55 cm corresponding to an increase in the maghemite-content. Similarly, the value of $B_{\text{hf, avg.}}$ shows that goethite dominate the iron oxides in the lower part of the profile, from about 50–60 cm, which can be seen from both the low average hyperfine field and the XRD-analysis.

The interpretation of the XRD-patterns in the natural burned site is made difficult, because of the low content of iron oxides (about 20 wt.% in the most iron-rich samples), and due to the high number of minerals in the samples (quartz, micas and silicates), making it difficult to identify the individual peaks and the width of these. The results from the Mössbauer analysis help in evaluating the amounts of the iron oxides with respect to each other, together with the determination of total free iron, extracted by dithionite–citrate–bicarbonate.

4 Conclusions

In the experimentally burned profile, four clearly distinct layers containing almost pure iron oxides could be identified. (1) A mixture of well crystalline hematite and maghemite, (2) a layer of maghemite, (3) a layer of hematite and (4) a layer dominated in goethite, like the unchanged parent material. The natural burned site shows some of the same characteristics as the experimentally burned site. A layer containing a mixture of maghemite and hematite has been identified at the top of the profile, followed by a layer with an increased content of maghemite. Below the maghemite containing layer some hematite is still present although in minor amounts. Goethite is present throughout the profile, probably due to continuous formation from ferrihydrite and Fe(II)-containing drain waters, or due to incomplete transformation. The result of this investigation suggests that burned soil profiles can be identified by the sequence above, even though the profile to some extent have been changed through soil forming processes.

Acknowledgement Ass. Prof. Mikkel Fougt Hansen, Technical University of Denmark is acknowledged for experimental help.

References

1. Nørnberg, P., Schwertmann, U., Stanjek, H., Andersen, T., Gunnlaugsson, H.P.: Clay Miner. **39**, 85 (2004)
2. Gunnlaugsson, J., Merrison, P., Mossin, L.A., Nørnberg, P., Sanden, J., Uggerhøj, E., Weyer, G.: Hyperfine Interact. **144/145**, 365 (2002)
3. Borggaard, O.K.: Soil Chemistry, 6th ed. Royal Veterinary and Agricultural University, Copenhagen, Denmark (2002)
4. Greenwood, N.N., Gibb, T.C.: Mössbauer Spectroscopy. Chapman & Hall, London (1971)
5. Mørup, S., Dumesic, J., Topsoe, H.: In: Cohen, R.L. (ed.) Application of Mössbauer Spectroscopy, vol. II, pp. 1–55, Academic, New York (1980)

6. Cornell, R.M., Schwertmann, U.: The Iron Oxides. Wiley, Weinheim, Germany (2003)
7. Gunnlaugsson, H.P.: Hyperfine Interact. (2006) (in press)
8. Schwertmann, U.: In: Stucki, J.W., Goodman, B.A., Schwertmann, U. (eds.) NATO Advanced Study Institute on Iron in Soils and Clay Minerals. Series C: Mathematical and Physical Sciences Col. 217, pp. 267–308. D. Reidel Publishing Comp., Netherlands (1988)
9. Bridges, E.M.: World Soils, 3rd ed. Cambridge University Press, Cambridge, UK (1997)

Hyperfine Interact (2005) 166:523–527
DOI 10.1007/s10751-006-9320-7

Backscattering Mössbauer spectroscopy of Martian dust

P. Bertelsen · M. B. Madsen · C. S. Binau · W. Goetz · H. P. Gunnlaugsson · S. F. Hviid · K. M. Kinch · G. Klingelhöfer · K. Leer · D. E. Madsen · J. Merrison · M. Olsen · S. W. Squyres · and the Athena Science Team

Published online: 27 October 2006

Abstract We report on the determination of the mineralogy of the atmospherically suspended Martian dust particles using backscattering ^{57}Fe Mössbauer spectroscopy on dust accumulated onto the magnets onboard the Mars Exploration Rovers. The spectra can be interpreted in terms of minerals of igneous origin, and shows only limited, if any, amounts of secondary minerals that may have formed in the presence of liquid water. These findings suggest that the dust has formed in a dry environment over long time in the history of the planet.

Key words Mars · dust · ^{57}Fe Mössbauer spectroscopy

1 Introduction

The Viking landers and the Pathfinder lander carried a number of instruments to investigate the magnetic properties of the soil and dust on Mars. These instruments

P. Bertelsen · M. B. Madsen (✉) · C. S. Binau · K. Leer · D. E. Madsen · M. Olsen
Niels Bohr Institute, University of Copenhagen, Juliane Maries Vej 30
2100 Copenhagen Oe, Denmark
e-mail: mbmadsen@fys.ku.dk

W. Goetz · S. F. Hviid
Max Planck Institut für Sonnensystemforschung, Katlenburg-Lindau, Germany

H. P. Gunnlaugsson · K. M. Kinch · J. Merrison
Institute for Physics and Astronomy, Aarhus University, Aarhus, Denmark

G. Klingelhöfer
Johannes Gutenberg-Universität Mainz, Mainz, Germany

S. W. Squyres
Department of Astronomy, Cornell University, Ithaca, NY, USA

consisted of small permanent magnets that accumulated magnetic dust particles suspended in the Martian atmosphere. Based on the magnetic and optical properties of the accumulated dust, the magnetic phase was tentatively identified as maghemite, γ-Fe_2O_3, present as cement in or stain on silicate agglomerates. However, it was not possible to exclude that the magnetic phase could be titanomagnetite ($Fe_{3-x}Ti_xO_4$) or titanomaghemite (γ-$Fe_{2-x}Ti_xO_3$) [1, 2] having been inherited directly from the bedrock (see e.g. discussion in [1] and references therein).

Revealing the properties and the process of formation of this magnetic material is of considerable interest and may give important clues in answering questions about the role of water in the soil forming processes on Mars [1]: Has the formation of the red soil on Mars taken place predominantly via oxidation of Fe(II) in aqueous solution eons ago, when Mars was warmer and wetter? Or is the soil forming process an ongoing process of breakdown and oxidation of surface rocks – possibly without an excessive role of water?

To contribute to the answer of this question the Mars Exploration Rovers (MER) were equipped with different permanent magnets that would attract and accumulate magnetic dust suspended in the atmosphere [3] for investigation with Mössbauer spectroscopy and with the APX-spectrometer for elemental composition.

Backscattering Mössbauer spectroscopy (b-MS) has successfully been applied on the MER mission, where various minerals in rocks and soils have been identified [4, 5]. The Mössbauer spectra of soils have been found to resemble relatively unweathered basaltic sand, which seems in variance with the characteristic reddish colour of the soils. A possible explanation for this is that the relative large sand particles – say ~ 50 μm – are mixed together with only few weight percentages of small sized dust particles (d ~ 3 μm). The dust particles give the soil its characteristic red colour, while having only minimal effect on the Mössbauer spectra. None of the instruments on the MER's are able to measure the dust component alone.

The magnets, however, accumulate the red dust suspended in the atmosphere that may then be the material that can give hints regarding the role of water in the surface processes. In this paper we shall describe the experiment, summarize the main findings based on Mössbauer spectroscopy and briefly discuss the consequences for the geology on Mars.

2 Experimental and samples

The right half of Figure 1a shows the position of the capture and filter magnets on the MERs. The rover is shown with the panoramic cameras (Pancam) on top of a mast and the robotic arm at the front. Figure 1a also shows the capture magnet and filter magnet, which are both located in front of the camera mast (left). During the mission the dust on the magnets has been studied using the panoramic camera (Pancam), the Mössbauer Spectrometer (MB), the Alpha Particle X-ray Spectrometer (APXS), and the Microscopic Imager (MI). Dust settles on a circular surface of high purity aluminum (45 mm in diameter), where the central part (25 mm in diameter) is magnetically active. The capture magnet was designed to

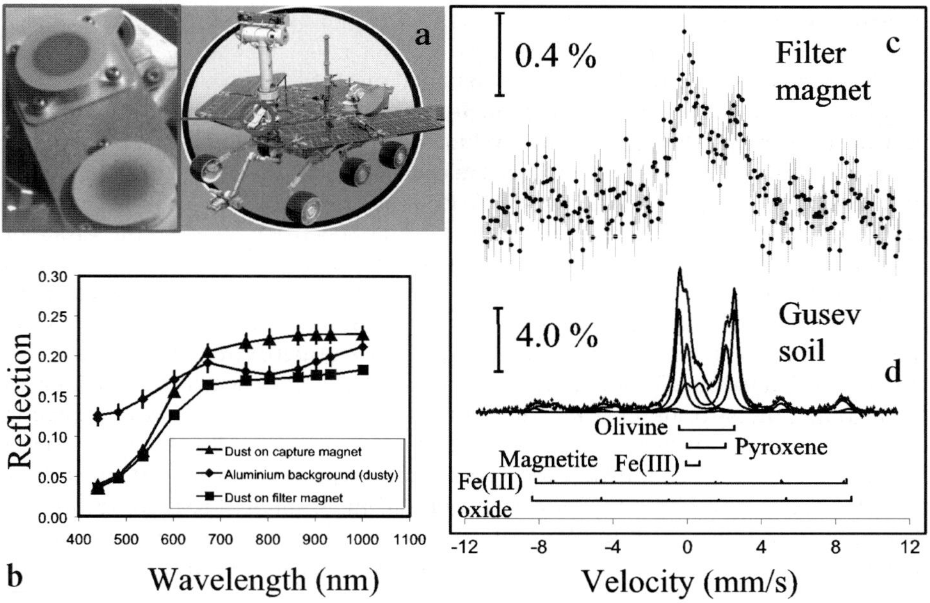

Figure 1 **a** The position of the capture and filter magnet on the Mars Exploration Rovers is indicated by the *red circle* on the rover. *Left* of this is shown a Pancam image of the capture magnet (*top*) and filter magnet (*bottom*) from the rover Spirit on sol 240. **b** Optical reflection spectra obtained of the targets indicated. **c** 14.4 keV Mössbauer spectrum obtained of the dust accumulated to the filter magnet on Spirit between sols 244 and 258 compared to a typical spectrum **d** of a sample of Gusev soil (Sol 74). The bar-diagram at the *bottom* shows the fitting components and their assignments.

be as strong as possible by maximizing the magnetic field B (0.46 T) and gradient ∇B (550 T m^{-1}). The strength of the filter magnet ($B = 0.20$ T and $\nabla B = 34$ T m^{-1}) was chosen such that it attracts mainly strongly magnetic dust particles (hence the term "filtering"). The left image, acquired 206 Sols (Mars days) after the landing of the Spirit rover, shows the reddish atmospheric dust, which has arrived on the two magnets.

Early in the missions, dust has accumulated onto the magnets at an almost constant rate as seen both optically and by the intensity of lines in APXS spectra. Later both rovers experienced dust clearing episodes, most probably due to strong winds [6]. Preliminary results [7] showed that the dust accumulating onto the filter magnet was darker than the dust accumulating to the capture magnet, pointing toward an igneous origin of the dust. The spectra of the dust on the two different magnets are shown in Figure 1b. The main part of the dust particles showed high magnetic susceptibility as seen in the pattern of dust on the so-called sweep magnet [7].

Mössbauer spectra of the material accumulating onto the magnets have been recorded several times during the missions. The temperature varies from 210 to 270 K during measurements. The analysis has been performed on spectra summed over all temperature bins. Calibration is done by the analysis of an internal instrument reference target as described in detail in [8].

3 Results and interpretation

Figure 1c shows a Mössbauer spectrum obtained on sols 244 through 258 of the dust attracted to the filter magnet on Spirit. In Figure 1d a spectrum of a typical soil (Spirit, sol 74) from the plains inside the Gusev crater is shown for comparison.

The spectrum of the dust on the filter magnet was recorded from sol 244 through sol 258 and over a total time interval of about 50 h. The dust layer is 1–2 mg/cm^2 thick while the sampling depth of the Mössbauer spectrometer is of the order of 50 mg/cm^2. The intensity of the spectrum is therefore low, especially for the magnetically ordered phases. From the striking similarity between the soil spectra published in [4, 5] and the dust spectrum, these spectra show that the doublets (paramagnetic Fe(II) in olivine ((Fe, Mg)$_2$[SiO$_4$]) and pyroxene (Fe, Mg, Ca, ...) [SiO$_3$], and Fe(III) in non-specified compounds) in the dust spectrum are similar to those found in spectra of soils of basaltic origin. In principle the magnetic components of the spectrum could be fitted almost equally well with two or three sextets. The scattered intensity in the velocity ranges around −8 and +9 mm/s show a broad scattering at −8 mm/s and narrower and more intense peak at +9 mm/s – this pattern is indicative of the Fe(II,III) nature in the mineral magnetite (Fe_3O_4). Magnetite is ubiquitous in the soil at Gusev and is therefore a likely candidate for this component. We have constrained the parameters of magnetite to those of the soil magnetite. The fit is improved by including a hematite component, which shows that the spectrum is consistent with the presence of hematite as a minor component of the dust. Based on the Mössbauer spectrum of the dust on the filter magnet, it is clear that the most probable candidate for the ferrimagnetic mineral in the airborne dust is magnetite. It is possible that the ferrimagnetic mineral is a somewhat oxidised variant of magnetite, be that non-stoichiometric magnetite or a solid solution between magnetite and maghemite. But maghemite alone (or maghemite and hematite) cannot be fitted to this spectrum with a reasonable result. It is also possible that the magnetite is isomorphically substituted with titanium. This element is enriched in the dust on the magnets compared to the soil [6]. Spectra obtained later in the missions have verified these interpretations [9].

It may be argued that the results obtained on the magnetically accumulated dust are not representative for the dust component. However, studies performed under simulated Martian conditions [10] verify that before dust clearing incidences, the dust accumulated onto the magnets is indeed representative of the true dust component. These studies suggest furthermore that the increased Fe(III) content of the Martian dust does not necessarily have to be due to interaction with liquid water.

We do not have access to analogue material of the Martian dust on Earth. Three micrometer particles of igneous composition are extremely unstable in the wet chemistry of our Earth, and in particular three micrometer particles of the mineral olivine would only last on the order of months in liquid water (estimate based on data from [11]).

4 Conclusions

The main results from Mössbauer spectroscopy on the MERs is that the dust collected on the magnets consists mostly of igneous minerals, and is not dominated by secondary iron oxide as we believed previously. The continuous breakdown of surface rocks generates soil with a broad particle size distribution. While the sand sized particles (>10 μm) form dunes and are not movable under current atmospheric conditions, the dust (<5 μm) can come into suspension and during dust storms is kept circulating planet wide. While this process has taken place, likely over geological time scales, water has not played any major role.

References

1. Knudsen, J.M., Madsen, M.B., Olsen, M., Vistisen, L., Koch, C.B., Mørup, S., Kankeleit, E., Klingelhöfer, G., Elanov, E.N., Khromov, V.N., Mukhin, L.M., Prilutski, O.F., Zubkov, B., Smirnov, G.V., Juchniewich, J.: Hyperfine Interact. **68**, 83 (1991)
2. Coey, J.M.D., Mørup, S., Madsen, M.B., Knudsen, J.M.: J. Geophys. Res. **95**, 14423 (1990)
3. Madsen, M.B., Bertelsen, P., Goetz, W., Binau, C.S., Olsen, M., Folkmann, F., Gunnlaugsson, H.P., Kinch, K.M., Knudsen, J.M., Merrison, J., Nørnberg, P., Squyres, S.W., Yen, A., Rademacher, J., Gorevan, S.: J. Geophys. Res. **108**, 8069 (2003)
4. Morris, R.V., Klingelhöfer, G., Bernhardt, B., Schröder, C., Rodionov, D.S., de Souza Jr., P.A., Yen, A., Gellert, R., Evlanov, E.N., Foh, J., Kankeleit, E., Gütlich, P., Ming, D.W., Renz, F., Wdowiak, T., Squyres, S.W., Arvidson, R.E.: Science **305**, 833 (2004)
5. Klingelhöfer, G., Morris, R.V., Bernhardt, B., Schröder, C., Rodionov, D.S., de Souza Jr., P.A., Yen, A., Gellert, R., Evlanov, E.N., Zubkov, B., Foh, J., Bonnes, U., Kankeleit, E., Gütlich, P., Ming, D.W., Renz, F., Wdowiak, T., Squyres, S.W., Arvidson, R.E.: Science **306**, 1740 (2004)
6. Madsen M.B., et al.: J. Geophys. Res. (2006) (manuscript in preparation)
7. Bertelsen, P., Goetz, W., Madsen, M.B., Kinch, K.M., Hviid, S.F., Knudsen, J.M., Gunnlaugsson, H.P., Merrison, J., Nørnberg, P., Squyres, S.W., Bell III, J.F., Herkenhoff, K.E., Gorevan, S., Yen, A.S., Myrick, T., Klingelhöfer, G., Rieder, R., Gellert, R.: Science **305**, 827 (2004)
8. Klingelhöfer, G., Morris, R.V., Bernhardt, B., Rodionov, D., de Souza Jr., P.A. Squyres, S.W., Foh, J., Kankeleit, E., Bonnes, U., Gellert, R., Schröder, C., Linkin, S., Evlanov, E., Zubkov, B., Prilutski, O.: J. Geophys. Res. **108**, 8067 (2003)
9. Goetz, W., Bertelsen, P., Binau, C.S., Gunnlaugsson, H.P., Hviid, S.F., Kinch, K.M., Madsen, D.E., Madsen, M.B., Olsen, M., Gellert, R., Klingelhöfer, G., Ming, D.W., Morris, R.V., Rieder, R., Rodionov, D.S., de Souza Jr., P.A., Schröder, C., Squyres, S.W., Wdowiak, T., Yen, A.: Nature **436**, 62 (2005)
10. Gunnlaugsson, H.P., Worm, E.S., Bertelsen, P., Goetz, W., Kinch, K., Madsen, M.B., Merrison, J.P., Nørnberg, P.: Hyperfine Interact. (2005) (submitted)
11. Stopar, J.D., Taylor, G.J., Hamilton, V.E., Browning, L., Pickett, D.: In: 6th Intl. Conf. on Mars (2003) (abstract no. 3151)

Hyperfine Interact (2005) 166:529–536
DOI 10.1007/s10751-006-9321-6

^{57}Fe Mössbauer spectroscopy and x-ray diffraction study of some complex metamict minerals

Dariusz Malczewski · Janusz E. Frąckowiak · Evgeny V. Galuskin

Published online: 26 October 2006

Abstract Metamict minerals are a class of natural amorphous materials which were initially crystalline but self-radiation damage mainly from alpha decays of ^{238}U and ^{232}Th series inside the structure can produce partially or fully amorphization (metamictization) of these minerals. This paper reports the results of ^{57}Fe Mössbauer spectroscopy, gamma-ray spectrometry and X-ray diffraction (XRD) study of some complex metamict minerals like: davidite-(La), gadolinite, steenstrupine-(Ce), vesuvianite and comparatively epidote. The absorbed α-dose for these minerals varies in wide range from 1.9×10^{14} α-decay/mg (epidote) to as high as 2.7×10^{16} α-decay/mg (steenstrupine). The Mössbauer spectra show decreasing IS values (except steenstrupine) for Fe^{2+} components with absorbed α-dose. Rather unexpected feature of these spectra is a noticeable decrease of the spectral line widths with increasing absorbed α-dose both Fe^{2+} and Fe^{3+} components in gadolinite, davidite and steenstrupine.

Key words davidite · steenstrupine · vesuvianite · Mössbauer spectroscopy · metamict state

D. Malczewski (✉) · E. V. Galuskin
Faculty of Earth Sciences, University of Silesia, Będzińska 60, 41-200 Sosnowiec, Poland
e-mail: malczews@us.edu.pl

E. Galuskin
e-mail: galuskin@us.edu.pl

J. E. Frąckowiak
Faculty of Computer and Materials Science, University of Silesia, Bankowa 12, 40-007 Katowice, Poland
e-mail: jfrack@us.edu.pl

1 Introduction and sample characterization

Metamict minerals incorporate of uranium and thorium that affect the crystal structure mainly by α-decay events. The mechanism for the transition from crystalline to amorphous (metamict) state is not quite understood but the progressive overlap recoil nuclei collision cascades from ^{238}U, ^{232}Th, ^{235}U and their daughter products is critical to the process [1]. Investigations of these materials are important because minerals containing large concentrations of uranium and thorium can serve as natural analogues for radiation effects in high-level nuclear waste form (HLW) [2].

Davidite (multiple oxides) is isostructural with the crichtonite group minerals having a structure based on a close-packed anion framework with the mixed stacking sequence ABCBCACABA and five cation sites M(1)–M(5) [3]. Investigated davidite crystals, massive grayish black with vitreous-metallic lustre, were found in granite pegmatite from Permian granitoid massif Bektau-Ata, Kazakhstan. Composition of davidite by EDS data is close to davidite-(La) with impurities of Mn and it responds to the common formula $(La,Ce,Ca,Th)(Y,U,Fe)(Ti,Fe,Mn)_{20}(O,OH)_{38}$, which was proposed by Gathehouse et al. [3].

Gadolinite $REE_2Fe^{2+}Be_2Si_2O_{10}$, (beryllosilicates) where REE means rare earth elements, yttrium, uranium and thorium, is commonly found in the fully or partially metamict state. The basic structure of gadolinite consists of sheets of SiO_4 and BeO_4 tetrahedra interconnected by layers of distorted octahedral iron and eight-coordinated REE [4,5]. The sample study here is from Ytterby (Sweden). It is fully metamict, pitch black and resembles a piece of glass.

Samples of steenstrupine (cyclosilicate with other anions) were collected in ussingite veins in Karnasurt Mountain (Lovozero alkaline Massif, Kola peninsular, Russia). Dark brown steenstrupine with characteristic resinous luster up to 1 cm in size associates with ussingite, vuonnemite, serandite, aegirine, nordite-(Ce), sphalerite. Veins with the steenstrupine belong to hyperagpaitic pegmatoid hydrothermalites [6]. Composition of the steenstrupine from Karnasurt Mountain was investigated by Pekov et al. [7], who concluded about high unhomogeneity of the composition and existence of solid solution between steenstrupine and thorosteenstrupine. The proposed common crystal chemical formula for the minerals of steenstrupine-thorosteenstrupine series [7] can be written as: $Na_{0-5}Ca_{1-3}(REE,Th)_6(Mn,Fe,Al,Ti)_{4-5}[Si_6O_{18}]_2[(Si,P)O_4]_6(OH,F,O)_x{\cdot}nH_2O$.

Metamict vesuvianites with high contents of U, Th and REE are very rare in nature and occur in pegmatites, nepheline sienite and skarns [8]. We investigate the metamict vesuvianite, which was found in pegmatite vein in contact with phlogopite skarn, Sludyanka, Baikal, Russia. The vesuvianite crystals up to 8–15 cm occur in granite pegmatite in association with microcline, plagioclase, garnet, quartz, galena, zircon and allanite. Two types of the vesuvianite zones were selected for investigation: early zones of the dark brown with glassy luster and late zones of the light brown mat vesuvianite. Vesuvianite in comparison with davidite and steenstrupine is meat in metamict state rarely. High-temperature vesuvianite with significant content of radioactive and rare earth elements crystallizes in *P4/nnc* group. Low-temperature vesuvianite has symmetry *P4/n* or *P4nc*, which connect with order–disorder phenomenon of cation in canal position [9,10]. Simple crystal chemical formula of vesuvianite (sorosilicate) may be show in the next manner: $X_{19}Y_{13}T_{0-5}Z_{18}O_{69}(O,OH,F)_9$ or $Ca_{19}(Al,Mg,Fe,Ti,Mn)_{13}(B,Al,E)_{0-5}Si_{18}O_{69}$

Table I Ages, Fe, U and Th contents and calculated α-doses of metamict samples studied by Mössbauer spectroscopy

	Davidite	Gadolinite	Steenstrupine	Vesuvianite (dark fraction)	Vesuvianite (light fraction)	Epidote
Age (Ma)	270(20)[a]	1795(2)[b]	362(17)[c]	447.3(24)[d]	447.3(24)[d]	280(12)[e]
Fe (wt.%)	19.0	10.5	1.8	3.9	3.9	24.2
U (wt.%)	1.47	0.10	0.89	0.10	0.23	0.02
Th (wt.%)	0.31	0.26	5.63	0.09	0.40	0.005
Calculated dose (α-decay/mg)[f]	$1.47(17)\times 10^{16}$	$1.06(3)\times 10^{16}$	$2.70(23)\times 10^{16}$	$1.81(14)\times 10^{15}$	$4.95(13)\times 10^{15}$	$1.94(11)\times 10^{14}$

[a] Permian granitoid massif

[b] [11]

[c] Age of the Lovozero complex

[d] [12]

[e] [13]

[f] Dose has been calculated from the equation: $D = 8 \times N_{238}(e^{t\lambda 238} - 1) + 7 \times N_{235}(e^{t\lambda 235} - 1) + 6 \times N_{232}(e^{t\lambda 232} - 1)$, where N_{238}, N_{235} and N_{232} are the present number of atoms of ^{238}U, ^{235}U and ^{232}Th per milligram, λ_{238}, λ_{235} and λ_{232} are the decay constants of ^{238}U, ^{235}U and ^{232}Th, t is the geologic age.

$(O,OH,F)_9$. Effect of Si-enrichment of metamict vesuvianite was noted for vesuvianite from the nepheline sienite of Alaska [8].

For comparison we took sample of crystalline epidote $Ca_2(Fe^{3+},Al)_3(SiO_4)(Si_2O_7)O(OH)$ (sorosilicates). Large dark green block, with very small impurities of U and Th was from cavern of Strzegom granite pegmatite, Sudetes Mts (Poland). Basic characteristics all six samples are given in Table I.

2 Experimental

The concentration of ^{232}Th, ^{238}U and ^{235}U were calculated for each sample based on gamma-ray activities of ^{228}Ac (^{232}Th), ^{226}Ra, ^{214}Pb, ^{214}Bi (^{238}U) and ^{235}U. Gamma-ray spectra were collected using characterized HPGe detector (30% efficiency with energy resolution 0.8 keV at 122 keV) and analyzed using the Genie 2000 v.3 software package.

The samples were powdered and prepared in the shape of a thin disc absorber. The Mössbauer transmission spectra were recorded at room temperature using a constant acceleration spectrometer, a multichannel analyzer with 1,024 channels and linear arrangement of ^{57}Co/Rh source (=50 mCi) absorber and detector. The spectrometer velocity was calibrated with a high purity α-Fe foil. X-ray powder diffraction (XRD) patterns were obtained using SIEMENS D5000 diffractometer in the Θ−Θ system and Cu Kα radiation in the scan mode with step size 0.02° and step time 5 s at each point.

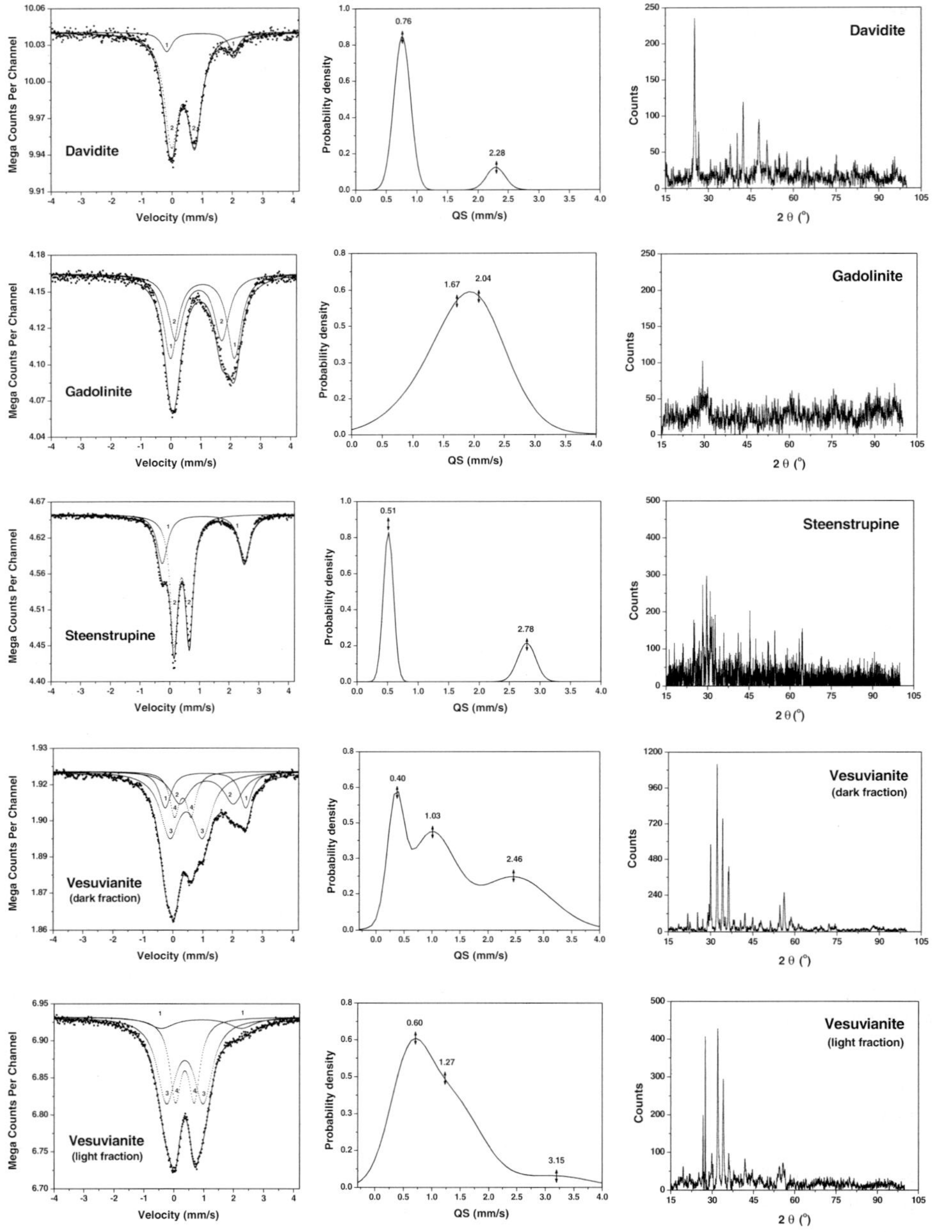

Figure 1 (*left*) Mössbauer spectra of metamict davdite, gadolinite, steenstrupine and vesuvianites. *Solid dots* – experimental data; *thick solid line* – fitted curve; *thin solid line* – Fe^{2+} doublets; *thin dotted line* – Fe^{3+} doublets. (*middle*) Corresponding quadrupole splitting distributions (QSD). (*right*) XRD patterns of the five metamicts samples.

Table II List of ^{57}Fe hyperfine fit parameters of all spectra (Figures 1 and 2)

Sample	Doublet no.	x^2	*IS* (mm/s)	*QS* (mm/s)	*Γ* (mm/s)	Assignment (CN)[a]	Intensity (%)
Davidite		1.0					
	1		0.942(13)	2.226(13)	0.223(19)	Fe^{2+} (6)	13.1(18)
	2		0.366(3)	0.766(6)	0.279(4)	Fe^{3+} (6)	86.9(16)
Gadolinite		2.7					
	1		1.064(6)	2.142(16)	0.285(11)	Fe^{2+} (6)	54.9(36)
	2		0.941(7)	1.547(22)	0.300(12)	Fe^{2+} (6)	45.1(35)
Steenstrupine		2.5					
	1		1.129(3)	2.770(3)	0.216(5)	Fe^{2+} (6)	36.2(7)
	2		0.397(2)	0.512(1)	0.151(1)	Fe^{3+} (6)	63.8(4)
Vesuvianite		1.1					
(dark	1		1.092(9)	2.696(13)	0.235(20)	Fe^{2+} (6)	16.2(31)
fraction)	2		1.132(11)	1.790(36)	0.382(36)	Fe^{2+} (6)	22.9(44)
	3		0.443(10)	1.088(15)	0.386(17)	Fe^{3+} (6)	45.3(36)
	4		0.331(7)	0.555(7)	0.188(16)	Fe^{3+} (5)	15.6(23)
Vesuvianite		1.2					
(light	1		0.925(32)	2.690(25)	0.514(42)	Fe^{2+} (6)	10.8(10)
fraction)	3		0.368(3)	1.227(14)	0.376(11)	Fe^{3+} (6)	56.0(43)
	4		0.368(2)	0.631(5)	0.232(8)	Fe^{3+} (5?)	33.2(33)
Epidote		2.0					
	1		1.136(6)	2.593(6)	0.234(10)	Fe^{2+} (6)	9.3(3)
	2		0.353(1)	2.035(1)	0.157(1)	Fe^{3+} (6)	90.7(2)

Isomer shift values IS are given relative to the α-Fe standard at room temperature. The uncertainties are given from fitting procedure using the MOS MOD and MEP software.

[a] Coordination number.

3 Results and discussion

The Mössbauer spectra of the investigated metamict samples with corresponding quadrupole splitting distributions and X-ray powder diffraction patterns are shown in Figure 1. The hyperfine parameters derived from fitting procedure for all six samples are summarized in Table II.

The Mössbauer spectrum of davidite is fitted to two quadrupole doublets assigned to Fe^{2+} and Fe^{3+} in octahedral positions. From crystalline structure of davidite results that these doublets correspond to Fe^{2+} and Fe^{3+} in distorted octahedral M(3) site. The quadrupole slitting distribution (QSD) of davidite has two components, which maxima agree well with the QS values of doublets no.1, and 2 in fitted Mössbauer spectrum (Figure 1, Table II). The XRD pattern shows high degree of amorphization (metamictization) of this mineral.

The main feature seen in the spectrum of metamict gadolinite is a presence of two broad asymmetric peaks due to overlap of numerous somewhat different positions of divalent iron ions. The corresponding QSD is continuous with central maximum at 1.84 mm/s and two weakly visible left and right side maxima at 1.67 mm/s and 2.04 mm/s, respectively. Based on this QSD the Mössbauer spectrum of this gadolinite can be fitted to two Fe^{2+} doublets, no. 1 and 2 (Figure 1, Table II). The doublet no. 1 with IS = 1.06 mm/s and QS = 2.14 mm/s represents the Fe^{2+} octahedra which have undergone an expansion during metamictization process

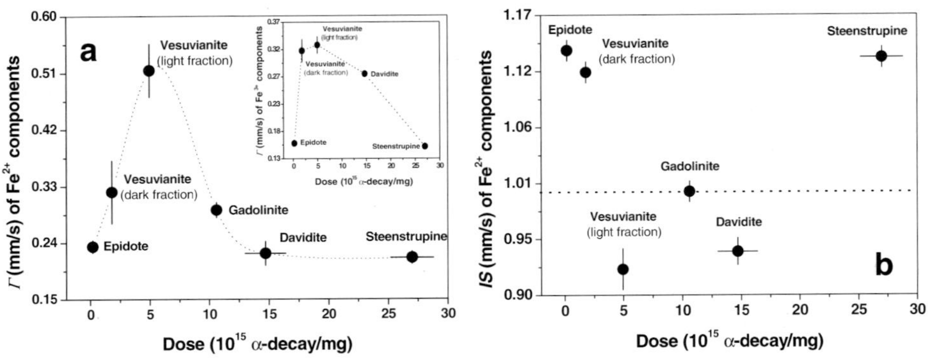

Figure 2 **a** Averaged line widths (Γ) of Fe^{2+} components as a function of absorbed α-dose. The *dotted line* is drawn to illustrate that for gadolinite, davidite and steenstrupine these line widths decrease with increasing α-dose; (*frame*) dose dependence of Γ for Fe^{3+} components. **b** Variation of isomer shift values with absorbed α-dose for Fe^{2+} components. The *dotted line* indicates IS = 1.

while the doublet no. 2 with IS = 0.94 mm/s and QS = 1.54 mm/s represents the Fe^{2+} octahedra which have undergone a contraction during metamictization. X-ray diffraction of gadolinite shows complete lack of long range structural order i.e., this mineral is fully metamict.

In the spectrum of steenstrupine two doublets are clearly separated (no. 1 and 2, Figure 1, Table II). The first of them with IS = 1.13 mm/s and QS = 2.77 mm/s corresponds to Fe^{2+} in octahedral site and the second narrow doublet, IS = 0.39 mm/s and QS = 0.51 mm/s, can be assigned to Fe^{3+} in the same octahedral site. From these IS and QS values may be inferred that the octahedral site is very regular and uniform over the whole structure. The result comes as surprise because this steenstrupine structure absorbed the highest α-dose (2.7×10^{16} α-decay/mg) among the other minerals presented here. No particular of lines broadening or similar effects in this Mössbauer spectrum are observed. The XRD pattern shows almost complete amorphization of the structure (sharp narrow reflects are from rare microcrystalline regions observed by HRTM). There is excellent agreement between the fitted values of QS and the two maxima in corresponding QSD.

The Mössbauer spectra of the two vesuvianite samples are complex similar to previous ^{57}Fe Mössbauer spectroscopy studies of vesuvianites [14,15]. The spectrum of dark vesuvianite can be fitted to a superposition of four doublets corresponding to Fe^{2+} (no. 1 and 2) and Fe^{3+} (no. 3 and 4, Figure 1, Table II). Based on the structure of vesuvianite [10] the values of hyperfine parameters of doublets no. 1 and 2 seem to be consistent with Fe^{2+} in octahedral Y(3) site occupied mainly by Al and Fe. It is expected that some part of Y(3) positions may be distorted due to α decays i.e., metamictization process. It can be a reason that we have observed the visible splitting of this site to two subsites represent by the two Fe^{2+} doublets with different isomer shift and quadrupole splitting values. On the other hand, this splitting may indicate that Fe^{2+} ions occupy more than one type of octahedral site in the vesuvianite structure. The wide doublet no. 3, IS = 0.44 mm/s and QS = 1.08 mm/s corresponds to Fe^{3+} in octahedral Y(3) site. The parameters of doublet no. 4, IS = 0.33 and QS = 0.55, allow us to assigned this to Fe^{3+} ions at fivefold coordinated Y'(1) site with possible influence of four O(9) oxygens at an average distance of 3.24

Å. The corresponding QSD consists of three components positioned at 0.40 mm/s, 1.03 mm/s and 2.48 mm/s. The distribution around 0.40 mm/s is connected with the narrow doublet no. 4 while the distribution with maximum at 1.03 mm/s with the doublet no. 3. The wide component with maximum at 2.46 mm/s reflects combined contribution of doublets no. 1 and 2. XRD pattern of dark vesuvianite shows still high crystallinity of this specimen.

Due to high absorbed α-dose into the structure of light vesuvianite its Mössbauer spectrum differs markedly from the spectrum of dark vesuvianite. The difference of absorbed α-dose between them is ca 3×10^{15} α-decay/mg. The spectrum of light vesuvianite was fitted to three doublets (Figure 1, Table II). The very wide doublet no. 1 with IS = 0.92 mm/s and QS = 2.69 mm/s represents the remains of Fe^{2+} into this structure at Y(3) site. The doublet no. 3 is assigned to Fe^{3+} ions in octahedral Y(3) site while the doublet no. 4 represents Fe^{3+} ions probably in still fivefold coordinated Y'(1) site. These Fe^{3+} doublets are characterized by slightly larger of QS values, lower (doublet no. 3) and slightly higher (doublet no. 4) of IS values in comparison to the fitted spectrum of dark vesuvianite. The corresponding QSD is more uniform than this for dark vesuvianite. This QSD has three broad components with maxima at 0.60 mm/s, 1.27 mm/s and 3.15 mm/s, and this distribution agree well with the parameters of the three fitted doublets in the Mössbauer spectrum. XRD pattern of the light vesuvianite shows decreasing crystallinity of this mineral.

The comparative spectrum of epidote, Table II, (the same silicate group as vesuvianite) is fitted to two narrow doublets assigned to Fe^{2+} (doublet no.1) and Fe^{3+} (doublet no. 2) ions in octahedral positions. The doublet no. 2 with large value of QS ~2 mm/s is characteristic for epidote and other minerals from epidote group like allanite [16] or stilpnomelane [17] and with greater QS for oxidized gadolinite [4,18]. The corresponding XRD pattern shows full crystallinity of this sample (published elsewhere).

The most striking feature visible in the spectra presented here is the unexpected relation between width of the spectral lines and absorbed α-dose. As can be seen in Figure 2a the line widths for Fe^{2+} components increase from epidote to light vesuvianite ($D = 4.9 \times 10^{15}$ α-decay/mg) and than rapidly decrease for gadolinite, davidite and steenstrupine. The future annealing experiments should determine if this variation is a characteristic feature of iron-bearing metamict minerals above a certain critical dose or it is a result of slight changes of coordinated octahedra some of these minerals during metamictization and consequently weak variability of the spectral lines. Similar behavior is also visible for Fe^{3+} components (Figure 2a right top). It is interesting to note that IS values of Fe^{2+} components (Figure 2b) lowered with absorbed α-dose except steenstrupine. For this mineral with the highest absorbed α-dose ($D = 2.7 \times 10^{16}$ α-decay/mg) the IS = 1.13 mm/s of Fe^{2+} doublet is nearly the same as for crystalline epidote. It means together with large value of QS = 2.77 mm/s that despite of amorphization of this structure the Fe^{2+} positions are solely octahedral and additionally very regular.

4 Conclusions

The present work shows that ^{57}Fe Mössbauer spectroscopy is a useful method for investigation of iron-bearing complex metamict minerals. It is characteristic for investigated minerals that line widths of Fe^{2+} doublets initially increase with absorbed α-dose in order of epidote, dark vesuvianite to light vesuvianite and then rapidly decrease for gadolinite, davidite and steenstrupine. The similar dependence is also noted for Fe^{3+} components except gadolinite, which does not contain trivalent iron in its structure. The IS values of Fe^{2+} doublets decrease with increasing absorbed α-dose except steenstrupine which Fe^{2+} doublet has nearly the same IS value as Fe^{2+} doublet in fully crystalline epidote. Annealing experiments of davidite, vesuvianite and steenstrupine are needed in order to explain the observed changes in Mössbauer spectra and consequently structural changes around Fe positions associated with metamictization process.

Acknowledgment Work supported by the State Committee for Scientific Research, Poland, trough grant no. 2P04D06229.

References

1. Ewing, R.C.: Nucl. Instrum. Methods Phys. Res., B **91**, 22 (1994)
2. Weber, W.J., Ewing, R.C., Catlow, C.R.A., Diaz de la Rubia, T., Hoobs, L.W., Kinoshita, C., Matzke, H., Motta, A.T., Nastas, M., Salje, E.K.H., Vance, E.R., Zinkle, S.J.: J. Mater. Res. **13**, 1434 (1998)
3. Gatehouse, B.M., Grey, I.E., Kelly, P.R.: Am. Mineral. **64**, 1010 (1979)
4. Ito, J., Hafner, S.S.: Am. Mineral. **59**, 700 (1974)
5. Miyawaki, R., Nakai, I., Nagashima, K.: Am. Mineral. **69**, 953 (1984)
6. Khomyakov, A.P.: Mineralogy of Hyperagpaitic Alkaline Rocks. Nauka, Moscow (1990)
7. Pekov, I.V., Ekimenkova, I.A., Kononkova, N.N.: Proc. RMS **6**, 35 (1997) (in Russian)
8. Eby, R.K., Janeczek, J., Ewing R.C., Ercit, T.S., Groat, L.A., Chakoumakos, B.F., Hawthorne, F.C., Rossman, G.R.: Can. Mineral. **31**, 357 (1993)
9. Armbruster, T., Gnos, E.: Am. Mineral. **85**, 563 (2000)
10. Galuskin, E.V., Armbruster, T., Malsy, A., Galuskina I.O., Sitarz, M.: Can. Mineral. **41**, 843 (2003)
11. Romer, R.L., Smeds, S.A.: Precambrian Res. **67**, 141 (1994)
12. Reznitsky, L.Z., Kotov, A.B., Salnikov, E.B., Vasiliev, E.P., Yakovleva, S.Z., Kovach, V.P., Fedoseenko, A.M.: Petrology **8**, 66 (2000)
13. Pin, Ch., Puziewicz, J., Duthou, J.L.: N. Jb. Miner. Abh. **160**, 71 (1989)
14. Tricker, M.J., Vaishnava, P.P., Manning, P.G.: J. Inorg. Nucl. Chem. **43**, 1169 (1981)
15. Kraczka, J., Żabiński, W.: Mineral. Pol. **34**, 37 (2003)
16. Dollase, W.A.: Z. Kristallogr. **138**, 41 (1973)
17. Malczewski, D., Popiel, E.S., Sitarek, A.: Nukleonika **49**(S3), 43 (2004)
18. Malczewski, D., Janeczek, J., Popiel, E.: Mol. Phys. Rep. **22**, 111 (1998)

Hyperfine Interact (2005) 166:537–542
DOI 10.1007/s10751-006-9322-5

KTB iron mineralogy at iridium deficient layers revealed by Mössbauer spectroscopy

Brajesh Pandey · Amita Tripathi · R. P. Tripathi · H. C. Verma

Published online: 14 November 2006

Abstract Mössbauer spectroscopic studies of two sedimentary layers from Petriccio section in Gubbio formation are presented. It is found that these layers have almost the same iron phases nanophase hematite and phyllosilicate and with the same compositions as those in the Cretaceous-Tertiary Boundary (KTB) in Petriccio. The two layers also show enhancement in the siderophile elements such as Cr, Ni, Fe, Ca etc. to that found in the KTB layer. The two layers are well separated from the KTB layer, 128 and 149 cm below this layer, and show no enhancement of iridium. From the data it is suggested that the two layers also represent some kind of impact by extraterrestrial (ET) bolides.

Key words Mössbauer spectroscopy · KT boundary · nanophase iron · iron mineralogy

1 Introduction

The 600 My history of life on our planet has seen several cases of severe stress when the number of living species dropped sharply over a small time interval on geological scale. The second most severe mass extinction, which occurred about 65 My ago, marks the end of Cretaceous period and beginning of Tertiary period. The cause of this extinction has been widely studied and debated [1–10]. Widespread Deccan volcanos erupted in western India during the same geological slot [7–11], and a possible impact by a large extraterrestrial bolide [1–5, 12–14], are two major possibilities on which scientists have focused their studies. Supplementary evidences

B. Pandey · H. C. Verma (✉)
Department of Physics, IIT Kanpur, Kanpur 208 016, India
e-mail: hcverma@iitk.ac.in

A. Tripathi · R. P. Tripathi
Department of Physics, JNV University, Jodhpur 342005, India

from the sedimentary rocks, both geochemical and paleontological have played important role to distinguish between the two possibilities.

The sedimentary rock layer corresponding to KTB usually distinguishes itself from the surrounding layers by its limonitic colour. This layer is typically 1–2 cm thick and shows a sudden enhancement of siderophile and platinum group elements over the background value of the underlying and overlying layers. The finding of Alvarez et al. that iridium concentration in the limonitic KTB layers in all the 50 sections [1, 2] studied was much higher (20–160 times) than that in the background, made the extraterrestrial origin of KT extinction a great favourite. Finding of ~200 km wide crater at Chicxulub [15], shocked quartz [16, 17], Nickel rich spinel [18, 19], abrupt change in the isotopic ratios such as $\delta^{13}C$ and $\delta^{18}O$ in the KTB layers, etc., are some of the supporting evidences for the impact scenario.

However, the actual mass extinction was caused due to the geothermal and geochemical conditions created in the atmosphere and on the earth's surface [20, 21] as a result of the impact. The billions of tons of ejecta must have gone up as vapour plume and chemical reactions at the high temperature must have changed the redox conditions. It is stipulated that the plume and the soot [6] released from wide spread fire blocked sunlight for an extended period, may be in kiloyears, bringing out a sharp drop in the temperature. Such extreme climatic changes are assumed to be responsible for the mass extinction.

Iron mineralogy of KTB sedimentary rocks from a number of sites has been studied by our group [22, 23] and Wdowiak's group [24, 25] using Mössbauer spectroscopy. While all the American and Asian KTB samples showed superparamagnetically relaxed Mössbauer spectra, the European sites samples including those from the Gubbio formation showed spectra representing collective magnetic excitations. Mössbauer spectra from all sections of Gubbio formation, i.e., Contessa, Bottaccione, Petriccio and Furlo, showed a well-split sharp sextet corresponding to hematite together with a strong doublet corresponding to phyllosilicate. From the temperature variation of hyperfine magnetic field (B_{hf}) corresponding to the hematite, Verma et al. [23] estimated the particle size of the hematite to be in the range of 16–27 nm.

Bhandari et al. [26] have reported the existence of two limonitic layers in the exposed section at Petriccio (in Gubbio formation) which are much below the recognized most iridium rich KTB layer. One of the layers is 128 cm below the well known KTB and the other is 149 cm below it. Correspondingly, the two samples are named P128 and P149. The boundary sample itself is named P0. The two layers P128 and P149 do not show any iridium enhancement but they are still special because several of the siderophile elements have high concentration in these layers. In this article we present the Mössbauer studies of samples from these limonitic layers and explore possible links of these limonitic sedimentary layers with the end Cretaceous events.

2 Experimental

The absorber for Mössbauer spectroscopy were prepared by pressing finely ground powders in a 12 mm diameter copper ring and fixing the open ends with transparent tapes. The spectra were recorded using a conventional constant acceleration

Table I Composition of different elements in P0, P128 and P149 samples, data taken from [26] (all values are in ppm except when mentioned otherwise)

Sample	Ir (ppb)	Cr	Ni	Co	Ba	Na	Fe (%)	Ca (%)	K (%)
P0	5.2	103.4	778	46.1	322	2294	3.51	7.6	2.53
P128	–	94.3	1240	44.8	661	3396	3.92	5.4	2.73
P149	–	90.0	1053	46.7	863	3558	3.16	6.0	2.05

Sample	La	Ce	Nd	Sm	Eu	Gd	Tb	Yb	Lu
P0	29.9	36.9	27.1	6.15	1.47	2294	5.5	1.65	0.23
P128	32.8	46.5	34.6	7.36	1.66	3396	5.7	2.44	0.33
P149	28.0	46.2	31.5	6.27	1.66	3558	6.7	2.10	0.31

Mössbauer spectrometer with ^{57}Co in Rh matrix as the Mössbauer source. Data were analysed by fitting an appropriate number of discrete doublets and sextets with Lorentzian peaks using a least-squares code developed at IIT Kanpur. The typical uncertainty in the isomer shift (IS) and quadrupole splitting (QS) are ~0.02 mm/s, whereas that for the hyperfine magnetic field (B_{hf}) is ~0.1 T. The relative areas under different component spectra have an uncertainty of ~2%.

3 Results

3.1 Age of the samples P128 and P149

Mukhopadhyay et al. [27] have used helium-3 as a constant-flux proxy of sedimentation rate and have calculated the rate of sedimentation of KTB clay at Gubbio to be 2.5 mm/kyr. The total deposition interval for KTB clay at Gubbio was obtained as 7.9±1.0 kyr. Though the sedimentation rate may not be the same for extended periods, one can estimate that P128 and P149 layers might have been formed several hundreds kiloyears prior to the formation of KTB layer.

3.2 Composition of elements

Table I gives the composition of some of the siderophile elements in P0, P128 and P149 samples as reported by Bhandari et al. [26]. P0 contains enormous iridium (5.2 ppb) over a non-detectable background. P128 and P149 layers are iridium deficient but have other elements such as Cr, Ni, Co, Ba, Na, etc. in much larger concentration than the background, almost the same as that at KTB. Iron too has highest concentration at P128 (3.92% as against 3.51% in P0). The first question that arises is whether these layers P128 and P149 which are widely separated from the accepted KTB Layer P0 are genetically related to it. To look into this question, we have studied the two samples using Mössbauer spectroscopy.

3.3 Mössbauer spectroscopy

From the Mössbauer studies of a number of KTB sections spread over American, European and Asian sites, it has been established that nanophase goethite and/or

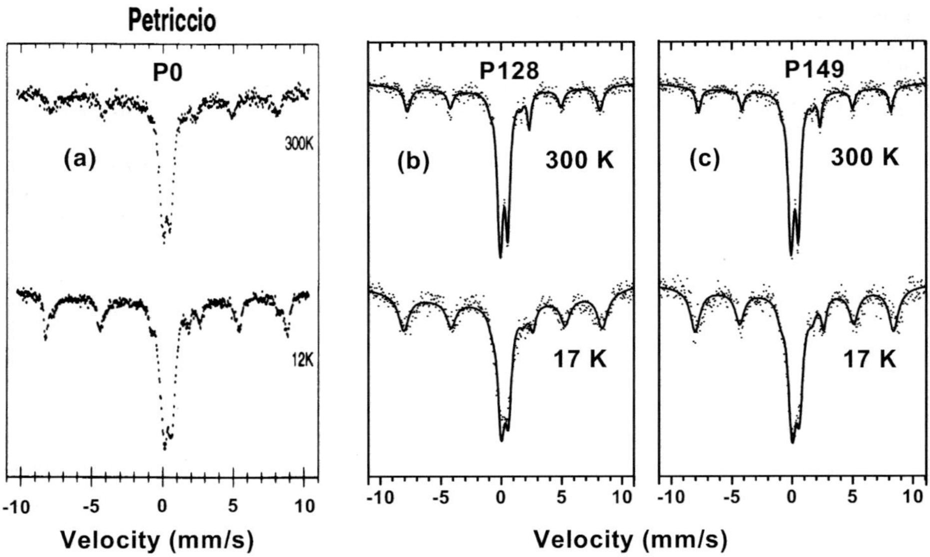

Figure 1 Mössbauer spectra of the Petriccio limonitic samples. **a** P0 (reproduce from [25] with permission), **b** P128 and **c** P149.

hematite without coexistence of their coarse size counterpart is a characteristic of KTB sediment [22–25]. This observation can further be extended to Permian-Triassic boundary (PTB) where evidences for impact are recently reported [28, 29]. The Mössbauer spectra of PTB samples from Spiti Valley in India showed nanophase goethite and hematite as the major iron containing phases [30]. It may be noted that the PTB layers were also limonitic, rich in siderophile elements but did not have iridium enhancement.

Figure 1 shows the Mössbauer spectra of samples from (a) the most iridium rich KTB layer in Petriccio, P0, (b) P128 and (c) P149 layers. The spectra for P0 are reproduced from [25] with permission. All three samples in Figure 1 give the spectra corresponding to the same iron phases, a six-line pattern corresponding to nanosize hematite, and two doublets, the major one corresponding to Fe^{3+}, most likely phyllosilicate, and the minor one corresponding to Fe^{2+}. In all the three, the ferric doublet is the strongest and the hematite is well split. Table II gives the Mössbauer parameter fitted to the spectra. We have included the parameters of KTB spectrum from Bottaccione section in the table. Bottaccione is also a well studied KTB section in the same Gubbio formation as the Petriccio and the Mössbauer spectra of the two are almost identical [23]. From Table II we see that all the three limonitic layers not only have the same iron phases, the relative areas corresponding to different components are also the same within experimental uncertainties. At room temperature the hematite shows B_{hf} somewhat smaller than the known values of 52.0 T in all the three cases. This is supposed to be due to smaller size of hematite particles [23, 31]. As the temperature is decreased, the value of B_{hf} and also the relative absorption area of the hematite component increase. This may be due to difference in the change of the recoil-free fraction of the phyllosilicate and hematite component as the temperature is changed. Another contribution to the increased

Table II Mössbauer parameter of B0, P128 and P149 layer

Sample	T (K)	IS (mm/s)	QS (mm/s)	B_{hf} (T)	Area (%)	Phase
Bottaccione (KTB)	RT*	0.34	0.60		53	Ferric doublet
		1.13	2.78		6	Ferrous doublet
		0.37	−0.22	49.9	41	Hematite
	12	0.38	0.55		45	Ferric doublet
		1.18	2.86		6	Ferrous doublet
		0.48	−0.08	52.1	49	Hematite
P128	RT	0.30	0.65		61	Ferric doublet
		1.28	2.26		11	Ferrous doublet
		0.37	−0.08	49.5	28	Hematite
	17	0.36	0.63		42	Ferric doublet
		1.48	2.43		9	Ferrous doublet
		0.43	−0.10	51.0	49	Hematite
P149	RT	0.29	0.64		64	Ferric doublet
		1.29	2.25		12	Ferrous doublet
		0.40	−0.10	49.5	24	Hematite
	17	0.36	0.66		42	Ferric doublet
		1.47	2.45		9	Ferrous doublet
		0.39	−0.09	51.0	49	Hematite

*[23]

magnetic area may come from the superparamagnetic fraction which gets split at low temperatures. The important point is that all the three samples show very similar temperature dependence too in the Mössbauer parameters.

The appearance of the same iron phases in P128 and P149 as in the KTB layer prompts us to think that these two limonitic layers are genetically related to the KTB layer P0. This is also supported by the enhanced concentration of the siderophile elements like Cr, Ni, Co, Ba, Na, Fe, etc. But then there is no iridium. If we assume that the P128 and P149 layers have originated from the KTB layer P0 due to some geological activity, the question arises what separated iridium from the sediment so selectively. As the Mössbauer parameters of the iron phases are characteristic of the extinction sites, most likely originating from impacts, and there is no explicit evidence of any hiatus, the another possibility is an impact other than that at KTB occurring prior to it. If so, these impacts were able to create the same geochemical climatic conditions as the KTB impact. This would also mean that the so-called KTB mass extinction started much earlier, several hundred thousand years ago. Indeed there are palentological evidences for extinction taking place over a long time.

References

1. Alvarez, L.W., Alveraz, W., Asaro, F., Michel, H.V.: Science **208**, 1095 (1980)
2. Alvarez, W., Alveraz, L.W., Asaro, F., Michel, H.V.: J. Geol. Soc. Am., Spec. Pap. **190**, 305 (1982)
3. Smith, J., Hertogen, J.: Nature **285**, 198 (1980)
4. Hsüv, J.K.: Nature **285**, 201 (1980)

5. Kyte, F.T., Zhou, Z., Wasson J.T.: Nature **288**, 651 (1980)
6. Wolbach, W.S., Lewis, R.S., Anders, E.: Science **230**, 167 (1985)
7. Rampino, M.R., Reynolds, R.C.: Science **219**, 495 (1983)
8. Officer, C.B., Drake, C.L.: Science **219**, 1383 (1983)
9. Officer, C.B., Drake, C.L.: Science **227**, 1161 (1985)
10. Officer, C.B., Hallam, A., Drake, C.L., Devine J.D.: Nature **326**, 143 (1987)
11. Hoffman, C., Feraud, G., Courtillot, V.: Earth Planet. Sci. Lett. **180**, 13 (2000)
12. Izett, G.A., Dalrymple, G.B., Snee, L.W.: Science **252**, 1539 (1991)
13. Marin, L.E., Quezada Muneton, J.M., Sharpton, V.L., Ryder, G., Schuraytz, B.C., Dalrymple, G.B.: Lunar Planet. Sci. **23**, 843 (1992)
14. Swisher, C.C.: Science **257**, 954 (1992)
15. Sharpton, V.L., Dalrymple, G.B., Marin, L.E., Ryder, G., Schuraytz, B.C., Urrutia-Fucugauchi, J.: Nature **359**, 819 (1992)
16. Bohor, B.F., Foord, E.E., Modreski, P.J., Triplehorn, D.M.: Science **224**, 867 (1984)
17. Alvarez, W., Claeys, P., Kieffer S.W.: Science **269**, 930 (1995)
18. Bohor, B.F., Foord, E.E., Ganapathy R.: Earth Planet. Sci. Lett. **81**, 57 (1986)
19. Robin, E., Bonte, P.H., Forget, L., Jehanno, C., Rocchia, R.: Earth Planet. Sci. Lett. **108**, 181 (1992)
20. Sutherland, F.L.: Earth-Sci. Rev. **36**, 1 (1994)
21. Bhandari, N.: Proc. Indian Acad. Sci. **107**, 251 (1998)
22. Bhandari, N., Verma, H.C., Upadhyay, C., Tripathi, A., Tripathi, R.P.: J. Geol. Soc. Am., Spec. Pap. **356**, 201 (2002)
23. Verma, H.C. Upadhyay, C., Tripathi, A., Tripathi, R.P., Bhandari, N.: Meteorit. Planet. Sci. **37**, 901 (2002)
24. Armendarez, L.P., Wdowiak, T.J., Agresti, D.G., Wade, M.L.: Lunar Planet. Sci. **29**, 1811 (1998)
25. Wdowiak, T.J., Armendarez, L.P., Agresti, D.G., Wade, M.L., Wdowiak, S.Y., Claeys, P., Izett, G.: Meteorit. Planet. Sci. **36**, 123 (2001)
26. Bhandari, N., Shukla, P.N., Castagnoli, G.C.: Palaeogeogr. Palaeoclimatol. Palaeoecol. **104**, 199 (1993)
27. Mukhopadhyay, S., Farley, K.A., Montanari, A.: Science **291**, 1952 (2001)
28. Becker, L., Poreda, R.J., Hunt, A.G., Bunch, T.E., Rampino, M.: Science **291**, 1530 (2001)
29. Basu, A.R., Petaev, M.I., Poreda, R.J., Jacobsen, B.S., Becker, L.: Science **302**, 1388 (2003)
30. Verma, H.C., Upadhyay, C., Tripathi, R.P., Shukla, A.D., Bhandari, N.: Hyperfine Interact. **142/143**, 357 (2002)
31. Morris, R.V., Agresti, D.G., Lauer H.V., Newcomb, J.A., Shelfer, T.D., Murali, A.V.: J. Geophys. Res. **94**, 2760 (1989)

Hyperfine Interact (2005) 166:543–547
DOI 10.1007/s10751-006-9323-4

Mössbauer study of glasses in meteorites: the D'Orbigny angrite and Cachari eucrite

Y. A. Abdu · I. Souza Azevedo · S. J. Stewart · A. López · M. E. Varela · G. Kurat · R. B. Scorzelli

Published online: 27 October 2006

Abstract Mössbauer spectroscopy measurements at room temperature (RT) and at liquid helium temperature (4.2 K) were carried out on bulk and glass samples from the D'Orbigny (angrite) and Cachari (eucrite) meteorites. The RT Mössbauer spectrum of the bulk sample of D'Orbigny shows the presence of Fe^{2+} in olivine and pyroxene and that of bulk Cachari contains only pyroxene. Very small amounts of Fe^{3+} are also present in the bulk samples, but are attributed to surface contamination. The RT spectra of the D'Orbigny and Cachari glasses are fitted with three doublets, which are assigned to Fe^{2+} at three different octahedral positions. No Fe^{3+} was detected in the glass samples. The spectra of the glasses measured at 4.2 K show the presence of relaxation effects. The results suggest a certain degree of structural ordering in these glasses.

Key words Mössbauer spectroscopy · glasses · meteorite · angrites · eucrites · relaxation effects

Y. A. Abdu · I. Souza Azevedo · A. López · R. B. Scorzelli (✉)
Centro Brasileiro de Pesquisas Físicas, Rua Dr. Xavier Sigaud 150,
22290-180 Rio de Janeiro, Brazil
e-mail: scorza@cbpf.br

S. J. Stewart
IFLP, CONICET, Depto. De Física, Facultad de Cs. Exactas,
Universidad Nacional de La Plata, Argentina

M. E. Varela
Depto. de Geologia, CONICET–UNS, San Juan 670, 8000 B. Blanca, Argentina

G. Kurat
Naturhistorisches Museum, Burgring 7, A-1010 Vienna, Austria

1 Introduction

The D'Orbigny meteorite belongs to the angrites, which is a small and rare group of achondrites. It was found in the Buenos Aires province of Argentina (1979) and is very rich in glasses, a feature that makes it an exceptional member of the angrites [1]. The Cachari meteorite, found in Argentina (1916), belongs to another group of achondrites known as eucrites. It is also an unusual meteorite as it contains appreciable amounts of glasses occurring mainly as dark glass veins [2]. Glasses in eucrites, as well as those in all achondrites, are believed to result from impact processes on the eucrite parent body. This is mainly because these glasses have practically the same composition as that of their respective bulk host rock. However, recent Secondary Ion Mass Spectrometry (SIMS) studies that compare glasses from eucrites with those from other achondrites proposed a different hypothesis on the origin of glasses envisaging a new role for silicate liquids in the genesis of meteorites [3].

^{57}Fe Mössbauer spectroscopy has been used successfully in studying Fe-bearing glasses, e.g. borate, phosphate, and silicate glasses [4]. For example, the technique has been used to determine Fe^{3+}/Fe^{2+} ratio in natural basalt glasses in order to reveal the oxygen pressure in the melt [5]. Here we report on the Mössbauer results on glasses from the D'Orbigny and Cachari meteorites in an attempt to contribute to the understanding of their possible origin(s).

2 Experimental

Glasses from D'Orbigny belong to the Glass Patches population described in [1]. Small chips of these glasses were separated under a binocular microscope and used for the preparation of the Mössbauer absorber. For Cachari, glasses cannot be separated by hand and therefore a double polished thin section of a glass-filled fracture was specially prepared for Mössbauer measurements. Mössbauer spectra were recorded in transmission geometry using a ^{57}Co(Rh) source. The spectrometer was calibrated with the RT spectrum of α-Fe. The low temperature spectra were collected using a liquid helium bath cryostat with the source and absorber being at 4.2 K.

3 Results and discussion

The room temperature (RT) Mössbauer spectra of the bulk samples of the D'Orbigny and Cachari meteorites are shown in Figure 1. The spectrum of D'Orbigny is fitted with four doublets resulting from the presence of Fe^{2+} in olivine (two outer doublets) and pyroxene (two inner doublets). On the other hand, the spectrum of Cachari consists of two strong doublets emanating from Fe^{2+} in pyroxene. Both meteorites contain very small amounts of Fe^{3+}. The Mössbauer results are consistent with the mineralogical and petrological studies reported for these meteorites [6, 7].

Figure 2 shows the RT Mössbauer spectra of glasses from the D'Orbigny and Cachari meteorites. The relevant hyperfine parameters are presented in Table I. The spectra can be well fitted with at least three overlapping doublets, which are

Figure 1 RT Mössbauer spectra of bulk samples of D'Orbigny and Cachari meteorites. The small absorption near the high velocity peak of the spectrum of bulk Cachari is ignored in the fitting.

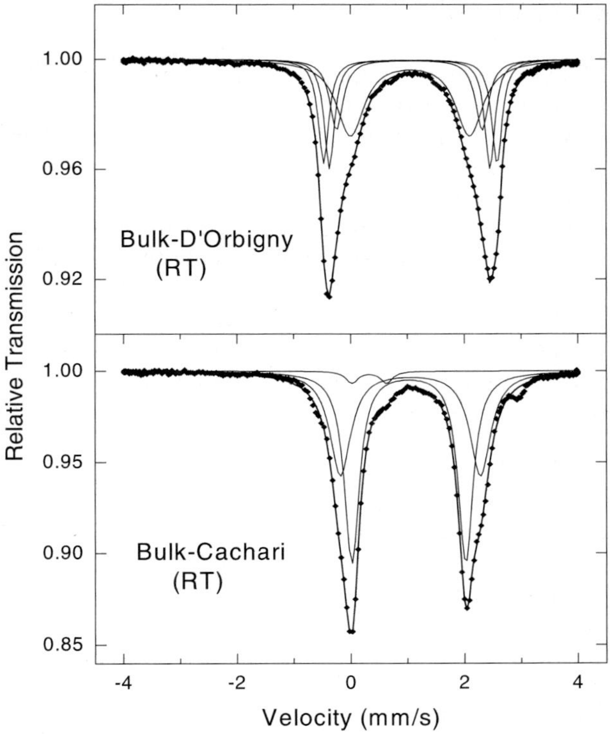

assigned to Fe^{2+} at three different octahedral sites A, B, and C. The quadrupole splitting of the doublets decreases from A to C, which indicates that distortion from octahedral symmetry is increasing in that direction (see Table I). In contrast to natural basalt glasses, which generally contain Fe^{3+} [4], the studied meteoritic glasses do not contain Fe^{3+}. The hyperfine parameters for sites A and B in the D'Orbigny and Cachari glasses are very close to the values obtained for the M1 and M2 sites in pyroxenes in these meteorites (e.g. D'Orbigny: IS = 1.15 mm/s, QS = 2.56 mm/s for M1; IS = 1.16 mm/s, QS = 2.09 mm/s for M2, Cachari: IS = 1.16 mm/s, QS = 2.46 mm/s for M1; IS = 1.13 mm/s, QS = 2.00 mm/s for M2).

The spectra of the D'Orbigny and Cachari glasses measured at 4.2 K are shown in Figure 3. They clearly indicate the presence of relaxation effects. A spectrum of pyroxene separated from D'Orbigny also exhibited a similar behaviour when measured at 4.2 K (unpublished data). This feature has been observed in natural orthopyroxenes at low temperatures, and was attributed mainly to spin–spin relaxation of Fe^{2+} ions [8]. The spectrum of D'Orbigny glass at 4.2 K shows a broader pattern compared to that of Cachari (see Figure 3). This implies a slower relaxation in the former and indicates a higher Fe^{2+} concentration in the D'Orbigny glass [9].

The similarity between the low temperature spectra of glasses from D'Orbigny and Cachari and those of natural and meteoritic pyroxenes, along with the RT hyperfine parameters suggest a certain degree of structural ordering in these glasses. Probably, cooling below the glass transformation temperature was slow enough to allow for elemental exchange processes like the replacement of Mg by Fe^{2+} [1].

Figure 2 RT Mössbauer spectra of D'Orbigny and Cachari glasses.

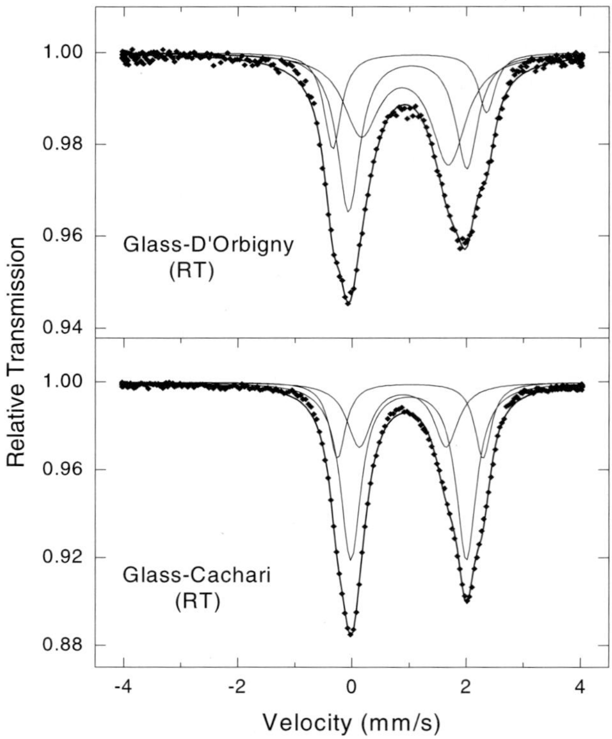

Table I Mössbauer hyperfine parameters of glasses from D'Orbigny and Cachari meteorites. A, B, and C are three octahedral sites occupied by Fe^{2+}

	IS (±0.02 mm/s)	QS (±0.02 mm/s)	*W* (±0.02 mm/s)	*A* (±5%)
D'Orbigny				
Fe^{2+} (A)	1.12	2.70	0.36	17
Fe^{2+} (B)	1.08	2.07	0.47	40
Fe^{2+} (C)	1.03	1.53	0.74	43
Cachari				
Fe^{2+} (A)	1.13	2.55	0.34	20
Fe^{2+} (B)	1.10	2.02	0.43	56
Fe^{2+} (C)	1.00	1.52	0.51	24

IS Isomer shift relative to α-Fe, *QS* quadrupole splitting, *W* line width, *A* relative area.

From the above discussion, it appears that glasses from D'Orbigny and Cachari are similar to each other. None of them appears to have any memory of a shock event but they have experienced some ordering, which could be related to a subsolidus elemental exchange process as identified for D'Orbigny silicates [1, 6]. Further variable temperature and in-field Mössbauer measurements are in progress

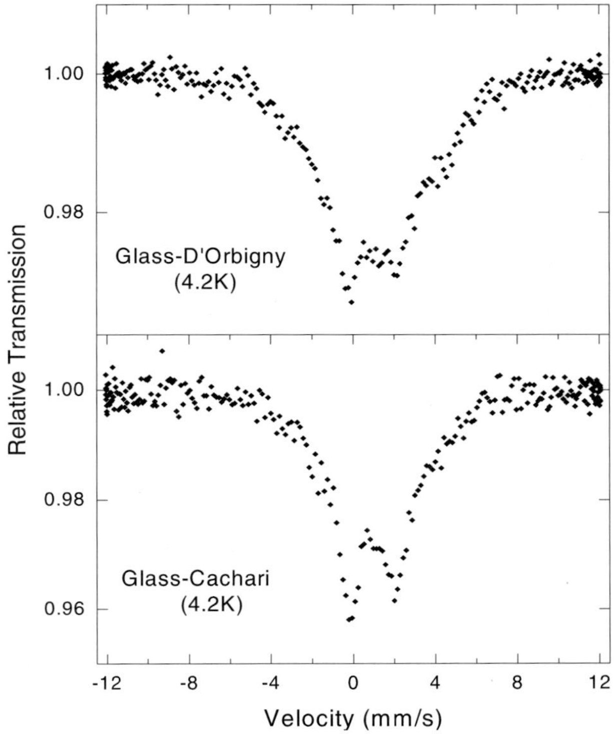

Figure 3 Mössbauer spectra (at 4.2 K) of glasses from D'Orbigny and Cachari meteorites.

in order to better determine Fe^{2+} distribution between the different octahedral sites in theses glasses, and to examine the temperature dependence of the relaxation.

References

1. Varela, M.E., Kurat, G., Zinner, E., Métrich, N., Brandstätter, F., Ntaflos, T., Sylvester, P.: Geochim. Cosmochim. Acta **67**, 5027 (2003)
2. Fredriksson, K., Kraut, F.: Geochim. Cosmochim. Acta **31**, 1701 (1967)
3. Varela, M.E., Kurat, G., Zinner, E., Brandstätter, F.: Geophys. Res. Abstr. **7**, 07425 (2005)
4. Dyar, M.D.: Am. Mineral. **70**, 304 (1985)
5. Helgason, O., Morup, S.: Hyperfine Interact. **70**, 985 (1992)
6. Kurat, G., Varela, M.E., Brandstätter, F., Weckwerth, R., Clayton, R., Weber, H.W., Schultz, L., Wäsch, E., Nazarov, M.: Geochim. Cosmochim. Acta **68**, 1901 (2004)
7. Mittlefehldt, D.W., Killgore, M., Lee, M.T.: Meteorit. Planet. Sci. **37**, 345 (2002)
8. Srivastava, K.K.P.: J. Phys. C **20**, 2161 (1987)
9. Gibb, T.C.: In: Principles of Mössbauer Spectroscopy, p. 150. Chapman & Hall, New York (1976)

Hyperfine Interact (2005) 166:549–554
DOI 10.1007/s10751-006-9329-y

Mössbauer spectroscopy on Mars: goethite in the Columbia Hills at Gusev crater

G. Klingelhöfer · E. DeGrave · R. V. Morris · A. Van Alboom · V. G. de Resende · P. A. De Souza, Jr · D. Rodionov · C. Schröder · D. W. Ming · A. Yen

Published online: 16 November 2006

Abstract In January 2004 the USA space agency NASA landed two rovers on the surface of Mars, both carrying the Mainz Mössbauer spectrometer MIMOS II. The instrument on the Mars-Exploration-Rover (MER) Spirit analyzed soils and rocks on the plains and in the Columbia Hills of Gusev crater landing site on Mars. The surface material in the plains have an olivine basaltic signature [1, 5] suggesting physical rather than chemical weathering processes present in the plains. The Mössbauer signature for the Columbia Hills surface material is very different ranging from nearly unaltered material to highly altered material. Some of the rocks, in particular a rock named Clovis, contain a significant amount of the Fe oxyhydroxide goethite, α-FeOOH, which is mineralogical evidence for aqueous processes because it is formed only under aqueous conditions. In this paper we describe the analysis of these data using hyperfine field distributions (HFD) and discuss the results in comparison to terrestrial analogues.

G. Klingelhöfer (✉) · D. Rodionov · C. Schröder
Institute Inorganic and Analytical Chemistry, Johannes Gutenberg-University Mainz, Staudinger Weg 9, 55128 Mainz, Germany
e-mail: klingel@mail.uni-mainz.de

E. DeGrave · A. Van Alboom · V. G. Resende
Department of Subatomic and Radiation Physics, NUMAT Division, University Gent, B-9000 Gent, Belgium

R. V. Morris · D. W. Ming
NASA Johnson Space Center, Houston, TX, USA

V. G. Resende
Departamento de Química, Universidade Federal de Ouro Preto, 35400 Ouro Preto, MG, Brazil

P. A. Souza
CVRD, Rio de Janeiro, RJ, Brazil

A. Yen
Jet Propulsion Laboratory, Pasadena, CA, USA

Key words MIMOS II · weathering · Mars · goethite · hematite · field distribution · Gusev crater · Mars-Exploration-Rovers · mineralogy

1 Introduction

In January 2004 the USA space agency NASA landed successfully two rovers on the surface of Mars. The Mars-Exploration-Rovers (MER) Spirit at the Gusev crater landing site and Opportunity at the Meridiani Planum landing site are both carrying our Mössbauer spectrometer MIMOS II mounted on an robotic arm (IDD) on the rovers [1–6]. The IDD instruments, in particular the Mössbauer spectrometer and the APXS (X-ray fluorescence for elemental analysis [8]) are used to determine the chemistry and mineralogy of rocks and soils. Both the rovers and instruments have been designed to operate for about 90 sols (1 sol = 1 Martian day) but were still completely functional and operating in September 2005 (close to 2 Earth years of continues operation) and expected to last for several more month. The performance of the Mössbauer spectrometers MIMOS II did not degrade, only the intensity of the Co57/Rh source, with an activity of about 340 mCi at launch in June 2003, decreased according to the half life.

2 Gusev crater landing site

During the first year of operation at the Gusev crater landing site the *Spirit* Mössbauer instrument analysed rocks and soils at Bonneville crater and the ejecta plains between Bonneville and the Columbia Hills before entering the hills.

2.1 The plains

As determined by MB already in the very first spectrum and confirmed by many others [1], the rocks and soil around Bonneville and in the cratered plains have a basaltic MB signature, and their Mössbauer spectra are dominated by an olivine doublet [1, 4, 5, 7]. A typical spectrum taken of the rock Adirondack is shown in Figure 1.

The presence of abundant olivine in rocks and in surrounding soil as determined by MB suggests that physical rather than chemical weathering processes currently dominate the plains at Gusev crater.

2.2 The Columbia Hills

The MB signature for the Columbia Hills is very different. Instead of relatively fresh olivine basalt, the MB signature is characteristic of highly altered rocks. In particular, one of the rocks, named *Clovis*, which was found in the West Spur region, contains the highest amount of the Fe oxyhydroxide goethite (Gt), α-FeOOH, of about 40% in area (see Figure 2) found to date in the Columbia Hills.

Detailed analyses of the data ([6] and references therein) using superpositions of two model-independent hyperfine-field distributions (HFD) with two ditto quadrupole-splitting distributions (QSD), have indeed yielded for the broad low-field component (lightly shaded in Figure 2) hyperfine parameters (see Table I) that are

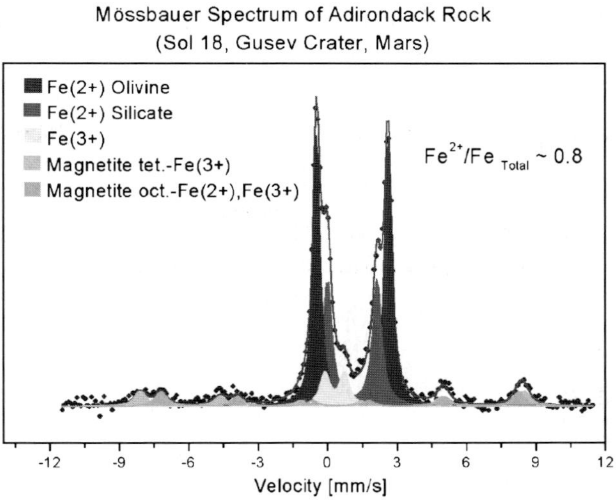

Figure 1 Mössbauer spectrum (14.4 keV; temperature range 220 to 280 K) of the rock Adirondack at the Spirit landing site. The data were taken at the as-is dusty surface (not yet brushed). The spectrum shows an olivine-basalt composition consisting of the minerals olivine, pyroxene, an Fe-3+ doublet, and non-stoichiometric magnetite.

in line with those of synthetic α-FeOOH having a particle size of 10 nm. In addition, the variations of the hyperfine parameters over the various temperature windows in which the Mars spectra have been collected, run more or less parallel to the respective variations observed for synthetic goethite. The ferric QSD resolved from the Clovis spectra may contain a (weak) contribution from goethite particles that are superparamagnetic in the temperature intervals measured on Mars, similar to the temperature variations as observed for the synthetic analogue. The fractional area taken up by the Gt component (see Table I, parameter RA) is fairly constant for the various ΔT windows. This implies that upon increasing the temperature, all of the Gt particles remain magnetic, instead of the smaller ones within the size-distributed conglomerates becoming superparamagnetic and thus giving rise to a quadrupole doublet that would contribute to the ferric doublet in the center of the spectrum. Like jarosite found at Meridiani Planum [3], the mineral goethite is a marker for aqueous processes as goethite forms only in the presence of water, and it contains the hydroxide anion.

The rock Clovis also contains a significant amount of hematite (see Table II and Figure 2). The temperature range of the measurements included the temperature where hematite (Hm) has its Morin transition. Therefore hematite was fitted with two sextets, one for the weakly ferromagnetic (*WF*) fraction of the hematite above the Morin transition, and a second one for the fraction of the antiferromagnetic (*AF*) spin state hematite below the Morin transition, well-known to co-exist in Hm powders or Al3+ substituted Hm; the isomer shifts δ for both contributions are very sharply defined for any Hm species encountered so far (natural or synthetic) and it is safe and justified to let their values fixed in the present iterations.

This behaviour indicates that the hematite present in these rocks has either a distribution in crystal sizes or Fe is substituted by other elements like Al. The derived hyperfine parameters for the Hm phase are listed in Table II. The difference between the AF and WF hyperfine fields, 8 kOe, is in excellent agreement with what is theoretically expected and experimentally observed for natural hematites sampled on Earth.

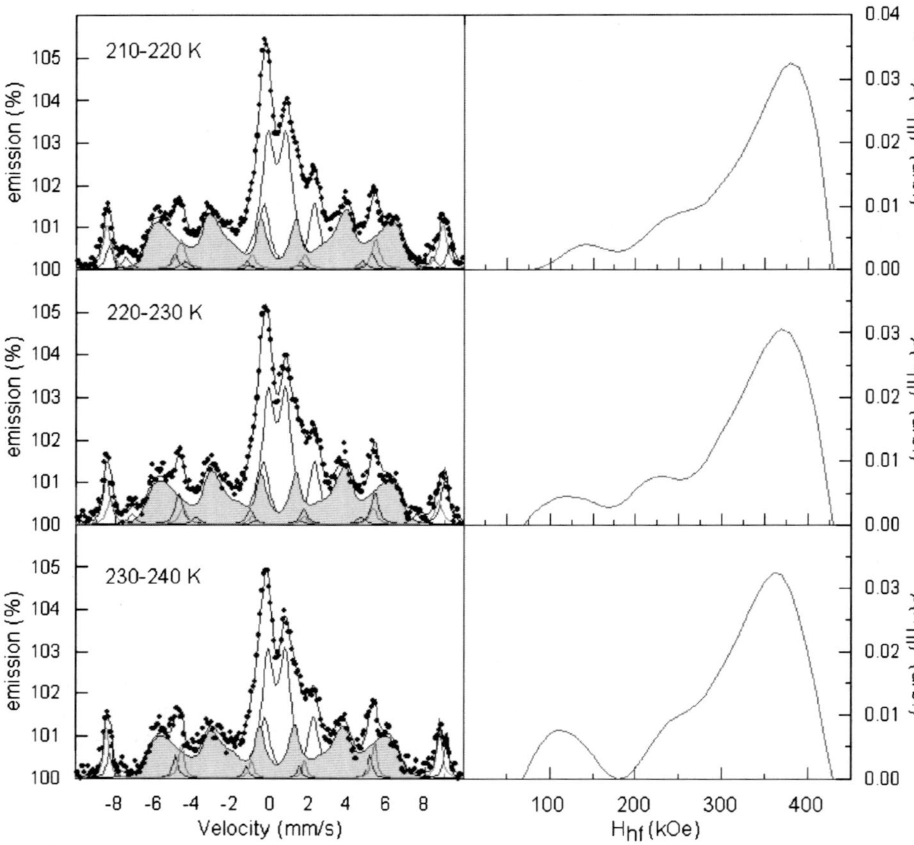

Figure 2 (*left*) Observed and fitted Mössbauer spectra (14.4 keV) acquired by MER Spirit at Gusev crater, rock 'Clovis' in the Columbia Hills, in three temperature windows as indicated. These spectra were fitted using a distribution for the relaxed goethite sextet and three discrete sites: a hematite sextet, an Fe3+ and a Fe2+ silicate doublet. The Fe3+ component possibly contains a contribution from the smaller goethite particles that behave superparamagnetic in the involved temperature window. (*right*) Corresponding hyperfine-field-distribution (HFD) profiles (in arb. units) of the goethite component calculated from the experimental data.

Besides the sextet components two doublet components are identified (see Figure 2), a ferric doublet and a ferrous one. All doublet parameters (see Table III), including RA remain almost unchanged upon passing thru the consecutive windows. This finding is in line with the general behaviour of isomer shift and quadrupole splitting of paramagnetic doublets: on changing the temperature, the variations, if any, are relatively small.

3 Summary

The MER Spirit Mössbauer spectrometer MIMOS II has identified different Fe-Oxides, silicates and other minerals and Fe-compounds (in total eight different phases) up to now at the Gusev landing site. The spatial distribution of the minerals

Table I Maximum-probability hyperfine field Hhf and some other relevant Mössbauer parameters (see caption Table II) of the MHFD fitted to the goethite component present in the Mars MS for the various ΔT "windows" (temperatures)

Window (K)	H_{hf}	δ	$2\varepsilon_Q$	RA
210–220	381	0.416*	−0.18	41
220–230	371	0.416*	−0.21	41
230–240	363	0.37	−0.19	44
240–260	354	0.37	−0.21	41

Table II Hyperfine parameters of the Hm phase present in the Mars MS for the various ΔT "windows"

Window (K)	H_{hf}	δ	$2\varepsilon_Q$	RA	r_{WF}
210–220	527	0.416*	−0.18	12	0.71
	529	0.416*	0.30*	5	
220–230	525	0.416*	−0.16	16	0.84
	521	0.416*	0.30*	3	
230–240	525	0.406*	−0.26	10	0.63
	523	0.406*	0.30*	6	
240–260	518	0.406*	−0.20*	16	

The quadrupole shifts ($2\varepsilon_Q$) and isomer shifts (δ, vs. α-Fe at RT) are given in millimeters per second, the hyperfine fields H_{hf} in kOe, the relative areas (RA) in percent of total spectrum. The parameter r_{WF} is the fractional contribution of the WF state to the total Hm fraction. Fixed values are marked by *.

Table III Fitted Mössbauer parameters for the ferrous and ferric quadrupole doublets: quadrupole splittings ΔE_Q and isomer shifts δ versus α-Fe, both in millimeters per second, and percentages of total area

Window (K)	ΔE_Q	RA	δ
Fe^{2+} doublet			
210–220	2.60	12	1.07
220–230	2.63	12	1.07
230–240	2.53	13	1.09
240–260	2.58	14	1.10
Fe^{3+} doublet			
210–220	0.94	27	0.43
220–230	0.90*	25	0.44
230–240	0.90	26	0.45
240–260	0.89	25	0.45

show that the plains of Gusev crater are mainly basaltic and relatively unaltered with an $Fe^{3+}/Fe_{total} < 0.2$. The Columbia hills on the other hand show a high degree of alteration. Only a few rocks are relatively unaltered. One of the highest degrees of alteration has been seen at the Clovis outcrop with a $Fe^{3+}/Fe_{total} \sim 0.8$. Another striking feature of Columbia Hill's outcrop is a variability of Fe-oxide phases for different locations. For example, Clovis has no magnetite component. Ebenezer, a different rock target in the hills, however, has substantial magnetite contribution in addition to hematite and goethite.

The rock Clovis contains a significant amount of goethite with about 40% of the total spectral area (see Figure 2). The mineral goethite is mineralogical evidence for the presence of H_2O/OH on Mars and for aqueous processes in the history of Mars.

Acknowledgement This work was funded by the German Space Agency DLR under contract 50QM99022, by the Fund for Scientific Research – Flanders, under grant G.0185.03, by CVRD Brazil, NASA MER project and NASA Johnson Space Center.

References

1. Morris, R.V., Klingelhöfer, G., Bernhardt, B., Schröder, C., Rodionov, D.S., de Souza Jr. P.A., Yen, A., Gellert, R., Evlanov, E.N., Foh, J., Kankleit, E., Gütlich, P. Ming, D.W., Renz, F., Wdowiak, T., Squyres, S.W., Arvidson, R.E. et al.: Science **305**, 833 (2004)
2. Klingelhöfer, G., Morris, R.V., Bernhardt, B., Rodionov, D., de Souza Jr. P.A., Squyres, S.W., Foh, J., Kankeleit, E., Bonnes, U., Gellert, R., Schröder, C., Linkin, S., Evlanov, E., Zubkov, B., Prilutski, O.: J. Geophys. Res. **108**, 8067 (2003)
3. Klingelhöfer, G., Morris, R.V., Bernhardt, B., Schröder, C., Rodionov, D.S., de Souza Jr. P.A., Yen, A., Gellert, R., Evlanov, E.N., Zubkov, B., Foh, J., Bonnes, U., Kankleit, E., Gütlich, P., Ming, D.W., Renz, F., Wdowiak, T., Squyres, S.W., Arvidson, R.E.: Science **306**, 1740 (2004)
4. Klingelhöfer, G.: In: Garcia, M., Marco, J.F., Plazaola, F. (eds.) Industrial Applications of the Mössbauer Effect. American Institute of Physics, Melville, NY (2005)
5. Morris, R.V., Klingelhöfer, G., Schröder, C., Rodionov, D.S., Yen, A., Ming, D.W., de Souza Jr. P.A., Fleischer, I., Wdowiak, T., Gellert, R., Bernhardt, B., Evlanov, E., Zubkov, B., Foh, J., Bonnes, U., Kankeleit, E., Gütlich, P., Renz, F., Squyres, S.W., Arvidson, R.: J. Geophys. Res. **111**, E02S12 (2006) doi: 10.1029/2005JE002560
6. Ming, D.W., Mittlefehldt, D.W., Morris, R.V., Golden, D.C., Gellert, R., Yen, A., Clark, B.C., Squyres, S.W., Farrand, W.H., Ruff, S.W., Arvidson, R.E., Klingelhöfer, G., McSween, H.Y., Rodionov, D.S., Schröder, C., de Souza Jr. P.A., Wang, A.: J. Geophys. Res. **111**, E02S12 (2006) doi: 10.1029/2005JE002560
7. Klingelhöfer, G. et al.: J. Geophys. Res. (2006), manuscript in preparation
8. Squyres, S.W. et al.: J. Geophys. Res. **108**, 8062 (2003), doi 10.1029/2003JE002121

Hyperfine Interact (2005) 166:555–560
DOI 10.1007/s10751-006-9334-1

Simulations of the magnetic properties experiment on Mars Exploration Rovers

H. P. Gunnlaugsson · E. S. Worm · P. Bertelsen · W. Goetz · K. Kinch · M. B. Madsen · J. P. Merrison · P. Nørnberg

Published online: 5 December 2006

Abstract We present some of the main findings from simulation studies of the Magnetic Properties Experiment on the Mars Exploration Rovers. The results suggest that the dust has formed via mechanical breakdown of surface rocks through the geological history of the planet, and that liquid water need not have played any significant role in the dust formation processes.

Key words dust on Mars · magnetite · olivine basalt

1 Introduction

The characteristic red colour of Mars is due to red dust sized (<5 μm in diameter [1]) particles, covering more dark/grey rocks units, and mixed together with the top-soil of basaltic origin [2]. Based on the optical and magnetic properties of the dust it has been suggested that it is composed of secondary ferric iron oxides [3], and that

H. P. Gunnlaugsson (✉) · E. S. Worm · K. Kinch · J. P. Merrison
Institute of Physics and Astronomy, Aarhus University, Ny Munkegade,
DK-8000 Århus C, Denmark
e-mail: hpg@phys.au.dk

P. Bertelsen · M. B. Madsen
Center for Planetary Science, University of Copenhagen, DK-2100 Copenhagen Ø, Denmark

W. Goetz
Max Planck Institute for Solar System Research, Max-Planck-Str. 2,
D-37191 Katlenburg-Lindau, Germany

K. Kinch
Department of Astronomy, Cornell University, 408 Space Sciences Building, Ithaca,
NY 14853-6801, USA

P. Nørnberg
Department of Earth Sciences, Aarhus University, Ny Munkegade,
DK-8000 Århus C, Denmark

identification of the mineralogy and chemical composition of the dust, and in particular the magnetic phase, could give information on the formation conditions [3–5]. For this reason, both Mars Exploration Rovers (MERs) Spirit and Opportunity [6] were equipped with a series of permanent magnets [7] that sample the dust suspended in the atmosphere for investigation with Mössbauer spectroscopy [8] and elemental analysis [9]. The magnets are of two types, the so-called capture magnet, designed to be as strong as possible and the so-called filter magnet, designed to filter and concentrate the more strongly magnetic particles from the dust.

The surprising result based mainly on Mössbauer spectroscopy is that the dust accumulated onto the magnets showed dominantly paramagnetic compounds identified as the rock forming minerals olivine and pyroxene with only moderate amounts of magnetite as the magnetic phase [10]. Around Spirit sol 300 (1 sol=1 Martian day) and Opportunity sol 200, dust clearing episodes, most probably related to very strong wind gusts, maybe even the passing of dust devils, resulted in a change in the composition of the dust by removal of the less magnetic particles and addition of particles with higher concentration of the magnetic phase [11]. Based on the simulation studies presented here, and as it will be discussed below, it seems evident that in the beginning of the missions, the accumulated dust was mostly characteristic for the true dust component in the atmosphere and got only sorted later due to strong dust clearing episodes. Analysis of the first spectra of the dust accumulated to the filter magnet on Spirit suggest a relatively high Fe(III) content, of the order of 30–40% of the total spectral area, while 15–25% is more common for soils at Gusev crater [2]. The elemental composition of the accumulated dust was not found to differ significantly from soil in general, but shows some characteristically different elemental ratios [10].

In order to understand these findings and how to interpret them, we have performed a series of studies under simulated Martian conditions ($P{\sim}6$–9 mbar CO_2 gas, $v{\sim}1$–10 m/s). Accumulation of dust particles onto magnets under such conditions has been found to differ significantly from terrestrial conditions [12] due to the difference in dynamics of dust particles caused by the different aerodynamic environment. In the study described here we used as a starting material a good Mössbauer analogue to the igneous rocks found on Mars [2], but the samples are neither good optical or particle size analogues to the Martian dust.

2 Experimental and samples

Transmission Mössbauer spectra were recorded at room temperature using 10–50 mCi ^{57}Co:Rh sources mounted on a conventional constant acceleration velocity drive. Velocities and isomer shifts are given relative to the centre of the spectrum of α-Fe at room temperature. Particle size distribution was measured using a laser diffraction instrument (Sympatec) mounted with a wet dispersion flow cell (QUIXEL).

Here are given the results of experiments using ~0.8 million years old olivine basalt from Grótta, Iceland. Studies of this basalt unit have shown that it is an excellent Mössbauer analogue to the unaltered volcanic rocks at Gusev Crater on Mars [13]. The samples were crushed in a mechanical grinder to form a powder with average particle size of ~20 μm suitable for wind tunnel experiments.

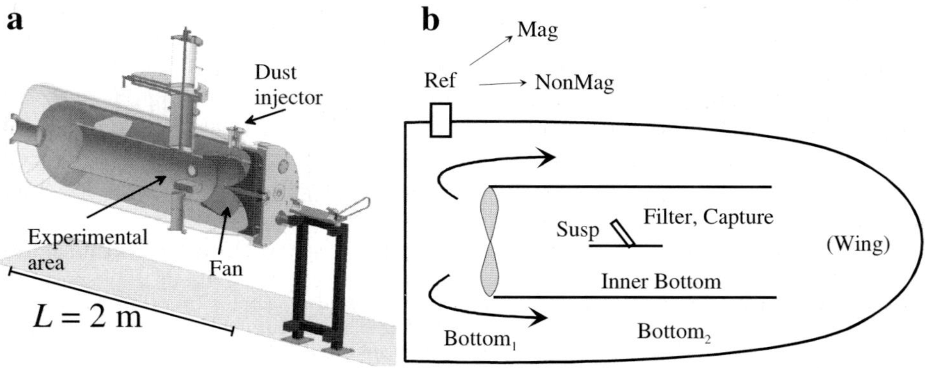

Figure 1 **a** Schematic illustration of the Aarhus wind tunnel. The tunnel is enclosed in a vacuum chamber, which can be evacuated to the desired pressure and the gas composition can be controlled. A fan causes wind flow and this can be regulated up to about 25 m/s with turbulence no higher than 16% at $v > 4$ m/s. A dust injection system allows dust to be put into suspension. **b** After an experiment, samples are gathered from different locations inside the wind tunnel system for further studies.

Wind tunnel experiments were performed in the Wind tunnel facilities at the Mars Simulation Laboratory at the University of Aarhus, Denmark [12] using wind speed of 5 m/s and 10 mbar air (similar aerodynamics as for 6 mbar CO_2 gas). A schematic drawing of the wind tunnel is shown in Figure 1.

After injection, the dust particles come into suspension and are circulated around the system for several minutes. Larger particles fall first out of suspension, and are mainly located at the bottom of the wind tunnel, but also on specific sample locations (wings) on the inner tube. Additionally a <2 μm separate was prepared by gravitational sedimentation after Stokes law in 0.002 M sodium pyrophosphate.

3 Results and discussion

Figure 2 shows representative Mössbauer spectra from the investigations.

The spectra can be analysed in terms of lines originating from paramagnetic ferrous iron assigned to olivine ((Fe, Mg)$_2$SiO$_4$, pyroxene ((Fe, Mg, ..)SiO$_3$, and mineralogically unspecific ferric iron (Fe(III)) and magnetic components assigned to titanomagnetite and ferric iron oxide, most likely titanomaghemite. The Mössbauer spectra of the dust that accumulated onto the magnets showed only minor differences (slightly more magnetic phases) from the spectra of dust accumulated from nonmagnetic sedimentation onto plates beside the magnets (cf. Figure 2), indicating that the dust accumulated onto the magnets is a fair analogue to the dust in general. This is despite the fact that the magnets capture well over an order of magnitude more material per unit area as compared to sedimentation to non-magnetic surfaces under the same geometrical conditions.

In general the experiment simulated well the results from the Magnetic Properties Experiment, in terms of relative capture rates of the two magnets, and optical variations over the surface of the magnets [14].

The data showed correlation between average particle size and Fe(III) content as illustrated in Figure 3.

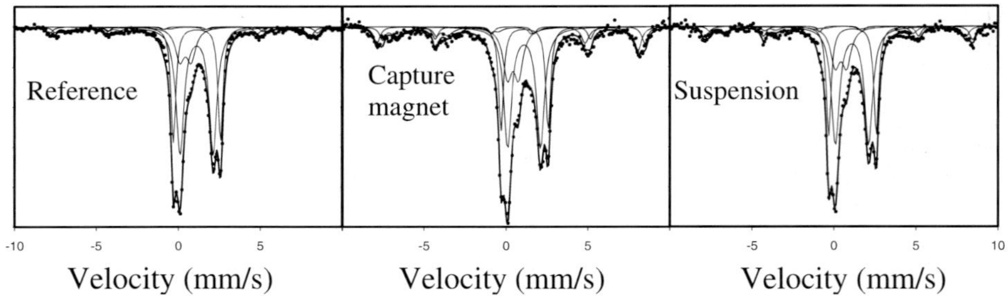

Figure 2 Room temperature Mössbauer spectra of representative samples from this study: *Left*: Reference sample injected into the wind tunnel. *Middle*: Sample accumulated onto the capture magnet. *Right*: Sample gathered from a plate beside the capture magnet.

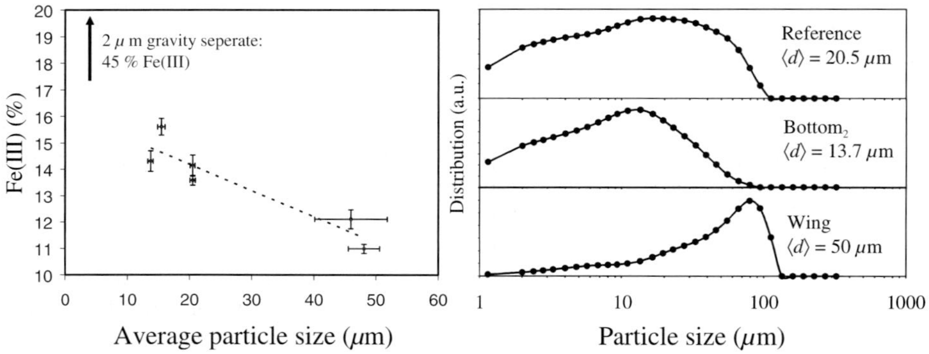

Figure 3 *Left*: Fe(III) content as determined from Mössbauer spectroscopy as a function of average particle size. *Right*: Particle size distribution shown for three different samples.

There are at least two phenomena that give rise to the observed behaviour: First, the oxidation of the sample during grinding into powder, where local heating in the break zones causes oxidation in the surface layers. The effect however is not sufficient to explain the whole tendency shown in the figure. In basaltic glass, where Fe(III) is probably mostly homogeneously distributed in the matrix, we observe only a moderate increase of Fe(III) as observed with CEMS (penetration depth ~0.25 μm, see e.g. [15]), typically Fe(III) is about 20% in the bulk and 27% in the CEMS measurement (H. P. Gunnlaugsson, unpublished data), and this cannot explain the extremely high oxidation state of the gravity separate. An alternative explanation is due to inhomogeneities in the material. Volumes representing weaknesses in the material can during grinding be crushed into finer powder on the average and these volumes are often characterised by higher oxidation state or less hard minerals. This is the effect observed in partially oxidised olivine basalt, where the olivine particles represent weaknesses in the material and surfaces containing olivine are overrepresented in CEMS measurements [16]. Both effects probably take place, and the crushing is associated with a change in the Na/Mg atomic ratio from 0.2 in the bulk sample to 0.38 in the 2 μm gravity separate which is a similar difference as observed on the magnets on Mars [10].

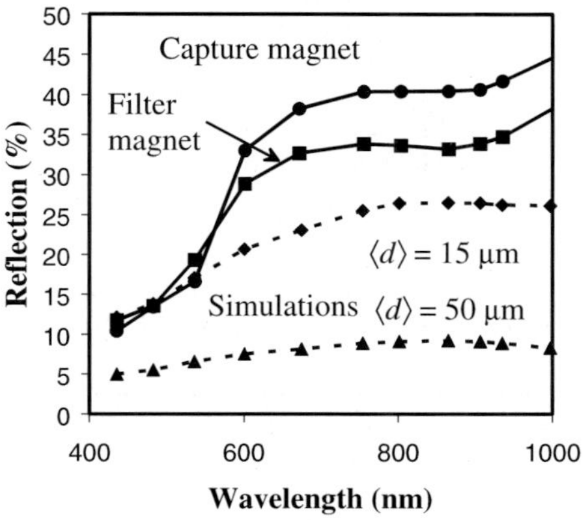

Figure 4 Optical reflection of the dust accumulated onto the Spirit magnets (*solid lines*) at sol 71 (modified from data from [17] by assuming incomplete covering) compared to two separates from the simulation experiments (*dashed lines*).

The colour of samples was found to depend significantly on the size distribution, the smaller samples, and in particular the samples accumulated onto the magnets show distinctly more reddish/brownish colour than the bulk or the samples with larger particle sizes (cf. Figure 4).

The smaller particles have clearly more reddish reflection, and the reason for this is probably twofold: The smaller fractions have higher oxidation state and relatively larger number of particles that have size close to the wavelength of red light resulting in increased reflection in the red end of the spectrum [18]. The effect observed in our simulation studies is not as strong as observed on Mars, possibly due to the fact that we have relatively larger particles. We were not able to obtain reliable optical data for the 2 µm separate due to contamination of bright coloured phosphorous compounds. All together, we can from this data offer – perhaps – a new explanation of the reddish colour of Mars.

It is noteworthy to add that the dust that has been shown to exists on Mars today is unstable on Earth. In particular the mineral olivine dissolves readily in water, and few micrometer particles would only withstand of the order of years in the wet chemistry of our atmosphere (estimate based on data from [19]).

4 Conclusions

Essential features of the results from the Magnetic Properties Experiment onboard the Mars Exploration Rovers on Mars can be reproduced in a simulated Martian dust environment. The results indicate that the reddish colour of the dust, at least some of the elemental composition, Mössbauer spectra and oxidation state of the dust, can be simulated without the involvement of water. The findings presented here suggest that the dust on Mars has formed during the geological history of the planet due to slow grinding of the surface rocks. While the larger, relatively darker particles are forming dunes, the dust that can be suspended in the atmosphere is kept in circulation, covering darker rock units and mixed with the top soil giving rise to the characteristic red colour of Mars: Without the extensive role of water.

References

1. Pollack, J.B., Colburn, D.S., Flasar, F.M., Kahn, R., Carlston, C.E., Pidek, D.: J. Geophys. Res. **84**, 2929 (1979)
2. Morris, R.V., Klingelhöfer, G., Bernhardt, B., Schröder, C., Rodionov, D.S., de Souza Jr., P.A., Yen, A., Gellert, R., Evlanov, E.N., Foh, J., Kankeleit, E., Gütlich, P., Ming, D.W., Renz, F., Wdowiak, T., Squyres, S.W., Arvidson, R.E.: Science **305**, 833 (2004)
3. Hargraves, R.B., Knudsen, J.M., Bertelsen, P., Goetz, W., Gunnlaugsson, H.P., Hviid, S.F., Madsen, M.B., Olsen, M.: Magnetic enhancement on Mars. J. Geophys. Res. **105**, 1819 (2000)
4. Knudsen, J.M., Madsen, M.B., Olsen, M., Vistisen, L., Koch, C.B., Mørup, S., Kankeleit, E., Klingelhöfer, G., Elanov, E.N., Khromov, V.N., Mukhin, L.M., Prilutski, O.F., Zubkov, B., Smirnov, G.V., Juchniewich, J.: Hyper. Interact. **68**, 83 (1991)
5. Madsen, M.B., Ågerkvist, P.D., Gunnlaugsson, H.P., Faurschou Hviid, S., Knudsen, J.M., Vistisen, L.: Hyper. Interact. **95**, 291 (1995)
6. Squyres, S.W., Arvidson, R.E., Baumgartner, E.T., Bell III, J.F., Christensen, P.R., Gorevan, S., Herkenhoff, K.E., Klingelhöfer, G., Madsen, M.B., Morris, R.V., Rieder, R., Romero, R.A.: J. Geophys. Res. **108**, 8062 (2003)
7. Madsen, M.B., Bertelsen, P., Goetz, W., Binau, C.S., Olsen, M., Folkmann, F., Gunnlaugsson, H.P., Kinch, K.M., Knudsen, J.M., Merrison, J., Nørnberg, P., Squyres, S.W., Yen, A., Rademacher, J., Gorevan, S.: J. Geophys. Res. **108**, 8069 (2003)
8. Klingelhöfer, G., Morris, R.V., Bernhardt, B., Rodionov, D., de Souza Jr., P.A., Squyres, S.W., Foh, J., Kankeleit, E., Bonnes, U., Gellert, R., Schröder, C., Linkin, S., Evlanov, E., Zubkov, B., Prilutski, O.: J. Geophys. Res. **108**, 8067 (2003)
9. Rieder, R., Gellert, R., Brückner, J., Klingelhöfer, G., Dreibus, G., Yen, A., Squyres, S.W.: J. Geophys. Res. **108**, 8066 (2003)
10. Goetz, W., Bertelsen, P., Binau, C.S., Gunnlaugsson, H.P., Hviid, S.F., Kinch, K.M., Madsen, D.E., Madsen, M.B., Olsen, M., Gellert, R., Klingelhöfer, G., Ming, D.W., Morris, R.V., Rieder, R., Rodionov, D.S., de Souza Jr., P.A., Schröder, C., Squyres, S.W., Wdowiak, T., Yen, A.: Nature **436**, 62 (2005)
11. Madsen, M.B. et al.: J. Geophys. Res. (2005), Manuscript in preparation
12. Merrison, J.P., Gunnlaugsson, H.P., Mossin, L., Nielsen, J., Nørnberg, P., Rasmussen, K., Uggerhøj, E.: Planet. Space Sci. **50**, 371 (2002)
13. Rasmussen, H., Gunnlaugsson, H.P., Tegner, C., Kristjánsson, L.: Hyperfine Interact. (in press, these proceedings). DOI 10.1007/s10751-006-9341-2 (2006)
14. Kinch, K.M., Merrison, J.P., Gunnlaugsson, H.P., Bertelsen, P., Madsen, M.B., Nørnberg, P.: Planet. Space. Sci. **54**, 28 (2006)
15. Gunnlaugsson, H.P., Bertelsen, P., Butz-Jørgensen, C., Klingelhöfer, G., Kuvvetli, I., Madsen, M.B., Merrison, J.P., Nørnberg, P., Rasmussen, H., Weyer, G.: On the merits of Conversion Electron Mössbauer Spectroscopy in geosciences. Hyperfine Interact. (in press, these proceedings). DOI 10.1007/s10751-006-9318-1 (2006)
16. Gunnlaugsson, H.P., Helgason, Ö., Kristjánsson, L., Nørnberg, P., Rasmussen, H., Steinþórsson, S., Weyer, G.: Phys. Earth Planet. Inter. **154**, 276 (2006)
17. Bertelsen, P., Goetz, W., Madsen, M.B., Kinch, K.M., Hviid, S.F., Knudsen, J.M., Gunnlaugsson, H.P., Merrison, J., Nørnberg, P., Squyres, S.W., Bell III, J.F., Herkenhoff, K.E., Gorevan, S., Yen, A.S., Myrick, T., Klingelhöfer, G., Rieder, R., Gellert, R.: Science **305**, 827 (2004)
18. Hapke, B.: In: Arvidsson, R.E., Rycroft, M.J. (eds.) Topics in Remote Sensing, vol. 3. Cambridge University Press (1993)
19. Stopar, J.D., Taylor, G.J., Hamilton, V.E., Browning, L., Pickett, D.: 6th International Conference on Mars (2003) abstract no. 3151

Hyperfine Interact (2006) 166:561–566
DOI 10.1007/s10751-006-9341-2

Magnetic properties of Martian olivine basalts studied by terrestrial analogues

H. Rasmussen · H. P. Gunnlaugsson · C. Tegner · L. Kristjánsson

Published online: 27 October 2006

Abstract Here we present a comparison study of terrestrial olivine basalt and relatively un-weathered basalt studied by the Mars Exploration Rover Spirit on Mars. The Mössbauer spectra of terrestrial olivine basalt exhibit some characteristics that can also be seen in the Mars spectra. The results from Mössbauer spectroscopy on Mars indicate that the olivine in the rocks has undergone alteration at high temperatures (600–1,000°C), a process known to give rise to anomalously magnetic rocks on Earth. This suggests that if the rocks at Gusev crater had solidified in an external magnetic field of terrestrial magnitude, these would have become highly magnetic enough to explain the presence of magnetic anomalies on Mars.

Key words Mars · olivine basalt · Mössbauer spectroscopy · magnetic anomalies

1 Introduction

Among the major discoveries of the Mars Global Surveyor mission is the finding of strong magnetic remanent anomalies [1]. Models suggest magnetic rocks tens of kilometers in thickness with remanent magnetisation up to $M_r \sim 20$ A/m [2], which is about an order of magnitude higher than average values for terrestrial mid-ocean ridge basalt (MORB). New results from the Mars Exploration Rovers (MER), in

H. Rasmussen (✉) · C. Tegner
Department of Earth Sciences, Aarhus University, C.F. Møllers Alle, DK-8000 Århus C, Denmark
e-mail: henrik.rasmussen@geo.au.dk

H. P. Gunnlaugsson
Institute of Physics and Astronomy, Aarhus University, Ny Munkegade, DK-8000 Århus C, Denmark

L. Kristjánsson
Science Institute, University of Iceland, Dunhaga 3, IS-107 Reykjavik, Iceland

particular by Mössbauer spectroscopy, suggest that olivine-rich basalt lavas may constitute a significant portion of the Martian crust, and yet remain difficult to detect by means of orbital remote sensing [3].

With the aim to better constrain the Mars data and geology we have studied olivine basalt (<0.8 Ma) from Iceland. The Mössbauer data from Iceland lavas has proved a good analogue to the relatively un-weathered volcanic rocks at Gusev crater on Mars [4].

The MER Mössbauer data [5] yields two spectra: one from backscattered 14.4 keV γ-rays and one from backscattered 6.4 keV X-rays. The sampling depth varies with the energy of the radiation. For typical volcanic rocks ($\rho \approx 3$ g/cm^3) the 14.4 keV γ-rays give information from ~100 μm beneath the surface, while the 6.4 keV X-rays give information from ~50 μm. A difference in these spectra therefore provides information on the depth of oxidation processes and/or mechanical properties of the sample.

The rock abrasion tool (RAT) [6] on the MERs instrument deployment arm can be applied to either brush the surface of a rock sample or mechanically remove the surface by a grinding tool (ratting) to look beneath the weathered/dust coated top layer.

2 Iceland samples and experiments

A total of 31 samples were collected from <0.8 Ma olivine basalt from Reykjavik and the Reykjanes peninsula, Iceland, hereafter called the HRI series. The samples were collected either by drilling or by hammering. Following detailed measurements of the remanent magnetisation of the samples, they were ground into a homogeneous powder for use in other experiments. Magnetic separates were obtained by placing the sample on a piece of paper and applying hand magnets to collect the most strongly magnetic particles.

Mössbauer spectra were recorded at room temperature using 10–50 mCi ^{57}Co:Rh sources moved relative to the sample on a conventional constant acceleration drive system. For conversion electron Mössbauer spectroscopy (CEMS) measurements, the bulk samples were placed on a double adhesive conduction tape and mounted as a cathode in a parallel plate avalanche detector. Velocities are given relative to α-Fe at room temperature.

Mössbauer spectra of samples from Mars spectra were acquired from the Planetary Data System (http://anserver1.eprsl.wustl.edu). Here we concentrate on the spectra obtained on relatively un-weathered rocks at Gusev crater, obtained before the rover found more weathered/oxidized rocks at Columbia Hills.

3 Results

Figure 1 shows representative spectra from Iceland. All the spectra were analysed in terms of eight spectral components, although no single spectrum contains more than seven components.

Symmetric quadrupole doublets were assigned to paramagnetic Fe(II) in olivine, $(Mg,Fe)_2[SiO_4]$, pyroxene, $(Mg,Fe,Ca\ldots)[SiO_3]$, ilmenite, $(FeTiO_3)$ and unspecific

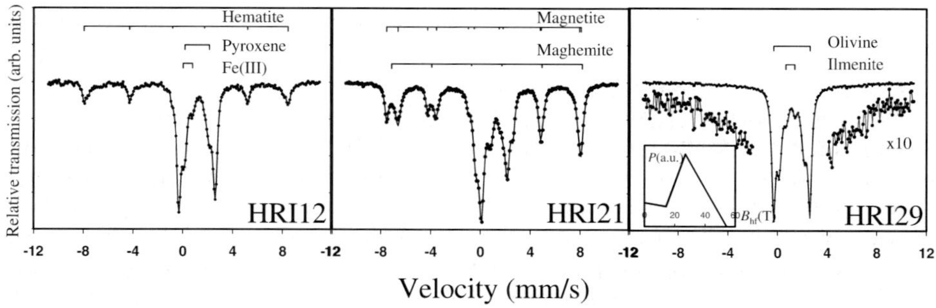

Figure 1 Room temperature Mössbauer spectra of olivine basalt samples from Iceland. The *solid lines* show the sum of the fitting components indicated with *bar diagrams* together with their assignments. The inserted figure in the HRI29 spectra shows the hyperfine distribution for titanomagnetite with $0.5<x<0.7$.

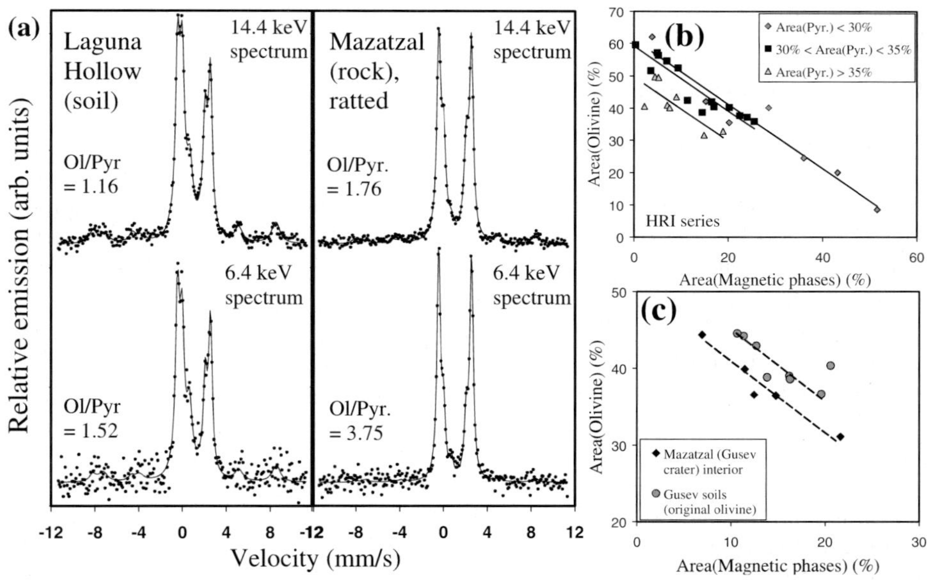

Figure 2 **a** Backscattering Mössbauer spectra from Mars of the targets and energy indicated. The spectral area ratio between olivine and pyroxene is indicated. **b** Correlation between the relative area of Fe in olivine and magnetic phases in spectra of the HRI series, binned after the pyroxene ratio. The *solid lines* are to guide the eye within each binning. **c** Correlation between the relative area of Fe in olivine and iron oxides for the Mars data following the procedures given in the text.

Fe(III). Simple magnetic hyperfine split sextets were assigned to hematite (α-Fe_2O_3) or maghemite (γ-Fe_2O_3) and stoichiometric magnetite, Fe_3O_4. Some spectra showed very broad features (cf. the spectrum of HRI29 in Figure 1). A magnetic hyperfine field distribution was applied in these cases, assigned to titanomagnetite ($Fe_{3-x}Ti_xO_4$) with $0.5<x<0.7$.

The spectra obtained from one sample (bulk, magnetic separates, CEMS measurements) were analysed simultaneously assuming the same hyperfine parameters of components in different amounts. The fitting procedure is described in more detail in [7].

Figure 2 shows representative Mössbauer spectra from Mars.

Mössbauer spectra from Mars (Figure 2a) have been analysed in terms of the same components as the spectra from Iceland, with the exceptions that ilmenite was not detected in significant amounts on Mars; hematite and maghemite were not differentiated and a single sextet was assigned to ferric iron oxide, probably a mixture of maghemite and hematite. The magnetic hyperfine field distribution for titanomagnetite was not applied, though misfits at $v\sim-4$ mm/s indicate the presence of this component in small amounts. The spectra from Mars are recorded at different temperatures, and were analysed simultaneously restricting the isomer shift of the components to follow the second order Doppler shift, and assuming that within the temperature range the quadrupole splitting could be described as linear in T. The fits obtained of the 14.4 keV spectra are in good agreement with the results presented in [4]. The spectra and analysis presented in Figure 2 are summed over all temperatures.

4 Discussion and conclusions

The Mössbauer spectra of olivine basalt from Iceland, show a negative correlation between the proportion of Fe in olivine and the proportion of Fe in magnetic phases [7] (cf. Figure 2b). The correlation depends on the proportion of Fe in pyroxene, however, it is important to note that no significant correlation is found between the proportion of Fe in pyroxene and olivine nor pyroxene and magnetic phases. The correlation between olivine and magnetic phases is attributed to partial oxidation of olivine that can lead to the formation of magnetic iron oxides imbedded in the original olivine host. The process can take place at both high (600–1,100°C) and low (<600°C) [8, 9] temperatures. In the former case, the magnetic iron oxides can form in a single domain state (<1 μm particles) above the Curie temperature giving rise to high magnetic remanence [10].

The correlation between the proportion of Fe in olivine and magnetic phases is not observed directly in the data from Mars (see e.g. the data from [4] based on the analysis of the 14.4 keV b-MS). The Mössbauer data from Mars shows a pronounced difference between the 14.4 keV spectra and the 6.4 keV spectra (cf. Figure 2a). The effect must be caused by variations in the mineralogy in the surface layer on the scale of tenths of micrometer. Generally, all rock targets, untreated surfaces, brushed or abraded, show the same characteristics. The difference can be characterised by the area ratio between olivine and pyroxene, showing enhancement of olivine in the surface layer of the rocks. For soils, the difference is less pronounced.

This characteristic difference is of importance for the conclusions that can be drawn from this work. First of all, the true bulk Mössbauer spectrum of the rock can be extrapolated from the measurements by subtracting the results of the analysis of the 6.4 keV spectra from the analysis of the 14.4 keV spectra, obtaining for the component labelled c: $A_{\text{int}}(c)=(A_{14.4\ \text{keV}}(c)-k{\cdot}A_{6.4\ \text{keV}}(c))(A_{14.4\ \text{keV}}(\text{tot})--k{\cdot}A_{6.4\ \text{keV}}(\text{tot}))$, where A is the area of the respective spectra. With k=0.48 the results shown in Figure 2c have been obtained for the rock target Mazatzal, where five measurements were performed. Also shown in the figure are results obtained from 14.4 keV spectra of soil samples where we have assumed that part of the Fe(III) component originates from olivine. These findings support the assignments of components and are at variance with suggestions that the spectra are dominated by hydrous iron sulphates [11].

The surface olivine enhancement observed in the Mars spectra resembles that observed in terrestrial samples when comparing CEMS (sensitive to ~0.25 μm) and transmission Mössbauer spectra of olivine basalt oxidised at high temperatures [7, 12]. In the latter the altered olivine crystals have lost their mechanical strength, and during sample grinding, they get overexposed in the surface layers [7]. In the Mars spectra the most probable explanation is that the ratting of the samples cuts through the weaker olivine crystals, while the more robust pyroxene crystals are removed as whole. The resulting surface layer will have higher olivine/pyroxene ratio than the rock as a whole. This observation indicates that the olivine crystals have been altered at high temperatures, a process known to lead to formation of highly magnetic rocks on Earth.

The fact that even untreated surfaces show the same effect, indicates that the weathering of these rocks is of mechanical nature.

The oxidation of the olivine in the rocks from Iceland is probably of low temperature origin. This is seen in the morphology of the olivine crystals [12] which shows more similarities with low temperature experiments [9] than high temperature experiments [8] and in the ratio of the olivine/pyroxene ratio between CEMS and transmission Mössbauer measurements (on the average 1.4 in spectra of the HRI series but up to 3 in highly magnetic olivine basalt [12]). The resulting remanence pr. weight percentage of highly magnetic phases of the HRI samples is likewise of the order of 1–4 A/m compared to ~20 A/m for highly magnetic olivine basalt [7]. If the rocks at Gusev crater had solidified in an external magnetic field of terrestrial magnitude (~50 μT) they would have acquired magnetic remanence of the order of ~40 A/m. Such rocks can easily explain the magnetic anomalies on Mars.

It has been shown, that olivine basalt can become an order of magnitude more magnetic than olivine poor basalt [7] due to high temperature oxidation of olivine during the solidification of the lava. The results from this work suggest that a major part of the magnetite found in the rocks at Gusev crater has formed in this way. For this to be true, the basalt must have had a high volatile content to drive the oxidation and this is corroborated by microscope images that show abundant vesicles and vugs, suggesting eruptive flows with high volatile content [13]. The rocks at the Gusev landing site have been reported to be considerably weaker physically than typical fine grained basalt [14] based on the grinding energy needed for the ratting. One possibility to explain this is that the weakening of the material as a whole is caused by weakening of the olivine component due to high temperature oxidation.

References

1. Acuña, M.H., Connerney, J.E.P., Ness, N.F., Lin, R.P., Michell, D., Carlson, C.W., McFadden, J., Anderson, K.A., Rème, H., Mazelle, C., Vignes, D., Wasilewski, P., Cloutier, P.: Science **284**, 790 (1999)

2. Purucker, M., Ravat, D., Frey, H., Voorhies, C., Sabaka, T., Acuña, M.: Geophys. Res. Lett. **27**, 2449 (2000)
3. McSween, H.Y., Wyatt, M.B., Gellert, R., Bell III, J.F., Morris, R.V., Herkenhoff, K.E., Crumpler, L.S., Milam, K.A., Stockstill, K.R., Tornabene, L., Arvidson, R.E., Bartlett, P., Blaney, D., Cabrol, N.A., Christensen, P.R., Clark, B.C., Crisp, J.A., Des Marais, D.J., Economou, T., Farmer, J.D., Farrand, W., Ghosh, A., Golombek, M., Gorevan, S., Greeley, R., Hamilton, V.E., Johnson, J.R., Joliff, B.L., Klingelhoefer, G., Knudson, A.T., McLennan, S., Ming, D., Moersch, J.E., Rieder, R., Ruff, S.W., Schroeder, C., de Souza Jr., P.A., Squyres, S.W., Wänke, H., Wang, A., Yen, A., Zipfel, J.: J. Geophys. Res. III, DOI 10.1029/2005JE002477 (2006)
4. Morris, R.V., Klingelhöfer, G., Bernhardt, B., Schröder, C., Rodionov, D.S., de Souza Jr., P.A., Yen, A., Gellert, R., Evlanov, E.N., Foh, J., Kankeleit, E., Gütlich, P., Ming, D.W., Renz, F., Wdowiak, T., Squyres, S.W., Arvidson, R.E.: Science **305**, 833 (2004)
5. Klingelhöfer, G., Morris, R.V., Bernhardt, B., Rodionov, D., de Souza Jr., P.A., Squyres, S.W., Foh, J., Kankeleit, E., Bonnes, U., Gellert, R., Schröder, C., Linkin, S., Evlanov, E., Zubkov, B., Prilutski, O.: J. Geophys. Res. **108**, 8067 (2003)
6. Gorevan, S.P., Myrick, T., Davis, K., Chau, J.J., Bartlett, P., Mukherjee, S., Anderson, R., Squyres, S.W., Arvidson, R.E., Madsen, M.B., Bertelsen, P., Goetz, W., Binau, C.S., Richter, L.: J. Geophys. Res. **108**, 8068 (2003)
7. Gunnlaugsson, H.P., Helgason, Ö., Kristjánsson, L., Nørnberg, P., Rasmussen, H., Steinþórsson, S., Weyer, G.: Phys. Earth Planet Inter. **154**, 276 (2006)
8. Haggerty, S.E., Baker, I.: Contrib. Mineral. Petrol. **16**, 233 (1967)
9. Baker, I., Haggerty, S.E.: Contrib. Mineral. Petrol. **16**, 258 (1967)
10. Hoye, G.S., Evans, M.E.: Geophys. J. R. Astron. Soc. **41**, 139 (1975)
11. Lane, M.D., Dyar, M.D., Bishop, J.L.: Geophys. Res. Lett. **31**, 19702 (2004)
12. Gunnlaugsson, H.P., Bertelsen, P., Butz-Jørgensen, C., Klingelhöfer, G., Kuvvetli, I., Madsen, M.B., Merrison, J.P., Nørnberg, P., Rasmussen, H., Weyer, G.: Hyp. Int. (2005), These proceedings
13. Herkenhoff, K.E., Squyres, S.W., Anderson, R., Archinal, B.A., Arvidson, R.E., Barrett, J.M., Becker, K.J., Bell III, J.F., Budney, C., Cabrol, N.A., Chapman, M.G., Cook, D., Ehlmann, B., Farmer, J., Franklin, B., Gaddis, L.R., Galuszka, D.M., Garcia, P.A., Hare, T.M., Howington-Kraus, E., Johnson, J.R., Johnson, S., Kinch, K., Kirk, R.L., Mae Lee, E., Leff, C., Lemmon, M., Madsen, M.B., Maki, J.N., Mullins, K.F., Redding, B.L., Richter, L., Rosiek, M.R., Sims, M.H., Soderblom, L.A., Spanovich, N., Springer, R., Sucharski, R.M., Sucharski, T., Sullivan, R., Torson, J.M.: J. Geophys. Res. III, DOI 10.1029/2005JE002574 (2006)
14. Arvidson, R.E., Anderson, R.C., Bartlett, P., Bell III, J.F., Blaney, D., Christensen, P.R., Chu, P., Crumpler, L., Davis, K., Ehlmann, B.L., Fergason, R., Golombek, M.P., Gorevan, S., Grant, J.A., Greeley, R., Guinness, E.A., Haldemann, A.F.C., Herkenhoff, K., Johnson, J., Landis, G., Li, R., Lindemann, R., McSween, H., Ming, D.W., Myrick, T., Richter, L., Seelos IV, F.P., Squyres, S.W., Sullivan, R.J., Wang, A., Wilson, J.: Science **305**, 821 (2004)

Hyperfine Interact (2005) 166:567–571
DOI 10.1007/s10751-006-9346-x

Mössbauer spectroscopy as a tool in astrobiology

Christian Schröder · Göstar Klingelhöfer ·
Brad E. Bailey · Hubert Staudigel

Published online: 27 October 2006

Abstract Two miniaturized Mössbauer spectrometers are part of the Athena instrument package of the NASA Mars Exploration Rovers, Spirit and Opportunity. The primary objectives of their science investigation are to explore two sites on the surface of Mars where water may once have been present, and to assess past environmental conditions at those sites and their suitability for life. Aqueous minerals – jarosite at Meridiani Planum, Opportunity's landing site, and goethite in the Columbia Hills in Gusev Crater, Spirit's landing site – were identified by Mössbauer spectroscopy, thus providing *in situ* proof of water being present at those sites in the past. The formation of jarosite in particular puts strong constraints on environmental conditions during the time of formation and hence on the evaluation of potential habitability. On Earth Mössbauer spectroscopy was used to investigate microbially induced changes in Fe oxidation states and mineralogy at the Loihi deep sea mount, a hydrothermal vent system, which might serve as an analogue for potential habitats in the Martian subsurface and the sub-ice ocean of Jupiter's icy moon Europa.

Key words Mössbauer spectroscopy · astrobiology · Mars · Europa · hydrothermal vent · biogeochemistry

C. Schröder (✉) · G. Klingelhöfer
Institut für Anorganische Chemie und Analytische Chemie, Johannes Gutenberg-Universität, Staudinger Weg 9, D-55128 Mainz, Germany
e-mail: schroedc@uni-mainz.de

G. Klingelhöfer
e-mail: klingel@mail.uni-mainz.de

B. E. Bailey · H. Staudigel
Institute of Geophysics and Planetary Physics, Scripps Institution of Oceanography, University of California San Diego, La Jolla, CA 92093-0225, USA

B. Bailey
e-mail: bebailey@ucsd.edu

H. Staudigel
e-mail: hstaudigel@ucsd.edu

1 Introduction

The field of astrobiology includes the study of the origin, evolution and distribution of life in the universe. Liquid water is the fundamental ingredient without which no life-form on Earth could exist. Mars is a prime target for the search for extinct or extant extraterrestrial life because its surface exhibits abundant evidence for the presence of liquid water in the past [1]. The NASA Mars Exploration Rover (MER) twins, Spirit and Opportunity, each carry an integrated suite of scientific instruments and tools called the Athena science payload. The primary objective of the Athena science investigation is to explore two sites on the Martian surface where water may once have been present, and to assess past environmental conditions at those sites and their suitability for life [2]. Part of that instrument suite are two miniaturized Mössbauer spectrometers MIMOS II [3], one on each rover. Fe-bearing minerals identified by Mössbauer spectroscopy help to classify rocks as either igneous, sedimentary, or metamorphic. The degree of oxidation, i.e. the Fe^{3+}/Fe_{Total} ratio measured by Mössbauer spectroscopy, as well as changes in the degree of oxidation and/or in Fe-bearing mineralogy between a rock's unaltered interior and its possibly weathered surface, offer clues on past and present environmental conditions [4].

The surface of Jupiter's moon Europa is covered by a thick ice sheet, which is underlain by a globe-spanning ocean. Because this ocean environment is isolated from sunlight as an energy source, deep sea hydrothermal vent systems on Earth – such as the Loihi Seamount – might serve as an analogue for possible habitats. Fe redox reactions can be utilized by chemolithotrophic and chemolithoautotrophic bacteria as an energy source. Mössbauer spectroscopy was used to investigate Fe oxidation states in basaltic glass exposed at the Loihi Seamount [5]. In an earlier study, Agresti et al. [6], investigating material from deep-sea smoker vents by Mössbauer spectroscopy, reported superparamagnetic material correlated with anaerobic bacteria found to thrive there.

2 Results and discussion

Gusev Crater was chosen as the landing site for MER Spirit, because the inflow channel Ma'adim Vallis entering the crater from the South suggests that it may once have contained a lake. However, Mössbauer spectra of angular-shaped rocks strewn across the plains surrounding the landing sites did not show a sedimentary signature. Instead, spectra of rocks and soil were dominated by a ferrous doublet whose hyperfine parameters are characteristic of the mineral olivine (Figure 1). Olivine shows the least resistance to chemical weathering of the common rock-forming minerals. Based on the ubiquitous presence of olivine in soil, Morris et al. [7] concluded that physical (i.e., non-aqueous) rather than chemical (i.e., aqueous) weathering processes currently dominate at Gusev Crater. Des Marais et al. [8] evaluated the potential for habitable environments in the Gusev Crater plains. Aqueous alteration of minerals in ultramafic rocks, such as olivine, can produce H_2, a near-universal source of energy and reducing power for microorganisms. Such reactions would also produce magnetite. Magnetite was identified in Mössbauer spectra of rocks and soils from the Gusev Crater plains (Figure 1). However, magnetite is also produced in the formation of igneous rocks. Des Marais et al. [8]

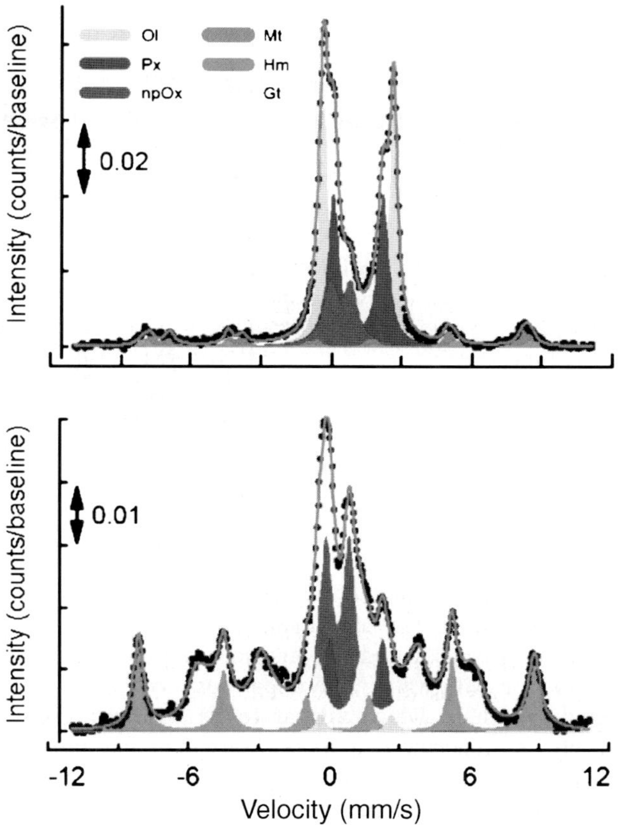

Figure 1 Typical Mössbauer spectra of basaltic soil at Gusev Crater (*top*), and of the rock Clovis in the Columbia Hills in Gusev Crater showing the highest abundance of goethite to date (*bottom*).

conclude, that although the sparsity of water puts up a barrier for life today, aquifers that were periodically recharged, for example during times of higher obliquity, might have sustained habitable conditions for geologically long periods of time. After traversing the plains surrounding its landing site, Spirit started ascending the Columbia Hills inside Gusev Crater on sol 157 of its mission. The Columbia Hills represent older terrain than the surrounding plains. Rocks encountered at the foot of the hills showed visible signs of physical and chemical weathering. In Mössbauer spectra of these rocks the mineral haematite was identified [9]. Haematite forms usually, though not exclusively, in the presence of water. In rocks showing a layered appearance further into the hills the mineral goethite was identified in Mössbauer spectra (Figure 1) [10], the first unequivocal *in situ* proof for the interaction of rocks and water in Gusev Crater. However, despite their layered appearance the goethite containing rocks do not represent sought-after lacustrine sediments [11].

Meridiani Planum was chosen as the landing site for MER Opportunity, because the area showed a clear signature of haematite in thermal emission spectra obtained from orbital observations [12]. The area surrounding the landing site is made up of basaltic soil underlain by finely layered outcrop rocks. Mössbauer spectra of the basaltic soil show the presence of olivine, and are identical to spectra obtained from soil in Gusev Crater [13, 14]. Contrary to the soil, Mössbauer spectra from the

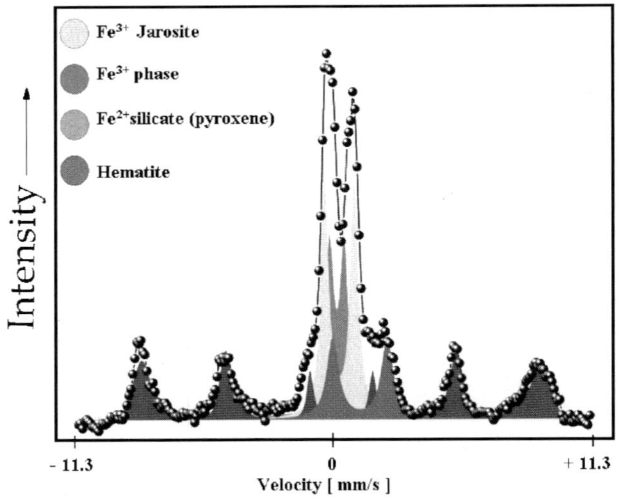

Figure 2 Typical Mössbauer spectrum of outcrop rocks at Meridiani Planum.

outcrop are dominated by a ferric doublet, whose hyperfine parameters are characteristic of the iron hydroxy sulfate mineral jarosite (Figure 2) [13]. That identification provided the first *in situ* proof, that these outcrop rocks formed in interaction with water. The uppermost layer of soil and outcrop is covered by mm-sized spherules, which appear to be weathering out of the outcrop. Mössbauer spectra of spherule-rich spots show, that they are composed mainly of haematite [13] and are thus the source of the haematite observed from orbit. The spherules likely formed diagenetically after deposition of the evaporite. Their spherical shape advocates another episode of water interaction during their formation. Based on the combined evidence from all MER instruments, Squyres et al. [15] concluded that the environmental conditions under which these sediments were laid down included episodic inundation by shallow surface water, evaporation, and desiccation, and that they were suitable for biological activity for some period of time in Martian history. Knoll et al. [16] provide an astrobiological perspective on Meridiani Planum. While Meridiani Planum may indeed have been a habitable place, high levels of acidity inferred from the presence of jarosite – jarosite forms at $pH < 3.5$ – would have posed a challenge for prebiotic chemical reactions thought to have played a role in the origin of life on Earth. Because sulfates and iron oxides can preserve detailed geochemical records of environmental history as well as chemical, textural and microfossil signatures of biological activity, Meridiani Planum is an attractive candidate for Mars sample return [16].

Several basaltic glass samples with varying Fe oxidation states were produced in order to detect the ability of microbial life to preferentially colonize chemically different substrates and subsequently metabolize the glass for various nutrients needed for survival. The Fe^{3+}/Fe^{2+} ratios were evaluated using Mössbauer spectroscopy [5]. Some of these basaltic glass samples were exposed to the iron-rich microbial mats and hydrothermal vent complexes at the Loihi Seamount for ~one year. Fe oxidation states were again evaluated after the exposure, showing a slight increase in Fe^{3+}/Fe^{2+} ratios. This increase may at least in part be due to strains of Fe-oxidizing bacteria that have been found on the exposed samples and have demonstrated their ability to use the basaltic glass as their only source of iron for growth [5].

3 Conclusion

Water is the most important requirement for all known life-forms. Fe-bearing minerals, goethite and jarosite in particular, identified by Mössbauer spectrometers onboard the Mars Exploration Rovers, Spirit and Opportunity, provided *in situ* proof for ancient aqueous and thus potentially habitable environments on Mars. Jarosite in particular adds acidity as a constraint to environmental conditions, which may have precluded prebiotic chemistry playing a role in the origin of life on Earth. Other minerals identified by Mössbauer spectroscopy such as olivine are a possible source of nutrients and energy.

Iron oxides can preserve signatures of biological activity, making Mössbauer spectroscopy an important tool to identify promising rocks for closer scrutiny by future dedicated biology missions to Mars, such as the planned European lander Exomars, or a Mars sample return mission.

Fe^{2+}–Fe^{3+} redox potentials are an energy source available to microorganisms at deep sea hydrothermal vent systems, which may be analogous environments compared to possible habitats in sub-ice oceans on the Jovian moon Europa. Monitoring changes in Fe^{3+}/Fe^{2+} ratios of basaltic glass samples exposed in such environments using Mössbauer spectroscopy can be used to determine the proportion of microbially induced redox reactions in such environments. The results discussed above show that Mössbauer spectroscopy is an important tool in the field of astrobiology.

Acknowledgements The development and realization of MIMOS II was funded by the German Space Agency DLR under contract 50QM99022. Parts of this work have been funded through the DFG priority programme "Mars and the Terrestrial Planets". The Agouron Institute has provided the funding for the Loihi portion of this work and has been a long time supporter of Astrobiology.

References

1. Carr, M.H.: Water on Mars. Oxford University Press, New York (1996)
2. Squyres, S.W., et al.: J. Geophys. Res. **108**(E12), 8062 (2003). doi: 10.1029/2003JE002121
3. Klingelhöfer, G., et al.: J. Geophys. Res. **108**(E12), 8067 (2003). doi: 10.1029/2003JE002138
4. Schröder, C., et al.: Planet. Space Sci. **52**, 997–1010 (2004)
5. Bailey, B.E., et al.: EOS Transactions AGU, 85(47), Fall Meeting Supplement (2004), Abstract B53C-1006
6. Agresti, D.G., et al.: Hyperfine Interact. **91**, 523–528 (1994)
7. Morris, R.V., et al.: Science **305**, 833–836 (2004)
8. Des Marais, D.J., et al.: Lunar Planet. Sci. **36**(No. 2353) (2005) (CD-ROM)
9. Schröder, C., et al.: Lunar Planet. Sci. **36**(No. 2309) (2005) (CD-ROM)
10. Morris, R.V., et al.: J. Geophys. Res. **111**, E02S13 (2006). doi: 10.1029/2005JE002584
11. Squyres, S.W., et al.: J. Geophys. Res. **111**, E02S11 (2006). doi: 10.1029/2005JE002562
12. Christensen, P.R., et al.: J. Geophys. Res. **105**, 9623 (2000)
13. Klingelhöfer, G., et al.: Science **306**, 1740–1745 (2004)
14. Yen, A.S., et al.: Nature **436**, 49–54 (2005). doi: 10.1038/nature03637
15. Squyres, S.W., et al.: Science **306**, 1709–1714 (2004)
16. Knoll, A.H., et al.: Earth Planet. Sci. Lett. **240**, 179–189 (2005). doi: 10.1016/j.epsl.2005.09.045

Hyperfine Interact (2005) 166:573–578
DOI 10.1007/s10751-006-9347-9

Octahedral cation partitioning in Mg,Fe^{2+}-olivine. Mössbauer spectroscopic study of synthetic $(Mg_{0.5}\ Fe^{2+}{}_{0.5})_2SiO_4$ (Fa_{50})

Mikhail Morozov · Christian Brinkmann · Michael Grodzicki · Werner Lottermoser · Gerold Tippelt · Georg Amthauer · Herbert Kroll

Published online: 5 December 2006

Abstract The high-temperature partitioning of Fe^{2+} and Mg between the two non-equivalent octahedral M1 and M2 sites in synthetic olivine (Fa_{50}) was studied by Mössbauer spectroscopy. Powder samples have been equilibrated in annealing experiments performed under reducing oxygen fugacity at temperatures between 500 and 800°C followed by rapid quenching in order to prevent redistribution of cations. M-site ordering with Fe^{2+} preferring M1, Mg preferring M2 sites increases continuously with rising equilibrium temperature. K_D values increase from 1.21 at 500°C to 1.48 at 750°C. The results are consistent with both room temperature as well as *in situ* high temperature single crystal X-ray diffraction experiments of Heinemann et al. [8, 9].

Key words Mössbauer spectroscopy · olivine · cation distribution

1 Introduction

Ordering phenomena in the crystal structure of minerals are traditionally in focus of scientific interest because they are related to both physical properties of crystals and thermodynamic properties of mineral assemblages [1]. In the case of olivine, $(Mg,Fe^{2+})_2SiO_4$ (space group *Pbnm*), Fe^{2+} and Mg occupy two non-equivalent 'octahedral' structural sites, denoted M1 with point symmetry $\overline{1}$ and M2 with point

M. Morozov (✉) · M. Grodzicki · W. Lottermoser · G. Tippelt · G. Amthauer
Department of Material Science, Division of Mineralogy, University of Salzburg, Hellbrunnerstr., 34, A-5020 Salzburg, Austria
e-mail: Gudrun.Riegler@sbg.ac.at

C. Brinkmann · H. Kroll
Institute of Mineralogy, University of Münster, Correnstr. 24, 48149 Münster, Germany

M. Morozov
Department of Mineralogy, Crystallography and Petrology, St. Petersburg Mining Institute (Technical University), 21st Line 2, 199106 St. Petersburg, Russia

symmetry *m* [2]. The differences in the crystallographic environment of the two M-sites as well as the differences in the chemical behaviour of Fe^{2+} and Mg potentially lead to cation partitioning between M1 and M2 sublattices. Unlike analogous silicates that also possess two sets of non-equivalent octahedral sites (e.g., pyroxenes, amphiboles, etc.), the M1 and M2 sites of olivine show less contrasting geometries which makes the prediction of the Fe^{2+},Mg distribution difficult. Deviations from the statistical distribution were experimentally confirmed for natural olivines since early structural X-ray and Mössbauer studies. A large amount of available data were reviewed recently [3] and can be summarised as follows: (1) Fe^{2+} predominantly occupies the M1 site in lunar and volcanic olivines, i.e., in crystals from rapidly cooled high-temperature rocks (2) Fe^{2+} prefers the M2 site in olivines from metamorphic rocks, which due to their slow cooling can attain equilibrium states down to lower temperatures. These observations suggest that the partitioning of Fe^{2+} and Mg between M1 and M2 sites is temperature-dependent.

Lastly, new attempts to recognise the ordering behaviour in olivine have been undertaken. The data obtained for natural and synthetic olivines with various intermediate compositions are of great curiosity because they retain the possibility of two unusual structural phenomena: (1) increasing M-site order at higher temperatures, (2) high-temperature crossover in the ordering trend separating two reversed ordered structures. However, the results published by different authors seem to be mutually incompatible. According to Artioli et al. [4] and Rinaldi et al. [5], *in situ* neutron diffraction experiments on natural Fa_{12} and Fa_{10} olivines reveal increasing order at temperatures up to about 900°C with Fe^{2+} preferring the M1 site followed by a crossover and reversed increasing order with Fe^{2+} preferring the M2 site up to 1,300°C. In partial agreement with these results, Heinemann et al. [6] reported for two natural Fa_{12} olivines that ordering increases at temperatures between 500 and 750°C with Fe^{2+} preferentially occupying the M1 site. A possible crossover at about 350°C leading to reversed ordering (Fe^{2+} at M2) at lower temperatures was also discussed. Completely incompatible results for a synthetic Fa_{50} olivine were presented by Redfern et al. [7] from *in situ* neutron diffraction experiments between 100 and 1,250°C. The authors presented a set of data that shows an almost uniformly increasing enrichment of Fe^{2+} in the M2 site with increasing temperatures between 600 and 1,250°C, whereas at temperatures lower than 600°C Fe^{2+} prefers the M1 site. It was supposed that annealed samples cannot be quenched fast enough to 'freeze' cation distributions corresponding to high T-conditions. These assumptions and results were finally disproved by Heinemann et al. [8, 9] who reported increasing order of Fe^{2+} at M1 with increasing temperatures from 525 to 900°C in various natural olivines (Fa_{12}, Fa_{25} and Fa_{48}) as revealed by single crystal XRD experiments of both quenched and *in situ* studied samples. We suggest that diffraction data should be completed by independent techniques (e.g. Mössbauer and other spectroscopic methods) which can verify the ordering behaviour in olivine. Therefore, a Mössbauer spectroscopic study is the aim of the present paper.

2 Experimental

Olivine powder samples (50 mol.% forsterite, 50 mol.% fayalite) were synthesized from stoichiometric mixtures of the oxides MgO, Fe_2O_3, and SiO_2 at high

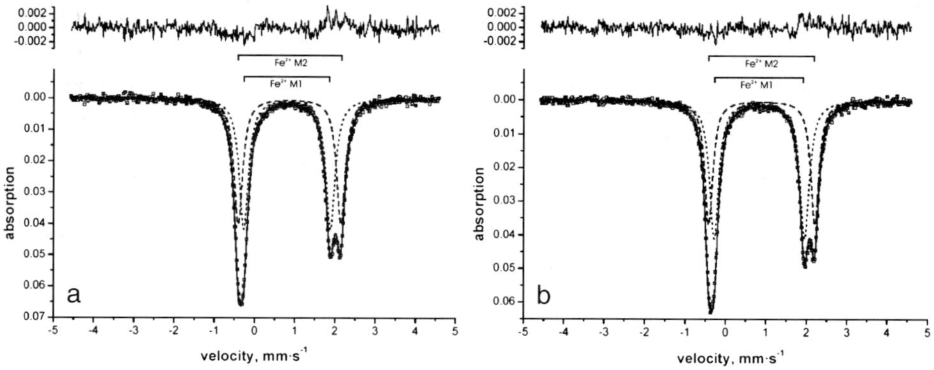

Figure 1 Mössbauer spectra of synthetic Fa_{50} olivines annealed at 500°C (**a**) and 725°C (**b**). $Fe^{2+}{}_{M1}$ – *dotted line*, $Fe^{2+}{}_{M2}$ – *dashed line*, total calculated spectrum – *solid line*, "residuals" above the spectra indicate the difference between the computed curve and the observed data. Spectra were taken at 300°C absorber temperature.

temperatures and oxygen fugacities controlled by CO/CO_2 gas mixtures. Samples were annealed at temperatures between 500 and 800°C in steps of 25°C for 64 to 2 days, depending on temperature, in order to ensure equilibration of the intra-crystalline cation distribution. After annealing, samples were rapidly quenched to avoid re-equilibration of cations.

Transmission ^{57}Fe Mössbauer experiments were carried out using a conventional Mössbauer spectrometer equipped with a furnace for absorber heating and a ^{57}Co/Rh single line source with an initial activity of 50 mCi. The spectrometer was operated in constant acceleration mode with a symmetrical triangular velocity shape and a multi-channel analyser with 1024 channels. The velocity range ($\pm$ 4 mm · s^{-1}) was calibrated with α-Fe at room temperature. Mössbauer doublets corresponding to the M1 and M2 sites were evaluated by a conventional least-square routine assuming Lorentzian lines. The total integrated intensity (area) of a doublet was taken as the amount of Fe^{2+} in the corresponding site. The error is about ± 1%.

Spectra taken with the absorber at room temperature exhibit a close overlap of the $Fe^{2+}{}_{M1}$ and $Fe^{2+}{}_{M2}$ doublets, respectively. Therefore they cannot be quantitatively evaluated with sufficiently high precision. Therefore, Mössbauer spectra were measured with the absorber at 300 C so that the high velocity peak is split up into 2 peaks (cf. Figure 2). This allows to better evaluate the spectra with regard to the exact determination of the doublet area, i.e. the Fe^{2+}-distribution. Measured and calculated spectra coincide with $\chi^2 < 1$. It is expected that at the absorber temperature of 300°C no cation exchange between M1 and M2 sites occurs within the measuring time of 5 days [10–12].

3 Results and discussion

In Figure 1 Mössbauer spectra of the samples heated at 500 and at 725°C, respectively, are displayed. The Mössbauer parameters of all the spectra taken are reported in Table I. The doublet with the smaller quadrupole splitting, i.e. the larger distortion of the corresponding octahedron, can be assigned to Fe^{2+} at M1, the other

Table I Mössbauer hyperfine parameters of Fe^{2+}(M1) and Fe^{2+}(M2) doublets (measured at 300°C) in synthetic olivines $(Mg_{0.5}Fe_{0.5})_2SiO_4$ equilibrated at different temperatures T_{eq}

T_{eq} (°C)	Fe^{2+} (M1)				Fe^{2+} (M1)			
	δ	ΔE_Q	Γ	X_{Fe}	δ	ΔE_Q	Γ	X_{Fe}
500	0.92	2.15	0.28	0.524	0.98	2.55	0.26	0.476
525	0.93	2.17	0.28	0.532	0.99	2.57	0.24	0.468
550	0.93	2.16	0.26	0.537	0.99	2.56	0.24	0.463
575	0.94	2.18	0.28	0.534	1.00	2.58	0.26	0.466
600	0.93	2.18	0.28	0.555	1.00	2.58	0.24	0.445
625	0.93	2.18	0.28	0.540	0.99	2.57	0.26	0.460
650	0.93	2.17	0.28	0.543	0.99	2.56	0.26	0.457
675	0.94	2.18	0.28	0.550	0.99	2.58	0.24	0.450
700	0.93	2.18	0.28	0.548	0.99	2.57	0.24	0.452
725	0.95	2.23	0.28	0.557	1.01	2.62	0.24	0.443
750	0.94	2.19	0.28	0.549	1.00	2.58	0.24	0.451
775	0.94	2.19	0.28	0.539	1.00	2.58	0.24	0.461
800	0.94	2.19	0.28	0.574	1.00	2.59	0.24	0.426

δ is the isomer shift (mm/s) related to α-Fe at room temperature, ΔE_Q is the quadrupole splitting (mm/s), Γ is the full width at half maximum (mm/s), X_{Fe} is the Fe^{2+} content calculated from the total integrated doublet intensities

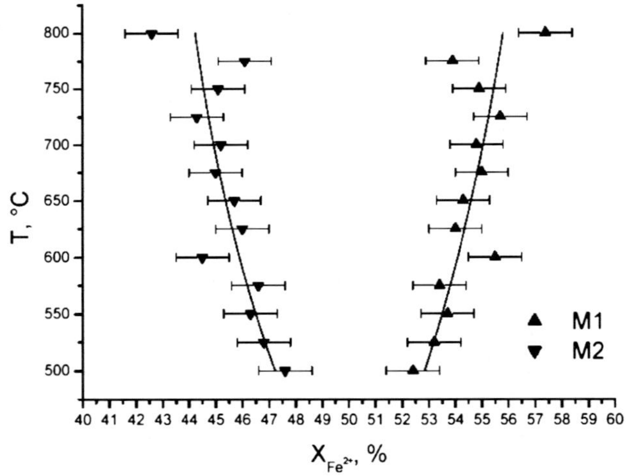

Figure 2 Area ratios X [%] of M1 and M2 in the temperature range from 500 to 800°C in synthetic Fa_{50} olivine determined by Mössbauer spectroscopy. *Solid curves* represent exponential fits.

one with the larger quadrupole splitting to Fe^{2+} at M2. This assignment is confirmed by electric field gradient and quadrupole splitting calculations for synthetic fayalite at moderate temperatures (from room temperature up to 300 C) which reveal the good agreement between the calculated and experimental Mössbauer spectroscopic data [13]. Fe^{3+} could not be detected in any of the spectra. Both isomer shift and quadruple splitting slightly increase with annealing temperature reflecting a rising difference in the geometries of the corresponding M-sites.

In Figure 2 the area ratios (%) X of Fe^{2+} at M1 and M2, respectively, are plotted as function of the heating temperature T. The systematic changes of the intensities

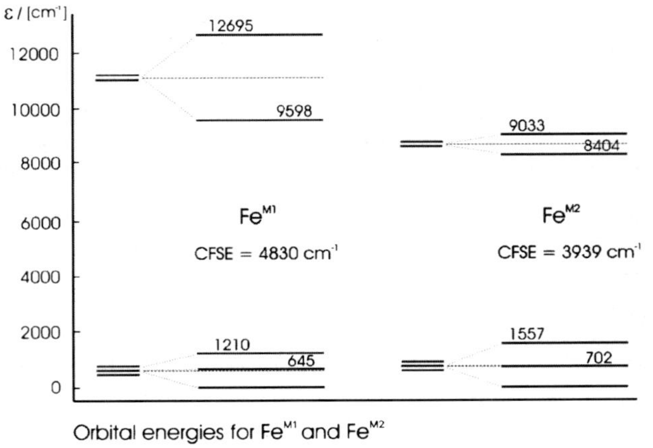

Figure 3 Crystal field stabilization energies CFSE of Fe^{2+} at M1 and M2 sites in olivines as evaluated by quantum mechanical calculations in the local spin density approximation.

of the doublets correspond to the temperature-dependent changes of the partitioning of Fe^{2+} and Mg between the M1 and M2-sites. The intensity (area) of the M1 doublet increases with increasing annealing temperature whereas that of M2 decreases. The site occupancies and intersite distribution coefficients reflect the increasing prevalence of Fe^{2+} on the M1 site and as a corollary the depletion of Fe^{2+} on the M2 site. Nevertheless, with exception of two outliers at 600 and 800°C the remaining trend corresponds to a 99.7% probability regression model. The drastic change of the area ratios in Figure 2 leads to the doubtless conclusion that between 500 and 750°C Fe^{2+} prefers the M1 site and that this tendency increases with temperature. It can be supposed that the angular distortion of the M1 octahedron becomes larger with temperature and thus controls the increasing incorporation of Fe^{2+} into the M1 site. It appears that the system could become completely disordered at 300–320°C so that the existence of a low-temperature crossover point cannot be excluded.

The increasing enrichment of Fe^{2+} at M1 with increasing temperature for synthetic Fa_{50} olivine studied in this work fully corresponds to recently published single crystal X-ray results for natural olivines Fa_{12}, Fa_{25} and Fa_{48} [6, 8, 9]. Recently reported *in situ* high temperature Raman experiments were carried out with natural and isotopically enriched olivines [14]. The behaviour of the vibrational mode of M2 cation mixed with external vibrations of SiO_4-tetrahedrà upon heating from 700 up to 1,000°C was explained by authors in terms of decrease in the concentration of Fe at M2 site. This conclusion is also fully consistent with results of the present work.

Some experimental points reported by [4] are also in a good agreement with the trend of this work. On the other hand, the values of the *in situ* neutron diffraction studies reported by [7] for synthetic Fa_{50} olivine are completely opposite to the results of this work and show a nearly uniform decrease of Fe^{2+} at M1 and increase of Fe^{2+} at M2 with increasing temperature and with a crossover point near 650°C. The results of [7] are also inconsistent with the results of recent *in situ* high temperature single crystal X-ray investigations by [8, 9] which confirm this study.

The results of the *in situ* neutron diffraction data published by [5] reveal a distribution similar to the one published by [7]. They observed progressive ordering of Fe^{2+} into the M2 site with a crossover at about 900°C for Fe-rich natural olivines [5]. This crossover temperature is situated out of the temperature range studied in the present work and also in [8, 9]. Hence, decisive evidence of the existence of ordering in olivine (Fe^{2+} in M2) at temperatures higher than 900°C has to be found by further investigations desirably involving independent experimental techniques such as Mössbauer spectroscopy. However, there is no obvious structural reason for a high-temperature transformation of olivine that leads to a crossover of the cation distribution behaviour.

This ordering of Fe^{2+} at M1 position with increasing temperature is quite unusual, and may be referred to the following: Transition metal cations such as Fe^{2+} prefer the site with the higher crystal field stabilisation energy (CFSE). As shown by optical absorption spectroscopy, the CFSE of Fe^{2+} at M1 is slightly larger than the CFSE of Fe^{2+} at M2 throughout the whole olivine solid solution [15]. This is confirmed by own electronic structure calculation in spin density approximation shown in Figure 3.

Acknowledgments This work was partially supported by DFG and ÖAD grants, and also by the Russian Ministry of Education (grants PD02-1.5-361 and E02-2.0-58), Administration of St. Petersburg, Russia (grant PD03-1.5-55) and BRHE (grant Y1-G-15-01).

References

1. Geiger, Ch. (ed.): EMU Notes in Mineralogy, vol. 3. Solid Solution in Silicate and Oxide Systems of Geological Importance, 466 p. Eötvös University Press, Budapest (2001)
2. Deer, W.A., Howie, R.A., Zussman, J.: 2nd ed., 919 p. Geological Society, London (2001)
3. Aikawa, N., Kumazawa, M., Tokonami, M.: Phys. Chem. Miner. **12**, 1 (1985)
4. Artioli, G., Rinaldi, R., Wilson, C.C., Zanazzi, P.F.: Am. Mineral. **80**, 197 (1995)
5. Rinaldi, R., Artioli, G., Wilson, C.C., McIntyre, G.: Phys. Chem. Miner. **27**, 623–629 (2000)
6. Heinemann, R., Staack, V., Fischer, A., Kroll, H., Vad, T., Kirfel, A.: Am. Mineral. **84**, 1400 (1999)
7. Redfern, S.A.T., Artioli, G., Rinaldi, R., Henderson, C.M.B., Knight, K.S., Wood, B.J.: Phys. Chem. Miner. **27**, 630–637 (2000)
8. Heinemann, R., Kroll, H., Kirfel, A., Barbier, B.: Mitt. Österr. Mineral. Ges. **148**, 159 (2003a) (in German)
9. Heinemann, R., Kroll, H., Kirfel, A., Barbier, B.: Eur. J. Mineral. **15**(Bh.1), 78 (2003b) (in German)
10. Morozov, M., Brinkmann, C., Kroll, H., Lottermoser, W., Tippelt, G., Amthauer, G.: Eur. J. Mineral. **13**(Bh.1), 127 (2001a)
11. Morozov, M., Brinkmann, C., Kroll, H., Lottermoser, W., Tippelt, G., Amthauer, G.: Mitt. Österr. Mineral. Ges. **146**, 202 (2001b) (in German)
12. Morozov, M., Brinkmann, C., Kroll, H., Lottermoser, W., Tippelt, G., Amthauer, G.: In: 18th general meeting of the International Mineralogical Association, 1–6 September 2002, Edinburgh, Scotland. Programme with Abstracts, 86 (2002)
13. Lottermoser, W., Steiner, K., Grodzicky, M., Jiang, K., Scharfetter, G., Bats, J.W., Redhammer, G., Treutmann, W., Hosoya, S., Amthauer, G.: Phys. Chem. Miner. **29**, 112 (2002)
14. Kolesov, B.A., Geiger, Ch.: Mitt. Österr. Mineral. Ges. **149**, 48 (2004)
15. Burns, R.G.: 2nd ed., XIII, 551 p. Cambridge University Press (1993)

Hyperfine Interact (2005) 166:579–584
DOI 10.1007/s10751-006-9324-3

Formation of iron oxides in soils developed under natural fires and slash-and-burn based agriculture in a monsoonal climate (Chittagong Hill Tracts, Bangladesh)

C. Bender Koch · O. K. Borggaard · A. Gafur

Published online: 13 December 2006

Abstract Fire-induced mineral transformations have been investigated in composite mineral grains separated from the coarse sand fractions (400–2,000 μm) from Ultisols developed in the Chittagong Hill Tracts (Bangladesh). Magnetic and colour based separation (into light brown, dark red, and magnetic, dark red classes) were used to select the grains that were studied by Mössbauer spectroscopy. Aluminium substituted goethite (α-FeOOH) dominates the light brown particles. Fire transform the goethite into a poorly crystalline hematite (α-Fe_2O_3) dominating in the dark red particles. In the dark red, magnetic grains a recrystallized hematite dominates, but small amounts of maghemite (γ-Fe_2O_3) are also present. The latter is indicated by comparing the line intensities in spectra measurement with and without an external magnetic field.

Key words goethite · hematite · maghemite · fire · magnetic minerals · external magnetic field

1 Introduction

Natural and man-controlled fires are recurrent events in many ecosystems. Among the more persistent signatures of fire conveyed to soils are charcoal and fire induced changes in the mineralogy. The latter originates in the fact that many secondary minerals in soils are susceptible to phase transformations at the temperatures reached during fires. These new formed minerals may contribute significantly to some soil properties such as texture, colour and magnetism (e.g. [1,2]). It is there-

C. Bender Koch (✉) · O. K. Borggaard
Department of Natural Sciences, KVL The Royal Veterinary and Agricultural University, Thorvaldsensvej 40, DK-1871 Frederiksberg C, Denmark
e-mail: cbk@kvl.dk

A. Gafur
Soil Resource Development Institute, Krishi khamar Sarak, Dhaka 1215, Bangladesh

fore of great importance to be capable of detecting, quantifying and understanding the distribution of these phases.

Ultisols developed in the monsoonal climate of the Chittagong Hill Tracts (CHT) in Bangladesh are currently subject to a slash-and-burn based agricultural system. Prior to this practice (i.e. essentially before 1870) the soils would also have experienced sporadic natural fires. The colour of the soils are dominated by yellowish brown horizons below a dark gray A horizon and neither reddening nor charcoal is evident in the soils [3,4].

We report here on Mössbauer investigations on the iron oxide present in different classes of pseudo-sand in these soils.

2 Experimental

Three Ultisols profiles sampled from the Upper, Middle and Lower parts of one hill slope were included in this study. Iron oxide in the clay fraction is dominated by aluminium substituted goethite and a very small amounts of a non-identified iron oxide throughout the profiles [4]. Further general information on the profiles can be found in Gafur et al. [3,4].

No primary mineral grains were found in the 400–2,000 μm fractions, that all consist of aggregates of secondary particles of clay, silt and fine sized sand. This sand fraction was dominated by particles of light brown colour, with a minor number of dark reddish grains. Some of the dark grains were susceptible to the presence of a permanent ferrite magnet, whereas no response was detected from any of the light brown coloured grains. These observations were used to depict a separation procedure. Firstly the dark magnetic grains were removed using a magnet and next the dark red particles were handpicked leaving the light brown particles.

Mössbauer spectra were measured using a conventional constant acceleration spectrometer and ^{57}Co in Rh as source. Samples were measured at room temperature and 80 K using liquid nitrogen bath cryostats. External magnetic fields were applied to enhance the response from different magnetic minerals in the sample and were applied perpendicular to the γ-ray direction. That is, natural ferrimagnetic minerals (e.g. maghemite and magnetite) are expected to show at least a partial increase in the relative intensity of lines 2 and 5, whereas a weakly ferromagnetic phase (e.g. hematite) is expected to show a minor decrease in the intensity of these lines. The intensities of the external magnetic fields applied in this work are comparatively small (<1 T) and thus no major line splitting is expected in magnetically ordered components (in particular due to the intrinsic high line widths of the components in the spectra), nor broadening of paramagnetic components. Solely line polarization effects and blocking of relaxation are expected. In this work we have used the simple ideal Lorentzian line shape in the fitting of the spectra and all doublets were constrained to have identical lines. Composite magnetically ordered components were fitted using only one sextet, the parameters representing the dominant component.

3 Results and discussion

Mössbauer spectroscopy revealed that aluminium-substituted goethite (α-FeOOH) was the dominating iron oxide in all horizons [4]. The Mössbauer spectra of the light

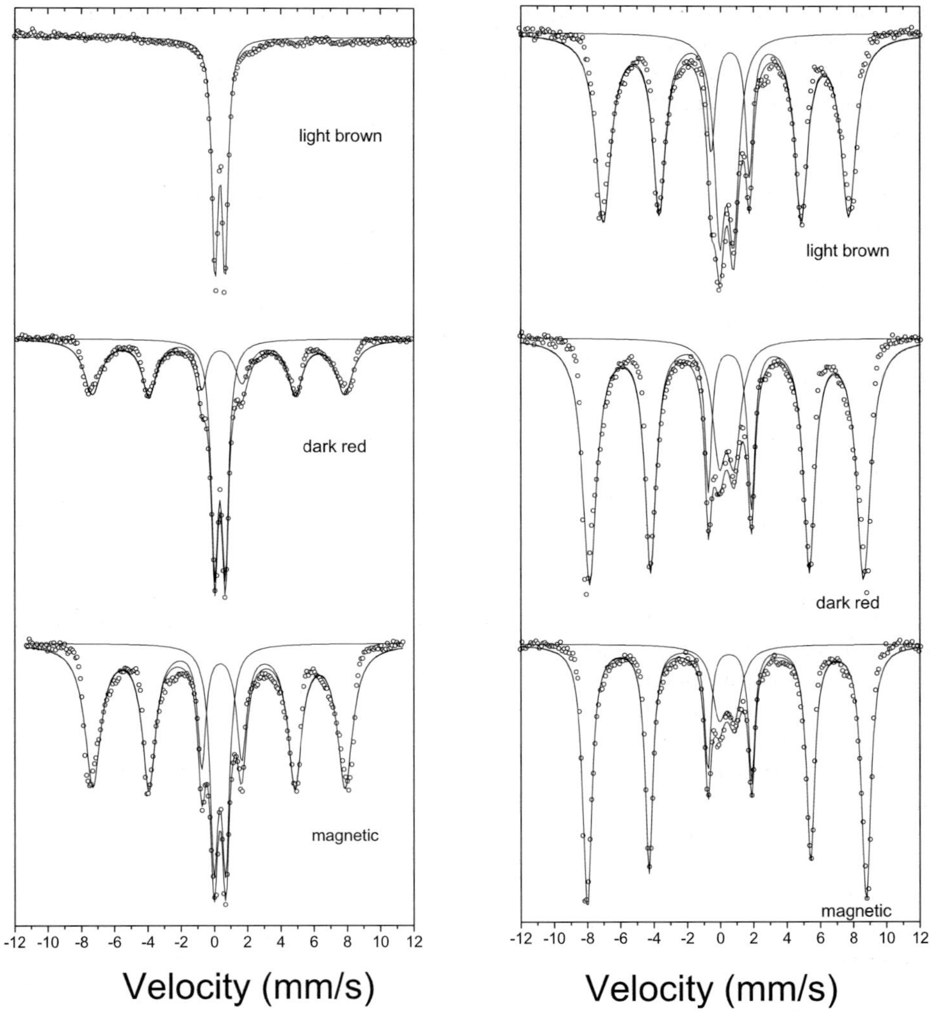

Figure 1 Mössbauer spectra of light brown, dark red, and magnetic grains from the 400–2,000 μm size fraction from the A horizon of the Middle profile measured at room temperature (*left*) and 80 K (*right*). The components of the fit and the sum of the components are shown as continuous lines.

brown, dark reddish, and magnetic fractions from the A horizon of the Middle profile is shown in Figure 1. At room temperature the spectra change from a dominantly paramagnetic spectrum of Fe(III) in the light brown sample, to a co-dominant sextet and doublet spectrum of the dark red sample, into a spectrum dominated by a sextet, but with significant amounts of doublet still in the magnetic sample. Lowering of the temperature to 80 K induces a substantial increase in the relative contribution of sextets to the spectra of all samples. The room temperature spectra of the light brown fractions consist essentially of one doublet due to (super)paramagnetic Fe(III). Cooling of the samples to 80 K induces magnetic splitting in approx. 34% of the iron with a magnetic hyperfine field of approx. 46 T. The hyperfine parameters of this component allow it to be identified as due to goethite [5]. The goethite from the light brown sand fractions has similar parameters and properties at 80 K to those found in the clay fraction. This implies that the goethite in the sand grains is highly substituted by aluminium similar to goethite in the clay fractions [4]. Most of the doublet in the 80 K spectrum is assigned to layer silicates, but part of the doublet may also be due to very small, defect goethite particles that order at approximately 40 K (spectra not shown). Because of the simple line shape used for fitting of the asymmetric lines in the spectra, the peak values of the hyperfine fields in the sextets from the room temperature spectra of the dark red samples will be slightly underestimated. A significant increase of the relative area of the sextet and a significantly better fit is obtained from the spectra measured at 80 K. The magnetic hyperfine field of approximately 46.6–47 T at room temperature increasing to 51.0–51.5 T at 80 K is indicative of hematite [5]. A slightly higher hyperfine field is observed for the magnetic grains with 47.1–47.3 T at room temperature and 51.9–52.0 T at 80 K. The sextet component in the spectra of the magnetic grains has significantly smaller line width in comparison to the spectra of the dark reddish grains (in particular at 80 K). Thus we may conclude that hematite is the dominant iron oxide in both the red and the magnetic sand grains. In comparison to pure hematite, the hyperfine field is reduced (from ca 53.5 T at 80 K) [5] and no Morin transition can be detected in the temperature range down to 80 K. This indicates both small size crystallites, and substitution and defects in the structure. Small size effects are also indicated from the increase of the sextet area on lowering the temperature from 298 to 80 K.

The magnetic susceptibility was measured on three classes from the sand fraction of the E horizon of the Upper profile. Values of 0.85, 1.05, and 84.7×10^{-3} SI units were found for the light brown, the red, and the red, magnetic fractions, respectively. Small iron oxide particles may not contribute to the bulk magnetic susceptibility due to rapid fluctuations in their magnetic moment at a given temperature, and substitutions may affect the magnetic structure. Differences in crystal sizes may thus in part explain why the red particles have a lower susceptibility in comparison to the red, magnetic particles. However, an increase by approximately two orders of magnitude, seems very high given the ratio of the superparamagnetic part in the spectra of both sand fractions (Figure 1). Rather, it is likely that the magnetic, red grains contains a ferrimagnetic oxide in addition to the hematite. Magnetite may be excluded based on the absence of its characteristic sextet due to the mixed valence site in the Mössbauer spectra, [5]. Maghemite would be the other possible candidate for a ferrimagnetic component. Maghemite and hematite have very similar Mössbauer parameters (including the magnetic hyperfine field – in particular if the hematite does not undergo a Morin transition) and either one is very difficult to detect in small amounts within mixtures. One clue to the presence of two magnetic components in the spectra of the magnetic grains is in fact indicated by the asymmetry of the second and fifth line, which in the case of pure hematite is expected to be symmetrical. A strategy for the identification of an oxide mixture using Mössbauer spectroscopy in external magnetic field, would ideally imply that specific components can be separated in the spectrum. However, application of high intensity fields causes significant smearing of spectral components and non-ideal line shape (e.g. [6]), adding to the difficulty in detecting the presence of a minor phase. Therefore, we have favoured an approach using a relatively small external magnetic field to induce polarization effects only, and use this as criteria for

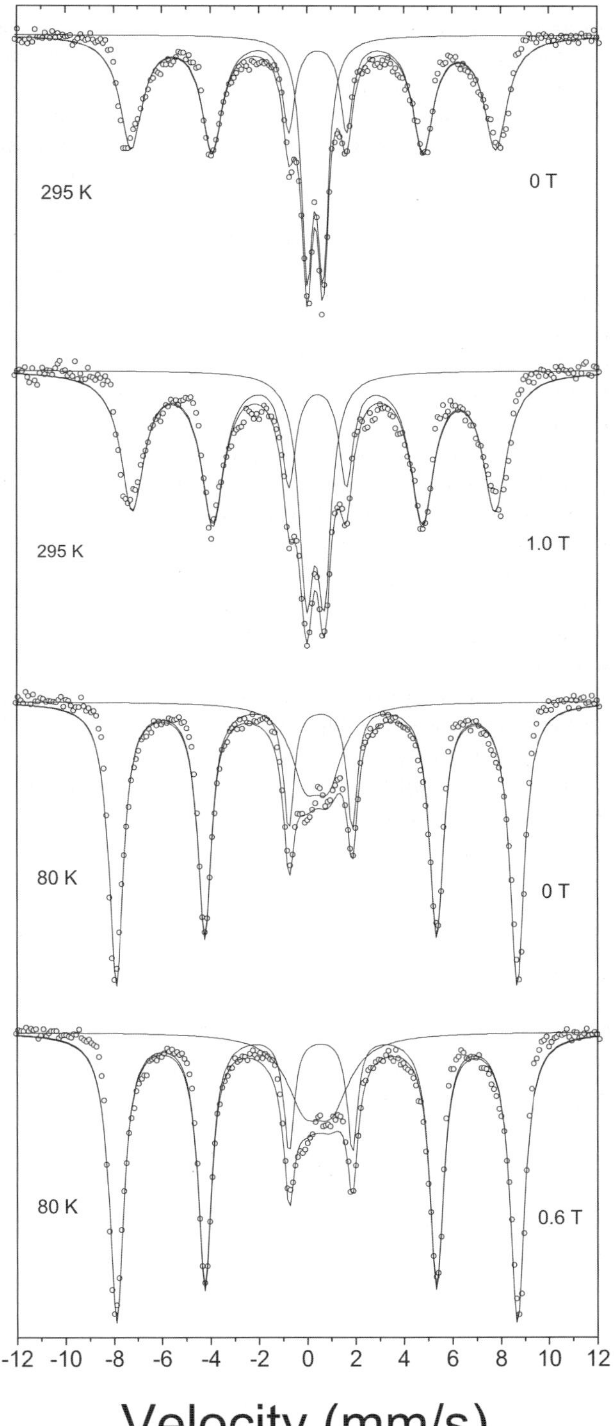

Figure 2 Mössbauer spectra of magnetic grains from the 400–2,000 μm size fraction from the A horizon of the Upper profile measured at room temperature and 80 K without and with an external magnetic field applied perpendicular to the γ-ray direction. The components of the fit and the sum of the components are shown as continuous lines.

identification. However, this approach also has its limitations, e.g. if defects and canting substantially changes the magnetic structure of a mineral.

The Mössbauer spectra of the magnetic grains from the A horizon of the Upper profile is shown in Figure 2. The hyperfine parameters of the room temperature and 80 K spectra are very similar to those of the magnetic grains from the Middle and Lower profiles. A comparison of the room temperature spectra measured with and without an external magnetic field reveals an enhancement of the intensity in lines 2 and 5. This is an indication of a component having ferrimagnetic spin structure in the sample. Ferrimagnetic particles exhibiting superparamagnetism at 298 K is essentially absent as evidenced by the identical ratio of doublet to sextet in the spectra measured with and without external field. As discussed above only the presence of hematite can be directly supported from the hyperfine parameters. Hematite has a weakly ferromagnetic structure and if any effect of the external field is expected, it would be a small decrease in the relative intensity of lines 2 and 5. This is contrary to what is actually observed at 298 K. Thus from the polarization behaviour alone it is suggested that maghemite is present in the red, magnetic samples. Since large crystallites of maghemite do not exhibit a phase transitions between 298 and 80 K it would also be expected to exhibit a polarization effect at 80 K, which is however not the case (Figure 2). We tentatively suggest that the polarization of the maghemite is inhibited, due to the highly aggregated state of the oxide particles. At room temperature maghemite may be polarized because the moments of the oxide particles behave independently. At temperatures where a significant number of neighbouring hematite particles order magnetically, the moment of the crystallites in the aggregates will interact and increase the resistance against polarization at low magnetic fields. This might have implications for the detailed process by which these oxides form, but further investigations are needed to understand this behaviour in detail.

References

1. Mullins, C.E.: J. Soil Sci. **28**, 223 (1977)
2. Ketterings, Q.M., Bigham, J.M., Leperche, V.: Soil Sci. Soc. Am. J. **64**, 1108 (2000)
3. Gafur, A., Borggaard, O.K., Jensen, J.R., Petersen, L.: Danish J. Geogr. **100**, 37 (2000)
4. Gafur, A., Bender Koch, C., Borggaard, O.K.: Soil Sci. **169**, 266 (2004)
5. Cornell, R.M., Schwertmann, U.: The Iron Oxides: Structure, Properties, Reactions, Occurrence and Uses. VCH, Weinheim (2003)
6. De Grave, E., da Costa, G.M., Bowen, L.H., Barrero, C.A., Vandenberghe, R.E.: Hyperfine Interact. **117**, 245 (1998)

Hyperfine Interact (2005) 166:585–591
DOI 10.1007/s10751-006-9325-2

Mössbauer studies of impactites from Huamalies province in Huanuco Region

A. Bustamante · S. Espinoza · G. Morales · R. B. Scorzelli

Published online: 2 November 2006

Abstract This report is about the X-ray diffraction and Mössbauer studies of three impactite samples denominated PMe-8, PMe-9 and PMe-11 from Huamalies Province in Huánuco Region, Peru. When terrestrial rocks are submitted to pressures higher than 60 GP, the majority is completely melted, forming a kind of glass called impactites. X-ray diffraction indicates the presence of quartz as the principal mineralogical phase in all samples. The ^{57}Fe Mössbauer spectra at room temperature of samples PMe-8 and PMe-9 show broadened spectra that were fitted using a distribution model. The most probable field of the magnetic component is 34 T, corresponding to the presence of small particles of goethite, confirmed by the 4.2 K spectrum. For the sample PMe-11, the MS showed the presence of well crystallized hematite.

Key words impactites · Mössbauer spectroscopy · hematite · goethite

1 Introduction

When terrestrial rocks are submitted to pressures higher than 60 GP, the majority are completely melted, forming a kind of glass called impactites. This report covers the Mössbauer studies of three impactite samples denominated PMe-8, PMe-9 and

A. Bustamante (✉) · S. Espinoza
Faculty of Physical Sciences, San Marcos University, P.O. Box 14-0149, Lima 14, Peru
e-mail: abustamanted@unmsm.edu.pe

G. Morales
Museo de Historia Natural(Museum of Natural History), Ave. Arenales 1251, Jesús Maria, Peru

R. B. Scorzelli
Centro Brasileiro de Pesquisas Físicas (Brazilian Center for Research in Physics), Rua Dr. Xavier Sigaud 150, Rio de Janeiro, Brazil

MAPA GEOLOGICO DEL CUADRANGULO DE LA UNION

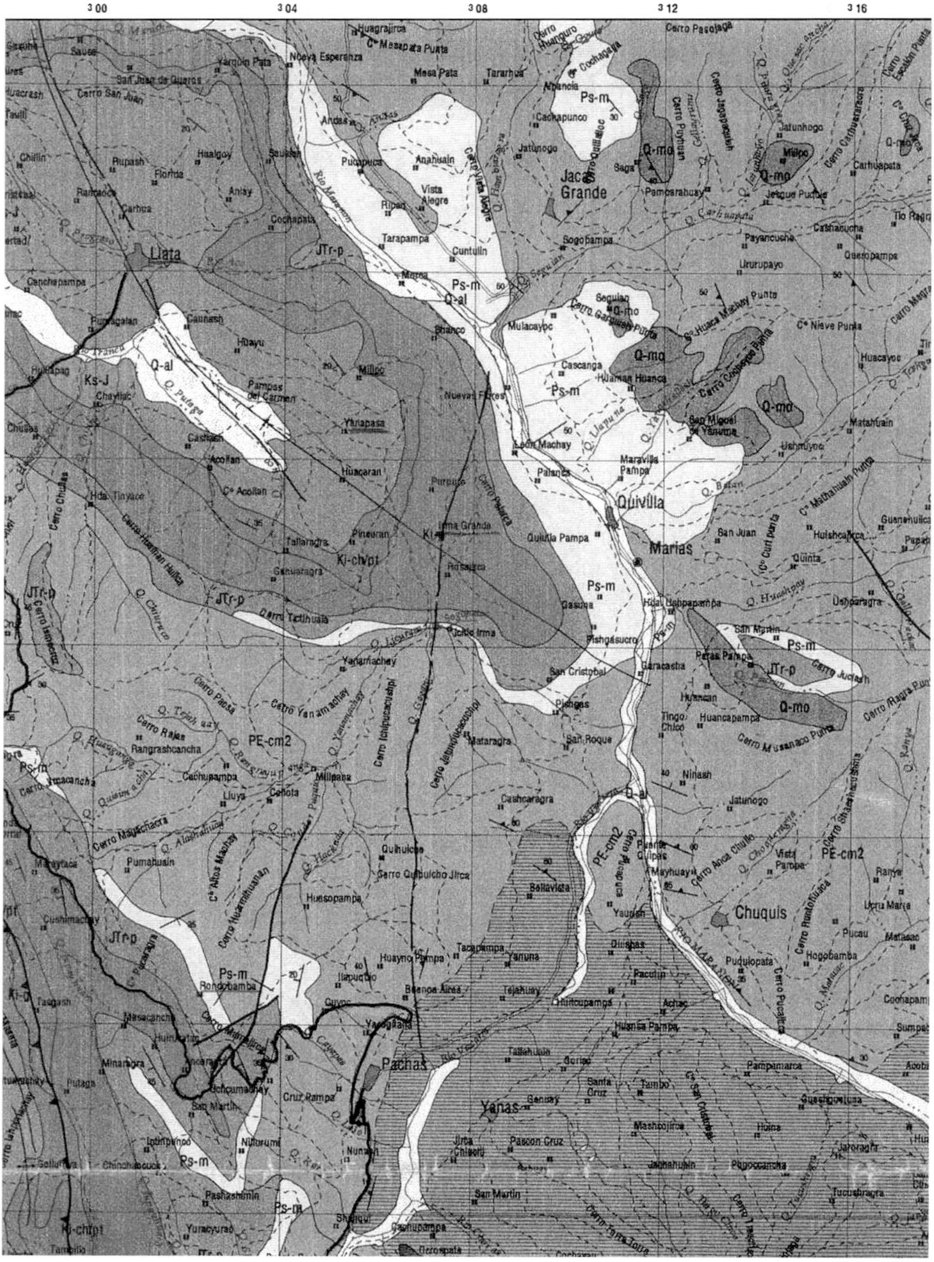

Figure 1 Location map of the meteoritic impact.

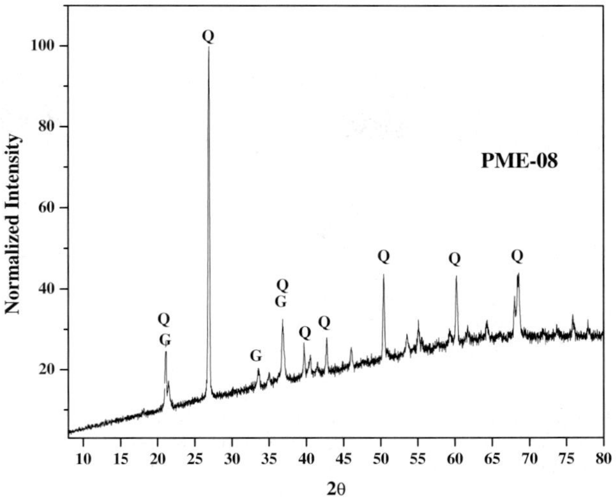

Figure 2 X-ray diffraction of sample PMe-8, Q = Quartz and G = Goethite.

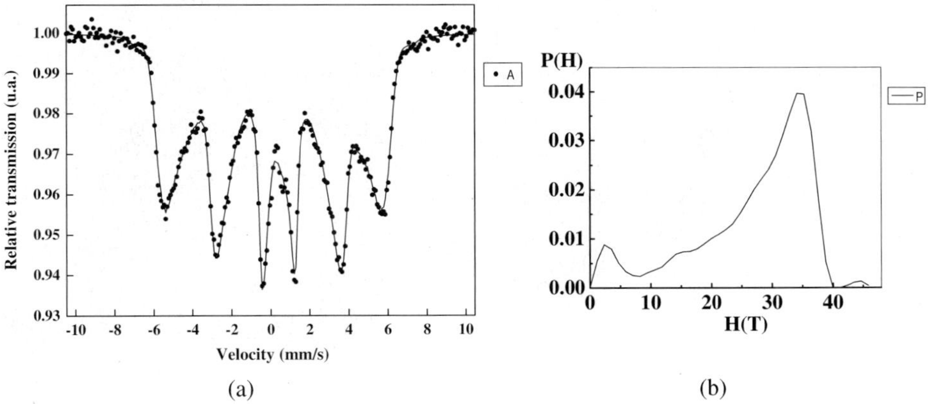

Figure 3 **a** RT Mössbauer spectra of PMe-8 sample and **b** the corresponding distribution of hyperfine fields $P(H)$.

PMe-11 from the Huánuco Region in Peru. The first geological and meteoritical place of interest in the region is the locality of Rondobamba, 20 km away from Llata, the capital of Huamalies Province (Figure 1), and 300 m from the right side of the road, where there is a depression with a crater 15 m in diameter and 5–7 m deep. The rock gathered has an elongated-pear shape about 15 cm long, with a smooth surface and dark brown color. After cutting it, the rock appears to have a silicone-like composition with quartz grains imbedded in a matrix of dark brown color (ferruginous). Sample PMe-09, from the depression of Yagapasa located 8 km southwest of Llata, has an approximately 50 m outside diameter and 15 m inside diameter. The surrounding rock of the crater is beige color limestone with

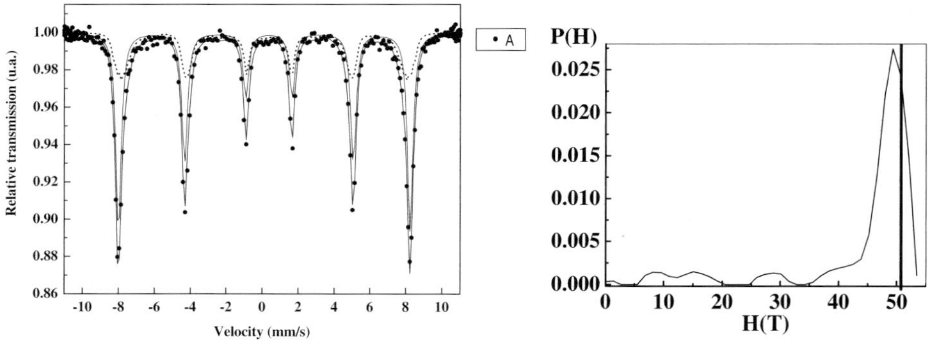

Figure 4 **a** 4.2 K Mössbauer spectra of PMe-8 sample showing the crystalline part (*continuous line*) and distribution part (*dotted line*) and **b** the corresponding distribution of hyperfine fields $P(H)$. The discrete H-value of the particular crystalline component is plotted by the *thick vertical line*.

Figure 5 Photomicrography PMe-9 sample (polarized light 10×).

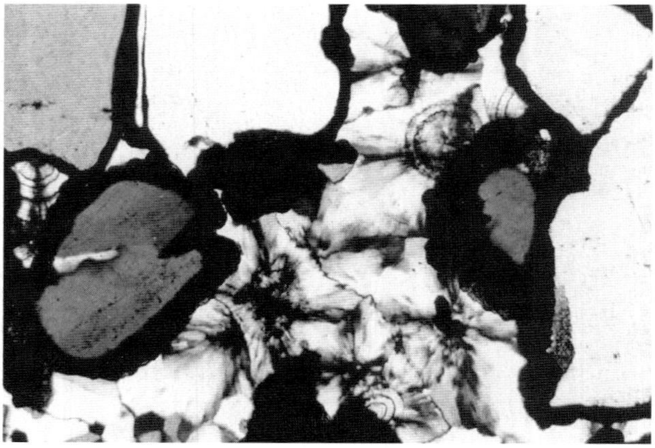

horizontal layers. The distance between the Yagapasa and Rondobamba craters is about 17 km. Sample PMe-11 was collected in Paugarhuilca, 1.5 km southwest of Llata.

2 Experimental

Room temperature (RT) and 4.2 K Mössbauer spectra (MS) were obtained at high velocities with standard transmission geometry using two spectrometers with 512 channels. The drives were operating in sinusoidal and triangular mode. Calibration was done using alpha-Fe foil. Isomer shifts were measured in relation with iron. Source and absorber were kept at the same temperature. NORMOS code [2] was used for the spectrum analysis. The X-ray powder data was collected between 8 and 80° with 0.02 step using a MINIFLEX RIGAKU using copper Kα radiation (λ = 1.54178 Å).

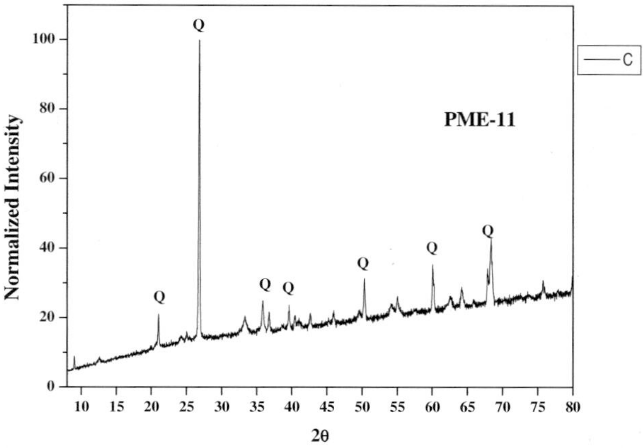

Figure 6 X-ray diffraction of sample PMe-11. *Q* = Quartz.

Figure 7 Photomicrography PMe-11 sample (polarized light 10×).

3 Results and discussion

3.1 PMe-8 and PMe-9 samples

The sample PMe-8 was collected in Rondobamba. The powder XRD pattern for sample PMe-8 (Figure 2) shows quartz as the principal mineral phase and goethite as a secondary phase, similar to PMe-9. The Mössbauer spectra at RT (Figure 3a) exhibit a sextet characteristic of the presence of an iron oxide or hydroxide, with broadened line width. The fitting of the spectra was done using a magnetic field distribution model. The most probable field of the magnetic component is 34 T, the average value of $BH_f = 26.96 \pm 0.08$ T and median value = 25.20 T, corresponding to the presence of small particles of goethite, confirmed by the 4.2 K spectrum (Figure 4a) [1]. Fitted Mössbauer parameters at 4.2 K show that the most probable field is 50.35 T, average value of $BH_f = 44.58 \pm 0.65$ T and median value = 43.17 T and the corresponding distribution of hyperfine fields $P(H)$ is shown in Figure 4b. The relative area is about 36%.

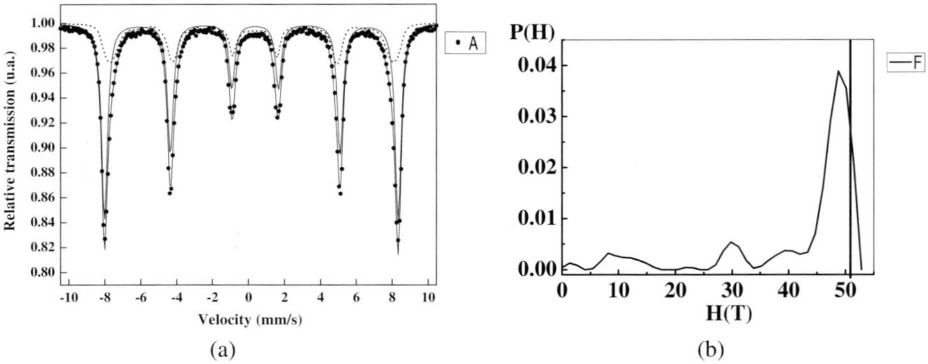

Figure 8 a RT Mössbauer spectra of PMe-11 sample and **b** the corresponding distribution of hyperfine fields *P*(*H*). Discrete *H*-value of the particular crystalline component (hematite) is plotted by the *thick vertical line.*

The discrete *H*-value of the particular crystalline component of goethite is plotted by the thick vertical line; the relative area is about 64%. The same features occur in sample PMe-9.

Under the microscope, dashed quartz grains with undulating (optical) extinction can be observed. This is due to a "shocked quartz" impact present in the samples PMe-09 and PMe-11; see Figure 5.

3.2 PMe-11 sample

The X-ray diffraction in Figure 6 indicates the presence of quartz as the principal mineralogical phase; another secondary phase is present but difficult to identify. In Figure 7 the presence of affluent reddish crystallized iron oxide particles is evident and its corresponding Mössbauer spectrum is shown in Figure 8. The best fitting of the spectra is obtained using a distribution field and one crystalline site. The most probable field of the magnetic component is 48.76 T, average value of $BH_f = 42.73 \pm 0.51$ T and median value = 47.81 T. The crystalline site corresponds to the presence of hematite ($Bh_f \cong 50.69$ T, $\varepsilon_Q \cong -0.110$ mm/s, IS = 0.38 mm/s) [3], confirmed by the 4.2 K measurements.

4 Conclusion

Samples PMe-8 and PMe-9 are composed mainly of quartz with goethite as a secondary phase, as detected by XRD and Mössbauer Spectroscopy, the latter confirming the presence of small particles. Sample PMe-11 also has quartz as the principal component; however the magnetic oxide detected by Mössbauer Spectroscopy corresponds to the presence of hematite.

Only in samples PMe-9 and PMe-11 was it possible, by optical microscopy, to observe the presence of "shocked quartz" due to impact.

Acknowledgement A.B.D would like to thank the CLAF for financial support at the CBPF, Brazil.

References

1. Lipka, J., Jensen, H.G., Knudsen, J.M., Madsen, M.B., Bentzon, M.D., Koch, C.B., Morup, S.: Hyperfine Interact. **70**, 965 (1992)
2. Brand, R.A.: Normos Programs, Angewandte Physik, Univ. Duisburg. Ecole des Mines, Nancy (1995)
3. Murad, E., Johnston, J.H.: In: Long, G.J. (ed.) Mössbauer Spectroscopy Applied to Inorganic Chemistry, vol. 1, p. 507. Plenum, New York (1987)

Hyperfine Interact (2005) 166:593–597
DOI 10.1007/s10751-006-9402-6

Mössbauer spectroscopy description of limonite from Taraco, in the Huancane Province of the Puno Region, Peru

A. Bustamante · J. Cabrera · V. Garcia · E. Urday · Y. A. Abdu · R. B. Scorzelli

Published online: 14 November 2006

Abstract Natural iron oxides are very common in nature and are the main components of many minerals, rocks and soils. There are a great variety of these minerals in Peru, especially in the Andes region. The mineral studied was extracted from the Taraco District in the Huancané Province of the Puno Region. The extracted sample is a yellowish mineral with very small particles which is called limonite. X-ray diffraction indicates the presence of goethite as the principal mineralogical phase and quartz as the secondary phase. The ^{57}Fe Mössbauer spectra at room temperature show broadened spectra that were fitted using a distribution model. The most probable field of the magnetic component is 21T, corresponding to the presence of small particles of goethite, confirmed by the 4.2 K spectrum. MS of the calcinated sample in air at 900°C show the presence of two hematite sextets, one related to bulk particles and another to surface particles. Chemical analysis using Scanning Electron Microscopy and Energy Dispersive X-ray Spectrometer indicates that O and Fe are the major components; other elements such as Al, Si, Mg and Ca are also present.

Key words Mössbauer spectroscopy · hematite · limonite · SEM/EDX

1 Introduction

Natural iron oxides are very common in nature and are the main components of many minerals, rocks and soils. There are many varieties of these minerals in Peru, especially in

A. Bustamante (✉)
School of Physical Sciences, San Marcos National University, Post Office Box 14.0149, Lima 14, Peru
e-mail: abustamanted@unmsm.edu.pe

J. Cabrera · V. Garcia · E. Urday
Electron Microscopy Center, Saint Augustine National University in Arequipa, N/N Indepencia Av., University Campus, Arequipa, Peru

Y. A. Abdu · R. B. Scorzelli
Centro Brasileiro de Pesquisas Físicas, Rua Dr. Xavier Sigaud 150, 22290-180 Rio de Janeiro, Brazil

Figure 1 X-ray diffractogram of natural limonite (*G* Goethite).

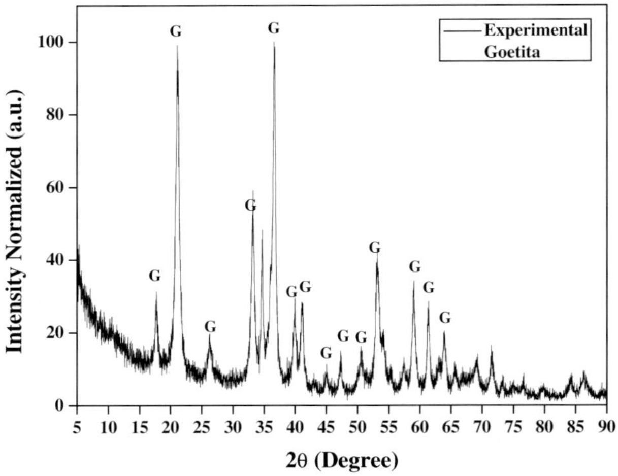

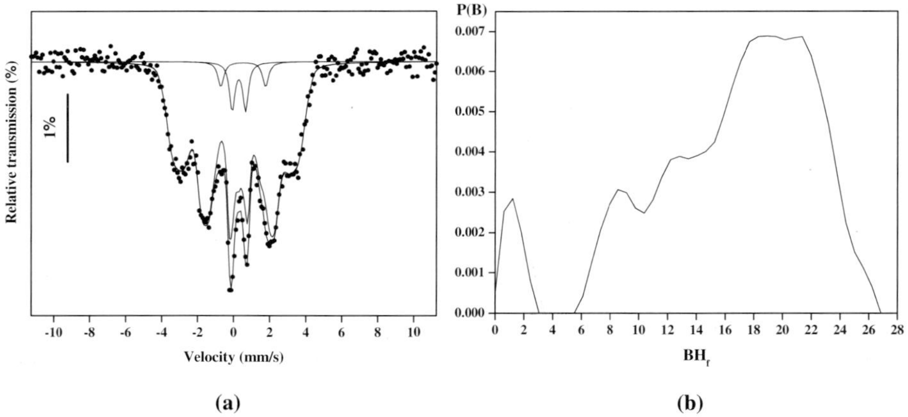

Figure 2 **a** Mössbauer spectra at RT of natural limonite and **b** its corresponding magnetic field distribution.

the Andes region. Limonite has many industrial applications, particularly as a highly concentrated pigment in paint, enamel and ceramics. It is also used in the manufacture of cement, mortar and glass. The mineral studied was extracted from the District of Taraco, Huancané Province, in the Puno Region. The extracted sample is a mineral of yellowish color called limonite. The main objective of this work was to identify and describe natural limonite using X-ray diffraction, Mössbauer spectroscopy, Scanning Electron Microscopy (SEM) and Energy Dispersive X-ray Spectrometer (EDX).

2 Experimental

The X-ray powder data was collected between 8 and 80° with 0.02 step using a SHIMADZU model 6000 and copper Kα radiation (λ=1.54178 Å). Room temperature

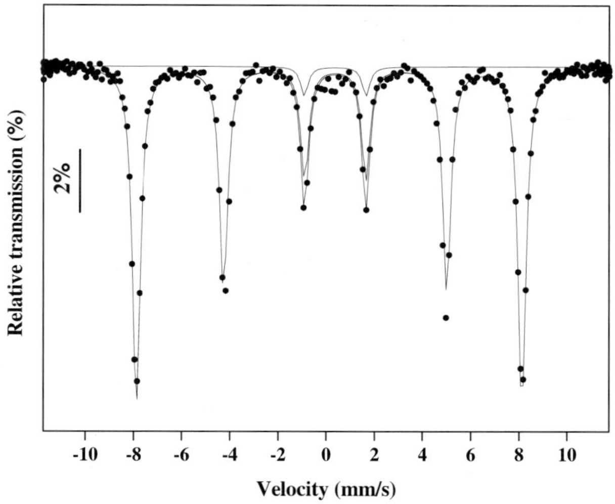

Figure 3 MS of natural limonite at 4.2 K.

Table I Mössbauer hyperfine parameters of natural limonite at RT and 4.2 K

Temp. (K)	IS (mm/s)	ΔQ (mm/s)	BH_f (T)	Γ (mm/s)	Area (%)
RT	0.307	−0.083	21		92
Phase 1	0.639	2.47	–	0.40	03
Phase 2	0.400	0.74	–	0.36	05
4.2	0.358	−0.177	49.62	0.40	96
Phase 1	0.474	2.55	–	0.37	04

(RT) and 4.2 K Mössbauer spectra (MS) were obtained at high velocities with standard transmission geometry using two spectrometers with 512 channels. The drives were operating in sinusoidal and triangular mode. Calibration was performed using α-Fe foil. Isomer shifts were measured relative to iron. Source and absorber were kept at the same temperature. NORMOS code [1] was used for the spectrum analysis. The microanalytical study was done using Scanning Electron Microscopy (SEM) with a Phillips model XL 20 microscope equipped with an Energy Dispersive X-ray Spectrometer (EDX).

3 Results and discussion

Limonite is not a true mineral but a mixture of similar hydrated iron oxide minerals. Limonite is mostly made up of goethite. Massive goethite and limonite can be indistinguishable. Limonite is formed mostly as sedimentary beds in/or near oxidized iron and other metal ore deposits [2]. Its chemical formula is FeO(OH) H_2O and it is used mainly as pigment and also to obtain iron [3].

Figure 1 shows the X-ray diffraction pattern that reveals the presence of goethite (α-FeOOH) as the main mineralogical phase and other phases, such as quartz, as secondary phases. SEM and EDX analysis showed that the main components are O and Fe, with variable amounts of Al and Si. The minor elements detected were Na, Mg, Al, Si, K and Ti,

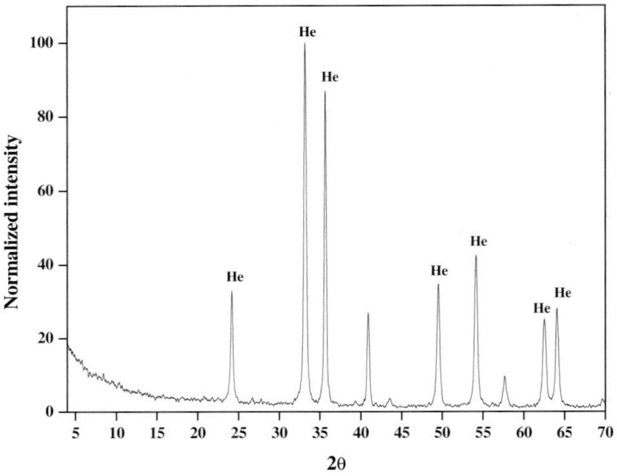

Figure 4 X-ray diffractogram of the sample after annealing in air at 900°C/3 h.

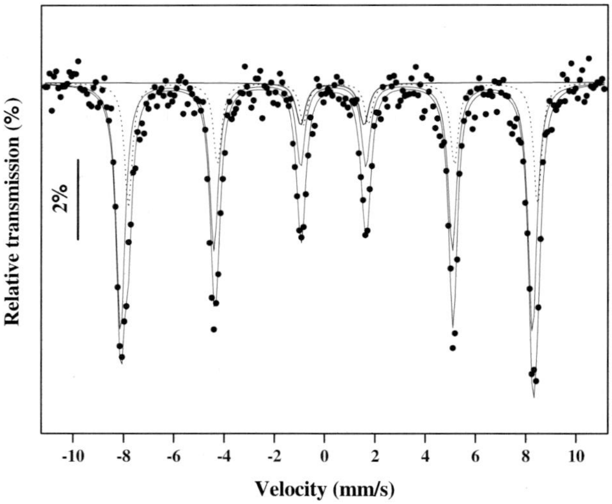

Figure 5 MS at RT of sample after annealing in air at 900°C/3 h.

indicating that the sample consists mainly of iron oxides with impurities of organic portions of clays. The result of a typical analysis of this mineral is Fe_2O_3 (89.86%) and H_2O (10.14%).

The Mössbauer spectra (MS) at room and liquid helium temperatures of natural Limonite from Puno are shown in Figures 2 and 3, respectively. The RT spectrum was fitted using a hyperfine magnetic field distribution, due to the ultrafine particle effect of very fine particles (~100 Å) and two paramagnetic doublets (see Table I), one of them (IS= 0.64 mm/s, QS=2.47 mm/s) associated with iron in an aluminum silicate phase and the second one (IS=0.40 mm/s, QS=0.74 mm/s) attributed to superparamagnetic iron oxides or hydroxides. The most probable magnetic field of the magnetic component is around 21T. In order to suppress the superparamagnetic effect, the sample was measured at 4.2 K

(Figure 3) and fitted using one sextet and one small paramagnetic doublet. The magnetic component has hyperfine parameters typical of goethite (IS=0.358 mm/s, $2\varepsilon_Q$= −0.253 mm/s, B_{hf}=49.62T) [4]. After a heat treatment at 500°C for 3 h, a change in color from yellow to brown was observed, indicating the presence of hematite and/or magnetite and maghemite (MS not showed). The MS and X-ray (Figure 4) of the sample calcinated in air at 900°C for 3 h indicate that all particles were crystallized hematite. The MS Figure 5 at RT was fitted using two sextets, one related to bulk particles (IS=0.322 mm/s, $2\varepsilon_Q$=−0.268 mm/s, B_{hf}=50.81T), another to surface particles (IS=0.499 mm/s, $2\varepsilon_Q$= −0.122 mm/s, B_{hf}=50.48T), and one paramagnetic phase [5].

4 Conclusion

Mössbauer spectroscopy and X-ray diffraction show that limonite is made up of ultrafine goethite particles. SEM/EDX analysis confirmed that iron and oxygen are the principal elements. MS for the sample calcinated at 900°C for 3 h consists of two types of crystallized hematite, one related to bulk and another to surface.

Acknowledgment A.B.D. wishes to thank the Centro Latinoamericano de Física/Latin American Center for Physics (CLAF) for financial support of the Brazilian Center for Research in Physics (CBPF), Brazil. This work was partially financed by CONCYTEC under contract No. 485-CONCYTEC-OAJ-2004.

References

1. Brand, R.A.: Laboratorium für Angewandre Physik, Universität Duisburg, Lotharstr 1, D-4100 Duisburg 1, Germany
2. http://mineral.galleries.com/minerals/oxides/limonite/limonite.htm
3. Diccionario Enciclopédico (Encyclopedic Dictionary) (ed.) p. 1599. Santillana (2000)
4. Stevens, J., Khasanov, A., Miller, J., Pollack, H., Li, Z.: Mössbauer Mineral Handbook. Mössbauer Effect Data Center, North Carolina, USA (1998)
5. Van Der Kraan, A.M.: Phys. Status Solidi, A **18**, 215 (1973)

Hyperfine Interact (2005) 166:599–603
DOI 10.1007/s10751-006-9326-1

The Mössbauer spectrum of synthetic hureaulite: $Fe_5^{2+}(H_2O)_4(PO_4H)_2(PO_4)_2$

Angel Bustamante · Enrico Mattievich · Helio S. de Amorim · Ivo Vencato · Máximo M. Silveira

Published online: 27 October 2006

Abstract The crystal structure of synthetic ferrous hureaulite, Fe_5^{2+} $(H_2O)_4$ $(PO_4H)_2(PO_4)_2$, was refined from single-crystal X-ray data. It is monoclinic, space group C2/c, with a=17.487(4), b=9.017(2), c=9.338(2) Å, β=96.27(3)°, V=1463.6(6) Å^3, Z=4 and D_{calc}=3.327 g/cm^3. This end member of the hureaulite series was crystallized under distinctly acidic conditions, by a method that gives perfect crystals, large enough for X-ray single crystal studies. The main feature of the hureaulite structure is that it has an equal number of normal $(PO_4)^{3+}$ and acid $(PO_4H)^{2+}$ tetradentate groups. These are centered on Fe^{2+} atoms and share corners with edge-linked octahedra, forming pentamer units. The five Fe^{2+} atoms are distributed on three distinct sites in these units. This can be directly observed in the Mössbauer spectrum at 295 K, which contains three doublets whose relative intensities correspond to the 1:2:2 distributions of crystallographic sites.

Key words iron phosphate · Hureaulite · Mössbauer Spectroscopy · X-ray refinement

1 Introduction

Several works appeared recently on the synthetic phosphate series named hureaulite, $Mn_{5-x}Fe_x(H_2O)_4(PO_4H)_2(PO_4)_2$, $0 < x < 5$, which in nature always has

A. Bustamante (✉)
Facultad de Ciencias Físicas, Universidad Nacional Mayor de San Marcos
Apto. Postal 14-0149 Lima-14, Peru
e-mail: abustamanted@unmsm.edu.pe

E. Mattievich · H. S. de Amorim · M. M. Silveria
Instituto de Física, Universidade Federal do Rio de Janeiro, P.O. Box: 68.528, 21945-970 Rio de Janeiro, RJ, Brazil

I. Vencato
Instituto de Física, Universidade Federal de Goiás, P.O. Box 131, 74001-970 Goiânia, GO, Brazil

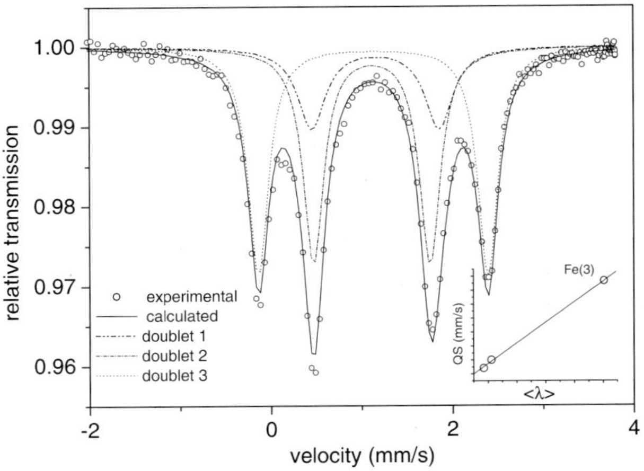

Figure 1 Mössbauer spectra of ferrous hureaulite at room temperature.

a high Mn content. Scientific interest was stimulated by the discovery that members of this series exhibit two types of magnetic behavior at low temperatures [1], particularly as it is easy to prepare as a microcrystalline powder with any composition between the manganous and ferrous end members [2, 3].

The mineral was discovered in 1825 by the ceramist Alluaud, in the pegmatite quarry of Huréaux, from which its name originates, in Haute Vienne, France [4, 5]. Mineralogists, chemists and physicists have tried to determine its properties, but as Moore [5] observed, "Few minerals have enjoyed more confusion in the literature! Names like 'baldafite,' 'wenzelite,' 'palaite,' 'pseudopalaite' and 'bastinite,' all proved to be hureaulite upon further investigation by various workers."

When von Klement et al. [6] presented their results on the synthesis of hureaulite-type crystals, they concluded that the solid solution of Fe^{2+} in synthetic manganous hureaulite appears limited to a maximum of 35% of substituted Fe^{2+} ions. This conclusion was accepted as experimental evidence to explain the natural paragenesis of hureaulite minerals, in which the Fe content appears to be limited to the ratio Fe/Mn = 1/3 [5]. It will be shown here that the conclusions of Klement et al. are incorrect.

2 Experimental

A 1,000 ml filtering flask was placed over a stirrer hot plate, in a N_2 atmosphere. A fresh solution of iron (II) sulfate was prepared from 7.5 g of reduced iron GR by adding sufficient water to cover the iron powder, heating to 100°C, and adding 9 ml of concentrated (98%) sulfuric acid GR. The mixture was stirred until the iron powder had completely dissolved. A solution of 3 g of ammonium acetate GR, CH_3COONH_4, in 300 ml of water was added to the iron sulfate solution. The ferrous phosphate was precipitated by adding, drop by drop, a solution of diammonium hydrogen phosphate GR, 21 g of $(NH_4)_2HPO_4$ dissolved in 300 ml of water (previously boiled and allowed to cool). Extra water was added to the filtering flask to bring the level up to the 1,000 ml mark. The acidity of the solution was then approximately pH = 4. The precipitate was maintained at 70°C and gently stirred for

Table I Mean distances (Å) between cations and oxygen atoms, quadratic elongations and variance of bond angles

Mean distances	Moore and Araki		This work	
M1–O	2.188		2.160	
M2–O	2.159		2.141	
M3–O	2.209		2.170	
P1–O	1.535		1.537	
P2–O	1.540		1.543	
Cation sites	$<\lambda>$	σ^2	$<\lambda>$	σ^2
M1 (octahedral)	1.0212	71.6	1.0242	81.7
M2 (octahedral)	1.0185	62.8	1.0179	59.2
M3 (octahedral)	1.0604	189.4	1.0617	187.3
P1 (tetrahedral)	1.00129	4.86	1.00158	6.35
P2 (tetrahedral)	1.00024	0.55	1.00028	0.73

M Metal (Fe, Mn); $<\lambda>$ quadratic elongation; σ^2 variance of bond angles.

3 days. After 3 days, the product was crystallized by raising the temperature to 96°C. Well-developed crystals are observed growing in the mother liquor between pH 3 and 6, from 70 to 100°C.

A colorless single crystal (0.13 × 0.10 × 0.05 mm^3) was mounted on a CAD-4 Enraf–Nonius diffractometer with graphite–monochromated MoK_α radiation and ω-2θ scan capability. The unit cell and the orientation matrix for the data collection were obtained by a least-squares fit of 25 centered reflections ($10.22 < \theta < 15.72°$). The structure was refined by full-matrix least-squares techniques on F^2 with SHELXL97 computer program [7]. The final refinement gave an R (on F) factor of 0.031, and wR (on F^2) = 0.076 for 147 refined parameters. In order to minimize the crystalline texture effect, the powdered crystals were mixed with sugar and fed laterally into the holder. The Mössbauer spectrum at 295 K, Figure 1, shows the least-square fitting of three symmetrical doublets. The lines of the external doublet (3) are well resolved, whereas those of the two internal doublets (2) and (1) are superimposed.

3 Results and discussion

The structure refinement confirms the relevant results of the crystal structure determination on a manganese rich (74%) natural sample, made by Moore and Araki [5]. There are three distinct Fe sites: Fe (1), Fe (2) and Fe (3) with proportions 1:2:2, respectively, which are fully occupied. The cations Fe^{2+} are octahedrally coordinated and these octahedra are linked by edges, forming pentamer groups according to the scheme Fe(2)–Fe(3)–Fe(1)–Fe(3)–Fe(2). These groups are linked by six corners, and share four equivalent oxygen sites and two other oxygen

Table II Mössbauer parameters of ferrous hureaulite at room temperature

Doublet	QS (mm/s)	IS (mm/s)	Width (mm/s)	Area (%)	Site assignment
1	1.38(2)	1.270(7)	0.365(5)	19.9(1.4)	Fe(1)
2	1.285(2)	1.226(2)	0.282(5)	39.1(1.2)	Fe(2)
3	2.533(2)	1.243(1)	0.275(4)	41.5(0.5)	Fe(3)

QS Quadrupole splitting; *IS* isomer shift with respect to Fe-metal at room temperature.

equivalent sites. The P cations are situated in two distinct tetrahedral sites, forming $P(2)O_4$ and $P(1)O_4H$ groups, that are linked to octahedra of different pentamers, completing the structure. One of the two independent phosphates, $P(1)O_3OH$, is an acid group. The mean Fe–O distances for each octahedron are presented in Table I with the corresponding values obtained by Moore and Araki [5]. The P(1)–O distances range from 1.519(4) to 1.568(4) Å and the P(2)–O distances from 1.523(4) to 1.556(4) Å. The P(1)–OH bond (1.568 Å) is considerable longer than both the average of the three shorter bonds (1.527 Å), and the average P(2)–O bond in the normal $P(2)O_4$ group (1.543 Å). Table I also presents the average quadratic elongation ($\langle\lambda\rangle$) and angular variance (σ^2), defined by Robinson et al. [8], calculated for octahedral and tetrahedral sites. There are small quantitative differences in both cases and a common tendency: the increasing site distortion follows the sequence M3 > M1 > M2, the same order observed in the mean M–O distances.

The values of the fitted Mössbauer parameters are displayed in Table II. The relative populations are in good agreement with the cation occupancy ratio 1:2:2; consequently, it is straightforward to attribute the less intense doublet (1) to Fe(1). Doublets (2) and (3), which have approximately equal intensities, cannot be unambiguously assigned to the crystallographic sites. However, they can be tentatively identified on the basis of their isomer shift (IS) values. For the same coordination number, sites with shorter average cation–anion distances generally correspond to higher electron densities at the cationic nuclei, and therefore have lower IS values. Since doublet (2) has a lower IS value than doublet (3), it can be attributed to site Fe(2), which has a shorter average Fe–O distance than Fe(3).

The main observed difference between these doublets is their quadrupole splitting (QS). As the octahedral distortion increases, so does the contribution of the crystal lattice to the electric field gradient and QS. On this basis, we attribute doublet 3 to the Fe(3) site.

Thus, the QS argument is consistent with the same Fe^{2+} site assignment as the IS argument. The relationship between QS and $\langle\lambda\rangle$ also confirms these site assignments. As suggested by the small insert in Figure 1, the relationship between the QS and $\langle\lambda\rangle$ is linear to a first approximation. This last correlation deserves more investigation, particularly with regard to other related ferrous phosphates. Shinno and Li [9] presented the relationship between $\langle\lambda\rangle$ and QS for chain silicates. In this case, for small values of $\langle\lambda\rangle$, the QS values increased almost linearly, reached a maximum at nearly 2.80 mm/s, and then decreased monotonically with increasing $\langle\lambda\rangle$.

Acknowledgement The authors thank the Chemistry Department of Federal University of Santa Catarina for the X-ray single crystal data collection. The research was funded by FUJB, FAPERJ, FINEP and CNPq.

References

1. Domingues, P.H., Moreira, L.F., Mattievich, E., Sarkissian, B.: J. Magn. Magn. Mater. **167**, 87 (1997)
2. Nomura, K., Ujihira, Y.: Journal of the Chemical Society of Japan **8**, 1352 (1982)
3. Moreira, L.F., Domingues, P.H., Mattievich, E.: J. Magn. Magn. Mater. **132**, 191 (1994)
4. Fisher, D.J.: Am. Mineral. **49**, 398 (1964)
5. Moore, P.B., Araki, T.: Am. Mineral. **58**, 302 (1973)
6. von Klement, R., Meinhold, M., Stang, H., Stenke, G.: Zeitschrift fuer Anorganische und Allgemeine Chemie **377**, 135 (1970)
7. Sheldrick, G.M.: SHELXL97. Program for the Refinement of Crystal Structures. University of Göttingen, Germany (1997)
8. Robinson, K., Gibbs, G.V., Ribbe, P.H.: Science **172**, 567 (1971)
9. Shinno, I., Li, Z.: Am. Mineral. **83**, 1316 (1998)

Hyperfine Interact (2005) 166:605–611
DOI 10.1007/s10751-006-9327-0

Invisible gold in Colombian auriferous soils

H. Bustos Rodriguez · D. Oyola Lozano · Y. A. Rojas Martínez · G. A. Pérez Alcázar · A. G. Balogh

Published online: 3 November 2006

Abstract Optic microscopy, X-ray diffraction (XRD), Mössbauer spectroscopy (MS), Electron microprobe analysis (EPMA) and secondary ions mass spectroscopy (SIMS) were used to study Colombian auriferous soils. The auriferous samples, collected from El Diamante mine, located in Guachavez-Nariño (Colombia), were prepared by means of polished thin sections and polished sections for EPMA and SIMS. Petrography analysis was made using an optical microscope with a vision camera, registering the presence, in different percentages, of the following phases: pyrite, quartz, arsenopyrite, sphalerite, chalcopyrite and galena. By XRD analysis, the same phases were detected and their respective cell parameters calculated. By MS, the presence of two types of pyrite was detected and the hyperfine parameters are: $\delta_1 = 0.280 \pm 0.01$ mm/s and $\Delta Q_1 = 0.642 \pm 0.01$ mm/s, $\delta_2 = 0.379 \pm 0.01$ mm/s and $\Delta Q_2 = 0.613 \pm 0.01$ mm/s. For two of the samples MS detected also the arsenopyrite and chalcopyrite presence. The mean composition of the detected gold regions, established by EPMA, indicated 73% Au and 27% Ag (electrum type). Multiple regions of approximately 200 × 200 μm of area in each mineral sample were analyzed by SIMS registering the presence of "invisible gold" associated mainly with the pyrite and occasionally with the arsenopyrite.

Key words invisible gold · Mössbauer spectroscopy · X-rays diffraction · secondary ions mass spectroscopy SIMS · pyrite · quartz · arsenopyrite

H. Bustos Rodriguez (✉) · D. Oyola Lozano · Y. A. Rojas Martínez
Departamento de Física, Universidad del Tolima, A. A. 546 Ibagué, Colombia
e-mail: hbustos@ut.edu.co

G. A. Pérez Alcázar
Departamento de Física, Universidad del Valle, A. A. 25360 Cali, Colombia

A. G. Balogh
Institute of Materials Science, Darmstadt University of Technology, Darmstadt, Germany

1 Introduction

The higher price of gold has contributed to a significant increase in the exploration and study of gold deposits. For most gold ores it is possible to recover until 90% of gold using conventional techniques such as gravity, flotation and cyanidation. However, the use of these and other techniques depends of many factors, such as: the mineral phase to which the gold is associated (visible or invisible), the type of gold present, the grain size, etc. [1]. Then it is very important to obtain a good characterization of the mineral gold ore to determinate the recovering technique. The term "invisible" arises from the fact that one cannot distinguish, by conventional microscopy, between gold chemically bound in the mineral and submicroscopic inclusions [2]. The question of the chemical form of "invisible" gold in sulfide and sulfarsenide minerals has also attracted much interest in the years following the report of chemically bound Au in sulfides [3, 4]. The ore genesis and metallurgical implications for the presence of invisible gold in minerals such as arsenopyrite are potentially very significant [5], and experimental techniques to determine the valence of gold in sulfides also are of current interest [6–8]. Experimental evidence, however, remains controversial because of low concentrations of gold, near the limits of resolution for conventional laboratory techniques, and its non-uniform distribution in sulfide minerals.

The intent of this paper is to complete the characterization of the Diamante gold mine, located in the municipality of Guachavez (Nariño) – Colombia. Previous studies showed that this is a precious metal deposit corresponding to a hydrothermal type composed principally by quartz and approximately 30% of sulfides like pyrite, sphalerite and arsenopyrite [9]. The gold in particular is associated with pyrite and quartz and secondarily with arsenopyrite, sphalerite and galena. The composition of gold, established by EPMA, indicated 73% Au and 27% Ag (electrum type).

In the current work, the results of the Colombian auriferous soil study are discussed and complemented by using XRD, MS and SIMS. Petrography analysis was made to each studied sample using optical microscopy.

2 Experimental

Thin polished sections of eight samples were prepared for petrography analysis and were marked as sample number 805700, 805695, 805694, 805691, F2L1S1, PD14163, PD14163R and PD10114. The degree of reflectivity of each of the phases present in the sample was used as identification method. For the petrography analysis a polarization microscopy was used with 5×, 10×, 20×, 50×, 100× objectives. XRD pattern of each sample (polished thin sections) was taken using a RIGAKU Rint 2100 diffractometer, and the Material Analysis Using Diffraction (MAUD) program [10] was used for the Rietveld refinament. ^{57}Fe Mössbauer Spectroscopy analyses was used in the polished thin sections of the mineral samples. The obtained Mössbauer spectra were fitted using the MOSFIT program [11] and a α-Fe foil was used as the calibration sample. Polished sections of 805700, 805694 and 805691 samples were prepared for the SIMS analysis. The SIMS analysis with direct O^{2+} ion were realized and the images of multiple regions of approximately 200×200 μm of area were obtained at CEMECA ims 5f with current 55.9 mA and voltage 12.47 keV (ion primary).

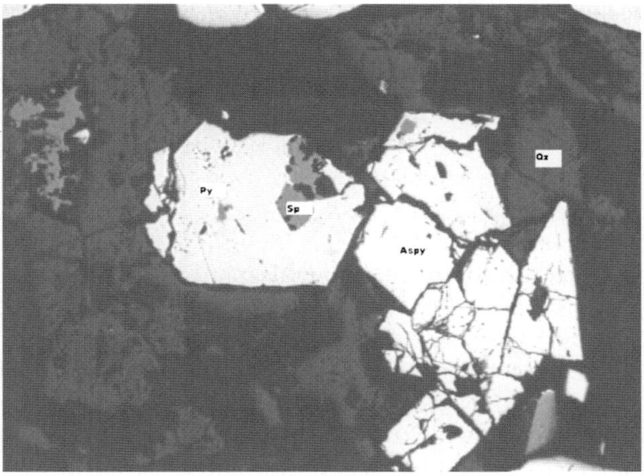

Figure 1 Photomicrograph of the 805701 mineral sample with optical camera, magnification 20× (*Aspy*, arsenopyrite; *Py*, pyrite; *Qz*, quartz; *Sp*, sphalerite).

3 Results and discussion

3.1 Petrography analysis results

In general, the petrography analysis of samples of the Diamante mine permits one to conclude that the principal phases are mainly the sulfides pyrite (cubic FeS_2), sphalerite (ZnS) and arsenopyrite (FeAsS). In the present investigation it was however found that pyrite is associated with arsenopyrite, and sometimes is associated with sphalerite, appearing as crystalline aggregates or isolated crystals. Arsenopyrite is commonly associated with the sphalerite and occasionally associated with pyrite. Sphalerite appears occasionally in isolated polycrystalline aggregates and normally associated with arsenopyrite and with pyrite. Chalcopyrite ($CuFeS_2$) is frequently associated with pyrite, arsenopyrite, sphalerite and freibergite [$(Ag, Cu, Fe)_{12}(Sb, As)_4S_{13}$]; sometimes found isolated in quartz. Galena (PbS) is associated with pyrite, arsenopyrite and sphalerite, and its occurrence is sporadic. Marcasite (orthorhombic FeS_2) is found generally included in pyrite, but its occurrence is very rare. It is generally associated with chalcopyrite, in accordance with the reported literature [12–14]. Typical microphotograph is shown in Figure 1, in which the sample 805701 is illustrated. It can be noted the association of pyrite with arsenopyrite, sphalerite and quartz.

The chemical analysis conducted by atomic absortion show that Zn, Fe, Cu, Pb, As, SiO2, S and Al appear as percent, whereas Au, Ag, Mn, Mg, Ni and Ca appear as ppm. The mean Au content is near 11 ppm and that of Ag is 200 ppm [9].

3.2 XDR results

Figure 2 shows the XRD patterns obtained for the 805694, 805691 and 805700 samples. From the Rietveld refinement conducted with the MAUD program α-quartz, β-quartz, pyrite and sphalerite were observed in the first sample; quartz and pyrite in the second and pyrite, quartz and gold in the third sample. A summary of the results obtained for these samples is presented in Table I. From this table it can

Figure 2 X-ray diffraction patterns of the 805694, 805691 and 805700 mineral samples.

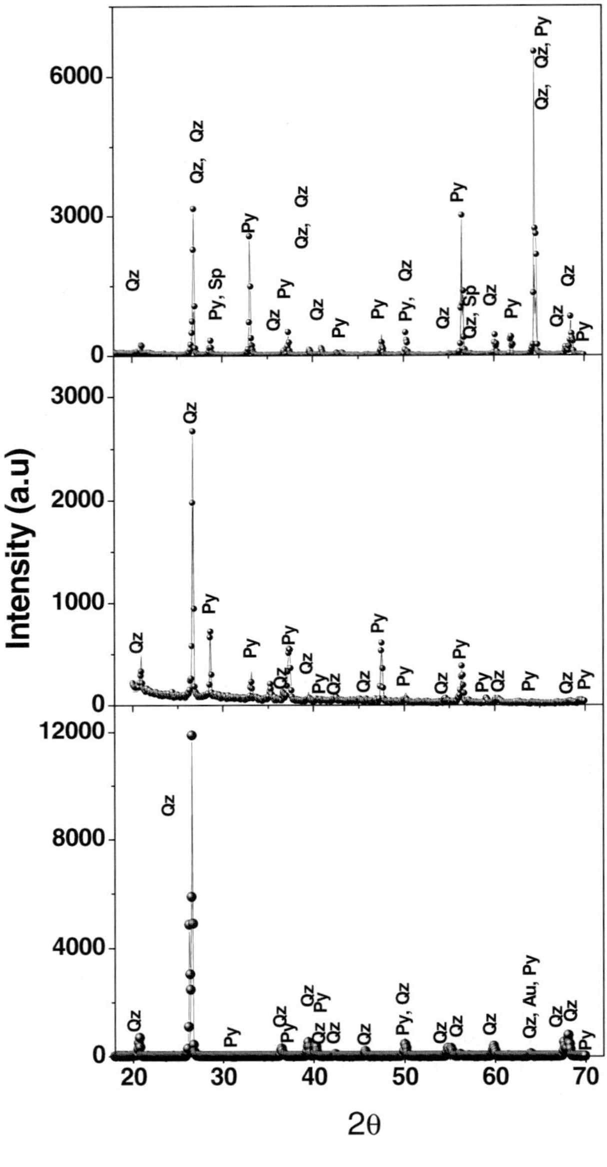

be seen that the different postulated phases from the petrography analysis were also detected by XRD, except the arsenopyrite phase, showing that its presence is under the detection limit of the XRD technique. The obtained XRD parameters are in accordance with those reported in the literature (http://webmineral.com).

3.3 Mössbauer spectroscopy results

Figure 3 shows typical Mössbauer spectra for the 805694, 805691, 805688 and F2L1S1 mineral samples. The fits show that the detected Fe phases are pyrite, in all

Table I Results of lattice parameter, grain size, volume and weight for all the studied samples, obtained by using MAUD program

Sample	Phases	Lattice parameters			Grain size (Å)	Fraction	
		a (Å)	*b* (Å)	*c* (Å)		volume	weight
805694	αQz	4.910		5.371	816.071	0.242	0.403
	Py	5.411			751.080	0.144	0.187
	Sp	5.423			499.619	0.004	0.004
	βQz	5.124		5.096	497.351	0.611	0.407
805691	Qz	4.897		5.411	501.318	0.806	0.686
	Py	5.400			509.294	0.194	0.314
805700	Py	5.429			3537.36	0.931	0.282
	Qz	4.919		5.411	500.021	0.063	0.711
	Au	4.098			500.001	0.006	0.007

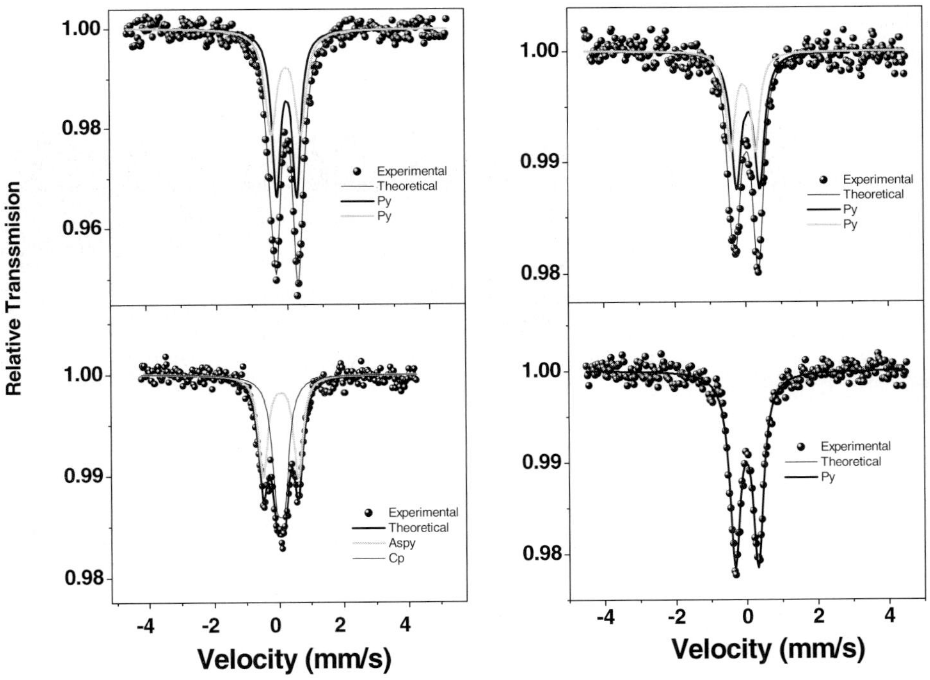

Figure 3 Mossbauer spectra of the 805694, 805691, 805688 and F2L1S1 mineral samples. The MS of the third sample shows the presence of arsenopyrite and chalcopyrite.

the samples, and arsenopyrite and chalcopyrite in two of them. Other Fe phases detected by petrography and XRD are present but not with the enough Fe abundance to be detected by MS. The other phases do not content Fe. The obtained hyperfine parameters are shown in Table II. It can be noted that in some cases it was necessary to fit with two doublets corresponding to two types of pyrite [15, 16].

Table II Hyperfine parameters obtained for 805691, 805694, 805695, 805696, 805700, PD1463 and F1L1S1 minerals samples, by fitting with MOSFIT program

Sample	δ (mm s^{-1}) ±0.02	Γ (mm s^{-1}) ±0.02	QS (mm s^{-1}) ±0.02	Area (%)	Phase
805688	0.353	0.170	1.060	49.54	Aspy
	0.344	0.184	0.211	50.46	Cp
805691	0.307	0.154	0.560	57.11	Py
	0.315	0.214	0.866	42.89	Py
805694	0.280	0.170	0.642	59.23	Py
	0.379	0.177	0.613	40.77	Py
805695	0.317	0.147	0.540	54.18	Py
	0.299	0.195	0.799	45.82	Py
805696	0.345	0.172	1.049	48.99	Aspy
	0.349	0.182	0.229	51.01	Cp
805700	0.368	0.171	0.617	60.34	Py
	0.234	0.151	0.665	39.66	Py
PD1463	0.314	0.153	0.630	100.0	Py
F2L1S1	0.32	0.180	0.613	100.0	Py

δ Isomer Shift, *QS*: Quadrupole Splitting, *A*: Relative Area, Γ: Width.

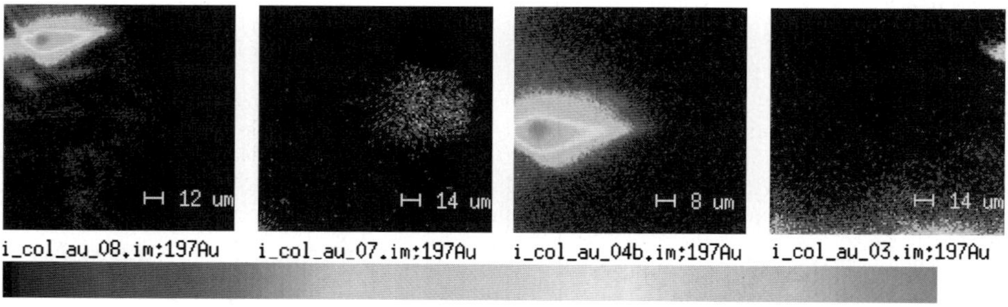

Figure 4 SIMS ion images for mineral sample 805700. In these areas the presence of invisible gold is registered (The *scale of colours* indicates the abundance of the content of the present mineral).

3.4 SIMS analysis results

Figures 4 and 5 show SIMS ion images on an area of 200 × 200 μm for the mineral samples 805700 and 805694, respectively. For the different areas the presence of invisible gold were registered with variable abundance (The colours scale is an indicative of the abundance). This indicates that the presence of the gold is not uniform in the studied mineral samples. In sample 805700 a general distribution is observed with little pyrite (FeS_2) and a bigger but more irregular distribution of arsenopyrite (FeAsS). Here the invisible gold is associated with pyrite. In sample 805691 a general distribution is observed with little pyrite (FeS_2) and a great arsenopyrite grain (FeAsS), and in the sample 805694 a general distribution is presented with little pyrite (FeS_2) and a great arsenopyrite grain (FeAsS). The invisible gold is associated with the pyrite.

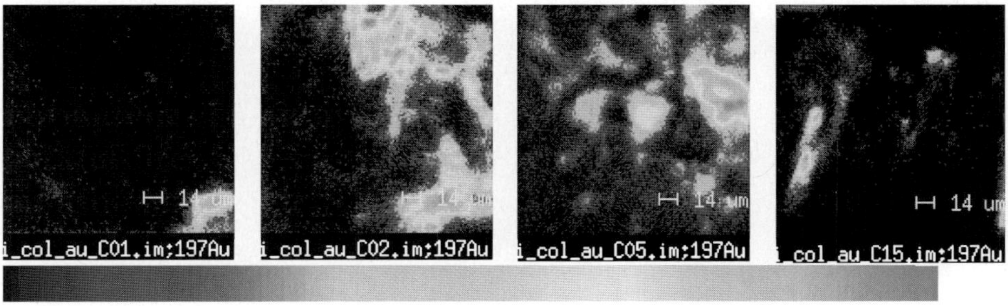

Figure 5 SIMS ion images for mineral sample 805694. In these areas the presence of invisible gold is registered (*Colours scale* indicates the abundance).

4 Conclusion

From the present study about Colombian auriferous soils of the Diamante mine, it can be conclude that the soils are rich in sulfides and register the presence of invisible gold associate mainly with pyrite and occasionally with arsenopyrite. Minerals such as pyrite, arsenopyrite, and sphalerite were the most frequently detected phases with pyrite commonly associated with the arsenopyrite and sphalerite.

Acknowledgements We are grateful to COLCIENCIAS and the Central Committee of Research of the University of Tolima, Colombia, for financial support; to the Mössbauer Laboratory of the University of Valle, Colombia, were the experimental work was conducted; and Institute for Materials Science, Darmstadt University of Technology, Darmstadt, Germany, for SIMS analysis of mineral samples.

References

1. Harris, D.C.: Miner. Depos. **25**, S3 (1990)
2. Cabri, L.J., Newville, M., Gordon, R.A., Darl, E., Sutton, J.R., McMahon, G., Tong Jiang, D.: Can. Mineral. **38**, 1265 (2000)
3. Marion, P., Regnard, J.R., Wagner, F.E.: C. R. Acad. Sci. Paris **302**, 571 (1986)
4. Wagner, F.E., Marion, P., Regnard, J.R.: S. Afr. Inst. Min. Metall. **2**, 435 (1986)
5. Mumin, A.H., Feet, M.E., Cryssoulis, J.L.: Miner. Depos. **29**, 445 (1994)
6. Li, J., Fing, D., Qi, J., Zhang, G.: Acta Geol. Sin. **69**, 67 (1995)
7. Simon, G., Huang, H., Penner-Hahn, J.E., Kesler, S.E., Kao, L.S.: Am. Mineral. **84**, 1071 (1999)
8. Wagner, F.E., Sawick, J.A., Frield, J., Mondonno, J.A., Harris, D.C., Cabri, L.J.: Can. Mineral. **32**, 189 (1994)
9. Molano, J.C., Mojica, J.: Seminario Internacional, p. 25. Ingeominas-Jica, Cali (1998)
10. Lutterotti, L., Scardi, P.: J. Appl. Crystallogr. **23**, 246 (1990)
11. Varret, F., Teillet, J.: MOSFIT Program
12. Friedl, J., Wagner, F.E., Sawicki, J.A., Harris, D.C., Mandarino, J.A., Marion, Ph.: Hyperfine Interact. **70**, 945 (1992)
13. Arehart, G.B., Chryssulis, S.L., Kesler, S.E.: Econ. Geol. **88**, 171 (1993)
14. Wu, X., Delbove, F., Touray, J.C.: Miner. Depos. **25**, S8–S12 (1990)
15. McCammon, C.: Mössbauer Spectroscopy of Minerals, p. 332. American Geophysical Union, Washington, DC (1995)
16. Stevens, J.G.: Mössbauer Mineral Handbook, p. 395. Asheville, North Carolina (1998)

Hyperfine Interact (2005) 166:613–617
DOI 10.1007/s10751-006-9328-z

Investigations of sulfide minerals with layered structure by Mössbauer spectroscopy methods

N. I. Chistyakova · V. S. Rusakov · T. V. Gubaidulina · S. V. Kozerenko

Published online: 27 October 2006

Abstract Samples obtained as a result of the tochilinite synthesis process were investigated by ^{57}Fe Mössbauer spectroscopy. Tochilinite, magnetite and troilite were obtained in investigated samples. The dependences of phase relative content on different content of Mg atoms in the initial mixture were received. It was found that Mg atoms stabilized tochilinite structure. It was shown that magnesium atom entering into tochilinite structure led to Fe atom redistribution in brucite-like layer. It was confirmed that in tochilinite structure equal number of sulfide and brucite-like layers conjugated.

Key words Mössbauer spectroscopy · tochilinite · mackinawite · brucite · troilite · magnetite

1 Introduction

Sulfide forming reactions are considered an important factor in the global carbon, sulfur and iron cycles in the Earth crust. Tochilinite is a hydroxide–sulfide with mixed layered crystal structure (chemical formula is $2Fe_{1-x}S{\cdot}1.67(Mg,Fe)(OH)_2$, $0.08 \le x \le 0.28$) [1]. The sizes of central triclinic cell of tochilinite are $a = 5.37$ Å, $b = 15.67$ Å, $c = 10.72$ Å, $\alpha = \gamma = 90°$, $\beta = 95°$; space group is C1. In the tochilinite structure mackinawite-like $Fe_{1-x}S$ nets interstratify with brucite-like $(Mg,Fe)(OH)_2$ nets. According to [1, 2] Fe^{2+} ions in mackinawite structure are in tetrahedral positions with sulfur neighborhood. Vacancies for Fe atoms exist in sulfide layer of tochilinite. It was shown in [3] that Fe^{2+} ions located in the low-spin state in sulfide

N. I. Chistyakova (✉) · V. S. Rusakov · T. V. Gubaidulina
Physics Department, M.V. Lomonosov Moscow State University,
Vorobyevy gory, 119992 Moscow, Russia
e-mail: nchistyakova@yandex.ru

S. V. Kozerenko
Vernadsky Institute of Geochemistry and Analytical Chemistry RAS,
Kosygin str. 19, 117975 Moscow, Russia

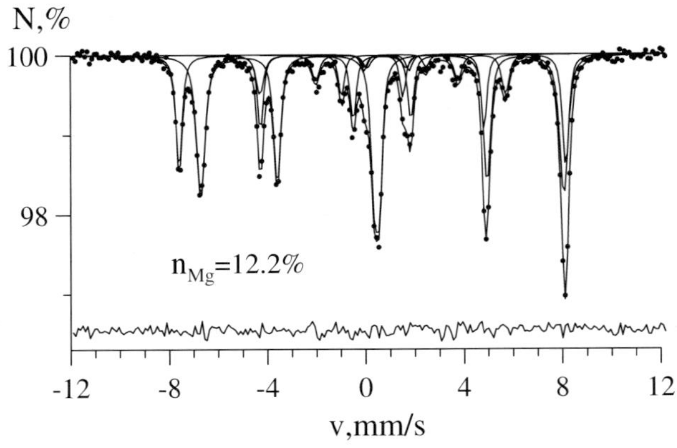

Figure 1 Mössbauer spectrum of investigated sample obtained at the relative content of Mg atoms in the initial mixture equal to n_{Mg} = 12.2%.

layer. In brucite $Mg(OH)_2$ and iron hydroxide $Fe(OH)_2$ structures divalent cations are located in the high-spin state in one crystallographic position with octahedral coordination of OH-groups [4].

2 Experimental

The samples for the Mössbauer investigation were obtained as a result of tochilinite synthesis under different Fe/Mg ratio in the initial mixture. The synthesis is realized in alkaline medium (pH = 11) by Fe (II)-hydroxide interaction with H_2S at the temperature equal to 160 °C. The content of iron in the initial mixture remained fixed (15 mmol). The relative content of Mg atoms in the initial mixture varied from 4.3 to 21.8%. Mössbauer investigations were carried out at room temperature on the spectrometer operated in constant acceleration mode and equipped with ^{57}Co-source in Rh-matrix. Model fitting processed the Mössbauer spectra.

3 Results and discussion

Mössbauer spectrum of one investigated samples is shown at the Figure 1. The model fitting and analyzes of experimental spectra are shown that besides tochilinite magnetite (Fe_3O_4) and troilite (FeS) are presented in samples. It was shown in [3] that tochilinite Mössbauer spectra was the superposition of three partial quadrupole doublets.

The hyperfine parameters of one doublet are close to the values of the mackinawite spectrum (the Mössbauer line shift δ^* = 0.43 ÷ 0.46 mm/s, the quadrupole shift ϵ = 0.08 ÷ 0.11 mm/s). The presence of this doublet is due to the presence of the sulfide layer in tochilinite. The appearance of the two other doublets is likely to be caused by the presence of iron in brucite-like layers (δ_1 = 1.13 ÷ 1.18 mm/s, ϵ_1 = 1.15 ÷ 1.29 mm/s, δ_2 = 0.88 ÷ 0.97 mm/s, ϵ_2 = 0.87 ÷ 0.95 mm/s). The hyperfine parameter values of subspectra, obtained as a result of the model fitting of our

* Mössbauer line shifts are relative to α-Fe.

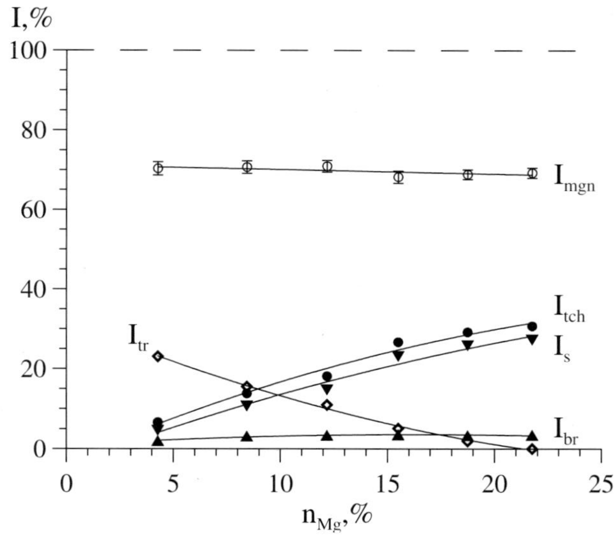

Figure 2 The dependencies of subspectrum relative intensities of obtained phases on Mg content in the initial mixture.

sample Mössbauer spectra, are close to early known data [3]. In the Mössbauer spectra of investigated samples subspectra of magnetite are two Zeeman sextets and the spectra of troilite is one sextet.

The relative intensities of subspectra of obtained phases as functions of Mg relative content in the initial mixture (n_{Mg}) were received (Figure 2). In the assumption of the equality of the Mössbauer effect probability for ^{57}Fe nuclei at different positions in investigated phases the relative intensities of subspectra are equal to Fe atom relative content in these phases.

It is apparently seen on the Figure 2 that the relative content of tochilinite (I_{tch}) in samples increases with growing of Mg atoms content in the initial mixture. In addition the relative content of troilite (I_{tr}) and magnetite (I_{mgn}) decreases and remains constant accordingly.

Under the practically absence of changing the relative intensities (I_{mgn}) of subspectra corresponding to magnetite with n_{Mg} increasing (Figure 2), the ratio of subspectrum intensities of Fe atoms in A- and B-positions in magnetite $I_B/I_A < 2$ for all samples at the same time. Therefore the assumption about entering of magnesium in magnetite in the synthesis process was proposed (the chemistry formula of magnetite is $(Fe^{3+})_A\lfloor Fe^{2+}_{1-z}Mg^{2+}_z Fe^{3+}\rfloor_B O_4$). The account of Mg atoms content in the magnetite was realized using known ratio for subspectrum intensities of Fe atoms in A- and B-positions in magnetite $\frac{I_B}{I_A} = \frac{2\cdot(1-z)}{(1+z)}$. We estimated the value of z, and it was obtained that $z = 0.06 \pm 0.02$ for all investigated samples.

The relative intensities of subspectra corresponding to tochilinite (I_{tch}) rise with an increase of Mg atom content in the initial mixture (Figure 2). The ratio of subspectrum intensities for brucite-like layer to tochilinite (I_{br}/I_{tch}) decreases (Figure 3), at the same time the relative intensity of subspectrum corresponding to sulfide layer rises with an increase of Mg atom content (Figure 3). Therefore we can conclude that Mg atoms enter into brucite-like layer during the synthesis process. The account of magnesium content in tochilinite was realized using the chemical formula of tochilinite and the relative contents of iron in sulfide and brucite-like layers. The

Figure 3 The ratio of subspectrum intensities corresponding to brucite-like and sulfide layers to the total intensity of tochilinite Mössbauer spectrum.

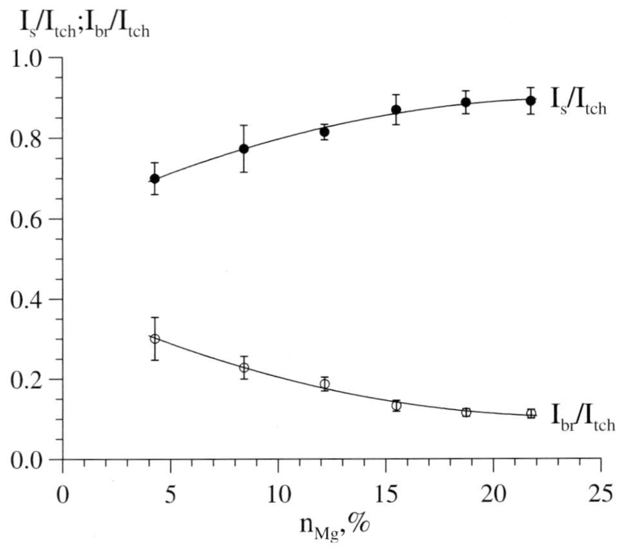

calculations have shown that magnesium content rise in tochilinite structure. In our opinion tochilinite growth is a result of Mg entering into the structure.

The total relative intensity of subspectra corresponding to brucite-like layer of tochilinite remains constant with entering of Mg atoms in the mineral structure (Figure 2). For Fe atom positions in brucite-like layer it is observed that the relative intensity of doublet with smaller values of hyperfine parameters decreases and with greater values – increases. It means that Mg atoms occupy only one of two nonequivalent positions in brucite-like layer, namely the crystallographic position with smaller hyperfine parameters of Mössbauer spectra. The Mg atoms replace the Fe atoms from these positions and occupy produced vacancies. Thereby with an increase of Mg atom content in the initial mixture the redistribution of Fe atoms between nonequivalent positions into brucite-like layer takes place.

With n_{Mg} increasing the correlated change of the relative intensities corresponding to troilite and tochilinite is observed (Figure 2). Since the magnesium stabilizes the tochilinite structure the rise of the tochilinite relative content in samples occurs due to a decrease of the troilite relative content.

The magnesium atom relative content in tochilinite can be evaluated as the difference between Mg content in the initial mixture and magnetite. In Figure 4 the abscissa axis is the ratio of Mg content in brucite-like layer and total content of cations (Fe^{2+} and Mg^{2+}) in the tochilinite.

The ordinate axis is the ratio of Fe content in brucite-like layer and total Fe content in tochilinite. In this figure the firm lines correspond to the curves calculated for different relations between number of sulfide (n_s) and brucite-like (n_{br}) layers in tochilinite structure. Apparently experimental data in the Figure 4 forms a group around the curve corresponding to the ratio n_s:n_{br} = 1:1. The small distinction between experimental and calculated data can be explained by the deficiency of iron in sulfide layer. According to [1, 2] the deficit of Fe in sulfide layer of tochilinite amounts from 8 to 28%. Dashed-line curves in the Figure 4 correspond to these values.

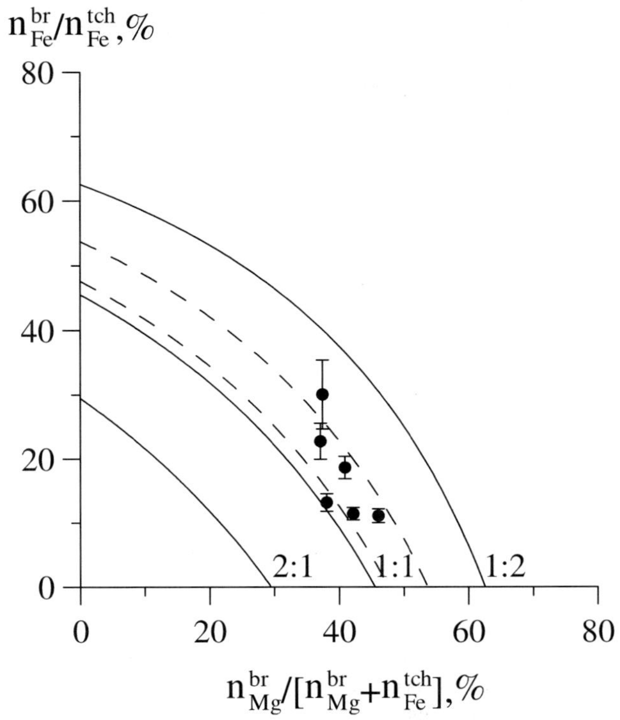

Figure 4 The relative content of Fe atoms in brucite-like layer as function of the Mg atoms relative content in brucite-like layer.

4 Conclusions

The series of samples obtained during a tochilinite synthesis process have been investigated by Mössbauer spectroscopy methods. Tochilinite, troilite and magnetite are found in synthesis products. The dependences of relative content of phases formed in the synthesis process under different magnesium content in the initial mixture were obtained. The stabilization of tochilinite phase with Mg entering into its structure is observed. It was shown that Mg atoms occupied only one of two nonequivalent positions into brucite-like layer with smaller values of hyperfine parameters of the Mössbauer spectrum. It was found that the redistribution of Fe atoms between nonequivalent positions into brucite layer arise with entering of Mg atoms into tochilinite structure. It was confirmed that in tochilinite structure the equal number of sulfide and brucite-like layers conjugated.

References

1. Organova, N.I., Drits, V.A., Dmitrik, A.L.: Kristallografiya **17**, 761 (1972)
2. Organova, N.I.: Crystal Chemistry of Incommensurable and Modulated Mixed-Layer Minerals. Nauka, Moscow (1989)
3. Rusakov, V.S., Chistyakova, N.I., Kozerenko, S.V., Fadeev, V.V.: Moscow Univ. Phys. Bull. **53**, 51 (1998)
4. Miyamoto, H.: Mater. Res. Bull. **11**, 329 (1976)

Hyperfine Interact (2005) 166:619–624
DOI 10.1007/s10751-006-9330-5

Clay fraction mineralogy of a Cambisol in Brazil

**A. S. Anastacio · J. D. Fabris · J. W. Stucki ·
F. S. Coelho · I. V. Pinto · J. H. M. Viana**

Published online: 17 November 2006

Abstract Clay minerals having a 2:1 (tetrahedral:octahedral sheet) structure may be found in strongly weathering soils only if the local pedo-climatic environment prevents them from further weathering to other minerals such as iron oxides. The clay minerals impart important chemical properties to soils, in part by virtue of changes in the redox state of iron in their crystal structures. Knowing the chemical nature of soil clays is a first step in evaluating their potential reactivity with other soil constituents and processes, such as the chemical decomposition of organic substrates to be potentially used in environmental remediation. The purpose of this work was to characterize the iron oxides and iron-bearing clay minerals from a B horizon of a Cambisol developed on tuffite in the State of Minas Gerais, Brazil, using chemical analysis, powder X-ray diffraction, Mössbauer spectroscopy, and thermal analysis. The iron oxides of this NaOH-treated clay-fraction were found to contain mainly maghemite (γFe_2O_3) and superparamagnetic goethite ($\alpha FeOOH$). Kaolinite ($Al_2Si_2O_5(OH)_4$), smectite, and minor portions of anatase (TiO_2) were identified in the CBD-treated sample.

Key words soil · iron oxides · smectite

A. S. Anastacio · J. D. Fabris (✉) · F. S. Coelho · I. V. Pinto
Department of Chemistry, Federal University of Minas Gerais, Campus – Pampulha, 31270-901 Belo Horizonte, Minas Gerais, Brazil
e-mail: jdfabris@ufmg.br

A. S. Anastacio · J. W. Stucki
Department of Natural Resources and Environmental Sciences, 1102 S. Goodwin Ave., Urbana, IL 61801, USA

J. H. M. Viana
Embrapa Milho e Sorgo, CP 285, CEP 35701-970 Sete Lagoas, Minas Gerais, Brazil

1 Introduction

Magnetic soils cover a significant land area (5%) in Brazil [5] and iron oxides such as magnetite (Fe_3O_4) and maghemite (γFe_2O_3) are responsible for the magnetic properties of these soils. Maghemite has a cation deficient spinel structure with a unit cell consisting of 21 1/3 Fe(III) atoms and 2 1/3 vacancies for every eight tetrahedral and 16 octahedral sites [2, 12]. Such magnetic soils are plentiful in the region named Alto Paranaíba in the State of Minas Gerais, Brazil, which were formed by Pre-Cambrian deposits that were later covered with sediments and pyroclastic material derived from a long extinct volcanic source.

Among the soils of this region is Cambisol (according to the FAO Soil Classification System), a soil class with an incipient B horizon that indicates either slow weathering conditions or recent formation [3]. They more commonly occur in low-land areas of the tuffite geodomain that overlies the Bambuí Group. Even though Cambisols are not actually typically widespread in this area, they do contain significant proportions of iron-bearing 2:1 silicates in their clay fraction. The purpose of this study was to characterize the constituent clay minerals and iron oxides of this Cambisol clay fraction in order to gain a better understanding of its mineralogy and potential reactivity with other soil components.

2 Materials and methods

The soil sample was collected from the B horizon of a Cambisol in an agricultural area of Patos de Minas, State of Minas Gerais, Brazil (18° 38′ 10 S, 46° 21′ 30 W). The soil material was disaggregated by hand, air dried, and sieved to obtain a fine-grain particle size (mean particle size, $\varphi < 2$ mm). The sample was dispersed in 1 mol L^{-1} NH_4OH and shaken, followed by wet sieving with a 0.053-mm sieve to remove the sand fraction. The silt and clay fractions were separated by centrifuge fractionation according to Stokes law [8]. The clay fraction was then treated either with NaOH to concentrate, or with CBD (citrate-bicarbonate-dithionite) to remove the iron oxides [11].

Chemical analysis of the CBD-treated sample was accomplished by fusing a 1:1 mixture of the sample with Na- and K-carbonate [9] in a Pt crucible and dissolving the residue in 50% (v/v) HCl, then analyzing the product for all elements except Na and K by plasma emission spectrometry (Spectroflame Analytical Instruments). A separate sample was digested with a concentrated mixture of $HCl+H_2SO_4+HF$, dried, and redissolved with concentrated HCl, then analyzed for Na and K by flame emission spectrometry.

Mössbauer spectra were collected at 298, 110, and 4 K in a transmission geometry using a $^{57}Co/Rh$ source, either in a CMTE spectrometer (model MA250) or a WEB Research spectrometer, with a drive system operating with a triangular waveform and a Janis Model SHI-850-5 cryostat, operating at temperatures varying from 4 to 298 K.

Powder X-ray patterns were collected at the Brazilian National Laboratory of Synchrotron Light Source, with a six-circle diffraction goniometer (Huber) equipped with a double Si (111) crystal monochromator and Cyberstar scintillation detector The wavelength of the synchrotron incident radiation was 0.1377 nm. Some samples were analyzed using a conventional theta-two theta diffractometer (Scintag

Table I Mössbauer parameters of 5 M NaOH-treated Cambisol at room temperature

Site	Mössbauer parameters				
	δ/mm s^{-1}	ε/mm s^{-1}	Δ/mm s^{-1}	B_{hf}/T	RA/%
Fe^{3+}	0.35(1)	–	0.48(2)	–	81
Mh	0.35(2)	−0.08(2)		49.3(9)	16
Fe^{2+}	1.08(1)	–	2.69(2)	–	3

δ: isomer shift relative to the αFe, ε: quadrupole shift, Δ: quadrupole splitting; B_{hf}: hyperfine field; RA: relative area, Mh: maghemite.

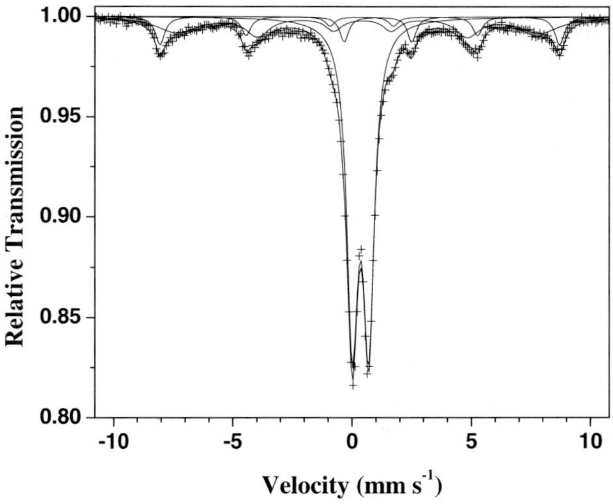

Figure 1 Mössbauer spectrum at 110 K of the 5 M NaOH-treated Cambisol sample.

PAD-V or Rigaku Geigerflex 2037) equipped with a cobalt tube and graphite diffracted beam monochromator (Rigaku only).

Differential thermal (DTA) and thermal gravimetric (TG) analyses of the CBD-treated samples were carried out simultaneously with a STA409EP Netzschi thermal analyzer, using approximately 20 mg of clay sample, at a heating rate of 10°C min^{-1}, up to 950°C in air. Alumina was the reference material.

3 Results and discussion

3.1 NaOH-treated sample

The room-temperature Mössbauer spectrum for the 5 mol L^{-1} NaOH-treated sample shows a six-line pattern and two doublets. The hyperfine parameters (Table I) indicate that the sextet is due to maghemite [4, 6] and the doublets are characteristic of octahedral Fe(II) in a 2:1 aluminosilicate and octahedral Fe(III) in kaolinite ($Al_2Si_2O_5(OH)_4$). The presence of maghemite and kaolinite was confirmed by powder X-ray diffraction; anatase was also identified in this clay fraction, as expected from previous reports of its occurrence in Oxisols of this geological

Table II Mössbauer parameters of the 5 M NaOH-treated Cambisol sample at 110 K

Site	Mössbauer parameter				
	δ/mm s^{-1}	ε/mm s^{-1}	Δ/mm s^{-1}	B_{hf}/T	RA/%
Fe^{3+}	0.356(1)	–	0.68(1)	–	62
Fe^{2+}	1.10(1)	–	2.83(2)	–	4
Mh	0.451(6)	−0.091(9)		51.66(5)	24
Gt	0.47(4)	–	−0.25	46.2(3)	10

δ: isomer shift relative to the αFe, $2\varepsilon_Q$: quadrupole shift, Δ: quadrupole splitting; B_{hf}: hyperfine field; RA: relative area, Mh: maghemite; Gt: goethite.

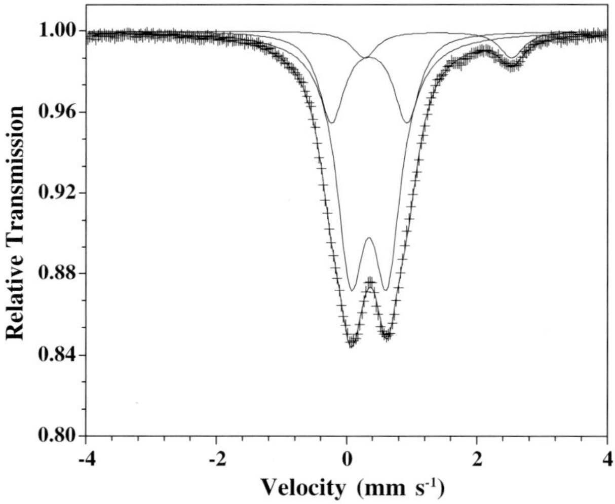

Figure 2 Mössbauer spectrum at 110 K of the CBD-treated Cambisol sample.

origin [1]. In the 110 K spectrum (Figure 1), the relative area of the central Fe(III) doublet decreased to 62%, compared to 81% at room temperature. The decreased doublet area at lower temperature was accompanied by the appearance of a second sextet assigned to goethite. This means that the particle size distribution for this goethite is sufficiently small to induce superparamagnetic behavior at room temperature, but at least part of the relaxation due to smaller particles is blocked at 100 K. The relative subspectral area of maghemite also increased from 16 to 24% as the temperature was decreased from room temperature to 110 K (Table II).

3.2 CDB-treated sample

The Mössbauer spectrum of the CBD-treated sample (Figure 2) at 110 K is essentially dominated by two paramagnetic doublets. The quadrupole distribution revealed two Fe(III) sites and one Fe(II) site (Table III). At 4 K, a small sextet still persists (data not shown). The hyperfine field of this residual sextet is consistent with goethite. The XRD reflection at d = 1.5 nm (Figure 2) is consistent with the presence of a 2:1 layer silicate, but no characteristic reflection can be attributed to any iron oxide; the presence of kaolinite may be confirmed by the 0.713 nm reflection, which disappeared by heating the sample at 550°C (Figure 3).

Table III Mössbauer parameters of the CBD-treated Cambisol sample

Site	Mössbauer parameters at 110 K		
	δ/mm s^{-1}	Δ/mm s^{-1}	AR/%
Fe^{3+}	0.459(4)	0.553(5)	68
Fe^{3+}	0.463(1)	1.15(9)	25
Fe^{2+}	1.40(4)	2.26(8)	7

δ: isomer shift relative to the αFe, Δ: quadrupole splitting; B_{hf}: hyperfine field; RA: relative area.

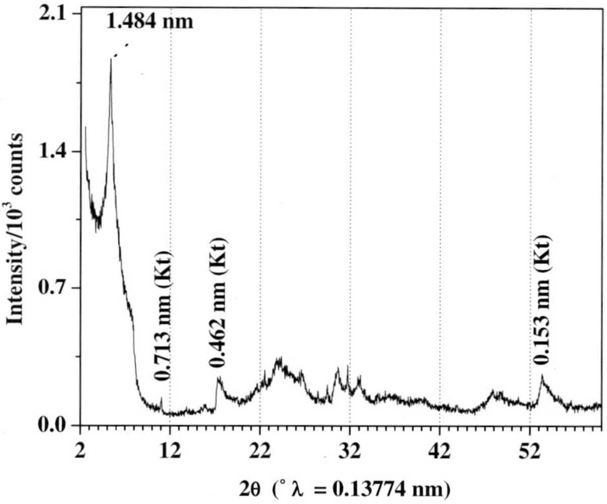

Figure 3 X-ray powder diffractogram of the CBD-treated Cambisol sample. *Kt* = kaolinite.

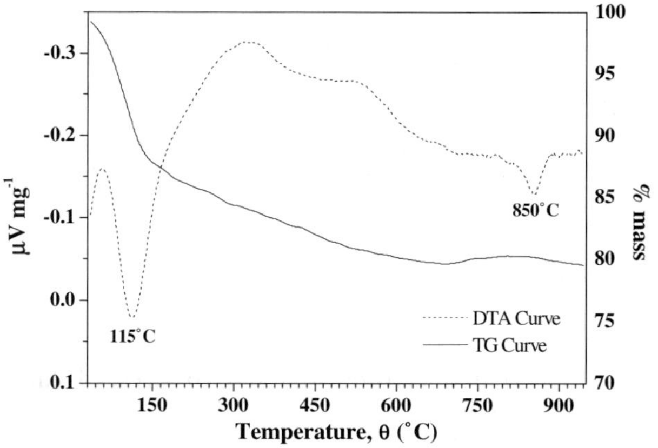

Figure 4 TG/DTA curve of the CBD-treated Cambisol sample.

Chemical analysis of the CBD-treated sample revealed high amounts of iron and magnesium, indicating some isomorphous substitution in kaolinite (iron for aluminum) and smectite (magnesium and/or iron for aluminum) (Table IV).

Table IV Chemical composition of the CBD-treated Cambisol sample

	Mass %										
Oxide	Al_2O_3	CaO	Cr_2O_3	Fe_2O_3	K_2O	LOI	MgO	Na_2O	P_2O_5	SiO_2	TiO_2	Total
	8.61	0.53	1.37	18.19	1.08	18.09	8.75	6.36	2.62	29.31	2.27	97.48

LOI: loss on ignition.

Results from thermal analyses of the CBD-treated sample revealed (Figure 4) an endothermic peak at 115°C due to loss of adsorbed water and a small shoulder at 680–690°C, due to the loss of structural water. The endothermic peak at 850°C is characteristic of the transformation of smectite into other silicate phases, such as quartz [7,10].

4 Conclusion

The mineralogy of the clay fraction of the studied Cambisol consists mainly of goethite, maghemite, kaolinite, smectite, and accessory amounts of anatase. Chemical treatment with CBD effectively removed the iron oxides, but some goethite apparently resisted dissolution by this treatment. The major silicate minerals in this clay fraction are kaolinte and a 2:1 clay of the smectite group.

Acknowledgements Work financially supported by CNPq, CAPES and FAPEMIG (Brazil). The Fulbright Commission (US) and Capes (Brazil) supported a visit of JDF to the University of Illinois at Urbana-Champaign (USA).

References

1. Carmo, D.N., Curi, N., Resende, M.: Caracterização e Gênese de Latossolos a Região do Alto Paranaíba (MG). Bras. Ciênc. Solo **8**, 235–240 (1984)
2. Coey, J.M.D.: Magnetic properties of soil iron oxides and clay minerals. In: Stucki, J.W., Goodman, B.A., Schwertmann, U. (eds.) Iron in Soils and Clay Minerals, pp. 397–466. Reidel, Dordrecht (1988)
3. Deckers, J.A., Nachtergale, F.O., Spaargaren, O.C.: FAO, Rome (1988)
4. Fabris, J.D., Coey, J.M.D., Qi, Q., da N. Mussel, W.: Characterization of Mg-rich maghemite form tuffite. Am. Mineral. **80**, 664–669 (1995)
5. Ferreira, S.A.D., Santana, D.P., Fabris, J.D., Curi, N., Nunes Filho, E., Coey, J.M.D.: Relações entre magnetização, elementos traços e litologia de duas seqüências de solos do Estado de Minas Gerais. Bras. Ciênc. Solo **18**, 167–174 (1994)
6. Goulart, A.T., de Jesus Filho, M.F., Fabris, J.D., Coey, J.M.D.: Multiple iron-rich spinel phaes and hematite in a magnetic soil developing on tuffite. Phys. Chem. Miner. **25**, 63–69 (1997)
7. Hatakeyama, T.: Handbook of Thermal Analysis. Wiley, New York (1998)
8. Jackson, M.L.: Soil Chemical Analysis – Advanced Course. Madison (1975) (published by the author)
9. Jeffery, P.G., Hutchison, D.: Chemical Methods of Rock Analysis. Pergamon, Oxford (1981)
10. Mackenzie, R.C.: Differential Thermal Analysis. Fundamental Aspects. Academic, London (1970)
11. Mehra, O.P., Jackson, M.L.: Iron Oxide Removal from Soils and Clays by a Dithionite–Citrate System Buffered with Sodium Bicarbonate. Clays Clay Miner. **7**, 317 (1960)
12. Murad, E., Jonston, J.H.: Iron oxides and oxyhydroxides. In: Long, G.J. (eds.) 629. Plenum, New York (1987)

Hyperfine Interact (2005) 166:625–630
DOI 10.1007/s10751-006-9331-4

Iron-containing atmospheric aerosols in the Chernobyl fallout

V. Rusanov · V. Gushterov · H. Winkler · A. X. Trautwein

Published online: 21 November 2006

Abstract Mössbauer spectroscopy was applied to determine the composition and the iron concentration in the atmospheric aerosols contaminated in Sofia, Bulgaria after the Chernobyl accident. The results confirm the major conclusion of the Kopcewiczs for Poland, i.e. that in the initial filters, collected during the contaminating fallout (30.04–05.05.1986), the iron concentration was highest, 3.69 $\mu g/m^3$ and that magnetite Fe_3O_4 was present. For the following days a change in the chemical composition including the presence of α-Fe_2O_3, α-FeOOH and γ-FeOOH as well as the absence of magnetite, was detected. Input of industrial iron contamination was negligible since the nearby steel plant had worked at minimum power due to official holidays. Unfortunately, Mössbauer spectroscopy studies only, do not allow a definite conclusion about an increase of the isotope abundance of ^{57}Fe in the Chernobyl fallout.

Key words Chernobyl accident · atmospheric aerosols · iron-containing compounds

1 Introduction

The Mössbauer effect can be successfully used for studies of the properties and the concentration of the iron in the atmospheric air [1–4]. Weathering processes inject in the form of fine particles large quantities of iron oxides, oxyhydroxides and iron containing clay and silicate minerals in the troposphere. Industrial air pollutions from steel production, coal combustion or highway traffic can change dramatically the iron concentration in the air. Under favourable conditions iron of extraterres-

V. Rusanov · V. Gushterov (✉)
Faculty of Physics, Department of Atomic Physics, University of Sofia, 5 James Bourchier Blvd., 1164 Sofia, Bulgaria
e-mail: viktor@phys.uni-sofia.bg

H. Winkler · A. X. Trautwein
Institut für Physik, Universität zu Lübeck, Ratzeburger Allee 160, D-23538 Lübeck, Germany

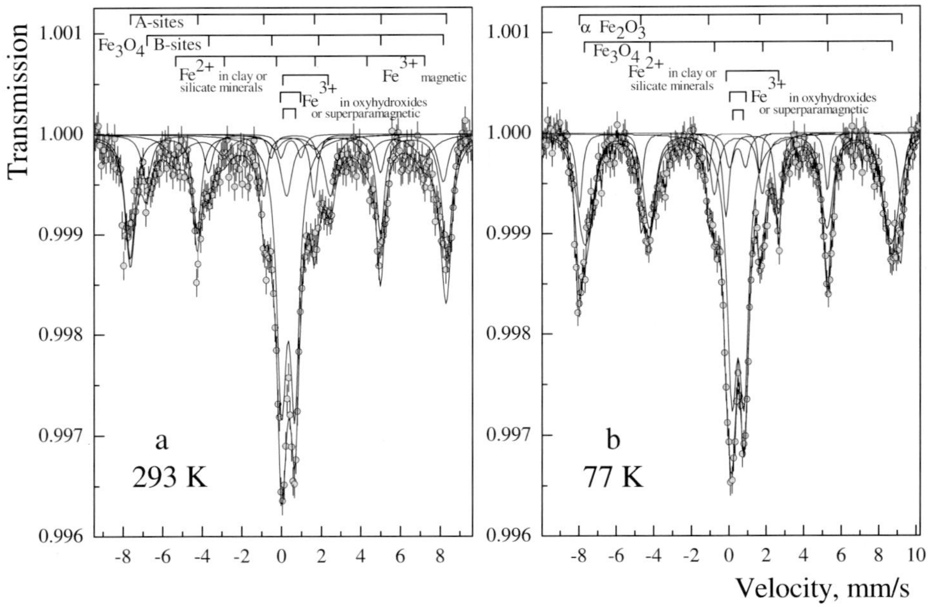

Figure 1 Mössbauer spectra of an air filter collected during 30.04–05.05.1986 taken at 293 and 77 K.

trial origin could be observed. During atmospheric nuclear bomb tests a change of the natural abundance of iron after the reaction $^{56}Fe(n,\gamma)^{57}Fe$ is expected.

After the Chernobyl accident on 26.04.1986, about 50 t of reactor fuel was spread out as radioactive fallout over Europe. In situ γ-ray measurements identified different radionuclides of reactor origin present in ground, soil, and grass area or in the air (by means of particle filters) [5]. Investigating iron-containing atmospheric aerosols collected before and during the Chernobyl accident the Kopcewiczs reported that air filters contained higher radioactivity, higher concentration of iron oxyhydroxides and also different chemical composition (more Fe_3O_4) compared to those collected before the accident [6]. In our contribution we present new measurements on air filters collected in Sofia, Bulgaria during the time of contaminating Chernobyl fallout and we compare them to similar air filters collected long after. Other possible sources of iron contamination are discussed.

2 Experimental details

Mössbauer measurements were performed at room and liquid nitrogen temperatures on several air filters collected in Sofia, Bulgaria during the time interval between 30.04 and 13.05.1986 and long after (1989–2001). In Bulgaria the maximum contamination of Chernobyl fallout was detected from 01.05 to 05.05.1986. The samples were taken from membrane filters used for air radioactivity monitoring by the Radiochemistry Lab of the Faculty of Chemistry, University of Sofia. The volume of the air pumped was recorded. A small part of the filters was prepared by dissolving in dichloromethane, CH_2Cl_2, evaporating and pressing in the form of pellets. The absorbers surface density was about 80 ÷ 100 mg/cm^2. Measurements

Table I Mössbauer parameters at 77 K of the detected iron-bearing compounds and Fe concentration in air for the different filters

Sample	IS (mm/s)	ΔE_Q (mm/s)	H_{hf} (T)	I_{rel} (%)	Fe (μg/m^3)
30.04–05.05.1986					
Fe_3O_4 + γ-Fe_2O_3	0.50(2)	−0.00(2)	50.2(5)	43.1(9)	3.69
α-Fe_2O_3	0.45(2)	0.33(2)	52.7(5)	12.6(9)	
Fe^{3+}	0.46–0.48(2)	0.65–0.85(2)	–	36.5(9)	
Fe^{2+}	1.25(2)	2.81(2)	–	7.8(9)	
06.05–07.05.1986					
α-Fe_2O_3	0.46(3)	0.41(3)	53.0(9)	9.4(9)	3.61
α-FeOOH	0.55(5)	−0.23(5)	48.2(9)	25.2(9)	
Fe^{3+}	0.48–0.52(2)	0.59–1.12(2)	–	54.6(9)	
Fe^{2+}	1.22(2)	2.85(2)	–	10.8(9)	
09.05–10.05.1986					
α-FeOOH	0.52(5)	−0.23(5)	49.6(9)	45.2(9)	0.79
Fe^{3+}	0.44–0.52(5)	0.53–1.01(5)	–	49.8(9)	
Fe^{2+}	1.2(1)	2.8(1)	–	5(2)	
Days with small air pollution (February 1992)					
α-Fe_2O_3	0.40(2)	0.33(2)	52.9(5)	24.2(9)	0.63
α-FeOOH	0.45(2)	−0.20(2)	50.6(5)	30.3(9)	
Fe^{3+}	0.44(2)	0.65(2)	–	30.5(9)	
Fe^{2+}	1.27(2)	2.78(2)	–	15.0(9)	
Days with high air pollution (July 1990)					
Fe_3O_4 + α-Fe_2O_3	0.45(2)	0.08(2)	51.8(9)	30.6(9)	5.0
α-FeOOH	0.58(5)	−0.30(5)	47.4(9)	26.4(9)	
Fe^{3+}	0.46(2)	0.66(2)	–	37.2(9)	
Fe^{2+}	1.26(2)	2.82(2)	–	5.8(9)	

were performed in transmission geometry using a standard constant acceleration spectrometer and ^{57}Co[Rh] source with activity of about 20 mCi. The isomer shifts were related to the α-Fe standard at room temperature. The filters with Chernobyl contaminations were compared to similar air filters collected long after the Chernobyl accident in days with large and small air pollutions. The concentration of Fe in air was calculated after the "area method" of Hafemeister and Brooks Shera [7].

3 Results and discussions

The first air filter was collected during the time interval 30.04–05.05.1986. These days were non-working days (May 1st and 2nd – official holidays, and May 3rd and 4th – weekend). The sampling time coincides with the maximum contaminating fallout due to the Chernobyl accident detected in Bulgaria. Measurements of this filter (Figure 1) confirmed the presence of large quantities of magnetite Fe_3O_4, mixtures of various oxyhydroxides mainly γ-FeOOH, a superparamagnetic α-Fe_2O_3, and Fe^{2+} in clay and silicate minerals. The shape of the magnetite spectra at room and liquid nitrogen temperatures changed considerably. This gave evidence that the magnetite particles had undergone a Verwey transition at about 120 K, which could happen only if they were larger than 20 nm [8, 9]. The iron concentration in air of

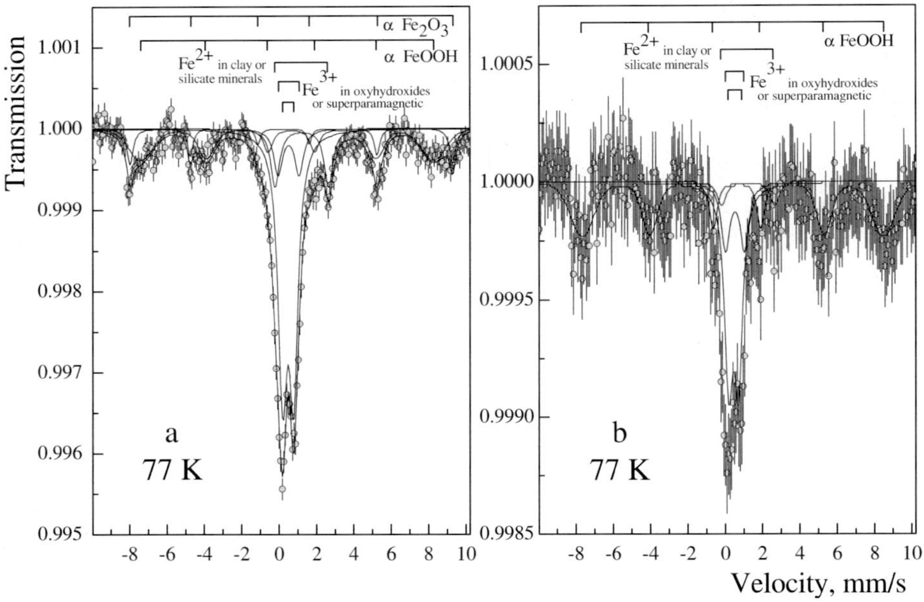

Figure 2 Mössbauer spectra of an air filter collected during **a** 06.05–07.05.1986 and **b** 09.05–10.05.1986 taken at 77 K.

3.69 μg/m^3 was high and typical for days with large air pollutions. The values of the Mössbauer parameters in Table I are slightly different from those for bulk materials [8, 9]. The differences are related with different particle size and particle size distribution.

In the filter from the following days, 06.05–07.05.1986, (Figure 2a; Table I), a change in the chemical composition including the presence of α-Fe_2O_3 and oxyhydroxides, as well as the absence of magnetite, were detected. The iron concentration remained high 3.61 μg/m^3. In the Mössbauer spectra of an air filter collected during the next days, 09.05–10.05.1986, the magnetically splitted component was considerably reduced at room temperature. Only small quantities of α-FeOOH and γ-FeOOH were detected (Figure 2b; Table I), and the measured iron concentration was extremely low, 0.79 μg/m^3, which is typical for days of small air pollution.

The measurements on consecutively collected filters (Figures 1 and 2), clearly showed that after the first wave of radioactive contaminations from 01.05 to 05.05.1986 the concentration of iron in the air dropped considerably. The overall pollution was not caused by this radioactive wave only. Similar effects were observed in the radioactivity of air in Sweden. Devell et al. reported about irregular oscillations of the air radioactivity in the days during and after the Chernobyl accident reaching differences by two orders of magnitude [5].

The results from the first air filter of Chernobyl fallout are compared to results from air filters collected long after the accident on days of high pollution (windy summer days in 1990) and low pollution (winter days after snowfall in 1992), (Figure 3; Table I). This comparison confirms the major conclusion of the Kopcewiczs [6]

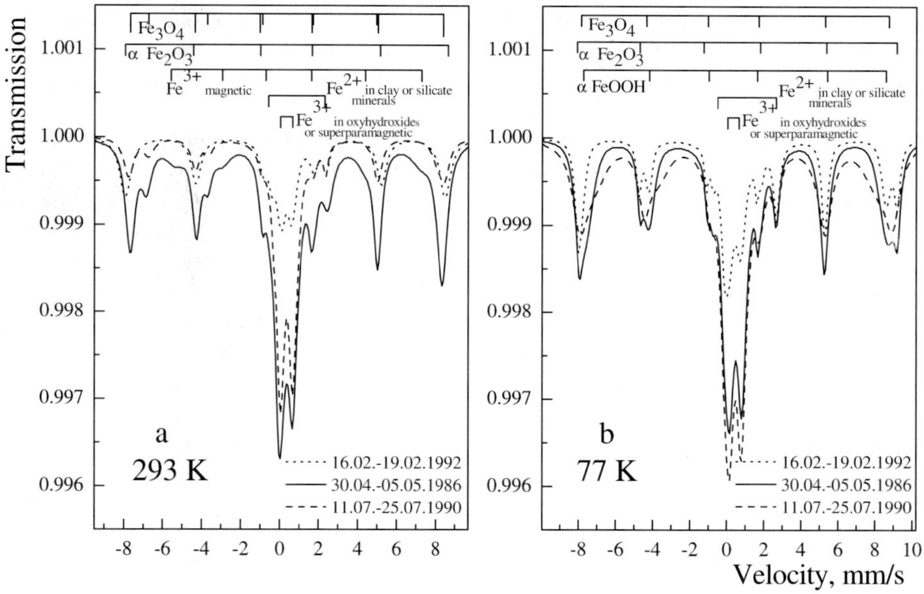

Figure 3 Comparison of fitted Mössbauer spectra of filters collected right after the Chernobyl accident (*solid line*) with those from filters collected long after it on days of high (*dashed line*) and low (*dotted line*) air pollution.

that the aerosols collected during the Chernobyl accident contain iron which is not completely oxidized to Fe^{3+} (magnetite, Fe_3O_4 and maghemite, γ-Fe_2O_3). Such contamination of air over Sofia is not unusual. In days of high air pollution we detected magnetite, which most probably originates from the steel production plant near Sofia. Nevertheless, the magnetite concentration in the Chernobyl filter is two to three times higher (Figure 3) and therefore should be a result of the pollution due to the accident.

For the calculation of air–iron concentration in this work the "area method" has been applied and the natural abundance of the Mössbauer isotope ^{57}Fe, 2.18%, was used. The question is still open: could have the nuclear reaction, $^{56}Fe(n,\gamma)^{57}Fe$, which has a small cross-section for thermal neutrons, 2.6 b, enriched the samples in ^{57}Fe. The matter has already been discussed by Kopcewicz et al. [1, 6]. In our opinion a conclusion about the enrichment of the samples in ^{57}Fe cannot be made by Mössbauer spectroscopy only, because enrichment in ^{57}Fe as well as larger concentration of iron lead to the same effect. Only mass-spectroscopic measurements could unambiguously prove a higher concentration of ^{57}Fe caused by neutron activation during the reactor operation.

4 Conclusion

The increase of iron-containing material in the air pollution as a result of the Chernobyl reactor accident was expected. First the Kopcewiczs have used Mössbauer spectroscopy for such studies of air filters from Warsaw [6]. Our results for air filters from Sofia confirm that for the initial filters, collected during the contaminating fallout (30.04–05.05.1986), iron concentration was highest and magnetite Fe_3O_4 was present. This is most likely due to the fact that during the accident large amounts of iron-construction materials were destroyed, evaporated and transported together with the radioactive fallout. The presence of magnetite particles of industrial origin in air pollutions in Sofia is not exceptional and originates mostly from the steel plant situated nearby. Nevertheless, the major Chernobyl fallout in Bulgaria happened during four consecutive holidays after May 1st, and we could assume that on these days the industrial plants had worked on minimal power. During these days the magnetite concentration in the air filter was 2–3 times higher compared to that from days with high air pollution in Sofia. This finding allows us to confirm the suggestion that the large part of the iron-containing contaminations was caused by release from the reactor accident. A definite conclusion about an increase of the isotope abundance of ^{57}Fe in the Chernobyl fallout cannot be based on Mössbauer spectroscopy only. Additional mass-spectroscopic measurements are needed.

Acknowledgements The authors are indebted to Mr V. Mavrodiev for providing the filters. Thanks are due to the Alexander von Humboldt Foundation, special program Stability Pact for South Eastern Europe (V. R. and A. X. T.).

References

1. Dzienis, B., Kopcewicz, M.: Tellus **XXV**, 213 (1973)
2. Kopcewicz, B., Kopcewicz, M.: Tellus **30**, 562 (1978)
3. Kopcewicz, B., Kopcewicz, M.: Hyperfine Interact. (C) **5**, 427 (2001)
4. Kopcewicz, B., Kopcewicz, M.: In: Spurny, K.R. (ed.) Analytical Chemistry of Aerosols, p. 185. Lewis, Boca Raton (1999)
5. Devell, L., Tovedal, H., Bergstöm, U., Appelgren, A., Chyssler, J., Andersson, L.: Nature **321**, 192 (1986)
6. Kopcewicz, B., Kopcewicz, M.: Hyperfine Interact. **139/140**, 657 (2002)
7. Hafemeiter, D.W., Brooks Shera, E.: Nucl. Instrum. Methods **41**, 133 (1966)
8. Murad, E., Johnston, J.H.: In: Long, G.L. (ed.) Mössbauer Spectroscopy Applied to Inorganic Chemistry, vol. 2, p. 507. Plenum, New York (1987)
9. Cornell, R.M., Schwertmann, U.: The Iron Oxides: Structure, Properties, Reactions, Occurrence and Uses, p. 146. VCH, Weinheim (1996)

Hyperfine Interact (2005) 166:631–636
DOI 10.1007/s10751-006-9332-3

Mössbauer study of chemical transformations in soil samples during thermomagnetic measurements

B. Kopcewicz · M. Kopcewicz · M. Jelenska · A. Hasso-Agopsowicz

Published online: 27 October 2006

Abstract Mössbauer spectroscopy was used in order to identify iron oxides responsible for soil magnetism and for changes of the shape of the magnetic susceptibility curves obtained in the thermomagnetic measurements. The chemical composition of the soil samples was transformed due to the thermal treatment through dehydration/dehydroxylation and than oxidation/reduction process. The largest content of the magnetite has been observed for samples annealed up to T_a = 700°C, which indicates that conversion process occurred under reduction conditions.

Key words soil annealing · soil pollution · soil susceptibility

1 Introduction

Recently, rock-magnetic measurements related to the concentration of ferromagnetic minerals were used to study the distribution of anthropogenic pollution in soil profiles. Especially convenient, as a simple cheap and fast method were the measurements of low-field magnetic susceptibility. A detailed environmental magnetic study has been carried out for the samples collected in the industrial region of Ukraine at Mariupole area, and, for comparison, at Homutovski steppe, relatively unpolluted region (the results were published elsewhere [1]). The magnetic susceptibility is highly sensitive to mineralogical changes, which take place during thermal treatment of soil samples. The studies of such changes can provide information concerning the magnetic mineral composition and the size of magnetic grains. In this paper the Mössbauer spectroscopy was used in order to identify the iron oxides

B. Kopcewicz (✉) · M. Jelenska · A. Hasso-Agopsowicz
Institute of Geophysics, Polish Academy of Sciences, Ks. Janusza 64, 01-452 Warsaw, Poland
e-mail: barbara@igf.edu.pl

M. Kopcewicz
Institute of Electronic Materials Technology, Wolczynska 133, 01-919 Warsaw, Poland

responsible for soil magnetism and to determine the chemical transformations of the soil samples during thermomagnetic measurements.

2 Experimental details

In order to verify the influence of industrial activity on soil composition the soil samples used in the present study came from two areas: Mariupol, a heavily polluted industrial region in south-east Ukraine (M4 profile) and Homutovski steppe, a relatively unpolluted area (H1 profile), situated 20 km from the northern coast of the Azov sea and about 80 km north-east of Mariupol. The sampling site of the M4 profile was situated close to the city. The soil samples (H1 and M4) were collected from two different layers: topsoil (depth 0–10 cm) and loess (depth 120–130 cm and of 100–110 cm for H1 and M4 profiles, respectively). The curves of magnetic volume specific susceptibility versus temperature (K(T)) were obtained by continuous exposure of the samples through heating cycles from room temperature (RT) up to 700°C and back to room temperature. Thermomagnetic measurements were done using KLY-3S Kappabridge. The heating rate was 11°C/min. at the applied field of 300 Am^{-1}. The samples for Mössbauer measurements were annealed up to $T_a = 250$, 400, 500, 600 and 700°C each time starting from room temperature. In this way we obtained four, for H1 sampling site, and five for M4 sampling site, annealed samples for both topsoil and loess layers. The samples were closed in a glass tube and heated and cooled in air.

The samples were used as absorbers in the Mössbauer measurements performed in the transmission geometry. The measurements were carried out using a conventional constant acceleration spectrometer at room and liquid nitrogen temperatures. The ^{57}Co in Rh source of activity about 20 mCi was used. The isomer shifts (δ) were related to the α-Fe standard.

3 Results

Figure 1 shows the changes of the magnetic susceptibility during continuous heating for sample H1 (Figure 1a) and during the heating and cooling K(T) curves for sample M4 (Figure 1b), both for topsoil layer. For samples collected at loess layers the K(T) curves look very similar: they are non-reversible in this heating–cooling cycle. Susceptibility curves obtained during heating lay always below the curves obtained during cooling which suggests the formation of secondary ferrimagnetic material. The temperature dependence of the susceptibility curves exhibit two peaks: a small one around $T = 250$°C and a larger one around $T = 500$°C (Figure 1a and b). Additionally, around $T = 450$°C the decrease of susceptibility was observed. The K(T) drops to zero at about $T = 580$°C, which indicates that the magnetite is a main contribution to the susceptibility. The aim of this study is to follow the changes at these transition points and explain the reason of their appearance. The Mössbauer measurements were performed at room temperature for soil samples annealed up to $T_a = 250$, 400, 500, 600 and 700°C.

The hyperfine parameters and the content of iron-containing minerals (spectral relative fraction of a component) were determined using the Mössbauer measure-

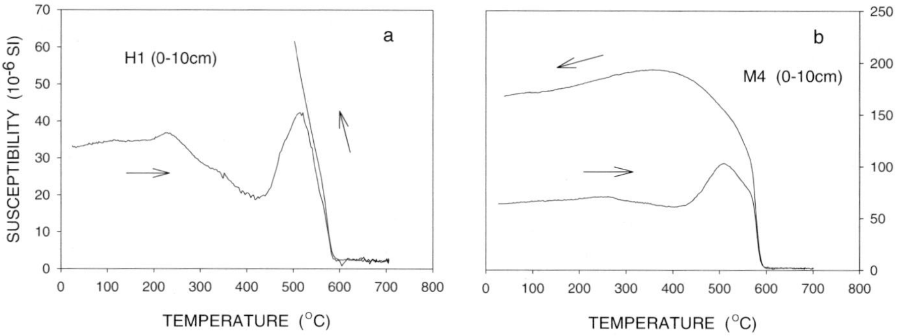

Figure 1 Temperature dependence of magnetic susceptibility for topsoil samples: H1 profile (**a**) and M4 profile (**b**).

ments for the original (as-collected) soils samples. Figure 2 shows, as an example, the Mössbauer spectra recorded for the topsoil sample M4(0–10 cm). The Mössbauer parameters obtained for all investigated original samples at room temperature revealed the presence of Fe^{3+}-bearing minerals (two paramagnetic doublets with isomer shift $\delta_1 = 0.36$ mm/s and quadrupole splitting $\Delta_1 = 0.53$–0.60 mm/s (mineral A) and $\delta_2 = 0.37$ mm/s and $\Delta_2 = 0.93$–1.0 mm/s (mineral B)) in addition to the Fe^{2+}-bearing silicate mineral (paramagnetic doublet with $\delta_3 = 1.08$ mm/s and $\Delta_3 = 2.74$ mm/s) as the main spectral components. Part of Fe^{3+} cations together with Fe^{2+} cations comes most probably from biotite. The paramagnetic doublets (A and B) may originate from Fe^{3+} cations in the oxyhydroxides (ferrihydrite, γ-FeOOH) and/or from iron-containing compounds in the form of ultra fine particles in superparamagnetic state. In the Mössbauer measurements performed at room temperature the poor crystallinity of ferrihydrite causes broad distributions of hyperfine parameters. In a more detailed analysis the QS spectral component related to the ferrihydrite phase may be described by two doublets with parameters: $\delta_1 = 0.37$ mm/s, $\Delta_1 = 0.6$ mm/s and $\delta_2 = 0.38$ mm/s, $\Delta_2 = 0.9$ mm/s [2]. Another oxyhydroxide, γ-FeOOH, exhibits also a broadened doublet with QS in the range of 0.55 to 0.7 mm/s [3]. The Mössbauer parameters obtained in the fitting procedure of our experimental data are very similar to those mentioned above [2, 3] but identification is not fully satisfactory because there are other paramagnetic Fe^{3+} species with very similar hf parameters. In order to verify the origin of the quadrupole doublets observed in the spectra the Mössbauer measurements were performed at $T = 80$ K (Figure 2b). The magnetically ordered compounds: α-Fe_2O_3 (hematite), γ-Fe_2O_3 (maghemite) and, for M4 topsoil layer only, Fe_3O_4 (magnetite) were identified in all samples studied In the case of M4(0–10 cm) sample the spectrum measured at 80 K (Figure 2b) shows a central unresolved doublet, due to the quadrupole splitting (QS) of much smaller intensity (the relative spectral area decreased by about 14%) than that obtained at 300 K (Figure 2a), and four six line subspectra due to magnetic hyperfine splitting. Beside the sextets with the hyperfine fields corresponding to hematite and magnetite that appeared already in the room temperature spectrum we observed additionally a six line spectrum (relative area of about 7%) with the Mössbauer parameters: $\delta = 0.47$ mm/s and $H_{hf} = 47.8$ T, which corresponds to α-FeOOH in the form of particles smaller than 20 nm [4]. The line

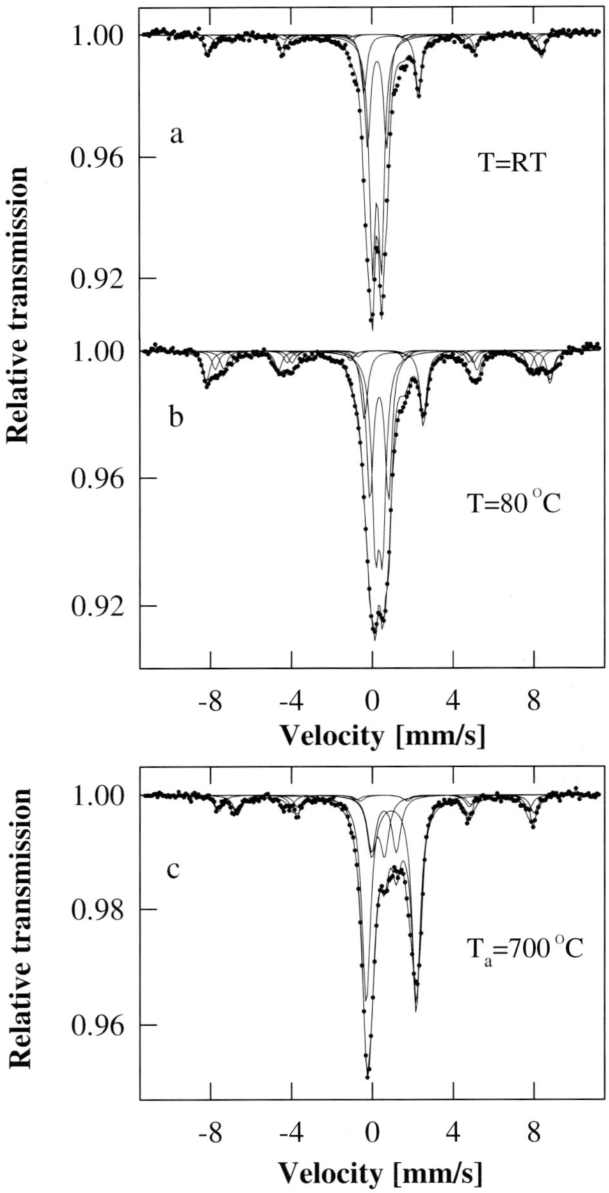

Figure 2 Mössbauer spectra for the M4 profile topsoil sample (as-collected form) measured at **a** room temperature (*RT*) and **b** liquid nitrogen temperature; **c** for sample annealed up to T_a = 700°C measured at RT.

intensities of the subspectra ascribed to hematite and magnetite in the low temperature measurements (Figure 2b) are higher then those at room temperature spectrum. The results suggest that the hematite, magnetite and goethite may appear partly in the form of ultra fine particles in the superparamagnetic state. The approximate mineral content was determined as the relative area (RA) of the Mössbauer spectral components obtained at room temperature corresponding to each compound. The chemical composition of soil samples was transformed during the thermal treatment. Figure 2c shows the Mössbauer spectrum of M4 (0–10 cm)

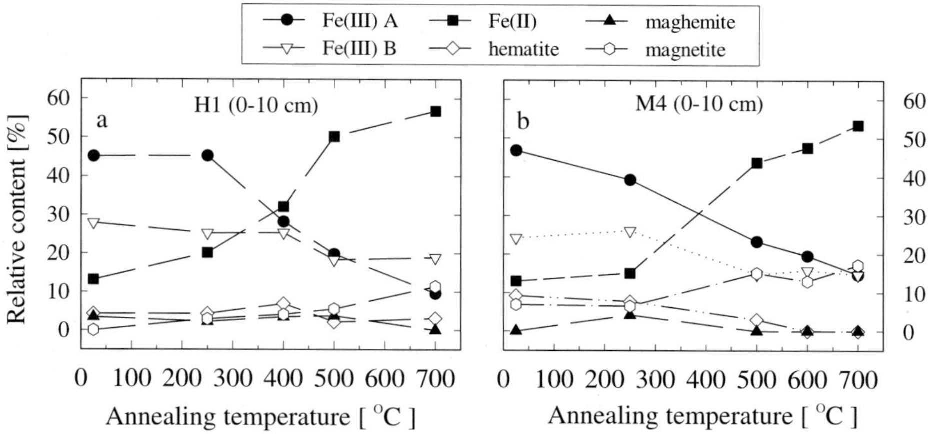

Figure 3 Changes of relative spectral fraction of iron compounds identified for annealed topsoil samples: H1 profile (**a**) and M4 profile (**b**).

sample annealed up to T_a = 700°C. The Mössbauer parameters (isomer shifts and quadrupole splittings) for Fe^{3+}-bearing minerals (A and B), which remained relatively stable up to 250°C increase after annealing at temperature higher than 250°C. It suggests that the process of mineral dehydration begun. The larger values of quadrupole splittings (Δ = 0.65 mm/s and Δ = 1.22 mm/s for A and B Fe^{3+}-bearing minerals, respectively) obtained for annealed samples are characteristic of a strongly distorted coordination.

The changes of relative content of iron-containing minerals for annealed H1(0–10 cm) and M4(0–10 cm) soil samples identified in Mössbauer experiment are shown in Figure 3. The small maximum in susceptibility heating curve at about T = 250°C (Figure 1a and b) may have different reasons: (1) It may arise from the reduction of hematite to magnetite during heating organic matter in the sample. Such interpretation is supported by the Mössbauer results obtained for H1 sample: the magnetite component appears in the Mössbauer spectrum obtained for the topsoil layer sample annealed up to 250°C (Figure 3a). For loess layer (absence of organic matter) the magnetite was not observed. Another reason for the susceptibility maximum at T = 250°C is (2) the formation of ferrimagnetic phases, e.g., γ-Fe_2O_3, during dehydration and dehydroxylation of ferrihydrite and oxyhydroxides [5]. Lepidocrocite phase dehydrates first to maghemite and then to hematite at temperature ranging from 275 to 410°C [3]. The actual transition temperature depends on the grain size and impurities which may stabilize the maghemite, resulting in a much higher transition temperature, up to T_a = 600°C. The increase of the relative area assigned to the maghemite component in the Mössbauer spectra (Figure 3b) support this kind of chemical transformation. Maximum of hematite spectral fraction was observed in the H1 samples annealed up to 400°C, both at topsoil and loess layers. This explains the minimum on the high-temperature susceptibility curve (Figure 1a). According to Florindo et al. [7] that decrease in susceptibility results from thermally induced conversion of maghemite to hematite. In the Mössbauer measurements for samples annealed above T_a = 500°C we observed a decrease of hematite content in all investigated samples.

The heating curves for all four investigated samples, H1(0–10 cm), H1(120–130 cm), M4(0–10 cm) and M4(100–110 cm), exhibit a secondary intense peak at about T_a = 500°C (Figure 1). The Mössbauer measurements performed for the annealed samples show that starting at temperature T_a = 500°C the content of the magnetite in the samples was continuously increasing and the relative change of the magnetite content in the original and annealed up to T_a = 700°C sample, both for topsoil and loess layers for H1 and M4 sampling sites, was very similar (about 10%). Also for M4 original topsoil sample in which a magnetite content was about 7%, the amount of the magnetite in the sample increased to about 17% after annealing to T_a = 700°C (Figures 2c and 3b). These data suggest that observed increase of magnetite content in the samples and related to this changes in the susceptibilities curves vs. temperature are caused by the formation of additional magnetically ordered phase in the chemical transformation during heating. It means that the temperature and atmosphere in which the thermomagnetic measurements were performed influence significantly the final composition of the investigated samples and, as a result of that, the susceptibility of the investigated samples. In our case the formation of additional magnetite phase and significant increase of Fe^{2+} cations content in the paramagnetic minerals indicates that the measurements of the susceptibility were performed under reduction conditions.

4 Conclusions

Mössbauer measurements, performed for soil samples annealed up to temperature indicated on thermomagnetic curve as a transition points (250, 400, 500, 600, 700°C), show the possible conversion between the different phases of iron oxide and iron oxyhydroxide systems. The annealing induced the chemical transformation of the samples, first a dehydration/dehydroxylation and than a reduction. The largest content of magnetite has been found for the samples annealed up to T_a = 700°C. Simultaneously to the process of the formation of additional magnetic phase a steady increase of Fe^{2+} cations content in the paramagnetic components was observed which indicates that transformation process occurred under reduction conditions.

References

1. Jeleńska, M., Hasso-Agopsowicz, A., Kopcewicz, B., Sukhorada, A., Tyamina, K., Kądziołko-Hofmokl, M., Matviishina, Z.: Geophys. J. Int. **159**, 104 (2004)
2. Murad, E., Johnston, J.H.: In: Long, G.L. (ed.) Mössbauer Spectroscopy Applied to Inorganic Chemistry, vol. 2, p. 507. Plenum, New York (1987)
3. Vanderberghe, R.E., DeGrave, E., Lanfuydt, C., Bowen, L.H.: Hyperfine Interact. **53**, 175 (1990)
4. Van der Kraan, A.M., Van Loef, J.J.: Phys. Lett., **20**, 614 (1966)
5. Cornell, R.M., Schwertmann, U.: The Iron Oxides, p. 367. Wiley, New York (2003)
6. Mullins, J.C.: J. Soil Sci **28**, 223 (1977)
7. Florindo, F., Zhu, B., Guo, B., Yue, L., Pan Y., Speranza, F.: J. Geophys. Res., **104**, 645 (1999)

Hyperfine Interact (2005) 166:637–642
DOI 10.1007/s10751-006-9333-2

Iron speciation of sediments from Kuhai Lake in China by Mössbauer spectroscopy

A. Kuno · M. Matsuo · Q. Wang · C. J. Zhang

Published online: 27 October 2006

Abstract The vertical distribution of iron species in the sediments collected from Kuhai Lake was investigated in this study. Their Mössbauer spectra were composed of one sextet and four doublets ascribable to one paramagnetic high-spin Fe^{3+}, two paramagnetic high-spin Fe^{2+} with different quadrupole splittings, and one diamagnetic low-spin Fe^{2+} that corresponds to pyrite (FeS_2). The distribution of pyrite was considered to reflect the climatic change during the past 2,000 years. Mössbauer spectroscopy was successfully applied to the analysis of the Kuhai Lake sediments so as to reconstruct the history of the past climate.

Key words Mössbauer spectroscopy · Kuhai Lake · brackish lake sediment · vertical distribution · iron speciation

1 Introduction

Kuhai Lake is an inland brackish lake in northwestern China. More than half of China's lakes are saline. Most salt lakes are in northwestern China. Tectonic movements have been of the greatest importance in lake origins, but aeolian activity and deflation have also played a role [1]. There have been few Mössbauer studies on

A. Kuno (✉) · M. Matsuo
Graduate School of Arts and Sciences, The University of Tokyo, 3-8-1 Komaba, Meguro, Tokyo 153-8902, Japan
e-mail: kuno@dolphin.c.u-tokyo.ac.jp

Q. Wang
Lanzhou Institute of Geology, Chinese Academy of Sciences, 324 West Donggang Road, Lanzhou 730000, China

C. J. Zhang
College of Earth Resources and Environment, Lanzhou University, Lanzhou 730000, China

sediments collected from salt lakes although a number of investigations have been carried out on marine and coastal sediments [2, 3]. In marine and coastal sediments, sulfate ion from seawater is reduced to hydrogen sulfide, which reacts with iron compounds to form iron sulfides. Therefore, the speciation of iron compounds in the sediments collected from salt lakes is geochemically important. The authors have earlier investigated the vertical distribution of iron species in the sediments collected from Qinghai Lake [4]. Comparing data from the two lakes gives a better understanding of the sedimentation process in brackish lakes and also of the past environmental changes in this area, which attracts attention from the viewpoint of environment. There have been few studies on Kuhai Lake whereas Qinghai Lake sediments are intensively studied in an attempt to reconstruct the history of the past climate traced by carbonate contents [5], fossil pigments [6] etc, since vertical sediment cores record the climatic change [7]. Mössbauer spectroscopy has the advantage of non-destructiveness and sensitivity to environmental change among such reconstruction studies. Chinese salt lakes are significant sites for paleoclimatic research. From these standpoints, ^{57}Fe Mössbauer spectroscopy and X-ray absorption fine structure (XAFS) were applied to analyze sediment samples collected from Kuhai Lake in order to investigate the vertical distribution of iron species.

2 Experimental

Kuhai Lake is located in the northeastern Qinghai-Tibetan plateau, 100 km to the south-southwest of Qinghai Lake. Kuhai Lake is an intermontane Quaternary basin with Triassic shale basement. A vertical sediment core of approximately 80 cm in length was subjected to the analysis. It is considered to record the environmental change during the past 2,000 years, basing on the estimated sedimentation rate (40 cm/ka). The collected samples were cut at every 1 cm length, freeze-dried, and ground into powder. Out of 80 samples from different depths, 22 samples were analyzed. Approximately 200 mg of each sample was mounted in a sample holder (16 mm in diameter, 1 mm thick). Absorbers thus prepared contained 5 mg/cm^2 of Fe, a quantity low enough so to allow relative peak areas to be used as a measure of the distribution of the corresponding Fe-bearing compounds, discarding differences of the f fractions. The spectra were measured with an Austin Science S-600 Mössbauer spectrometer using a 1.11 GBq ^{57}Co/Rh source at 293 and 78 K. The curve fitting of the resulting spectra was performed with a personal computer, assuming that the spectra were composed of peaks with Lorentzian line shapes. Half-widths and peak areas within each quadrupole doublet were constrained to be equal. Isomer shifts were expressed with respect to the centroid of the spectrum of metallic iron foil. Pyrite (FeS_2) peak positions were constrained (using values of δ=0.307 mm/s, ΔE_Q=0.610 mm/s at 293 K and δ=0.405 mm/s, ΔE_Q=0.629 mm/s at 78 K) as in Manning and Ash [8], since the peak positions of paramagnetic high-spin Fe^{3+} and pyrite containing diamagnetic low-spin Fe^{2+} are in very close proximity. In addition to the measurements in the velocity range of −10.00 to 10.00 mm/s, the spectra were also measured in the velocity range of −3.33 to 3.33 mm/s for a better resolution. The X-ray absorption fine structure (XAFS) measurements were made using synchrotron radiation with a Si(111) double crystal monochromator at beam

Figure 1 Mössbauer spectra of the sediments collected from Kuhai Lake at 293 K.

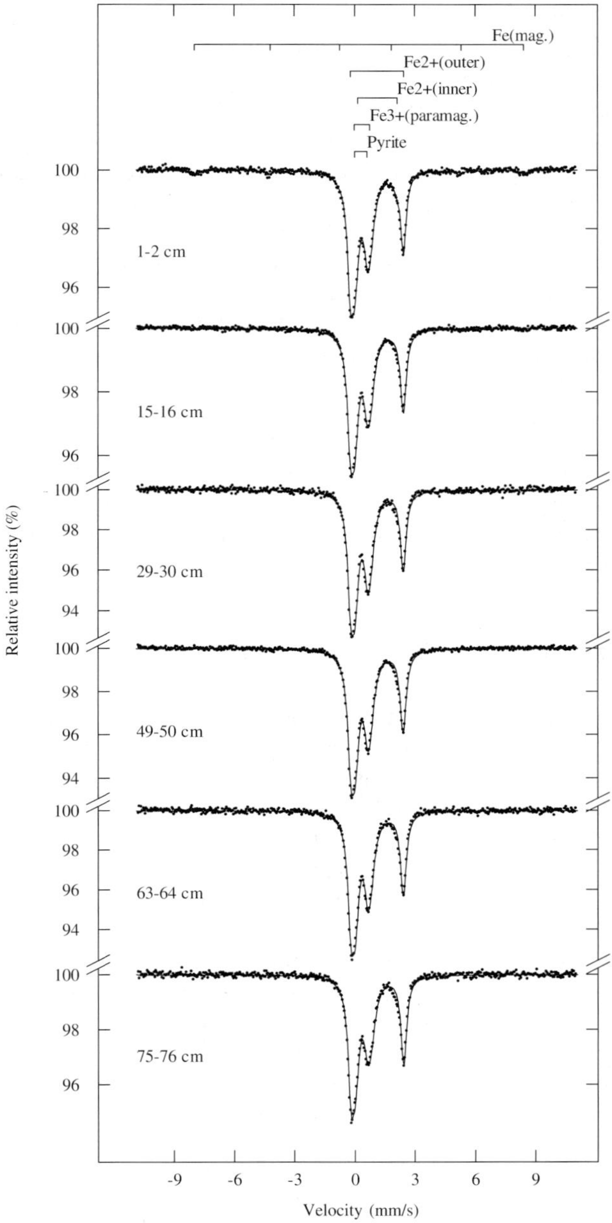

line 12C [9] of Photon Factory, Tsukuba, Japan. Constituent minerals were checked by X-ray diffraction (XRD). As a measure of organic matter content, the total organic carbon (TOC) and the loss on ignition (LOI) between 105 and 550°C were measured. The $CaCO_3$ content and $\delta^{13}C$ were also measured.

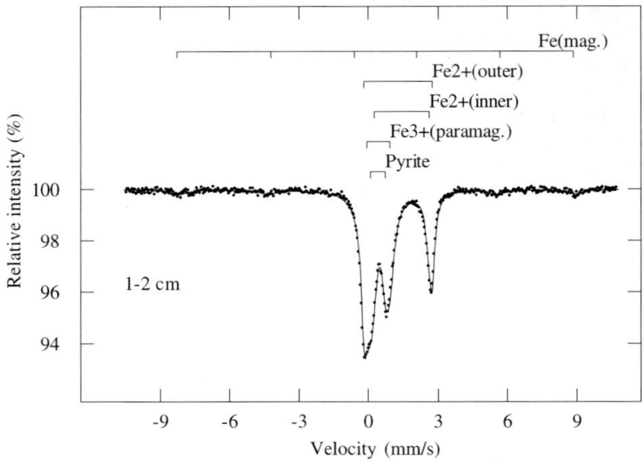

Figure 2 Mössbauer spectrum of the Kuhai Lake sediment collected from the depth of 1–2 cm at 78 K.

Table I Mössbauer parameters of the Kuhai Lake sediment collected from the depth of 1–2 cm measured at 293 K

Species	Area (%)	δ (mm/s)	Δ (mm/s)	Γ (mm/s)	H_i (T)
Fe^{3+}(paramag.)	43.7(7)	0.38(1)	0.80(1)	0.54(1)	
Fe^{2+}(outer)	30.9(4)	1.13(1)	2.67(1)	0.31(1)	
Fe^{2+}(inner)	3.9(2)	1.19(1)	2.04(3)	0.26(4)	
Pyrite	15.5(8)	0.31	0.61	0.37(2)	
Fe(mag.)	6.0(4)	0.41(2)	−0.21(4)	0.43(7)	50.8(2)

The errors in the least significant figure are given in parentheses.

3 Results and discussion

Figure 1 shows a composite of six Mössbauer spectra out of 22 spectra of the sediments collected from different depths measured at 293 K. As shown at the top of the spectra, four doublets and one sextet were separated. The fit curves shown as solid lines in Figure 1 well threaded the observed spectra, and the chi-squared values were sufficiently small, which justifies the validity of the applied model. The relative peak areas of the components in the spectra measured at 78 K (Figure 2) were almost the same as those at 293 K, as it would be expected in case that the sediments did not contain fine particles showing a doublet at 293 K and a sextet at 78 K by a superparamagnetic effect [10]. In addition, this observation supports the accuracy of the determined ratio of the components. Pyrite was also confirmed by the XRD patterns, in which the peaks of quartz and feldspar predominated. In the XRD patterns of some samples, relatively large peaks of calcite were detected. Table I shows the Mössbauer parameters of the sediment collected from the depth of 1–2 cm measured at 293 K. They did not vary much throughout the core. An incipient Mössbauer sextet was detected only for the samples from the top of the sediment. The value of the corresponding hyperfine field was typically about 49 T, which is assignable to magnetic high-spin Fe^{3+}, presumably in an admixture of ferric hydroxides or ferric oxides with various degrees of hydration. The paramagnetic

Figure 3 Vertical distribution of iron species in the sediments collected from Kuhai Lake: *open circle*, paramagnetic Fe^{3+}; *shaded square*, paramagnetic Fe^{2+} (outer); *open square*, paramagnetic Fe^{2+} (inner); *shaded circle*, pyrite; *open triangle*, magnetic components.

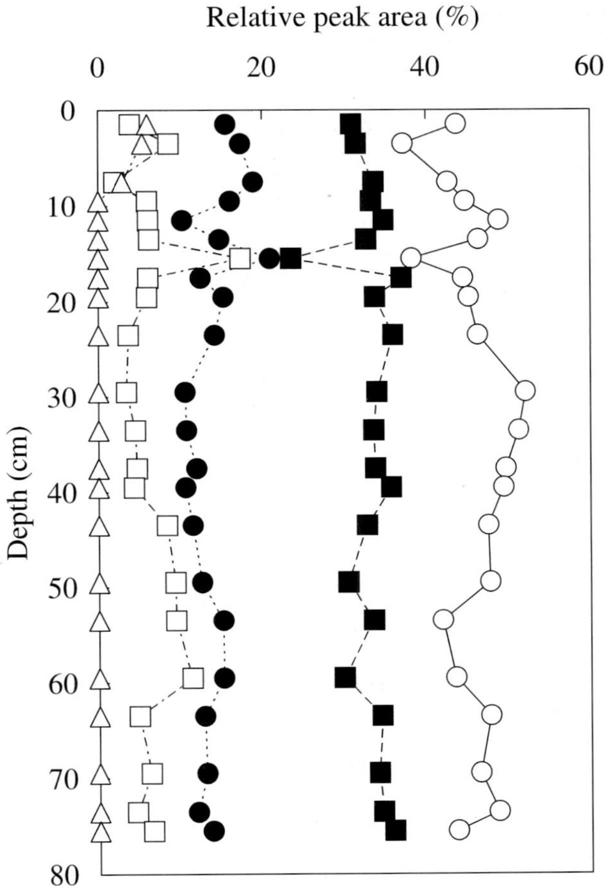

high-spin Fe^{3+} would originate from clay minerals and/or hydrated ferric oxides [8]. Two doublets with larger isomer shifts were attributable to paramagnetic high-spin Fe^{2+}. The doublet with a larger quadrupole splitting shown as Fe^{2+}(outer) in Figure 1 was a major component and accompanied by a small amount of the other doublet with a smaller quadrupole splitting shown as Fe^{2+}(inner). The Mössbauer parameters of the paramagnetic Fe^{2+} showing an inner doublet were highly similar to those of siderite ($FeCO_3$) or ferrous carbonate minerals, in which iron is partly substituted by other metals. The paramagnetic Fe^{2+} showing an outer doublet would be ascribable to clay minerals such as chlorite, which was confirmed by the XRD patterns.

Figure 3 shows the vertical distribution of relative peak areas of these species in the sediments. The Fe K-edge X-ray absorption near-edge structure (XANES) of the samples was also consistent with the Mössbauer results; Fe K-XANES of samples with higher Fe^{3+}/Fe^{2+} ratio shifted to a higher energy side. The magnetically ordered components detected only in the surface sediments would be formed by the oxidation of Fe^{2+} with dissolved oxygen in the lake water. This sextet was not detected in the Qinghai Lake sediments [4]. This would be due to the difference in the lake size. In deep lakes such as Qinghai Lake, the thermal circulation rate is low and the bottom water lacks dissolved oxygen [1]. In the case of the Kuhai Lake, results indicated that the influence was restricted within 10 cm from the surface. The vertical distribution of pyrite showed a positive correlation with the organic matter content measured by LOI and TOC, which increases under warm climate [11]. This observation is reasonable because one of the factors controlling sulfate reduction is temperature that strongly affects the activity of sulfate reducing bacteria. A high content of organic matter produces reducing environment, which also favors the activity of sulfate reducing bacteria. A maximum pyrite proportion at the depth of 16 cm from the surface suggested a relatively warmer climate around 400 years B.P. The past climate revealed by Mössbauer spectroscopy was consistent with the $CaCO_3$ content, $\delta^{13}C$ data and other reconstruction studies [12, 13]. The results were consistent with the Qinghai Lake sediments [4], which means that the paleoclimatic change in the Kuhai Lake and Qinghai Lake regions are almost the same even though the existence of magnetic components was different due to the lake size.

4 Conclusions

Mössbauer spectroscopy was applied to sediments collected from Kuhai Lake in China, to clarify the vertical distribution of iron species. One sextet and four doublets ascribable to one paramagnetic high-spin Fe^{3+}, two paramagnetic high-spin Fe^{2+}, and one diamagnetic low-spin Fe^{2+} that corresponds to pyrite were separated in their spectra. The distribution of pyrite suggested climatic changes during the past 2,000 years. Compared to Qinghai Lake, the surface sediments of Kuhai Lake contained the magnetic component, presumably formed by the oxidation of Fe^{2+} with dissolved oxygen in the lake water.

References

1. Williams, W.D.: Hydrobiologia **210**, 39 (1991)
2. Matsuo, M., Takano, B., Sugimori, K.: Il Nuovo Cimento D **50**, 757 (1996)
3. Kuno, A., Matsuo, M., Takano, B.: Hyperfine Interact. C **3**, 328 (1998)
4. Kuno, A., Zheng, G.D., Matsuo, M., Takano, B., Shi, J.A., Wang, Q.: Hyperfine Interact. **141/142**, 321 (2002)
5. Wang, S.M., Li, J.R.: Chin. Sci. Bull. **36**, 1364 (1991)
6. Han, X.Z., Shen, H., Wang, S.M., Yu, Y.S.: Chin. J. Oceanol. Limnol. **10**, 320 (1992)
7. Zheng, G.D., Takano, B., Kuno, A., Matsuo, M.: Appl. Geochem. **16**, 1201 (2001)
8. Manning, P.G., Ash, L.A.: Can. Miner. **17**, 111 (1979)
9. Nomura, M., Koyama, A.: In: Hasnain, S.S. (ed.) X-ray Absorption Fine Structure, p. 667. Ellis Horwood, Chichester (1991)
10. Kündig, W., Bömmel, H., Constabaris, G., Lindquist, R.H.: Phys. Rev. **142**, 327 (1966)
11. Huang, Q., Sun, N.: Chin. Sci. Bull. **34**, 1457 (1989)
12. Kelts, K., Zao, C.K., Lister, G., Qing, Y.J., Hong, G.Z., Niessen, F., Bonani, G.: Eclogae Geol. Helv. **82**, 167 (1989)
13. Wang, W., Chen, K.: China Earth Sci. **1**, 233 (1990)

Hyperfine Interact (2005) 166:643–649
DOI 10.1007/s10751-006-9335-0

Effect of $FeCl_3$ and acetone on the structure of Na–montmorillonite studied by Mössbauer and XRD measurements

Andrea Komlósi · Ernő Kuzmann · Zoltán Homonnay · Noémi M. Nagy · Shiro Kubuki · József Kónya

Published online: 21 November 2006

Abstract ^{57}Fe Mössbauer spectroscopy, X-ray diffraction, X-ray fluorescence spectroscopy and infrared spectroscopy were used to study the effect of $FeCl_3$ and acetone on the structure of a Na–bentonite. XRD indicated the incorporation of Fe^{3+} ions into the interlayer space since the basal lattice spacing of montmorillonite increased to 1.6 from 1.24 nm after treatment with $FeCl_3$ dissolved in acetone. Interlayer Na^+ ions could be exchanged to Fe^{3+}. Magnetically split Mössbauer subspectra with internal magnetic fields 41 and 46 T at 74 K, were associated with two main Fe^{3+} microenvironments within the interlayer regions. The resultant Fe–montmorillonite was successfully applied as a catalyst in the preparation of 1,1-diacetates from aromatic aldehydes and acetic acid anhydride.

Key words bentonite · montmorillonite · Mössbauer spectroscopy · XRD

1 Introduction

Nowadays it is popular to use cation-exchanged clays as catalysts in organic reactions [1–6], since they are harmless, reusable and they can be easily separated from the starting materials and the products. It was claimed formerly, that

Shiro Kubuki is on leave from Ube National College of Technology, Japan.

A. Komlósi · N. M. Nagy · J. Kónya
Isotope Laboratory, Department of Colloid Chemistry,
Debrecen University, Debrecen, Hungary

E. Kuzmann (✉) · Z. Homonnay
Research Group for Nuclear Methods in Structural Chemistry,
Hungarian Academy of Sciences, Pázmány P. s. 1/a, Budapest 1117, Hungary
e-mail: kuzmann@ludens.elte.hu

Z. Homonnay · S. Kubuki
Department of Nuclear Chemistry, Eötvös Loránd University,
Pázmány P. s. 1/a, Budapest 1117, Hungary

Fe–montmorillonite acts as a Brönsted-acid, when Fe is in the lattice of the clay, while it acts as a Lewis-acid, when Fe is situated in the interlayer space of the clay [7–11]. The preparation of Fe–montmorillonite is rather difficult because of the hydrolysis of Fe^{3+}. Iron can be present in two different oxidation states and can be situated both in the lattice and in the interlayer space. When an aqueous solution of an Fe-salt is stirred with montmorillonite suspension in the presence of air to exchange the cation in the interlayer space, Fe^{III}–oxyhydroxides or oxides could be formed, at least partially when the starting Fe-salt contains Fe^{II} (e.g. [3, 5, 12–16]). Li et al. [16] have reported simple preparation of 1,1-diacetates using Fe^{III}–montmorillonite derived from K10 montmorillonite, but the structure of the catalyst was not examined. In our recent paper [17] 1,1-diacetates were prepared from aromatic aldehydes catalysed by Zn–montmorillonite obtained from Ca–montmorillonite by several ion exchange procedures.

Our purpose was to study the structural changes of Na–montmorillonite, as a catalyst in the preparation of 1,1-diacetates under mild circumstances, due to its treatment with acetone and with $FeCl_3$ dissolved in acetone. For these investigations ^{57}Fe Mössbauer spectroscopy XRD, XRF, IR spectroscopy, chemical analysis and catalytic activity tests were used. In this report the Mössbauer and XRD results will be emphasized.

2 Experimental

Na–bentonite from Valea Chioarului (Kővár, Romania) was used as starting material. The elemental analysis of the mineral gave 73.2% SiO_2, 14.2% Al_2O_3, 1.15% Fe_2O_3, 0.82% CaO, 2.24% MgO, 0.4% TiO_2, 2.68% Na_2O, and 1% K_2O. Adapting the procedure of Li et al. [16] Fe–montmorillonite was prepared by impregnation of Na–bentonite with a solution of anhydrous $FeCl_3$ (Reanal, Hungary) in acetone (Spektrum 3D, Hungary). Three different bentonite to $FeCl_3$ ratios were applied, 1.5:1, 3:1, and 6:1. In each case, the solvent was evaporated under reduced pressure, the remaining material was dried in desiccator. The effect of acetone was studied by stirring 1 g of the starting material in acetone (5 cm^3), then the solvent was evaporated and the clay was also dried in desiccator.

The samples were characterised before and after treatments by powder XRD, using a computer controlled powder diffractometer (DRON-2). The measurements were performed with a goniometer speed of 1 cm^{-1}, with β-filtered CoK_α radiation (λ = 1.7890), in the range of 2Θ = 3–110° at room temperature. The powder diffraction patterns were evaluated by "EXRAY" peak searching computer program (Klencsár, 1998), which gave cell parameters.

^{57}Fe Mössbauer spectra of the powdered samples were recorded in transmission geometry with a conventional Mössbauer spectrometer (Wissel). The γ-rays were provided by a 3×10^9 Bq ^{57}Co/Rh source. The measurements were performed at 300 K and in some cases at 78 K in a temperature-controlled cryostat (Leybold, Germany). Isomer shifts are given relative to α-iron. The Mössbauer spectra were analyzed by least-square fitting of Lorentzian lines by the help of the MOSSWINN code [18].

In addition to Mössbauer and XRD studies, X-ray fluorescence (XRF) and infrared (IR) spectra of all samples were also measured, and the catalytic activity of

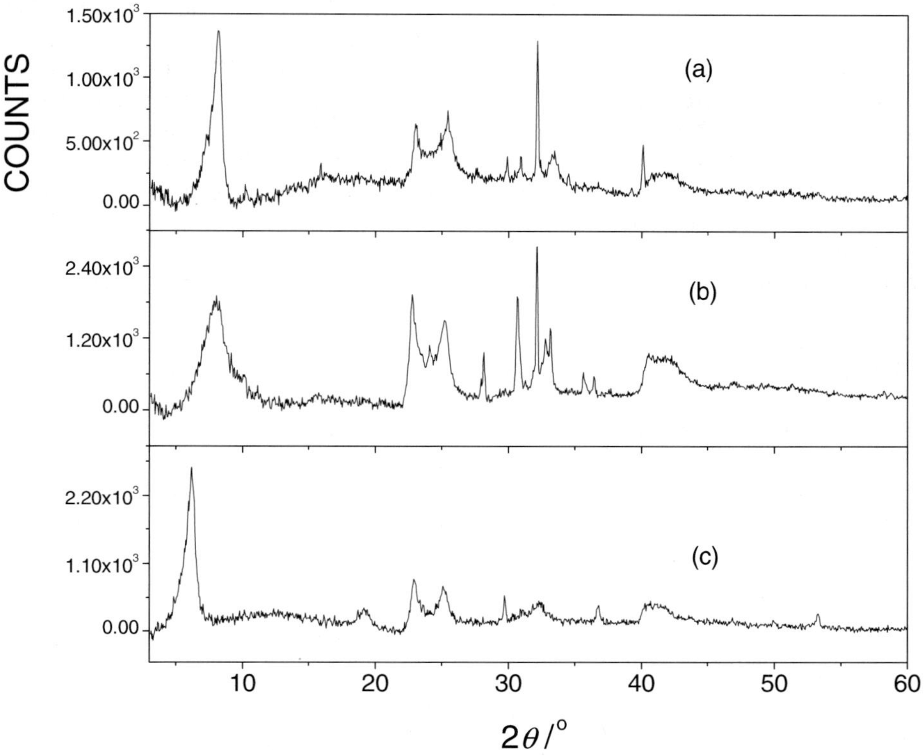

Figure 1 X-ray diffractograms of bentonites. Untreated (a) treated with acetone (b) and with $FeCl_3$ in acetone (c).

the $FeCl_3$-treated bentonite for the formation of 1,1-diacetates from aromatic aldehydes was investigated as well. These results will be published in details elsewhere.

3 Results and discussion

XRF analysis of the samples showed, that the Fe content of the resultant bentonite sample increased from 3.4 (for original bentonite) to 7.07 to 9.7 to 12.2 m/mass% with increasing amount of $FeCl_3$ added in acetoneous solution.

XRD of the original Na–bentonite from Kővár (Figure 1a) exhibits mainly cristobalite phase besides the montmorillonite phase with a small amount of quartz, illite and kaolinite. The basal lattice spacing of montmorillonite in the original Na–bentonite was observed to be 1.26 nm, which is consistent with other data found in literature for Na–montmorillonite (e.g. [15]). The treatment with acetone did not change the basal lattice spacing of the montmorillonite (Figure 1b), but the relative intensity of the main peak decreased and small new peaks arose. When the acetone contained $FeCl_3$ as well, the corresponding XRD (Figure 1c) showed significant changes, namely, the basal lattice spacing became 1.6 nm, which can be explained by the incorporation of Fe^{3+} ions into the interlayer space. On the other hand, there

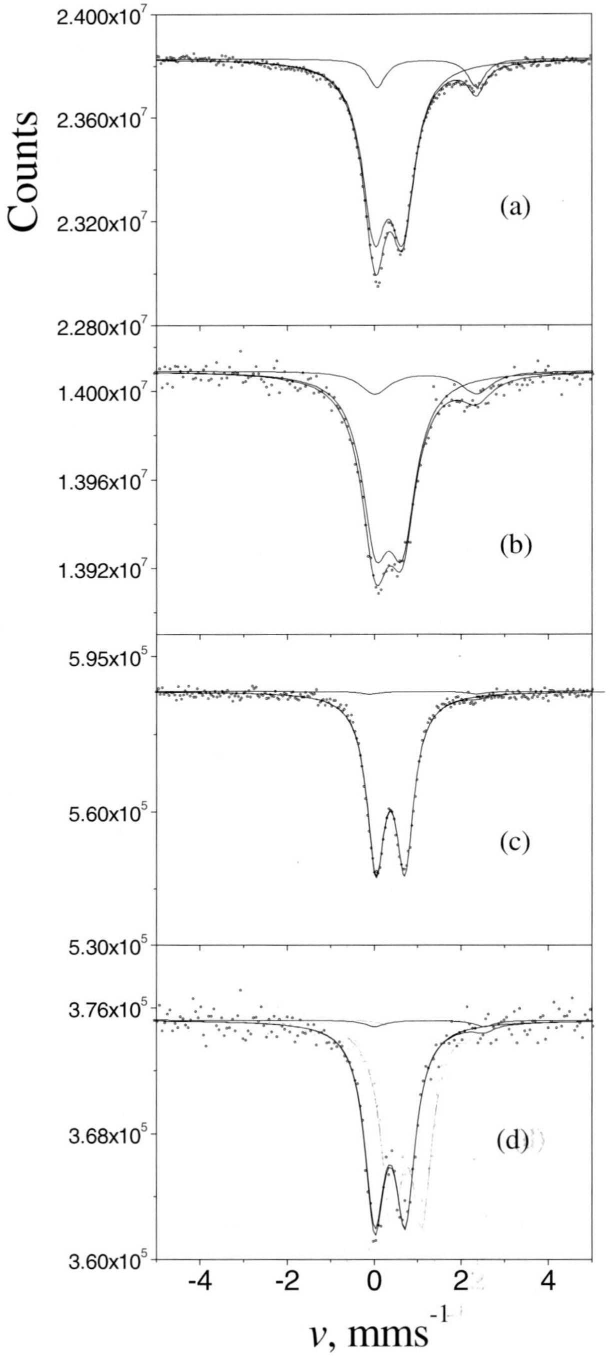

Figure 2 Room temperature Mössbauer spectra of bentonite originated from Kővár (a) and after its treatment with acetone (b) as well as with $FeCl_3$ in acetone with $FeCl_3$ to montmorillonite ratio 1:3 (c) and 1:6 (d).

Table I Mössbauer parameters of bentonites

Sample	T (K)	δ_1 (mms^{-1})	Δ_1 (mms^{-1})	A_1 (%)	δ_2 (mms^{-1})	Δ_2 (mms^{-1})	A_2 (%)	δ_3 (mms^{-1})	B_3 (T)	A_3 (%)	δ_4 (mms^{-1})	B_4 (T)	A_4 (%)
Na-bentonite	293	0.34	0.64	88.4	1.21	2.28	11.6						
Na-bentonite	88	0.44	0.70	89.7	1.34	2.35	10.3						
Na-bentonite treated with acetone	293	0.34	0.63	87.0	1.20	2.31	13.0						
Na-bentonite treated with $FeCl_3$ in acetone 1.5:1	293	0.35	0.63	96.7	1.23	2.31	3.3						
Na-bentonite treated with $FeCl_3$ in acetone 3:1	293	0.36	0.66	97.3	1.22	2.32	2.7						
Na-bentonite treated with $FeCl_3$ in acetone 6:1	293	0.36	0.69	97.8	1.24	2.31	2.2						
Na-bentonite treated with $FeCl_3$ in acetone 3:1	74	0.47	0.53	26.3	1.40	2.41	2.7	0.48	41	32.1	0.47	46	37.9

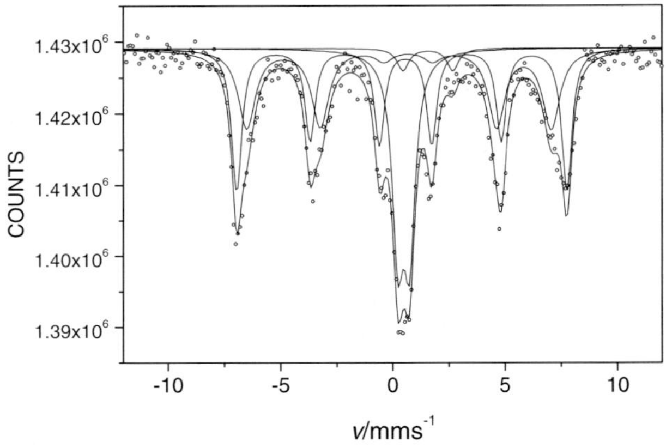

Figure 3 Mössbauer spectrum, recorded at 74 K, of bentonite after its treatment with with $FeCl_3$ in acetone with $FeCl_3$ to acetone.

was no significant change in the X-ray diffractograms of montmorillonite treated with acetoneous $FeCl_3$ solution as a function of the mass proportion of $FeCl_3$ relative to montmorillonite.

Figure 2 shows typical room temperature Mössbauer spectra of bentonite originated from Kővár (Figure 2a) and after its treatment with acetone (Figure 2b) as well as with $FeCl_3$ in acetone with different $FeCl_3$ to montmorillonite ratios (Figure 2c and d). The spectra were evaluated for two doublets. The Mössbauer parameters are shown in Table I.

δ_1 and Δ_1 values reveal Fe^{3+} while values of δ_2 and Δ_2 reflect Fe^{2+} microenvironments. The Mössbauer parameters obtained for the original Na–bentonite correspond well to those found earlier for montmorillonites [19]. They can mainly be associated with octahedrally coordinated Fe ions in the lattice. No magnetically split spectrum parts appeared in the spectrum of the original bentonite recorded at the temperature of liquid nitrogen.

After the treatment with acetone no significant changes can be observed in the Mössbauer spectrum. This treatment did not change the Fe^{2+}/Fe^{3+} ratio compared to that of original Na–bentonite.

In the case of the samples treated with acetoneous $FeCl_3$ solution, however, the most striking observable change in the spectra is a considerable decrease of the relative occurrence of the Fe^{2+} subspectrum. The decrease in the ratio of Fe^{2+}/Fe^{3+} upon this treatment can be explained by the high Fe^{3+} uptake of the samples. The change in the relative occurrence of Fe^{2+} as a function of the $FeCl_3$ to montmorillonite ratio (Table I) correlates with the iron content of the samples determined by XRF analysis. The Mössbauer pattern of the Fe^{3+} microenvironments, newly introduced by the treatment, can hardly be distinguished from those in the original montmorillonite on the basis of the room temperature Mössbauer spectra. However, new magnetically split components (two sextets) appear in the Mössbauer spectrum of the $FeCl_3$ treated samples recorded at around the temperature of liquid nitrogen (Figure 3). Although the Mössbauer parameters of the sextets (Table I) are not so far from those characteristic for microenvironments in Fe^{3+} oxyhydroxides, we propose to attribute these sextets to two different main microenvironments of Fe^{3+} incorporated within the interlayer regions of the montmorillonite lattice. This is strongly supported by the concomitant increase of the lattice spacing as indicated by XRD results. On the other hand, a similar magnetic component was assigned to Fe^{3+} species intercalated between the layers in iron exchanged montmorillonite clays by Berry et al. [12].

The catalytic activity of the $FeCl_3$-treated bentonite for the formation of 1, 1-diacetates from aromatic aldehydes was studied. The reaction scheme is the following:

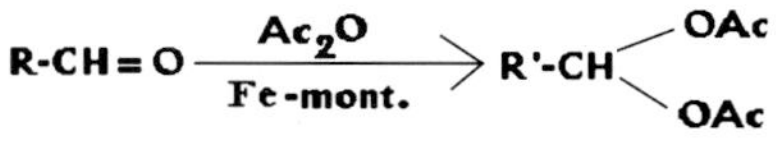

It was established that bentonite samples treated with $FeCl_3$ in acetone are efficient catalysts, in good agreement with, for example, ref. [16]. It was also found that the same reactions occur about twice faster with our Fe–montmorillonite than with those catalysed by Zn-montmorillonite [17].

We can conclude that the preparation of Fe–montmorillonite by reacting montmorillonite with $FeCl_3$ in acetone was successful and Fe^{3+} could be introduced into the interlayer space. The resulting clay can be used as a catalyst in the acylation of aldehydes.

Acknowledgement This study was supported by the Hungarian Science Foundation (OTKA T043687).

References

1. Cseri, T., Békássy, S., Figueras, F., Rizner, S.: J. Mol. Catal., A Chem. **98**, 101–107 (1995)
2. Pai, S.G., Baipai, A.R., Deshpande, A.B., Samant, S.D.: J. Mol. Catal., A Chem. **156**, 233–243 (2000)
3. Shrigadi, N.B., Shinde, A.B., Samant, S.D.: Appl. Catal., A Gen. **252**, 23–25 (2003)
4. Singh, D.U., Samant, S.D.: J. Mol. Catal., A Chem. **223**, 111–116 (2004)
5. Pillai, U.R., Sahle-Demessie, E.: Appl. Catal., A Gen. **245**, 103–109 (2003)
6. Zhang, Z.-H., Li, T.-S., Fu, Ch.-G.: J. Chem. Res. (S) 174–175 (1997)
7. Freeman, F., Karchevski, E.M.: J. Chem. Eng. Data **22**, 355–357 (1977)
8. Oláh, G.A., Mehrotra, A.K.: Synthesis 962–963 (1983)
9. Scriabine, I.: Bull. Soc. Chim. Fr. 1194–1198 (1961)
10. Kochhar, K.S., Bal, B.S., Deshpande, R.P., Rajadhyaksha, S.N., Pinnick, H.W.: J. Org. Chem. **48**, 1765–1767 (1983)
11. Aggarwal, V.K., Fonquerna, S., Vennal, G.P.: Synlett 849–850 (1998)
12. Berry, F.J., Hayes, M.H.B., Jones, S.L.: Inorg. Chim. Acta **122**, 19–24 (1986)
13. Pálinkó, I., Lázár, K., Hannus, I., Kiricsi, I.: J. Phys. Chem. Solids **57**, 1067–1072 (1996)
14. Green-Pedersen, H., Pind, N.: Colloid Surf., A. Physicochem. Engin. Asp. **168**, 133–145 (2000)
15. Kozai, N., Adachi, Y., Kawamura, S., Inada, K., Kozaki, T., Sato, S., Ohashi, H., Ohnuki, T., Banba, T.: J. Nucl. Sci. Technol. **38**, 1141–1143 (2001)
16. Li, T.-S., Zhang, Z.-H., Gao, Y.-J.: Synth. Commun. **28**(24), 4665 (1998)
17. Nagy, N.M., Jakab, M.A., Kónya, J., Antus, S.: Appl. Clay Sci. **21**, 213–216 (2002)
18. Klencsár, Z., Kuzmann, E., Vértes, A.: J. Radioanal. Nucl. Chem. **210**, 105–118 (1996)
19. Stevens, J.G. (ed.): Mössbauer Effect Reference and Data Index (MERDI). Interscience, New York (1958–2003)

Hyperfine Interact (2005) 166:651–655
DOI 10.1007/s10751-006-9336-z

Analysis of surface layers of an iron axe from Early Medieval Age

Marcel Miglierini · Július Dekan · Danica Staššíková-Štukovská

Published online: 27 October 2006

Abstract An ancient axe from Early Medieval Age which was unearthed on a territory of the archaeological excavation site near the village Borovce (Slovakia) is investigated. The surface of the axe was supposed to be covered with a sort of glassy system. Mössbauer effect measurements performed in backscattering and transmission geometry upon original as well as powdered pieces of the axe confirmed that the surface layers of the axe consist of goethite, lepidocrite, akageneite, and/or magnetite. After temperature treatment at 400°C/100 min hematite and maghemite were found; these are products of thermal decomposition of the above mentioned iron phases. These findings ruled out the hypothesis that the surface covering substance was put on the axe by intention.

Key words Mössbauer spectra · corrosion · oxidation · aging

PACS 33.45.+x · 81.65.−b · 81.65.Mq · 81.40.Cd

1 Introduction

Archaeological excavations near Borovce (Slovakia) have revealed a set of iron axes in different graves dated from 8th to 10th century. The grave no. 221 in which socially highly ranked individual of the ancient society was buried, was of a particular interest. One of the work of art found in the grave was an axe of an ancient Slovak type. Its surface seemed to be decorated with non-ferrous metals or enamel featuring smooth shape and a light brownish colour which was referred to as

M. Miglierini (✉) · J. Dekan
Department of Nuclear Physics and Technology, Slovak University of Technology, Ilkovičova 3, 812 19 Bratislava, Slovakia
e-mail: marcel.miglierini@stuba.sk

D. Staššíková-Štukovská
Institute of Archaeology, Slovak Academy of Sciences, Akademická 2, 949 21 Nitra, Slovakia

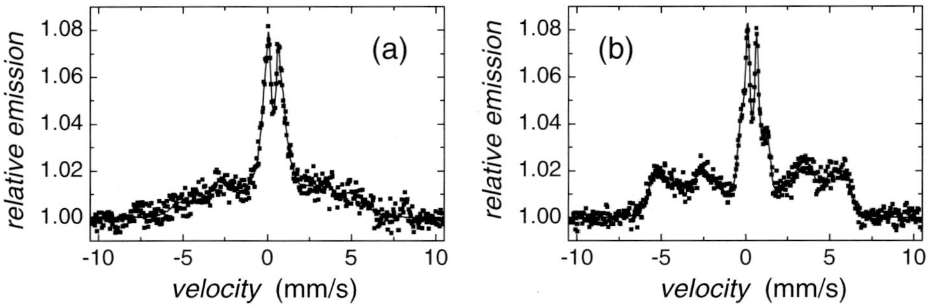

Figure 1 Backscattering Mössbauer spectra taken from the inner (**a**) and outer (**b**) surface of sample A.

"an applied glass system" [1]. It is assumed that this particular iron axe was not used for "ordinary," i.e. working or fighting purposes. Instead, a hypothesis was put forward according to which this axe has served as a symbol of a social status of its holder.

The aim of this work is to apply Mössbauer spectrometry to provide analytical evidence on the nature of surface cover of the axe from Borovce.

2 Experimental details

After the excavation, the axe was cleaned from sediments by distilled water and mechanically brushed. It was not conserved by conventional procedures. Instead it is kept in the original state for the sake of continuing investigations.

We have analyzed specimens taken from the surface of the iron axe. ^{57}Fe Mössbauer spectrometry was employed using a ^{57}Co(Rh) source. Calibration was performed by a α-Fe foil and isomer shift values are given relative to the Mössbauer spectrum of α-Fe taken at room (300 K) temperature. Evaluation of hyperfine parameters was done by the help of the CONFIT fitting program [2].

Three sets of samples were investigated. Sample A was a compact piece of surface covering substance about $30 \times 20 \times 5$ mm^3 large, which has accidentally disintegrated from the spike of the investigated axe. It belonged to what is called in [1] "an applied glassy system." Sample A was taken from the side of the axe which was in direct contact with the floor of the grave (the axe laid this side downwards). Sample B was prepared from one piece of a surface fragment about $15 \times 5 \times 2$ mm^3 in size originating from the opposite side of the axe as sample A. It was mechanically separated into its inner and outer parts thus providing two batches of sample B. Each part was ground into fine powder. Sample C consisted of tiny debris from all over the surface that were smashed into powder.

Both surfaces (the inner and the outer one) of sample A were measured in backscattering geometry at room temperature as this sample was not allowed to be powdered because of its historical value. Samples B and C were measured in transmission geometry at room (RT) and at liquid nitrogen (LNT) temperature. Some of this material was exposed to heat treatment in the air at 400°C for 100 min aimed to induce thermal transformations.

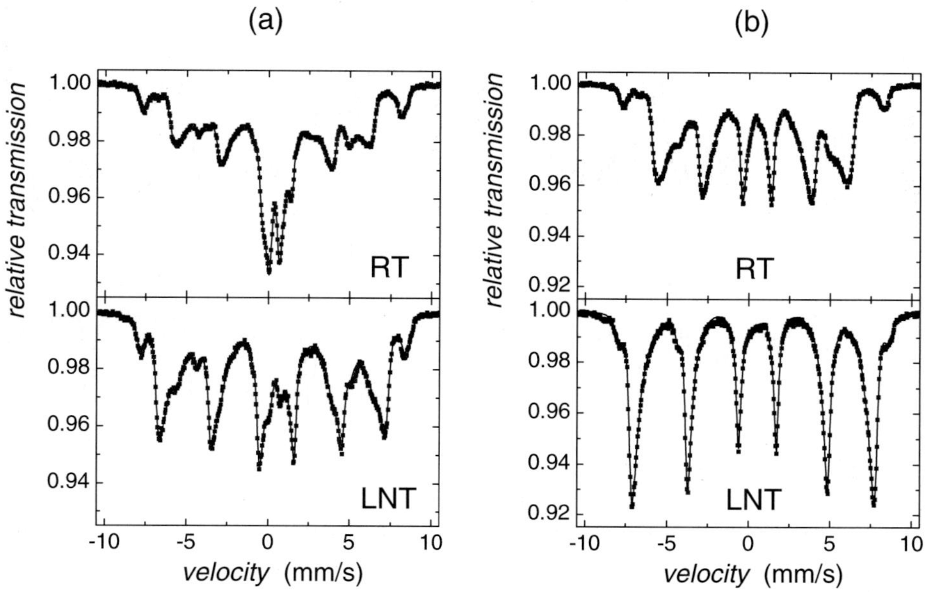

Figure 2 Mössbauer spectra of the inner (**a**) and outer (**b**) parts of sample B taken at room (*RT*) and liquid nitrogen (*LNT*) temperature.

3 Results and discussion

Mössbauer spectrum in Figure 1a recorded in backscattering geometry from the inner part of sample A shows the presence of two doublets that corresponds to lepidocrocite and akageneite. Magnetically split components are not very well resolved and probably represent goethite. Mössbauer spectrum in Figure 1b taken from the outer surface of sample A exhibits only one doublet of lepidocrocite. The magnetically split components are assigned to goethite. The differences in Mössbauer spectra of both samples stem from the fact that the outer part of sample A was in direct contact with the surrounding soil whereas the inner part was originally attached to the body of the axe.

Transmission Mössbauer spectra in Figure 2 recorded from two batches of sample B, i.e. the inner and the outer parts, revealed the presence of magnetite and goethite. In addition, a quadrupole doublet with the parameters characteristic for Fe^{II} in coordination 6 was observed. Relative fraction of this doublet, which is not completely identified so far, is significantly higher in the inner part of sample B. As its contents decreases after measurements at LNT it can also partly represent small superparamagnetic particles. No traces of lepidocrocite were revealed in sample B. The inner part of sample B was originally associated with the body of the axe, i.e. with its basic material and the outer part was actually the very surface of the axe which was not in a direct contact with the floor of the grave.

Sample C was prepared from tiny debris. The part of the axe they are coming from could not be unambiguously identified and in this respect these debris provide a sort of "integral" or overall information about the surface of the axe. Nevertheless, corrosion products as goethite, lepidocrocite, and magnetite were recognized. Sample *C* was heat treated at 400°C for 100 min in air and subsequently measured

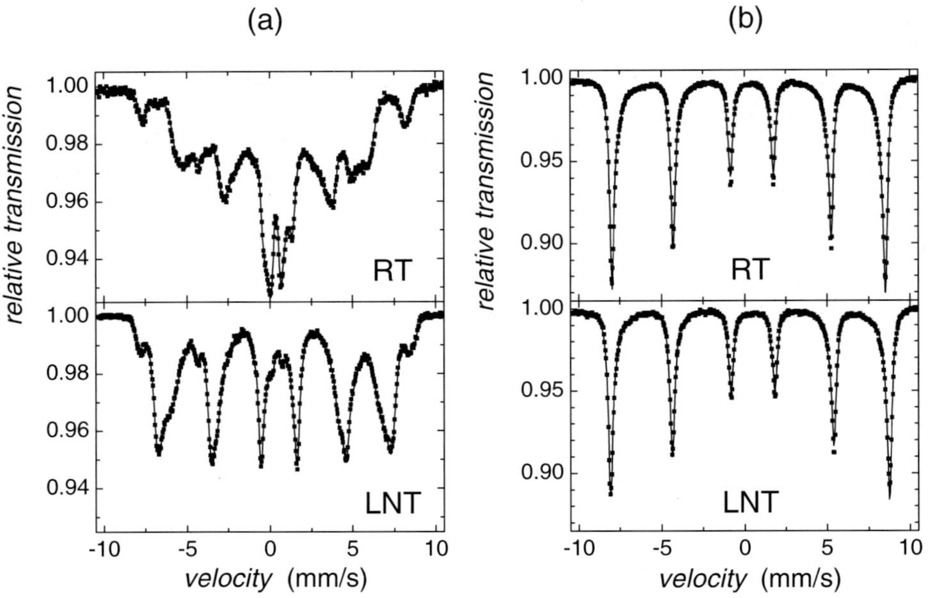

Figure 3 Mössbauer spectra of sample C before (**a**) and after (**b**) thermal treatment at 400°C/100 min taken at room (*RT*) and liquid nitrogen (*LNT*) temperature.

in transmission experiment. The existence of hematite and maghemite was confirmed in the resulting Mössbauer spectra. Hematite is the end product of dehydration reaction of goethite and maghemite whereas the latter originates from thermal decomposition of lepidocrocite and magnetite (Figure 3) [3].

Similar results were reported by Gauzzi et al. [4] who have studied powder samples of an ancient Etruscan axe. They have reported only goethite with very distorted and substituted structure as a principal crystallographic phase which was identified in their sample. In addition, a non-magnetic component with a super-paramagnetic character was also revealed. No traces of ancient metal were detected.

Our samples which are not as old as the one studied by Gauzzi [4] show presence of several corrosion products. Together with goethite that dominates, lepidocrocite, akageneite, and magnetite were also identified. A small amount of Fe^{II} was found in sample B.

4 Conclusion

The main goal of our study was to find out whether a glassy system was applied to the surface of the axe, which might support the hypotheses about the origin of this axe [1]. Taking the results obtained from Mössbauer effect data into account we cannot confirm the presence of "an applied glassy system." We can conclude that the investigated samples of the Early Medieval Age axe from the grave no. 221 in Borovce demonstrate the incidence of corrosion products. Samples taken from the part of the axe which was facing the floor of the grave (outer surface of sample A) demonstrate the occurrence of goethite and lepidocrocite, only. Small amounts of magnetite were found in the other samples (B and C) in addition to the aforementioned corrosion products. This demonstrates differences in alternation processes which have an effect upon this ancient archaeological artifact before and after its excavation.

As the Mössbauer spectrometry is sensitive only to materials containing at least fractions of atomic percent of iron we shall continue this research by examining other parts of the surface as well as the body of the axe. To support the results obtained by Mössbauer spectrometry also additional analytical methods will be employed.

Acknowledgement This work was partially supported by the grants VEGA 2/5060/25 and KEGA 3/100503.

References

1. Staššíková-Štukovská, D., Brziak, P.: In: Proc. Archaeometallurgy in Karpaty region, Herl'any, Slovak, (1994)
2. Kučera, J., Veselý, V., Žák, T.: Mössbauer Spectra Convolution Fit for Windows 98/2K, ver. 4.161 (2004)
3. Cornell, R.M., Schwertmann, U.: The Iron Oxides. Wiley-VCG, Weinheim (2003)
4. Gauzzi, F., Verdini, B., Principi, G., Gupta, R., Raccanelli, A., Russo, U.: In: Ortalli, I. (ed.) Proc. Int. Conf. on Application of the Mössbauer Effect, Part II, vol. 50, pp. 777–780. Societa Italiana di Fisica, Bologna (1996)

Hyperfine Interact (2005) 166:657–664
DOI 10.1007/s10751-006-9337-y

Mössbauer studies of Jarosite, Mikasaite and Yavapaiite, and implication to their Martian counterparts

Kiyoshi Nomura · Masuo Takeda · Takeshi Iiyama · Hitoshi Sakai

Published online: 27 October 2006

Abstract In order to confirm Mössbauer spectra of Jarosite, unidentified Fe^{3+} species (Fe3D3) and Fe_2O_3 found on Martian surface, some Jarosite specimens produced on Earth were characterized by ^{57}Fe Mössbauer spectroscopy. For comparison, anhydrous ferric sulfate minerals, Mikasaite $(Fe^{3+}{}_{1.56}Al_{0.44})(SO_4)_3$, and Yavapaiite, KFe $(SO_4)_2$ were also measured. The thermal decomposition products of Jarosite and their hydration products were also traced by scattering Mössbauer spectra (CEMS and XMS). These data are useful for the identification of hydrous and anhydrous ferric sulfate phases on Mars, including Fe3D3.

Key words Mössbauer spectra · CEMS · XMS · Jarosite · thermal decomposition · hydration · Mars

1 Introduction

Mini Mössbauer spectrometer (MIMOS II) loaded on Opportunity, one of the two rovers sent to Mars by NASA identified abundant Jarosite $[(K, Na, X)(Fe^{3+}, Al)_3(SO_4)_2(OH)_6]$, in the outcrop of the Meridiani Planum at Eagle crater: the first strong evidence for the existence of liquid water on Mars [1]. Unidentified

K. Nomura (✉)
School of Engineering, The University of Tokyo, 7-3-1 Hongo, Bunkyo-ku, Tokyo 113-8656, Japan
e-mail: k-nomura@t-adm.t.u-tokyo.ac.jp

M. Takeda · T. Iiyama
Faculty of Science, Toho University, 2-2-1 Miyama, Funabashi, Chiba 274-8510, Japan

H. Sakai
Okayama University, Okayama, Japan

octahedral high spin Fe^{3+} phase (Fe3D3) and magnetic components were also detected together with Jarosite from the Martian rocks.

In order to elucidate the structure and bonding of Jarosite produced on Mars and Earth, and to see if any difference exists between them, we measured ^{57}Fe Mössbauer spectra of Jarosite from Chino mine and Aso mine, Japan, one from South Australia, and one USA. While Japanese and S. Australian specimens are microcrystalline aggregates, the one from USA is coarse crystalline.

While Jarosite was found by Opportunity, there has been no mentioning on anhydrous ferric sulfate phases in the Martian rocks so far studied by MIMOS II. Considering the extremely dry surface conditions on Mars, this is rather as surprising as the finding of hydrous ferric sulfate. In order to assess the presence of anhydrous ferric sulfate on Mars, it is essential to have the Mössbauer parameters of such phases. However, Yavapaiite [KFe $(SO_4)_2$] is only the mineral whose Mössbauer parameters has been reported [2]. Here, we also report the Mössbauer spectra of Mikasaite $(Fe^{3+}{}_{1.56}Al_{0.44})(SO_4)_3$, from Mikasa, Hokkaido, Japan [3], as well as Yavapaiite, KFe $(SO_4)_2$, from United Verde mine, Jerome, Yavapai County, Arizona, USA [4]. Furthermore, Mössbauer spectra of thermal decomposition products of Jarosite were also measured, as they should contain some anhydrous ferric sulfate phases. These results will be discussed with relevance to the unidentified Fe3D3 phase.

2 Experimental

All the samples used were analyzed for their compositions by an electron probe microscopic analyser (EPMA). Thermal gravimetric and differential thermal analyses were carried out using a Jarosite sample. Transmission Mössbauer spectra were obtained by a conventional constant acceleration mode using a 25 mCi γ source of ^{57}Co at Toho University. Scattering Mössbauer spectra were measured by a homemade dual gas counter obtained simultaneously for conversion electron Mössbauer (CEMS) and X-ray Mössbauer spectroscopy (XMS) [5] at the University of Tokyo. CEMS can provide us the characterization of thin surface layer within ca. 100 nm in thickness. The probing depth by XMS is about 10 μm. The areas and widths of doublet peaks in CEMS and XMS spectra were constrained to be equal for analysis of Mössbauer spectra. The isomer shift values were referred to αFe. All the samples for transmission Mössbauer spectroscopy were first ground to coarse grains to make sure of their purity under magnifying lenses. Impurities, if any, were removed by handpicking. The samples were then pulverized to fine powder and mixed with appropriate amounts of glucose.

3 Results and discussions

It was confirmed from EPMA that the Japanese and S. Australian specimen are K-rich Jarosite, whereas a crystalline specimen from USA is Na-rich one. The results of the transmission Mössbauer spectroscopy are listed in Table I. Isomer shifts (δ) of Jarosite increased with decreasing temperatures, that is, from 0.37 mm/s

Table I Mössbauer parameters of Jarosite, Mikasaite and Yavapaiite

δ ($mm\ s^{-1}$)	Δ ($mm\ s^{-1}$)	2Γ ($mm\ s^{-1}$)[1]	Measur. temp. (K)
Jarosite, Aso mine			
0.38	1.25	0.28, 0.28	290
0.43	1.26	0.29, 0.29	193
0.49	1.26	0.32, 0.33	77
Jarosite, Chino mine			
0.38	1.26	0.30, 0.31	290
0.43	1.27	0.30, 0.30	193
0.48	1.28	0.30, 0.31	77
Jarosite, USA			
0.37	1.29	0.25, 0.25	290
0.43	1.30	0.25, 0.26	193
0.48	1.30	0.27, 0.28	77
Jarosite, South Australia			
0.38	1.15	0.35, 0.35	290
0.43	1.16	0.28, 0.29	193
0.48	1.16	0.29, 0.31	77
Jarosite, Meridiani P. Mars [1]			
0.39	1.22		200–280
Mikasaite			
0.47	0.34	0.41, 0.41	290
0.47	0.34	0.41, 0.41	193
0.56	0.36	0.44, 0.44	77
Yavapaiite			
0.48	0.32	0.28, 0.28	290
0.52	0.32	0.27, 0.27	193
0.57	0.33	0.27, 0.29	77

at room temperature to 0.49 mm/s at 77 K, although the quadrupole splitting (Δ) stayed almost constant within 0.02 mm/s. It is notable that the Na-rich Australian Jarosite exhibits Δ slightly smaller than that of K-rich Jarosite. This result is consistent with the previous results that Δ values decrease in the order of K > Na > H_2O for Jarosite [6]. The values of δ and Δ of the Martian Jarosite are 0.39 and 1.22 mm/s, respectively, as also listed in Table I. Considering the average temperature at Martian surface during the measurements was 240 K, they are well within the range for the terrestrial ones at temperatures between 290 and 193 K.

As shown in Figure 1 and Table I, Mikasaite and Yavapaiite have nearly the same values of δ and Δ, reflecting their similar octahedral coordination of oxygen around Fe^{3+}. Our results on Yavapaiite at 290 K are in close agreement with δ=0.48 mm/s and Δ=0.305 mm/s reported by Lane et al. [2]. The values of δ (0.47 mm/s at room temperature) is unusually high compared to normal hexacoordinated Fe^{3+}, although at present we have no clear interpretation for this behavior. On the other hand, the values of Δ are much smaller than Jarosite, reflecting the highly symmetrical oxygen coordination around Fe^{3+}. Detailed comparison of the Mössbauer spectra of these two anhydrous ferric sulfates with the spectra of Jarosite-bearing Martian rocks reported by [1] indicates no significant contribution of this type of ferric sulfate phase in the rocks. Considering the extremely dry, oxidative surface condition of Mars, non-existence of anhydrous ferric sulfate phase

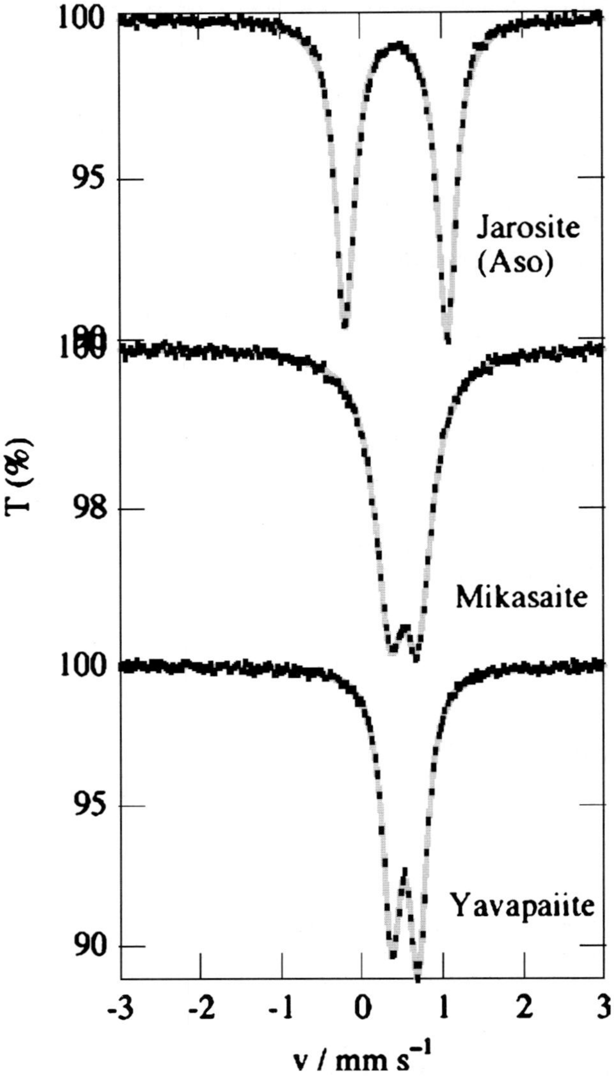

Figure 1 Mössbauer spectra of Jarosite, Mikasaite and Yavapaiite, measured at 193 K.

is rather surprising. Search for anhydrous ferric sulfate, which are potentially important in the sulfur cycle of Mars, should be continued in future exploration of Mars.

The CEMS and XMS spectra of Jarosite (Chino) are shown in Figure 2. Fitting, as two doublets were better than as one doublet. The widths of two doublets were constrained to be equal. At room temperature, δ and Δ of the major doublet and the minor doublet were δ=0.37 mm/s, Δ=1.28 mm/s, and δ=0.38 mm/s, Δ=0.7 mm/s, respectively. It was confirmed that heating at 200 °C for 2 h did not change Jarosite.

TGA and DTA curves showed that the gravity decreased at two steps at 423 and 676 °C. Both are corresponding to the endothermic peaks of K Jarosite (Aso). Jarosite became from yellow to dark yellow by heating. The first endothermic peak is due to dehydration of Jarosite, and the second one due to the decomposition of the sulfate units. The change of gravity was not observed at the exothermic peak of

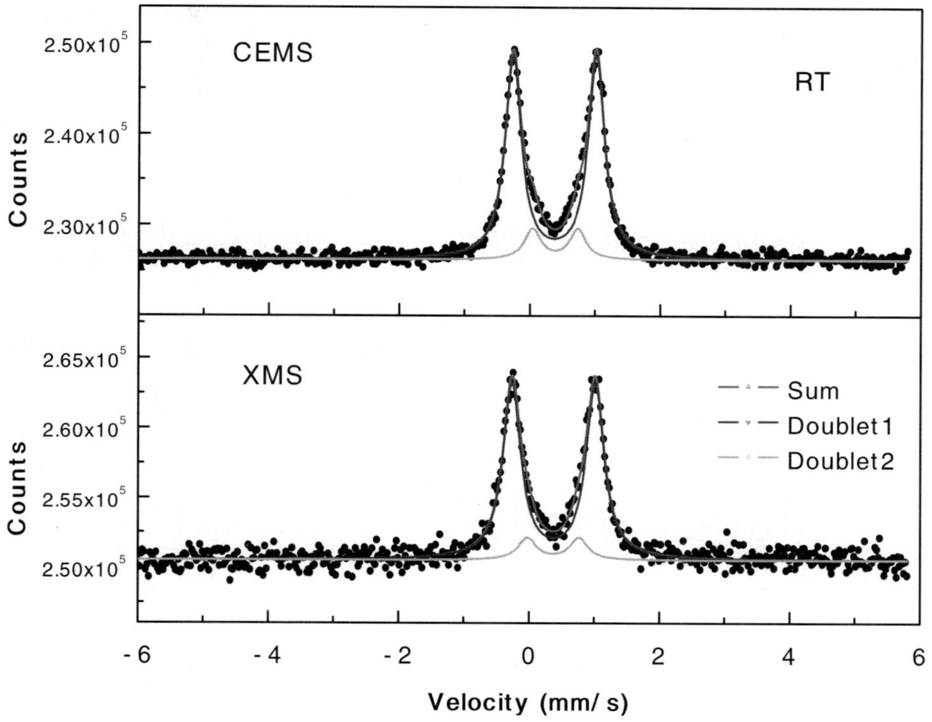

Figure 2 CEMS and XMS spectra of Jarosite (Chino).

CEMS	δ (mm/s)	Δ (mm/s)	Γ (mm/s)	Int. (%)
Doublet1	0.38 (1)	1.28 (1)	0.30 (1)	87.0
Doublet2	0.39 (1)	0.69 (2)	0.30 (1)	13.0
XMS				
Doublet1	0.37 (1)	1.28 (1)	0.33 (1)	89.0
Doublet2	0.37 (2)	0.79 (7)	0.33 (1)	11.0

506 °C, where the dehydrated phase may be changed in structure without mass change. The decomposition process is considered to be the following:

$$K_2[Fe_3(SO_4)_2(OH)_6]_2 \rightarrow K_2SO_4 \cdot Fe_2(SO_4)_3 + 2Fe_2O_3 + 6H_2O \qquad (423^\circ C) \qquad (1)$$

$$K_2SO_4 \cdot Fe_2(SO_4)_3(amorphous) \rightarrow K_2SO_4 \cdot Fe_2(SO_4)_3(crystalline) \qquad (506^\circ C) \qquad (2)$$

$$K_2SO_4 \cdot Fe_2(SO_4)_3 \rightarrow K_2SO_4 + Fe_2O_3 + 3SO_3 \qquad (676^\circ C) \qquad (3)$$

CEMS and XMS spectra of the Jarosite heated at 490 °C for 2 h are shown in Figure 3. The black spot's area increased on the surface by heating. The area intensity of doublet 2 (Figure 2) (δ=0.37 mm/s, Δ=0.87 mm/s) increased in both

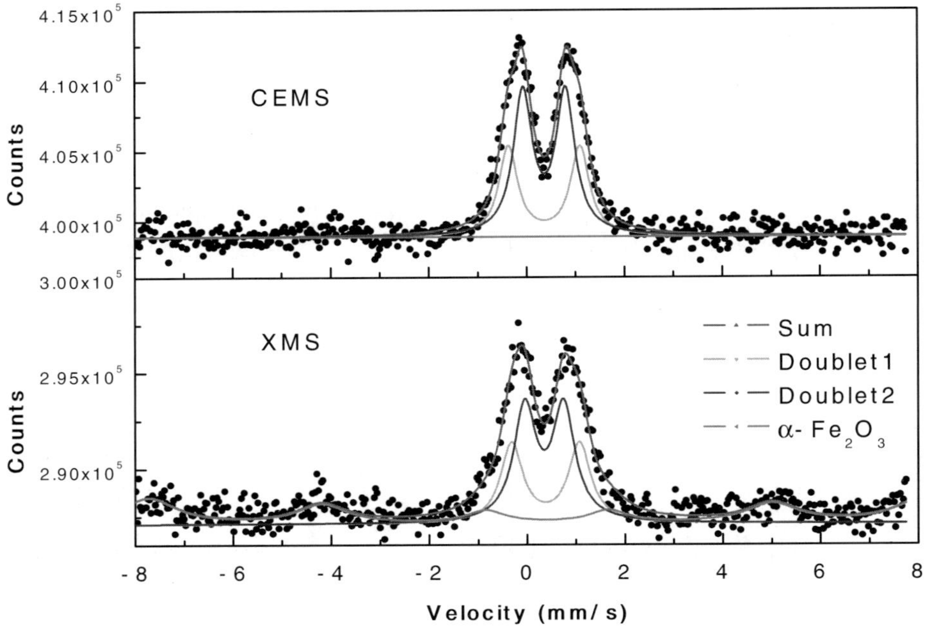

Figure 3 CEMS and XMS spectra of Jarosite heated at 490 °C for 2 h.

CEMS	δ (mm/s)	Δ (mm/s)	Γ (mm/s)	B_{hf} (T)	Area Int (%)	Ratio
Doublet1	0.36 (1)	1.46 (3)	0.47 (1)	–	38.7	1.0
Doublet2	0.37 (1)	0.87 (2)	0.47 (1)	–	61.2	1.58
XMS						
Doublet1	0.38 (1)	1.38 (5)	0.55 (3)	–	24.6	1.0
Doublet2	0.35 (1)	0.79 (3)	0.55 (3)	–	35.1	1.42
Magnetic	0.31 (5)	−0.19 (9)	1.4 (3)	49.3 (4)	40.3	

CEMS and XMS spectra and the magnetic sextet (δ=0.31 mm/s, Δ=−0.19 mm/s, B_{hf}=50 T) was observed in addition to the doublets especially in XMS spectra.

These results indicate that, by heating, stable anhydrous KFe^{3+} sulfate was formed on the surface and hematite was produced inside with the KFe^{3+} sulfate. When the transmission Mössbauer spectra of the heated sample at 490 °C for 2.5 h were measured at RT and 77 K, the magnetic components appeared strongly at 77 K although they were not clearly observed at room temperature. The intensity ratio of magnetic to paramagnetic components was different among Jarosite products. It is considered that the difference depends on the particle sizes of Fe_2O_3 produced by decomposition of Jarosite. They might be affected by impurity in each sample.

Given (1) to represent the thermal decomposition process at 420 °C, a phase of $KFe(SO_4)_2$ is responsible for doublet 2. Mössbauer parameters of Fe3D3 (δ= 0.36 mm/s, and Δ=0.64 mm/s at 200–280 K [1]) are close to those of doublet 2, implying that Fe3D3 might be a ferric sulfate phase similar in composition and structure to $KFe(SO_4)_2$ phase. The $KFe(SO_4)_2$ phases cannot be Yavapaiite, be-

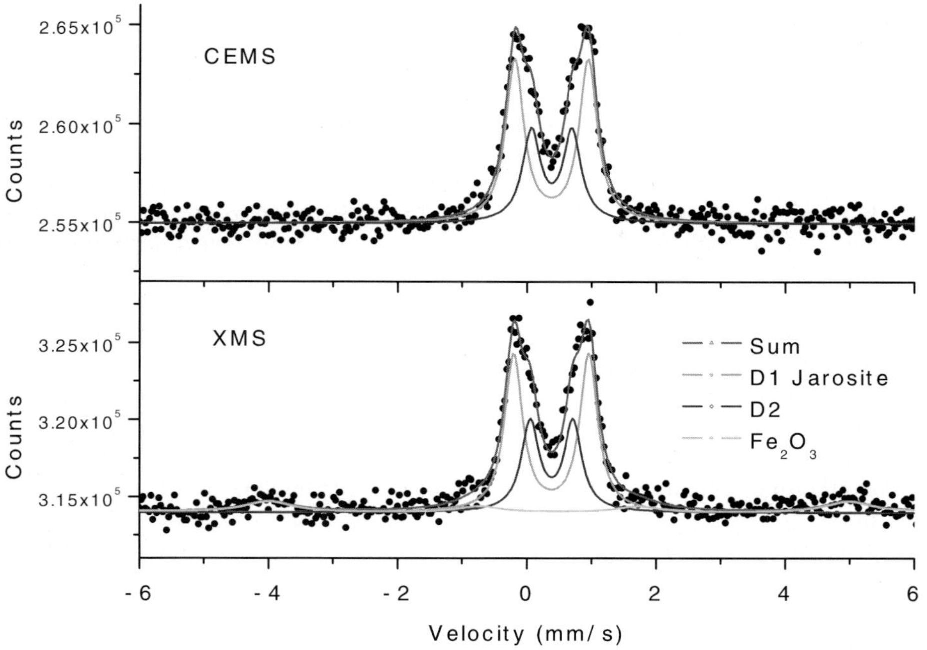

Figure 4 CEMS of Jarosite kept in high humidity air at room temperature for 13 days after heated at 490 °C for 2 h.

	δ (mm/s)	Δ (mm/s)	Γ (mm/s)	Int.(%)
CEMS				
D1	0.38	1.14	0.34	64.2
D2	0.38	0.63	0.34	35.8
XMS				
D1	0.38	1.17	0.33	46.8
D2	0.38	0.66	0.33	26.6

Remaining Mag. Components: Fe_2O_3.

cause of the quite difference in their Mössbauer parameters. Further study is needed to identify the $KFe(SO_4)_2$ phase in the decomposition products and its relevance to Fe3D3 phase.

After 2 h heating at 490 °C, the sample was kept in high humidity atmosphere at room temperature for 13 days and was subjected to Mössbauer measurements. The results are shown in Figure 4. The peak intensity of Jarosite doublets increased again. QS value of Jarosite produced was smaller than that of original Jarosite. This suggests that the reproduced Jarosite is hydronium Jarosite, with K being fixed as K_2SO_4 formed by the thermal decomposition.

The Mössbauer parameters of the thermal decomposition products of Jarosite at 480 °C for 2 h are different from those at 510 °C. The difference may be ascribed to the exothermic phase change that occurs at temperature above 480 °C and modifies the structure of ferric sulfate phase produced by decomposition at 480 °C.

4 Conclusions

From the foregoing results, we conclude as follows:

1. The Mössbauer parameters (δ and Δ) of the Martian Jarosite reported by [1] are in good agreement with those of the terrestrial samples of the present study.
2. Anhydrous ferric sulfate minerals, Mikasaite and Yavapaiite exhibit significantly larger δ and smaller Δ than Jarosite. The Mössbauer spectra of the Martian rocks so far studied by [1] do not seem to have any significant contribution from these phases.
3. The unidentified octahedral Fe3D3 at Martian surface may be attributable to an anhydrous ferric sulfate phase or phases similar in structure and composition to those formed by thermal decomposition of Jarosite.
4. Δ of K-rich Jarosite is slightly greater than that of Na-rich one, supporting the previous finding that it decreases in the order K>Na.
5. Thermal decomposition products of Jarosite formed below and above the exothermic temperature have the Mössbauer parameters different from each other.
6. Thermal decomposition products of Jarosite, when exposed to wet atmosphere at room temperature, is hydrated to form H_3O-Jarosite.
7. The thermal decomposition experiments are not related to the identification of Jarosite on Mars.

We only want to say that the thermal decomposition of Jarosite produced phases, which have the similar Mössbauer parameters to the unidentified Fe3D3 phase on Mars. But we have no intention to say that the Fe3D3 phase was formed by the similar process on Mars.

Acknowledgements We thank Mr S. Iio and Mr S. Ohostuka, in The University of Tokyo, for their measurements of the scattering Mössbauer spectra and electron probe microanalysis. Sample specimens of Mikasaite, Jarosite (Aso) and Jarosite (Chino) are gifts from Prof. H. Miura, Hokkaido Univ.; Dr M.Aoki, Geol. Surv. Japan and Dr T. Suzuki, Nikkou Mining and Metals Co., LTD, respectively, to whom our thanks are due. We also thank Dr W. P. Gates, Primary Industries Research, Australia for his providing us with the S. Australia Jarosite.

References

1. Klingelhoefer, G., et al.: Science **306**, 1740 (2004)
2. Lane, M.J., et al.: Lunar Planet. Sci. **XXXVI**(1442), (2005)
3. Miura, H., Hirama, T.: Mineral. Mag. **58**, 649 (1994)
4. Hutton, C.O.: Am. Mineral. **44**, 1105 (1959)
5. Johnson, J.H.: Geochim. Cosmochim. Acta **41**, 539 (1977)
6. Nomura, K., et al.: Spectrochim. Acta, Part B: Atom. Spectrosc. **59**, 1259 (2004)

Further Reading

1. Hrynkiewicz, A.Z., Kubisz, J., Kulgawezuk, D.S.: J. Inorg. Nucl. Chem. **27**, 2513 (1965)
2. Takano, M., Shinjo, T., Kimiya, M., Takada, T.: J. Phys. Soc. Jpn. **25**, 902 (1968)

Hyperfine Interact (2005) 166:665–670
DOI 10.1007/s10751-006-9338-x

Mössbauer spectroscopy of ordinary chondrites: an analysis of the metal phases

E. V. Zhiganova · M. I. Oshtrakh · O. B. Milder · V. I. Grokhovsky · V. A. Semionkin · A. V. Mezentsev

Published online: 3 November 2006

Abstract Mössbauer spectroscopy was used for the analysis of the metal part in various ordinary chondrites. One and two magnetic sextets were used to fit the metal subspectra and the main Fe–Ni–Co phases with different nickel and/or cobalt concentrations were proposed in different chondrites.

Key words Mössbauer spectroscopy · ordinary chondrites · metal phases

1 Introduction

Mössbauer spectroscopy is a useful tool to study meteorites and especially ordinary chondrites due to their chemical and phase composition. Ordinary chondrites usually contain metal phases (Fe–Ni–Co alloys), troilite FeS, olivine $(Fe, Mg)_2SiO_4$, pyroxene $(Fe, Mg)SiO_3$, and sometimes oxidized iron as ferric oxides or hydrous oxides. Ordinary chondrites were related to three groups: H, L and LL. These groups were characterized by different total iron content and different metal iron contents. The metal iron content in these groups of ordinary chondrites varied as follows: 15–19 wt.% (H), 4–10 wt.% (L) and 1–3 wt.% (LL). Metal grains in ordinary chondrites may contain various phases such as α-Fe(Ni,Co) or kamacite with Ni content up to 7.5 wt.%, α_2-Fe(Ni,Co) – distorted phase supersaturated in Ni or so-called martensite with Ni content in the range of ~15–25 wt.%, α′-Fe(Ni,Co) or kamacite resulted from decomposition of α_2-Fe(Ni,Co) with Ni content in the range of ~7–15 wt.%, and γ–Fe(Ni,Co) or taenite with Ni content in the range of 25–45 wt.%. Low metal content in ordinary chondrites makes the determination of

E. V. Zhiganova · M. I. Oshtrakh (✉) · V. I. Grokhovsky · A. V. Mezentsev
Faculty of Physical Techniques and Devices for Quality Control, Ural State Technical University – UPI, Ekaterinburg 620002, Russian Federation
e-mail: oshtrakh@mail.utnet.ru

O. B. Milder · V. A. Semionkin
Faculty of Experimental Physics, Ural State Technical University – UPI, Ekaterinburg 620002, Russian Federation

its phase composition from Mössbauer spectra difficult. Assumptions about the presence of kamacite and taenite in ordinary chondrite were made [1]. It was difficult to distinguish the Fe–Ni phases in Mössbauer spectra of ordinary chondrites due to low total metal content and especially due to lower partial content of each Fe–Ni phase. For instance, the presence of ~7% of kamacite and ~2% of tetrataenite (ordered FeNi phase) was found in New Halfa meteorite by Mössbauer spectroscopy [2]. Visual evaluation of the symmetry of the most negative velocity absorption peak was used to determine the presence of taenite contribution in [3], however, this evaluation should not be used for spectra analysis in general. In this work, we tried to analyze the Mössbauer spectra of the H and L chondrites to distinguish the main metal phases.

2 Materials and methods

We studied equilibrated ordinary chondrites from H and L chemical groups: Tzarev L5, Farmington L5, Mbale L5, Kunashak L6, Zvonkovoe H6 and Vengerovo H5. All chondrites are falls with low weathering grades some of which have experienced relatively intense shock processing. All samples were studied by metallography and Mössbauer spectroscopy. Samples for metallography were sections polished with diamond and etched by Nital (2 vol.% HNO_3, balance ethyl alcohol) for 40–60 sec. Samples of ordinary chondrites for Mössbauer measurements were prepared in the form of powder. Images of ordinary chondrites sections were obtained using a Neophot 32 reflected light microscope with a Luminera LU205C digital camera. The area which was explored for each meteorite was not lower than 50 mm^2. The content of the metal phases was evaluated using the SIMAGIS Research^{TM} software. Chemical analyses of metal phases were made by Philips SEM 535 with EDAX using a 30 kV accelerating voltage and a 30 μA current. It should be noted that Co concentration could not be accurately determined by this technique. Mössbauer spectra of the sample powders were measured at room temperature with the constant acceleration computerized high precision, sensitive and stable spectrometer that was a part of the multi-dimension parametric Mössbauer spectrometer SM-2201 [4]. The noise of the velocity signal of the spectrometer was 1.5×10^{-3} mm/s, the drift of zero velocity point was $\pm2.6\times10^{-3}$ mm/s, the nonlinearity of the velocity signal was 0.01%, the harmonic distortion factor was 0.005% for the frequency band in the range of 0–1120 Hz. The 0.3×10^{9} Bq ^{57}Co(Cr) source was used at room temperature. Mössbauer spectra were measured in transmission geometry with moving absorber and computer fitted with the least squares method using Lorentzian line shape. Mössbauer parameters isomer shift δ, quadrupole splitting ΔE_Q, magnetic hyperfine field H_{eff}, line width Γ and subspectrum area S were determined. The values of isomer shift are given relative to α-Fe at 295 K.

3 Results and discussion

Mössbauer spectra of ordinary chondrites are shown in Figure 1. These spectra showed usual chondrite spectra consisted of pyroxene, olivine, troilite, metal phase, and oxidized iron in some cases (see [1–3]). All spectra were fitted with two or three

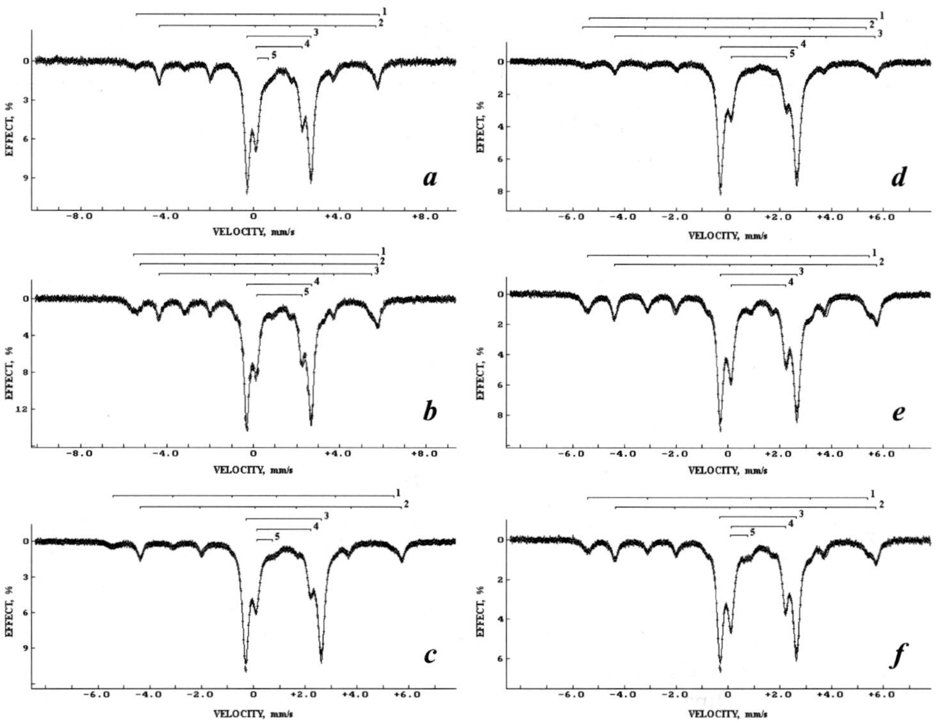

Figure 1 Mössbauer spectra of ordinary chondrites: Tzarev L5 (**a**) Farmington L5 (**b**) Mbale L5 (**c**) Kunashak L6 (**d**) Zvonkovoe H6 (**e**) and Vengerovo H5 (**f**). 1–5 are the results of the better fit (see Table I). T=295 K.

doublets and two sextets. However, Mössbauer spectra of Farmington L5 and Kunashak L6 were better fitted with the addition of the third sextet. The results of the better fit are given in Table I. The Mössbauer parameters of pyroxene, olivine, troilite and oxidized iron were similar to those published in the literature [1–3]. Samples of meteorites Mbale L5, Zvonkovoe H6 and Vengerovo H5 contained metal phase subspectrum which was fitted using one sextet with H_{eff} at about 335–337 kOe. Therefore, this metal phase may be considered as α-Fe(Ni,Co). The sample of meteorite Tzarev L5 contained a metal phase subspectrum which was fitted using one sextet, however, the value of H_{eff} was higher (348 kOe). It is well known that meteorite Tzarev L5 was shock-metamorphosed with reheating [5] and, therefore, higher value of H_{eff} for metal phase in meteorite Tzarev L5 may be related to structural and/or concentration variations of the initial metal alloy. As a contrary, samples of meteorites Farmington L5 and Kunashak L6 contained metal phases subspectra which were better fitted using two sextets. However, lager and smaller values of H_{eff} for corresponding sextets were slightly different for these meteorites. The value of H_{eff} for sextet 1 in the Mössbauer spectrum of meteorite Farmington L5 was the same, within the experimental error, as that of meteorite Tzarev L5. It was interesting that meteorite Farmington was also shock-metamorphosed with reheating [6, 7].

Photomicrographs of metal grains of studied chondrites are shown in Figure 2. Metal grains for meteorites Mbale L5, Zvonkovoe H5 and Vengerovo H5

Table I Mössbauer parameters of ordinary chondrites

Sample	Γ^a (mm/s)	δ (mm/s)	ΔE_Q (mm/s)	H_{eff} (kOe)	S (%)	Compound[b]
Tsarev	0.275±0.076	0.110±0.038	0.126±0.038	348.6±1.2	6.06	Fe-Ni (1)
L5	0.379±0.076	0.764±0.038	−0.226±0.038	311.0±1.2	18.42	Troilite (2)
	0.304±0.076	1.186±0.038	2.957±0.038	–	45.56	Olivine (3)
	0.333±0.076	1.183±0.038	2.119±0.038	–	23.50	Pyroxene (4)
	0.457±0.076	0.435±0.038	0.519±0.038	–	6.46	Fe^{3+} (5)
Farmington	0.332±0.076	0.094±0.038	0.036±0.038	350.2±1.2	9.34	Fe-Ni (1)
L5	0.300±0.076	0.120±0.038	0.262±0.038	341.2±1.2	6.74	Fe-Ni (2)
	0.425±0.076	0.712±0.038	−0.313±0.038	305.8±1.2	16.12	Troilite (3)
	0.311±0.076	1.183±0.038	2.991±0.038	–	44.11	Olivine (3)
	0.374±0.076	1.192±0.038	2.103±0.038	–	23.70	Pyroxene (4)
Mbale	0.563±0.064	0.014±0.032	−0.041±0.032	336.2±1.0	8.12	Fe-Ni (1)
L5	0.287±0.064	0.768±0.032	−0.183±0.032	312.0±1.0	14.47	Troilite (2)
	0.296±0.064	1.163±0.032	2.923±0.032	–	52.69	Olivine (3)
	0.302±0.064	1.165±0.032	2.092±0.032	–	19.42	Pyroxene (4)
	0.424±0.064	0.440±0.032	0.606±0.032	–	5.30	Fe^{3+} (5)
Kunashak	0.420±0.064	0.139±0.032	0.120±0.032	343.9±1.0	5.26	Fe-Ni (1)
L6	0.400±0.064	−0.083±0.032	−0.102±0.032	338.5±1.0	3.67	Fe-Ni (2)
	0.420±0.064	0.762±0.032	−0.216±0.032	310.9±1.0	12.12	Troilite (3)
	0.289±0.064	1.181±0.032	2.941±0.032	–	58.90	Olivine (4)
	0.317±0.064	1.189±0.032	2.096±0.032	–	20.05	Pyroxene (5)
Zvonkovoe	0.464±0.064	0.025±0.032	−0.021±0.032	336.9±1.0	19.63	Fe-Ni (1)
H6	0.284±0.064	0.778±0.032	−0.178±0.032	313.1±1.0	16.94	Troilite (2)
	0.288±0.064	1.182±0.032	2.950±0.032	–	40.68	Olivine (3)
	0.312±0.064	1.183±0.032	2.095±0.032	–	22.75	Pyroxene (4)
Vengerovo	0.429±0.064	0.013±0.032	−0.028±0.032	335.0±1.0	12.19	Fe-Ni (1)
H5	0.307±0.064	0.773±0.032	−0.185±0.032	311.7±1.0	17.88	Troilite (2)
	0.293±0.064	1.169±0.032	2.937±0.032	–	43.97	Olivine (3)
	0.278±0.064	1.168±0.032	2.087±0.032	–	22.26	Pyroxene (4)
	0.489±0.064	0.443±0.032	0.675±0.032	–	6.69	Fe^{3+} (5)

[a] Γ values for sextets were given for the first and sixth lines.

[b] Numbers given in parenthesis corresponded to numbers of components in Figure 1.

(Figure 2a–c) showed that the major part of metal was located in the region marked K (~90, ~97 and ~97%, respectively) whereas in two samples, small quantity of another metal regions, marked T, were observed. Metallography of these grains evidenced that the K region was α-Fe(Ni,Co) kamacite whereas the T region was γ-Fe(Ni,Co) taenite. The results of the chemical analysis of K regions in meteorites Mbale L5, Zvonkovoe H5 and Vengerovo H5 showed the presence of ~6.2 wt.% of Ni and ~0.5 wt.% of Co, ~5.9 wt.% of Ni and ~0.6 wt.% of Co, and ~5.6 wt.% of Ni and ~0.7 wt.% of Co, respectively. Therefore, Mössbauer spectra of these chondrites were fitted using one sextet for metal phase and obtained values of H_{eff} corresponded to usual α-Fe(Ni,Co). Determination of the paramagnetic singlet correspondent to taenite was not possible due to extremely low content of this phase in comparison with other paramagnetic phases in chondrites. Metal grain of meteorite Tzarev L5 (Figure 2d) showed the presence of the main metal regions marked P that were determined as plessite (mixture of α′-Fe(Ni,Co) and γ-Fe(Ni,Co) phases). In case of meteorite Tzarev L5 metal grains consisted of a mixture of ~60%

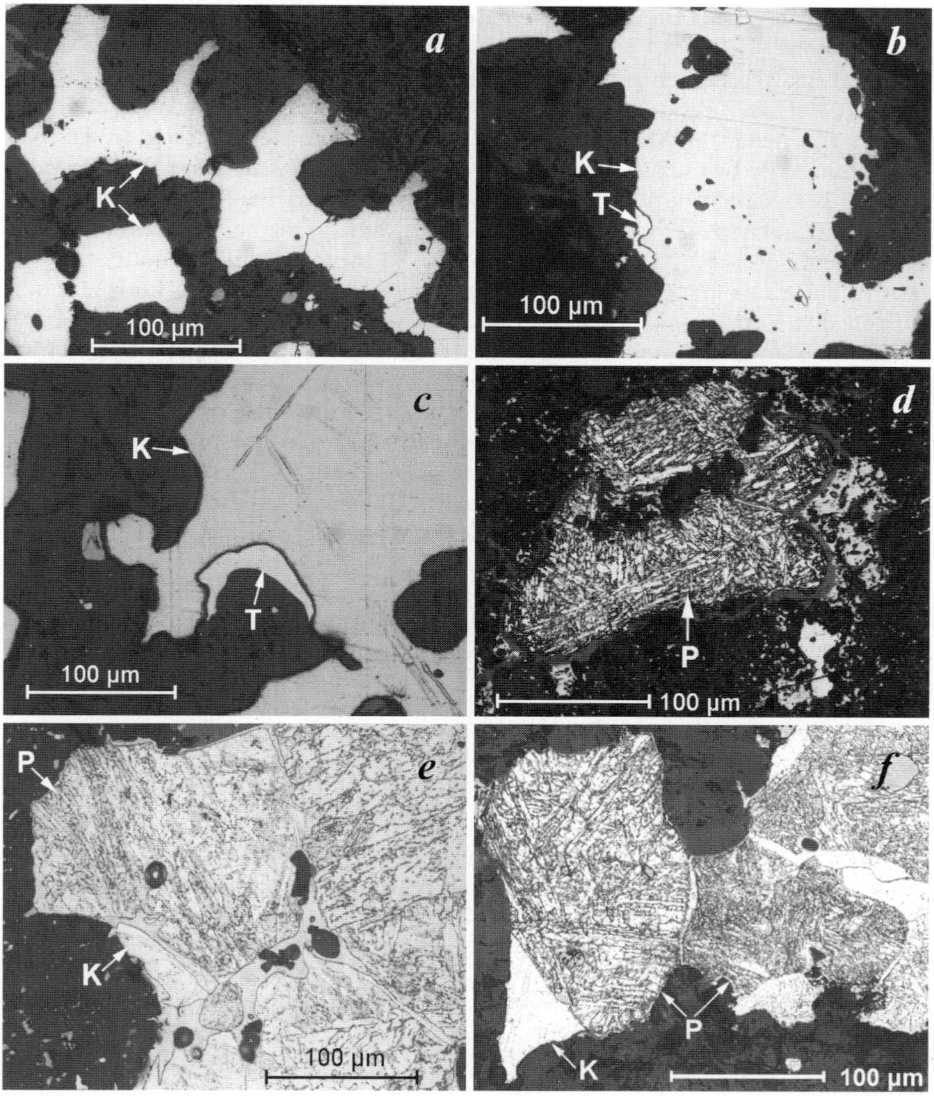

Figure 2 Digital photomicrographs of metal grains in ordinary chondrites: Mbale L5 (**a**) Zvonkovoe H6 (**b**) Vengerovo H5 (**c**) Tzarev L5 (**d**) Farmington L5 (**e**) and Kunashak L6 (**f**). K, T and P – various metal regions (explanation is given in the text).

of α'-Fe(Ni,Co) and small quantity of γ-Fe(Ni,Co) in plessite as well as ~25% of α_2-Fe(Ni,Co) and ~10% of α-Fe(Ni,Co). Chemical analysis demonstrated the presence of ~11.4 wt.% of Ni and ~0.8 wt.% of Co in the main α'-Fe(Ni,Co) phase. Therefore, metal in meteorite Tzarev L5 was characterized by one sextet in Mössbauer spectrum with a higher value of H_{eff}. The presence of small quantities of other Fe–Ni–Co phases determined by metallography, unfortunately, cannot be resolved in Mössbauer spectra of the bulk chondrite samples. Metal grain of meteorite Farmington L5 (Figure 2e) contained two main regions marked P and K. Regions

P (~70%) in meteorite grains represented plessite α'-Fe(Ni,Co)+γ-Fe(Ni,Co) while regions K (~25%) represented α-Fe(Ni,Co) kamacite. The results of the chemical analysis demonstrated the presence of ~11 wt.% of Ni and ~1.1 wt.% of Co in α'-Fe(Ni,Co) and ~7.6 wt.% of Ni and ~0.9 wt.% of Co in α-Fe(Ni,Co). Therefore, metal in Mössbauer spectrum of meteorite Farmington L5 was characterized by two sextets with higher values of H_{eff}. We supposed that the value of 350.2 kOe should be related to α'-Fe(Ni,Co) whereas the value of 341.2 kOe should be related to α-Fe(Ni,Co) probably due to higher Co concentration. Contribution of α_2-Fe(Ni,Co) and γ-Fe(Ni,Co) to the spectrum was not possible to resolve due to its low content. Metal grain of meteorite Kunashak L6 (Figure 2f) contained two main regions marked P – α'-Fe(Ni,Co)+γ-Fe(Ni,Co) plessite (~70%) and K – α-Fe(Ni,Co) kamacite (~30%). Metal components in Mössbauer spectrum of meteorite Kunashak L6 were characterized by two sextets with different values of H_{eff}, however, negative δ value for the second sextet gives rise to some doubts (see [8]). The chemical analysis of meteorite Kunashak L6 was not made. Nevertheless, higher value of H_{eff} (343.9 kOe) may be related to α'-Fe(Ni,Co) while lower value of H_{eff} (338.5 kOe) may be related to α-Fe(Ni,Co). We pointed out that meteorites Tzarev L5, Farmington L5 and Kunashak L6 Mössbauer spectra which were fitted using sextets with higher values of H_{eff} were reheated chondrites after intensive shock-events. On the basis of these data we can suppose that variation of H_{eff} may be related to variation of Ni and/or Co concentrations in Fe–Ni–Co phases in ordinary chondrites similar to iron meteorites (see [9, 10]).

4 Conclusion

The results obtained showed that metal grains in ordinary chondrites might contain various metal phase compositions: one or at least two main Fe–Ni–Co phases with different Ni and/or Co concentrations which were characterized by one or two sextets in Mössbauer spectra of ordinary chondrites with different values of the magnetic hyperfine field. However, for further detailed analysis, Mössbauer measurements with higher velocity resolution and study of extracted metal grains are required.

Acknowledgement This work was supported in part by the grant of the Russian Foundation for Basic Research #06-08-00705-a.

References

1. Dunlap, R.A.: Hyperfine Interact. **110**, 209 (1997)
2. Abdu, A.A., Ericsson, T.: Meteorit. Planet. Sci. **32**, 373 (1997)
3. Verma, H.C., Rawat, A., Paliwal, B.S., Tripathi, R.P.: Hyperfine Interact. **142**, 643 (2002)
4. Irkaev, S.M., Kupriyanov, V.V., Semionkin, V.A., Sokolov, M.M.: British Patent No. 10745 (7 May, 1987)
5. Migdisova, L.F., Zaslavskaya, N.I., Ivanov, A.V., Grokhovsky, V.I.: Lunar Planet. Sci. **13**, 518 (1982)
6. Smith, B.A., Goldstein, J.I.: Geochim. Cosmochim. Acta **41**, 1061 (1977)
7. Lambert, P., Lewis, C., Moore, C.B.: Lunar Planet. Sci. **12**, 583 (1981)
8. Grandgean, F., Long, G.J., Hautot, D., Whitney, D.L.: Hyperfine Interact. **116**, 105 (1998)
9. Grokhovsky, V.I., Oshtrakh, M.I., Milder, O.B., Semionkin, V.A.: Bull. Russ. Acad. Sci., Ser. Phys. **69**, 1522 (2005)
10. Oshtrakh, M.I., Milder, O.B., Grokhovsky, V.I., Semionkin, V.A.: Hyperfine Interact. **158**, 365 (2004)

Hyperfine Interact (2005) 166:671–677
DOI 10.1007/s10751-006-9339-9

Mössbauer spectroscopy of iron meteorite Dronino and products of its corrosion

V. I. Grokhovsky · M. I. Oshtrakh · O. B. Milder · V. A. Semionkin

Published online: 3 November 2006

Abstract The results of the first study of iron meteorite Dronino and products of its corrosion by Mössbauer spectroscopy are presented. Mössbauer parameters for iron phase, sulfide inclusions, products of terrestrial corrosion and iron oxides in concretion are determined and analyzed.

Key words Mössbauer spectroscopy · iron meteorite · metal phase · iron corrosion · concretion

1 Introduction

Dronino is a huge iron meteorite shower which recently occurred in Russia. The Dronino meteorite was classified as ungrouped iron with about 9.3 wt% of Ni. The terrestrial age of meteorite Dronino was evaluated to be more than 1,000 years, therefore, its fragments were weathered for a long time. Samples of meteorite Dronino were collected at a depth of 0.5–2.0 m in clay sand. The fragments of meteorite Dronino found were rusted and covered by several centimeter-thick ferric hydrous oxides. Non-oxidized fragments of meteorite Dronino contained numerous iron sulfide inclusions. In addition, many concretions were found near meteorite fragments. Recently, the metal phase of meteorite Dronino was studied for the first time by Mössbauer spectroscopy to compare it with some other iron meteorites [1, 2]. In this work, we have considered the results of a Mössbauer study of meteorite Dronino and the products of its corrosion as well as of its concretions.

V. I. Grokhovsky · M. I. Oshtrakh (✉)
Faculty of Physical Techniques and Devices for Quality Control, Ural State Technical University—UPI, Ekaterinburg 620002, Russian Federation
e-mail: oshtrakh@mail.utnet.ru

O. B. Milder · V. A. Semionkin
Faculty of Experimental Physics, Ural State Technical University—UPI, Ekaterinburg 620002, Russian Federation

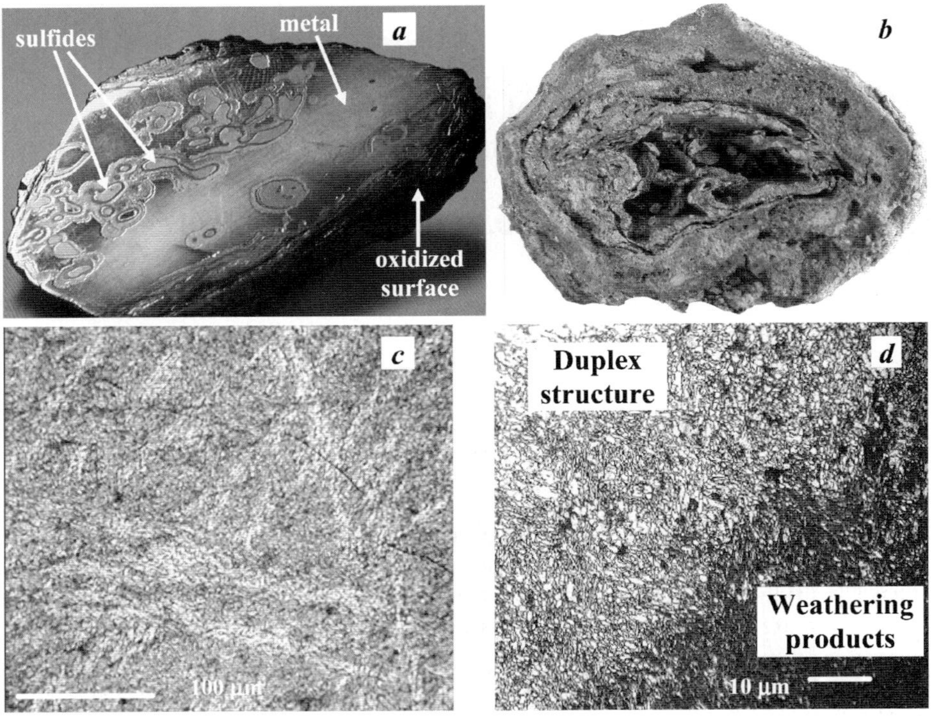

Figure 1 Photographs of the sections of fragment of meteorite Dronino (**a**) and concretion (**b**); optical micrographs of typical regions of Dronino: oriented bands formed by mixture of α-Fe(Ni) and α_2-Fe(Ni) phases (**c**); duplex structure of α-Fe(Ni) and α_2-Fe(Ni) with weathering products which replaced kamacite at the area surrounded sulfide nodule (**d**).

2 Materials and methods

Samples of metal, sulfide, surface corrosion products from meteorite Dronino as well as friable internal layers of concretion and solid external component of concretion were prepared as powder. Mössbauer spectra of the sample powders were measured at room temperature with the constant acceleration computerized high precision, sensitive and stable spectrometer that was a part of the multidimension parametric Mössbauer spectrometer SM-2201 [3]. The noise of velocity signal of the spectrometer was 1.5×10^{-3} mm/s, the drift of zero velocity point was $\pm2.6\times10^{-3}$ mm/s, the nonlinearity of velocity signal was 0.01%, the harmonic distortion factor was 0.005% for the frequency band in the 0–1120 Hz range. The 0.3×10^9 Bq ^{57}Co(Cr) source was used at room temperature. Mössbauer spectra were measured in transmission geometry with moving absorber and computer fitted with the least squares method using Lorentzian line shape. The Mössbauer parameters, i.e. isomer shift δ, quadrupole splitting ΔE_Q, magnetic hyperfine field H_{eff} and subspectrum area S were determined. The values of isomer shift are given relative to α-Fe at 295 K. Additionally, in order to determine the phase compositions, a section of the meteorite was studied by optical microscopy and electron probe microanalysis using scanning electron microscope Philips 30XL with EDAX. X-ray diffraction

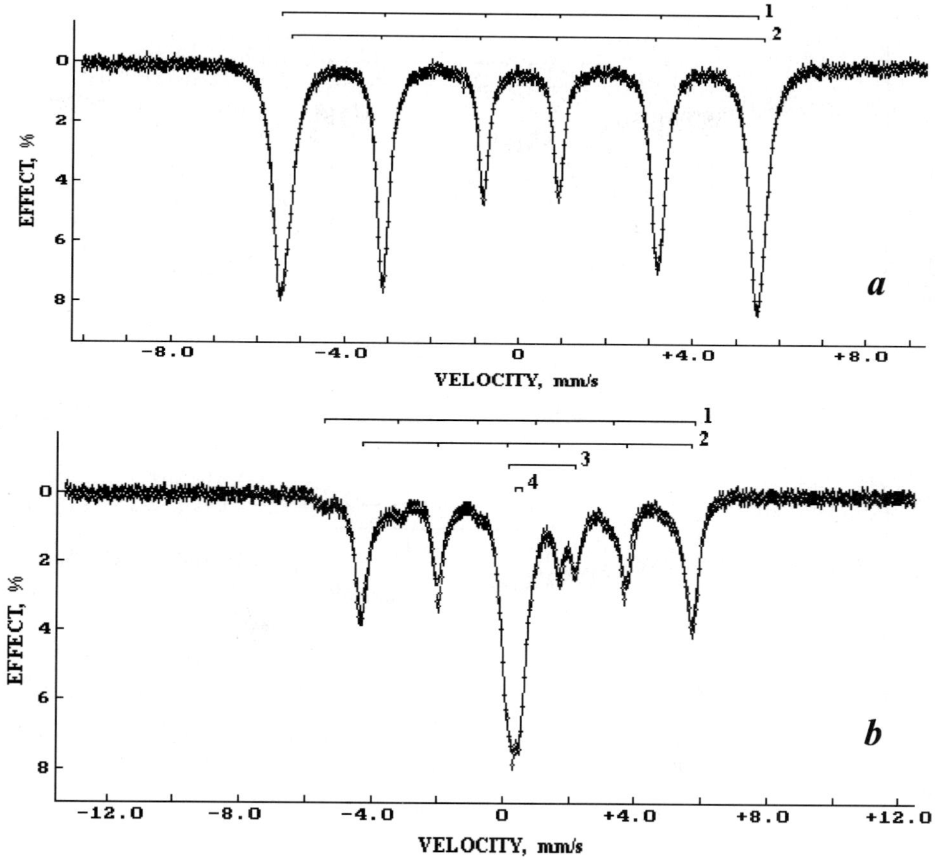

Figure 2 Mössbauer spectra of meteorite Dronino: **a** metal phase, *1* α-Fe(Ni,Co), *2* α_2-Fe(Ni,Co); **b** sulfides, *1* Fe(Ni,Co), *2* FeS, *3* high spin Fe^{+2}, *4* high spin Fe^{+3}. T=295 K.

patterns of the sample powders were measured using DRON-3M diffractometer with $FeK\alpha_1$ radiation.

3 Results and discussion

Photographs of the sections of meteorite Dronino and concretion are shown in Figure 1a and b. The microscopic analysis demonstrated that the metal part of meteorite Dronino has a complicated plessite-like structure, two examples of which are shown in Figure 1c and d. The Ni content in two Fe(Ni,Co) phases was determined by the electron probe microanalysis to be ~7.0 and ~26.5 wt%, respectively. Lower volume fraction of the phase with high Ni concentration was determined to be 12% in the oriented bands and 23% in the other areas by the quantitative image analysis.

Mössbauer spectra of meteorite Dronino (metal and sulfide), products of meteorite corrosion and samples of concretion are shown in Figures 2–4. The

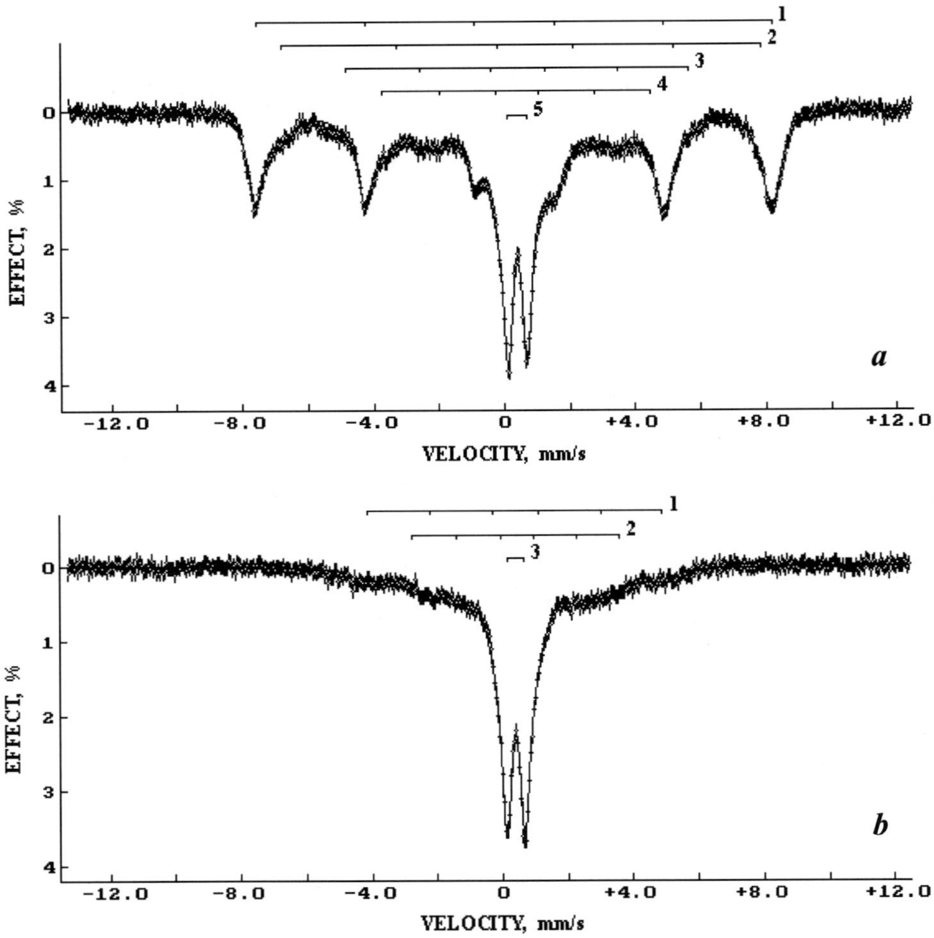

Figure 3 Mössbauer spectra of meteorite Dronino corrosion products: **a** internal zone of oxidation, *1* and *2* γ-Fe_3O_4, *3–5* β-FeOOH; **b** external zone of oxidation, *1–3* β-FeOOH. T=295 K.

results of spectra fitting are given in Table I. Mössbauer spectrum of metal consisted of two sextets, the parameters of which were related to α-Fe(Ni,Co) (kamacite)—component 1, and to α_2-Fe(Ni,Co)—component 2 [1, 2]. Comparison of the Mössbauer parameters of Dronino metal with those of some other iron meteorites showed several differences. For instance, H_{eff} for the α-Fe(Ni,Co) phase in Dronino was higher than those in meteorites Chinga, Sikhote-Alin and Bilibino. This fact was discussed in [1, 2] and was related to the presence of ~0.7 at.% of Co in Dronino kamacite leading to an increase in H_{eff}. It was interesting to see that relative areas of metal phases in Mössbauer spectrum correlated with phase area fractions determined by the quantitative image analysis. Mössbauer spectrum of Dronino sulfides was fitted using four components which are two sextets and two doublets. Minor sextet 1 had Mössbauer parameters which can be related to the Fe(Ni,Co) phase with higher value of H_{eff} than that in kamacite. It is possible that Ni and Co concentrations in this phase were higher than those in usual kamacite. The largest part of iron (component 2) was in the form of FeS. The Mössbauer parameters of

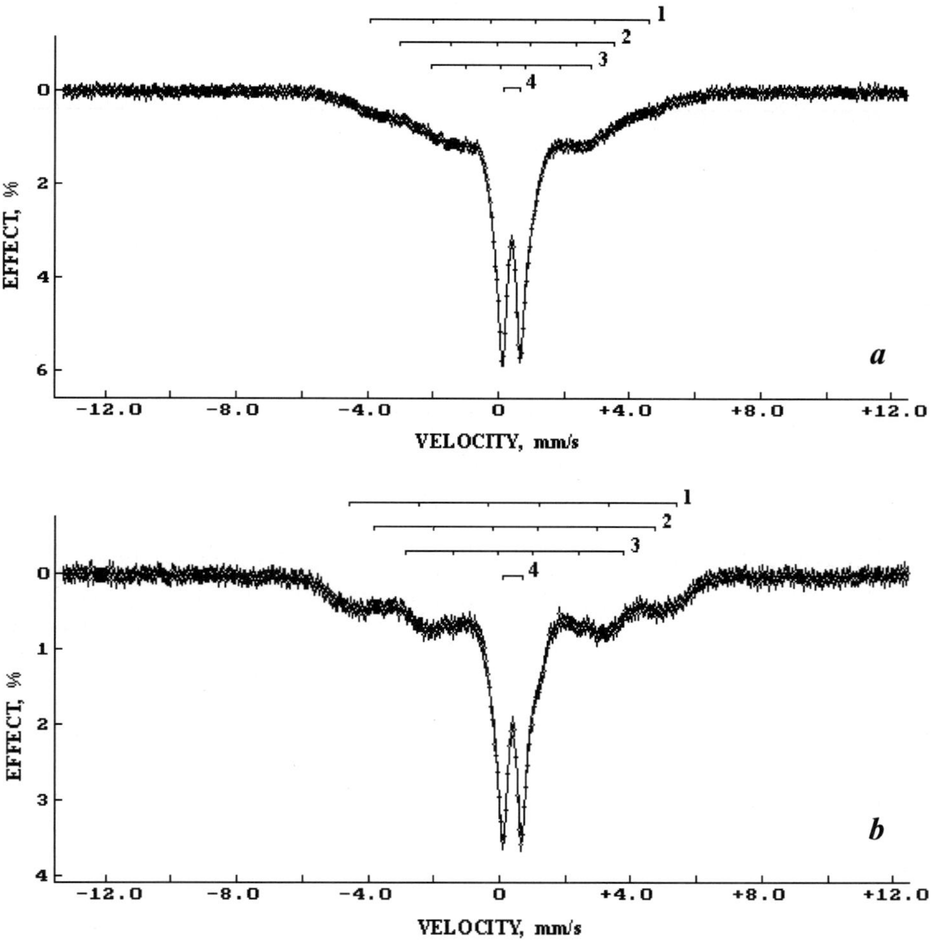

Figure 4 Mössbauer spectra of concretion: **a** friable internal layers of concretion, *1–4* β-FeOOH; **b** solid component of concretion, *1–4* β-FeOOH. T=295 K.

this component were the same as those for troilite found in ordinary chondrites [4]. Component 3 had Mössbauer parameters which characterized the high spin ferrous compound while component 4 had Mössbauer parameters which characterized the high spin ferric compound with very small quadrupole splitting. Comparison with the Mössbauer parameters of well known ferric compounds (see [5]) allowed us to suppose that this compound may be related to Co containing pentlandite. Unfortunately, it was not possible to determine the appropriate compound corresponding to component 4. Mössbauer spectra of corrosion products were different. Spectrum of internal zone of oxidation consisted of five components. Components 1 and 2 had values of H_{eff} correspondent to tetrahedral and octahedral sites in γ-Fe_3O_4 whereas the intensities of spectral lines had inverse ratio compared to the Mössbauer spectrum of natural γ-Fe_3O_4 [6]. However, it was recently shown [7] that 0.4 wt% Ni doping of magnetite led to a decrease in Fe occupation of octahedral sites and Mössbauer spectrum had ratio of line intensities similar to our results. Therefore, components 1 and 2 were related to γ-Fe_3O_4 doped with Ni

Table I Mössbauer parameters of meteorite Dronino and products of its corrosion (T=295 K)

Sample	δ (mm/s)	ΔE_Q (mm/s)	H_{eff} (kOe)	S (%)	Compound[a]
Metal phase	0.021±0.038	−0.078±0.038	339.6±1.2	80	α-Fe(Ni,Co) (1)
	0.066±0.038	−0.218±0.038	337.3±1.2	20	α_2-Fe(Ni,Co) (2)
Sulfides	0.114±0.050	0.131±0.050	348.6±1.6	6	Fe(Ni,Co) (1)
	0.773±0.050	−0.162±0.050	309.0±1.6	50	FeS (2)
	1.135±0.050	2.037±0.050	–	15	High spin Fe^{+2} (3)
	0.435±0.050	0.229±0.050	–	29	High spin Fe^{+3} (4)
Internal zone of oxidation	0.321±0.050	−0.004±0.050	489.6±1.6	39.5	γ-Fe_3O_4 (1)
	0.742±0.050	−0.417±0.050	454.7±1.6	10.5	γ-Fe_3O_4 (2)
	0.426±0.050	−0.021±0.050	324.6±1.6	12	β-FeOOH (3)
	0.394±0.050	−0.020±0.050	254.8±1.6	11	β-FeOOH (4)
	0.398±0.050	0.584±0.050	–	27	β-FeOOH (5)
External zone of oxidation	0.382±0.050	−0.021±0.050	278.7±1.6	28	β-FeOOH (1)
	0.410±0.050	−0.021±0.050	197.0±1.6	28	β-FeOOH (2)
	0.392±0.050	0.549±0.050	–	44	β-FeOOH (3)
Friable internal layers of concretion	0.378±0.050	−0.113±0.050	264.0±1.6	27	β-FeOOH (1)
	0.350±0.050	−0.178±0.050	202.8±1.6	29	β-FeOOH (2)
	0.390±0.050	−0.033±0.050	152.2±1.6	19	β-FeOOH (3)
	0.392±0.050	0.529±0.050	–	25	β-FeOOH (4)
Solid component of concretion	0.400±0.050	−0.033±0.050	309.0±1.6	21	β-FeOOH (1)
	0.451±0.050	−0.033±0.050	265.7±1.6	26	β-FeOOH (2)
	0.459±0.050	−0.033±0.050	204.9±1.6	26	β-FeOOH (3)
	0.385±0.050	0.577±0.050	–	27	β-FeOOH (4)

[a] Numbers in parentheses correspond to components in Figures 2–4.

(α-Fe(Ni,Co) in Dronino contained ~7.0 wt% of Ni). Other components (3–5) were related to ferric hydrous oxides in magnetic and paramagnetic states possibly depending on various sizes of microparticles (see [6, 8]). It is well known that Néel temperatures for α-FeOOH is about 400 K, for β-FeOOH about 295 K, for δ-FeOOH about 440 K and for ε-FeOOH about 570 K [6]. Therefore, we can suppose that β-FeOOH is the most appropriate compound with various particle sizes that was in the superparamagnetic state at room temperature, i.e. the temperature of the beginning of the magnetic phase transition. Spectrum of external zone of oxidation consisted of five components too; however, its parameters were related to β-FeOOH in magnetic and paramagnetic states only on the basis of the above-mentioned reasons. The difference between the values of H_{eff} may be related to smaller particle sizes from external zone of oxidation. The presence of γ-Fe_3O_4 and β-FeOOH in magnetic and paramagnetic states were also observed in corrosion products of meteorites Chinga, Sikhote-Alin and Bilibino [2]. Mössbauer spectra of concretion consist of four components, the parameters of which were related to β-FeOOH in magnetic and paramagnetic states. Higher values of H_{eff} for solid components of concretion in comparison with those of friable internal layers of concretion may be a result of larger particles in solid component.

X-ray diffraction study of the metal part powder of meteorite Dronino showed the presence of broadened lines of the bcc α-Fe(Ni,Co) phase. On the other hand, there were no lines of the fcc taenite phase. We supposed that broadening of the bcc α-Fe(Ni,Co) phase lines may be a result of the presence of the bcc α_2-Fe(Ni,Co) phase that correlates with Mössbauer data and metallography. In this case the α_2-Fe(Ni,Co) phase contained high Ni concentration and this phase may be considered as martensite. An analysis of the X-ray diffraction pattern of sulfide powder showed that the major part of the powder contained troilite with small amounts of pentlandite $(Fe, Ni)_9S_8$, pyrite $(Fe, Ni)S_2$, chromite $FeCr_2O_4$, kamacite α-Fe(Ni,Co) and taenite γ-Fe(Ni). However, these data were not fully in line with the Mössbauer results possibly due to a lower content of some compounds observed by X-ray diffraction and difficulties in the minor spectral components resolution. X-ray diffraction patterns of both products of surface corrosion showed that ferric hydrous oxide was the main iron containing component in the samples while quantity of γ-Fe_3O_4 was larger in sample of internal zone of oxidation. X-ray diffraction patterns of concretion samples showed that ferric hydrous oxide was the main iron containing component also.

4 Conclusion

Mössbauer study of meteorite Dronino and products of its corrosion showed that metal consisted of two Fe(Ni,Co) phases, sulfide inclusions consisted of troilite and possibly pentlandite, internal zone of oxidation contained magnetite and akaganeite whereas external zone contained akaganeite only. Study of concretions formed as a result of ferric hydrous oxides re-sedimentation due to ground water transport of its particles or due to diffusion of its molecules from meteorite Dronino oxidation external zone demonstrated the presence of akaganeite particles with different sizes in solid component and in friable internal layers of concretion.

Acknowledgment This work was supported in part by the grant of the Russian Foundation for Basic Research # 06-08-00705-a.

References

1. Oshtrakh, M.I., Milder, O.B., Grokhovsky, V.I., Semionkin, V.A.: Hyperfine Interact. **158**, 365 (2004)
2. Grokhovsky, V.I., Oshtrakh, M.I., Milder, O.B., Semionkin, V.A.: Izv. Russ. Acad. Sci., Ser. Phys. **69**, 1522 (2005), Bull. Russ. Acad. Sci., Phys. **69**, 1710 (2005)
3. Irkaev, S.M., Kupriyanov, V.V., Semionkin, V.A., Sokolov, M.M.: British Patent 10745, 7 May, 1987
4. Forder, S.D., Bland, P.A., Galazka-Friedman, J., Urbanski, M., Gontarz, Z., Milczarek, M., Bakun-Czubarow, N.: Hyperfine Interact. (C) **5**, 405 (2002)
5. McCammon, C.: In: Mineral Physics and Crystallography, p. 332. American Geophysical Union, New York (1995)
6. Murad, E., Johnston, J.H.: In: Long, G.J. (ed.) Mössbauer Spectroscopy Applied to Inorganic Chemistry, vol. 2, p. 507. Plenum Publishing Corporation, New York (1987)
7. Mohapatra, M., Pandey, B., Upadhyay, C., Anand, S., Das, R.P., Verma, H.C.: J. Magn. Magn. Mater. **295**, 44 (2005)
8. Bowen, L.H., Weed, S.B.: In: Herber, R.H. (ed.) Chemical Mössbauer Spectroscopy, p. 217. Plenum Publishing Corporation, New York (1984)

Hyperfine Interact (2005) 166:679–685
DOI 10.1007/s10751-006-9340-3

Mössbauer effect in natural strunzite, ferristrunzite and ferrostrunzite

Antoine Van Alboom · Eddy De Grave

Published online: 16 November 2006

Abstract The temperature dependence of the ferric and ferrous hyperfine fields in natural samples of strunzite, ferristrunzite and ferrostrunzite is determined by Mössbauer spectroscopy between 4.2 K and their magnetic transition temperatures (T_N), i.e. 50.5±0.5 K, 43.0±0.5 K and 44.0±0.5 K respectively, which are determined by Mössbauer thermoscanning. Two dominating magnetically split ferric subspectra were consistently present in all of the samples and are related to the Fe(1) and Fe(2) sites in the crystallographic structure, but an unambiguously assignment to a specific site is not possible. The difference between the corresponding hyperfine fields is very small. In the strunzite sample these fields are well defined and rather weakly dependent of temperature. In the other samples the corresponding hyperfine fields are more distributed especially at higher temperatures (below T_N). The relative contribution in the spectra of the third magnetic ferric component differs strongly between the samples and is assigned to ferric ions at the Mn site. At the lowest temperatures applied, its hyperfine field exceeds all other field values, but it decreases rather rapidly with increasing temperature, in so far that the corresponding spectral lines make a crossover with the lines of the other ferric subspectra. The magnetically split spectra of ferrostrunzite consist additionally of a ferrous magnetic component, which could be successfully analysed by introducing two magnetically split ferrous subspectra, which strongly overlap with each other but also with the ferric components. At higher temperatures in the magnetic region

A. Van Alboom (✉)
Physics, Department of Industrial Sciences BME-CTL, Hogeschool Gent,
Schoonmeersstraat 52, 9000 Gent, Belgium
e-amil: toon.vanalboom@hogent.be

E. De Grave
NUMAT, Department of Subatomic and Radiation Physics, University of Gent,
Proeftuinstraat 86, 9000 Gent, Belgium

E. De Grave
e-mail: eddy.degrave@Ugent.be

all subspectra overlap more and in the case of ferri- and ferrostrunzite the ferric hyperfine fields were distributed over a wider range.

Key words strunzite · ferristrunzite · ferrostrunzite · Mössbauer · liquid helium · hyperfine field

1 Introduction

Strunzite, ferristrunzite and ferrostrunzite are three isostructural minerals with triclinic crystallographic structure (P1) which is already described by several authors [1–3]. The ideal chemical formulas are respectively $Mn^{2+}Fe^{3+}{}_2(PO4)_2(OH)_2{\cdot}6(H_2O)$, $Fe^{3+}Fe^{3+}{}_2(PO4)_2(OH)_2{\cdot}[(H_2O)_5(OH)]$ [4] and $Fe^{2+}Fe^{3+}{}_2(PO4)_2(OH)_2{\cdot}6(H_2O)$ [5]. In strunzite the iron is distributed over two equivalent sites Fe(1) and Fe(2), of which Fe(1) is somewhat more distorted than Fe(2). In ferristrunzite the manganese is substituted by Fe^{3+}, in ferrostrunzite by Fe^{2+}. To this position is further referred as to the Mn site.

2 Experimental

Mössbauer spectra (MS) were collected with standard equipments (see [4–6] for details). All center-shift (δ) values quoted hereafter are relative to α-Fe at room temperature (RT). The numerical analyses were always carried out assuming Lorenztian line shaped absorption lines. Three programs were used for analysing the MS. Two simple fit programs were used when spectra were composed of some simple magnetic split subspectra. With the first program (SFP1) simple (sub)spectra were reproduced using the first order approximation of the nuclear interaction hamiltonian (NIH) and besides doublet spectra it is applicable on ferric magnetic split spectra (sextets) for which the linepositions are given by ordinary linear combinations of δ, the quadrupole shift ($2\varepsilon_Q$) and the hyperfine field (H_{hf}). It has the advantage that line area ratios can be determined more or less independently of each other following the ratio 3:x:y or coupled following the ratio 3:2:1 (ideal powder sample) from outer to middle to inner lines. With the second program (SFP2) the position of the absorption lines of a simple magnetic component is more generally determined by the diagonalization of the NIH and depends on δ, the quadrupole splitting (ΔE_Q), H_{hf}, the asymmetry parameter $\left(\eta = \frac{V_{xx}-V_{yy}}{V_{zz}}\right)$, the polar ($\Omega$) and the azimuthal angle (Ψ) of H_{hf} with respect to the principal axis (V_{zz}) of the electric field gradient (EFG) tensor. For ferric magnetically split subspectra, the position of the absorption lines is usually given in terms of $2\varepsilon_Q$ instead of ΔE_Q, both related by $2\varepsilon_Q = \frac{1}{2}\Delta E_Q(3\cos^2\Omega - 1)$. Additionally to the given parameters, the line width (Γ: full width at half maximum height) and the relative area (RA) for each subspectrum can be determined by fitting the experimental MS.

A distribution fit program (DFP) was used if subspectra themselves were the superposition of a number of simple magnetic split subcomponents related to a distribution of hyperfine fields. In this case one width parameter was used in order to calculate all the absorption lines of the same subspectrum. Only model independent distributions of hyperfine fields [7] were used.

The magnetic order–disorder transition temperatures (T_N) of the samples were determined by Mössbauer thermoscanning. Therefore the temperature variation of the γ-ray transmission through the absorber was recorded as a function of temperature during a fixed time interval with the source at zero velocity [8].

The samples used in this study were described formerly in [4–6], where their composition can also be found.

3 Results and discussion

The magnetic order–disorder temperatures of the three samples were determined as 50.5±0.5 K, 43.0±0.5 K and 44.0±0.5 K for respectively strunzite, ferristrunzite and ferrostrunzite by Mössbauer thermoscanning.

At 4.2 K the MS of all samples were successfully reproduced by a superposition of a small number of simple magnetic split subspectra using a simple fit program. At higher temperatures (T) below T_N considerable asymmetric broadening of the absorption lines was observed, which increased with increasing T. DFP was used for fitting these MS.

The dominant sextet absorption present in all samples at 4.2 K, consists of two sextet contributions S1 and S2, with slightly different hyperfine fields and are assigned to Fe^{3+} at the Fe(1 or 2) and Fe(2 or 1) sites in the structure, but further unambiguously assignment cannot be done. The additional ferric magnetic split component (S3) with a remarkably higher hyperfine field, which is resolved in the MS at 4.2 K of ferristrunzite [4] and of strunzite [5] and at 5 K in ferrostrunzite, is assigned to Fe^{3+} at the Mn position. SFP1 is used for reproducing the MS at 4.2 K for ferristrunzite and strunzite. In strunzite a doublet component S4 at small velocities, of which the RA is approximately 2%, was additionally resolved. In the MS of ferrostrunzite at 5 and at 10 K two ferrous magnetic split components (S5 and S6) were resolved besides S1 and S2 and SFP2 was used in these cases. The ΔE_Q values of the two ferrous magnetic components (S5 and S6) were fixed on the values at 50 K just above the magnetic transition temperature, 3.04 and 2.52 mm/s respectively [6]. The RA of both ferrous components were also fixed on their proportion determined from the MS at 50 K.

The results, given at all other temperatures, are obtained by DFP.

The T-dependence of the hyperfine fields of the ferric magnetically split components in ferristrunzite and strunzite is remarkably different. While $H_{hf,S1}$ and $H_{hf,S2}$ show nearly the same T-dependence, $H_{hf,S3}$ decreases more rapidly than $H_{hf,S1}$ and $H_{hf,S2}$ with increasing T. As a consequence $H_{hf,S3}$ becomes smaller than that of the other ones when temperature increases above about 15 K. Between 8 K and 15 K S3 overlaps strongly with S1 and S2, with the consequence that the parameters for S3 cannot be determined adequately for these T. Above 15 K, S3 was sufficiently resolved again but considerable and increasing asymmetric broadening of the absorption lines of one or more ferric subspectra was observed with increasing T. Because of the always nearly total overlap of S1 and S2, the RA results for S1 and S2 show some, but plausible fluctuation, except at 4.2 K in ferristrunzite where both subspectra are sufficiently separated. Also above 15 K, extra absorption was determined in the MS at small velocities, of which the RA increases with increasing T. It is found that it is best taken into account by an

Table I Results of the Mössbauer parameters for ferristrunzite (FISTR), strunzite (STR) and ferrostrunzite (FOSTR) below their respective T_N

Sample	T		χ^2		H_{hfmin}	H_{hfmax}	#div	δ	ΔE_Q	$2\varepsilon_Q$	η	Ω	Γ	$\Delta\Gamma$	H_{max}	RA
FISTR	4.2	DFP	938	S1	430	520	18	0.508		−0.624			0.250		487	0.330
				S2	430	520	18	0.510		0.096			0.250		481	0.376
				S3	520	570	10	0.502		−0.227			0.250		548	0.293
	21	DFP	687	S1	300	450	30	0.508		−0.469			0.330		403	0.217
				S2	300	450	30	0.508		0.140			0.330		407	0.229
				S3	100	350	25	0.498		0.177			0.435		261	0.377
				S7	0	30	6	0.508		0.755			0.542		1	0.177
STR	4.2	DFP	773	S1	440	520	16	0.520		−0.420			0.308		486	0.424
				S2	440	520	16	0.520		0.100			0.308		484	0.426
				S3	520	570	10	0.508		0.160			0.376		542	0.128
				S4	0	0	0	0.313	0.270				0.599		0	0.023
	20	DFP	718	S1	360	470	22	0.508		−0.110			0.277		448	0.426
				S2	360	470	22	0.508		0.180			0.277		448	0.423
				S3	50	300	25	0.508		0.440			0.280		188	0.133
				S4	0	0	0	0.313	0.270				0.600		0	0.017
FOSTR	5	SFP2	845	S1	–	–	–	0.510		−0.292			0.314	15	482	0.360
				S2	–	–	–	0.510		0.095			0.314	15	479	0.360
				S3	–	–	–	0.518		0.098			0.254	0	552	0.025
				S5	–	–	–	1.366	3.04		0.08	74	0.983	0	216	0.109
				S6	–	–	–	1.366	2.52		0.31	81	0.983	0	285	0.146
	22	DFP	544	S1	350	455	23	0.516		−0.175			0.39		446	0.333
				S2	350	455	23	0.516		0.06			0.39		444	0.332
				S5				1.372	3.04		0.20	79	0.634		89	0.103
				S6				1.372	2.52		0.16	89	0.634		92	0.163
				S7	0	100	20	0.516	0.755				0.355		75	0.069

δ, ΔE_Q, $2\varepsilon_Q$ and Γ are given in mm/s, $\Delta\Gamma$ (field dependent broadening), H_{hfmin} (lower limit of the H_{hf} distribution range), H_{hfmax} (higher limit of the H_{hf} distribution range), H_{max} (H_{hf} with maximum probability) in kOe, Ω in °, T in K. #div is the number of divisions of the H_{hf} range. Typical errors are: 0.5 K on T, 0.005 mm/s on δ, 0.005 mm/s on ΔE_Q and $2\varepsilon_Q$, 0.10 on η, 0.005 mm/s on Γ, 5° on Ω, 1 kOe on $\Delta\Gamma$, 5 kOe on H_{max}.

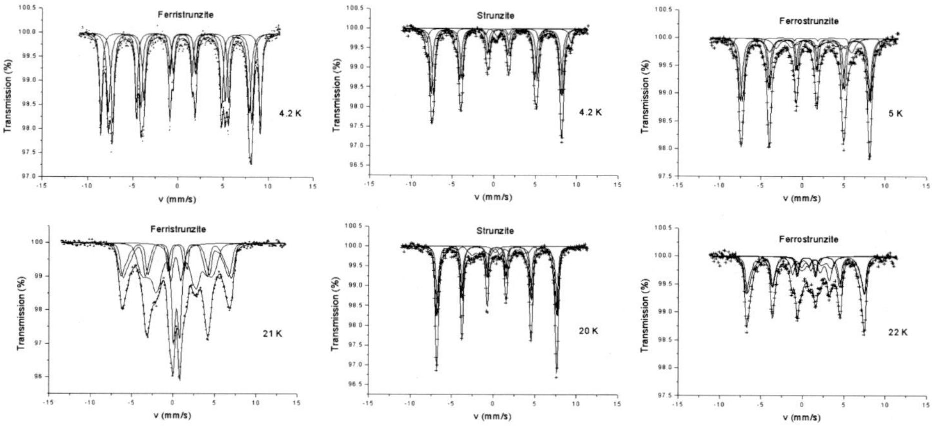

Figure 1 Experimental and fitted MS for ferristrunzite, strunzite and ferrostrunzite at selected temperatures. The parameters are shown in Table I.

additional hyperfine field distributed component (S7) with a small range at low values, or a doublet. Above 15 K the RA of S1 and S2 decreases substantially especially in ferristrunzite. S7 has to be considered as composed of the collapsing S3 at the one side, and collapsed parts of S1 and S2 at the other side. In ferrostrunzite S3 can only resolved at 5 K.

The quadrupole splitting of component S7 in ferristrunzite and ferrostrunzite was determined as ≈0.75 mm/s, which is in good agreement with the value of the ferric quadrupole doublet absorption at maximum probability above T_N [4, 6]. In ferristrunzite $2\varepsilon_{Q,S1}$ and $2\varepsilon_{Q,S2}$ have always a different sign, and vary between −0.63 and −0.47 mm/s for S1 and 0.098 and 0.14 mm/s for S2. Most probably these fluctuations are a consequence of the overlap of S1 and S2, by which it is not possible to determine these values more precisely. That the RA of S3 above 15 K is larger than the value at 4.2 K, is also most probably an artifact of the fit. Above 15 K, the overlap of S1, S2 and S3 is in such a way that S1, S2 and S3 are not sufficiently resolved and apparently RA_{S3} is overestimated by the fitting routine, whereas RA_{S1} and RA_{S2} are somewhat underestimated. Remarkably is the change in sign of $2\varepsilon_{Q,S3}$ below 8 K with respect to T higher than 15 K.

Because the sign of $2\varepsilon_Q$ is related to the orientation of the local hyperfine field to the principal axis of the EFG, the difference in sign of $2\varepsilon_{Q,S1}$ and $2\varepsilon_{Q,S2}$ indicates a difference in the orientation at Fe(1) and Fe(2) sites. For the same reason the difference in sign of $2\varepsilon_{Q,S3}$ below 8 K and above 15 K should be interpreted as a change in the direction of the local hyperfine field with respect to the principal axis of the EFG.

The effect of increasing RA of S7 and decreasing RA for the other components with increasing T above 15 K, is interpreted in the context of relaxation of the magnetic moments, by which the magnetic moments stay shorter and shorter in their equilibrium positions when T increases. When the relaxation time is sufficiently shorter than the nuclear Larmor precession time (of the order 10^{-7} s), then the absorption is of a doublet type. In the opposite case when the magnetic moments stay long enough in their equilibrium state, one observes a magnetically split spectrum. In the intermediate case, the absorption manifests oneself as the

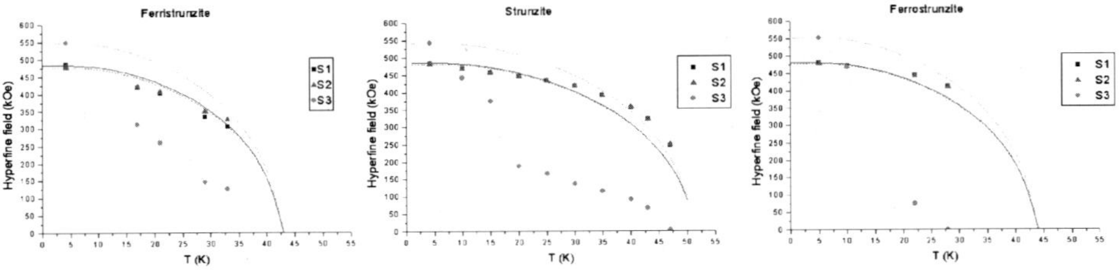

Figure 2 T-dependence of the hyperfine fields in ferristrunzite, strunzite and ferrostrunzite. The lines are calculated Brillouin curves (*full line*, S1; *dashed line*, S2; *dotted line*, S3) for following parameters: 2.5 for spinvalues, the fields at lowest temperatures for saturation field values and the T_N values as determined from thermoscanning.

superposition of a continuum of hyperfine fields between 0 kOe (doublet situation) and a certain maximum value.

So the S7 component has to be associated to Fe^{3+} for which the relaxation time is already short enough to get a more doubletlike absorption. That the relaxation phenomenon is spread over a relatively wide temperature range means that the magnetic exchange interaction with neighbouring magnetic ions, which is responsible for the orientation of the magnetic moments, is not just as strong at each individual site, and is most weakly for Fe^{3+} at Mn positions.

For comparison with the results in [4] and [5], obtained with SFP1, the MS at 4.2 K of ferristrunzite and strunzite are also analysed with DFP. The fitting results for the respective samples at some selected T are given in Table I with an indication of the used program. Figure 1 contains the corresponding experimental and fitted MS.

In order to interpret the experimental H_{hf} data and because H_{hf} is in good approximation proportional to the sublattice magnetisation related to the different ferric sites, Brillouin curves were calculated, which describe the T-dependence of the magnetisation within the molecular field approximation [9]. These curves are determined by the mean spin value of the probe-ions (2.5 for ferric ions), the saturation value of the magnetisation at low temperatures (here: the hyperfine field at lowest T) and the corresponding magnetic transition temperature (here: the T_N value as determined from thermoscanning). Figure 2 shows the experimental T-dependences of H_{hf} and the calculated Brillouin curves for the respective sites and samples. Its clear that the molecular field model gives a raw description of the T-dependence of $H_{hf,S1}$ and $H_{hf,S2}$, but it is totally deficient for $H_{hf,S3}$. The $H_{hf,S3}$ T-dependence is remarkable. In strunzite $H_{hf,S3}$ (with maximum probability) decreases strongly when T increases from 0 to about 18 K, after which it decreases less fast when T further increases up to 44 K. A less pronounced analog tendence is determined for $H_{hf,S3}$ in ferristrunzite. In ferrostrunzite it is clearly demonstrated that different sublattices have different magnetic transition temperatures and that the T_N values as determined from thermoscanning give an overall value for the whole sample and not the value corresponding to all sublattices.

References

1. Frondel, C.: N. Jb. Miner. Mh., 222 (1957)
2. Fanfani, L., Tomassini, M., Zanazzi, P., Zanzari, A.R.: Tschermak's Mineral. Petrogr. Mitt. **25**, 77 (1978)

3. Peacor, D.R., Dunn, P.J., Simmons, W.B.: N. Jb. Miner. Mh., 524 (1983)
4. Vochten, R., De Grave, E.: N. Jb. Miner. Mh., 176 (1990)
5. Vochten, R., De Grave, E., Van Springel, K., van Haverbeke, L.: N. Jb. Miner. Mh., 11 (1995)
6. Van Tassel, R., De Grave, E.: N. Jb. Miner. Mh., 207 (1992)
7. Vandenberghe, R.E., De Grave, E., de Bakker, P.M.A.: Hyperfine Interact. **83**, 29 (1994)
8. Chambaere, D., De Grave, E.: J. Magn. Magn. Mater. **42**, 263 (1984)
9. Morrish, A.H.: The Physical Principles of Magnetism. Wiley, New York (1965)

Hyperfine Interact (2005) 166:687–691
DOI 10.1007/s10751-006-9342-1

Mineral associations in coal and their transformation during gasification

F. B. Waanders · A. Govender

Published online: 26 October 2006

Abstract The principle aim of this investigation was to determine the effect that minerals and mineral associations in dense medium coal fractions have on the ash fusion temperature (AFT) of coal, where the mineral matter associated with coal undergoes transformations during gasification. Samples from three coal sources used by Sasol for gasification were acquired and a comprehensive characterisation on all dense medium fractions was conducted, including proximate analyses, AFT, ash oxide analyses, XRD and Mössbauer spectroscopy. From the proximate analyses, the ash content was the highest for the higher density fractions, with an accompanying decrease in fixed carbon content, consistent with the XRD and Mössbauer analyses. From the ash oxide analyses, it was evident that at higher relative densities more Fe_2O_3 and SiO_2 were present than in the lower density samples. From XRD analyses, the low density fractions contained calcite and dolomite. Pyrite and quartz were found in the higher density fractions whereas kaolinite occurred in all density separated fractions. From the different techniques it follows that with the Fe-content high in high density fractions, the AFT was low. When Ca and other basic oxide levels were abundant in low density fractions, the AFT was low and when the SiO_2/Al_2O_3 ratio was high in high density fractions the AFT increased.

Key words ash fusion temperature · minerals · coal · Mössbauer spectroscopy · proximate analyses

F. B. Waanders (✉)
School of Chemical and Minerals Engineering, North West-University, Potchefstroom 2520, South Africa
e-mail: chifbw@puk.ac.za

A. Govender
Sasol Technology (Pty) Ltd, Box 1, Sasolburg, South Africa

1 Introduction

Coal is fundamentally composed of the fossilised remains of plant debris which have undergone progressive physical and chemical alteration through geological time and is thus not a homogeneous material, as it is composed of a number of microscopic organic and inorganic constituents, which occur together in various proportions or associations [1]. The term “mineral matter” usually applies to all the inorganic non-coal material occurring in coal [2]. The most abundant mineral groups found in South African coals are clays, carbonates, sulphides and quartz with pyrite the most abundant Fe-mineral present [1]. The behaviour of these minerals during gasification is of interest to the operation of the gasifier, the selection of coals and the utilisation of the gasification ash. Heat induced transformations due to a low ash fusion temperature (AFT) leads to agglomeration of the ash particles with agglomerate sizes varying from tiny particles to lumps larger than 100 mm.

The principle aim of this investigation was to determine the effect that minerals and mineral associations have on the AFT of coal. The AFT analyses determine the average temperature at which the bulk mineral composition starts to become soft or obtain liquid properties, as well as to what extent agglomeration can occur in the gasifier where the agglomeration inside the gasifier can cause various undesirable operational conditions [3]. In a gasifier the coal enters the gasifier through the top while reactant gases (steam and oxygen) are introduced at the bottom. The reactant gases flow upwards through the spaces between the coal lumps. As the coal descends it is first dried and devolatilised by the heat of the rising gas. The devolatilised coal then enters a gasification zone and is finally burnt to ash which is removed at the bottom of the gasifier.

The AFT of a coal sample can be used, amongst others, as a measure of the expected behaviour of the ash bed during gasification, which in turn can be used in the gasification process to optimally operate the gasifier.

2 Experimental

Representative run-of-mine coal samples from three sources used by Sasol for gasification were acquired. These samples were then density separated with relative densities ranging from 1.4 g cm^{-3} to 2.2 g cm^{-3}. A comprehensive characterisation on all dense medium fractions was conducted, including proximate analyses, AFT, ash oxide analyses, XRD and Mössbauer spectroscopy.

For the proximate analyses to determine ash, moisture and fixed carbon of the coal the SABS 924, ISO 1171 and ISO 562 standard procedures were followed, whilst for the ash oxide analyses the ASTM D3682 standard was followed. The XRD analyses were performed on a Siemens D500 X-ray goniometer equipped with a Cu-tube, variable slit and secondary graphite monochromator for mineral composition determination.

The Mössbauer spectra were obtained with the aid of a Halder Mössbauer spectrometer, capable of operating in conventional constant acceleration mode using either a proportional counter filled with Xe-gas to 2 atm or a backscatter-type gas-flow detector. The spectrometer was calibrated using α-Fe as the reference and a 50 mCi ^{57}Co source, plated into a Rh-foil, was used to produce the γ-rays. The

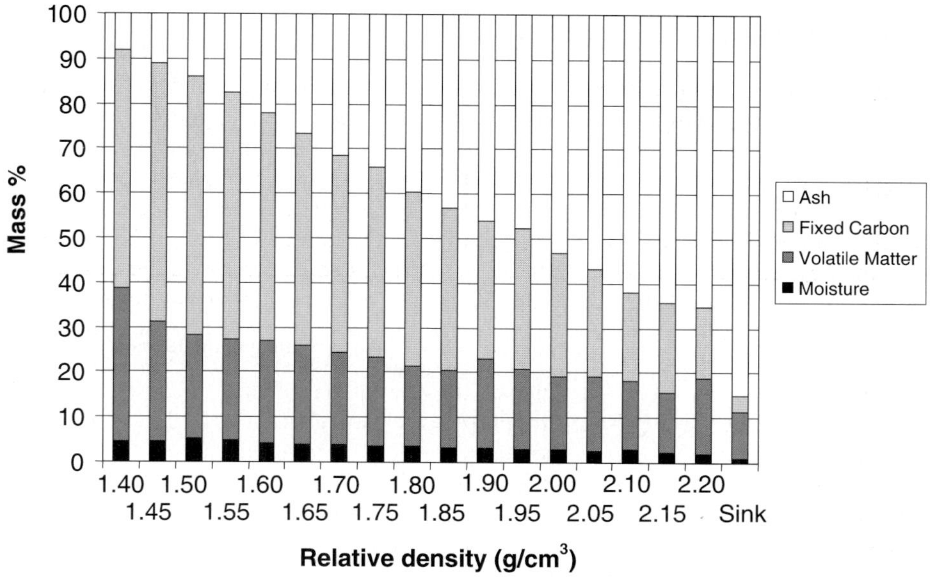

Figure 1 Proximate analyses of density separated coal sample 1.

samples were analysed at room temperature and data collected in a multi-channel analyser (MCA) to obtain a spectrum of count rate against source velocity. A least-squares fitting program was used and by superimposing Lorentzian line shapes, the isomer shift, quadrupole splitting and/or hyperfine magnetic field of each constituent was determined with reference to the centroid of the spectrum of a standard α-iron foil at room temperature. The amount of each constituent present in a sample was determined from the areas under the relevant peaks.

3 Results and discussion

The proximate analyses on the three coals showed a similar decrease in fixed carbon and increase in ash content when moving to higher densities, consistent with the XRD analyses and the results are shown in Figure 1.

From the ash oxide analyses the following conclusions can be drawn. The basic oxide distribution typically shows an interesting trend that explains why the AFT first increases and then decreases across the densities. The lower density samples show high concentrations of CaO and MgO and very low concentrations of Fe_2O_3. These concentrations then decrease as the densities get higher. Therefore, less fluxing agents are present, hence the AFT would increase. But from RD 1.90 onwards, more Fe_2O_3 is present together with small amounts of CaO and MgO and this would once again push down the AFT. Therefore, it is the presence and interaction of a combination of minerals that results in lower AFT values.

From the XRD analyses, the low density fractions contained CaO and MgO, whilst SiO_2 and the clay minerals predominantly reported in the higher density fractions with kaolinite found in all density separated fractions. The XRD analyses

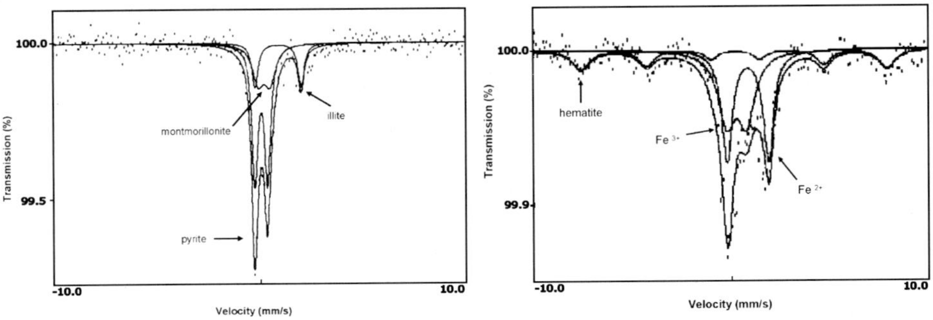

Figure 2 On the *left* the Mössbauer spectrum of a DMS sample containing pyrite, montmorillonite and illite, whilst on the *right* the spectrum of an ash sample.

Table I Room temperature Mössbauer parameters of various coal samples

Sample	Mineral	δ (mm s^{-1}) ±0.01	Δ (mm s^{-1}) ±0.01	*B* (T)	Relative area (%)
Coal sample 1	Pyrite	0.27	0.58	–	100
Coal sample 2	Pyrite	0.28	0.64	–	100
Coal sample 3	Pyrite	0.26	0.62	–	100
1.4 g cm^{-3} fraction	Pyrite	0.28	0.57	–	100
Sink fraction	Pyrite	0.25	0.60	–	55
	Illite	1.05	2.21	–	18
	Montmorillonite	0.38	0.57	–	27
Fly ash	Fe^{2+}	0.93	2.11	–	27
	Fe^{3+}	0.38	1.01	–	38
	Hematite	0.34	−0.08	49.8	35

δ=Isomer shift relative to α-iron, Δ=Quadrupole splitting, *B*=magnetic hyperfine field

also showed that the Fe predominantly reported in the higher density fractions. In the majority of samples, pyrite was the abundant Fe-containing mineral, whilst in the 2.2 g cm^{-3} sink fraction, clay (illite and/or montmorillonite) and quartz were also found.

Typical Mössbauer spectra for the sink fraction and for an ash sample are shown in Figure 2 and the Mössbauer parameters of the three run-of-mine coal samples, two different dense medium separated fractions and the ash are shown in Table I. The Mössbauer parameters are in accordance with those found in literature [4].

The AFT was determined to serve as a measure of the expected behaviour of the ash in the coal during gasification and a typical result for coal sample 1 is shown in Figure 3. It is evident that at the intermediate relative densities the AFT was the largest and furthermore the value is dependant on the mineral phase present if the data are compared with the other analytical techniques used in this investigation.

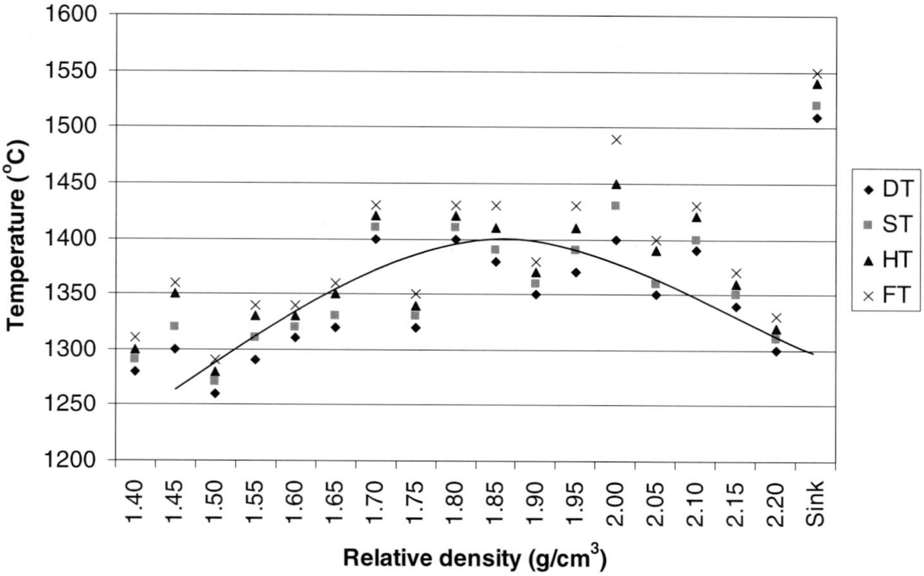

Figure 3 The AFT of the various DMS fractions of one of the coal samples with DT=initial deformation temperature, ST=softening temperature, HT=hemisphere temperature and FT=flow temperature.

4 Conclusions

From the different techniques, the following can be highlighted for the samples investigated.

(1) A high Fe-content in high density fractions it will reduces the AFT.
(2) The Ca and other basic oxide levels are high in low density fractions and reduce the AFT.
(3) The SiO_2/Al_2O_3 ratio is high in high density fractions and increase the AFT.

It is thus evident that the presence and interaction of a combination of minerals result in changes in the AFT values and further studies are conducted to determine which mineral combinations result in the highest possible AFT value.

Acknowledgements The authors wish to thank Sasol and the North West University for financial support.

References

1. Falcon, R.M.S., Snyman, C.P., In: An Introduction to Coal Petrography: Atlas of Petrographic Constituents in the Bituminous Coals of Southern Africa, p. 19. (1986)
2. Bryers, R.W., Taylor, T.E.: J. Eng. Power **98**, 528 (1976)
3. Seggiani, M.: Fuel **78**, 1121 (1999)
4. Stevens, J.G., Khasanov, A.M., Miller, J.W., Pollak, H., Li, Z. (eds.): Mössbauer Mineral Handbook, Mössbauer Effect Data Centre. University of North Carolina, Asheville, USA (1998)

Hyperfine Interact (2005) 166:693–698
DOI 10.1007/s10751-006-9343-0

Mössbauer study of two different aged rock types in the Vredefort structure, South Africa

F. B. Waanders · M. C. Brink · A. A. Bisschoff

Published online: 14 November 2006

Abstract The Vredefort Structure in South Africa was recently inscribed into the list of World Heritage sites as the oldest and largest recognised impact structure on earth. Due to upliftment of more than 30 km of the Archaean basement core and subsequent exposure of the deeply eroded central portion of the crater a unique opportunity exists to study rocks of the crust, especially those that have undergone recrystallization due to various thermal metamorphic events over time. Two rock types occurring in the central uplifted part of the impact structure were studied. One was a typical Archaean iron formation of sedimentary origin. The other rock studied, adjacent to this much older rock, is a homogeneous, medium grained recrystallized norite of immediately pre- or post-impact age, indicating a possible mafic igneous activity related to impact. Microscopy, XRD and Mössbauer analyses were performed on both samples.

Key words eulite · eulysite · norite · metamorphic alteration · Mössbauer spectroscopy

1 Introduction

In the 1930s, the inner circular portion of the Vredefort impact crater, South Africa, was included into a world wide list of structures that possibly owed their formation to meteoritic impacts. It is now widely accepted as the oldest (2,023±4 Ma) and largest (≈380 km diameter) recognized impact structure on earth (http://www.unb.ca.passe/ImpactDatabase) and during the World Heritage Committee meeting held during July 2005 in Durban, South Africa, [1], the Vredefort Structure was inscribed

F. B. Waanders (✉) · M. C. Brink · A. A. Bisschoff
School of Chemical and Minerals Engineering, North-west University, Potchefstroom, North West 2520, South Africa
e-mail: chifbw@puk.ac.za

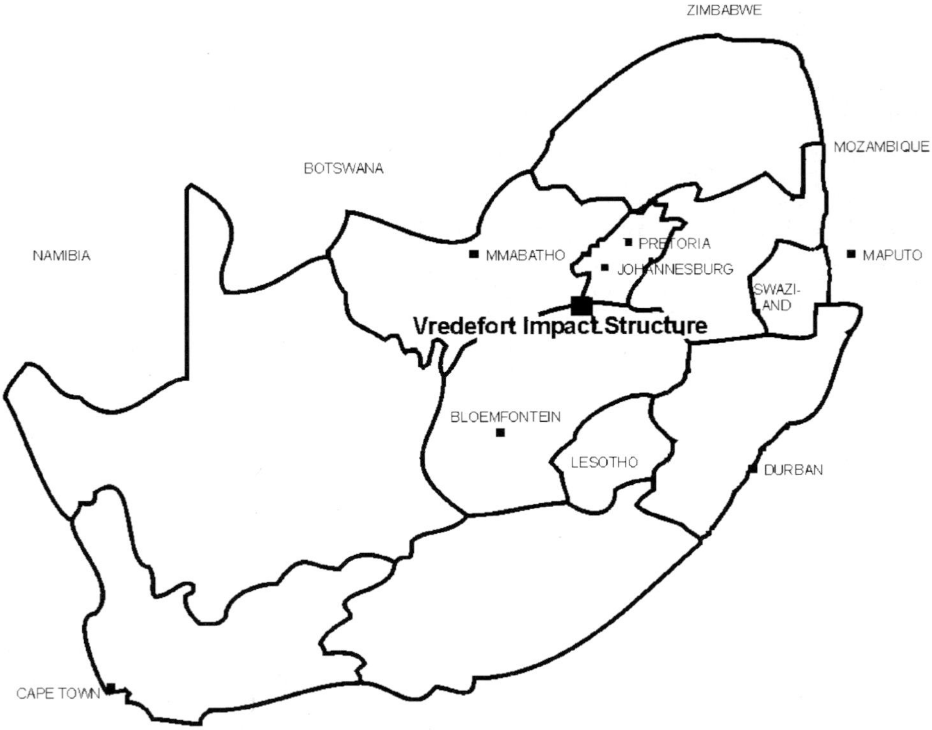

Figure 1 Locality map of the Vredefort Impact Structure found in South Africa.

into the list of natural heritage sites of the world. The locality of the Vredefort Structure is shown in Figure 1. The development of the structure in the first few seconds after impact can be deciphered from the deformation of the rock masses that were displaced. The thick layers of mainly sedimentary rock, some high in iron oxide content, were broken during impact. These contain recognisable marker beds, which can now be mapped, and a reconstruction can be made of the way large rock blocks moved relatively to each other during and immediate after impact.

Due to upliftment of more than 30 km of the Archaean basement core and subsequent exposure of the deeply eroded central portion of the crater a unique opportunity exists to study not only the sedimentary rocks from the inner circular portion of the Vredefort Structure, but also rocks of the Archaean basement, especially those that have undergone recrystallization due to various thermal metamorphic events over time.

2 Experimental

During field work in the central portion of the Structure two different aged rocks were found and samples taken for investigation. The one rock type studied, is a typical Archaean iron formation of sedimentary origin dated as being ≈3.5 Ga old [2]. Since the initial sampling of the Archaean iron formation, three more confined localities have been discovered, of which one, not close to the dyke mentioned

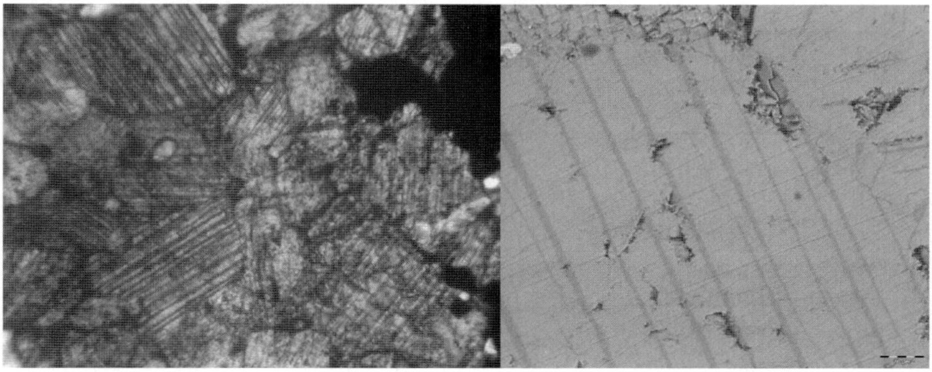

Figure 2 On the *left* an optical photo of the eulysite with clinoeulite as *grey areas* between the *white bars* of ferroaugite with the *black in the right top* being magnetite. On the *right* a SEM micrograph to show the intergrowth of ferroaugite (*dark grey thin lines*) with the *light grey thick lamellae* of clinoeulite.

below, was also analysed in the present investigation. In the fresh state the rocks are a very dark green to black colour with brown weathering crusts with unusually high density due to its high content of magnetite. The other rock, adjacent to the one locality of the much older rock, is a homogeneous, medium grained 0.5 m-wide dyke of recrystallized norite of immediately pre- or post-impact age and dated to be 2,019±2 Ma old [4], indicating a possible mafic igneous activity related to impact.

Optical and SEM microscopy and Mössbauer analyses were performed on both samples. A FEI Quanta 200 ESEM Scanning Electron Microscope fitted with an Oxford Inca 400 energy dispersive X-ray spectrometer (EDS) was used for the SEM analyses of the samples. The Mössbauer spectra were obtained with the aid of a Halder Mössbauer spectrometer, capable of operating in conventional constant acceleration mode using either a proportional counter filled with Xe-gas to 2 atm. or a backscatter-type gas-flow detector. A 50 mCi ^{57}Co source, plated into a Rh-foil, was used to produce the γ-rays. The samples were analysed at room temperature and data collected in a multi-channel analyser (MCA) to obtain a spectrum of count rate against source velocity. A least-squares fitting program was used and by superimposing Lorentzian line shapes, the isomer shifts, quadrupole splitting and/or hyperfine magnetic field of each constituent was determined with reference to the centroid of the spectrum of a standard α-iron foil at room temperature. The amount of each constituent present in a sample was determined from the areas under the relevant peaks.

3 Results and discussion

Petrographic observation under the microscope showed that the Archaean iron stone, or eulysite, consists of between 10 and 70% magnetite, depending on the locality. In addition, quartz (≈5%) and feldspar (≈5%) are present with the balance an intergrowth of clinoeulite and ferroaugite. Typical optical and SEM micrographs are shown in Figure 2.

Table I Mössbauer parameters of the different mineral species found in each sample

Sample	Mineral	δ ±0.02	Δ ±0.02	H (T)	Rel. area (%)
Eulysite 1	Magnetite				
	Sextet 1	0.21	−0.01	48.4	23
	Sextet 2	0.65	−0.01	45.3	46
	Eulite				
	Doublet 1	1.05	2.39	–	4
	Doublet 2	1.29	2.33	–	6
	Ferroaugite				
	Doublet 1	0.97	2.11	–	12
	Doublet 2	0.95	2.06	–	1
	Doublet 3	0.39	0.66	–	8
Eulysite 2	Hematite	0.32	−0.13	51.2	9
	Eulite				
	Doublet 1	1.01	2.83	–	34
	Doublet 2	1.30	2.21	–	23
	Ferroaugite				
	Doublet 1	1.07	2.27	–	18
	Doublet 2	0.95	2.06	–	4
	Doublet 3	0.39	0.66	–	12
Norite	Magnetite/hematite	0.28	0.00	49.6	33
	Hypersthene	1.11	2.09	–	56
	Ferroaugite	0.59	0.43	–	8
	Illmenite	0.97	0.73	–	3

δ=Isomer shift relative to α-iron and Δ=Quadrupole splitting, both in mm.s^{-1}, H=magnetic hyperfine field.

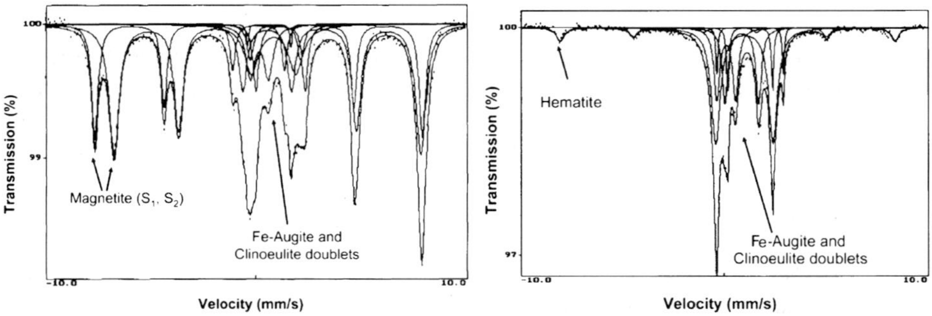

Figure 3 On the *left* a Mössbauer spectrum of the eulysite containing a large amount of magnetite, whilst on the *right* a spectrum of a sample containing no magnetite.

The rocks were altered to a ferropigeonite during metamorphic alteration as they were initially subjected to an extended period of static alteration, followed by hot metamorphism of the rocks immediately prior to the Vredefort event. Resulting, after cooling down to 850°C, the Fe-pigeonite exsoluded ferroaugite parallel to the (001) crystallographic orientation of the pigeonite before inversion to the orthopyroxene (orthoeulite) and finally forming an intergrowth of rare clinopyroxene, clinoeulite and ferroaugite together with magnetite, quartz, feldspar and other minor minerals to form the eulysite [3].

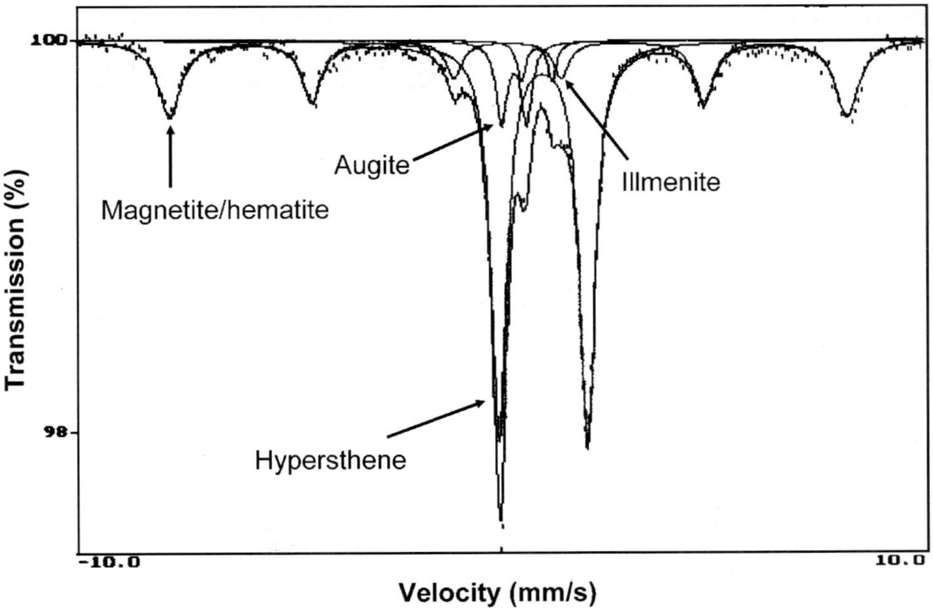

Figure 4 Mössbauer spectrum of the norite.

Mössbauer spectroscopy revealed the presence of the main constituents in the eulysite as being the intergrowth of ferroaugite and eulite with magnetite and the determined Mössbauer parameters correlating with those found in literature [5], shown in Table I (Figure 3).

The norite consists mainly of hypersthene, augite and feldspar with minor minerals such as magnetite/hematite, ilmenite, and zircons and the representative spectrum is shown in Figure 4 with the resulting Mössbauer parameters shown in Table I.

4 Conclusion

The small dykes and pod-like intrusions of Archaean age contain clinoeulite with exsolution lamellae of ferroaugite indicative that the parent phase of this intergrowth was ferropigeonite, requiring a minimum temperature in the order of ≈850°C. The norite magma, not related to the Archaean rocks of the area, intruded into the country rock that were at the elevated temperatures. After crystallization of the pigeonite and feldspar, at temperatures well above the 850°C, further cooling was inhibited and sufficient time was available for the pigeonite to exsolve augite and invert to orthopyroxene, indicative that the rocks of the impact centre were at abnormally high temperatures shortly before and during impact, in line with observations made elsewhere in the impact area.

Acknowledgements The authors wish to thank the North West University, Potchefstroom Campus, for financial support. The help of Dr L Tiedt of the Laboratory for Electron Microscopy is greatly appreciated.

References

1. IUCN Evaluation of Nominations of Natural and Mixed Properties to the World Heritage List, Document WHC-05/29.COM/INF.8B.2 prepared by IUCN – The world Conservation Union, p. 143 (May 2005)
2. Hart, R.J., Welke, H.J., Nicolaysen, L.O.: J. Geophys. Res. **86**, 10663–10680 (1981)
3. Schreyer, W.: J. Petrol. **24**(1), 26–47 (1983)
4. Moser, D.E.: Geology **25**, 7–10 (1997)
5. Stevens, J.G., Khasanov, A.M., Miller, J.W., Pollak, H., Li, Z. (eds.). In: Mössbauer Mineral Handbook. Mössbauer Effect Data Centre, University of North Carolina, Asheville, USA, p. 527 (1998)

Hyperfine Interact (2005) 166:699–703
DOI 10.1007/s10751-006-9344-z

Volume and surface oxidation of basalt

D. J. M. Burkhard · T. Scherer

Published online: 27 October 2006

Abstract Natural, Fe^{2+}-rich basalt glass (quenched lava) was heat treated as glass pieces and glass powder in air, in 6.0 Ar and in a 9×10^{-6} mbar vacuum below temperatures of significant crystallization to access volume and surface oxidation by ^{57}Fe Mössbauer spectroscopy. While no oxidation occurs upon heating in vacuum, the amount of Fe^{3+} formed in powder (surface oxidation) is about 10 times higher than in pieces (volume oxidation), and surface oxidation is of the same order in air and Ar. This effect is assigned to chemisorption of water or CO_2. Crystalline basalt, investigated by wet chemistry, includes five glass pieces treated above T of crystallization in air and in 6.0 Ar, and three lava samples of increasing depth up to 9 cm of lava lobes. The high Fe_2O_3 of all these crystalline samples is explained as a stabilization of Fe^{3+} due to the change of the local electronic environment in the course of crystallization; volume oxidation therefore appears to be independent on the environmental atmosphere.

Key words basalt · oxidation kinetics · surface oxidation · volume oxidation · ^{57}Fe Mössbauer spectroscopy

1 Introduction

Basalt glass/melt may experience at least three modifications upon heat-treatment/cooling, which include crystallization [1–3], cation diffusion to the free surface

D. J. M. Burkhard (✉)
Institute for Technical Chemistry (ITC-WGT), Research Center Karlsruhe, 76021 Karlsruhe, Germany
e-mail: Dorothee.Burkhard@itc-wgt.fzk.de

D. J. M. Burkhard
Institute for Mineralogy and Geochemistry, University of Karlsruhe, 76128 Karlsruhe, Germany

T. Scherer
Institute for Nanotechnology (INT), Research Center Karlsruhe, 76021 Karlsruhe, Germany

Table I Wet-chemical analyses of glass SL02, obtained by quenching, and products heated in Ar and air

Sample	SL02	SL-22	SL-11	SL-19	960/70.5	990/70.5	TT	TM	LT
Medium		Ar	Ar	Ar	Air	Air			
T (°C)		880	1000	1030	960	990			
Time (h)		48.2	93	48	70.5	70.5			
% FeO	9.92	6.28	5.78	3.96	4.01	1.92	4.32	1.58	2.12
% Fe_2O_3	1.53	5.58	6.14	8.16	8.10	10.42	7.76	10.8	10.2

TT, TM, and LT are samples from lava lobes (Figure 1)

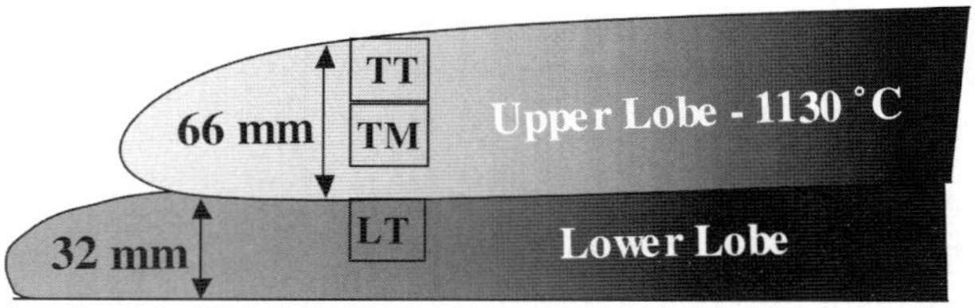

Figure 1 Upper and lower lava lobe with 1-cm block-samples, TT, TM, LT.

(Mg, Ca, or Fe, depending on temperature [1, 4, 5], and oxidation [1, 5, 6]. Although oxidation is a well-observed phenomenon, its mechanism is often not understood; for instance, iron-bearing silicate glasses such as basalt glass, may oxidize upon heat treatment, of relevance for industries that use these materials for heat insulation; oxidation is also observed with lava on earth or other planets. However, the process or mechanism of oxidation is a matter of discussion, e.g. [1, 4, 5]. To contribute to the comprehension of oxidation, we considered heat-treated natural glassy and crystalline basalt with Mössbauer spectroscopy and wet chemistry, respectively.

2 Sample description and experimental details

Basalt glass, obtained by quenching lava from Puu Oo-Kupaianaha on Kilauea, Hawaii, has a tholeiitic composition and is rich in Fe^{2+}, 93% of total Fe, (Table I). Further details and characterizations are in [1–3, 5–7].

Glass pieces (0.8 cm in diameter) and powder (50 μ size, on average) were heat-treated in air and in 6.0 Ar at various temperatures and periods of times, below temperatures of significant crystallization, i.e. in air <900°C [1, 3], in Ar <800°C [6], and at 760°C in 9×10^{-6} mbar in vacuum. Mössbauer spectroscopy applied ^{57}Co source embedded in a 6 μm Rh matrix, a triangular voltage, 1,024 channels, and a calibration to 25 μm α-Fe-foil. Spectra were fitted with two doublets for Fe^{2+} and one doublet for Fe^{3+} [1] using the program NORMOS. Glass pieces treated at T>920°C and basalt samples across lava lobes from Puu Oo (Figure 1, Table I) are crystalline and were investigated wet-chemically according to [8].

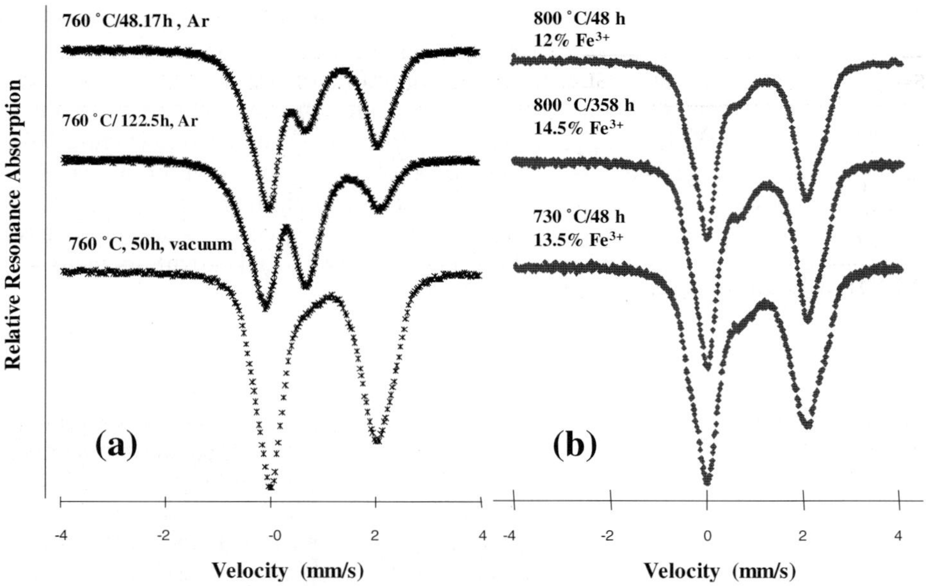

Figure 2 Some Mössbauer spectra of heat treated basalt glass; **a** treatment at 760°C shows an increase in Fe^{3+} with time, no oxidation is seen upon treatment in vacuum. **b** Heat treatment of pieces shows no systematic change of Fe^{3+} with time or temperature.

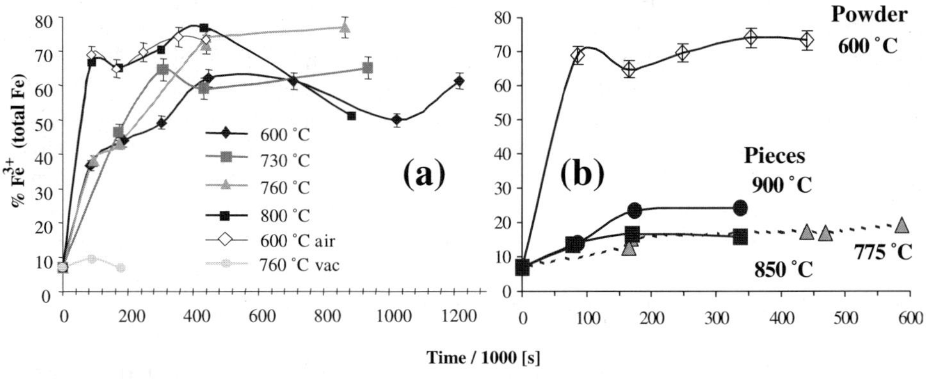

Figure 3 Oxidation rates **a** of powder in Ar, compared to vacuum and air, and **b** of pieces and powder in air, from [1].

3 Results and discussion

Temperature and time-dependent treatment of powder (examples are in Figure 2) and subsequent evaluation of the increase of Fe^{3+} show a saturation of Fe^{3+} that is independent of whether heating occurred in air or in Ar whereby the increase in Fe^{3+} occurs faster in air (Figure 3). No significant oxidation is observed upon treatment in vacuum (Figures 2a, 3). Glass pieces show no time or temperature dependence if treated in Ar, however some increase with time upon treatment in air (Figure 3b). This means that surface oxidation is hampered only in vacuum, and occurs otherwise independent of the atmosphere. Whereas volume oxidation is minor in air or insignificant in Ar, if treated below *T* of crystallization (Figure 2b), a significant increase in Fe^{3+} is derived if *T* of treatment is at or above *T* of crystallization (Table I). Volume oxidation increases with the degree of crystallization in air and Ar to a similar degree. Likewise, the degree of oxidation increases with depth from the free surface of lava lobes (TT,TM,LT), i.e., with an increasing degree of crystallization (Table I).

Surface oxidation Despite the oscillation of Fe^{3+}-increase in Ar, suggesting some slower kinetics compared to air, the comparable saturation in Ar and air suggests a comparable availability of oxygen. The small amount of residual oxygen in Ar (fO_2 of 10^{-6} to 10^{-7}, vs. air, $10^{-0.68}$) is unlikely to be actively involved noticed already by [9]. The presence of an atmosphere is, however, required, since no oxidation is taking place in vacuum. Our observations are consistent with a model of chemisorption of water or CO_2, i.e., adsorption of such oxygen-bearing species at the time the surface was exposed to air, and reaction of Fe^{2+} to Fe^{3+} upon heating [6].

Volume Oxidation Volume oxidation is unlikely related to oxygen migration into the sample, based on stable $\delta^{18}O$ values upon heating [1]. Volume oxidation has been related to hydrogen escape (e.g., [10]), however, this effect, if active at all, appears to be too small [5]. The observed volume oxidation at temperatures of crystallization, occurring in the laboratory upon heating and in the field upon cooling, implies a relation between oxidation and crystallization. We suggest that the change of glass structure upon crystallization specifically the local environment of iron, is responsible for oxidation; Fe^{3+} is stabilized with the basicity of a silicate glass, or the number of non-bridging oxygens per tetrahedron (NBO/T) [11, 12]. Kilauea basalt glass has NBO/T of 0.82 [2] where pyroxene for example per definition has NBO/T of 2. A local stabilization of Fe^{3+} in relation to pyroxene crystallization is therefore likely, also in view of the detection of Fe^{3+} in pyroxene by the Mössbauer milliprobe, correlating with smallest increase in Na_2O [13]. If significant variations of Fe^{3+} are to be expected upon crystallization, the redox state of basalt rocks may be significantly overestimated compared to the melts they originate from. Mechanisms of surface and volume oxidation suggested herein could be of relevance even in oxygen-poor atmospheres, as in extraterrestrial environments.

Acknowledgements We thank D. Lattard and G. Partzsch, University of Heidelberg, for help and access to the Mössbauer Laboratory.

References

1. Burkhard, D.J.M.: J. Petrol. **42**, 507 (2001)
2. Burkhard, D.J.M.: Contrib. Mineral. Petrol. **142**, 724 (2002)
3. Burkhard, D.J.M.: Eur. J. Mineral. **17**, 675 (2005)

4. Cooper, R.F., Fanselow, J.B., Poker, D.B.: Geochim. Cosmochim. Acta **60**, 3253 (1996)
5. Burkhard, D.J.M.: Geol. Soc. Am. Spec. Pap. **396**, 67 (2005)
6. Burkhard, D.J.M., Scherer, T.: J. Non-Cryst. Solids **352**, 241 (2006)
7. Burkhard, D.J.M.: Mineral. Mag. **69**, 103 (2005)
8. Heinrichs, H., Hermann, A.G.: Praktikum der analytischen Geochemie. Springer, Berlin Heidelberg New York, 669 (1990)
9. Virgo, D., Luth, R.W., Moats, M.A., Ulmer, G.C.: Geochim. Cosmochim. Acta. **52**, 1781 (1988)
10. Haggerty, S.E.: In: Oxide Minerals, Rev. Mineral. 3 D. Rumbe. (ed.) Mineral. Soc. Am. Washington DC (1976)
11. Duffy, J.A.: J. Non-Cryst. Solids **196**, 45 (1996)
12. Duffy, J.A.: J. Non-Cryst. Solids **297**, 275 (2002)
13. Sobolev, V.N., McCammon, C.A., Taylor, L.A., Snyder, G.A., Sobolev, N.V.: Am. Mineral. **84**, 78 (1999)

Hyperfine Interact (2005) 166:705–708
DOI 10.1007/s10751-006-9345-y

Investigation of near-source basaltic glasses using ^{57}Fe Mössbauer spectroscopy

G. R. Fern · T. A. Mather · D. M. Pyle

Published online: 3 November 2006

Abstract Two basaltic glass samples have been studied, one from the Masaya volcano in Nicaragua and one from the Villarrica volcano in Chile using ^{57}Fe Mössbauer spectroscopy. The iron(II) to iron(III) ratio has been determined for each sample. The 80 K data shows some separation of the sites and is fitted allowing for discrete doublets.

Key words basaltic glass · Mössbauer spectroscopy

1 Introduction

Basaltic glasses are of interest as they are the major constituents of many of the products (e.g., Pele's hair and scoria) from persistently basaltic volcanism. Such basaltic volcanism contributes to volcanic sulfur emissions to the atmosphere. Such volcanic sulfur emissions released during eruption and via continuous degassing, are estimated to be ~6.7–10.5 Tg s year^{-1} [1–3] accounting for a significant quantity of natural sulfur emissions.

Near-source or 'primary' sulfate aerosol emission has also been observed in some volcanoes [4]. This aerosol accounts for a small proportion of the sulfurous emission but is important locally and is concentrated in the fine (<1 μm diameter) aerosol and appears to have been formed by more rapid processes than those operating in plumes during transport downwind [4]. The mechanism of formation of this sulfate aerosol remains unknown. Although a small contribution to the total global sulfur budget, such particle emissions may constitute an important source of aerosol on

G. R. Fern (✉)
Wolfson Centre for Materials Processing, Brunel University, Uxbridge, Middlesex, London UB8 3PH, UK
e-mail: george.fern@brunel.ac.uk

T. A. Mather · D. M. Pyle
Department of Earth Sciences, University of Cambridge, Downing Street, Cambridge CB2 3EQ, UK

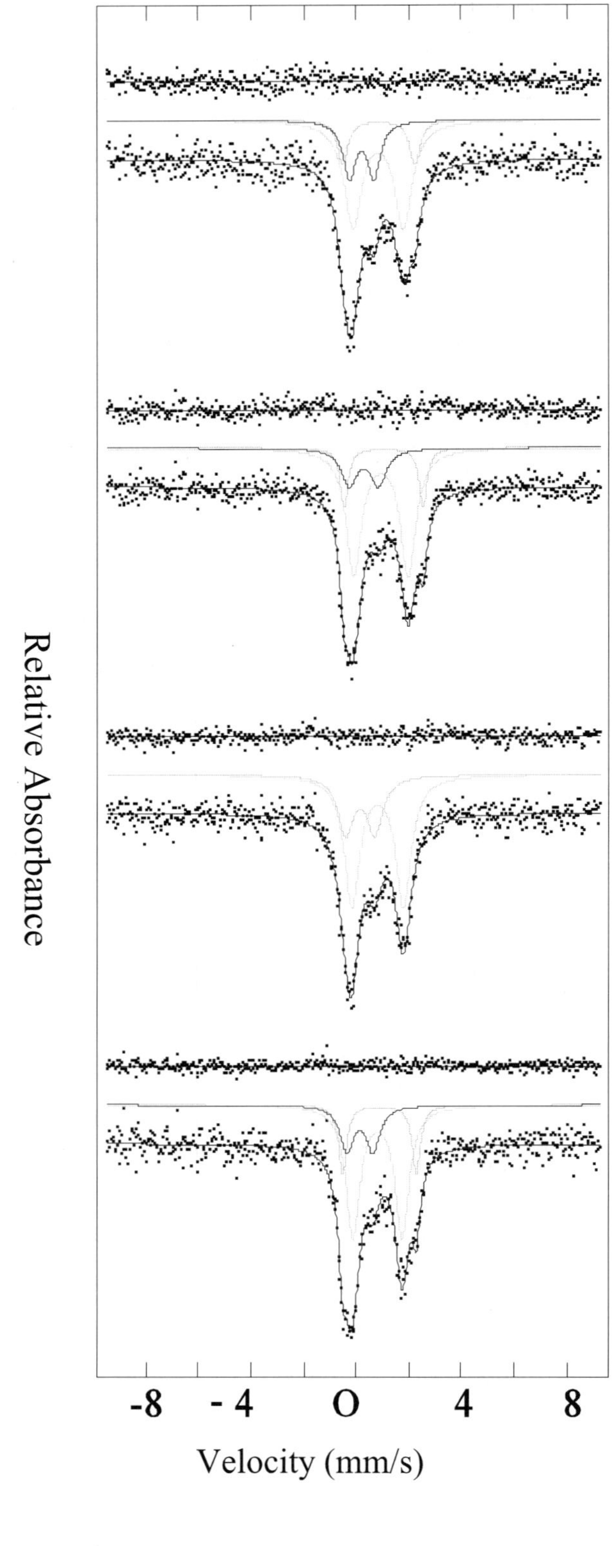

Figure 1 Mössbauer Spectroscopic data for (*from top the bottom*) Masaya at 80 K, Villarrica at 80 K, Masaya at 298 K, Villarrica at 298 K.

Table I Mössbauer spectroscopic data for basaltic glass from the Masaya volcano in Nicaragua and Villarrica Volcano, Chile

T (K)	Sample	Site	IS (mm/s)	QS (mm/s)	FWHM (mm/s)	Area (%)
80	Masaya	Site 1 (Fe^{2+})	1.057(17)	1.903(58)	0.815(48)	61.5(66)
		Site 2 (Fe^{2+})	1.084(16)	2.737(56)	0.841(106)	15.2(54)
		Site 3 (Fe^{3+})	0.435(19)	0.929(42)	0.583(54)	23.3(17)
80	Villarrica	Site 1 (Fe^{2+})	1.157(7)	2.063(19)	0.708(23)	64.5(21)
		Site 2 (Fe^{2+})	1.257(8)	2.994(16)	0.356(29)	16.2(20)
		Site 3 (Fe^{3+})	0.506(19)	1.111(54)	0.715(47)	19.3(9)
298	Masaya	Site 1 (Fe^{2+})	1.036(6)	1.946(14)	0.714(20)	69.0(19)
		Site 3 (Fe^{3+})	0.375(12)	1.104(26)	0.720(49)	31.0(17)
298	Villarrica	Site 1 (Fe^{2+})	1.034(06)	1.860(20)	0.637(26)	60.7(29)
		Site 2 (Fe^{2+})	1.106(9)	2.794(21)	0.373(34)	18.7(20)
		Site 3 (Fe^{3+})	0.367(20)	0.991(46)	0.624(67)	20.6(16)

IS=Isomer shift relative to α iron, QS=Quadrupole splitting, the figure in brackets relates to the error. FWHM=Full width at half height.

local and regional scales often at elevated altitude and to otherwise relatively unpolluted areas. The sulfur and oxygen isotopic composition of atmospheric sulfate can be used to elucidate the oxidation pathways that were involved in its production. Here, data on the oxidation state of iron in basaltic glass samples from the source magma associated with these near-source volcanic sulfate aerosol emissions are presented. The Fe(II)/Fe(III) ratio of the magma can be used to yield information on sulfur speciation and possible degassing mechanisms in the magma.

2 Experimental

Samples of Pele's hair (fine glass shards formed during bubbles bursting on a lava lake surface and transported up in the hot gas stream) from Masaya and fresh scoria from Villarrica were ground to a fine powder and mounted in aluminium disks and placed in a cryostat either at 298 or 80 K. The data was allowed to accumulate for approximately 8 days due to their relatively low iron content of 13.5 wt.% for the Masaya sample and 11.5 wt.% for the Villarrica sample (from electron microprobe analysis). The data was then fitted using the NORMOS fitting program written by R.A. Brand.

3 Results and discussion

Fitting of glasses can often be carried out using a distribution of sites [5]. However in this instance it was found that two Fe^{2+} environments were required to give good fitting statistics and therefore a simpler fitting model has been employed in accordance with a recent Mössbauer spectroscopic study of basaltic glasses which investigated the Fe^{3+}/Total Fe ratio [6].

The residual analysis of the Villarrica sample (See Figure 1) indicated that using two iron environments for the Fe^{2+} sites at room temperature was satisfactory

[1] See reference [5], page 382, Table 2.

whereas the Masaya sample fitted well to a single Fe^{2+} environment. The IS for the Masaya sample doesn't show as much change with temperature as the Villarrica sample which must be due to the overlap of the two Fe^{2+} sites.

The Mössbauer spectroscopic data is useful for not only elucidating the Fe^{2+}/Fe^{3+} ratio but also the extent of disorder of the glass is indicated. The coordination environment is alluded to by comparison with other work.[1] This suggests that both the iron(II) and iron(III) sites are sited in a six coordinate environment.

The large degree of distortion of the iron environment is evident in the large quadrupole splitting parameters of site 2 shown in Table I. This indicates that the Fe^{2+} sites present are in quite different local environments. The FWHM for site 1 are large for a single Fe^{2+} environment which is expected for such a glass system and is therefore within the range of acceptance for this study. The most important parameters are the IS and relative proportions of Fe^{2+} to Fe^{3+} which have been determined and will be useful in the further understanding of the formation processes of near-source sulfate aerosol from Masaya and Villarrica [7].

References

1. Stoiber, R.E., Williams, S.N., Huebert, B.: J. Volcanol. Geotherm. Res. **33**, 1 (1987)
2. Andres, R.J., Kasgnoc, A.D.: J. Geophys. Res. **103**, 25251–25261 (1998)
3. Halmer, M.M., Schmincke, H.-U., Graf, H.F.: J. Volcanol. Geotherm. Res. **115**, 511 (2002)
4. Mather, T.A., Tsanev, V.I., Pyle, D.M., McGonigle, A.J.S., Oppenheimer, C., Allen, A.G.: J. Geophys. Res. **109**, D21303 (2004)
5. McCammon, C.A.: EMU Notes Mineral. **6**, 369–398 (2004), Chapter 9
6. Wilke, M., Partzsch, T.G.M., Bernhardt, R., Lattard, D.: Chem. Geol. **220**, 143 (2005)
7. Mather, T.A., McCabe, J.R., Kumar Rai, V., Thiemens, M.H., Pyle, D.M., Heaton, T.H.E., Sloane, H.J., Fern, G.R.: J. Geophys. Res. **111**(D18), D18205 (2006)

Author Index to Volumes 165/1–4 and 166/1–4 (2005)

Abdelmoula, M., 391
Abdu, Y. A., 543, 593
Alfroj, D., 409
Aïssa, R., 391
Alboom van, A., 549, 679
Amthauer, G., 573
Anastacio, A. S., 619
Antonyuk, L. P., 303
Arthur, J., 3
Asahi, K., 357
Aubin, V., 489
Augustin, C. O., 231

Babalola, O. J., 279
Bailey, B. E., 567
Balaji, G., 153
Balan, L., 101
Balogh, A. G., 605
Balogh, J., 49
Ban, S., 95
Barčová, K., 203
Barone, G., 17, 299
Bartlett, A., 373
Bartonkova, H., 221
Basak, S., 153
Basset, J.-M., 55
Bauminger, E. R., 285
Becker, K. D., 81
Bender Koch, C., 455, 505, 579
Bergmann, I., 81
Berry, F. J., 363, 449
Bertelsen, P., 511, 523, 555
Bettati, S., 279
Bhattacharyya, S., 147, 153
Bhatia, S. C., 339
Billaud, D., 101
Binau, C. S., 523
Bingham, P. A., 135
Bisschoff, A. A., 693
Black, C. T., 239
Borggaard, O. K., 579
Boufas, S., 345
Boukheddaden, K., 37
Böttger, L. H., 17, 299
Brink, M. C., 693
Brinkmann, C., 573
Brown, D. E., 499
Bürck, U. V., 141
Burkhard, D. J. M., 699
Bustamante, A., 585, 593, 599
Bustos Rodriguez, H., 605
Butz-Jørgensen, C., 511
Byakov, V. M., 437

Cabrera, J., 593
Candy, J.-P., 55
Cardelino, B. H., 339
Carney, E., 333
Cashion, J. D., 5, 161
Caurant, D., 489
Chasteen, N. D., 333
Chistyakova, N. I., 613
Chumakov, A. I., 17, 295
Codjovi, E., 37
Coelho, F. S., 619
Constantini, J.-M., 489
Costa, B. F. O., 107
Croci, S., 279

Dailly, A., 101
Datas, L., 113
Das, R. K., 153
da Costa, G. M., 113
Degmová, J., 121
de Amorim, H. S., 599
de Grave, E., 113, 549, 679
de Resende, V. G., 113, 549
De Souza, Jr, P. A., 549

Dedushenko, S. K., 367
Dekan, J., 651
Dénès, G., 345, 373, 379
Dias Filho, J. H., 247
D'Orléans, C., 61
Ducourett, A., 489
Duda, P., 227

Espinoza, S., 585
Estournès, C., 61, 215, 475

Fabris, J. D., 619
Fan, C., 273
Fanciulli, M., 69
Fedotov, A., 127
Fedotova, J., 127
Feldhoff, A., 81
Fern, G. R., 705
Ferrow, E. A., 505
Fischer, L., 55
Florez, J. M., 253
Fodor, F., 289
Forder, S. D., 135
Franz, H., 141
Frąckowiak, J. E., 529
Friedman, A., 285
Fukagawa, M., 429

Gafur, A., 579
Gajbhiye, N. S., 147, 153
Galazka-Friedman, J., 285
Galuskin, E. V., 529
Galyametdinor, Y., 385
Gancedo, J. R., 449
Garcia, V., 593
Garitaonandia, J. S., 161
Gaspar, A. B., 385
Génin, J.-M.R. 1, 391
Gich, M., 203
Goetz, W., 523, 555
Gona, M. N., 167
Goikolea, E., 161
Gorria, P., 161
Gorun, S. M., 333
Goujon, A., 37
Gourier, D., 489
Govender, A., 687
Grabski, J., 227
Greaves, T. L., 5
Greig, D., 167
Grodzicki, M., 573
Grokhovsky, V. I., 665, 671
Gubaidulina, T. V., 613
Gunnluagsson, H. P., 511, 517, 523, 555, 561
Gushterov, V., 625
Gütlich, P., 385

Haase, W., 385
Hand, R. J., 135
Hantash, J., 373
Hashino, H., 483
Haslbeck, S., 89
Hasso-Agopsowicz, A., 631
Helgason, Ö., 517
Heitjans, P., 81
Herber, R. H., 351
Homonnay, Z., 429, 643
Horn, R., 333
Hoya, H., 95
Hviid, S. F., 523

Iijima, S., 397
Iiyama, T., 657
Ikeda, K., 403
Inoue, K., 469
Insausti, M., 161
Iriyama, T., 483
Iwunzie, M., 499

Jablonska, A., 127
Jelenska, M., 631
Jirásková, Y., 75
Jumas, J.-C., 1, 55, 101

Kákay, A., 195
Kalai Selvan, R., 231
Kalinowska, J., 183, 189
Kalinin, J., 127
Kameda, D., 357

Kamikawa, T., 429
Kamnev, A. A., 289, 303
Kaňuch, T., 75
Kaptás, D., 49
Károly Varga, L., 195
Katada, M., 409
Kawakami, N., 429
Kawasaki, T., 417
Kemény, T., 49
Keune, W., 175
Khasanov, A. M., 209
Kilcoyne, S. H., 167
Kinch, K. M., 523
Kinch, K., 555
Kiss, L. F., 49
Kitazawa, T., 417
Klingelhöfer, G., 511, 523, 549, 567, 599
Kobayashi, Y., 357, 403, 425
Kochuparampil, J., 379
Kojima, M., 397
Kojima, N., 403
Komlósi, A., 643
Kónya, J., 643
Kopcewicz, B., 631
Kopcewicz, M., 183, 189, 631
Kovács, K., 289
Kozerenko, S. V., 613
Koziorowski, D., 285
Kristjánsson, L., 561
Kroll, H., 573
Ksenofontov, V., 385
Kubo, M. K., 357, 425
Kubuki, S., 429, 643
Kulik, T., 183
Kulikov, L. A., 303, 437
Kunev, B., 61, 215
Kuncser, V., 175
Kuno, A., 637
Kurat, G., 543
Kurmoo, M., 61, 475
Kurth, D. G., 465
Kuvvetli, I., 511
Kuzmann, E., 195, 289, 303, 429, 643

Ladriere, J., 443
Lakatos-Varsányi, M., 195
Lakshmi, N., 261
Latuch, J., 189
Lančok, A., 203
LaManna, G., 17, 299
Lavaysierre, A., 135
Lee, J., 333
Leer, K., 523
Le Caër, G., 107
Li, Y., 273
Lindner, A., 465
Lippens, P.-E., 1
Litterst, F. J., 81
Liu, W., 273
López, A., 543
Lottermoser, W., 573
Loureiro, J. M., 107
Lumpkin, G. R., 363

Mabe, D. R., 209
Madsen, D. E., 523
Madsen, M. B., 511, 523, 555
Maksimov, Y. V., 25
Malakheeva, L. I., 321
Malczewski, D., 529
Manova, E., 61, 215
Mantovan, R., 69
Marco, J. F., 449
Martinec, P., 221
Martinek, K.-P., 89
Mashlan, M., 75, 221
Mather, T. A., 705
Mattievich, E., 599
Matsue, H., 425
Matsumoto, N., 397
Matsuo, M., 637
May, L., 499
Mazo-Zuluaga, J., 253
Meaz, T. M., 455
Menzel, M., 465
Merazig, H., 345
Merrison, J. P., 511, 555
Merrison, J., 523
Merodiiska, T., 267
Mezentsev, A. V., 665
Miglierini, M., 75, 651
Mikó, A., 195

Milder, O. B., 231, 321, 665, 671
Ming, D. W., 549
Minguez, P., 161
Mitov, I., 61, 215
Miyazaki, J., 357
Miyoshi, H., 357
Mizutani, F., 397
Molins, E., 203
Morales, G., 585
Morin, S., 55
Morita, T., 469
Morozov, M., 573
Morris, R. V., 549
Muntasar, A., 345, 373, 379
Murray, C. B., 239

Nagae, D., 357
Nagy, F., 195
Nagy, N. M., 643
Nakada, M., 417
Nakashima, S., 469
Nedkov, I., 267
Nguyen, N., 489
Ningthoujam, R. S., 153
Nishida, T., 429
Nishimura, T., 417
Nishizumi, T., 429
Niwa, O., 397
Niznansky, D., 475
Nomura, K., 657
Nonaka, H., 357, 425
Nørnberg, P., 511, 517, 555
Nowik, I., 351

Oates, G., 363
O'Brien, S. P., 239
Ohkubo, N., 95
Ohya, T., 309
Oldfield, P., G., 373
Oleniacz, J., 227
Oliver, F. W., 499
Olivier-Fourcade, J., 55, 101
Olsen, M., 523
Omatsuzawa, R., 483
Ortalli, I., 279
Oshtrakh, M. I., 231, 309, 313, 321, 327, 665, 671
Oyola Lozano, D., 605

Pandey, B., 537
Paneva, D., 61, 215
Paniago, R., 247
Papaefthymiou, G. C., 239, 333
Parak, F., 279
Paulsen, H., 17, 295, 299, 495
Pechousek, J., 221
Pérez Alcázar, G. A., 605
Perfiliev, Y. D., 303, 367, 437
Petrov, L., 61
Pfannes, H.-D., 247
Pieper, J., 175
Pinto, I. V., 619
Pizarro, R., 161
Plazaola, F., 161
Popovici, M., 203
Prokopenko, P. G., 321
Pyle, D. M., 705

Rasmussen, H., 511, 561
Ravi, N., 339
Redl, F. X., 239
Rehspringer, J.-L., 61, 215, 475
Reiman, S., 385
Ren, X., 449
Rentschler, E., 385
Renz, F., 465
Restrepo, J., 253
Rodionov, D., 549
Roig, A., 203
Roisin, E., 55
Rojas Martinez, Y. A., 605
Rüffer, R., 17, 295
Ruby, C., 391
Rusakov, V. S., 613
Rusanov, V., 625
Ryan, D. H., 499

Saeki, M., 417
Sahoo, B., 175

Saito, T., 483
Sakai, H., 657
Sakai, Y., 425, 483
Sakazaki, I., 483
Sandstrom, R. L., 239
Sato, M., 309
Savii, C., 203
Scherer, T., 699
Schmidt, C., 17, 295
Schneider, R., 101
Schneeweiss, O., 221
Schröder, C., 549, 567, 599
Schünemann, V., 17, 295
Scorzelli, R. B., 543, 585, 593
Sekizawa, K., 95
Semionkin, V. A., 231, 321, 665, 671
Šepelák, V., 81
Senterre, V., 443
Seredyuk, M., 385
Sergueev, I., 17, 141, 299
Seto, M., 403
Sharma, R. K., 261
Shenoy, G., 3
Shimada, K., 357
Shoji, H., 425
Silveria, M. M., 599
Silvestri, A., 17, 299
Sitek, J., 121
Sitnikov, A., 127
Slavov, L., 267
Smirnov, G., 3
Souza Azevedo, I., 543
Spiga, S., 69
Squyres, S. W., 523
Staššíková-Štukovská, D., 651
Staudigel, H., 567, 599
Stepanov, S. V., 437
Stevens, J. G., 209
Stewart, S. J., 543
Stievano, L., 89
Stucki, J. W., 619
Studer, F., 489
Sunatsuki, Y., 397
Suzdalev, I. P., 25
Suzuki, K., 161
Suzuki, S., 483
Suwalka, O., 261
Svito, I., 127
Szlachta, K., 285

Takahashi, Y., 95
Takano, Y., 95
Takase, K., 95
Takeda, J., 309
Takeda, M., 657
Taneja, S. P., 499
Tegner, C., 561
Temst, K., 49
The Athena Science Team, 523
Thünemann, A. F., 465
Tippelt, G., 573
Trautwein, A. X., 17, 295, 299, 495, 625
Tripathi, A., 537
Tuchagues, J.-P., 495
Tsoncheva, T., 215

Ueno, H., 357
Urday, E., 593
Uzio, D., 55

Valenti, C., 279
Vandenberghe, R. E., 267
Van Haesendonck, C., 49
Varela, M. E., 543
Varret, F., 37
Vencato, I., 599
Vendelboe, A. L., 517
Venugopalan, K., 261
Verma, H. C., 537
Vertes, A., 289, 303
Viana, J. H. M., 619
Vieira Junior, D. S., 247
Viescas, A. J., 333
Vilminot, S., 475
Vincze, I., 49
Vůjtek, M., 75

Waanders, F. B., 687, 693
Wagner, F. E., 89

Wang, Q., 637
Wegner, P., 17, 295
Weissmüller, J., 153
Wellenreuther, G., 141
Weyer, G., 511
Whittle, K. R., 363
Willmann, P., 101
Winkler, H., 17, 295, 495, 625
Wolny, J. A., 17, 299, 495
Worm, E. S., 555
Wynter, C. I., 499

Yamada, K., 469
Yamada, Y., 357, 425
Yen, A., 549
Yin, M., 239
Yoshimi, A., 357
Yuryeva, E. I., 327

Zaleski, A., 127
Závěta, K., 203, 475
Zbořil, R., 75, 221
Zhang, C. J., 637
Zhao, G., 333
Zhiganova, E. V., 665
Zhizhin, M. G., 367
Zhu, S., 273
Zych, W., 227